*Facets of Genetics*

*Readings from*

**SCIENTIFIC AMERICAN**

# *Facets of Genetics*

*Selected and Introduced by*

**Adrian M. Srb**
*Cornell University*

**Ray D. Owen**
*California Institute of Technology*

**Robert S. Edgar**
*University of California, Santa Cruz*

W. H. Freeman and Company
*San Francisco*

Each of the SCIENTIFIC AMERICAN articles in *Facets of Genetics* is available as a separate Offprint. For a complete list of approximately 700 articles now available as Offprints, write to W. H. Freeman and Company, 660 Market Street, San Francisco, California 94104.

---



Printed in the United States of America

Library of Congress Catalog Card Number: 78–99047

Standard Book Number: 7167 0950–3 (cloth);
7167 0949–X (paper)



10 9 8 7 6 5 4 3

# *Preface*

A textbook is traditionally impersonal. No matter how broad its author's knowledge and understanding, he usually finds himself writing mostly about other people's work, and in a language that his readers readily identify as textbookese. Readers of *Scientific American,* on the other hand, are regularly treated to short, well-illustrated articles by scientists who are reporting with enthusiasm the results of their own research and observations in the field of their major competence—and in language that is not like that of a textbook at all.

We have selected the thirty-five *Scientific American* articles in this book from a rich mosaic of material published over the past two-and-a-half decades, samplings from a field that over that period has experienced a burst of vitality and progress. We left out of our selection several excellent pieces of research and writing, in part because we did not want the book to become so large as to be unwieldy and in part because some of the choices open to us were already so well known, already so thoroughly incorporated into textbook material, that their marginal utility in still another publication seemed questionable. Furthermore, at one end of its spectrum genetics now shades imperceptibly into the relatively new field of molecular biology; we have left out much relevant material we felt to be preponderantly biochemical or biophysical, and have chosen instead to present at that end of the spectrum of our subject only those "molecular" articles that illustrate primarily the application of genetic approaches, techniques, and reasoning.

We doubt that any other discipline currently spreads over so great an array of materials as genetics, sheds light on life at so many levels, and yet displays the same essential techniques and concepts throughout. The articles in this book range from the humanity of a British king to the genetic definition of the triplet code by which the nucleic acids control amino acid sequences in proteins. The collection is divided into five parts, each beginning with an introduction intended to provide some degree of integration and focus for the articles. Each illustrates how genetics illuminates our study of life at a different level: Part I is concerned with the processes of inheritance; Part II, with the nature of genetic material and how it works in cells; Part III, with the development of individual organisms; Part IV, with the diversity of living things that has occurred in the course of evolution; and Part V, with our efforts toward understanding and serving mankind.



ADRIAN M. SRB
RAY D. OWEN
ROBERT S. EDGAR

*November 1969*

# *Contents*

## I. *The Elements of Inheritance*



## II. *The Nature of the Gene*



## III. *From Gene to Organism*

## IV. *Genetics and Evolution*



## V. *Genetics and Man*



---

*Note on cross-references*

Cross-references within the articles are of three kinds. A reference to an article included in this book is noted by the title of the article and the page on which it begins; a reference to an article that is available as an offprint but is not included here is noted by the article's title and offprint number; a reference to a SCIENTIFIC AMERICAN article that is not available as an offprint is noted by the title of the article and the month and year of its publication.

*Facets of Genetics*

# I
# *The Elements of Inheritance*

# I

# *The Elements of Inheritance*

## INTRODUCTION

Although the science of genetics was born in 1866, it is really a creature of this century. The pioneering work of Gregor Mendel accumulated dust on library shelves until 1900, when his notions of the mechanics of heredity were discovered anew. Since 1900, genetics has prospered and established for itself a central place in biology. The primary concern of genetics is the blueprint of living organisms—its mode of transmission from generation to generation and the means by which it becomes manifest through the development of the organism. Conceptually, the gene soon emerged as the elementary unit of this blueprint. Here was the biologist's atom, an elementary particle endowed with great complexity and exhibiting an essential, mysterious feature of life—the power of self-duplication.

The period since 1950 has been a time of spectacular growth in our understanding of the nature of the gene and the details of the mechanism of inheritance. This growth can in large part be attributed to two factors, the successful application of biochemical and biophysical techniques to the "gene problem," and the exploitation of microorganisms, such as bacteria and viruses, for the study of the fundamental problems of genetics—the mechanism of reproduction and function of the genes. *Molecular Genetics* is the name often given to this field, which encompasses the studies of the genetic apparatus of bacteria and their viruses. Workers in this field are as apt to use radioactive tracers as mutations to tackle the problem of their immediate concern.

Our present understanding of the chemical nature of the gene stems principally from the elucidation of the structure of deoxyribose nucleic acid (DNA), the chemical substance of which genes are made. This understanding of DNA structure led almost directly to an appreciation of how the genes reproduce and eventually to our present picture of the role genes play in influencing the synthesis of the vast array of proteins that, to a large degree, determine the nature and activities of cells.

The first article in this collection, "The Gene," by Norman H. Horowitz, was written in 1956, only two years after Watson and Crick announced their solution for the structure of DNA. This article was chosen to begin this collection because in many ways it is an introduction to the articles that follow. Horowitz skillfully sets the stage for the subsequent triumphs of molecular genetics by describing the basic Mendelian principles, the concept of the gene, its location in the chromosome, and the major problems concerning gene function and reproduction—problems that were soon to be clarified.

Although it is true that genes are composed of DNA, in the living organism the vehicle for transmission is not the gene but the chromosome, a complex organelle containing large amounts of DNA and a variety of proteins. The second article, "The Duplication of Chromosomes," by J. Herbert Taylor, shows that the DNA residing in the chromosomes reproduces and is parceled out to the daughter cells just as expected on the basis of the notions of Watson and Crick for DNA duplication. Even though Taylor's article was written as long ago as 1958, the organization of the DNA and protein within the chromosomes and the detailed mode of reproduction of this organelle are still unknown. The model proposed by Taylor has now become one of a large family of speculations.

The intractability of chromosome structure to analysis is perhaps one motive for looking in detail at the mode of reproduction of the genetic material of less complex creatures—for example, bacteria. The beautiful and painstaking work of John Cairns, described in his article "The Bacterial Chromosome," reveals a single circular DNA molecule of prodigious length. As Cairns points out, new and fascinating problems arise in comprehending the reproduction of the genetic apparatus even of the simple bacterium.

During the first half of this century the bacteria were of little interest to geneticists. Although their simplicity might have proved attractive for the study of basic problems, they appeared to lack a sex life. Sexual reproduction is the crucial feature of an organism that must be exploited by the geneticist if he is to analyze the properties of genes and study their interactions through recombination and its consequences. Actually, bacteria turn out to have a most bizarre and varied sex life. The first recognized mode of gene exchange among bacteria is described in the article by Rolin D. Hotchkiss and Esther Weiss, "Transformed Bacteria." The phenomenon of transformation has in fact been known since 1928; it played a crucial role in the proof that the genetic material was DNA. Perhaps an even more unusual sexual mechanism is the phenomenon of transduction, described in " 'Transduction' in Bacteria" by Norton D. Zinder. Here a virus plays the role of Eros carrying genes from one bacterial host to another!

Many bacterial viruses display an exquisitely complex and intimate relationship with their hosts. This situation has been revealed largely through the ingenious genetic studies of François Jacob and Elie L. Wollman, described in the article "Viruses and Genes." Their work led not only to the understanding of the mechanism of conjugation in bacteria but also to the discovery that the DNA of temperate viruses physically inserts itself into the host DNA, becoming in fact a part of the genetic apparatus of the cell. From these studies it became clear that in bacteria there exists a veritable "zoo" of virus-like creatures, leading only semi-autonomous lives, slipping by various means from cell to cell, and occasionally insinuating themselves into the genetic apparatus of the host. One such "episome" of present medical concern is the infectious drug resistance factor described in the article by Tsutomu Watanabe.

This rich and varied sex life of bacteria is not only of intrinsic interest, with implications for possible genetic mechanisms in higher organisms, but has also provided a variety of powerful and versatile tools for tackling other problems of genetic interest, such as gene function.

The last article in this section, "Genes Outside the Chromosomes," by Ruth Sager, shows that higher organisms with proper chromosomes, meiosis, and "mendelizing factors" are far from free of such unconventional genetic creatures as episomes. Her studies demonstrate the existence in cells of genes other than those in the chromosomes. DNA has now been found in mitochondria, chloroplasts, and other cellular organelles. The role that such nonchromosomal genetic material plays in the life of cells is an exciting new problem, which is at present under intensive study.

These articles show quite well the present vigor of this area of genetics. It is now seen to include a menagerie of strange creatures to tame and understand. Meanwhile, the familiar chromosome still awaits elucidation of its structure.

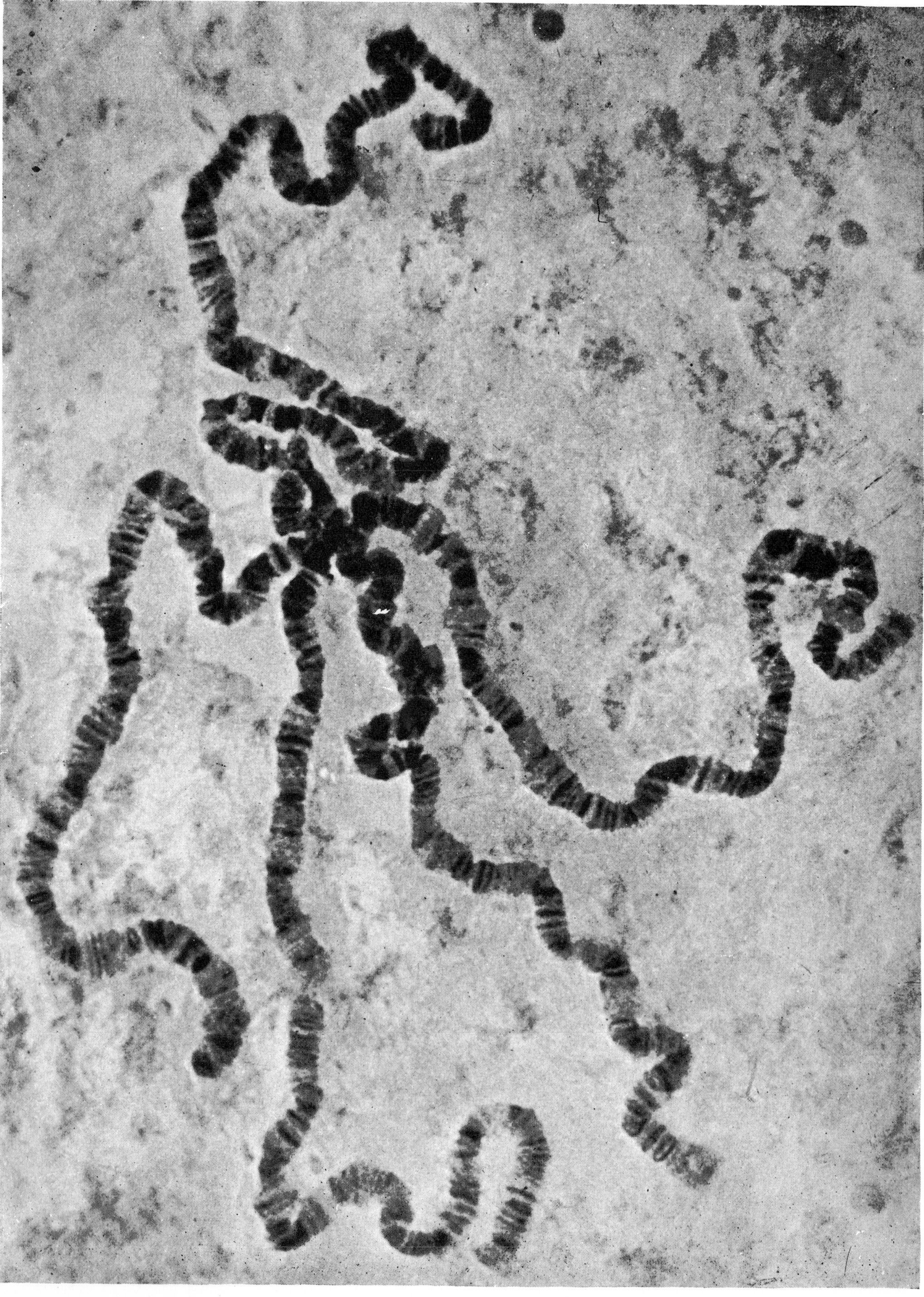

# 1

# The Gene

NORMAN H. HOROWITZ
*October 1956*

The distinguished theoretical physicist Erwin Schrödinger has called the science of genetics "easily the most interesting of our days." No branch of science in the 20th century has contributed more to man's understanding of himself and the living world in general. Genetic discoveries have provided new insights into such fundamental problems as the origin of life, the structure of living matter and evolution; and they have yielded practical benefits over a broad range of human concerns, from plant and animal breeding to the investigation of disease. Genetics, in short, has become the theoretical backbone of biology.

The central concept of genetics is the gene, the elementary unit of inheritance. This article presents an account of the development of the gene concept and of modern researches into the nature and mode of action of genes.

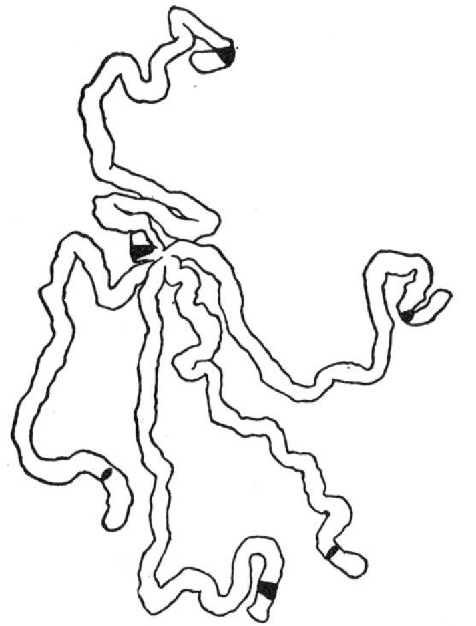

**CHROMOSOMES in the cells of the salivary glands of *Drosophila*, sometimes called the vinegar fly (*opposite*), are useful to geneticists because they are large and easy to "read." On the diagram above are marked the loci of certain identifiable features in the chromosomes which are found in association with heritable characteristics, especially color, in the eye of the vinegar fly.**

Almost everyone knows that the science of genetics was founded by a monk named Gregor Mendel, who carried on plant breeding experiments as a hobby in the garden of his monastery at Brünn in Moravia (now Brno, Czechoslovakia). The story of how Mendel's remarkable paper, published in 1866 in the proceedings of a provincial natural history society, was ignored by his contemporaries and was rescued from oblivion by three separate investigators in the year 1900, 16 years after Mendel's death, is one of the most dramatic incidents of modern science. Many historians of biology have speculated on why so far-reaching an advance was disregarded at the time it was made, only to be enthusiastically welcomed 35 years later. The interment of Mendel's paper is commonly attributed to the fact that he was an amateur and to the obscurity of the journal in which it was published. But Charles Darwin also was an amateur, and the standing of amateurs in science was still strong in the 19th century. As for the journal—the *Proceedings of the Brünn Society for the Study of Natural Science*—it was regularly received by the leading research centers of Europe, and Mendel is known to have called his paper to the attention of Carl K. von Nägeli, one of the leading botanists of the day. A more likely answer to the puzzle is that Mendel's contemporaries were incapable of appreciating the significance of his discoveries. His conclusions were essentially of an abstract nature, based on numerical data obtained by counting the various kinds of offspring produced by crosses of pea plants. Because chromosomes and many other aspects of the biology of reproduction were unknown in Mendel's day, his paper must have seemed to his contemporaries arbitrary and formalistic—mere numerology.

It is curious to reflect that although Mendel is justly entitled to acclaim for his discovery, the development of genetics would have been the same even if he had never lived. His three rediscoverers—Carl Correns in Germany, Hugo de Vries in Holland, Erich Tschermak in Austria—were led to his paper only after they had independently arrived at the same conclusions. Some geneticists believe that Correns, Tschermak and de Vries were greatly aided in the interpretation of their experiments by Mendel's brilliant paper, but this cannot be proved by the historical record.

## "The Elementary Game"

The essence of Mendel's discovery was that hereditary traits—or "characters"—are independent of one another and each is transmitted as a separate unit from a parent to the offspring: in other words, the organism is a mosaic of distinct and independent qualities. The units of inheritance are distributed in families and populations according to the laws of independent events—*i.e.*, the laws of chance. This view contrasts with the older idea that the characteristics of the parents are blended in the offspring, as one might blend two liquids by mixing them together (a misconception which still persists in the expressions "full-blooded," "half-blood" and the like). Mendel showed that there is no blending or dilution of individual characters in a hybrid. The expression of a given character may disappear, but the hybrid

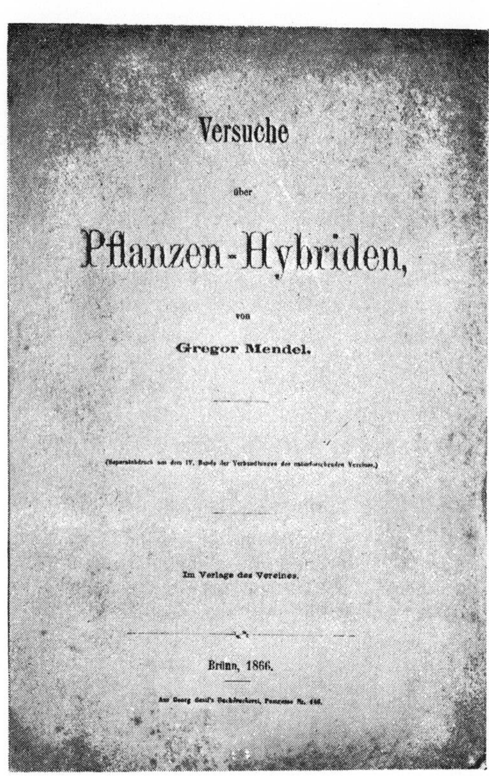

Versuche

über

Pflanzen-Hybriden,

von

Gregor Mendel.

Im Verlage des Vereines.

Brünn, 1866.

**GREGOR MENDEL published his first comprehensive report on his investigations in 1866. Copies of his paper were circulated to most of the university libraries of Europe and to many contemporary scientists, yet its significance went unrecognized until 1900.**

carries the character as a "recessive" unit, and it may emerge in later generations. Mendelian inheritance is essentially atomistic, the heritable qualities of the organism behaving as if they were determined by irreducible particles (we now call them genes). Mendel was not concerned with the nature of the particles but with showing that inheritance can be understood in these terms and with working out the rules of transmission of traits.

An individual is composed of thousands of heritable characteristics. Each character may take one of several possible forms. For example, there are three principal forms, called A, B and O, of a gene for human blood type. Every person carries a pair of the genes, the possible combinations being AA, AO, BB, BO, AB and OO. A and B are "dominant" and O is "recessive" to them, so that the blood of a person with the combination AO, for instance, shows the properties of the blood group A.

Mendel's laws can be understood in terms of a game of chance in which the genes are represented by counters. There are two players (the parents) and each is provided with a pair of counters for each of the thousands of hereditary characters (*e.g.*, AA or AO or the like for blood group). To play the game each player selects at random one counter from each of his pairs and puts it in a pile in the middle of the table. In the end the pile will contain just as many counters as were originally held by each player, half of them contributed by each player. The pile of counters represents the genetic endowment of an offspring of the two players. In principle this is all there is to the game of Mendelian inheritance. We shall call it the "elementary game."

Mendel's great achievement was to recognize that this simple game—that is, the random separation and reuniting of pairs of inheritance determiners in the germ cells—would provide an orderly explanation of the seemingly unsystematic results of his experiments. Mendel found the evidence for the separation of determiners in a statistical analysis of pea-plant offspring: the characters tended to occur in certain orderly proportions among the members of successive generations as the determiners were shuffled and reshuffled. Nowadays the separation of genes (technically called "segregation") can be demonstrated in a direct way by experiments with some of the lower plants, such as the red bread mold *Neurospora* [see "The Genes of Men and Molds," by George W. Beadle; SCIENTIFIC AMERICAN, September, 1948]. For example, the segregation of a certain gene pair results in exactly equal numbers of black and white spores in every mature set of eight spores produced by this fungus. The Lysenkoists in the U.S.S.R. used to attack Mendelian genetics on the ground that it was based on statistics, which, for reasons not explained, they aimed to eliminate from biology. It was a moment of some dramatic interest, therefore, when, at the International Botanical Congress in Stockholm in 1950, a Portuguese scientist rose to ask a Soviet speaker how he explained the nonstatistical demonstration of segregation in Neurospora and similar organisms. It appears from the official account of the meeting that the Soviet delegate was unaware of these demonstrations.

## Enter the Chromosome

In the years following the rediscovery of Mendel's laws at the turn of the century, the new science of genetics advanced rapidly. Investigators soon showed that the Mendelian rules of inheritance applied to animals as well as plants. The most important advance came when T. H. Morgan and his group at Columbia University, working with the vinegar fly *Drosophila*, discovered that the genes are material particles carried in the chromosomes of the cell nucleus. They were led to this discovery by their finding that genes are not altogether independent, as Mendel had thought, but tend to be transmitted in groups. In terms of the elementary game, we would say that the choice of counters is not entirely free: when one counter is selected, there is a tendency for certain other counters to be selected also, as if they were linked by a weak physical bond. Morgan and his students A. H. Sturtevant, C. B. Bridges and H. J. Muller found that genes in the same chromosome (where they are arranged like beads on a string) are transmitted sometimes as an intact group, sometimes not. That is to say, a pair of chromosomes may exchange segments of their strings of genes, forming new chromosomes which consist in part of one and in part of the other—a process known as "crossing over." By grouping genes in chromosomes and yet allowing them some freedom to change their lodgings, nature reconciles two conflicting requirements of inheritance and evolution. On the one hand, total disorganization of the genes in a cell would make the reproduction of cells exceedingly difficult. The tiny vinegar fly has something of the order of 10,000 genes. If they were loose in the cell nucleus, like buckshot, the problem of passing them on in exactly equal number to every daughter cell would be formidable. The problem is reduced to manageable proportions by the fact that the genes are grouped in four pairs of chromosomes: thus the cell has only eight objects to cope with, instead of 10,000. On the other hand, if the genes were forever bound in the same chromosomes, the organism would lack the flexibility for recombination of genes which is essential for evolutionary development. The situation is saved by the fact that genes may cross over from one chromosome to its partner when the germ cells are formed.

The Mendelian theory led to a new understanding of the biological significance of sex. It showed that the sexual method of reproduction provides an elaborate lottery which serves the function of recombining genes in new ways, thus permitting living things to explore a practically limitless range of possible variations. If each of 10,000 genes determining the make-up of a species of organism existed in only two different forms, the number of different gene combinations possible would be $3^{10.000}$ As we have seen, some genes are known to occur in more than two forms. The practically infinite number of possible combinations provides a vast reservoir of potential variability upon which the species can draw for its evolutionary

BOTANIQUE. — *Sur la loi de disjonction des hybrides.* Note de M. **Hugo de Vries**, présentée par M. Gaston Bonnier.

« D'après les principes que j'ai énoncés ailleurs (*Intracellulare Pangenesis*, 1889), les caractères spécifiques des organismes sont composés d'unités bien distinctes. On peut étudier expérimentalement ces unités soit dans des phénomènes de variabilité et de mutabilité, soit par la production des hybrides. Dans le dernier cas, on choisit de préférence les hybrides dont les parents ne se distinguent entre eux que par un seul caractère (les monohybrides), ou par un petit nombre de caractères bien délimités, et pour lesquels on ne considère qu'une ou deux de ces unités en laissant les autres de côté.

» Ordinairement les hybrides sont décrits comme participant à la fois des caractères du père et de la mère. A mon avis, on doit admettre, pour comprendre ce fait, que les hybrides ont quelques-uns des caractères simples du père et d'autres caractères également simples de la mère. Mais quand le pere et la mère ne se distinguent que sur un seul point, l'hybride ne saurait tenir le milieu entre eux; car le caractère simple doit être considéré comme une unité non divisible.

» D'autre part l'étude des caractères simples des hybrides peut fournir la preuve la plus directe du principe énoncé. L'hybride montre toujours le caractère d'un des deux parents, et cela dans toute sa force; jamais le

HUGO DE VRIES, in Holland, discovered Mendel's paper after he had performed the same experiments and come to the same conclusion. His first publication was in March, 1900.

**19. C. Correns: G. Mendel's Regel über das Verhalten der Nachkommenschaft der Rassenbastarde.**

Eingegangen am 24. April 1900.

Die neueste Veröffentlichung HUGO DE VRIES': „Sur la loi de disjonction des hybrides"[1]), in deren Besitz ich gestern durch die Liebenswürdigkeit des Verfassers gelangt bin, veranlasst mich zu der folgenden Mittheilung.

Auch ich war bei meinen Bastardirungsversuchen mit Mais- und Erbsenrassen zu demselben Resultat gelangt, wie DE VRIES, der mit Rassen sehr verschiedener Pflanzen, darunter auch mit zwei Mais rassen, experimentirte. Als ich das gesetzmässige Verhalten und die Erklärung dafür — auf die ich gleich zurückkomme — gefunden hatte, ist es mir gegangen, wie es DE VRIES offenbar jetzt geht: ich habe das alles für etwas Neues gehalten[2]). Dann habe ich mich aber überzeugen müssen, dass der Abt GREGOR MENDEL in Brünn in den sechziger Jahren durch langjährige und sehr ausgedehnte Versuche mit Erbsen nicht nur zu demselben Resultat gekommen ist, wie DE VRIES und ich, sondern dass er auch genau dieselbe Erklärung gegeben hat, soweit das

CARL CORRENS, in Germany, saw de Vries' paper and wrote in April, 1900: "The same thing happened to me." He thought he had something new but then found Mendel's work.

**26. E. Tschermak: Ueber künstliche Kreuzung bei Pisum sativum[1]).**

Eingegangen am 2. Juni 1900.

Angeregt durch die Versuche DARWIN's über die Wirkungen der Kreuz- und Selbstbefruchtung im Pflanzenreiche, begann ich im Jahre 1898 an *Pisum sativum* Kreuzungsversuche anzustellen, weil mich besonders die Ausnahmefälle von dem allgemein ausgesprochenen Satze über den Nutzeffect der Kreuzung verschiedener Individuen und verschiedener Varietäten gegenüber der Selbstbefruchtung interessirten, eine Gruppe, in welche auch *Pisum sativum* gehört. Während bei den meisten Species, mit welchen DARWIN operirte (57 gegen 26 bezw. 12), die Sämlinge aus einer Kreuzung zwischen Individuen derselben Species beinahe immer die durch Selbstbefruchtung erzeugten Concurrenten an Höhe, Gewicht, Wuchs, häufig auch an Fruchtbarkeit übertrafen, verhielt sich bei der Erbse die Höhe der aus der Kreuzung stammenden Pflanzen zu jener der Erzeugnisse von Selbstbefruchtung wie 100:115. DARWIN erblickte den Grund dieses Verhaltens in der durch viele Generationen sich wiederholenden Selbstbefruchtung der Erbse in den nördlichen Ländern. In Anbetracht

ERICH TSCHERMAK, in Austria, was busy on "the second correction of my own paper" when he saw de Vries' and Correns' reports. His abstract was published in June, 1900.

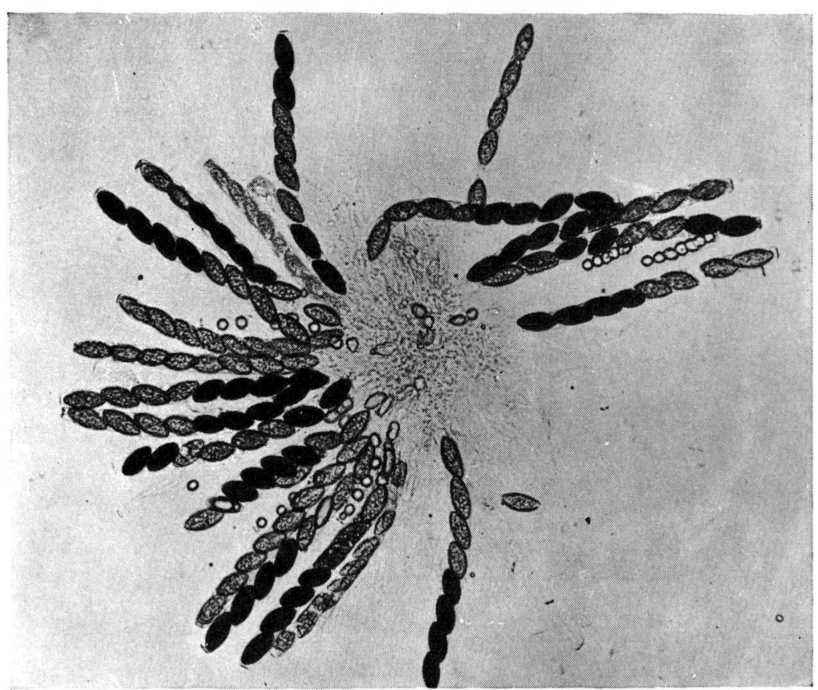

**NEUROSPORA SPORES** reflect crossing of dark- and light-colored strains. Spore groups have four light and four dark spores, just as pea plants bear peas in fixed ratios (*see opposite*).

progress. This is true of all species that reproduce sexually, from microbes to man.

## Self-Duplication

Let us turn to the genes themselves. How are genes reproduced in the cell? How do they act in controlling heredity? What are they made of?

The genes are, of course, self-reproducing. In this they are like the cell or an organism as a whole, such as a bacterium. Bacteria, as we know, arise only from pre-existing bacteria. We can prepare a broth that contains all of the raw materials needed for the production of bacteria, and we can provide all the necessary environmental conditions—acidity, temperature, oxygen supply and so on—but if we fail to inoculate the broth with at least one bacterial cell, then no bacteria will ever be produced in it. The situation is the same with respect to gene production, the only difference being that we cannot prepare an artificial broth for growing genes: the only medium in which they are known to multiply is in the living cell itself. Indeed, it appears that the genes are the only self-replicating elements in a cell; all the other components of cells apparently are produced, directly or indirectly, by the activities of genes.

Just what is involved in the process of self-duplication as we have defined it? One way to explore this question is to look into certain chemical reactions which seem to parallel reproduction by a living organism. An interesting case in point is pepsin, a gastric enzyme which is important for the digestion of proteins. As a catalyst, pepsin acts upon pepsinogen, a protein found in the wall of the stomach, and the product of its breakdown of pepsinogen is pepsin itself. Thus pepsin in the formal sense is self-duplicating: it acts upon the appropriate substance to produce a molecule exactly like itself. Moreover, its production of pepsin from pepsinogen over a period shows a curve of increase like the growth curve of a population of organisms. In other words, the equations for the production of pepsin and the production of cells are the same. The question now is: Does this formal similarity reflect a similarity in the mechanism of duplication?

A number of years ago Roger M. Herriott, Quentin R. Bartz and John H. Northrop at the Rockefeller Institute for Medical Research carried out the following interesting experiment. They added pepsin obtained from chickens to pepsinogen prepared from pigs, and *vice versa*. Their purpose was to learn whether the pepsin produced would depend on the species of pepsin or on the species of pepsinogen. If pepsin behaved like a living organism, the pepsin formed should be the same as the pepsin added, regardless of the source of the pepsinogen, just as the species of bacteria obtained from a culture depends on the species inoculated, not on the nutrients supplied. But the results were just the opposite. Swine pepsin reacting with chicken pepsinogen produced only chicken pepsin, and the mixture of chicken pepsin with swine pepsinogen yielded swine pepsin.

It follows that pepsin is not strictly self-duplicating. The product is determined not by the pepsin but by the material on which it acts. This is not the case with living organisms, as we have seen, and neither is it the case with genes. The reproduction of genes is a *copying* process: they copy themselves when they multiply, and if a gene happens to mutate to a new form, the new type reproduces itself in its mutant version. No such copying process has been found in simple chemical reactions.

## Mutation

The mutation of genes has been investigated very extensively by experimental work with X-rays and other radiations. This exploration began in 1927, when Muller, working with Drosophila, and L. J. Stadler, working independently with barley, discovered that treatment of cells with X-rays speeded up the rate of mutation. The alteration of the genes undoubtedly is due to a chemical change, which is caused by the ionization of atoms (*i.e.*, removal of electrons) by the radiation. The main conclusion drawn from the many experiments with radiation is that a single ionization, in the right place, suffices to cause a gene mutation. This conclusion is particularly interesting because it suggests that the gene is a single molecule. Other ideas have strong champions—among them the theory that the only real unit is the chromosome, a kind of supermolecule—but there is little doubt that at the present time the gene-molecule theory provides the most satisfactory general account of the properties and behavior of genes.

From a practical point of view this

**MENDELIAN LAWS are illustrated on the opposite page. At top, a plant bearing smooth yellow peas is mated with one bearing wrinkled green peas. In the first generation (*middle*) the plants bear only smooth yellow peas because smooth and yellow are dominant. In the next generation (*bottom*) the plants bear smooth yellow, wrinkled yellow, smooth green and wrinkled green in the approximate ratios 9:3:3:1.**

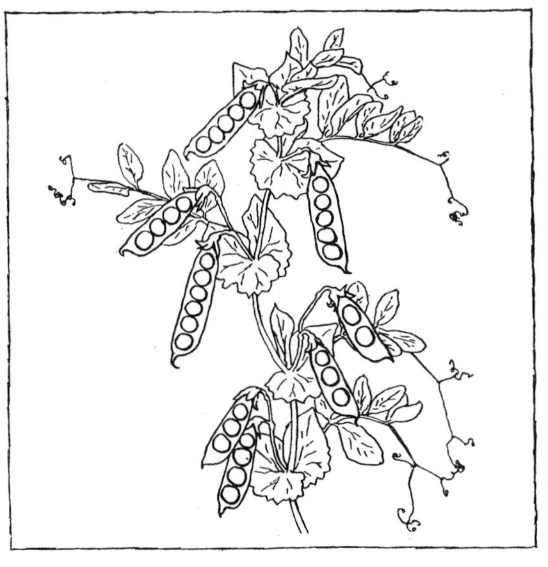

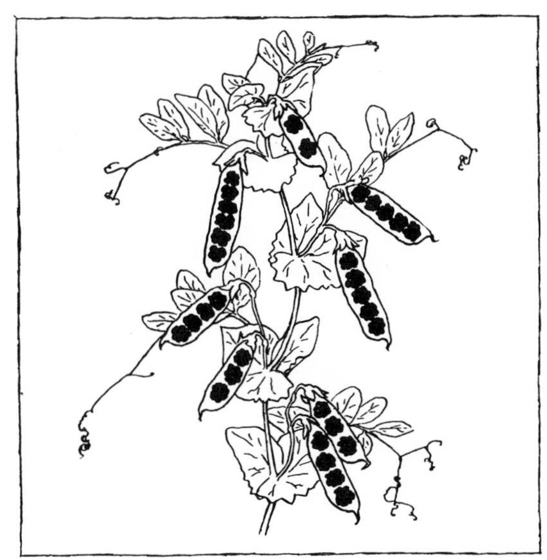

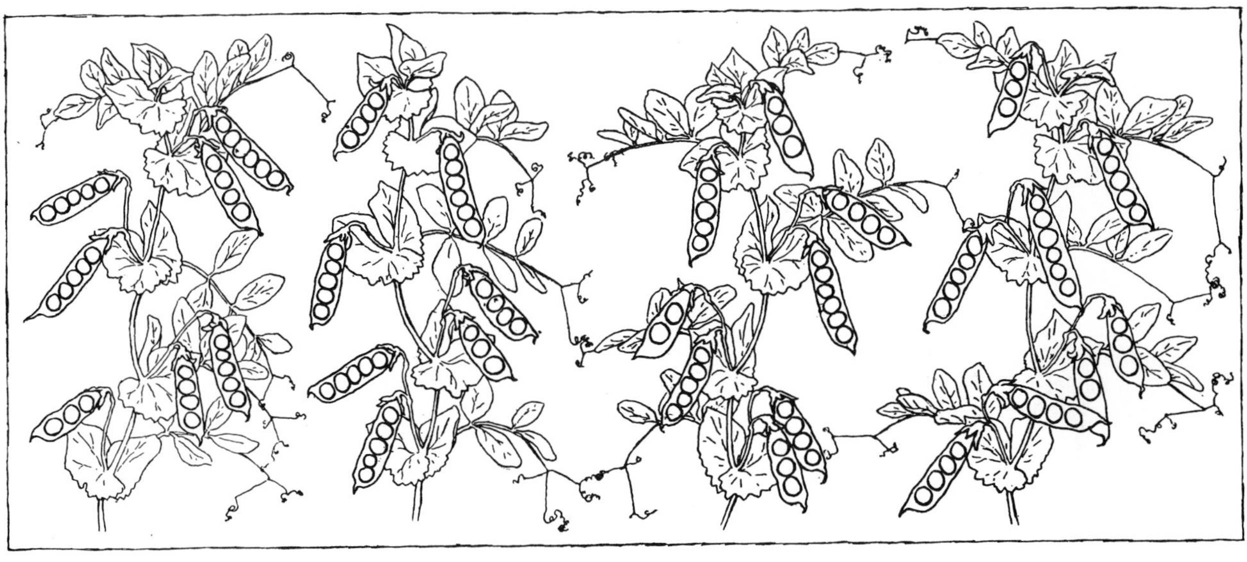

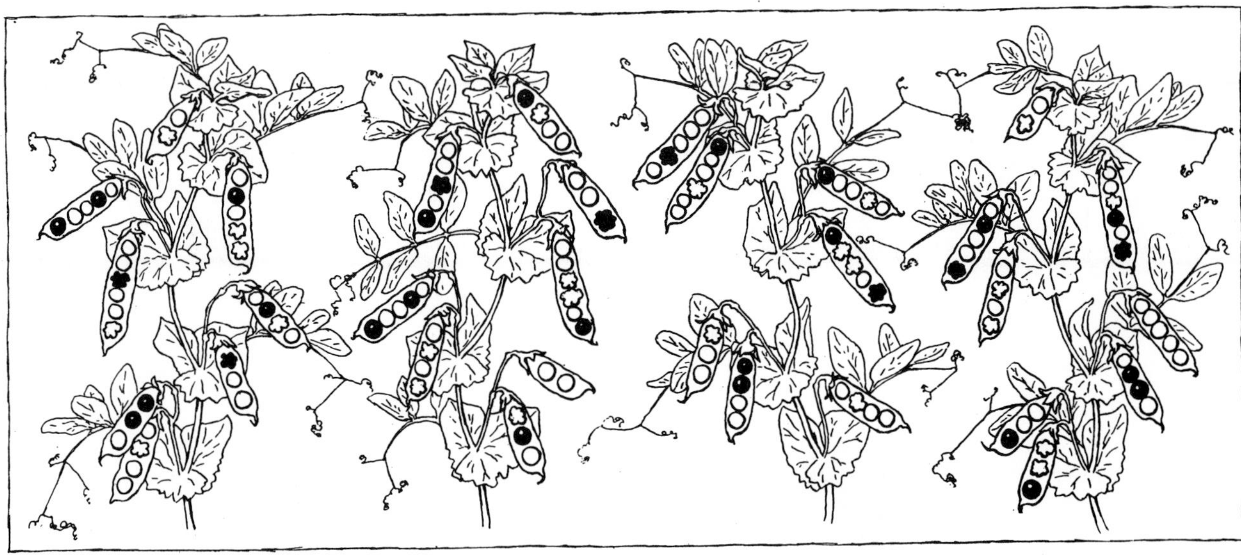

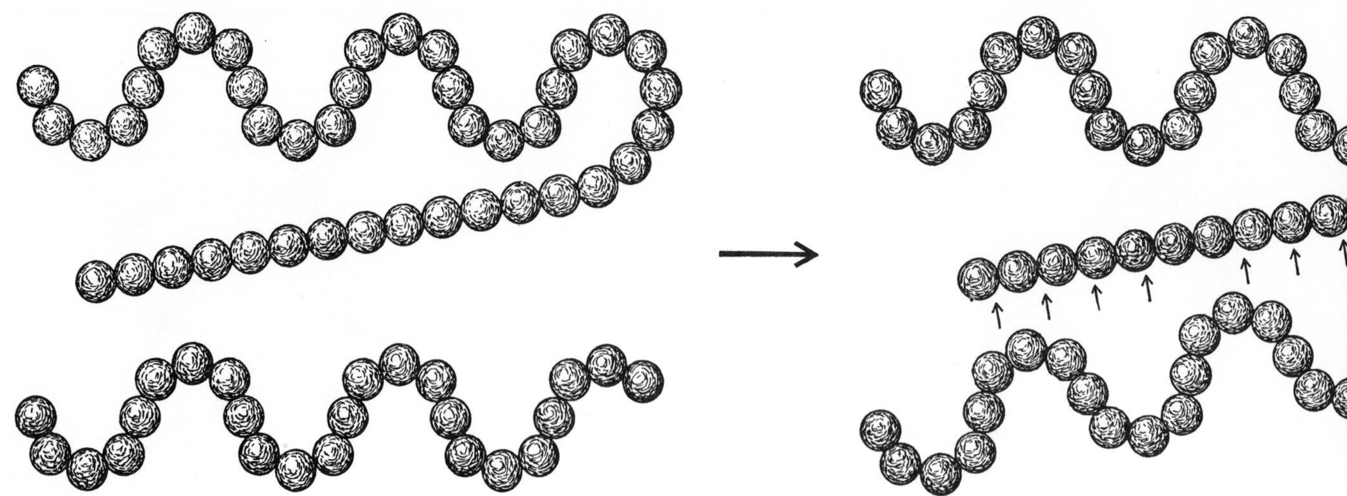

**SELF-DUPLICATION OF PEPSIN,** an example of autocatalytic reaction in the biosynthesis of proteins, is shown in three steps in this highly schematic diagram. In the first stage (*at left*) a molecule of pepsinogen, the precursor of pepsin, appears at top and a molecule of pepsin below. The beads represent the amino acid units out of which pepsins are made. As indicated, pepsinogen is

interpretation has an important bearing on the possible harmful effects of atomic and other man-made radiations. If the gene is a single molecule, mutable by a single quantum of radiation, then even the smallest exposure to radiation may produce a mutation: in other words, there is no "safe" dose. Experiments bear out this view. Over a wide range of X-ray dosages the frequency of mutation in Drosophila is directly proportional to the number of ionizations, with no indication of a threshold below which mutations are not induced. It is possible that a tolerance level might be found at doses lower than have yet been tested, but such a threshold would be very difficult to detect, because as the dose decreases, we approach the natural, spontaneous rate of mutation, which acts as a background "noise" to obscure the small additional effect of the radiation. Clearly it would be rash to base one's hopes or the national policy on the chance that a threshold exists. The only reasonable course is to assume that no amount of ionizing radiation, however small, is without an effect on the genes. Knowing that gene mutations are irreparable and for the most part harmful, we must weigh this hazard as best we can against the expected benefits of X-rays and other uses of ionizing radiation.

## Genes and Enzymes

The sensitivity of genes to radiation brings us to our second question: How do genes act on the cell? A gene mutation can sterilize the cell or permanently alter all of its descendants. Considering that this profound effect is triggered by an almost infinitesimal change in the gene—a single ionization—we must conclude that the genes function in a far-reaching way. That is to say, they act not merely as enzymes (which themselves have profound effects in the cell, determining the rate and direction of its chemical activities) but as catalysts for the production of enzymes.

This idea occurred to the early workers in genetics, but techniques for exploring it have not been available until fairly recently. By now it has won strong support, as the result of the pioneer experiments of George W. Beadle and E.

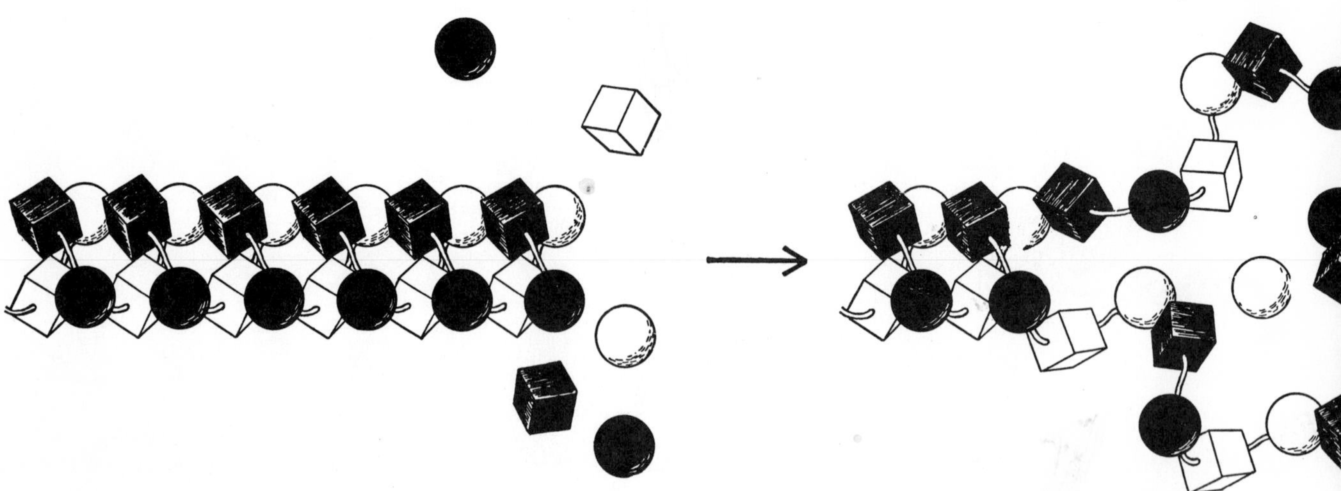

**SELF-DUPLICATION OF NUCLEIC ACID,** which is thought to be the ultimate genetic material, is shown in three stages in this diagram. In the first stage (*at left*) the nucleic acid is seen as a structure of two helices coiled about one another, with four different nucleotides (represented by cubes and spheres) arranged in complementary order. In the second stage (*center*) the structure

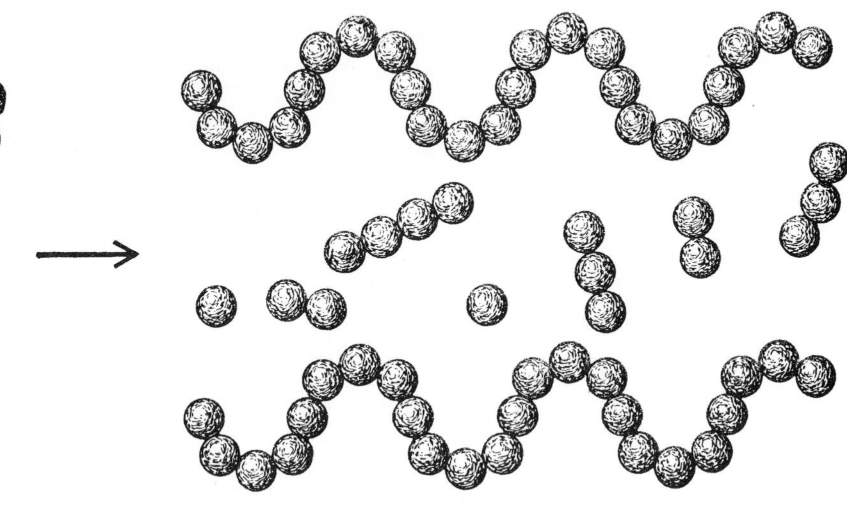

**a more complex molecule, incorporating the structure of pepsin itself plus a "tail." In the second stage (*center*) the two molecules enter into reaction with one another. In the third stage the tail has been separated from pepsinogen and there appear two pepsin molecules.**

L. Tatum on Neurospora and of many other studies of microorganisms, notably the colon bacillus, *Escherichia coli.* Hundreds of mutations have been produced in these organisms and in each case the effect of the mutation is to abolish the organism's ability to make some essential chemical, for example, a vitamin or an amino acid. The mutation usually blocks just one step in the series of reactions required to make the vitamin or amino acid. It evidently interferes with the production of a single specific enzyme: all the other enzymes involved in catalyzing the series of reactions are apparently unaffected.

In order to account for this selectivity, it is necessary to assume that the structure of the enzyme is related in some way to the structure of the gene. By a logical extension of this idea we arrive at the concept that the gene is a representation—a blueprint, so to speak—of the enzyme molecule, and that the function of the gene is to serve as a source of information regarding the structure of the enzyme. It seems evident that the synthesis of an enzyme—a giant protein molecule consisting of hundreds of amino acid units arranged end-to-end in a specific and unique order—requires a model or set of instructions of some kind. These instructions must be characteristic of the species; they must be automatically transmitted from generation to generation, and they must be constant yet capable of evolutionary change. The only known entity that could perform such a function is the gene. There are many reasons for believing that it transmits information by acting as a model, or template.

If the template theory is correct, a mutant gene may produce a mutant enzyme—an enzyme whose structure and properties are changed in some way. A systematic search for mutations of this sort has been started recently, and already several interesting examples have been found.

In our laboratory at the California Institute of Technology we have been studying an enzyme of Neurospora which converts the amino acid tyrosine into melanin—a black pigment widely distributed in nature (it is the black pigment of hair, skin and of the ink of the squid). We find that this enzyme, tyrosinase, may occur in either of two different forms in Neurospora. The forms differ only in their stability toward heat. At a temperature of 138 degrees Fahrenheit, for example, one form is reduced to half of its original activity in three minutes; the other in 70 minutes. Our experiments show that this difference in stability is inherited in a simple Mendelian way—*i.e.*, it is controlled by a single gene. One form of the gene causes the organism to produce tyrosinase which is comparatively stable to heat; the other yields unstable tyrosinase. It is interesting that the forms of the enzyme produced by the two strains of Neurospora are exactly alike in every detail, as far as we have tested them, except in stability to heat. This fact indicates that the genetic control of enzyme structure is exceedingly fine-grained, permitting the separate alteration, as in this case, of a single feature of that structure.

Another example of a gene which influences the structure of a protein is one affecting the hemoglobin of human blood. There is a mutant which is known as the sickle-cell gene, because it leads to production of a form of hemoglobin that causes the red blood cells to take a sickle shape. Linus Pauling and a group of his co-workers at the California Institute of Technology found that the sickle-cell hemoglobin molecule has a different electric charge from normal hemoglobin. A very interesting feature of the sickle-cell mutation, from the evolutionary point of view, is the fact that it apparently confers resistance to malaria [see "Sickle Cells and Evolution," by Anthony C. Allison; SCIENTIFIC AMERICAN Offprint 1065].

The discovery of structural mutations of proteins is gratifying but is only one step toward a proof of the template theory: to prove conclusively that genes do

**uncoils, freeing the components of each helix for attachment to free nucleotide units diffused in the nearby environment. In the third stage (*right*) each helix has bound nucleotide units to itself, thus beginning the formation of two complete new nucleic acid molecules.**

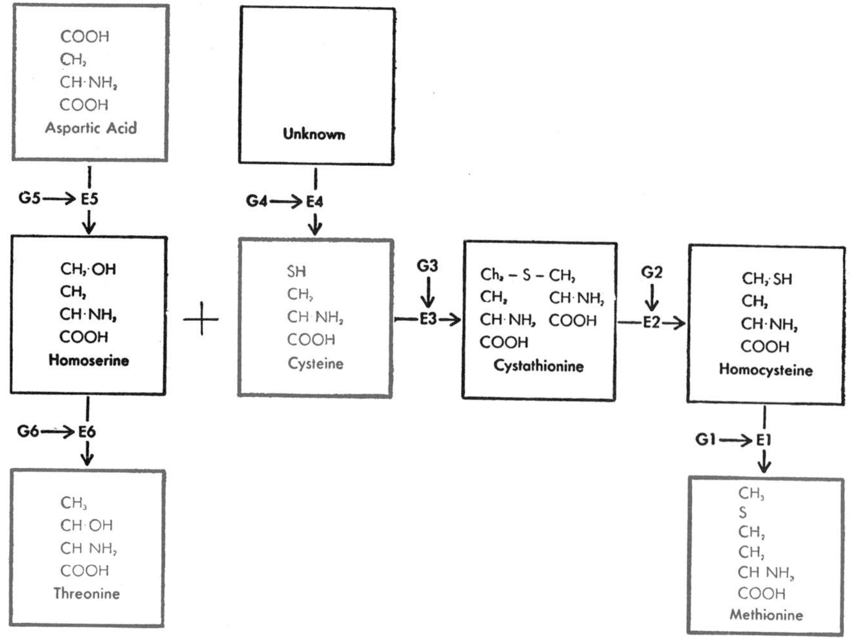


**SYNTHESIS OF AMINO ACIDS** may be mediated by many genes; absence of any one gene will stop the cycle. Each gene (G) catalyzes a specific enzyme (E) which in turn catalyzes one step in the reaction. Compounds bóxed in red lines are stable end-product amino acids.

in fact act as templates, it would have to be demonstrated that every specific property of a protein can be modified by a gene mutation. Experiments along these lines are being pursued actively in several laboratories.

## DNA and RNA

Final answers to our first two questions—how genes reproduce themselves and how they function—will not be obtained until we have found the answer to the third: What are they made of? It may seem strange, considering the great power of modern methods of chemical analysis, that the chemical composition of the genetic material in the chromosomes should still be something of a mystery. The explanation is quite simple: no one has ever isolated a gene from any animal or higher plant—at least so far as anyone is aware, for we have no way so far of recognizing a gene after it has been removed from the cell.

Nevertheless, we do have some definite ideas about the chemical nature of genes. There is very good evidence that the genetic material of some bacteria and viruses consists of nucleic acid, and there is some reason to believe that this is also true in higher organisms. It has been known for a long time that desoxyribonucleic acid (DNA) is a prominent constituent of the chromosomes; this fact marked it for special attention as possible genic material. A number of studies of bacteria and viruses confirm that it does indeed play a genetic role.

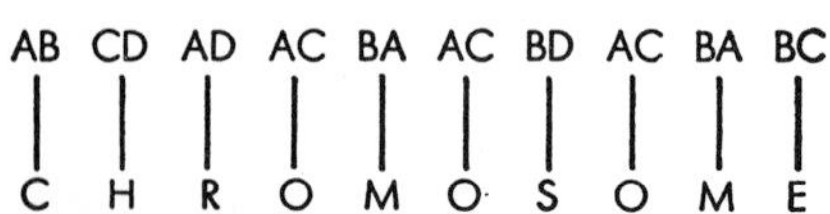


**FOUR-LETTER CODE** suggests one way in which the four nucleotide components of nucleic acid, here designated as A, B, C, D, may control the synthesis of the 20 different amino acids of which proteins are made.

Some years ago a substance with gene-like properties was extracted from heat-killed cells of *Pneumococcus*, the pneumonia organism. Strains of Pneumococcus grown in the presence of this substance acquired hereditary characteristics of the particular strain from which it was derived; the characteristics included virulence, resistance to drugs, the ability to synthesize certain enzymes and so on. The transformations were permanent: they were passed on from generation to generation of the bacteria. Moreover, the substance appeared to be subject to mutation. Eventually Oswald T. Avery, Colin M. MacLeod and Maclyn McCarty of the Rockefeller Institute identified the transforming agent as DNA. More recently a new series of transforming agents, also varieties of DNA, has been found in another species of bacteria, *Hemophilus influenzae*.

In the realm of the viruses, there have been two definite identifications of nucleic acid as genetic material. A. D. Hershey and Martha Chase at the Cold Spring Harbor Biological Laboratory have found that DNA plays a genetic role in a bacterial virus which attacks the bacterium *E. coli*. Heinz Fraenkel-Conrat at the University of California identified the genetic substance of the tobacco mosaic virus as ribonucleic acid (RNA).

All the available evidence thus points toward nucleic acid as the ultimate genetic material. Naturally its chemical structure has come in for a great deal of attention. F. H. C. Crick and J. D. Watson at the University of Cambridge have proposed a structure for DNA which not only accounts for many of its known physical and chemical properties but also seems capable of accounting for the properties of a gene [see "The Structure of the Hereditary Material," by F. H. C. Crick; SCIENTIFIC AMERICAN Offprint No. 5]. According to their scheme DNA is composed of two close-fitting, complementary chains, each chain consisting of a long series of nucleotides in linear order. There are only four kinds of nucleotides in DNA, but since the number of nucleotide molecules per chain is of the order of 10,000, the number of possible arrangements is very large indeed. Replication of the molecule is thought to involve separation of the two complementary chains, each of which then acts as a template for the synthesis of a new partner.

The idea of the two-stranded structure seems to have a firm basis. However, it is not easy to see how this model can account for template action by genes. The specificity of the Watson-Crick structure rests on the sequence of nucleotides in each chain, suggesting that the genetic information is coded on a linear tape in an alphabet of four symbols. Mutation would consist of the rearrangement, deletion or substitution of parts of this coded message. The difficulty is that whereas the nucleic acid alphabet contains only four symbols (corresponding to the four different nucleotides), the protein alphabet contains 20 or more (corresponding to the 20-odd kinds of amino acids). There is no known mechanism at the present time for translating instructions from the nucleotide code into the amino acid code. But this difficulty may not be insuperable [see "Information Transfer in the Living Cell," by George Gamow; SCIENTIFIC AMERICAN, October, 1955].

Thus for the first time we have a definite working hypothesis as to the structure of the gene. There is, however, a puzzling feature about the present situation. The experiments on the tobacco mosaic virus clearly show that RNA is capable of performing a genetic function. But RNA does not usually act in this way, as far as can be determined. It is found chiefly in the cytoplasm of cells (*i.e.*, outside the nucleus), and genetic experimentation with animals has failed to show any regular mechanism of inheritance in the cytoplasm. Hereditary mechanisms do exist in the cytoplasm of plant cells (for example, in connection with the production of chlorophyll) but they are of minor significance compared to the chromosomal mechanism.

Possibly the RNA that controls heredity in the tobacco mosaic virus (and other plant viruses) is of a different kind from that found in the cytoplasm of animal cells. Such a difference could explain why RNA acts like a gene in one situation and apparently not in the other. But this possibility cannot be tested at the present time, because the chemistry of RNA is still relatively unknown.

## The Origin of Life

A general article on the gene ought to make at least some mention of what bearing all this may have on the problem of the origin of life. Probably no question in biology has a wider appeal than this one—especially among nonbiologists. Historically the basic difficulty has been to define "life." Up to the 17th century the most primitive forms of life known were worms, fleas, scorpions and the like, and there was a notion that these creatures originated spontaneously from decaying organic matter. This idea was demolished in 1668 by the Italian physician Francesco Redi, when he showed that no maggots developed in meat shielded from egg-laying insects. But it was reborn at another level when, a few years later, Anton van Leeuwenhoek discovered bacteria. They seemed so small and rudimentary that many people were convinced they must be on the dividing line between living and nonliving matter. Actually bacteria are just as complex as any cell of our own bodies, and their spontaneous origin from nonliving material is not much more likely than the spontaneous generation of scorpions.

Nowadays many biologists and biochemists tend to regard the question of how life started as essentially meaningless. They view living and nonliving matter as forming a continuum, and the drawing of a line between them as arbitrary. Life, on this view, is associated with the complex chemical paraphernalia of the cell—enzymes, membranes, metabolic cycles, etc.—and no one can say at what point it begins. Geneticists are apt to take a different view. If genes are required to produce enzymes, then life began only when they began.

We can imagine the spontaneous origin of some chemical substance capable of reproducing itself, of mutating and of directing the production of specific catalysts in its environment. It would not be long before this substance, trying out new molecular arrangements by blind mutation, began to evolve along lines favored by natural selection. In time all the complexity that is now associated with living matter might well develop.

It may be objected that an unstated assumption is hidden in this theory: namely, that the gene arose in an environment which was already prepared to supply all the materials needed for its multiplication and other chemical activities. But if this is an objection, it holds for any theory which supposes that life began in some accidental combination of chemicals. The material of life as we know it could have come into being only in a complex chemical environment.

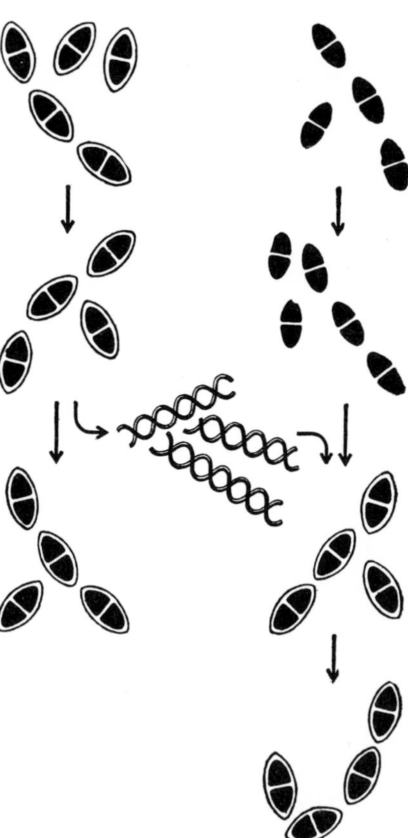

**PNEUMOCOCCUS is genetically transformed by mixing cells of one strain with nucleic acid from another. The first strain (*right*) thereupon assumes the characteristics which mark the second strain (*left*).**

# 2

# *The Duplication of Chromosomes*

J. HERBERT TAYLOR
*June 1958*

In the search for the secret of how living things reproduce themselves, geneticists have recently focused on the substance called DNA (deoxyribonucleic acid). DNA seems pretty clearly to be the basic hereditary material—the molecule that carries the blueprints for reproduction. F. H. C. Crick and J. D. Watson have found that the molecule consists of two complementary strands, and they have developed the theory that it duplicates itself by a template process, each strand acting as a mold to form a new partner [see "Nucleic Acids," by F. H. C. Crick; SCIENTIFIC AMERICAN Offprint 54].

The problem now is: How does this model fit into the larger picture of chromosomes? For half a century we have known that chromosomes, the rodlike bodies found in the nucleus of every cell, are the carriers of heredity. They contain the genes that pass on the hereditary traits to offspring. As each cell divides, the chromosomes duplicate themselves, so that every daughter cell has copies of the originals. How is the reproduction of chromosomes related to the reproduction of DNA? The question is being pursued by two approaches: from DNA up toward chromosomes and from chromosomes down toward the molecular level. This article will report some experimental studies of the behavior of chromosomes which have suggested a general model of the mechanism of reproduction.

Chromosomes take their name from the fact that they readily absorb dyes and stand out in strong color when cells are stained [*see photographs on the opposite page*]. They become visible under the microscope shortly before a cell is ready to divide. At that time each chromosome consists of a pair of rods side by side. When the cell divides, the two members of the pair (called chromatids) separate, and one chromatid goes to each daughter cell.

In the new cell the chromatid disappears. Then as this cell approaches division, each chromatid reappears, now twinned with a new partner. It has made a copy of itself for the destined daughter cell. There are two possible ways it may have done this: (1) by staying intact (even though invisible in the microscope) and acting as a template, or (2) by breaking down and generating small

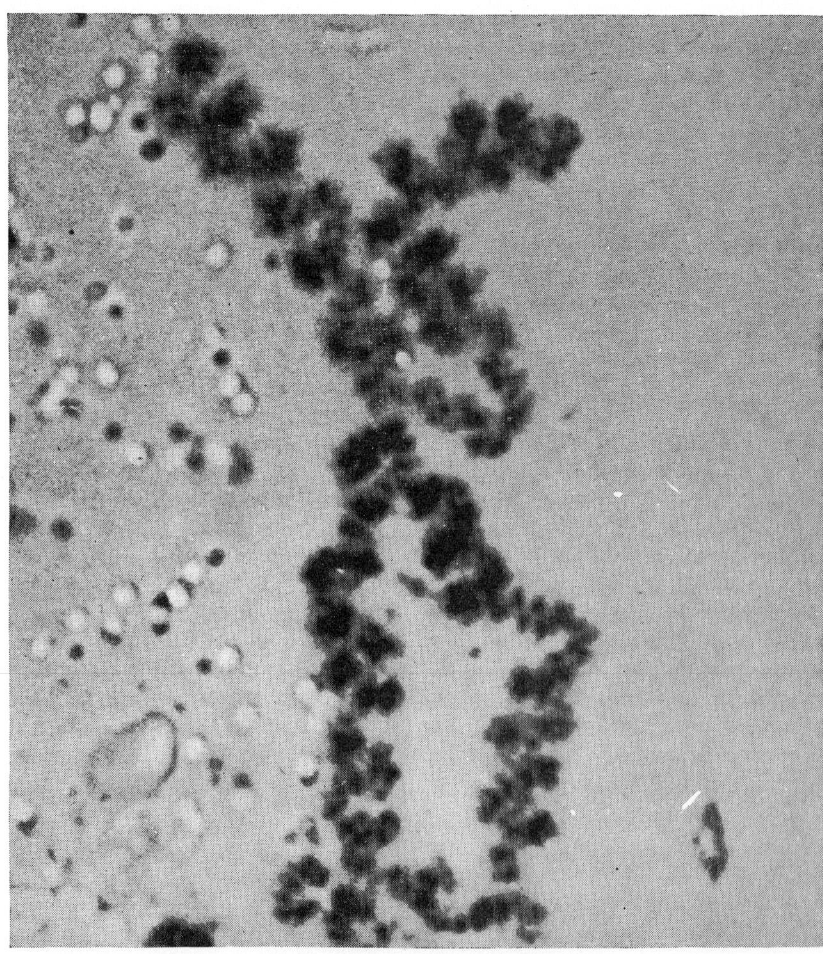

**DOUBLE COILED STRUCTURE of a chromosome can be seen in this photomicrograph, the magnification of which is about 4,000 diameters. Chromosome was partly unwound by treatment with dilute potassium cyanide. The chromosome is from a cell of the Easter lily.**

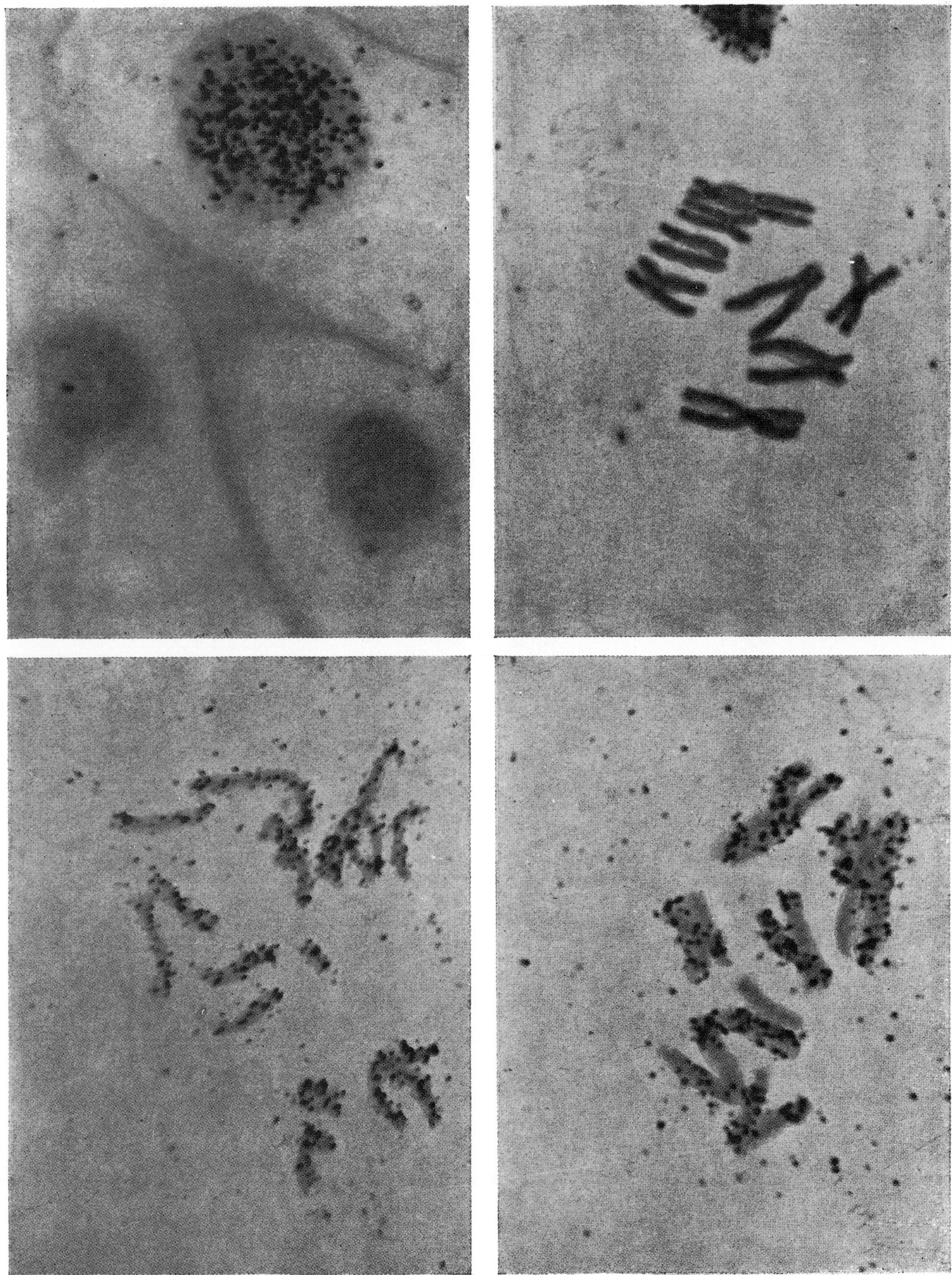

**CHROMOSOMES** of Bellevalia, a plant of the lily family, are tagged with radioactive thymidine in experiment on duplication. Radiation from the thymidine strikes photographic film placed over the cells, producing black specks. The upper nucleus in the photomicrograph at top left has taken up the tracer material but its chromosomes have not yet become visible. Chromosomes at top right completed their duplication before the cells were placed in radioactive solution and are not tagged. Those at bottom left duplicated once in radioactive solution. Both members of each pair are labeled. The chromosomes at bottom right duplicated once in radioactive solution and once after cells were removed. Only one member of each pair is tagged, except where segments crossed.

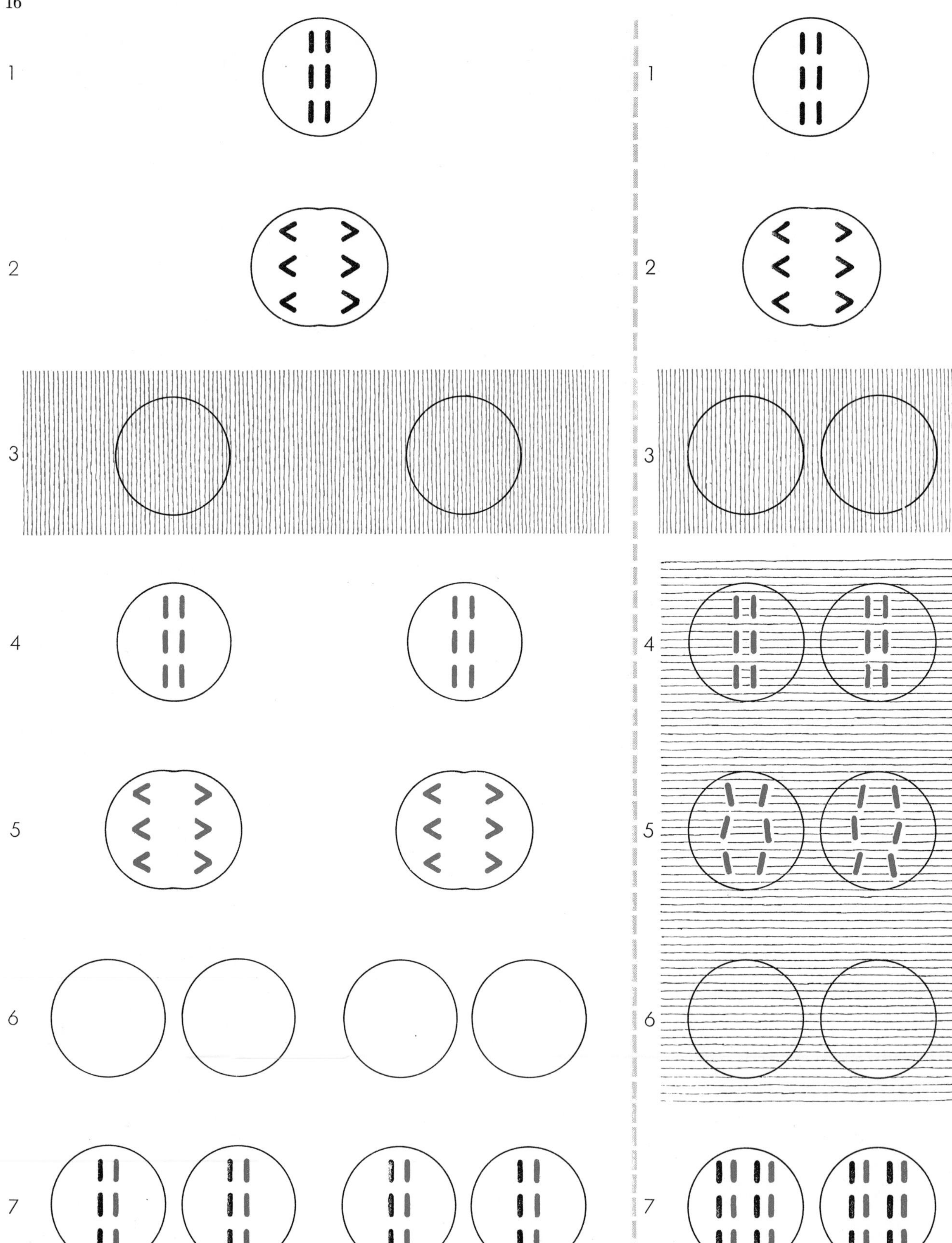


**DUPLICATION CYCLES** in tracer experiment are diagrammed schematically. Cells to left of the vertical broken line are allowed to divide normally. Those to the right are prevented from dividing when they are placed in colchicine (*black shading*), but their chromosomes continue to duplicate. Black rods represent unlabeled chromatids; colored rods, labeled chromatids. Colored shading indicates radioactive thymidine solution. The empty circles represent stage when chromatids are invisible and duplicating themselves.

units which then reassemble themselves in the form of the original chromatid.

It has recently become possible to resolve this question by means of radioactive tracers. When cells grow in a medium containing thymidine, a component of DNA, all of the thymidine is taken up by the chromosomes; none of it is built into any other part of the cell. Thus if we label thymidine with radioactive atoms (the radioisotopes of hydrogen or carbon), we can follow the transmission of the material through successive replications of the chromosomes. For our own experiments, which I conducted in collaboration with Walter L. Hughes and Philip S. Woods of the Brookhaven National Laboratory, we chose radiohydrogen (tritium) as the tracer. This substance makes it possible to distinguish a radioactive chromatid from a nonradioactive one lying next to it. To localize the radioactivity we use the technique of autoradiography. The cells are squashed flat on a glass slide and covered with a thin sheet of photographic film. Radioactive emanations from the cells produce darkened spots on the film. The emissions from radioactive carbon are fairly penetrating and therefore darken a comparatively wide area of the film; we selected tritium instead because its emissions travel only a short distance—so short that we can narrow down the source to a single chromosome or part of a chromosome.

Our first experiment followed the fate of the thymidine through one duplication of labeled chromosomes. In order to control the situation so that we could identify newly formed chromosomes we treated the cells with colchicine—a drug which prevents cells from dividing but allows chromosomes to go on duplicating themselves. This enabled us to sequester the new chromosomes within the original cells and to tell how many generations had been produced. The cells we studied were those in the growing roots of plants, cultured in a solution containing tritium-labeled thymidine.

We found, to begin with, that in cells that had taken up this thymidine (*i.e.*, produced a new generation of chromosomes preparatory to division), all the chromosomes were labeled, and radioactivity was distributed equally between the two chromatids of each chromosome. This might suggest that the new chromosomes had been formed from a mixture of materials generated by a breakdown of the original chromatids. But when we

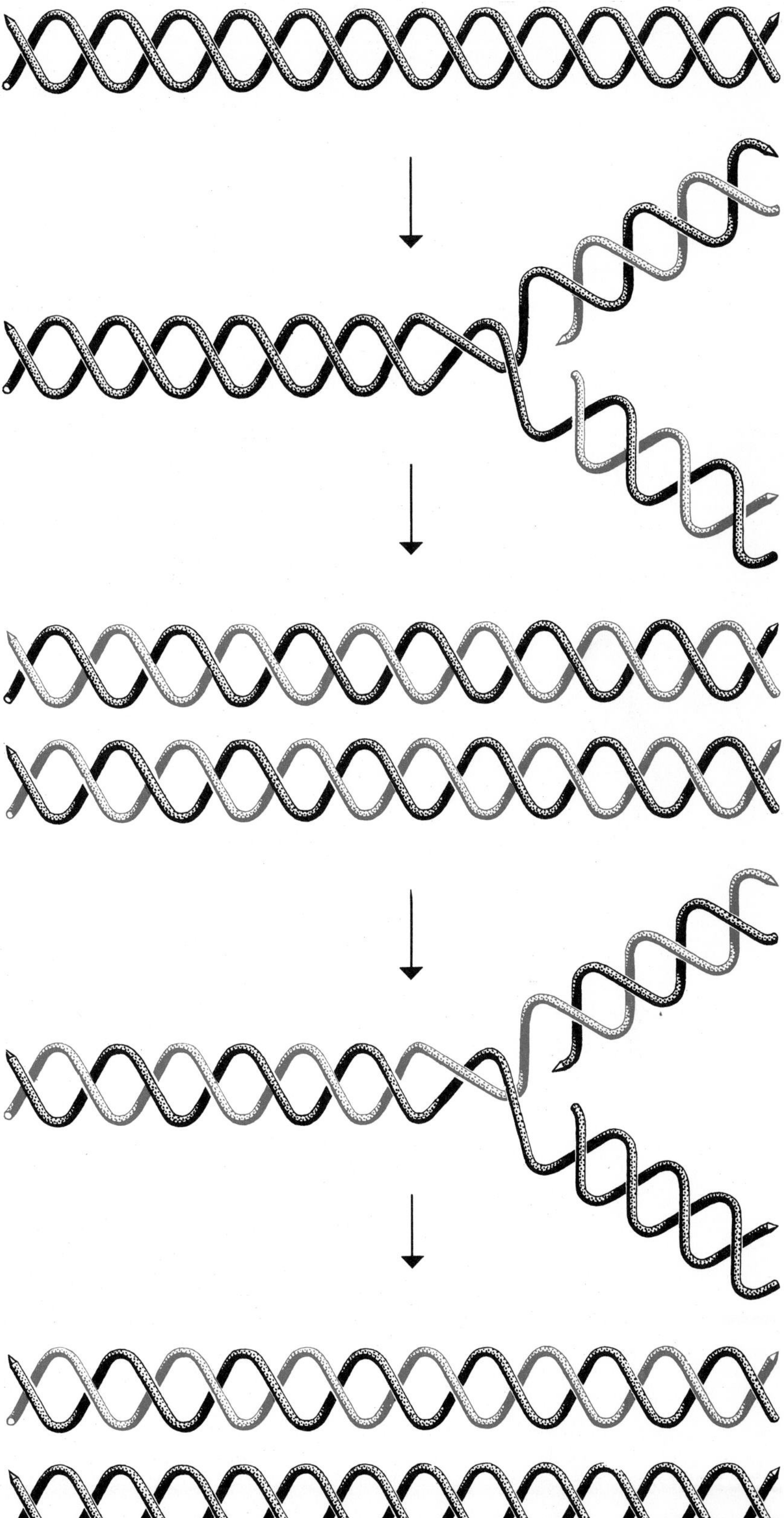

**DNA MOLECULES** consist of two complementary chains wound around each other in a double helix. When they duplicate, they unwind and each chain builds itself a new partner. Shown here are two cycles of duplication. The first cycle takes place in radioactive solution, producing two labeled chains (*colored helixes*). When a labeled molecule duplicates itself again in nonradioactive solution, only one of its descendants contains a labeled chain.

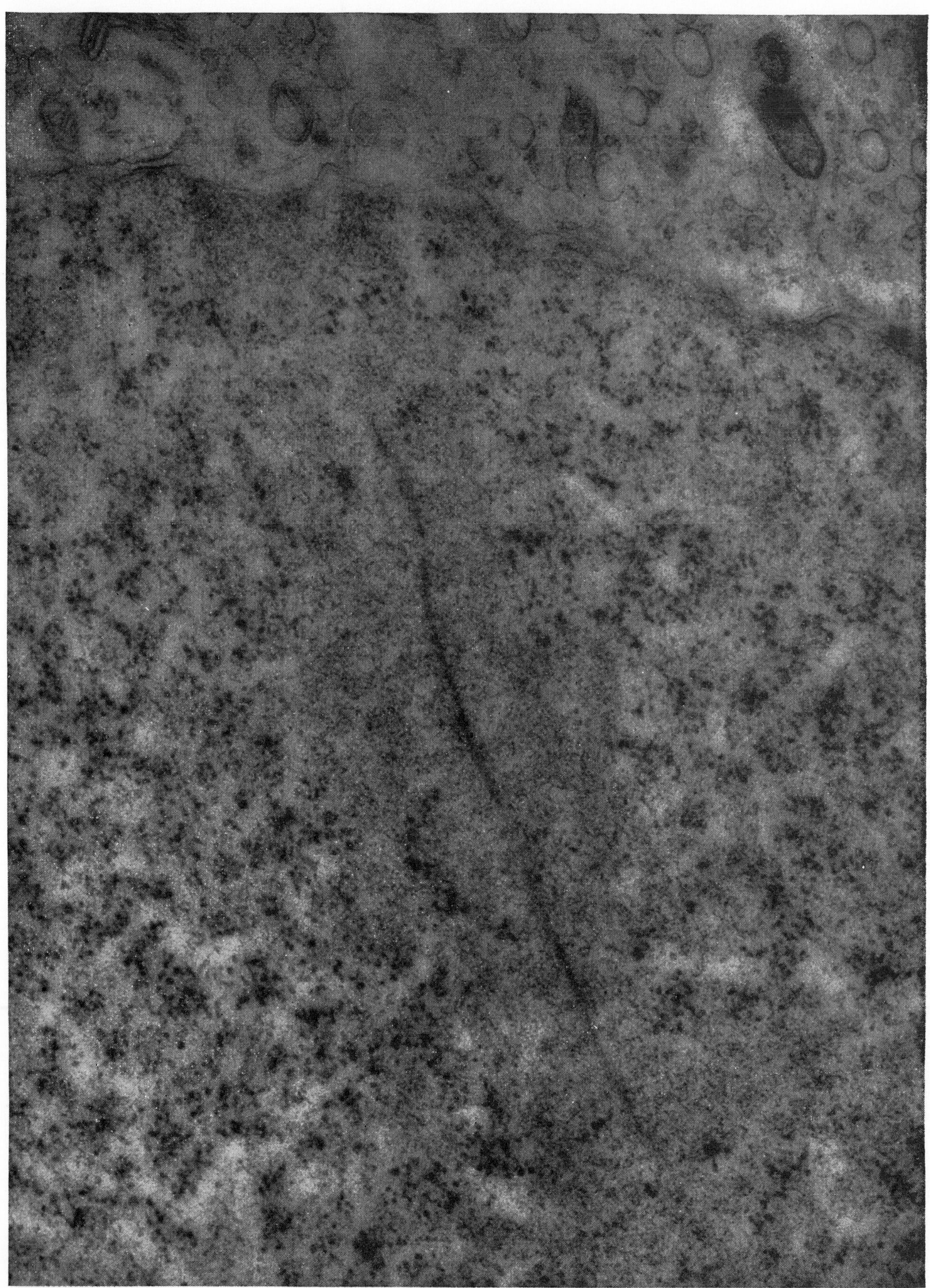

CHROMATID appears as a linear structure running diagonally down the middle of this electron micrograph, made by Montrose J. Moses of the Rockefeller Institute for Medical Research. The magnification is some 50,000 diameters. The line may represent the central column in the author's models and the fuzzy material surrounding the line may be DNA strands running perpendicularly outward.

followed root cells through a second generation of chromosome reproduction, where the second generation was synthesized in a medium containing nonradioactive thymidine, we found to our delight that in each new doubled chromosome one chromatid was labeled and the other was not!

What might this mean? The simplest and most likely answer was that a chromatid itself consists of two parts, each of which remains intact and acts as a template. In the radioactive medium each of the original chromatids, after splitting in two, builds itself a radioactive partner. Therefore all the new chromosomes are labeled. Now when the labeled chromatids split again to produce a second generation, half of the strands are labeled and half are not. In a nonradioactive medium all of them will build unlabeled partners. As a result, half of the newly formed chromatids will be partly labeled, half will have no label at all [*see diagrams on page 16*].

Our picture of the chromatid as a two-part structure fits very well with what we know about the DNA molecule and with the Crick-Watson theory. DNA too is a double structure, consisting of two complementary helical chains wound around each other. And some of our recent experiments indicate that the two strands of a chromatid are complementary structures. It is tempting, therefore, to suppose that a chromatid is simply a chain of DNA. But when we consider the question of scale, we realize that the matter cannot be so simple. If all the DNA in a chromatid formed a single linear chain, the chain would be more than a yard long, and its two strands would be twisted around each other more than 300 million times! It seems unthinkable that so long a chain could untwist itself completely, as the chromatid must each time it generates a new chromosome. Furthermore, the chromatid has the wrong dimensions to be a single DNA chain. When fully extended, it is about 100 times thicker and only one 10,000th as long as the linear DNA chain would be.

Under a high-power microscope we can see that the chromatid is a strand of material tightly wound in a helical coil—in fact, so strongly wound that the coil itself often winds up helically, like a coiled telephone cord which is twisted into a series of secondary kinks [*see photograph on page 14*]. But beyond this the optical microscope cannot resolve details of the chromatid's structure. Assuming that it is made up of pieces of DNA as its

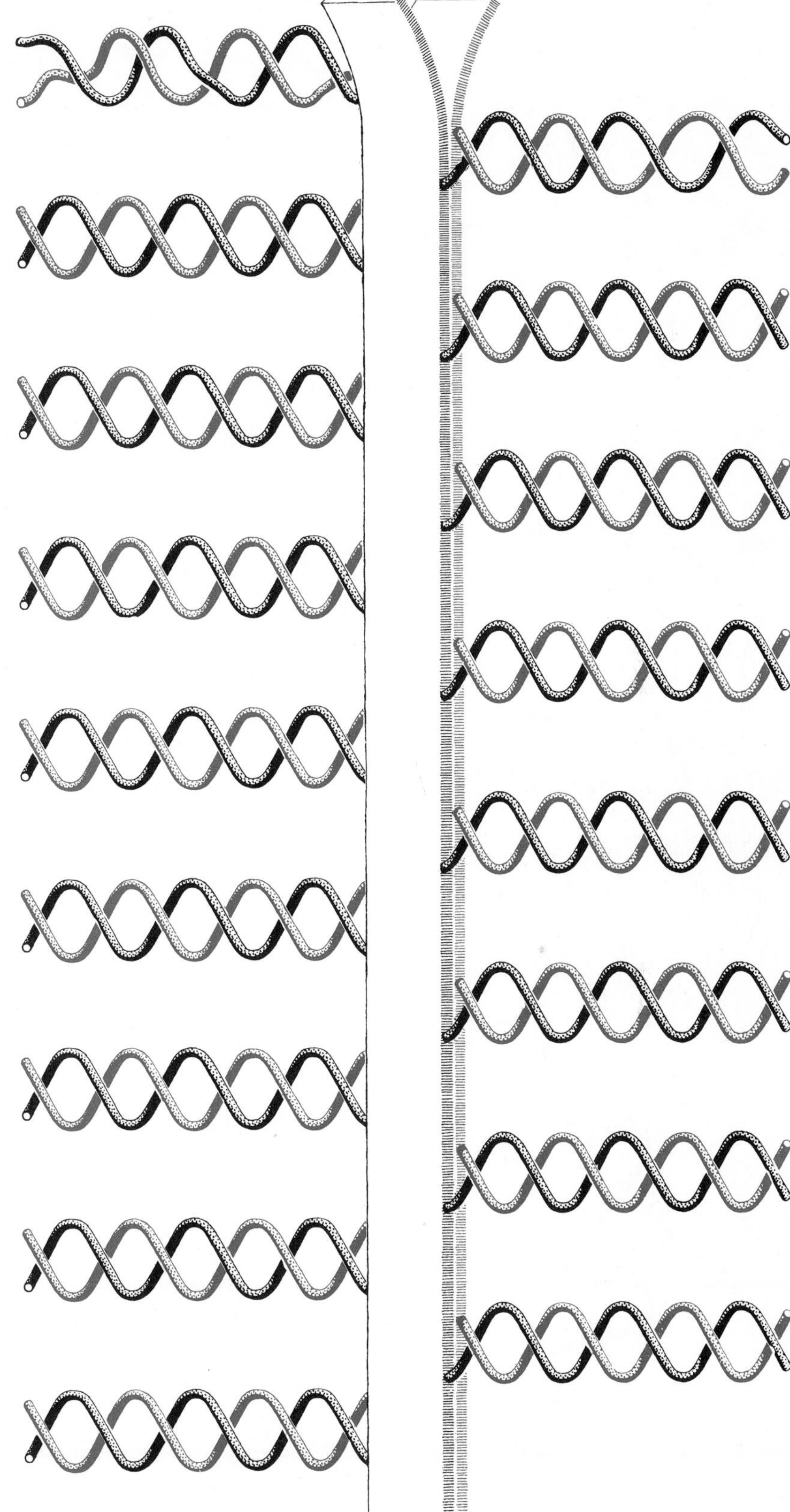

**RIBBON MODEL of a chromatid consists of a two-layered central column to which DNA molecules are attached. One chain of each molecule is anchored to the front layer and the other to the back layer. When the chromatid duplicates, the central ribbon peels apart, unwinding the DNA molecules. Each half of the structure then builds itself a new partner.**

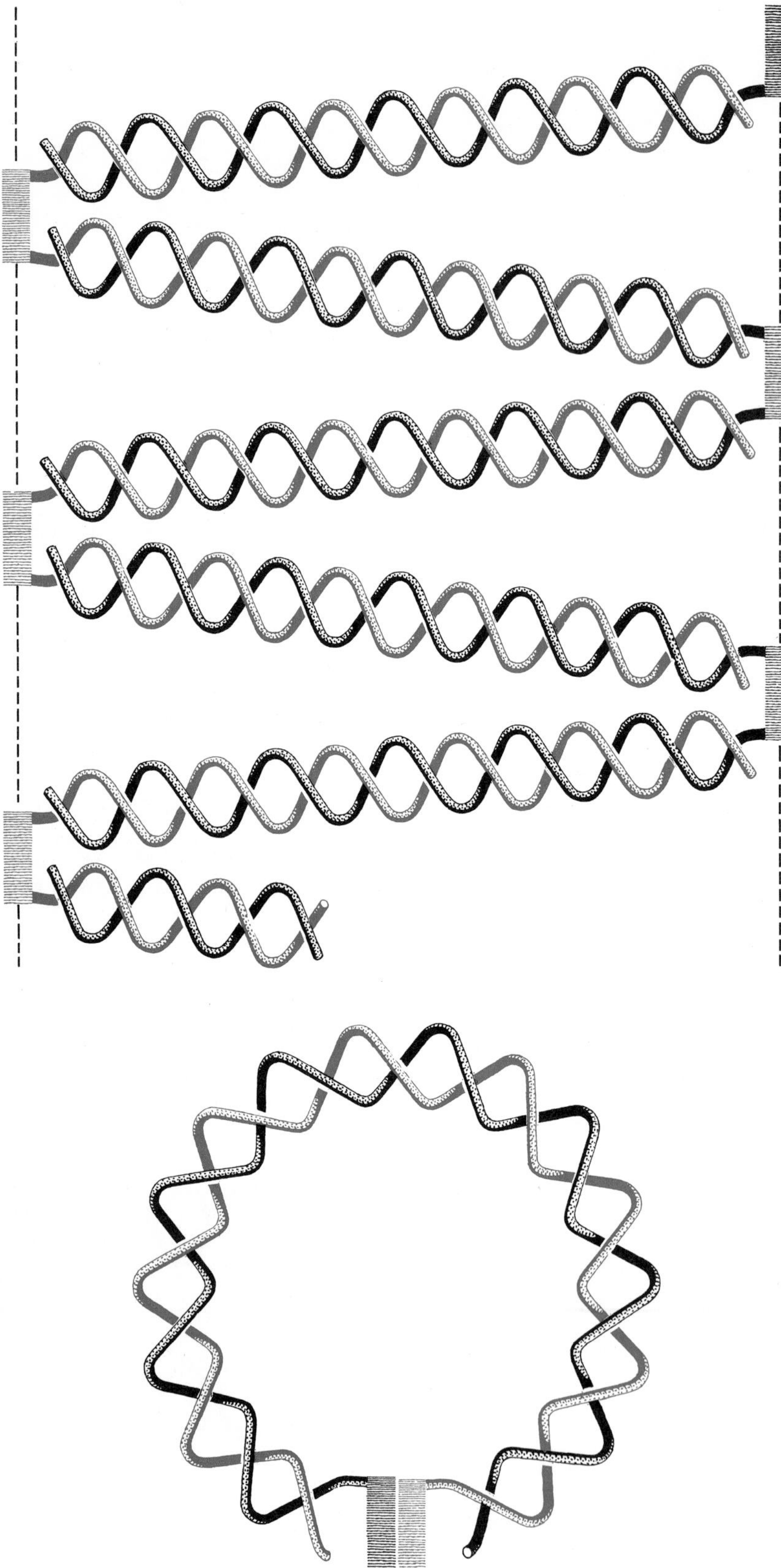

ANOTHER MODEL of the chromatid places its molecules of DNA between two columns, with one chain of each molecule attached to the left-hand column and the other to the right-hand column. Shaded rectangles represent structural material of the columns; broken lines indicate bonds which may include calcium. When the chromatid becomes visible, the columns come together so that the structure appears as in lower drawing when viewed end on.

basic replicating units, what sort of model can we imagine to explain its construction?

We know that chromosomes contain protein. So as a start we may picture the chromatid as a long protein backbone with DNA molecules branching out to the sides like ribs [*see diagram on page 19*]. Because the chromatid splits in two, we visualize the backbone as a two-layered affair whose layers can separate. The ends of the two strands of a DNA molecule are attached to these layers: one strand to one layer, the complementary strand to the other [*see diagram*]. As the layers peel apart, they unwind the strands. The unwinding strands promptly begin to build matching new strands for themselves. Eventually the new strands also assemble a new backbone, and the original chromatid is thus fully duplicated.

This model seems to have all the necessary mechanical specifications except one. The genes in a chromosome are arranged in a fixed linear order. Here the DNA segments, with one end waving freely about, are not so arranged. To meet this objection Ernst Freese of Harvard University has suggested a slightly different model which joins the free ends of the DNA molecules so that they form a definite sequence. Instead of one spine there are two, with the DNA segments crossing between them somewhat like the rungs of a ladder [*see diagram at the left*]. The spines may consist of blocks of protein joined by flexible bonds involving calcium atoms. The DNA rungs zigzag so that they march up the ladder, and the points on the rungs thus have a sequential order.

Now we can suppose that the calcium bonds give the structure considerable flexibility, allowing it to fold and coil on itself. The two spines may come together and so form a long tube [*see lower drawing at left*]; the tube may then coil into a tight helix. Replication in this model is accomplished by a stretching of the chain and unwinding of the DNA strands, each of which has one free end, as the diagrams show.

Recently Montrose J. Moses of the Rockefeller Institute for Medical Research and Don W. Fawcett of Cornell Medical College, using the electron microscope, have obtained pictures of chromatids which do indeed show a spine structure with DNA branches [*see photograph on page 18*]. It appears that we are beginning to penetrate down to the detailed mechanisms of the duplication of life.

# 3

# *The Bacterial Chromosome*

JOHN CAIRNS

*January 1966*

The information inherited by living things from their forebears is inscribed in their deoxyribonucleic acid (DNA). It is written there in a decipherable code in which the "letters" are the four subunits of DNA, the nucleotide bases. It is ordered in functional units—the genes—and thence translated by way of ribonucleic acid (RNA) into sequences of amino acids that determine the properties of proteins. The proteins are, in the final analysis, the executors of each organism's inheritance.

The central event in the passage of genetic information from one generation to the next is the duplication of DNA. This cannot be a casual process. The complement of DNA in a single bacterium, for example, amounts to some six million nucleotide bases; this is the bacterium's "inheritance." Clearly life's security of tenure derives in large measure from the precision with which DNA can be duplicated, and the manner of this duplication is therefore a matter of surpassing interest. This article deals with a single set of experiments on the duplication of DNA, the antecedents to them and some of the speculations they have provoked.

When James D. Watson and Francis H. C. Crick developed their two-strand model for the structure of DNA, they saw that it contained within it the seeds of a system for self-duplication. The two strands, or polynucleotide chains, were apparently related physically to each other by a strict system of *complementary* base pairing. Wherever the nucleotide base adenine occurred in one chain, thymine was present in the other; similarly, guanine was always paired with cytosine. These rules meant that the sequence of bases in each chain inexorably stipulated the sequence in the other; each chain, on its own, could generate the entire sequence of base pairs. Watson and Crick therefore suggested that accurate duplication of DNA could occur if the chains separated and each then acted as a template on which a new complementary chain was laid down. This form of duplication was later called "semiconservative" because it supposed that although the individual parental chains were conserved during duplication (in that they were not thrown away), their association ended as part of the act of duplication.

The prediction of semiconservative replication soon received precise experimental support. Matthew S. Meselson and Franklin W. Stahl, working at the California Institute of Technology, were able to show that each molecule of DNA in the bacterium *Escherichia coli* is composed of equal parts of newly synthesized DNA and of old DNA that was present in the previous generation [*see lower right figure on page 23*]. They realized they had not proved that the two parts of each molecule were in fact two chains of the DNA duplex, because they had not established that the molecules they were working with consisted of only two chains. Later experiments, including some to be described in this article, showed that what they were observing was indeed the separation of the two chains during duplication.

The Meselson-Stahl experiment dealt with the end result of DNA duplication. It gave no hint about the mechanism that separates the chains and then supervises the synthesis of the new chains. Soon, however, Arthur Kornberg and his colleagues at Washington University isolated an enzyme from *E. coli* that, if all the necessary precursors were provided, could synthesize in the test tube chains that were complementary in base sequence to any DNA offered as a template. It was clear, then, that polynucleotide chains could indeed act as templates for the production of complementary chains and that this kind of reaction could be the normal process of duplication, since the enzymes for carrying it out were present in the living cell.

Such, then, was the general background of the experiments I undertook beginning in 1962 at the Australian National University. My object was simply (and literally) to look at molecules of DNA that had been caught in the act of duplication, in order to find out which of the possible forms of semiconservative replication takes place in the living cell: how the chains of parent DNA are arranged and how the new chains are laid down [*see top illustration on page 23*].

Various factors dictated that the experiments should be conducted with *E. coli*. For one thing, this bacterium was known from genetic studies to have only one chromosome; that is, its DNA is contained in a single functional unit in which all the genetic markers are arrayed in sequence. For another thing, the duplication of its chromosome was known to occupy virtually the entire cycle of cell division, so that one could be sure that every cell in a rapidly multiplying culture would contain replicating DNA.

Although nothing was known about the number of DNA molecules in the *E. coli* chromosome (or in any other complex chromosome, for that matter), the dispersal of the bacterium's DNA among its descendants had been shown to be semiconservative. For this and other reasons it seemed likely that the

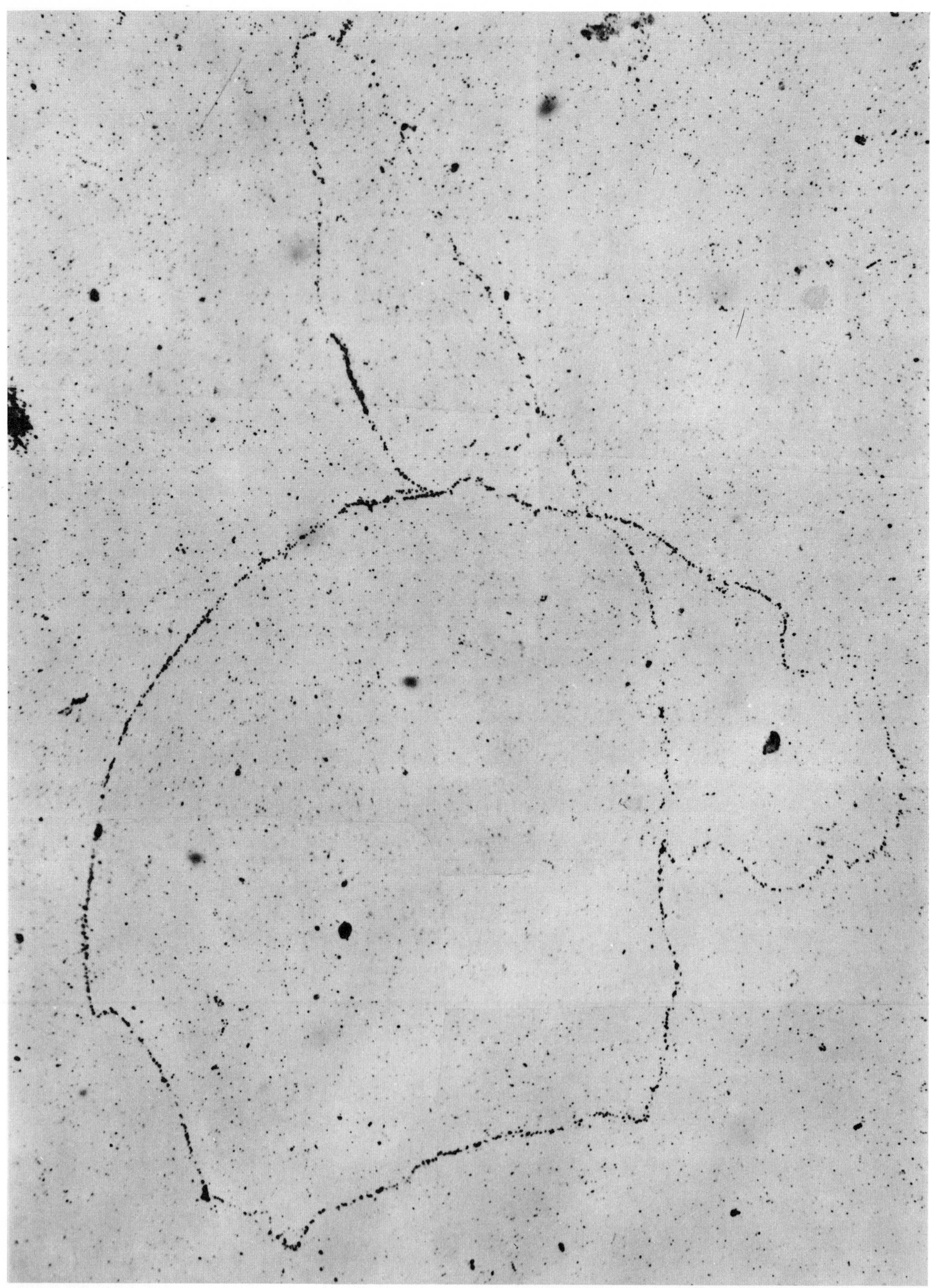

AUTORADIOGRAPH shows a duplicating chromosome from the bacterium *Escherichia coli* enlarged about 480 diameters. The DNA of the chromosome is visible because for two generations it incorporated a radioactive precursor, tritiated thymine. The thymine reveals its presence as a line of dark grains in the photographic emulsion. (Scattered grains are from background radiation.) The diagram on the opposite page shows how the picture is interpreted as demonstrating the manner of DNA duplication.

bacterial chromosome would turn out to be a single very large molecule. All the DNA previously isolated from bacteria had, to be sure, proved to be in molecules much smaller than the total chromosome, but a reason for this was suggested by studies by A. D. Hershey of the Carnegie Institution Department of Genetics at Cold Spring Harbor, N.Y. He had pointed out that the giant molecules of DNA that make up the genetic complement of certain bacterial viruses had been missed by earlier workers simply because they are so large that they are exceedingly fragile. Perhaps the same thing was true of the bacterial chromosome.

If so, the procedure for inspecting the replicating DNA of bacteria would have to be designed to cater for an exceptionally fragile molecule, since the bacterial chromosome contains some 20 times more DNA than the largest bacterial virus. It would have to be a case of looking but not touching. This was not as onerous a restriction as it may sound. The problem was, after all, a topographical one, involving delineation of strands of parent DNA and newly synthesized DNA. There was no need for manipulation, only for visualization.

Although electron microscopy is the

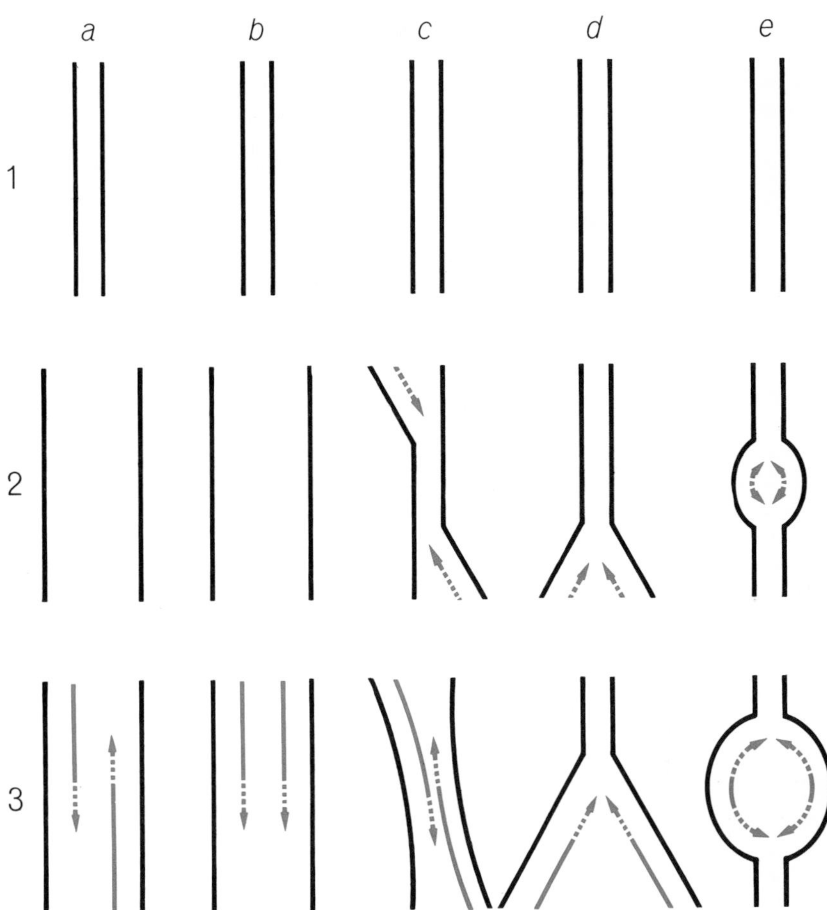


**DUPLICATION** could proceed in various ways (*a–e*). In these examples parental chains are shown as black lines and new chains as colored lines. The arrows show the direction of growth of the new chains, the newest parts of which are denoted by broken-line segments.

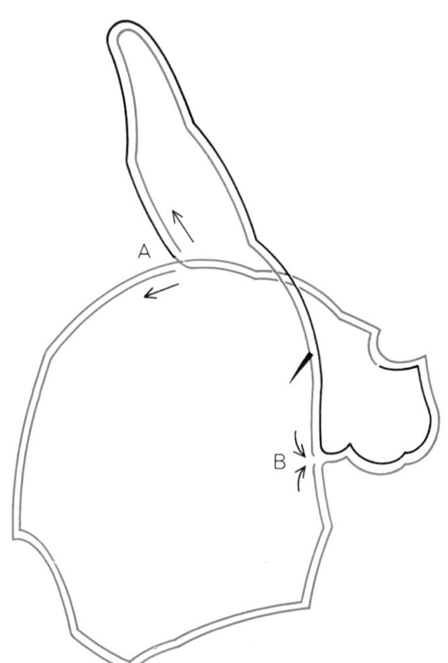


**INTERPRETATION** of autoradiograph on opposite page is based on the varying density of the line of grains. Excluding artifacts, dense segments represent doubly labeled DNA duplexes (*two colored lines*), faint segments singly labeled DNA (*color and black*). The parent chromosome, labeled in one strand and part of another, began to duplicate at *A;* new labeled strands have been laid down in two loops as far as *B.*

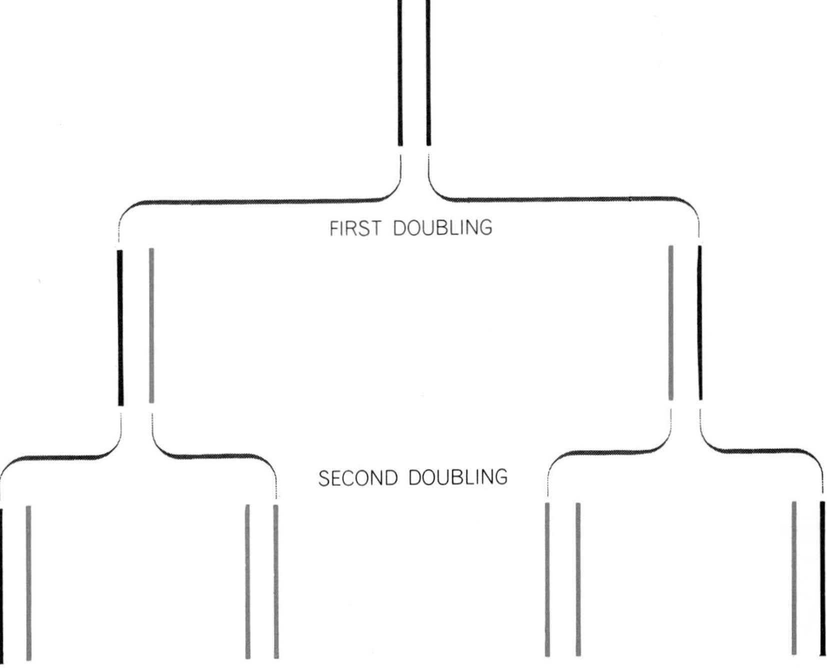


**SEMICONSERVATIVE DUPLICATION** was confirmed by the Meselson-Stahl experiment, which showed that each DNA molecule is composed of two parts: one that is present in the parent molecule, the other comprising new material synthesized when the parent molecule is duplicated. If radioactive labeling begins with the first doubling, the unlabeled (*black*) and labeled (*colored*) nucleotide chains of DNA form two-chain duplexes as shown here.

obvious way to get a look at a large molecule, I chose autoradiography in this instance because it offered certain peculiar advantages (which will become apparent) and because it had already proved to be the easier, albeit less accurate, technique for displaying large DNA molecules. Autoradiography capitalizes on the fact that electrons emitted by the decay of a radioactive isotope produce images on certain kinds of photographic emulsion. It is possible, for example, to locate the destination within a cell of a particular species of molecule by labeling such molecules with a radioactive atom, feeding them to the cell and then placing the cell in contact with an emulsion; a developed grain in the emulsion reveals the presence of a labeled molecule [see "Autobiographies of Cells," by Renato Baserga and Walter E. Kisieleski; SCIENTIFIC AMERICAN Offprint 165].

It happens that the base thymine, which is solely a precursor of DNA, is susceptible to very heavy labeling with tritium, the radioactive isotope of hydrogen. Replicating DNA incorporates the labeled thymine and thus becomes visible in autoradiographs. I had been able to extend the technique to demonstrating the form of individual DNA molecules extracted from bacterial viruses. This was possible because, in spite of the poor resolving power of autoradiography (compared with electron microscopy), molecules of DNA are so extremely long in relation to the resolving power that they appear as a linear array of grains. The method grossly exaggerates the apparent width of the DNA, but this is not a serious fault in the kind of study I was undertaking.

The general design of the experiments called for extracting labeled DNA from bacteria as gently as possible and then mounting it—without breaking the DNA molecules—for autoradiography. What I did was kill bacteria that had been fed tritiated thymine for various periods and put them, along with the enzyme lysozyme and an excess of unlabeled DNA, into a small capsule closed on one side by a semipermeable membrane. The enzyme, together with a detergent diffused into the chamber, induced the bacteria to break open and discharge their DNA. After the detergent, the enzyme and low-molecular-weight cellular debris had been diffused out of the chamber, the chamber was drained, leaving some of the DNA deposited on the membrane [*see illustration below*]. Once dry, the membrane was coated with a photographic emulsion sensitive to electrons emitted by the tritium and was left for two months. I hoped by this procedure to avoid subjecting the DNA to appreciable turbulence and so to find

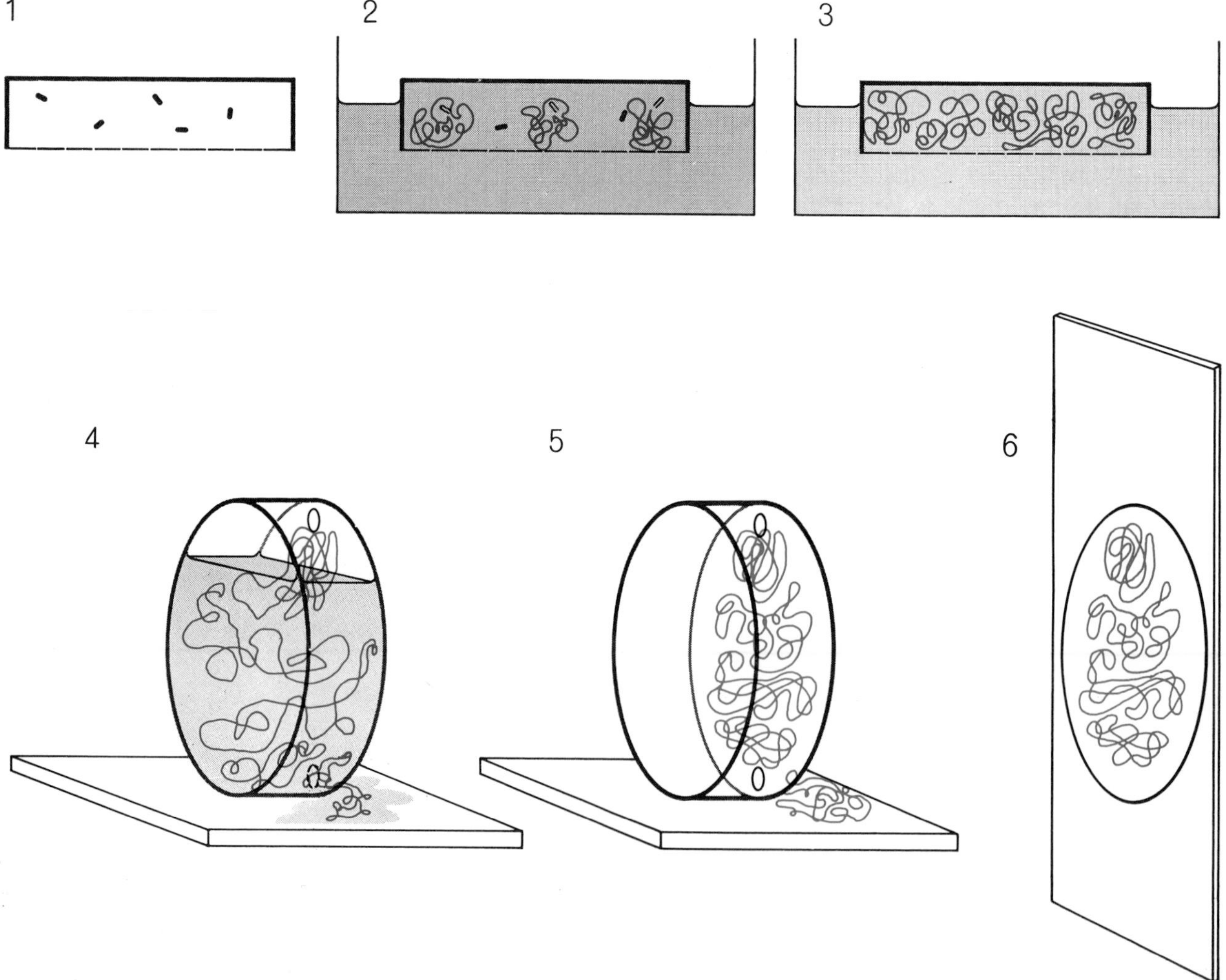


**AUTORADIOGRAPHY EXPERIMENT begins with bacteria whose DNA has been labeled with radioactive thymine. The bacteria and an enzyme are placed in a small chamber closed by a semipermeable membrane (*1*). Detergent diffused into the chamber causes the bacteria to discharge their contents (*2*). The detergent and cellular debris are washed away by saline solution diffused through the chamber (*3*). The membrane is then punctured. The saline drains out slowly (*4*), leaving some unbroken DNA molecules (*color*) clinging to the membrane (*5*). The membrane, with DNA, is placed on a microscope slide and coated with emulsion (*6*).**

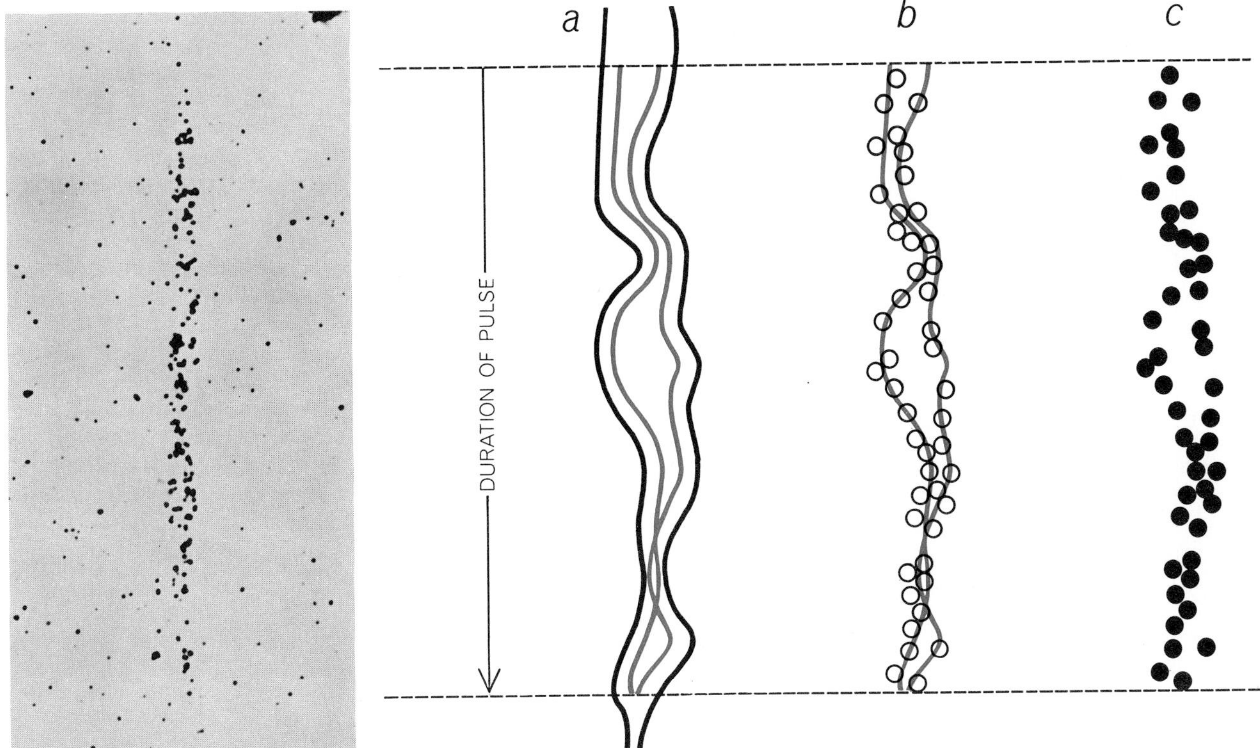


**DNA synthesized in *E. coli* fed radioactive thymine for three minutes is visible in an autoradiograph, enlarged 1,200 diameters, as an array of heavy black grains (*left*). The events leading to the autoradiograph are shown at right. The region of the DNA chains synthesized during the "pulse-labeling" is radioactive and is shown in color (*a*). The radioactivity affects silver grains in the photographic emulsion (*b*). The developed grains appear in the autoradiograph (*c*), approximately delineating the new chains of DNA.**

some molecules that—however big—had not been broken and see their form. Inasmuch as *E. coli* synthesizes DNA during its entire division cycle, some of the extracted DNA should be caught in the act of replication. (Since there was an excess of unlabeled DNA present, any tendency for DNA to produce artificial aggregates would not produce a spurious increase in the size of the labeled molecules or an alteration in their form.)

It is the peculiar virtue of autoradiography that one sees only what has been labeled; for this reason the technique can yield information on the history as well as the form of a labeled structure. The easiest way to determine which of the schemes of replication was correct was to look at bacterial DNA that had been allowed to duplicate for only a short time in the presence of labeled thymine. Only the most recently made DNA would be visible (corresponding to the broken-line segments in the upper illustration on page 23), and so it should be possible to determine if the two daughter molecules were being made at the same point or in different regions of the parent molecule. A picture obtained after labeling bacteria for three minutes, or a tenth of a generation-time [*at left in illustration above*], makes it clear that two labeled structures are being made in the same place. This place is presumably a particular region of a larger (unseen) parent molecule [*see diagrams at right in illustration above*].

The autoradiograph also shows that at least 80 microns (80 thousandths of a millimeter) of the DNA has been duplicated in three minutes. Since duplication occupies the entire generation-time (which was about 30 minutes in these experiments), it follows that the process seen in the autoradiograph could traverse at least 10 × 80 microns, or about a millimeter, of DNA between one cell division and the next. This is roughly the total length of the DNA in the bacterial chromosome. The autoradiograph therefore suggests that the entire chromosome may be duplicated at a single locus that can move fast enough to traverse the total length of the DNA in each generation.

Finally, the autoradiograph gives evidence on the semiconservative aspect of duplication. Two structures are being synthesized. It is possible to estimate how heavily each structure is labeled (in terms of grains produced per micron of length) by counting the number of exposed grains and dividing by the length. Then the density of labeling can be compared with that of virus DNA labeled similarly but uniformly, that is, in both of its polynucleotide chains. It turns out that each of the two new structures seen in the picture must be a single polynucleotide chain. If, therefore, the picture is showing the synthesis of two daughter molecules from one parent molecule, it follows that each daughter molecule must be made up of one new (labeled) chain and one old (unlabeled) chain—just as Watson and Crick predicted.

The "pulse-labeling" experiment just described yielded information on the isolated regions of bacterial DNA actually engaged in duplication. To learn if the entire chromosome is a single molecule and how the process of duplication proceeds it was necessary to look at DNA that had been labeled with tritiated thymine for several generations. Moreover, it was necessary to find, in the jumble of chromosomes extracted from *E. coli*, autoradiographs of unbroken chromosomes that were disentangled enough to be seen as a whole. Rather than retrace all the steps that led, after many months, to satisfactory pictures of the entire bacterial chromosome in one piece, it is simpler to present two sample autoradiographs and explain how they can be interpreted and what they reveal.

The autoradiographs on page 22 and at the right show bacterial chromosomes in the process of duplication. All that is visible is labeled, or "hot," DNA; any unlabeled, or "cold," chain is unseen. A stretch of DNA duplex labeled in only one chain ("hot-cold") makes a faint trace of black grains. A duplex that is doubly labeled ("hot-hot") shows as a heavier trace. The autoradiographs therefore indicate, as shown in the diagrams that accompany them, the extent to which new, labeled polynucleotide chains have been laid down along labeled or unlabeled parent chains. Such data make it possible to construct a bacterial family history showing the process of duplication over several generations [*see illustration on next page*].

The significant conclusions are these:

1. The chromosome of *E. coli* apparently contains a single molecule of DNA roughly a millimeter in length and with a calculated molecular weight of about two billion. This is by far the largest molecule known to occur in a biological system.

2. The molecule contains two polynucleotide chains, which separate at the time of duplication.

3. The molecule is duplicated at a single locus that traverses the entire length of the molecule. At this point both new chains are being made: two chains are becoming four. This locus has come to be called the replicating "fork" because that is what it looks like.

4. Replicating chromosomes are not Y-shaped, as would be the case for a linear structure [*see "d" in upper illustration on page 23*]. Instead the three ends of the Y are joined: the ends of the daughter molecules are joined to each other and to the far end of the parent molecule. In other words, the chromosome is circular while it is being duplicated.

It is hard to conceive of the behavior of a molecule that is about 1,000 times larger than the largest protein and that exists, moreover, coiled inside a cell several hundred times shorter than itself. Apart from this general problem of comprehension, there are two special difficulties inherent in the process of DNA duplication outlined here. Both have their origin in details of the structure of DNA that I have not yet discussed.

The first difficulty arises from the opposite polarities of the two polynucleotide chains [*see illustration on page 28*]. The deoxyribose-phosphate backbone of one chain of the DNA duplex has the sequence $-O-C_3-C_4-C_5-O-P-O-C_3-C_4-C_5-O-P-\ldots$ (The $C_3$, $C_4$ and $C_5$ are the three carbon atoms of the deoxyribose that contribute to the backbone.) The other chain has the sequence $-P-O-C_5-C_4-C_3-O-P-O-C_5-C_4-C_3-O-\ldots$

If both chains are having their complements laid down at a single locus moving in one particular direction, it follows that one of these new chains must grow by repeated addition to the $C_3$ of the preceding nucleotide's deoxyribose and the other must grow by addition to a $C_5$. One would expect that two different enzymes should be needed for these two quite different kinds of polymerization. As yet, however, only the reaction that adds to chains ending in $C_3$ has been demonstrated in such experiments as Kornberg's. This fact had seemed to support a mode of replication in which the two strands grew in opposite directions [*see "a" and "c" in upper illustration on page 23*]. If the single-locus scheme is correct, the problem of opposite polarities remains to be explained.

The second difficulty, like the first, is related to the structure of DNA. For the sake of simplicity I have been representing the DNA duplex as a pair of chains lying parallel to each other. In actuality the two chains are wound helically around a common axis, with one complete turn for every 10 base pairs, or 34 angstrom units of length (34 ten-millionths of a millimeter). It would seem, therefore, that separation of the chains at the time of duplication, like separation of the strands of an ordinary rope, must involve rotation of the parent molecule with respect to the two daughter molecules. Moreover, this rotation must be very rapid. A fast-multiplying bacterium can divide every 20 minutes;

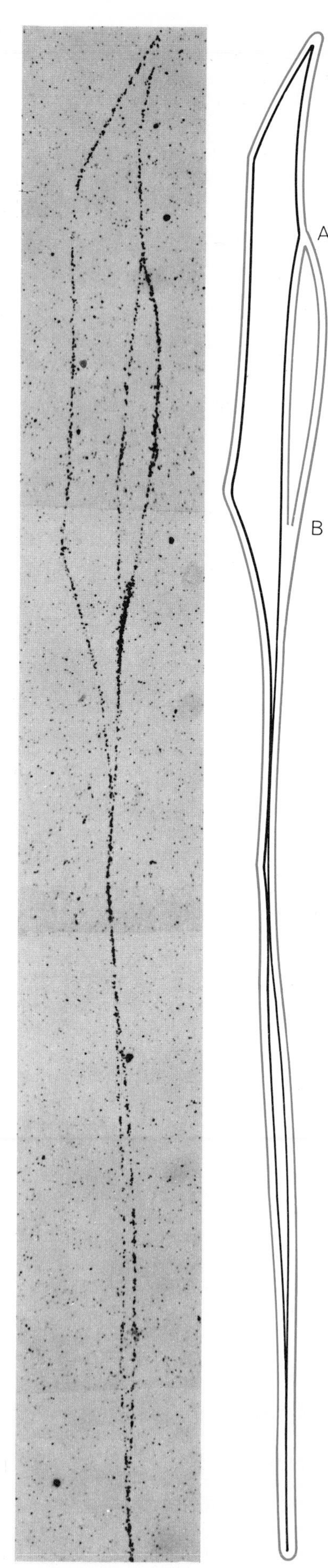


COMPLETE CHROMOSOME is seen in this autoradiograph, enlarged about 370 diameters. Like the chromosome represented on pages 22 and 23, this one is circular, although it happens to have landed on the membrane in a more compressed shape and some segments are tangled. Whereas the first chromosome was more than halfway through the duplication process, this one is only about one-sixth duplicated (*from A to B*).

during this time it has to duplicate—and consequently to unwind—about a millimeter of DNA, or some 300,000 turns. This implies an average unwinding rate of 15,000 revolutions per minute.

At first sight it merely adds to the difficulty to find that the chromosome is circular while all of this is going on. Obviously a firmly closed circle—whether a molecule or a rope—cannot be unwound. This complication is worth worrying about because there is increasing evidence that the chromosome of *E. coli* is not exceptional in its circularity. The DNA of numerous viruses has been shown either to be circular or to become circular just before replication begins. For all we know, circularity may therefore be the rule rather than the exception.

There are several possible explanations for this apparent impasse, only one of which strikes me as plausible.

First, one should consider the possibility that there is no impasse—that in the living cell the DNA is two-stranded but not helical, perhaps being kept that way precisely by being in the form of a circle. (If a double helix cannot be unwound when it is firmly linked into a circle, neither can relational coils ever be introduced into a pair of uncoiled circles.) This hypothesis, however, requires a most improbable structure for two-strand DNA, one that has not been observed. And it does not really avoid the unwinding problem because there would still have to be some mechanism for making nonhelical circles out of the helical rods of DNA found in certain virus particles.

Second, one could avoid the unwinding problem by postulating that at least one of the parental chains is repeatedly broken and reunited during replication, so that the two chains can be separated over short sections without rotation of the entire molecule. One rather sentimental objection to this hypothesis (which was proposed some time ago) is that it is hard to imagine such cavalier and hazardous treatment being meted out to such an important molecule, and one so conspicuous for its stability. A second objection is that it does not explain circularity.

The most satisfactory solution to the unwinding problem would be to find some reason why the ends of the chromosome actually *must* be joined together. This is the case if one postulates that there is an active mechanism for unwinding the DNA, distinct from the mechanism that copies the unwound

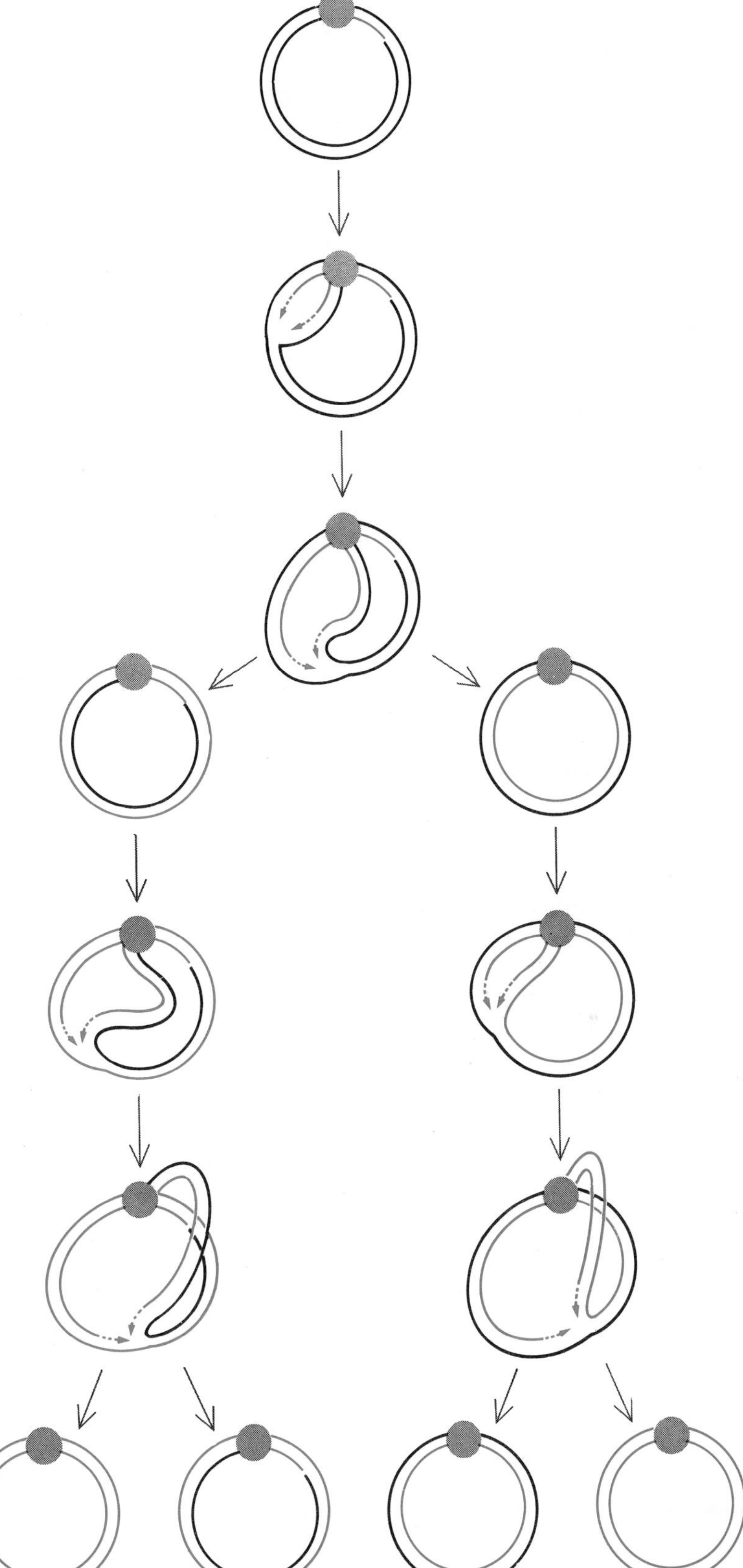

BACTERIAL DNA MOLECULE apparently replicates as in this schematic diagram. The two chains of the circular molecule are represented as concentric circles, joined at a "swivel" (*gray spot*). Labeled DNA is shown in color; part of one chain of the parent molecule is labeled, as are two generations of newly synthesized DNA. Duplication starts at the swivel and, in these drawings, proceeds counterclockwise. The arrowheads mark the replicating "fork": the point at which DNA is being synthesized in each chromosome. The drawing marked *A* is a schematic rendering of the chromosome in the autoradiograph on page 22.

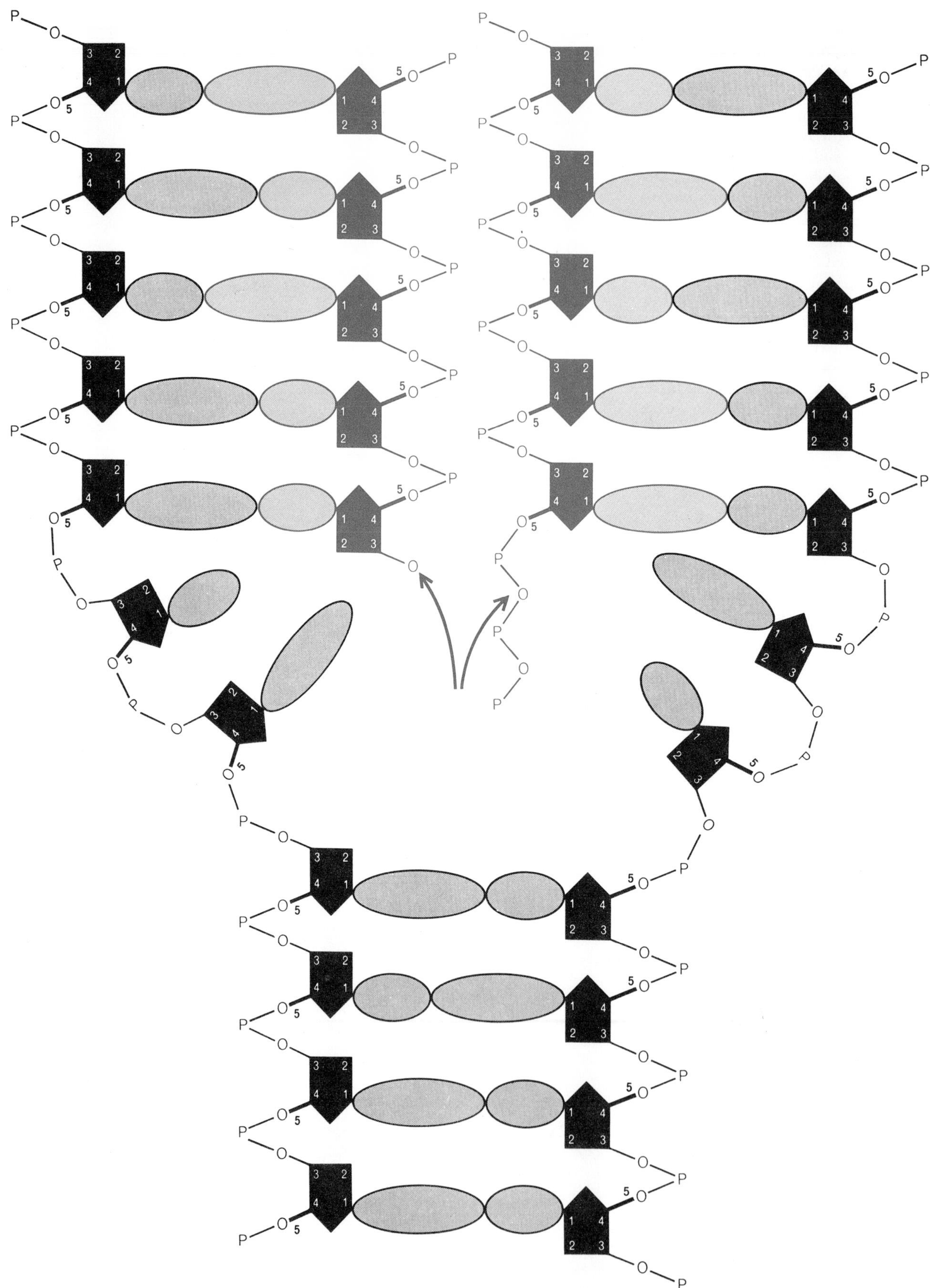

**OPPOSITE POLARITIES** of the two parental chains of the DNA duplex result in opposite polarities and different directions of growth in the two new chains (*color*) being laid down as complements of the old ones during duplication. Note that the numbered carbon atoms (*1 to 5*) in the deoxyribose rings (*solid black*) are in different positions in the two parental chains and therefore in the two new chains. As the replicating fork moves downward, the new chain that is complementary to the left parental chain must grow by addition to a $C_3$, the other new chain by addition to a $C_5$, as shown by the arrows. The elliptical shapes are the four bases.

chains. Now, any active unwinding mechanism must rotate the parent molecule with respect to the two new molecules—must hold the latter fast, in other words, just as the far end of a rope must be held if it is to be unwound. A little thought will show that this can be most surely accomplished by a machine attached, directly or through some common "ground," to the parent molecule and to the two daughters [*see illustration below*]. Every turn taken by such a machine would inevitably unwind the parent molecule one turn.

Although other kinds of unwinding machine can be imagined (one could be situated, for example, at the replicating fork), a practical advantage of this particular hypothesis is that it accounts for circularity. It also makes the surprising—and testable—prediction that any irreparable break in the parent molecule will instantly stop DNA synthesis, no matter how far the break is from the replicating fork. If this prediction is fulfilled, and the unwinding machine acquires the respectability that at present it lacks, we may find ourselves dealing with the first example in nature of something equivalent to a wheel.

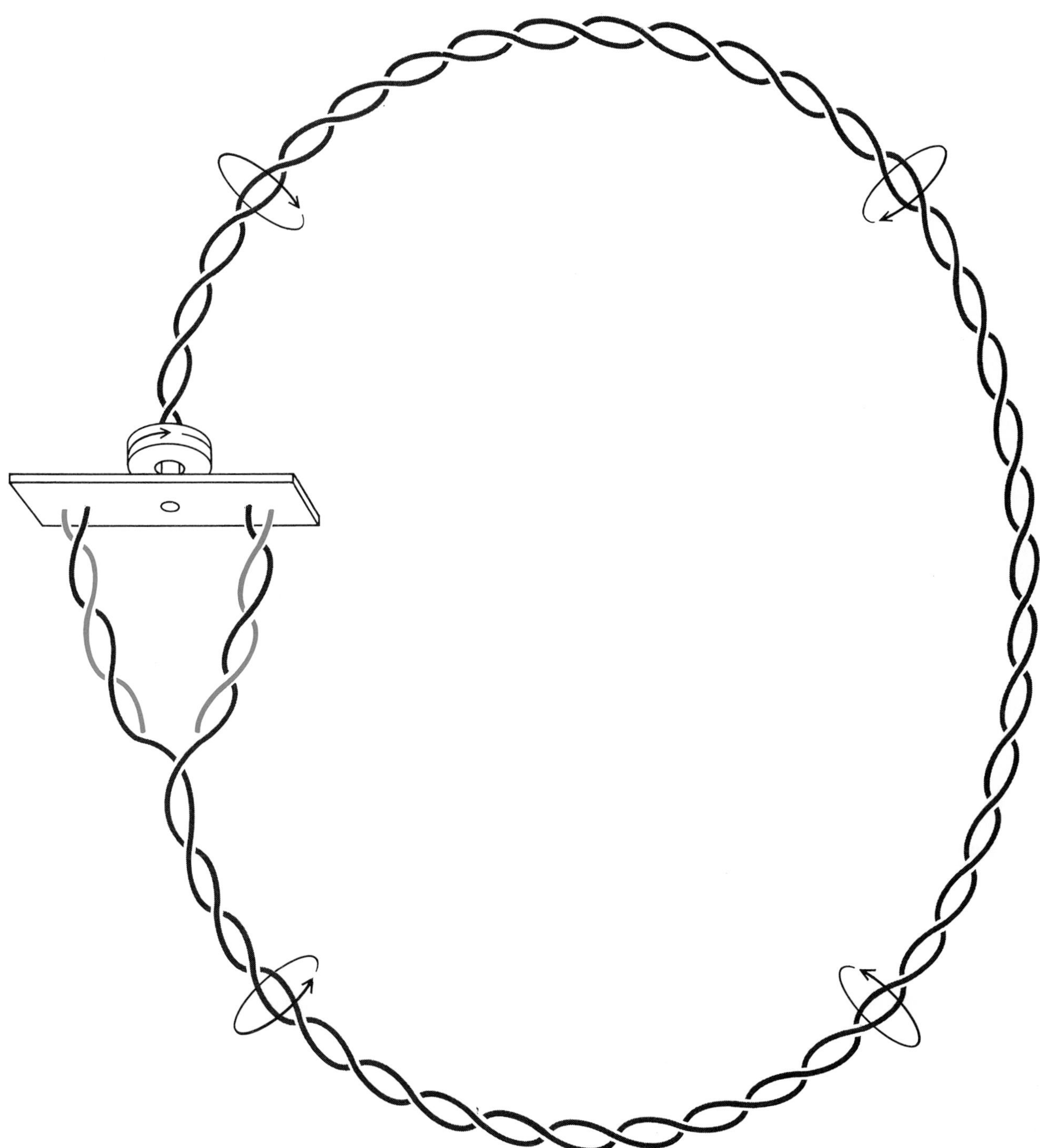

**POSSIBLE MECHANISM for unwinding the DNA double helix is a swivel-like machine to which the end of the parent molecule and also the ends of the two daughter molecules are joined. The torque imparted by this machine is considered to be transmitted along the parent molecule, producing unwinding at the replicating fork. If this is correct, chromosome breakage should halt duplication.**

# 4

# *Transformed Bacteria*

ROLLIN D. HOTCHKISS and
ESTHER WEISS

*November 1956*

If man reproduced his kind the way bacteria do, a grown man at 25 would more or less abruptly become two young men in his own exact image. These two in turn would "divide" in another 25 years, so that after 50 years there would be four young men indistinguishable from the original ancestor. A rather large family could eventually be built up by this process, but all its members would be monotonously alike in appearance, abilities, temperament and vigor. The same would be true of every family. It would be entirely male or entirely female: the two sexes would be aloof from each other. There would be families of burly, competitive athletes, and others made up exclusively of gray-eyed introverts liking nothing better than to write sad poems on the haunting loveliness of subdivision.

After about 20 generations (500 years) we could expect an occasional "mutant" individual to show up, differing from the million other individuals of his family in some one trait such as eye color. In a community so ordered, a means of transferring these rare traits at will from one family to another would have dramatic value indeed. Actually such transformations can be accomplished with bacteria. It is possible to transfer mutations of a particular kind and thereby make controlled studies of heredity.

Only in the last 10 years or so have bacteria become prominent in the study of genetics. Until a little more than a decade ago it was supposed that a bacterial cell did not have a nucleus or the elaborate genetic apparatus characteristic of higher forms of life. The general feeling was that bacteria were simple enough not to need so cumbersome a system for passing along their hereditary characteristics. But in 1944 C. F. Robinow, a British bacteriologist, found signs of nuclei in bacteria. Since then bacteria have gained in complexity the more they have been studied, and their mechanisms have been found to parallel those of cells of higher organisms. The bacterial genetic system has proved to be one of the most fascinating of all to bacteriologists and geneticists alike. Undoubtedly some of the appeal of bacteria for geneticists lies in their rapid rate of multiplication. What other laboratory subject provides a new generation every half-hour? There is no problem of eyestrain involved, either. The tiny creatures take only a few hours to grow to a many-celled colony which is easily visible in the test tube or on a gelatin-like agar plate, and they are much more manageable than the classical fruit flies. When a bacterium achieves a new characteristic, it quickly produces a whole colony of the new type, which can often be readily identified.

Such a new characteristic may turn up in perhaps one in a few million cells in a growing population. The few mutant bacteria may be swamped out as they attempt to grow in the enormous competitive population. Occasionally, however, the environment may favor the rare individuals of a newly arising type, so that they outgrow the usual type. A bacterial geneticist is always on the lookout for these natural events, and he can make detection of them easier by providing a selective environment which allows only certain mutants to grow. For example, if bacteria are put upon agar plates containing penicillin, only the mutants that have resistance to the drug will grow to produce colonies. From a population of a hundred million cells, usually about five to 10 such colonies will emerge, indicating that something like one cell in 10 million of the original population became, mysteriously and suddenly, penicillin-resistant.

The technique of transformation developed in the laboratory permits us to take matters more into our own hands. We can introduce a specific hereditary characteristic into a strain of bacteria by treating the cells with an extract from killed bacteria of a related strain which possess the characteristic in question. The procedure does not create new traits but transfers traits already present in the donor bacteria. In effect we are "robbing Peter to pay Paul."

The transforming material is desoxyribonucleic acid (DNA)—the type of substance now so well known as a fundamental constituent of the chromosomes of higher plants and animals. DNA from the donor cells seems to enter into the recipient bacterial cell and, like a gene, direct part of the cell's internal mechanism. The cell even learns to make more of the directing substance itself. How this controlling substance carries its specific instructions is still a mystery: our most sensitive chemical analyses cannot distinguish any differences between DNA varieties responsible for different traits. But each variety must be chemically distinctive, because the recipient cells repeatedly respond in the same predictable way to a particular extract.

Pneumococci, the pneumonia germs, have been extensively studied by means of laboratory transformation. From about 30 million pneumococcal cells, killed and broken down by sodium desoxycholate (a substance found in bile), we can extract one microgram of purified DNA. This can be preserved for years as a white precipitate in alcohol.

The pneumococci we have used to test the transforming DNA are Type II strains which have lost the ability to produce the sugar capsule that makes the bacteria virulent. On agar plates these strains normally grow as shiny pinpoint colonies. In a broth containing a small amount of antiserum (made from the blood of an animal inoculated with pneumococci) they grow in chains and clumps which settle to the bottom as separate white colonies. By diluting samples of the culture and counting the colonies that develop, we can make quantitative studies of what happens when bacteria are transformed.

More than 25 specific characteristics have been transferred to this strain from various mutants. They include every sort of trait we can observe: acquisition by the bacterium of various types of capsule coating, development of resistance to drugs, formation of certain types of colonies, and so on. Usually only one trait is passed on to any particular cell, even if the DNA preparation is from a strain having two or three identifiable characteristics. For example, if a million pneumococci are treated with a tenth of a microgram of a DNA which carries three traits—resistance to penicillin and streptomycin and formation of a coat of Type III—some 50,000 of the million may be transformed; of these about 49,000 will acquire only one of the three traits, 800 may have two, and only four will take all three.

The DNA preparation evidently does not carry a complete package of the donor's traits into a recipient cell but only some part of the package. This part is not as small as a gene, however. Certain traits seem to travel together as if they were linked. For example, the DNA factors responsible for the ability of pneumococci to utilize mannitol (a form of sugar found in manna) as a source of energy and for the ability to resist streptomycin tend to be coupled: 20 per cent of cells transformed by a DNA carrying these two markers will show both characteristics.

Experiments in transmission of the mannitol-utilization trait illustrate an-

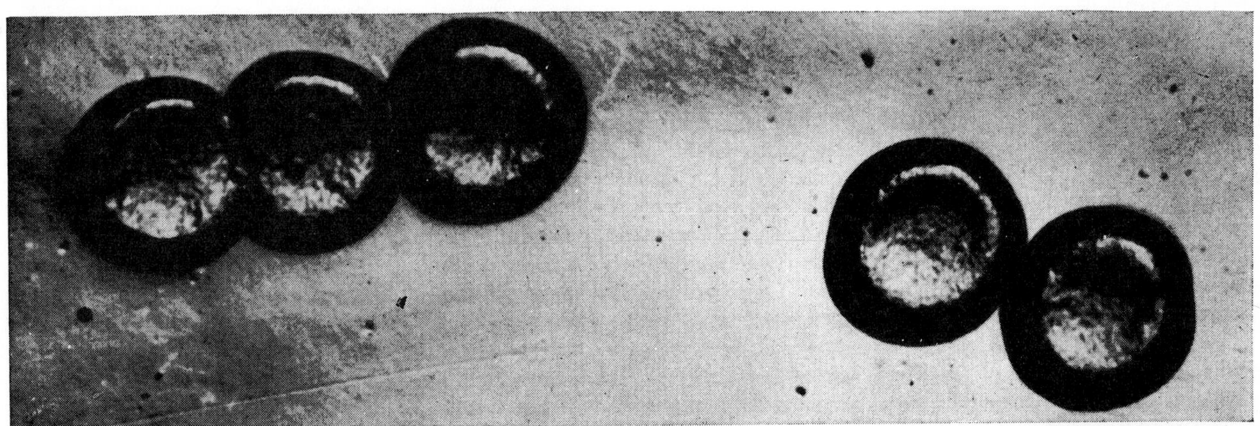

**DIFFERENT STRAINS OF PNEUMOCOCCI** grow on the surface of agar in colonies of characteristic form. At the top are the small colonies of pneumococci without capsules. At the bottom are the larger colonies of encapsulated Type III pneumococci. The colonies of Type III produce a sticky sugar; they are larger because they absorb moisture from the agar. If desoxyribonucleic acid (DNA) is removed from pneumococci of Type III and added to the nonencapsulated pneumococci, the latter will acquire capsules.

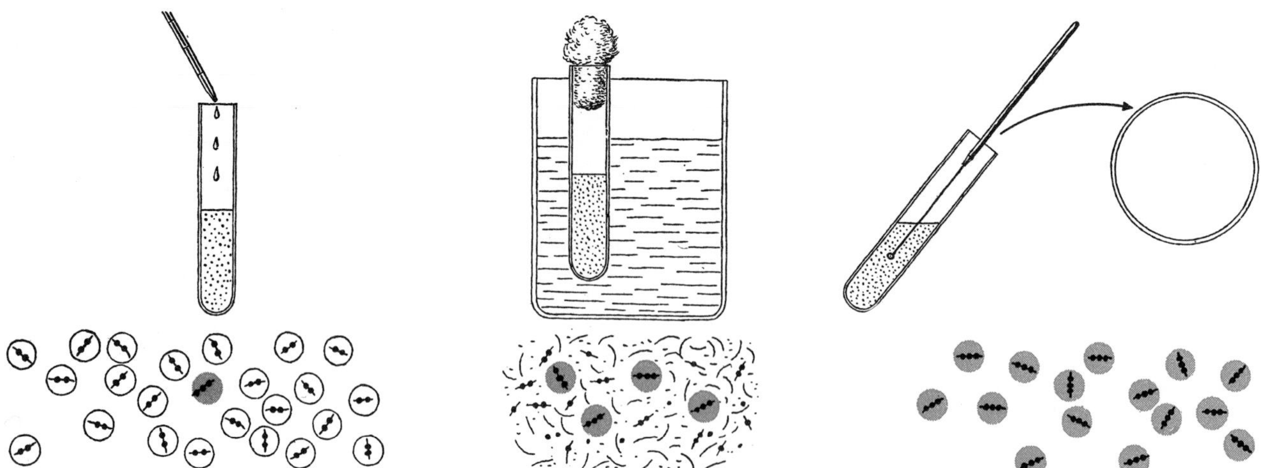

**PREPARATION OF DNA** from a strain of pneumococci resistant to streptomycin is traced in the drawings at the top of this and the next three pages. First, a broth containing streptomycin is inoculated with pneumococci. One cell (*color*) is a rare mutant resistant to streptomycin. Second, the broth is incubated overnight at 37 degrees centigrade. The mutant multiplies and makes the broth

other important point: namely, that DNA may carry a hereditary trait as a latent ability, regardless of whether or not the ability is developed. Like children who display their innate musical talent only after they have taken some piano lessons, the talented strains of pneumococci do not exhibit their ability to utilize mannitol until they have been exposed to this sugar for a short time. To metabolize it they have to learn to make an enzyme which oxidizes mannitol.

Under the most ideal conditions we have devised up to now, it has been possible to transmit a new trait to 17 per cent of the treated cells. Many factors influence the yield obtained. Foremost of these is the capacity of the recipient strain itself. Some strains seem to be transformed more readily than others, and many seem altogether incapable of responding to DNA. Another factor is the concentration of DNA present: one half of a millionth of a gram per cubic centimeter is an optimal concentration, and one 10,000th of that amount will have some effect. The length of exposure to DNA also is important. We think that the bacteria are most susceptible to transformation just after cell division. After pneumococci have been cooled to a growth-arresting temperature, so that all start out "in step" in the division cycle when they are rewarmed, transformations are exceptionally numerous. About 15 minutes after the rewarming the cells abruptly lose their susceptibility, and just at this time very few cells will be dividing. Still another important factor is the composition of the medium in which the cells grow before and during DNA exposure.

After acquiring the new DNA, a cell multiplies more slowly than the others for a time and is at a disadvantage in growth until its transformation has been completed. Some of the transformed cells are likely to survive in any event, but the percentage transformed is easier to observe if the conditions are adjusted to favor them. For example, in an experiment in which the transformation makes the cells resistant to penicillin, placing the bacteria in a penicillin broth will kill off all but the transformed cells. One must be careful, however, not to challenge the bacteria with the selective agent too soon. The transformed cells require 30 to 60 minutes to manifest their new drug resistance, and it takes still longer for the cells to set up the mechanism necessary to duplicate the new DNA. During this time the new DNA is beginning to perform its genelike functions in the cells.

A transformed pneumococcus remains susceptible to further transformations, even by the same DNA if the DNA

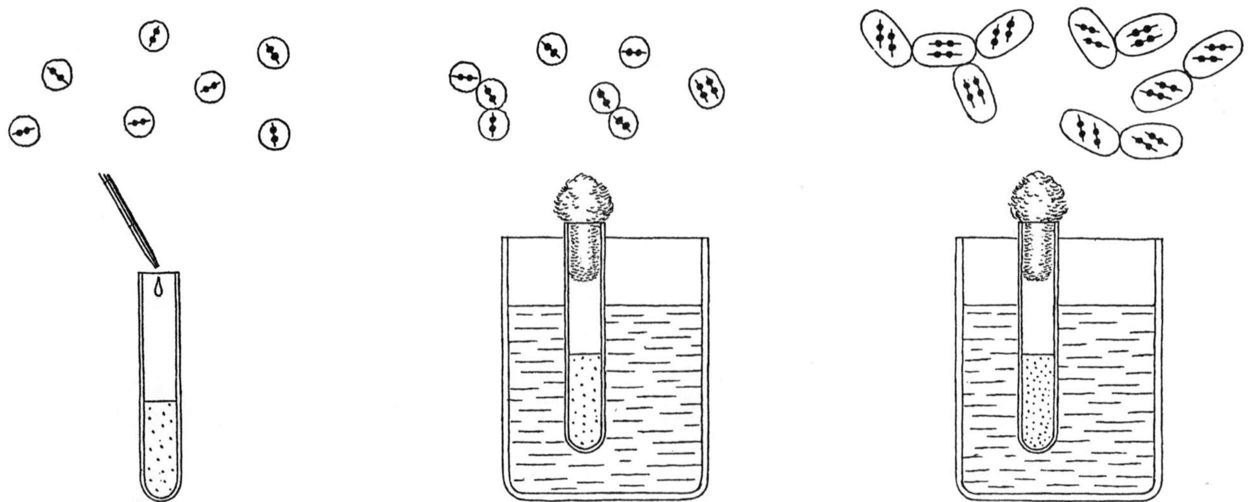

**TRANSFORMATION OF PNEUMOCOCCI** to a strain resistant to streptomycin from a nonresistant strain is illustrated at the bottom of this and the next three pages. First, young pneumococci are added to a rich broth containing serum albumin. Second, the tube is incubated for three hours at 37 degrees C.; the cells multiply. Third, the tube is cooled to 25 degrees for 20 minutes. This arrests

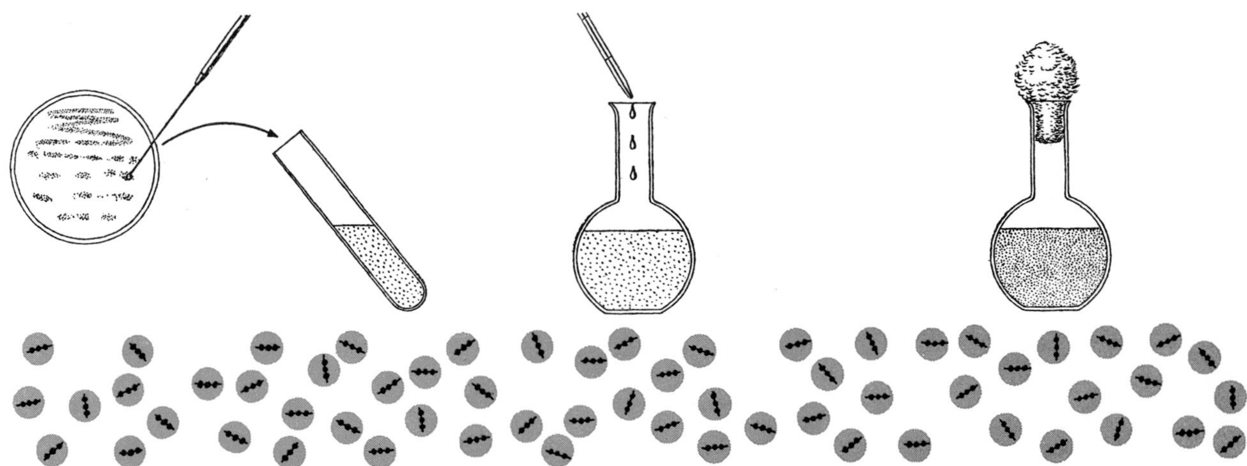

**turbid; the other cells are killed by the streptomycin. Third, a bit of turbid broth is removed and spread on an agar medium. Fourth, one of the colonies on the agar plate is transferred to a tube of fresh broth. Fifth, a flask of broth is inoculated with the mutant strain. Sixth, the flask is incubated overnight until the culture is full-grown. This sequence is continued at the top of the next page.**

carries more than one trait. Indeed, a particular trait, such as resistance to penicillin, may be developed by a series of stepwise mutations rather than by a single transformation. Beginning with pneumococci that survived exposure to low concentrations of penicillin, we submitted them to successively higher concentrations until we had a mutant strain which was resistant to 30 units of penicillin per 100 milliliters of culture. We then administered the DNA from this strain to pneumococci which were fully sensitive to penicillin. None of the sensitive cells acquired as much resistance to penicillin as the donor strain possessed, and most of those transformed could not resist more than five units of the drug. When these were again treated with the DNA of the highly resistant donor, some became resistant to 12 units of penicillin. It took several such steps to produce transformants able to resist 30 units. We think these indicate the number of spontaneous mutations that must have taken place in the original evolution of the mutant strain.

Experiments with streptomycin produced the same stepwise development of resistance. However, in a large population of cells an occasional cell spontaneously acquired a high level of resistance in just one step, and the DNA from this mutant produced equally resistant cells in a single transformation. Evidently in this case one mutation of a single genetic unit modified the DNA so that it could effect the entire transformation in one step.

Another kind of phenomenon emerges when pneumococci without a capsule are treated with a DNA which confers the ability to form a capsule of Type III. Normally the DNA effects this transformation in one step, but at times it produces cells with intermediate varieties of sugar capsules. Colonies of these cells are smaller than those of the full Type III. Harriett Ephrussi-Taylor, formerly of the Rockefeller Institute for Medical Research and now at the University of Paris, has done many experiments with two such varieties of cells. They seem to differ from each other and from normal Type III only in the quantity of capsule material they produce. If DNA from both varieties is mixed in a single culture, large, juicy colonies characteristic of the normal Type III cells will appear on the agar plates. Dr. Ephrussi-Taylor concluded that the two kinds of DNA could combine in a single recipient bacterium to yield the normal DNA of Type III cells, and this conclusion was strongly supported by other experiments.

The more the action of DNA is studied, the clearer it becomes that each organism's DNA is biologically distinc-

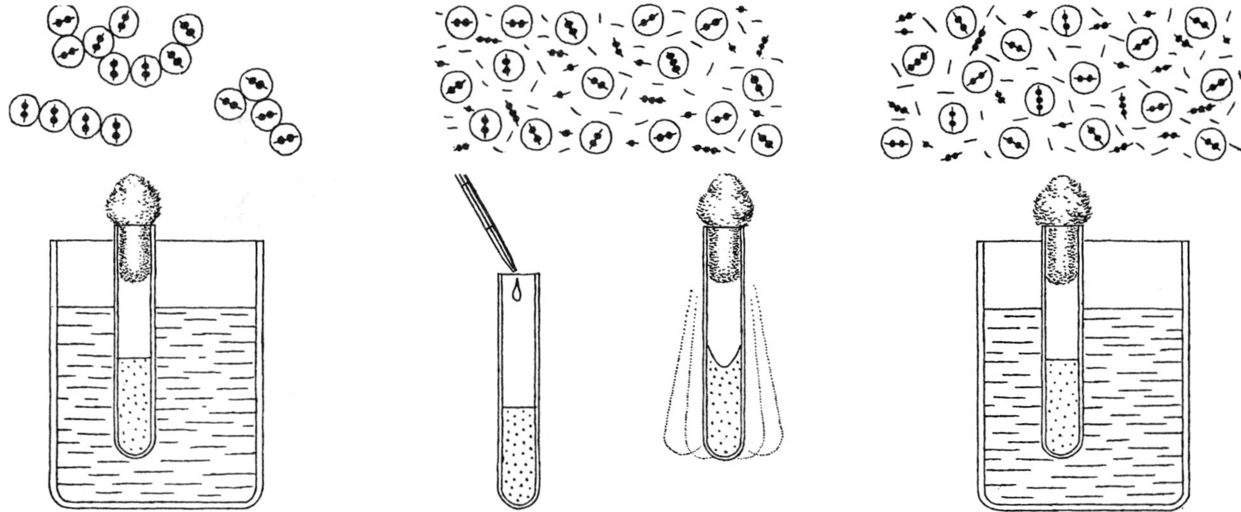

**the activities of the cells at the same point in their cycle of division. Fourth, the culture is rewarmed to 37 degrees. Now all the cells start dividing at the same time. Fifth, DNA removed from a resistant strain is added to the culture. The DNA spreads through the broth, and some of the cells react with it. Sixth, the culture is reincubated for five minutes; the cells continue to react with DNA.**

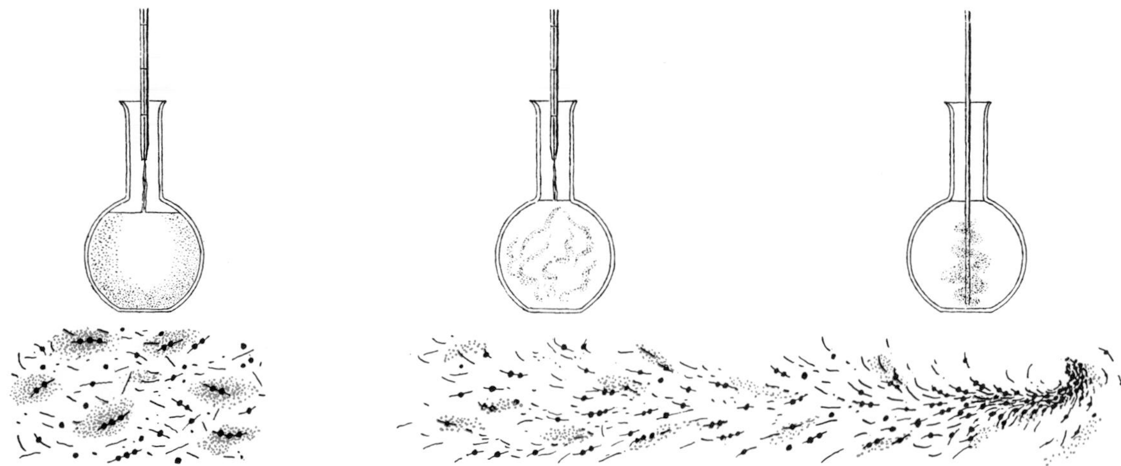

PREPARATION OF DNA IS CONTINUED from the drawings at the top of the preceding two pages. First, sodium desoxycholate is added to the flask containing the full-grown culture of pneumococci. This kills and breaks up the cells; the broth clears. Second, alcohol is added to the cleared flask. This causes the DNA to precipitate in threads. Third, the threads are collected from the flask

tive, even though we cannot detect any chemical difference. Attempts have been made to bring about transformations in pneumococci by injecting DNA from species of bacteria distantly related to them. The attempts have not succeeded. The foreign DNA may enter the cell, but it does not produce any detectable change in the cell's traits. The machinery of the cell apparently recognizes something unusual about the foreign DNA and makes no genetic response to it, so far as we can determine. However, the cell's incorporation of the bogus DNA prevents it from reacting freely with a suitable DNA. (In this respect it behaves something like a fertilized egg: the egg, having accepted one spermatozoon, repels all others.) It seems that the various kinds of DNA are sufficiently alike to penetrate the outer defenses of the bacterial cell. Even DNA from thymus-gland cells of the calf will react with pneumococci, inhibiting their transformation by an appropriate DNA. Indeed, a foreign DNA can compete on about equal terms with pneumococcal DNA for entry into a susceptible pneumococcus. But only the native DNA seems capable of producing a genetic effect on the bacterium.

The experiments in transforming bacteria go back to a discovery made in 1928 by Fred Griffith, an English bacteriologist. Pneumonia was then one of the most challenging problems in medical research (the "miracle" drugs having not yet been discovered), and the pneumococcus was as popular a subject of study as the viruses are now. Griffith injected into mice pneumococci without capsules, which are not virulent, together with killed pneumococci of a virulent type (Type III capsules). To his surprise, the tissues of the mice were soon teeming with live, virulent pneumococci of Type III. Since the dead cells could not have come to life, it became evident that their material must somehow have transformed the nonencapsulated bacteria into virulent germs which had the ability to make Type III capsules.

Griffith's discovery was followed up by workers at the Rockefeller Institute, under the great and inspiring leadership of the late Oswald T. Avery. One cannot fail to note that the study of pneumococcal transformation since the initial discovery has been carried out essentially by a single "school," consisting of Avery's students and their followers. This school, now in its second generation and widely spread, can trace its lines of descent as accurately as those of the bacteria it studies.

By 1944 Avery, with Colin M. MacLeod and Maclyn McCarty, had identi-

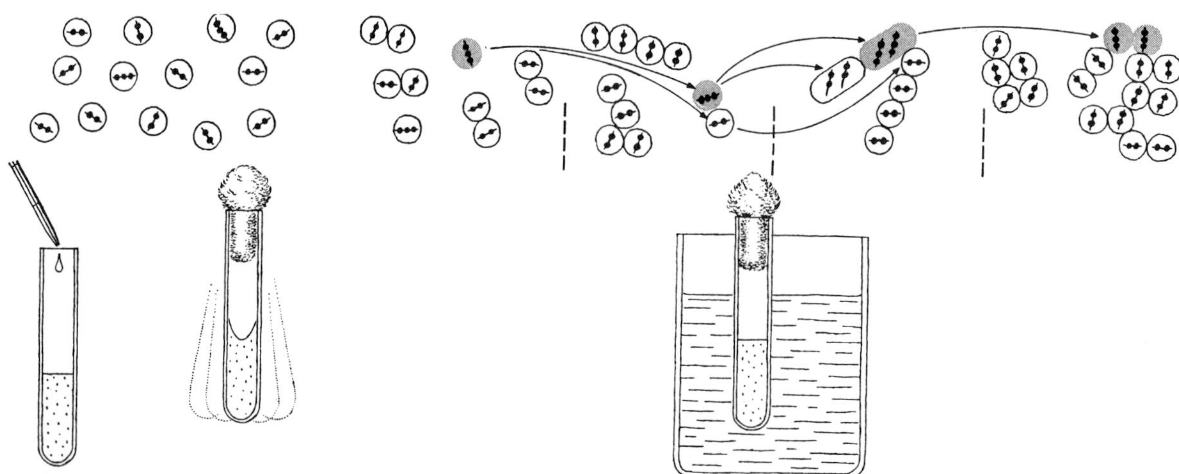

TRANSFORMATION OF PNEUMOCOCCI IS CONTINUED from the preceding two pages. First, desoxyribonuclease is shaken in the culture, which destroys the DNA that is not inside the cells. Second, the culture is reincubated for two hours. After about 45 minutes some cells show the new trait. These cells are slow to divide; for several generations the new DNA is passed on to only one daughter

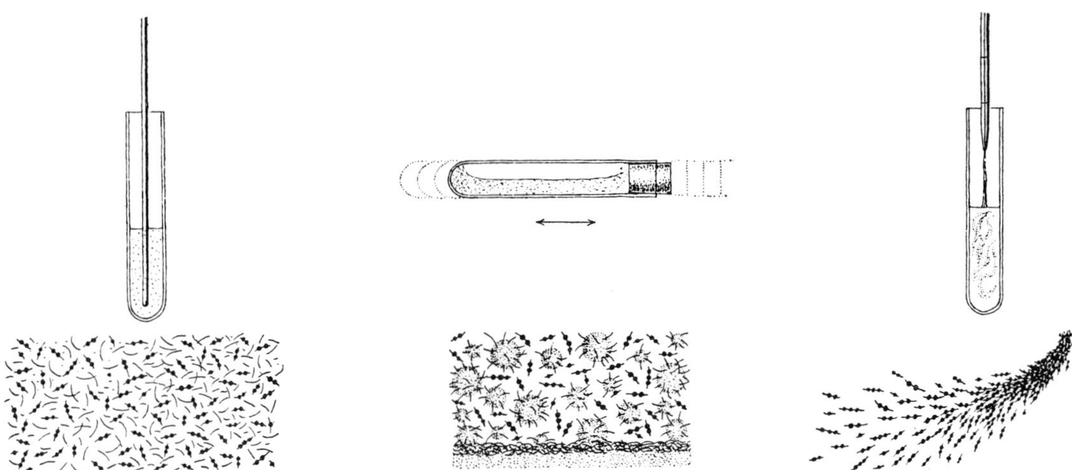

by winding them on a glass rod; this separates the DNA from most of the other substances in the debris of the broken cells. Fourth, the DNA is dissolved in a salt solution. Fifth, impurities are removed by adding various solvents to the solution and shaking it; the solvents tend to precipitate the impurities in particles. Sixth, alcohol is added to the purified DNA, causing it to form threads again.

fied the substance responsible for the transformation of pneumococci as DNA. There followed a long series of transformation experiments, not only on pneumococci but also on other species of bacteria. The germ once thought to be the cause of influenza (*Hemophilus influenzae*) has been studied extensively by Hattie Alexander and her associates at Columbia University, and their findings largely parallel those on pneumococci. Other investigators have reported success in transforming strains of *Escherichia coli*, *Shigella paradysenteriae* and meningococci with DNA.

Over the last 30 years there has been an impressive accumulation of evidence that nucleic acids play a central role in the hereditary mechanism of all living creatures [see "The Chemistry of Heredity," by A. E. Mirsky; Scientific American, Offprint No. 28]. The transformation work has had a decisive part in that vast investigation. This lead has generated a great number of exciting genetic experiments with animal and plant cells, bacteria and viruses, as the many recent articles on the subject in Scientific American have made plain. In the nucleic acids biologists at last have definite chemical substances which embody the properties of the somewhat hypothetical units long known as genes. Biochemists have not been slow to take up the challenge to explore the structure of the nucleic acids for the key to the machinery of heredity [see "The Structure of the Hereditary Material," by F. H. C. Crick, Scientific American Offprint 5; and "The Gene," by Norman H. Horowitz, page 5 in this book].

Happily, in the mid-20th century we can feel that we are on the threshold of still more exciting discoveries. We can expect to learn new kinds of facts about heredity in the coming years, and also to find theories which will unify the facts. The transformation of bacteria is one of our most promising laboratory tools for further discovery. It means that we can interbreed organisms by transferring a comparatively small and simple genetic unit—much simpler than the intricate apparatus of chromosomes involved in other genetic systems. The simplicity of this process gives many possibilities for controlled manipulation and variation. Transforming agents may be added or withheld at will, used in various concentrations or in combination with other materials, pretreated with chemicals, modified or damaged by exposure to acid, heat or radiant energy. The outcome of the transformations with DNA so treated should do much to throw light upon the still mysterious processes set in motion by the fascinating entities that we call genes.

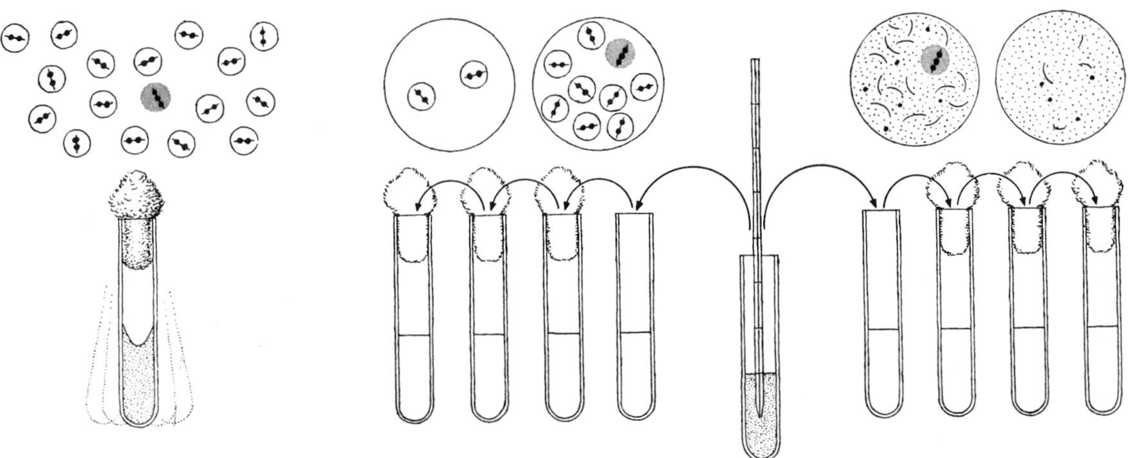

cell. Then one cell duplicates the DNA and passes it on to both daughter cells, which continue the duplication. Third, the tube is shaken to break up clumps of cells. Fourth, samples of the culture are placed in tubes, some of which (*right*) contain streptomycin. The streptomycin kills the untransformed cells. Later, colonies are visible in the tubes. In higher dilutions they can be counted.

# 5

# "Transduction" in Bacteria

NORTON D. ZINDER
*November 1958*

Viruses first made their existence known as especially tiny germs that cause disease in animals and plants. Then it was discovered that a virus can multiply itself only inside the living cell; the new viruses are released, and the cell dies. During the time it is inside, the virus vanishes in the biochemical system of its host, so that its activities there have been largely obscure. Bit by bit, however, the life story of the virus is being pieced together. As the parts fall into place, the virus is assuming a new identity. From the point of view of the biologist the germ has become a valuable ally in the exploration of the life processes of the cell.

Geneticists have found viruses particularly useful as a means of getting at the mechanism of heredity. The typical virus is a bit of genetic material encapsulated in a protein coat. As such it may affect the genetics of its host in important ways. This was first observed in the case of viruses that infect bacteria. At times these viruses cause a latent infection; they do not kill their host but become a part of its genetic apparatus. The latent virus then acts like a bit of the bacteria's genetic material and induces new traits in its host. In this role it may be reproduced through many generations along with the bacteria's own genes before it resumes its existence as a separate virus. It is such a latent virus infection which causes the normally innocuous diphtheria bacillus to make its lethal toxin.

This article is concerned with a discovery which implicates the virus even more deeply in the genetic processes of its host. It now appears that a bacterial virus can carry the bacteria's own genes from cell to cell. Like a disease carrier, it infects one bacterium with hereditary material picked up from another.

This "transduction" of bacterial heredity was discovered by accident and good luck in an investigation that was at first not concerned with viruses at all. The discovery occurred during an attempt to induce sexual mating in the bacteria which cause a disease in mice resembling typhoid fever in man. Mating is a rare process in bacteria. A bacterium ordinarily multiplies simply by dividing into two cells, each of which usually has the same genetic constitution as the other. In 1946, however, Joshua Lederberg and Edward L. Tatum (then at Yale University) found that a bacterium in one strain of the bacillus *Escherichia*

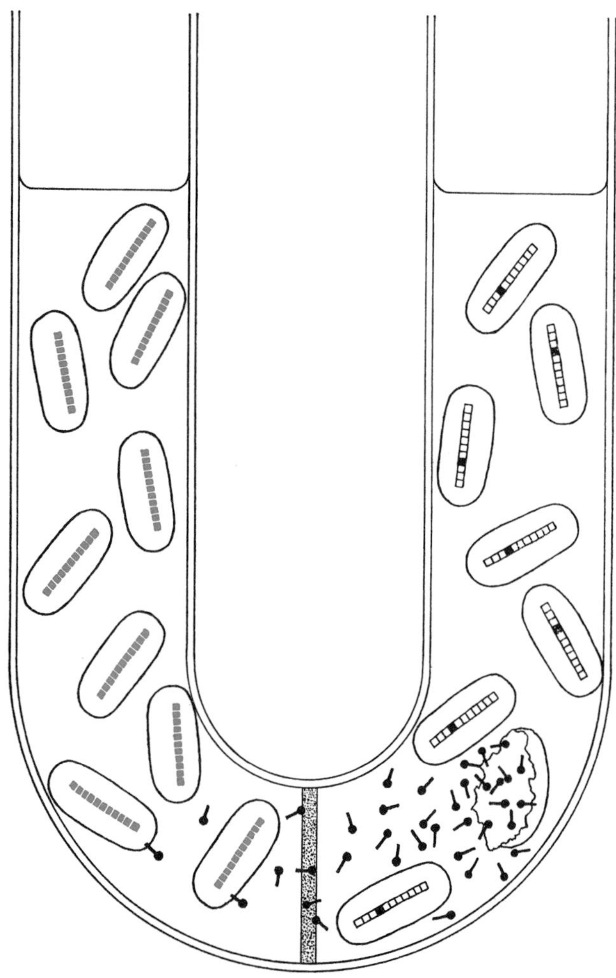

**U-TUBE EXPERIMENT led to discovery of transduction. At bottom of tube was a filter. On left side of filter was one strain of the bacterium *Salmonella typhimurium*; on right side, another strain. The strain at right harbored "latent" virus (*small black square on***

*coli* could mate under certain conditions with a bacterium in another strain of the same species, and thus give rise to bacteria with some characteristics of both strains. There seemed to be no reason why *E. coli* alone among bacteria should have the ability to mate, so in 1949 Lederberg and I (then at the University of Wisconsin) undertook to induce mating in another species.

For our first experiment we chose two strains of the mouse-typhoid bacterium, *Salmonella typhimurium*. Each strain lacked the capacity to synthesize a particular amino acid needed for its growth, but was able to make the amino acid which the other strain could not produce. If mating occurred, some of the offspring should be able to synthesize both factors; they could then be isolated by transfer to an agar medium that did not contain either of the two amino acids. On such a medium the parent strains would not be able to proliferate, but the new cells would form visible colonies in a few hours. Accordingly we mixed cells of the two parent strains and spread them on the selective agar. Colonies of the new type of cells appeared; apparently we had succeeded in mating *Salmonella*.

To be sure that the new cells were the product of mating, we tried to mate strains that were distinguished from each other by more than one trait. The offspring of bacterial mating may combine genes from their two parents in any proportion. We could therefore expect to find a variety of new cells, some more like one parent and some more like the other. We were surprised to find, however, that the offspring of these matings all resembled one parent, except for the trait by which they were isolated and which was supplied by the other parent. The transfer of only one trait at a time was not consistent with the idea of a mating process. Furthermore, one of the two strains always acted as the donor, and the other as the recipient, of the genetic trait.

We considered first the simplest alternative: the change was merely a random mutation. But this possibility had to be rejected because the change appeared much more frequently in mixed cultures of the parent strains than in unmixed cultures. Nor could it be an increase in the mutation rate in the mixed cultures, for the new trait of the recipient strain was always related to traits of the donor strain. We concluded that we had stumbled upon an instance of bacterial "transformation." In this process the genetic traits of one strain of bacteria are transformed by contact with the genetic material of another strain. No contact between the cells is necessary. In familiar instances of transformation, in fact, the cells of the donor strain are dead, and their genetic material—the deoxyribonucleic acid, or DNA, contained in their chromosomes—is released into the culture medium. The recipient strain incorporates some of this free DNA into its chromosomes, and thus acquires traits of the dead strain [see "Transformed

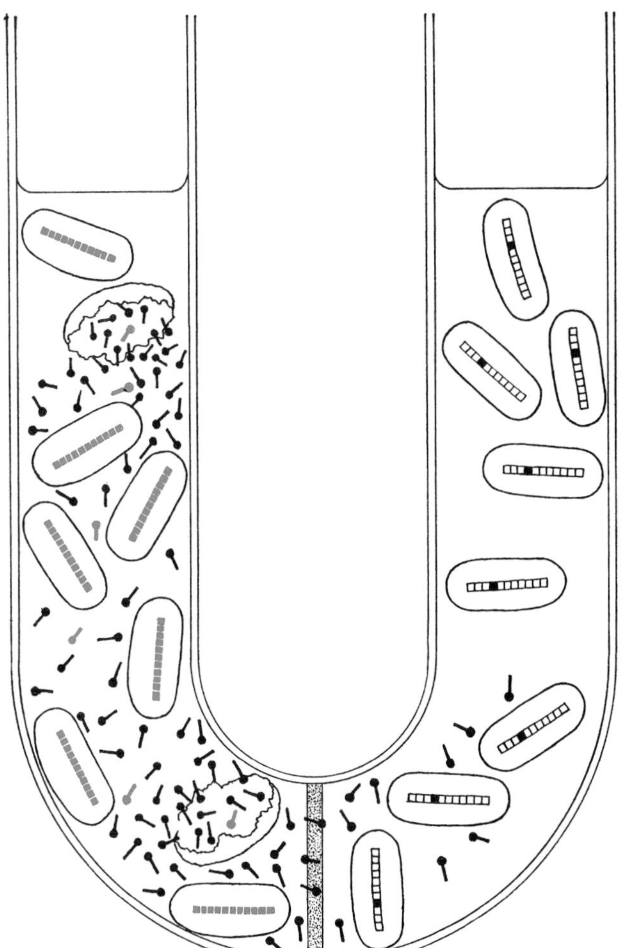

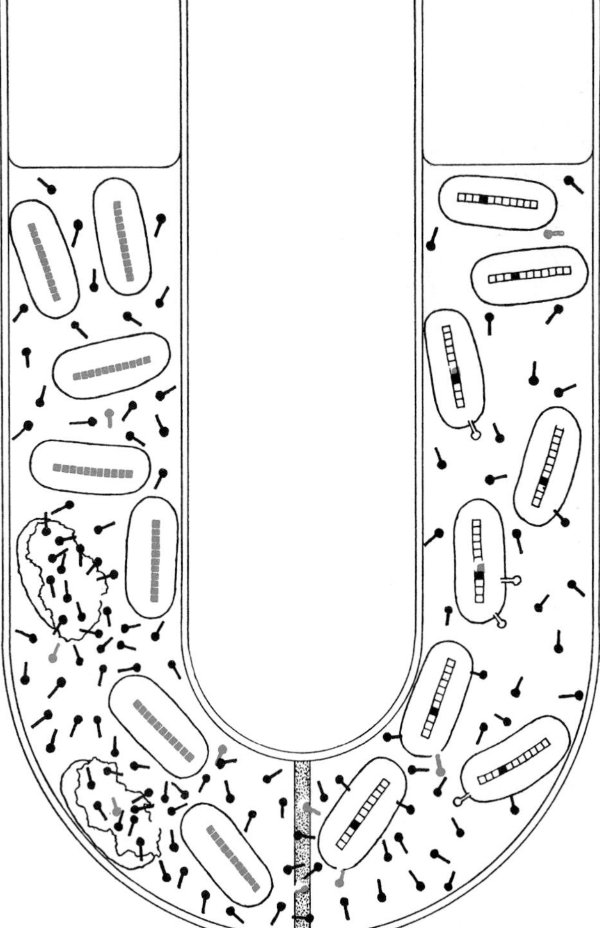

*schematic chromosome*). Occasionally one of these bacteria released live viruses (*tadpole-shaped objects*) which passed through filter (*first drawing*). In strain at left the viruses multiplied rapidly (*second drawing*). Some of the viruses (*colored*) carried bits of genetic material of the strain at left back through the filter to become part of genetic material of the strain at right (*third drawing*).

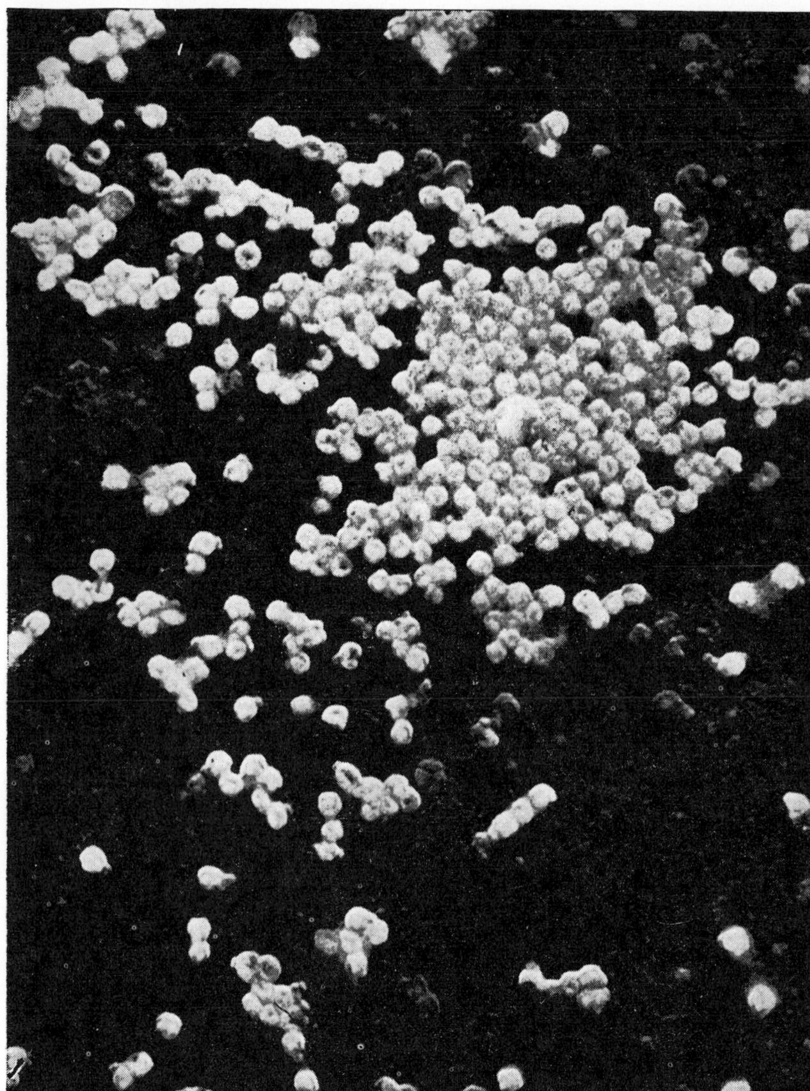

**VIRUS PARTICLES used for transduction of *Salmonella* are enlarged some 60,000 diameters in this electron micrograph made by Keith R. Porter of the Rockefeller Institute.**

Bacteria," by Rollin D. Hotchkiss and Esther A. Weiss, which begins on page 30 in this book]. Though well-established in certain species of bacteria, transformation had never been observed in *Salmonella*.

To test the transformation hypothesis, we now grew our two *Salmonella* strains in a specially constructed U-shaped tube. One strain was grown in one arm of the tube; the other strain, in the other arm. Between the two arms was a filter which prevented mating contact between the two strains [*see illustration on pages 36 and 37*]. To promote the exchange of substances secreted by the two strains, we gently flushed the nutrient broth from one side of the tube to the other. When the two strains were transferred to a selective medium, some offspring of the recipient strain showed a new trait picked up from the donor strain. Transformation seemed to be the answer. We then performed a more conventional transformation experiment, treating the recipient culture with pure DNA extracted from the donor strain. This, unexpectedly, had no effect at all. With our hypothesis now shaken, we went back to the U-tube. This time, with the idea of eliminating the possibility of transformation, we added an enzyme which destroys free DNA. In spite of the presence of the enzyme, genetic changes appeared just as before. We were obliged to discard transformation as well as mating, and to look for another explanation.

What could be happening in the U-tube? The donor bacterium passed its genetic material to the recipient in pieces small enough to pass through the filter, but the pieces were not damaged by the DNA-destroying enzyme. Both parents had to be present for the transfer to take place, but they did not have to be in direct contact. We wondered whether the donor perhaps gave off something other than DNA which could affect the heredity of the recipient. To test this idea, we added broth filtered from a culture of donor cells to a culture of recipient cells. No heritable changes resulted. However, when we grew the two strains together and filtered the broth, this fluid did produce changes in a fresh culture of recipient cells. At this point we realized there were two steps in the process: first the recipient had to produce something to stimulate the donor, and then the donor could send genetically active material back to the recipient.

Now we made a further discovery: once a culture of donor cells had been stimulated by exposure to fluids from the recipient strain, the fluid from this culture could stimulate other donor cells. The stimulating material was somehow reproduced in the donor bacteria. This reproductive capacity made us think of bacterial viruses. Viruses are small enough to pass with ease through the filter in the U-tube, and the protein coat of the virus protects its DNA from the enzyme which destroys free DNA.

The most familiar bacterial viruses are so destructive that an infected culture virtually disappears before your eyes. Obviously we would have noticed a virus of this type immediately. The "temperate" virus that causes a latent infection is harder to detect. Killing off only a small fraction of the cells, a latent infection may scarcely change the density of a culture. But occasionally one of the latent viruses regains its original form, multiplies in the cell and bursts forth to invade others. As a result a little free virus is always present in a culture of bacteria harboring a latent infection.

Sure enough, when we looked for viruses in cultures of donor *Salmonella* which had been treated with stimulating fluids, we found large numbers of virus particles. Further experiments indicated that virus activity and stimulating activity went hand in hand, and confirmed the fact that the virus was the stimulating agent.

It was but one step further to the theory that the viruses also acted as carriers of genetic material from donor to recipient bacteria. We hesitated to take this step, for the theory implied too much. It suggested, for example, that the

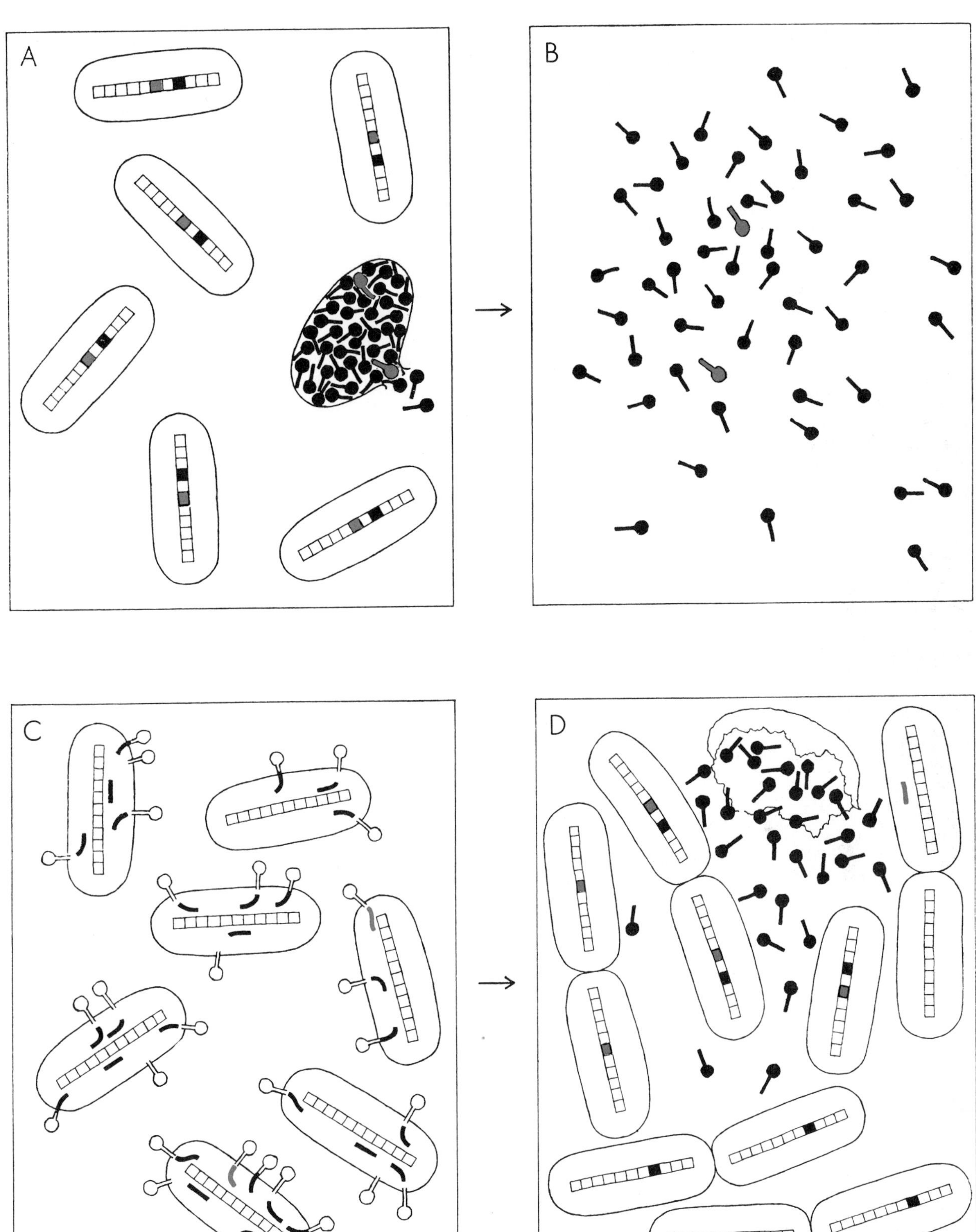


TRANSDUCTION requires first (A) a culture of donor cells which is infected with virus. Liquid filtered from this culture contains viruses (B), some of which carry genetic material from the bacteria (*colored*). Another strain of bacteria can be infected with this virus (C). Some of them (D) then produce more virus (*top*), some acquire genes of the donor bacteria (*colored square*), some acquire the genes of latent virus (*black square*), and some receive both. Occasionally a cell takes in a bacterial gene but does not incorporate it into its chromosome (*upper right*); gene is not duplicated when cell divides, so only one daughter cell receives it.

viruses of human diseases may carry genetic material from one host cell to another. But no other theory could explain the compelling evidence that the bacterial virus and the carrier of the genetic material from the donor to the recipient bacteria were identical in physical, chemical and biological properties.

The *Salmonella* mystery was now easily resolved. One of our two strains (the recipient) carried a temperate virus as a latent infection. The second strain (the donor) was especially susceptible to this virus. When the two were mixed, an occasional infective virus developed in the recipient strain and invaded a cell of the donor strain. When the offspring viruses erupted from the dying cell, most of them had genetic cores of virus DNA synthesized from the substance of the bacterial cell. But some of them incorporated particles of *Salmonella* DNA unchanged in their genetic cores. The normal viruses went on to cause infections in other bacterial cells, active or latent depending upon the strain of the bacteria. Those that carried bacterial DNA and invaded the recipient strain of bacteria brought about entirely different consequences. Instead of killing their host, they simply disappeared. The DNA they brought with them took its place in the chromosomes of the recipient cell and modified the nutritional or other characteristics of the cell's offspring.

It was pure chance that one of the strains chosen for our studies contained a latent virus, and that another was susceptible to this virus. We had certainly not expected to encounter a new genetic mechanism, and, considering the many factors that had to be in harmony, the discovery was extremely fortuitous.

In some ways the transduction of genetic traits by a virus closely resembles latent infection by a virus. The difference between the two processes lies in the nature of the genetic material carried by the virus, whether it is DNA picked up from a bacterial chromosome or true virus DNA. The DNA in a virus of a given strain has the same composition as the DNA in other viruses of the same strain. In latent infection this DNA will always take the same station in the host cell's chromosome, and will induce the same change in the characteristics of the host cell. The traits associated with latent virus—a new synthetic capacity or a change in the cell wall—appear in every cell harboring the virus. If the bacterial cells lose the virus genetic material, as evidenced by the disappearance of free virus from the culture, they simultaneously lose the trait associated with the virus.

On the other hand, the bacterial DNA carried by the virus may vary in composition, and each kind of DNA may be capable of producing a different trait. The properties of this DNA do not depend on the virus at all, but only on the bacterium from which it came. In the virus's new host, the bacterial DNA takes a station in the chromosome corresponding to the position it had occupied in the chromosome of the previous host. Only a very few of the cells in a culture will gain traits by transduction, but once incorporated such traits remain even after the strain loses the virus DNA.

Actually the transduction of a new trait to a bacterium is not always accompanied by a latent infection. Conversely a virus can take the latent form in a bacterium without establishing a transduction. Whether one virus particle can produce both effects is uncertain; a given cell is usually invaded by more than one virus. It is quite possible that a virus must lose some of its own DNA in order to acquire bacterial DNA, and as a result it may no longer be able to produce an active infection.

The piece of DNA that is picked up by the virus must be very small indeed. It was thought for a time that this fragment might correspond to a single gene unit, since a virus particle appeared to transfer just one bacterial trait at a time. In the course of our work, however, we discovered several traits which the virus regularly transfers in pairs. The piece of DNA carried by a virus must therefore be large enough for at least two genes, the two presumably lying adjacent or closely linked in the chromosome.

Studies of the swimming ability of

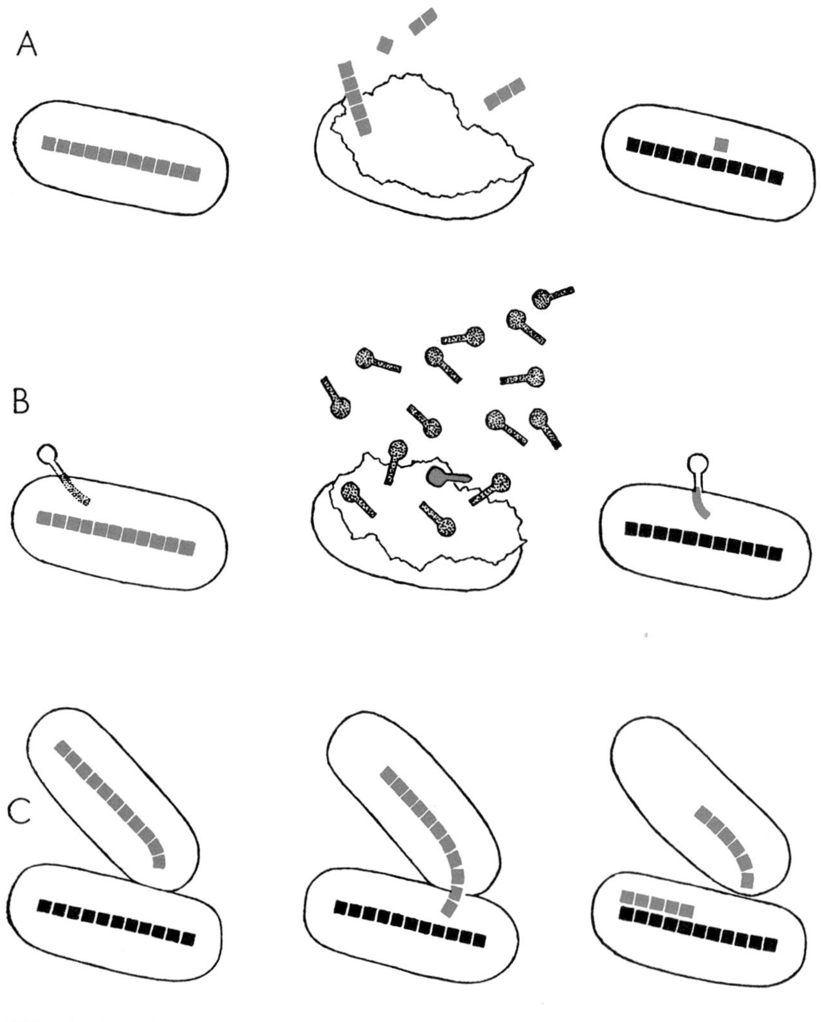

A

B

C

**THREE METHODS** of transferring genetic material from one bacterial cell to another are known. In "transformation" (A) the cells are disrupted to free the genetic material (*colored*), a small bit of which may then enter another cell. In transduction (B) viruses which can infect the cells carry genetic material from one to another. In mating (C) the material is passed by direct contact of cells, and a greater quantity can be transferred at one time.

*Salmonella* led us to the first pair of linked genes. Some *Salmonella* strains have whiplike tails (flagella) with which they can swim through liquids or semi-solid gelatins; others have no tails and cannot move. The tailless bacteria stay put and grow into colonies wherever they are placed, but the others swim off and spread in a cloudy swarm throughout the culture medium. The swimming strains are of different types, distinguished by the proteins of which their tails are made. We found that the ability to swim depends on two genes—one to determine whether or not the tail is made, and the other to determine the specific protein of which it is made. A tailless cell may already have a dormant gene for a kind of tail protein; if it acquires a gene for tail-making by transduction, it will make a tail of the protein type determined by its own gene. Almost as often, we found, the tailless cell will pick up a new gene for tail protein along with the tail-making gene; as a result it produces a tail of the donor's type instead of its own. The new genes introduced by transduction push out the corresponding genes in the bacterial chromosome and take their place.

The linkage between genetic traits revealed by transduction offers a clue to the linkage of the genes in the structure of the chromosomes. Transduction thus promises to be a useful tool in the important task of "mapping" chromosomes.

Bruce Stocker, a British bacteriologist, has demonstrated an abortive mode of transduction. The transducing DNA does not, in this case, replace a gene in the chromosome of the recipient cell, but by its very presence it induces a new trait. When the cell divides, however, the new gene is not duplicated, and only one of the daughter cells exhibits the trait. Stocker made this interesting discovery in an investigation of the swimming trait, when the recipient cells produced a trail of colonies rather than a spreading swarm. The colonies along the trail consisted entirely of tailless cells. But the trail had been laid by the swimming daughter cell which moved on to the next site after each division.

It seems possible to move any heritable trait from one cell to another by transduction. We have succeeded in transducing almost every trait we can reliably detect by experiment, including drug resistance, motility factors and antigenic factors. Transduction has been demonstrated in many kinds of *Salmonella*, in related species and in mating strains of *E. coli*. But ironically no one has succeeded in mating *Salmonella*.

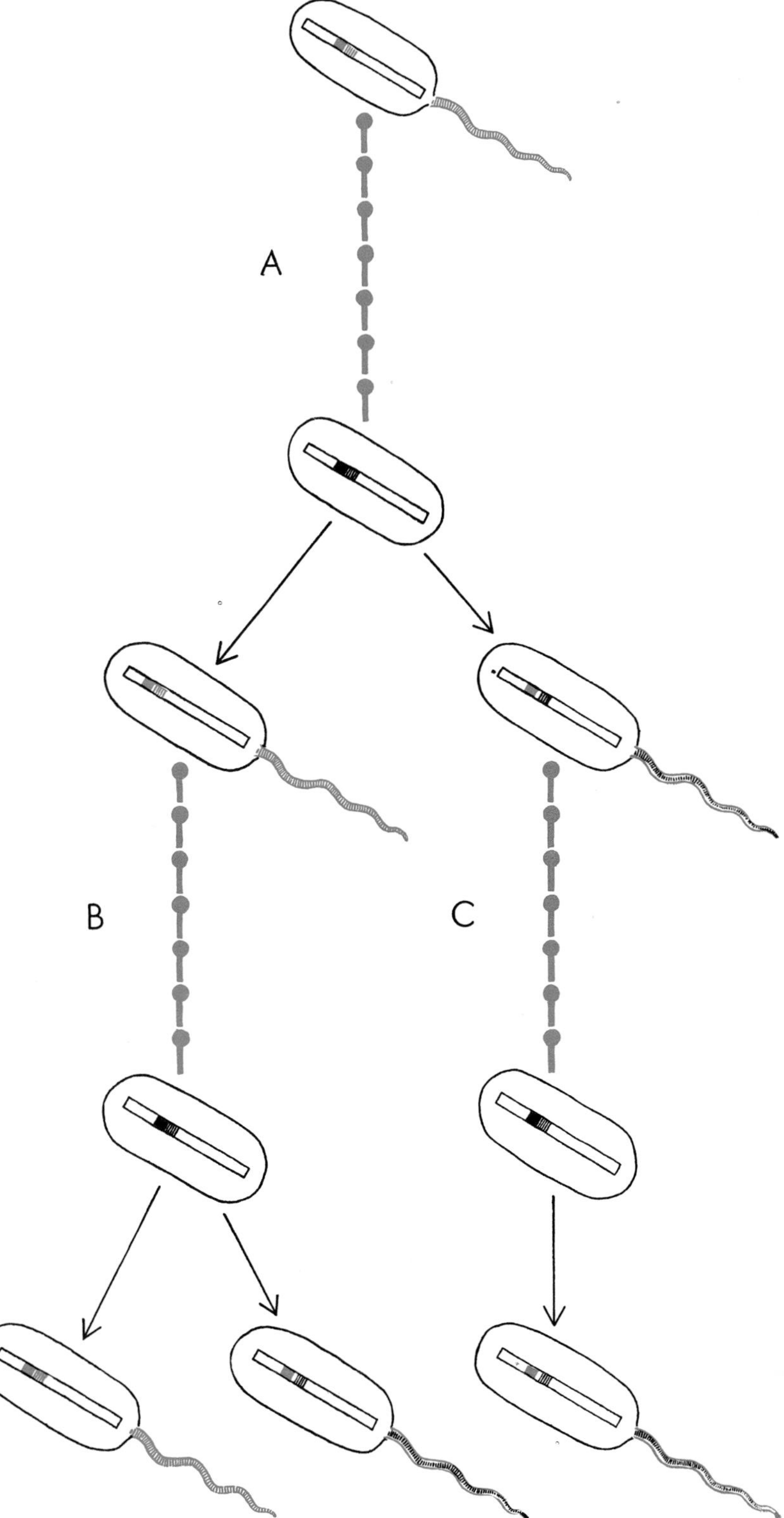


**TRANSDUCTION TO MOTILITY** revealed that two genes, presumably linked in the bacterial chromosome, can be carried together in a virus. Virus (A) from a culture of tailed cells (*top*) was used for transduction of untailed cells. Two types of motile cells resulted: some (*right center*) had received only the gene for tail-making and made tails of the protein type for which they already had a latent gene; an equal number (*left center*) had tails like the donor because they had also received a tail-protein gene. Transductions of untailed cells were repeated with virus from each of the new types of cells. One (B) gave rise to two types of tailed cells; the other (C) produced only one type of tailed cell.

# 6

# *Viruses and Genes*

FRANÇOIS JACOB and
ELIE L. WOLLMAN
*June 1961*

Almost everyone now accepts the unity of the inanimate physical world. Physicists do not hesitate to extrapolate laboratory results obtained with a small number of atoms to explain the source of the energy produced by stars. In the world of living things a comparable unity is more difficult to demonstrate; in fact, it is not altogether conceded by biologists. Nevertheless, most students of bacteria and viruses are inclined to believe that what is true for a simple bacillus is probably true for larger organisms, be they mice, men or elephants.

Accordingly we shall be concerned here with seeking lessons in the genetic behavior of the colon bacillus (*Escherichia coli*) and of the still simpler viruses that are able to infect the bacillus and destroy it. Viruses are the simplest things that exhibit the fundamental properties of living systems. They have the capacity to produce copies of themselves (although they require the help of a living cell) and they are able to undergo changes in their hereditary properties. Heredity and variation are the subject matter of genetics. Viruses, therefore, possess for biologists the elemental qualities that atoms possess for physicists. When a virus penetrates a cell, it introduces into the cell a new genetic structure that interferes with the genetic information already contained within the cell. The study of viruses has thus become a branch of cellular genetics, a view that has upset many old notions, including the traditional distinction between heredity and infection.

For a long time geneticists have worked with such organisms as maize and the fruit fly *Drosophila*. They have learned how hereditary traits are transmitted from parents to progeny, they have discovered the role of the chromosomes as carriers of heredity and they have charted the results of mutations—the events that modify genes. Complex organisms, however, multiply too slowly and in insufficient numbers for the high-resolution analyses needed to clarify such problems as the chemical nature of genes and the processes by which a gene makes an exact copy of itself and influences cellular activity. These detailed problems are most readily studied in bacteria and in viruses. Within the space of a day or two the student of bacteria or bacterial viruses can grow and study more specimens than the fruit-fly geneticist could study in a lifetime. An operation as simple as the mixing of two bacterial cultures on a few agar plates can provide information on a billion or more genetic interactions in which genes recombine to form those of a new generation.

It is the events of recombination, together with mutations, that model and remodel the chromosomes, the structures that contain in some kind of code the entire pattern of every organism. In recent years geneticists and biologists have clarified the nature of the hereditary message and have gained some clues as to what the letters of the code are. The primary, and perhaps the unique, bearers of genetic information in all forms of life appear to be molecules of nucleic acid. In living organisms, with the exception of some of the viruses, these long-chain molecules are composed of deoxyribonucleic acid (DNA). In all plant viruses and in some animal viruses the genetic substance is not DNA but its close chemical relative ribonucleic acid (RNA). DNA molecules are built up of hundreds of thousands or even millions of simple molecular subunits: the nucleotides of the four bases adenine, thymine, guanine and cytosine. These subunits, in an almost infinite variety of combinations, seem capable of encoding all the characteristics that all organisms transmit from one generation to the next. RNA molecules, which are somewhat shorter in length and not so well understood, act similarly for the viruses in which RNA is the genetic material.

Ultimately the role of the genes—the words of the hereditary message—is to specify the molecular organization of proteins. Proteins are long-chain molecules built up of hundreds of molecular subunits: the 20 amino acids. The sequence of nucleotides in the nucleic acid that contains the hereditary message is thought to determine the sequence of amino acids in the protein it manufactures. This process involves a "translation" from the nucleic-acid code into the protein code through a mechanism that is not yet understood.

## The Bacterial Chromosome

Before considering viruses as cellular genetic elements, we shall summarize the present knowledge of the genetics of the bacterial cell. In bacteria the hereditary message appears to be written in a

**SCORES OF VIRUSES of the strain designated $T_2$ are attached to the wall of a colon bacillus in this electron micrograph. The viruses are fastened to the bacterial wall by their tails, through which they inject their infectious genetic material. (Walls of the cell collapsed when the specimen was dried by freezing. "Shadowing" with uranium oxide makes objects stand out in relief.) The electron micrograph was made by Edouard Kellenberger of the University of Geneva. The magnification is 70,000 diameters.**

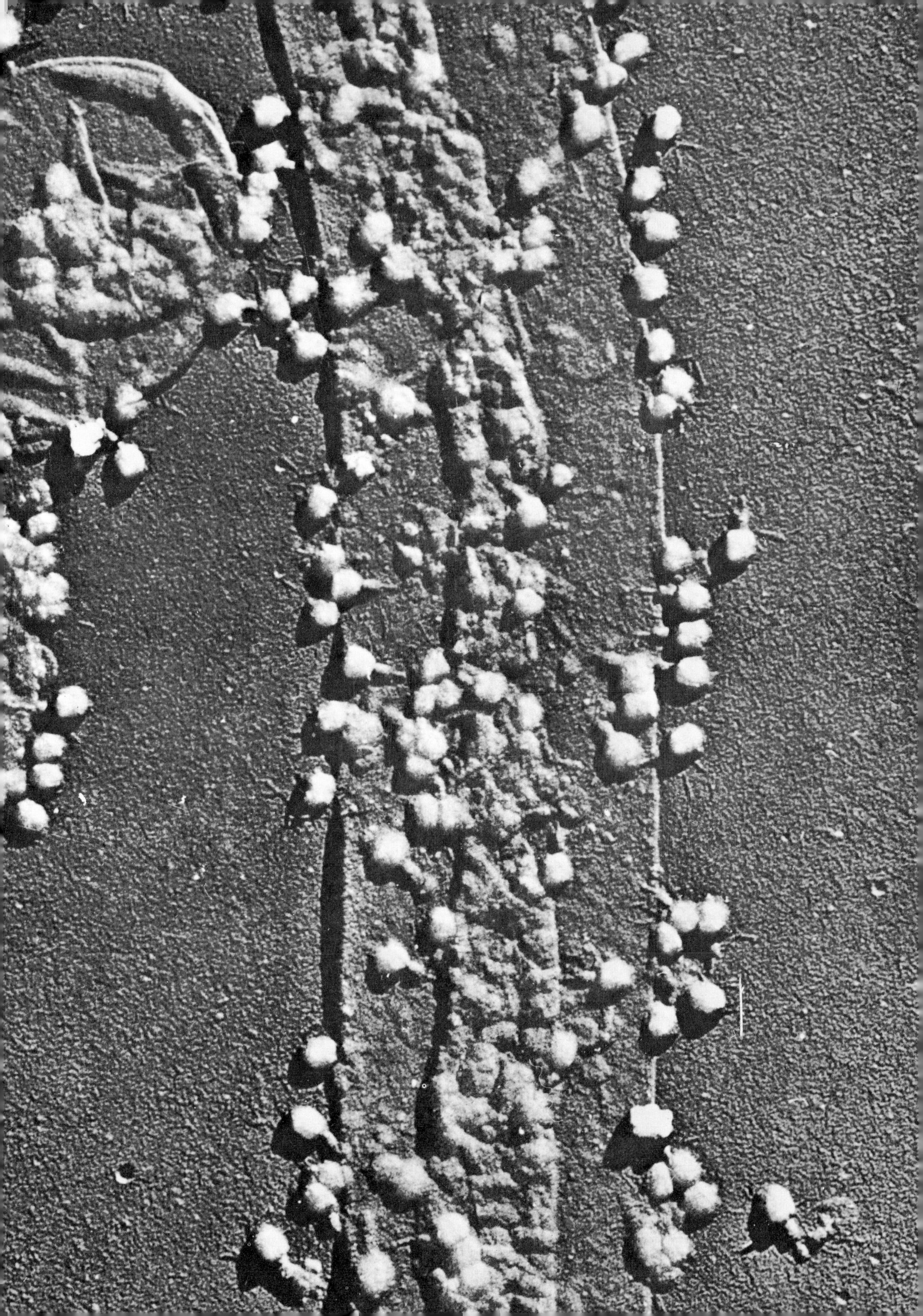

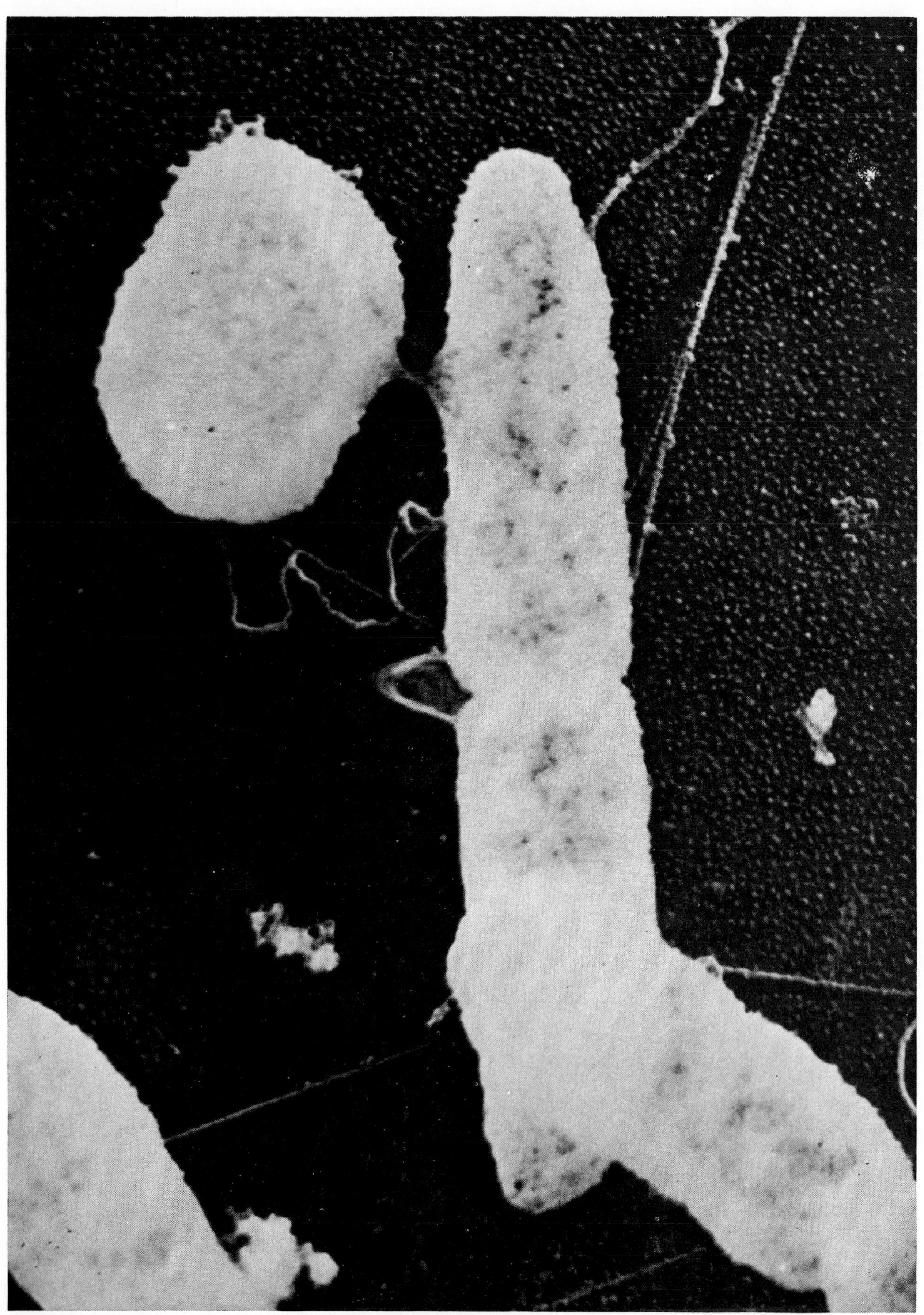

**CONJUGATING BACTERIA** conduct a transfer of genetic material. Long cell (*right*) is an *Hfr* "supermale" colon bacillus, which is attached by a short temporary bridge to a female colon bacillus (*see illustration on next two pages*). This electron micrograph, shown at a magnification of 100,000 diameters, was made by Thomas F. Anderson of the Institute for Cancer Research in Philadelphia.

single linear structure, the bacterial chromosome. For the study of this chromosome an excellent tool was discovered in 1946 by Joshua Lederberg and Edward L. Tatum, who were then working at Yale University. They used the colon bacillus, which is able to synthesize all the building blocks required for the manufacture of its nucleic acids and proteins and therefore to grow on a minimal nutrient medium containing glucose and inorganic salts. Mutant strains, with defective or altered genes, can be produced that lack the ability to synthesize one or more of the building blocks and therefore cannot grow in the absence of the building block they cannot make. If, however, two different mutant strains are mixed, bacteria like the original strain reappear and are able to grow on a minimal medium.

Lederberg and Tatum were able to demonstrate that such bacteria are the result of genetic recombination occurring when a bacterium of one mutant strain conjugates with a bacterium of another mutant strain. Further work by Lederberg, and by William Hayes in London, has shown that the colon bacillus also has sex: some individuals act as males and transmit genetic material by direct contact to other individuals that act as recipients, or females. The difference between the two mating types may be ascribed to the fertility factor (or sex factor) *F*, present only in males. Curiously, females can easily be converted into males; during conjugation certain types of male, called $F^+$, transmit their sex factor to the females, which then become males.

## The Chromosome "Essay"

Our own work at the Pasteur Institute in Paris has shed light on the different steps involved in bacterial conjugation and on the mechanism ensuring the transfer of the chromosome from certain strains of male, called *Hfr*, to females. When cultures of such males and of females are mixed, pairings take place between male and female cells through random collisions. A bridge forms between the two mating bacteria; one of the chromosomes of the male (bacteria have generally two to four identical chromosomes during growth) begins to migrate across the bridge and to enter the female. In the female, portions of the male chromosome have the ability to recombine with suitable portions of one of the female chromosomes. The chromosomes may be compared to written essays that differ only by a few letters, or a few words, corresponding to the mutations. Portions of the two essays may become paired, word for word and letter for letter. Through the process known as genetic recombination, which is still very mysterious and challenging, fragments of the male chromosome, which can be anything from a word or a phrase up to several sentences, may be exactly substituted for the corresponding part of the female chromosome. This process gives rise to a complete new chromosome that contains a full bacterial essay in which some words from the male have replaced corresponding words from the female. The new chromosome is then replicated and transmitted to the daughter cell.

Perhaps the most remarkable feature of bacterial conjugation is the way in which the male chromosome migrates across the conjugation bridge. For a given type of male the migration always starts at the same end of the chromosome, which, if we represent the bacterial chromosome by the letters of the alphabet, we can call *A*. Then, with the chromosome proceeding at constant speed, it takes two hours before the other end, Z, has penetrated the female. After the mating has begun, conjugation can be interrupted at will by violently stirring the mating mixture for a minute or so in a blender. The mechanical agitation does not kill the cells but it disrupts the bridge and breaks the male chromosome during its migration. The fragment of the male chromosome that has entered the female before the interruption is still functional and has the ability to provide words or sentences for a chromosome [*see illustration on pages 46 and 47*]. If conjugation is mechanically interrupted at various intervals after the onset of mating, it is found that any gene carried by the male chromosome, from *A* to Z, enters the female at a precise time. We have therefore been able to draw two kinds of detailed chromosome map showing the location of genes. One map, the conventional kind, is based on the observed frequency of different sorts of genetic recombination; the second is a new kind of map reflecting the time at which any gene penetrates the female cell. The latter can be compared to a road map drawn by measuring the times at which a car proceeding at a constant speed passes through various cities.

Finally, the mode of the male chromosome's migration has provided a unique opportunity for correlating genetic measurements with chemical measurements of the chromosome. In collaboration with Clarence Fuerst, who is now working at the University of Toronto, we have grown male bacteria in a medium containing the radioactive isotope phosphorus 32, which is incorporated into the DNA of the bacterial chromosome. The labeled bacteria are then frozen and kept in liquid nitrogen to allow some of the radioactive atoms to disintegrate. At various times samples are thawed and the labeled males are then mated with unlabeled females. The experiments show that the radioactive disintegrations sometimes break the chromosomes. If the break occurs between two markers, say *E* and *F*, the head part, *ABCDE*, is transferred to the female, but the tail part, *FGHIJKLMNOPQRSTUVWXYZ*, is not. Therefore the greater the number of phosphorus atoms between the *A* extremity of the chromosome and a given gene, the greater the chance that a break will prevent this gene from being transferred to the female. It is thus possible to draw a chromosomal map showing the location of the genes in terms of numbers of phosphorus atoms contained in the chromosome between the known genes. When we compare this map with those obtained by genetic analysis or by mechanical interruption, we find that for a given type of male all three maps are consistent.

In some types of male mutant the genetic characters have the same sequence along the chromosome but the character injected first differs from one mutant to another. The characters can also be injected either in the forward direction or in the backward direction, that is, from *A* to Z or from Z to *A*, with the alphabet capable of being broken at any point. These observations can be explained most simply by assuming that all the genetic "letters" of the colon bacillus are arranged linearly in a ring and that the ring can be opened at various points by mutation. It seems, furthermore, that the opening of the ring is a consequence of the attachment of the sex factor to the chromosome. The ring opens at precisely the point where the factor *F*, which is free to move, happens to affix itself. A cell with the *F* factor affixed to the chromosome is called an *Hfr* male, or "supermale," because it enhances the transmission of chromosomal markers. *Hfr* stands for "high frequency of recombination." When the chromosome is opened by the *F* factor, one of the free ends initiates the penetration of the chromosome into the female, carrying the sequence of characters after it. The other end carries the sex factor itself and is the last to enter the female. The sex factor has other remarkable properties and we shall bring it back into our story later.

The long-range objective of such studies is to learn how the thousands of genes strung along the chromosome control the molecular pattern of the bacterial cell: its metabolism, growth and division. These processes imply precise regulatory mechanisms that maintain a harmonious equilibrium between the cellular constituents. At any time the bacterial cell "knows" which components to make and how much of each is needed for it to grow in the most economical way. It is able to recognize which kind of food is available in a culture medium and to manufacture only those protein enzymes that are required to get energy and suitable building blocks from the available food.

At the Pasteur Institute, in collaboration with Jacques Monod, we have recently found new types of gene that determine specific systems of regulation. Mutants have been isolated that have become "unintelligent" in the sense that they cannot adjust their syntheses to their actual requirements. They make, for example, a certain protein in large amounts when they need only a little of it or even none at all. This waste of energy decreases the cells' growth rate. It seems that the production of a particular protein is controlled by two kinds of gene. One, which may be called the structural gene, contains the blueprint for determining the molecular organization of the protein—its particular sequence of amino acid subunits. Other genes, which may be called control genes, determine the rate at which the information contained in the structural gene is decoded and translated into protein. This control is exercised by a signal embodied in a repressor molecule, probably a nucleic acid, that migrates from the chromosome to the cytoplasm of the cell. One of the control genes, called the regulator gene, manufactures the repressor molecule; thus it acts as a transmitter of signals. These are picked up by the operator gene, a specific receiver able to switch on or off the activity of the adjacent structural genes. Metabolic

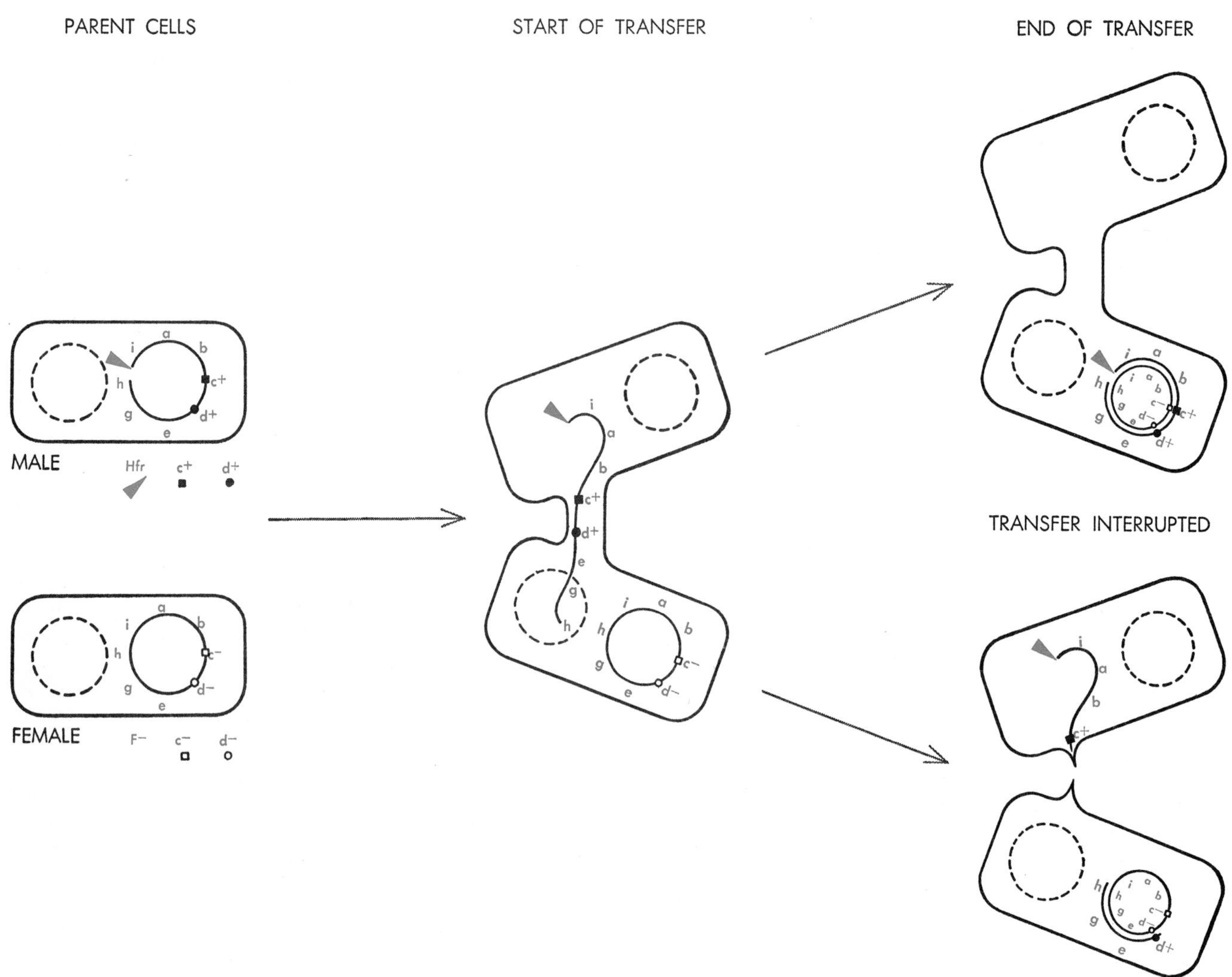


**CHROMOSOMAL TRANSFER** provides a primitive sexuality for colon bacillus. The bacterial chromosome, which appears to be ring-shaped, carries genetic markers (*designated by letters*), the presence or absence of which can be determined by studying cell's nutritional requirements. When the sex, or *F*, agent is attached to the chromosome, opening the ring, the cell is called an *Hfr* supermale. Two markers, labeled $c^+$ and $d^+$ when present and $c^-$ and $d^-$ when absent, can be traced from parents to daughter cells. When male and female cells conjugate, one of the male chromosomes (there are usually several, all identical) travels through the bridge.

products can interfere with the signals, either activating or inactivating the proper repressor molecules and thereby initiating or inhibiting the production of proteins.

Within the bacterial cell, then, there exists a complex system of transmitters and receivers of specific signals, by means of which the cell is kept informed of its metabolic requirements and enabled to regulate its syntheses. The bacterial chromosome contains not only a series of blueprints for the manufacture of individual molecular components but also a plan for the co-ordinated production of these components.

Let us now turn to the events that take place when a bacterial virus of the strain designated $T_2$ infects the colon bacillus. A $T_2$ virus is a structure shaped like a tadpole; by weight it is about half protein and half DNA. The DNA is enclosed in the head, the outside of which is protein; the tail is also composed of protein. The roles of the DNA and the protein in the infective process were clarified in 1952 by the beautiful experiments of Alfred D. Hershey and Martha Chase of the Carnegie Institution of Washington's Department of Genetics in Cold Spring Harbor, N. Y. By labeling the DNA fraction of the virus with one radioactive isotope and the protein fraction with another, Hershey and Chase were able to follow the fate of the two fractions. They found that the DNA is injected into the bacterium, whereas the protein head and tail parts of the virus remain outside and play no further role. Electron micrographs reveal that the tail provides the method of attachment to the bacterium and that the DNA is injected through the tail. The Hershey-Chase experiment was a landmark in virology because it demonstrated that the nucleic acid carries into the cell all the information necessary for the production of complete virus particles.

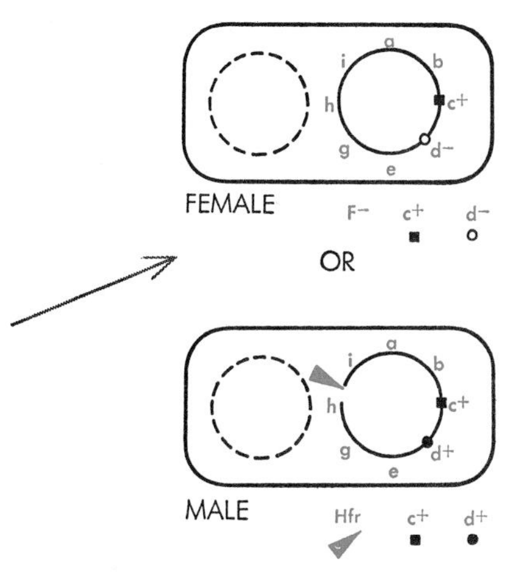


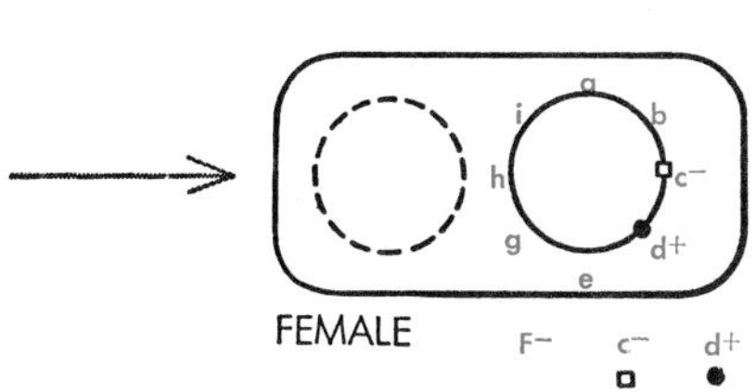


**If transfer is complete, daughter cells may be male or female and carry any marker of the male. If transfer is interrupted, daughters are all female and can carry only those markers passed before bridge was broken.**

## How Viruses Destroy Bacteria

A bacterium that has been infected by virus DNA will break open, or lyse, within about 20 minutes and release a new crop of perhaps 100 particles of infectious virus, complete with protein head and tail parts. In this brief period the virus DNA subverts the cell's chemical facilities for its own purposes. It brings into the cell a plan for the synthesis of new molecular patterns and the cell faithfully carries it out. The infected cell creates new protein subunits needed for the virus head and tail, and filaments of nucleic acid identical to the DNA of the invading particle. These pools of building blocks pile up more or less at random, and in excess amounts, inside the cell. Then the long filaments of virus DNA suddenly condense and the protein subunits assemble around them, creating the complete virus particle. The whole process can be compared to the occupation of one country by another; the genetic material of the virus overthrows the lawful rule of the cell's own genetic material and establishes itself in power.

A virus can therefore be considered a genetic element enclosed in a protein coat. The protein coat protects the genetic material, gives it rigidity and stability and ensures the specific attachment of the virus to the surface of the cell. As André Lwoff of the Pasteur Institute has pointed out, viruses can be uniquely defined as entities that reproduce from their own genetic material and that possess an apparatus specialized for the process of infection. The definition excludes both the cell and the specialized particles within the cell that serve its normal functions.

Another important criterion of viral growth is that of unrestricted synthesis. Infection with a virus is a sort of molecular cancer. The replication of the genetic material of the virus and the synthesis of the viral building blocks do not appear to be subject to any control system at all.

## Lysogenic Bacteria

When a $T_2$ virus infects a bacterium, it forces the host to make copies of it and ultimately to destroy itself. Such a virus is said to be virulent, and when it is inside the cell, reproducing itself, it is said to be in the vegetative state.

There are, however, other bacterial viruses, called temperate viruses, which behave differently. After entering a cell the genetic material of a temperate virus can take two distinct paths, depending on the conditions of infection. It can enter the vegetative state, replicate itself and kill the host, just as a virulent virus does. Under other circumstances it does not replicate freely and does not kill the host. Instead it finds its way to the bacterial chromosome, anchors itself there and behaves like an integrated constituent of the host cell. Thereafter it will be transmitted for years to the progeny of the bacterium like a bacterial gene. We know that the bacterial host has not destroyed the invading particle, because from time to time one of the daughter cells in the infected line will break open and yield a crop of virus particles, as it would if it had been freshly attacked by a virulent virus. When the virus is in the subdued and integrated state, it is called a provirus. Bacteria carrying a provirus are called lysogenic, meaning that they carry a property that can lead to lysis and death.

Lysogeny was discovered in the early 1920's, soon after the discovery of the bacterial virus itself, and it remained a profound mystery for some 25 years. The mystery was explained by the fine detective work of Lwoff and his colleagues [see "The Life Cycle of a Virus," by André Lwoff; SCIENTIFIC AMERICAN, March, 1954]. Lwoff found that when he exposed certain types of lysogenic bac-

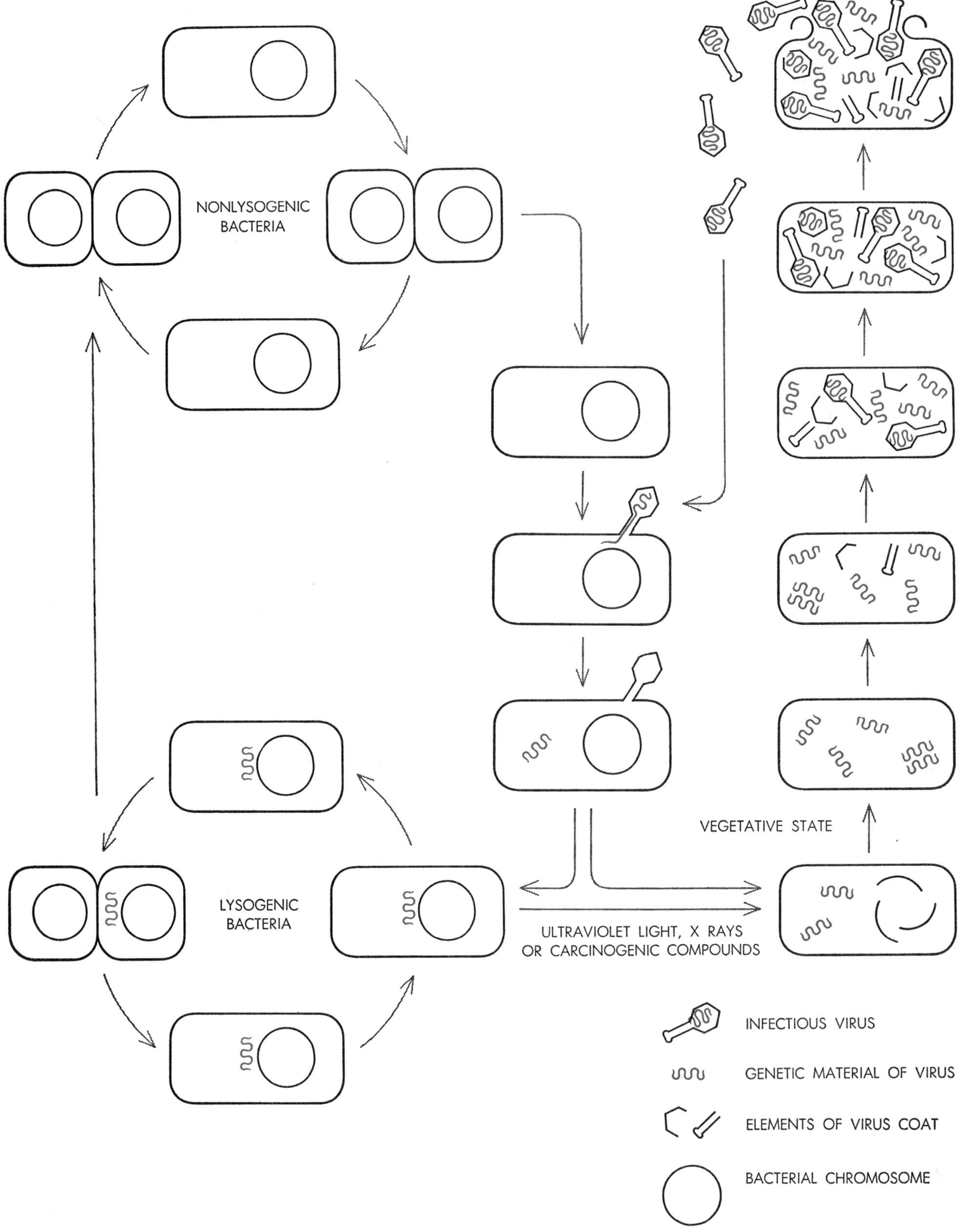


**LIFE CYCLE OF BACTERIAL VIRUS** shows that, for the bacterium attacked, infection and death are not inevitable. After the genes of the virus (*color*) enter a cell descended from a completely healthy line (*top left*), the cell may take either of two paths. One (*far right*) leads to destruction as the virus enters the vegetative state, makes complete copies of its infective self and bursts open the cell, a process called lysis. The other path leads to the so-called lysogenic state, in which the viral genes attach themselves to the bacterial chromosome and become a provirus; the cell lives. Exposure to ultraviolet light, however, can dislodge the provirus and induce the vegetative state. The provirus is sometimes lost during cell division, returning the cell to the nonlysogenic state.

teria to ultraviolet light, X rays or active chemicals such as nitrogen mustard or organic peroxides, the whole bacterial population would lyse within an hour, releasing a multitude of infectious virus particles. When a provirus is thus activated, or "induced," it leaves the integrated state and enters the vegetative state, eventually destroying the cell [*see illustration on opposite page*].

To determine the position of the provirus inside the host cell, we can apply the method of interrupting the sexual conjugation of bacteria that carry a provirus and are therefore lysogenic. In this way we can correlate the location of the provirus with that of known characters on the bacterial chromosome. Each of 15 different types of provirus takes a particular position at a specific site on the bacterial chromosome. Only one is an exception; it seems free to take a position anywhere. In the proviral state the genetic material of the virus has not become an integral part of the bacterial chromosome; instead it appears to be added to the chromosome in an unknown but specific way. However it may be hooked on, the genetic material of the virus is replicated together with the genetic material of the host. It behaves like a gene, or rather as a group of genes, of the host.

## Nonviral Effects of Provirus

The presence of this apparently innocuous genetic element, the provirus, can confer on the lysogenic bacteria that harbor it some new and striking properties. It is not at all obvious why some of these properties should be related to the presence of a provirus. As one example, diphtheria bacilli are able to produce diphtheria toxin only if the bacilli carry certain specific types of provirus. The disease diphtheria is caused solely by this toxin.

In other instances the presence of a provirus is responsible for a particular type of substance coating the surface of a bacterium. The substance can be identified by various immunological tests (typically by noting if a precipitate forms when a certain serum is added). The nonlysogenic strain, carrying no provirus, will bear a different substance. In such cases the genes of the virus are responsible for hereditary properties of the host. They can scarcely be distinguished from the genes of the bacterium.

The most striking property the provirus confers on its bacterial host is immunity from infection by external viruses of the same type as the provirus. When

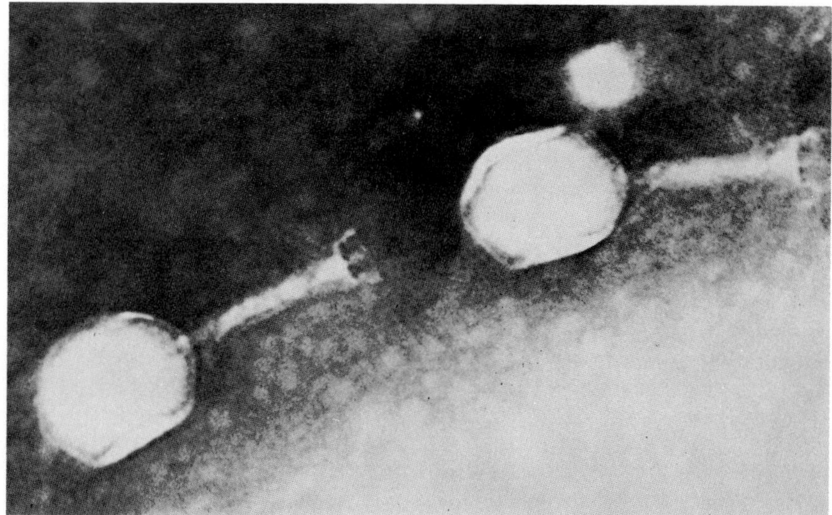

**INTACT $T_2$ VIRUS has polyhedral head membrane and a curious pronged device at the end of its tail. The magnification is 200,000 diameters. This electron micrograph and the two below were made by S. Brenner and R. W. Horne at the University of Cambridge.**

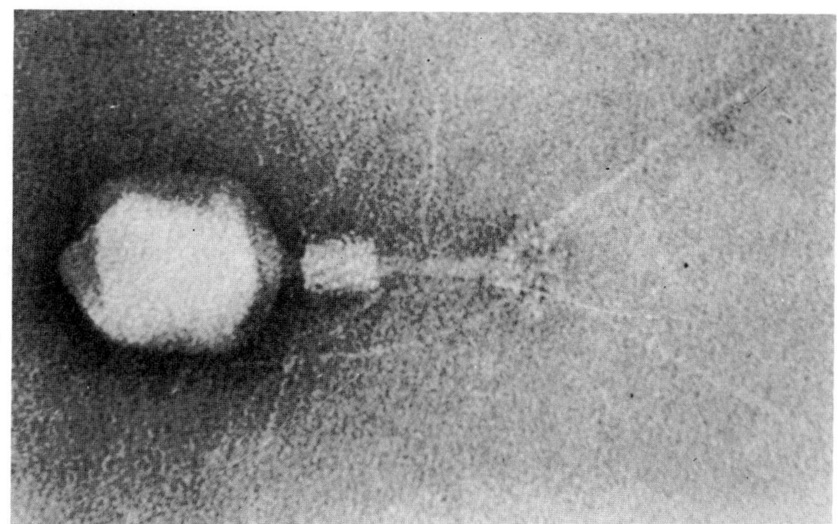

**"TRIGGERED" $T_2$ VIRUS results from exposure to a specific bacterial substance that causes contraction of the tail sheath (*stubby cylinder*) and discharge of viral genes.**

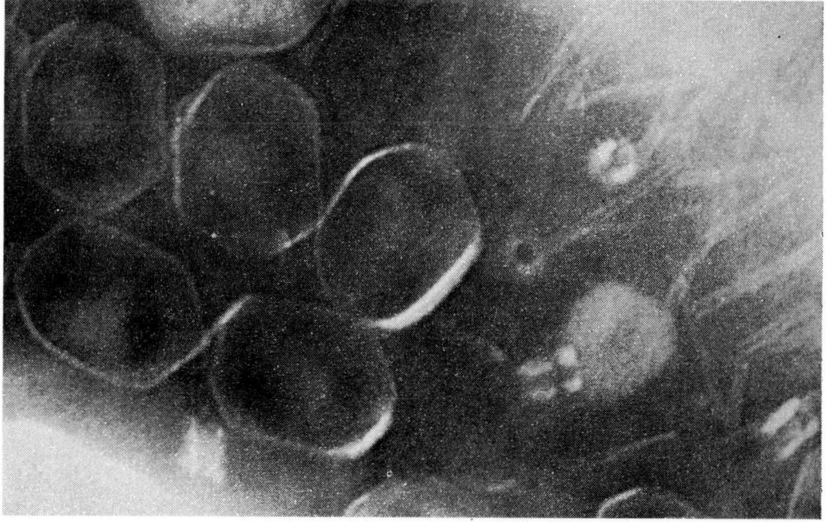

**ISOLATED $T_2$ PARTS can be found still unassembled if host cell is forced to burst open before synthesis of virus particles is complete. Parts include head membranes and tails.**

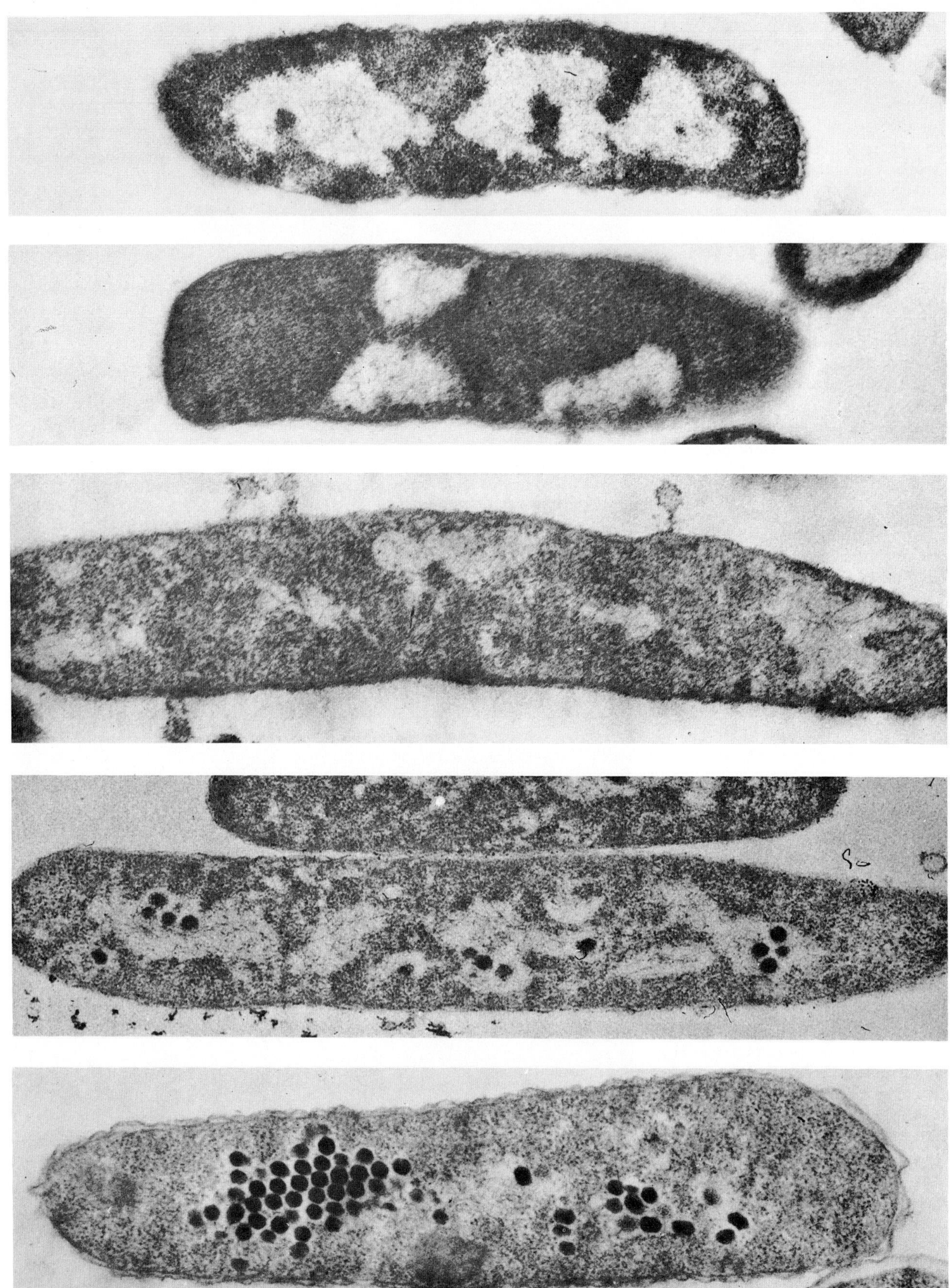

**GROWTH OF $T_2$ VIRUS** inside bacterial host is revealed in a striking series of electron micrographs by Kellenberger. Top picture shows the colon bacillus before infection. Four minutes after infection (*second from top*) characteristic vacuoles form along the cell wall. Ten minutes after infection (*third from top*) the virus has reorganized the entire cell interior and has created pools of new viral components. Twelve minutes after infection (*fourth from top*) new virus particles have started to condense. Thirty minutes after infection (*bottom*) more than 50 fully developed $T_2$ viruses have been produced and the cell is about ready to burst open.

lysogenic cells are mixed with such viruses, the virus particles adsorb on the cell and inject their genetic material into the cell, but the cell survives. The injected material is somehow prevented from multiplying vegetatively and is diluted out in the course of normal bacterial multiplication.

In the past two years we have attempted to learn more about the mechanism of this immunity. It seems clear that the mere attachment of the provirus to the host chromosome cannot account for the immunity of the host. The provirus must do something or produce something. We have evidence that the immunity is expressed by a substance or factor not tied to the chromosome. Remarkably enough, the system of immunity appears to be similar to the cellular systems already described that regulate the synthesis of protein in growing bacteria. It seems that the provirus produces a chemical repressor capable of inhibiting one or several reactions leading to the vegetative state. Thus immunity can be visualized as a specific system of regulation, involving the transmission of signals (repressors), which are received by an invading virus particle carrying the appropriate receptor.

## Transduction

The close association that may take place between the genetic material of the virus and that of the host becomes even more striking in the phenomenon of transduction, discovered in 1952 by Norton D. Zinder and Lederberg at the University of Wisconsin [see " 'Transduction' in Bacteria," by Norton D. Zinder, which begins on page 36 in this book]. They found that when certain proviruses turn into infective viruses, thereby killing their hosts, they may carry away with them pieces of genetic material from their dead hosts. When the viruses infect a host that is genetically different, the genes from the old host—the transduced genes—may be recombined with the genes of the new host. The transduction process seems able to move any sort of gene from one bacterial host to another.

Lysogeny and transduction therefore represent two complementary processes. In lysogeny the genes of the virus become an integral part of the genetic apparatus of the host and replicate at the pace of the host's chromosome. In transduction genes of the host become linked to the genes of the virus and can replicate at the unrestricted viral pace when the virus enters the vegetative state.

Viruses, like all other genetic ele-

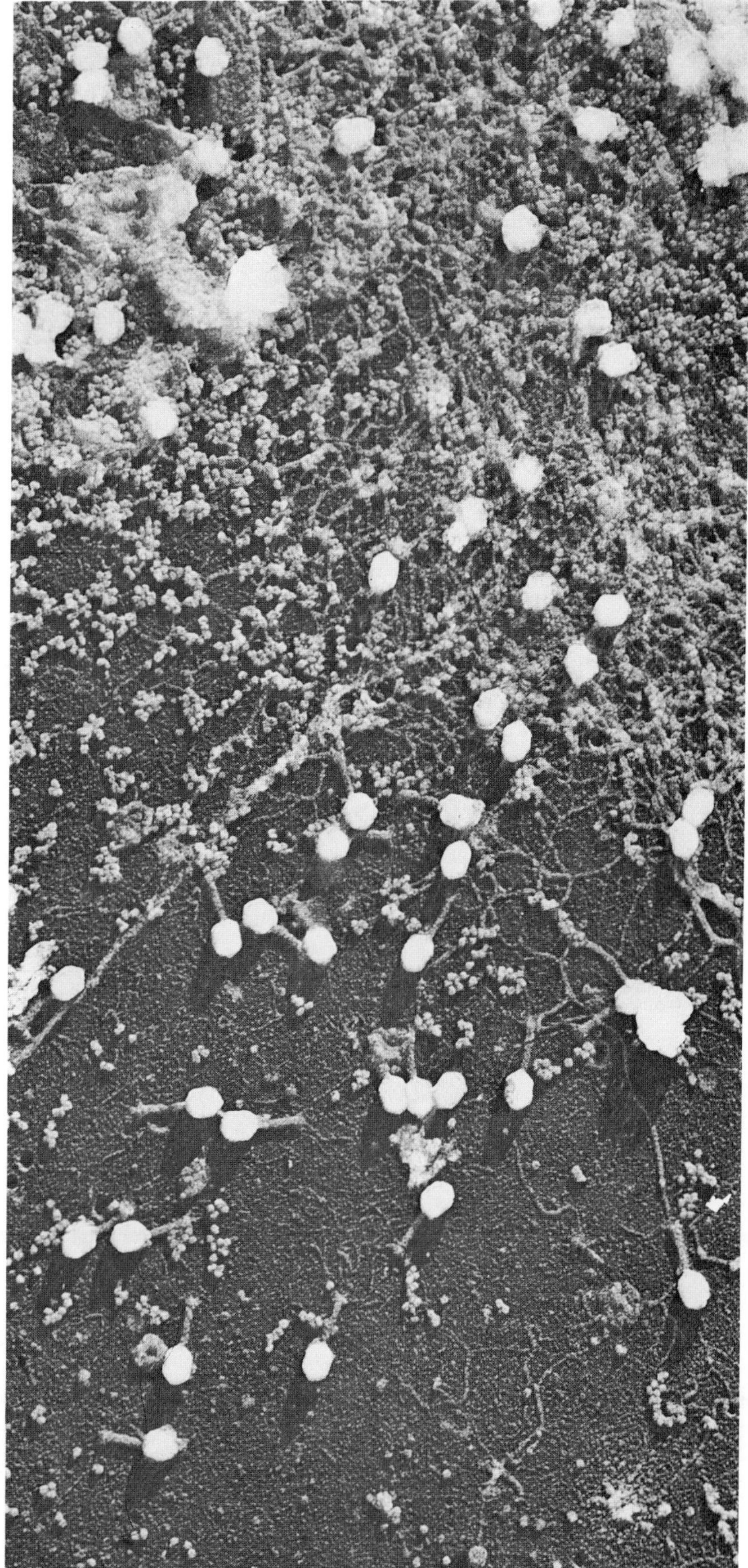

DEATH OF A BACTERIUM occurs when $T_2$ virus particles, having multiplied inside their host (*see sequence on opposite page*), dissolve the walls of the bacterial cell and spill out—a phenomenon called lysis. Viruses are the large white objects; the other matter is cellular debris. The electron micrograph (magnification: 50,000 diameters) is by Kellenberger.

ments, can undergo mutations, and these produce a variety of stable, heritable changes. The mutations of particular interest are those that prevent the formation of mature, infectious virus particles. Lysogenic bacteria in which such mutations have taken place are called defective lysogenic bacteria. These bacteria hereditarily perpetuate a mutated provirus, which is perfectly able to replicate together with the host's chromosome. If these cells are exposed to ultraviolet radiation, which activates the provirus, we observe that the defective lysogenic cells die without releasing any infectious viruses. Examination of such bacteria usually shows that virus subunits have started to appear inside the cell but have failed to reach maturity [*see illustrations on pages 54 and 55*]. Evidently some essential step in the formation of

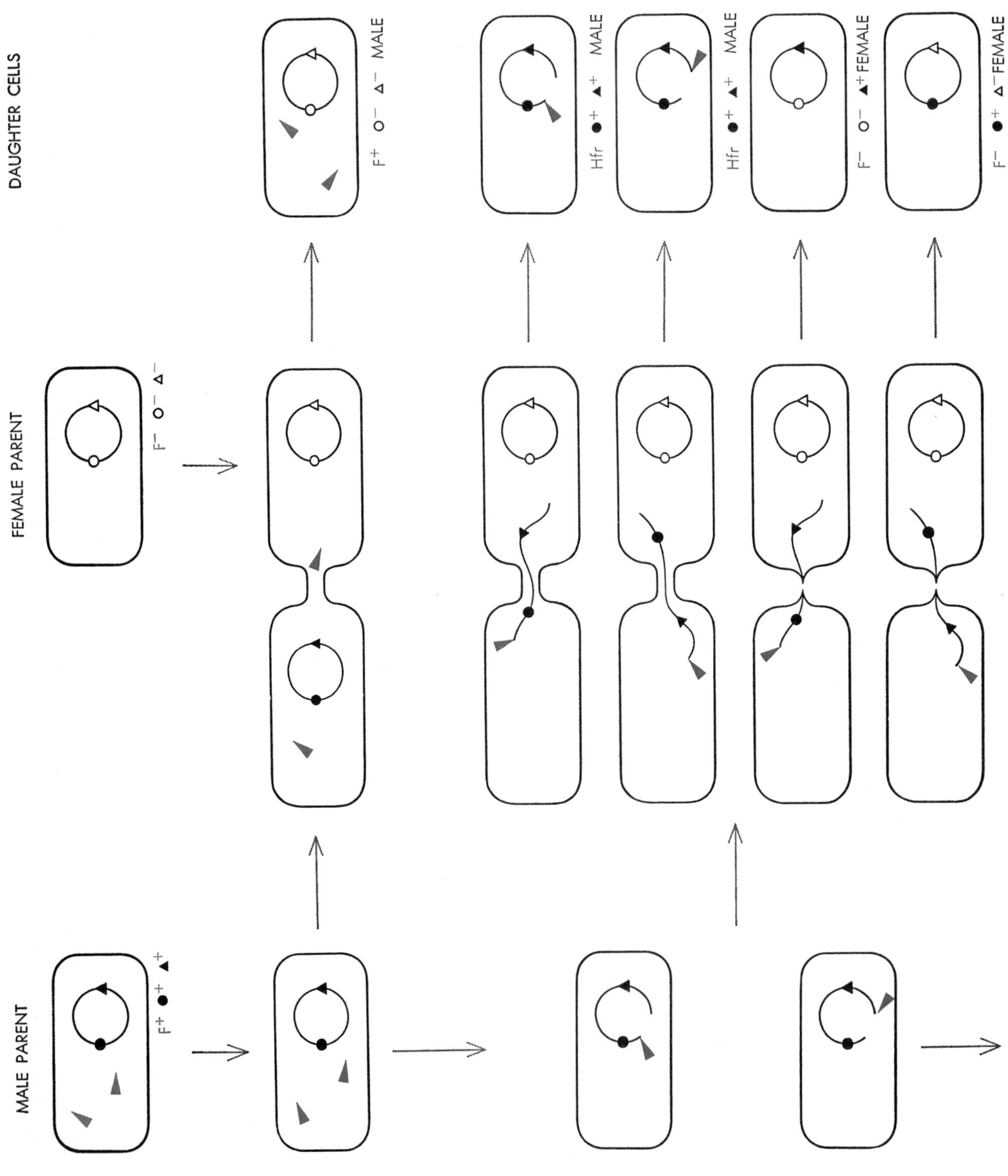

an infectious virus has been blocked by the mutation.

Just as we can study how other kinds of mutation block biochemical pathways associated with cell nutrition, we can try to identify the biochemical blockages that keep the provirus from multiplying normally. When a defective provirus turns to the vegetative state, some viral components begin to appear but the process halts. By using various biological tests, together with electron microscopy, we try to establish how far the process has gone. We have been able to identify two ways in which the process is halted and to relate the blockage to two main groups of viral genes.

One group of genes is concerned with the autonomous reproduction of the genetic material of the virus. The DNA of the provirus, which was able to repli-

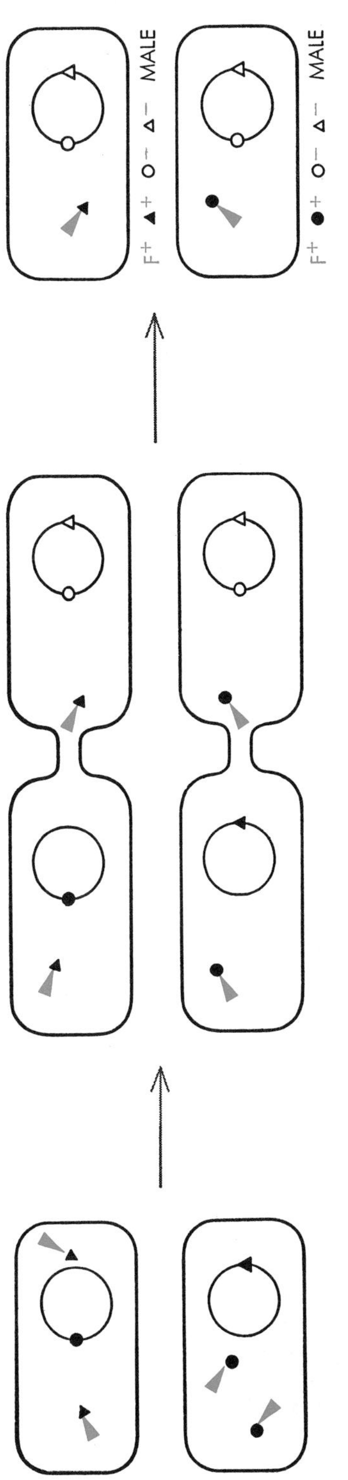


**F, OR SEX, AGENT**, indicated by colored wedge, is a versatile and busy "broker" in genes. It can be attached to the bacterial chromosomes (*integrated*) or unattached (*nonintegrated*) and can alternate between the two states. When nonintegrated, it usually transmits only itself when bacteria conjugate (*top sequence*). When integrated, it opens chromosome ring and is the last marker transferred in conjugation (*middle sequence*). Daughters may inherit markers in combinations other than those shown. When *F* agent leaves integrated state (*bottom*), it may remove a marker and transfer it (*sex-duction*).

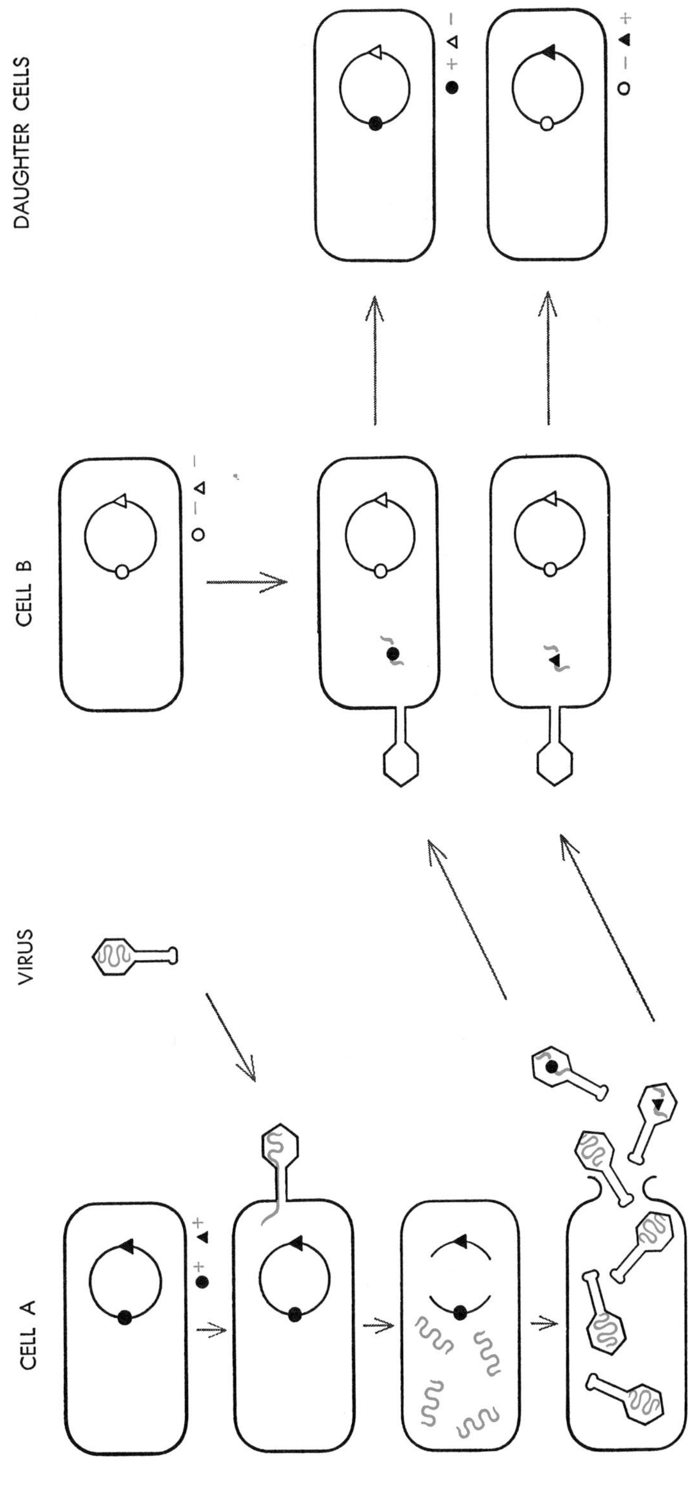


**TRANSDUCTION** is similar to sex-duction and was discovered earlier. In transduction the agent for transferring bacterial genes is a virus particle rather than an *F* agent. The virus injects its genes (*color*) into bacterial cell *A* and the genes create new copies of the virus. Occasionally the new virus particles so formed enclose a few genes from the chromosome of the bacterial host along with a few viral genes. These imperfect viruses are able to inject their contents into another cell (*cell "B"*) but are unable to destroy it. In this way genes (*solid black shapes*) can be transferred from cell *A* to the daughters of cell *B*.

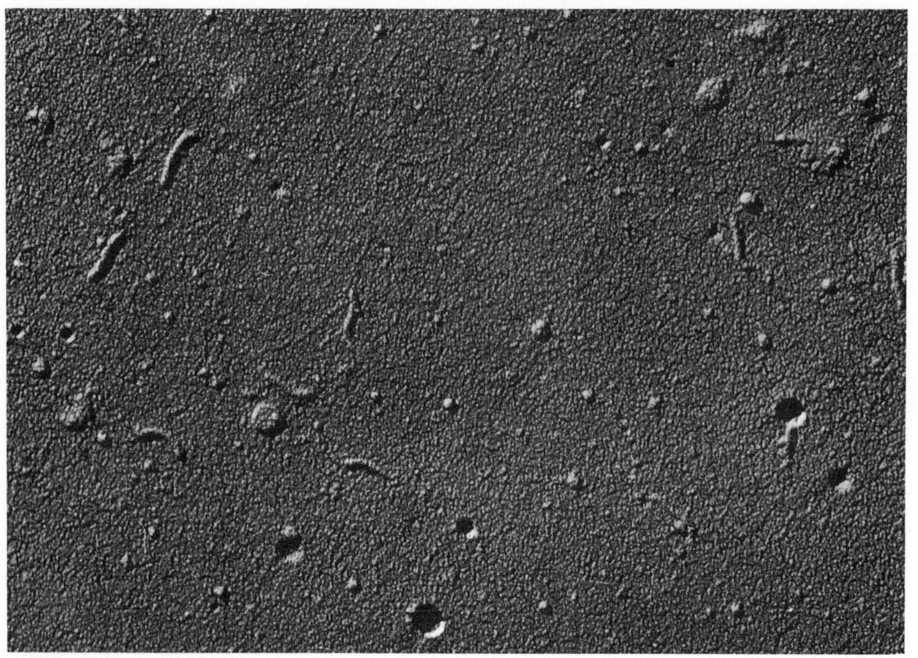

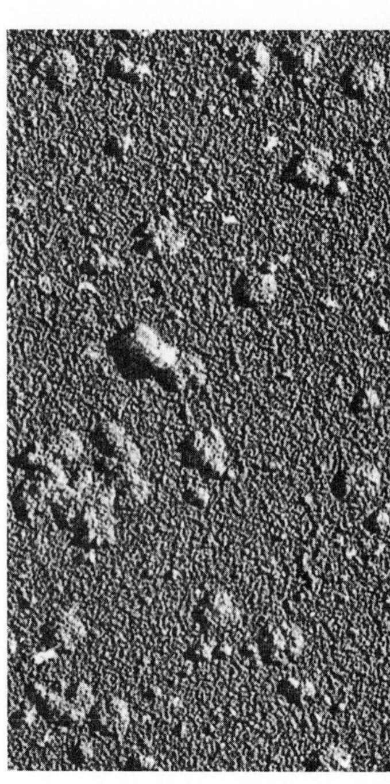

**INCOMPLETE VIRUS PARTICLES are created by defective proviruses (*see illustration below*). The electron micrograph at left shows virus heads and tails that remain unassembled because of some defect. Occasionally (*right*) only heads can be found. Electron**

cate when attached to the host chromosome, becomes unable to replicate on its own. A second group of genes is involved in the manufacture of the protein molecules that provide the coat and infectious apparatus of a normal virus. We have examples in which there is plenty of viral DNA, and many components of the coat material, but one or another essential protein is missing.

This study leads us to conclude that what distinguishes the genetic material of a virus from genetic elements of other types is that the virus carries two sets of information, one of which is necessary for the unrestricted multiplication of the viral genes and the other for the manufacture of an infectious envelope and traveling case.

The concept of a virus as it has emerged from the study of bacterial vi-

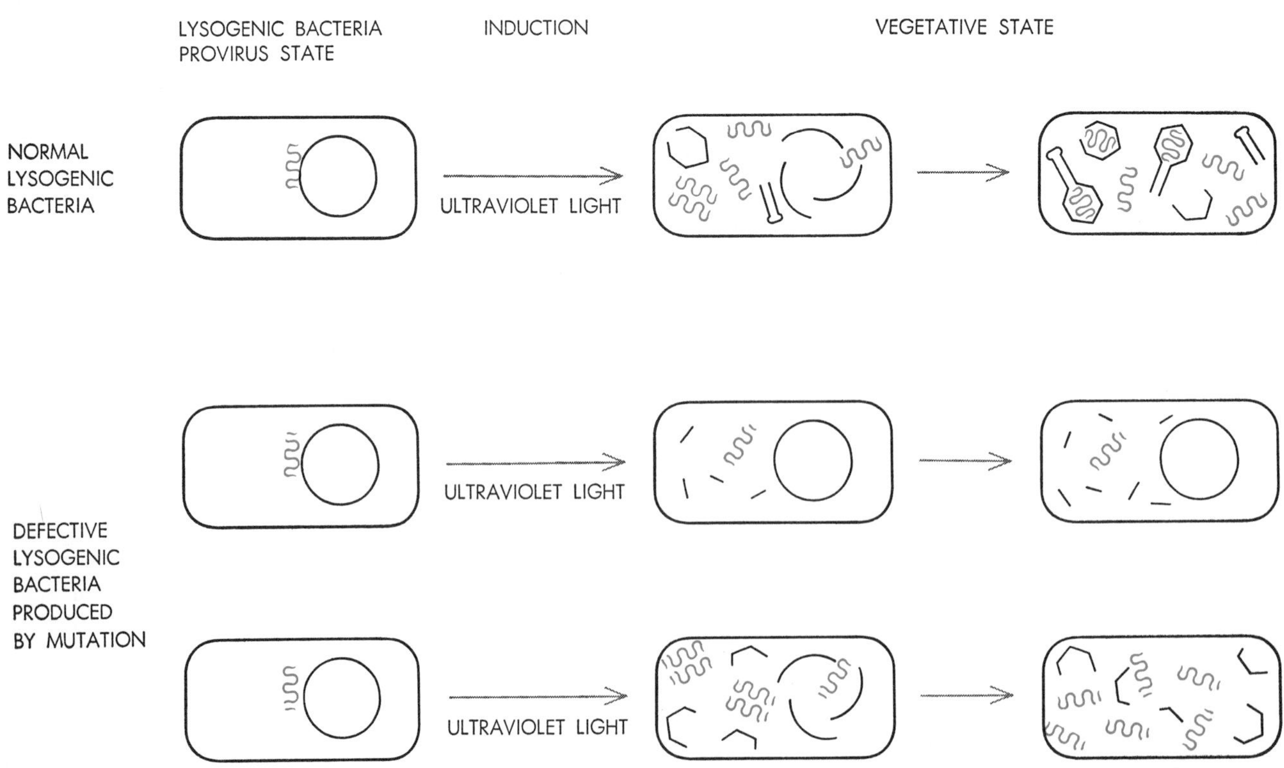


**DEFECTIVE LYSOGENIC BACTERIA appear as mutations among normal lysogenic bacteria. Upon induction with ultraviolet light a normal provirus (*color, top left*) leaves the bacterial chromosome, replicates, produces infectious virus particles and kills its host. When defective proviruses are induced, the host cell may also be killed, but no infectious viruses appear at lysis. In**

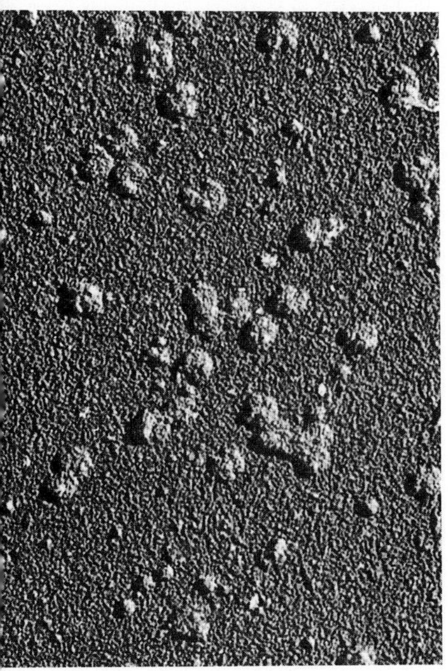

micrographs (magnification: 57,000) were made by Kellenberger and W. Arber.

ruses is far more complex and more fascinating than the concept that prevailed only a decade ago. As we have seen, a virus may exist in three states; the only thing common to the virus in the three states is that it carries at all times much the same genetic information encoded in

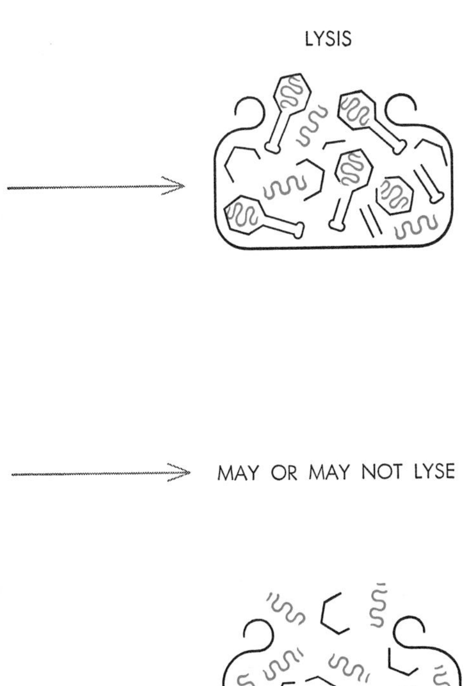


some cases (*middle*) the viral genes fail to replicate. In others (*bottom*) they replicate but the jacketing components are defective.

DNA. In the extracellular infectious state the nucleic acid is enclosed in a protective, resistant shell. The virus then remains inert like the spore of a bacterium, the seed of a plant or the pupa of an insect. In the vegetative state of autonomous replication the genetic material is free of its shell, overrides the regulatory mechanism of the host and imposes its own commands on the synthetic machinery of the cell. The viral genes are fully active. Finally, in the proviral state the genetic material of the virus has become subject to the regulatory system of the host and replicates as if it were part of the bacterial chromosome. A specific system of signals prevents the genes of the virus from expressing themselves; complete virus particles are therefore not manufactured.

## The Concept of the "Episome"

Less than a decade ago there was no reason to doubt that virus genetics and cell genetics were two different subjects and could be kept cleanly apart. Now we see that the distinction between viral and nonviral genetics is extremely difficult to draw, to the point where even the meaning of such a distinction may be questionable.

As a matter of fact there appear to be all kinds of intermediates between the "normal" genetic structure of a bacterium and that of typical bacterial viruses. Recent findings in our laboratory have shown that phenomena that once seemed unrelated may share a deep identity. We note, for example, that certain genetic elements of bacteria, which we have no reason to class as viral, actually behave very much like the genetic material of temperate viruses. One of these is the fertility, or *F*, factor in colon bacilli; in the so-called *Hfr* strains of males the *F* agent is attached to one of various possible sites on the host chromosome. In the males bearing the *F* agent designated $F^+$ the agent is not fixed to the chromosome and so it replicates as an autonomous unit. It bears one other striking resemblance to provirus. The integrated state of the *F* factor excludes the nonintegrated replicating state, just as a provirus immunizes against the vegetative replication of a like virus.

Another genetic agent resembling provirus is the factor that controls the production of colicines. These are extremely potent protein substances that are released by some strains of colon bacillus; the proteins are able to kill bacteria of other strains of the same or related species. The colicinogenic factors also seem to exist in two alternative states: integrated and nonintegrated. In the latter state they seem able to replicate freely and eventually at a faster rate than does the bacterial chromosome. Bacteria that lack these genetic elements—*F* agents and colicinogenic factors—cannot, so far as we know, gain them by mutation but can only receive them (by sexual conjugation, for example) from an organism that already possesses them. They may replicate either along with the chromosome or autonomously. Such genetic elements, which may be present or absent, integrated or autonomous, we have proposed to call "episomes," meaning "added bodies" [*see illustration on page 56*].

The concept of episomes brings together a variety of genetic elements that differ in their origin and in their behavior. Some are viruses; others are not. Some are harmful to the host cell; others are not. The important lesson, learned from the study of mutant temperate bacterial viruses, is that the transition from viral to nonviral, or from pathogenic to nonpathogenic, can be brought about by single mutations. We also have impressive evidence that any chromosomal gene of the host may be incorporated in an episome through some process of genetic recombination. During the past year, in collaboration with Edward A. Adelberg of the University of California, we have shown that the sex factor, when integrated, is able to pick up the adjacent genes of the bacterial chromosome. Then this new unit formed by the sex factor and a few bacterial genes is able to return to the autonomous state and to be transmitted by conjugation as a single unit. This process, in many respects similar to transduction, has been called sex-duction [*see illustration at left on pages 52 and 53*].

Do episomes exist in organisms higher than bacteria? We do not know; but if we accept the basic unity of all cellular biology, we should be confident that the answer is yes and that mice, men and elephants must harbor episomes. So far the great precision and resolution that can be achieved in the study of bacterial viruses cannot be duplicated for more complex organisms. There is, nevertheless, evidence for episome-like factors in the fruit fly and in maize. There have been reports of two viruses in the fruit fly, transmitted through the egg to the offspring, which may exist either as nonintegrated or as integrated elements. Although it does not seem that the virus is actually located on the chromosome in the latter state, the resemblance to provirus is striking. Barbara McClintock, of

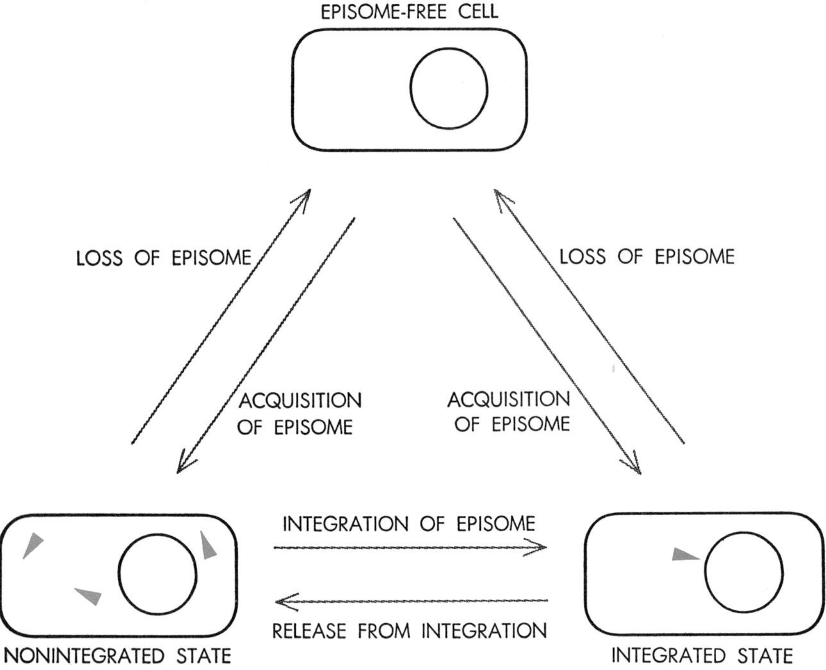


**CONCEPT OF THE "EPISOME," as put forward by the authors, describes a genetic element, such as the *F* agent, that may be either attached to the chromosome or unattached. When integrated, it replicates at host's pace; nonintegrated, it replicates autonomously.**

the Carnegie Institution of Washington's laboratory at Cold Spring Harbor, has discovered in maize "controlling elements" that are able to switch a gene off or on. (A gene responsible for a reddish color in corn may be switched on and off so fast that a single kernel may turn out speckled.) The controlling elements in maize are not always present, but when they are, they are added to specific chromosomal sites and can move from one site to another or even from one chromosome to another. These elements, therefore, act like episomes.

The discovery of proviruses and episomes has brought to light a phenomenon that biologists would scarcely have considered possible a few years ago: the addition to the cell's chromosome of pieces of genetic material arising outside the cell. The bacterial episomes provide new models to explain how two cells that otherwise possess an identical heredity can differ from each other. The episome brings into the cell a supplementary set of instructions governing additional biochemical reactions that can be superimposed on the basic metabolism of the cell.

The episome concept has implications for many problems in biology. For example, two main hypotheses have been advanced for the origin of cancer. One assumes that a mutation occurs in some cell of the body, enabling the cell to escape the normal growth-regulating mechanism of the organism. The other suggests that cancers are due to the presence in the environment of viruses that can invade healthy cells and make them malignant [see "The Polyoma Virus," by Sarah E. Stewart; SCIENTIFIC AMERICAN Offprint No. 77]. In the light of the episome concept the two hypotheses no longer appear mutually exclusive. We have seen that proviruses, living peacefully with their hosts, can be induced to turn to the vegetative, replicating state by radiation or by certain strong chemicals—the very agents that can be used to produce cancer experimentally in mice. If defective, the provirus will not even make viral particles. Malignant transformation involves a heritable change that allows a cell to escape the growth control of the organism of which it is a part. We can easily conceive that such a heritable change may result from a mutation of the cell, from an infection with some external virus or from the action of an episome, viral or not. Thus in the no man's land between heredity and infection, between physiology and pathology at the cellular level, episomes provide a new link and a new way of thinking about cellular genetics in bacteria and perhaps in mice, men and elephants.

# 7
# *Infectious Drug Resistance*

TSUTOMU WATANABE
*December 1967*

The advent of sulfonamide drugs and antibiotics brought with it the promise that bacterial disease might be brought under control, but that promise has not been fulfilled. Although many infections respond dramatically to chemotherapy, tuberculosis, dysentery and typhoid fever continue to be endemic in many parts of the world; cholera and plague erupt periodically; staphylococcal infections persist in the most advanced medical centers. One major reason is that the disease organisms have developed resistance to the drugs.

Until recently it was assumed that the appearance of drug-resistant bacteria was the result of a predictable process: the spontaneous mutation of a bacterium to drug resistance and the selective multiplication of the resistant strain in the presence of the drug. In actuality a more

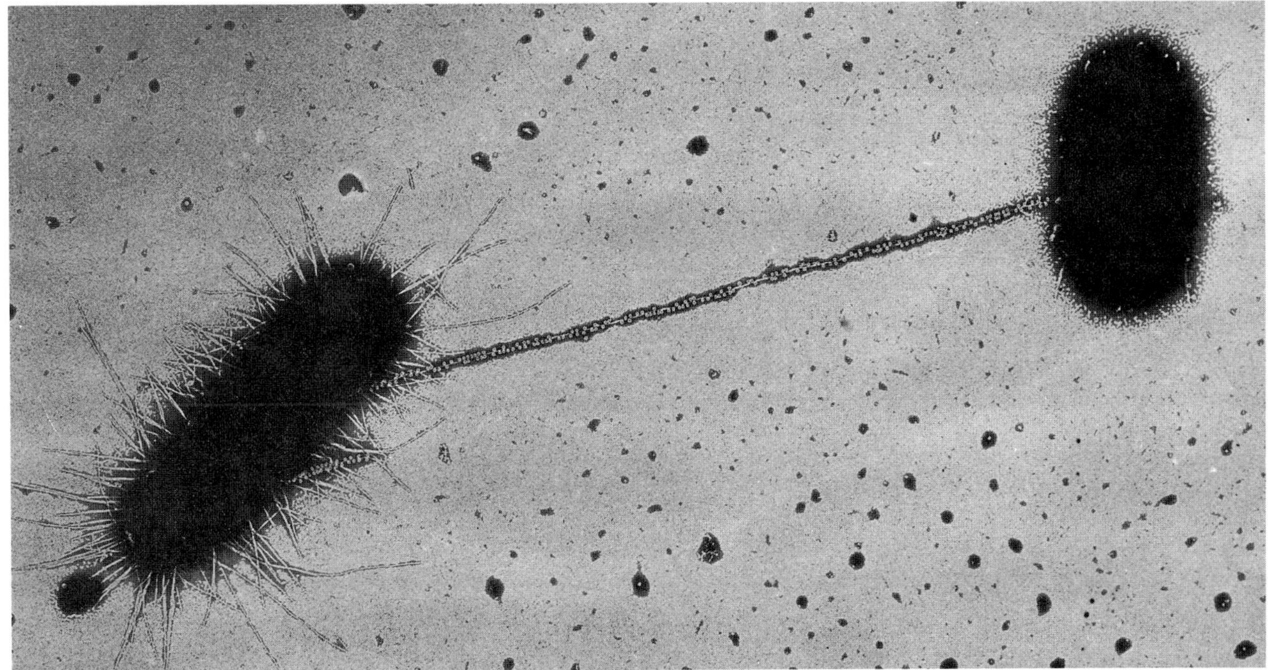

**R FACTOR, the particle that imparts infectious drug resistance, is transferred from one bacterial cell to another by conjugation. The various forms of conjugation are thought to be effected by way of thin tubules called pili. In this electron micrograph made by Charles C. Brinton, Jr., and Judith Carnahan of the University of Pittsburgh a male *Escherichia coli* cell (*left*) is connected to a female bacterium of the same species by an *F* pilus, which shows as a thin white line in the negatively stained preparation. Numerous spherical bacterial viruses, or phages, adhere to the *F* pilus. The cells have been magnified about 20,000 diameters.**

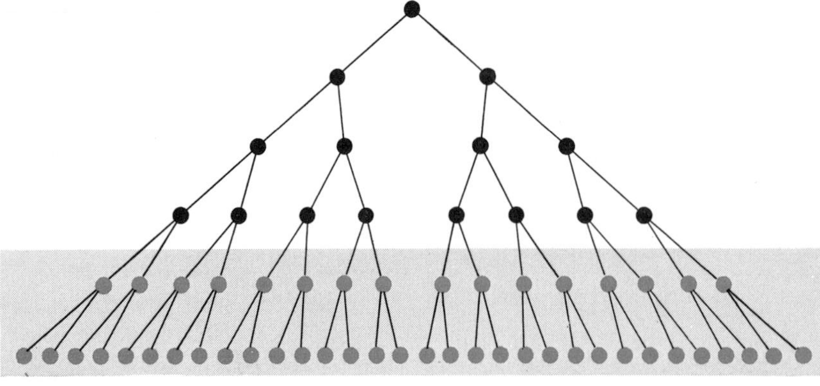

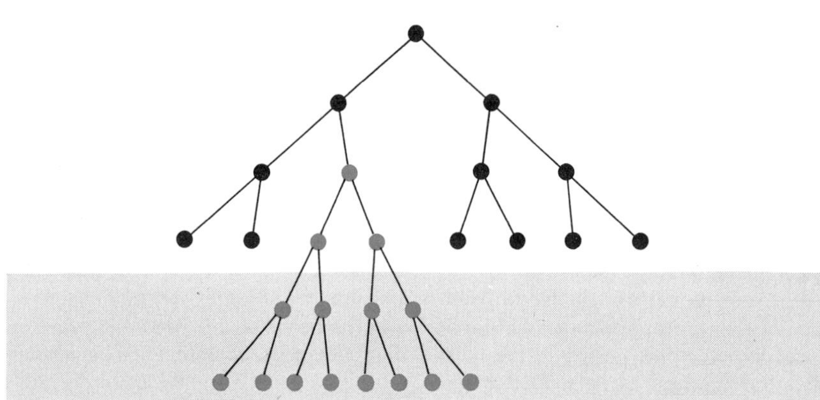

**DRUG RESISTANCE involves a change in the genetic material of a bacterial cell. The change from drug-sensitive cell (*black*) to drug-resistant cell (*color*) is not induced by the presence of the drug (*light color*), as was once thought (*top*). It is the result of a spontaneous mutation that gives rise to cells that survive in the drug environment (*bottom*).**

ominous phenomenon is at work. It is called infectious drug resistance, and it is a process whereby the genetic determinants of resistance to a number of drugs are transferred together and at one stroke from a resistant bacterial strain to a bacterial strain, of the same species or a different species, that was previously drug-sensitive, or susceptible to the drug's effect. Infectious drug resistance constitutes a serious threat to public health. Since its discovery in Japan in 1959 it has been detected in many countries. It affects a number of bacteria, including organisms responsible for dysentery, urinary infections, typhoid fever, cholera and plague, and each year it is found to confer resistance to more antibacterial agents. (What may be a related form of transmissible drug resistance has been discovered in staphylococci and may be responsible for "hospital staph" infections.) Quite aside from its importance to medicine, the study of infectious drug resistance is making significant contributions to microbial genetics by illuminating the complex and little understood relations among viruses, genes and the particles called episomes that lie somewhere between them.

If an antibacterial drug is added to a liquid culture of bacteria that are sensitive to the drug, after a while all the cells in the culture are found to be resistant to the drug. Once it was thought that the drug must somehow have induced the resistance. What has actually happened, of course, is that a few cells in the original culture were already resistant; these cells survive and their daughter cells multiply when the sensitive majority of bacteria succumb to the drug [*see illustration above*]. The resistance was not induced by the drug but was the result of a spontaneous mutation. Bacteria, like higher organisms, have chromosomes incorporating the genetic material, and from time to time a gene—perhaps one controlling drug resistance—undergoes a mutation. The mutation of a drug-sensitivity gene occurs only once in 10 million to a billion cell divisions, and when it occurs it alters a cell's sensitivity to one particular drug or perhaps two related drugs.

In 1955 a Japanese woman recently returned from Hong Kong came down with a stubborn case of dysentery. When the causative agent was isolated, it turned out to be a typical dysentery bacillus of the genus *Shigella*. This shigella was unusual, however. It was resistant to four drugs: sulfanilamide and the antibiotics streptomycin, chloramphenicol and tetracycline. In the next few years the incidence of multiply drug-resistant shigellae in Japan increased, and there were a number of epidemics of intractable dysentery.

The familiar process of mutation and selection could not explain either this rapid increase in multiple resistance or a number of other findings concerning the dysentery epidemics. For one thing, during a single outbreak of the disease resistant shigellae were isolated from some patients and sensitive shigellae of exactly the same type from other patients. Even the same patient might yield both sensitive and resistant bacteria of the same type. Moreover, the administration of a single drug, say chloramphenicol, to patients harboring a sensitive organism could cause them to excrete bacteria that were resistant to all four drugs. Then it was found that many of the patients who harbored drug-resistant shigellae also harbored strains of the relatively harmless colon bacillus *Escherichia coli* that were resistant to the four drugs. It was impossible, on the other hand, to obtain multiple resistance in the laboratory by exposing sensitive shigellae or *E. coli* to any single drug; multiply resistant mutants could be obtained only after serial selections with each drug in turn, and these mutants, unlike the ones taken from sick patients, multiplied very slowly.

Taken together, these characteristics of the resistant shigellae suggested to Tomoichiro Akiba of Tokyo University in 1959 that resistance to the four drugs might be transferred from multiply resistant *E. coli* to sensitive shigellae within a patient's digestive tract. Akiba's group and a group headed by Kunitaro Ochiai of the Nagoya City Higashi Hospital thereupon confirmed the possibility by transferring resistance from resistant *E. coli* to sensitive shigellae—and from resistant shigellae to sensitive *E. coli*—in liquid cultures. Other investigators demonstrated the same kind of transfer in laboratory animals and eventually in human volunteers. Clearly a new kind of transferable drug resistance had been discovered. What, then, was the mechanism of transfer? There were three known mechanisms of genetic transmission in bacteria that had to be considered as possibilities.

One was transformation, which involves "naked" deoxyribonucleic acid (DNA), the stuff of genes. DNA can be extracted from a donor strain of bacteria and added to a culture of a recipient strain; some of the extracted genes may "recombine," or replace homologous genes on chromosomes of the recipient bacteria, thus transferring a mutation from the donor to the recipient [*see top illustration below*]. In this way, for example, streptomycin-sensitive bacteria can become streptomycin-resistant.

Transformation occurs in a number of different bacteria, and it can occur spontaneously as well as experimentally. Because only small fragments of DNA are taken up by bacteria in transformation, however, it is seldom that more than two different drug-resistance genes are transferred together. It requires optimal laboratory conditions, moreover, for transformation to occur at a significant frequency, and such conditions are not likely to prevail in nature.

Another mechanism of gene transmission is transduction, in which genes are

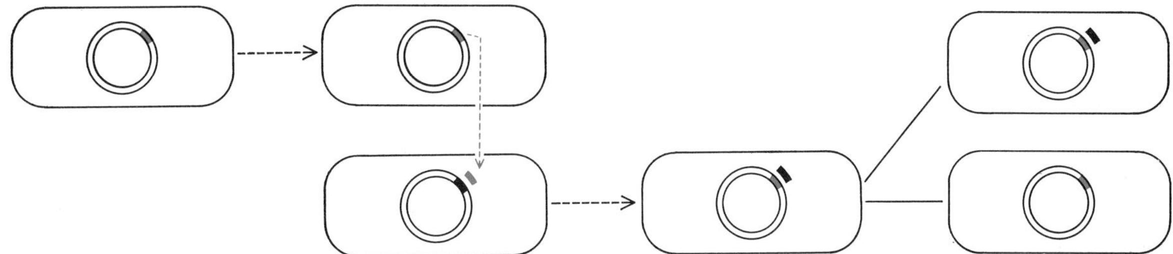

**TRANSFORMATION is a form of genetic transmission in which deoxyribonucleic acid (DNA) extracted or excreted from a donor cell (*top*) enters a recipient cell (*bottom*) and is incorporated into its chromosome. In this way a mutated gene (*color*) controlling resistance to a drug may be transferred to a drug-sensitive cell, replacing a homologous gene, which is unable to replicate and dies out.**

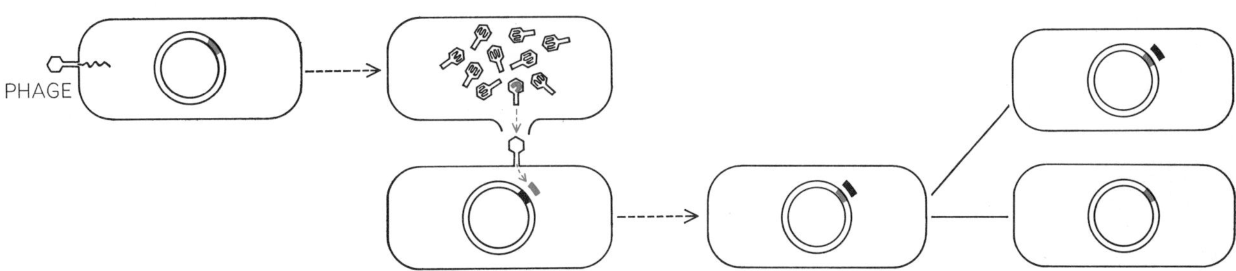


**TRANSDUCTION is effected by phage, or bacterial virus. Phage DNA enters a cell (*left*) and directs the synthesis of new phage, killing the cell (*second from left*). A bit of bacterial DNA (*color*), perhaps a mutated gene that causes drug resistance, may be incorporated inside a newly formed phage, be carried to a sensitive cell (*bottom*) and "recombine," or replace a gene on the chromosome.**

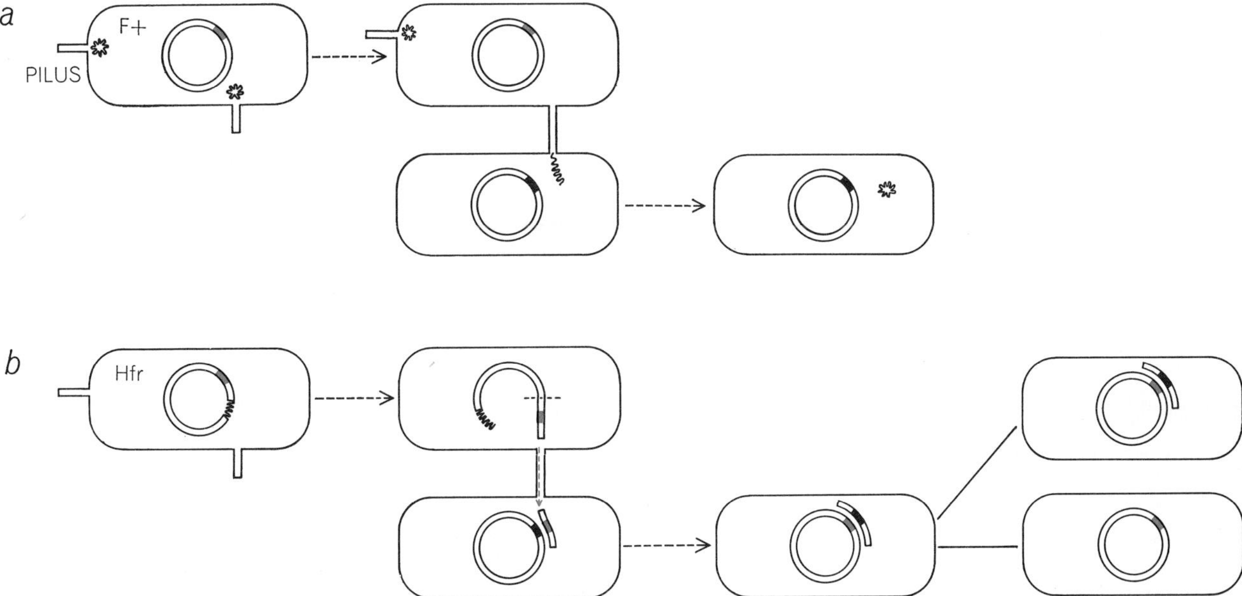


**SEXUAL MATING is a form of conjugation. If a fertility factor (*F*) is in the cytoplasm of a male ($F^+$) cell (*a*), it is transferred alone through a pilus to a female ($F^-$) cell. In an *Hfr* cell (*b*) the *F* is incorporated in the chromosome. Cell-to-cell contact causes part or all of the chromosome, perhaps including a mutation for drug resistance (*color*), to pass to a female cell and recombine.**

carried from one bacterial cell to another by infecting phages, or bacterial viruses. Transduction occurs when a phage, reproducing inside a cell by taking over the cell's synthesizing machinery, incorporates a bit of the bacterial chromosome within its protein coat "by mistake." When the phage subsequently infects a second cell, the bacterial genes it carries may recombine with homologous genes on the second cell's chromosome. The phage in effect acts as a syringe to bring about what in transformation is accomplished by the movement of naked DNA [*see middle illustration on preceding page*]. Transduction takes place in a variety of bacteria, but at a very low frequency. Genes for resistance can be transduced like other genes, but it is unlikely that more than two resistance genes could be transferred together because the small transducing phage can carry only a short segment of bacterial chromosome.

The third type of genetic transmission in bacteria is conjugation: a direct contact between two cells during which genetic material passes from one cell to the other. Transfer by conjugation occurs primarily from male to female cells of certain groups of bacteria. The male bacteria carry a fertility factor, the *F* factor, that is ordinarily located in the cytoplasm of the cell but may become integrated into the chromosome. When the *F* is cytoplasmic, the male cells are called $F^+$. In such cells the *F* is readily transferred to female ($F^-$) cells by conjugation, but it is transferred alone. When the *F* factor is integrated into the bacterial chromosome, it serves to "mobilize" the chromosome. That is, the chromosome, which in bacteria forms a closed loop, opens and portions of it can pass by conjugation to a female cell, recombine with the female chromosome and thereby endow the female bacterium with traits from the male. Because this transfer occurs with a high frequency in male cells with an integrated *F*, such cells are called *Hfr*, for "high frequency of recombination" [*see bottom illustration on preceding page*].

The *F* factor is what is generally called an episome: a genetic element that may or may not be present in a cell, that when present may exist autonomously in the cytoplasm or may be incorporated into the chromosome, and that is neither essential to the cell nor damaging to it. An episome is something like a virus without a coat; indeed, some bacterial viruses can become "temperate" and exist as harmless episomes inside certain bacterial cells [see "Viruses and Genes," by François Jacob and Elie L. Wollman, which begins on page 42 in this book].

Until recently the actual route of transfer was not known. In 1964 Charles C. Brinton, Jr., of the University of Pittsburgh and his colleagues proposed that the *F* factor or the *F*-mobilized chromosome passes from one cell to the other through a thin tubular appendage, the *F* pilus, that is formed on both $F^+$ and *Hfr* cells by the presence of the *F* factor. Another kind of pilus, the Type 1 pilus, is seen on female cells as well as male cells, but the two can be distinguished: the *F* pilus is the site of infection by certain phages, and so the phages cluster along the *F* pili, marking them clearly in electron micrographs [*see top and middle illustrations on page 63*].

If a male chromosome transferred to a female cell by conjugation carries drug-resistance genes, these genes may be incorporated into the female chromosome. Experiments with sexual mating showed that drug-resistance genes are in fact sometimes scattered along bacterial chromosomes. Rather long segments—sometimes the entire length—of the chromosome can be transferred in sexual mating, and so it is possible for several resistance genes to be transferred in a single mating event.

In 1960 we took up the study of the resistant shigellae in my laboratory at the Keio University School of Medicine. It soon became clear that the mechanism of transfer of multiple resistance was not transformation, because sensitive strains were not made resistant by DNA extracted from the resistant bacteria. It was not transduction, because it could not ordinarily be effected by cell-free filtrates of the resistant cultures.

There was strong evidence that some form of conjugation must be responsible. Microscopic examination of a mixed culture of sensitive and resistant bacteria revealed pairing between the different kinds of cells. When a mixed liquid culture was agitated in a blender to break off any cell-to-cell contact, and the culture was then diluted to prevent further pairing, the transfer of resistance ceased. If the mechanism of resistance transfer was conjugation, however, it was not the familiar process of sexual mating. For one thing, it occurred between $F^-$ cells. Moreover, two observations showed that unlike the transmission of traits by sexual mating the transfer did not involve the chromosome itself. First, we noted that known chromosomal traits of certain strains, such as the inability to synthesize particular substances, were not usually transferred along with the drug-resistance traits. Second, we noted that the recipient cells became resistant immediately after the transfer occurred, whereas chromosomal drug resistance is ordinarily expressed only after the original drug-sensitivity genes have been lost in the course of cell division through the process known as segregation.

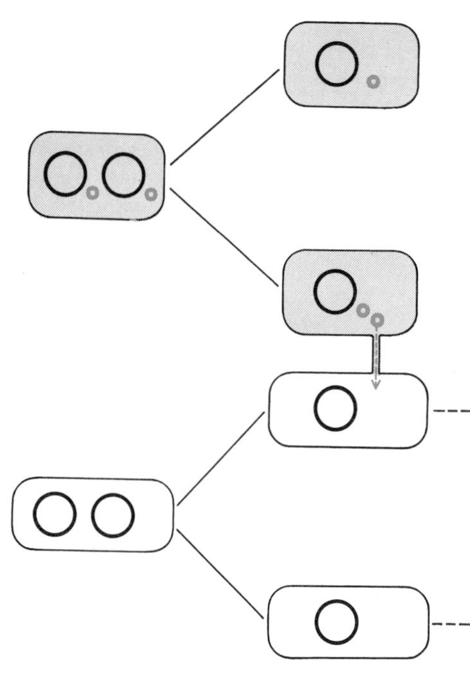

**INFECTIOUS DRUG RESISTANCE, another form of conjugation, involves transfer of the *R* (resistance) factor. A cell of a re-**

We concluded that the factor responsible for infectious drug resistance was an extrachromosomal element, which we called the *R* factor (for "resistance"). A number of experiments have confirmed the cytoplasmic nature of these factors. They are obtained by bacteria only by infection from other *R*-factor-carrying cells, never by spontaneous mutation. They can be eliminated from cells by treatment with acridine dyes; *F* factors can be eliminated in the same way when they are in the cytoplasm of $F^+$ cells but not when they are incorporated into the

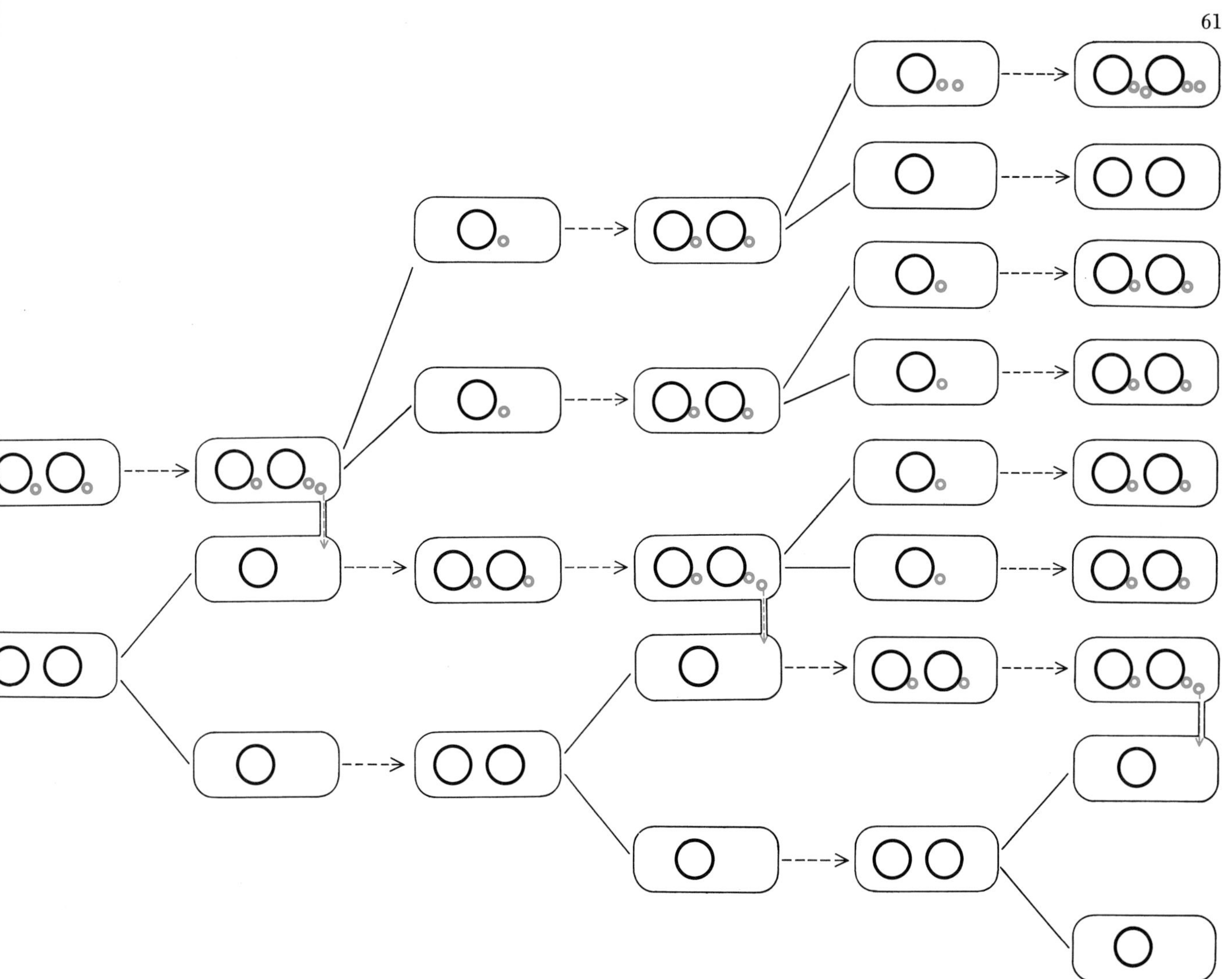

sistant strain (*light color*) comes in contact with one of a sensitive strain (*white*); one of its *R* factors (*color*) replicates and a copy passes through a pilus to the sensitive recipient. The procedure is repeated as cells come in contact. In the course of cell division an *R* factor is sometimes lost. The diagram is highly schematic; the actual sequence of replication and transfer is not established.

chromosome of *Hfr* cells. Finally, consider what happens when one adds a small number of bacteria with *R* factor to a culture of drug-sensitive cells. There is a rapid increase in the relative number of drug-resistant cells; in 24 hours or so the culture is almost completely resistant. This must be owing to the rapid infectious spread of *R* factors to the once sensitive bacteria, because it occurs at a much faster rate than the overall growth of the culture [*see top illustration on next page*]. Since chromosome replication is synchronized with cell division, the *R* factor must be replicating faster than the chromosomes and must therefore replicate outside the chromosome, in the cytoplasm.

Although the *R* factor is usually located in the cytoplasm, in rare instances it is integrated into the chromosome, and when that happens it is transferred together with some chromosomal genes. Such behavior suggests that the *R* factor, like the *F* factor, is episomal in nature. Both of them may be of selective advantage to the cells in which they exist, the *F* factor by making for genetic variability and the *R* factor of course by providing drug resistance. When they are not providing an advantage, they at least do the host cells no harm; they are symbionts rather than harmful parasites. Their behavior is similar to that of a temperate, or nonvirulent, phage, and it may be that both are descended from bacterial viruses. Unlike viruses, they cannot exist at all outside the cell; they are obligatory intracellular symbionts with even less biological function than viruses, which are usually considered to be on the borderline between living and nonliving matter.

There is a further major point of similarity between the *F* and the *R* factor, and that is their method of transfer. In London, Naomi Datta of the Royal Postgraduate Medical School, A. M. Lawn of the Lister Institute of Preventive Medicine and Elinor Meynell of the Medical Research Council observed in 1965 that most *R* factors induce the formation of pili that are shaped like *F* pili and attract the same phages as *F* pili: apparently they *are F* pili [*see bottom illustration on page 63*]. When bacteria that have such pili and are able to transmit multiple resistance are severely agitated in a blender, the pili are sheared off. Such "shaved" cells are unable to transfer the *R* factor; later, when the *F* pili have been regenerated, the cells are once again infectious. It now appears that both *R* factors and any chromosomal genes mobilized by *R* factors are transferred by the *F* pili or another closely related kind of pili.

The big difference between the transfer of *F* factors and the transfer of *R* factors is in the frequency with which they occur. In a mixed culture of male

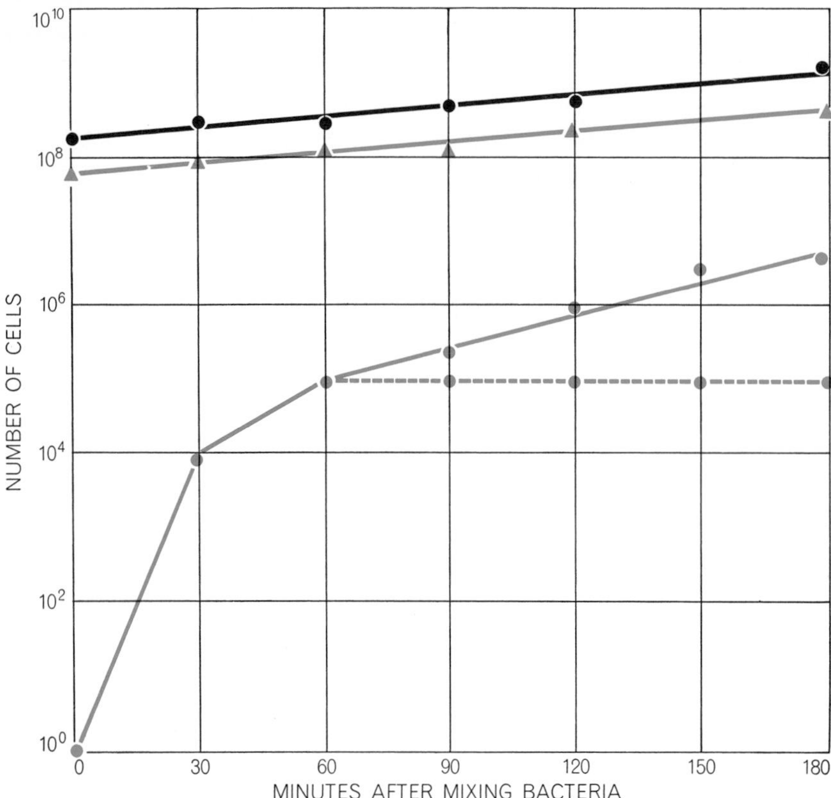


**TIME COURSE of *R*-factor transfer is shown. Equal volumes of cells of a donor *E. coli* strain infected with *R* factors (*triangles*) and of an initially sensitive recipient strain (*black dots*) were mixed. Sampling at intervals traced the increase in the number of resistant *E. coli* (*colored dots*). After one hour some of the culture was removed, agitated to break off cell-to-cell contact and diluted to prevent pairing. In the diluted culture there was no increase in the number of resistant cells (*broken line*), indicating that conjugation was the mechanism of transfer. The data are from David H. Smith of the Harvard Medical School.**

and female bacteria the transfer of nearly 100 percent of the *F* factors or *F*-mobilized chromosome, as the case may be, occurs within an hour. In a culture of drug-resistant (donor) and drug-sensitive (recipient) bacteria, on the other hand, only 1 percent or less of the donor cells transfer their *R* factors in an hour. The low frequency of transfer is due to the relative scarcity of cells with *F* pili in a culture of bacteria carrying *R* factors. Bacteria that have newly acquired the *R* factor, on the other hand, can transfer it at a very high frequency—almost 100 percent. (If this were not the case, *R* factors could hardly multiply so rapidly in a newly infected culture.) They lose this high competence after several cell-division cycles. The explanation seems to be that most *R* factors form a "repressor" substance that somehow inhibits the formation of *F* pili. Cells that are newly infected with such *R* factors contain no repressor, and so *F* pili are initially induced at a high frequency. Later, as the repressor accumulates, the formation of the pili is inhibited.

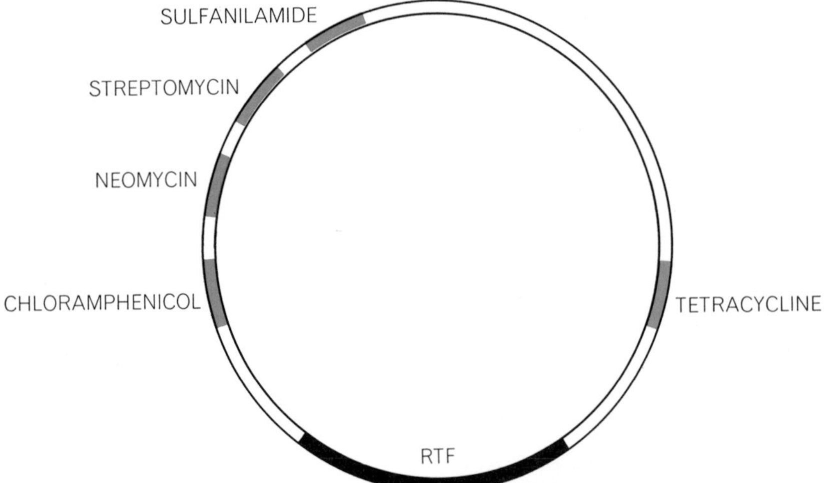


**MAP OF *R* FACTOR, still tentative, shows a closed loop. There are five determinants (*color*) of resistance to five different drugs. There is also a determinant, the resistance-transfer factor (*black*), that controls the ability of the *R* factor to replicate and be transferred.**

It is now possible to describe what happens when bacteria with the *R* factor come into contact with a population of drug-sensitive bacteria [*see illustration on preceding two pages*]. A few *R* factors are transmitted by conjugation from donor cells bearing pili into the cytoplasm of recipient cells, which immediately become resistant. The transfer process is repeated from cell to cell, and the normal process of cell division also contributes to the rapid proliferation of multiple resistance in the recipient population. From time to time an *R* factor is lost. Both the rate of transfer and the rate of loss vary in different strains of bacteria and *R* factors, thus accounting in part for the fact that naturally occurring multiple resistance is much more common in some bacteria that are susceptible to infectious drug resistance than in others.

For several years we have been seeking to map the various elements of an *R* factor as one maps the genes of a chromosome. To do this we capitalize on the fact that although *R* factors are not normally transferred by transduction, it is possible to transduce them under carefully controlled conditions. If we grow large phages in a culture of bacteria with *R* factors, a few of the phages pick up entire *R* factors and are capable of transferring them to recipient cells. If we use small phages, there is room for only part of the *R* factor to be incorporated inside their protein coats and transduced. Some of the transduced particles impart drug resistance but lack the ability to replicate or to be transferred by conjugation; others lack determinants of one or more of the multiple drug resistances. By calculating the frequency with which various segments of the *R* factor are transduced together, we can determine their relative distance from one another and so visualize the structure of the *R* factor we are studying.

The map is not yet conclusive, but we think the factor is circular and that it has a segment—the resistance-transfer factor, or RTF—that controls replication and transferability, as well as segments determining resistance to each of five types of drug [*see bottom illustration on opposite page*]. We have suggested that the *R* factors originate when a resistance-transfer factor picks up resistance genes from some bacterial chromosome and that the two then form a single episomal unit. E. S. Anderson and M. J. Lewis of the Central Public Health Laboratory in London have advanced a different view. They consider that the resistance-transfer factor and the set of resistance determinants exist as two separate units, which on occasion become associated to form *R* factors.

Since *R* factors are self-replicating units, carry genetic information and can recombine with bacterial chromosomes, it is safe to assume that they are composed of DNA. This is confirmed by the fact that *R* factors, like nucleic acids in general, are inactivated by ultraviolet radiation and by the decay of incorporated radioactive phosphorus. At the Walter Reed Army Institute of Research and at Keio, Stanley Falkow, R. V. Citarella, J. A. Wohlhieter and I were able to isolate the DNA of *R* factors by density-gradient centrifugation. A first attempt to separate *R*-factor DNA from that of *E. coli* was unsuccessful, suggesting that the densities of the two DNA's are very similar. We then selected as the host cell the bacterium *Proteus mirabilis,* which was known to have a DNA of unusually low density and to be subject to infectious drug resistance.

When DNA extracted from *Proteus* carrying the *R* factor is centrifuged in a solution of cesium chloride, two satellite bands of DNA appear in addition to the band characteristic of the bacterial DNA [*see illustration on next page*]. These bands disappear if the *Proteus* loses its *R* factors spontaneously or if they are eliminated by the acridine dye treatment, and so we conclude that the bands do represent the *R*-factor DNA. Analysis of this fraction by column chromatography shows that it is typical double-strand DNA. It is possible that *R* factors contain components other than DNA, but this is not likely in view of the fact that entire factors are transduced and transducing phages incorporate only DNA.

The original finding that infectious drug resistance affected four unrelated drugs implied that some factor was altering the cell membrane, reducing its permeability and thereby barring all the drugs from their normal sites of action inside the cell. The finding that there are separate resistance determinants for the various drugs, however, indicated that

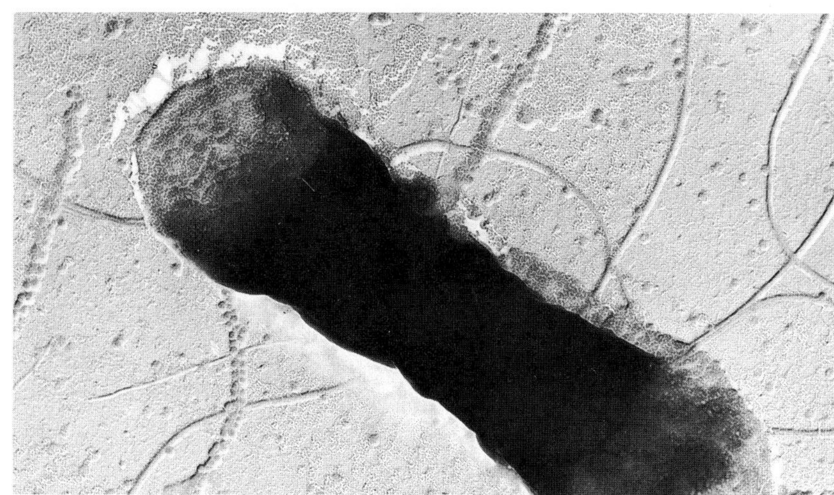

**MALE BACTERIUM, an *E. coli* infected with phage, has *F* pili. They are thin fibers, here hidden below the spherical phage particles. The thin fibers without phages are Type 1 pili and the thick fibers are flagella. The preparation has been enlarged about 30,000 diameters.**

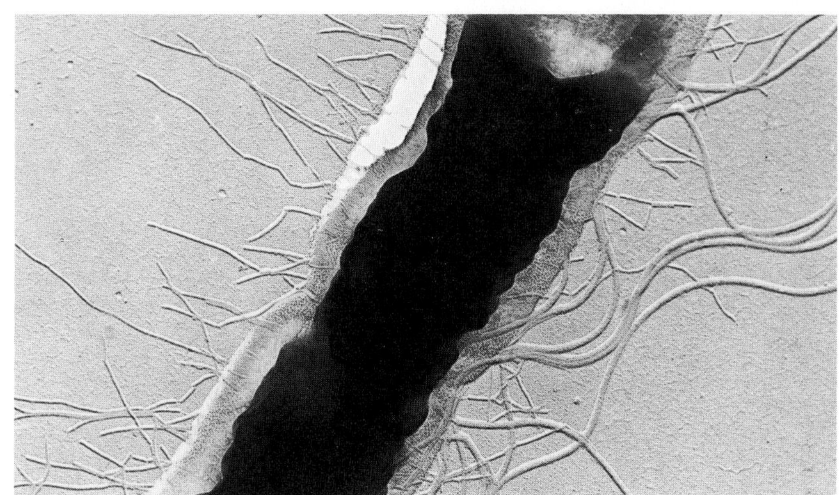

**FEMALE *E. COLI,* which lacks the *F* factor, also lacks *F* pili. It does have both the Type 1 pili and flagella, which are organelles of locomotion for the cell. The electron micrograph, like the others on this page, was made by Toshihiko Arai in the author's laboratory.**

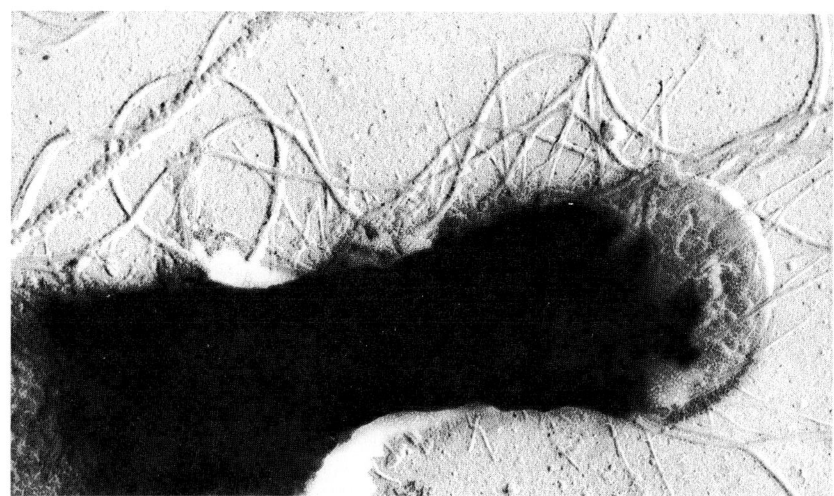

**_E. COLI_ WITH *R* FACTOR, although a female cell, does carry an *F* pilus, the phage-covered fiber at top left. It also has Type 1 pili and flagella. The most common type of *R* factor initially induces the formation of *F* pili, but it also tends to repress them later.**

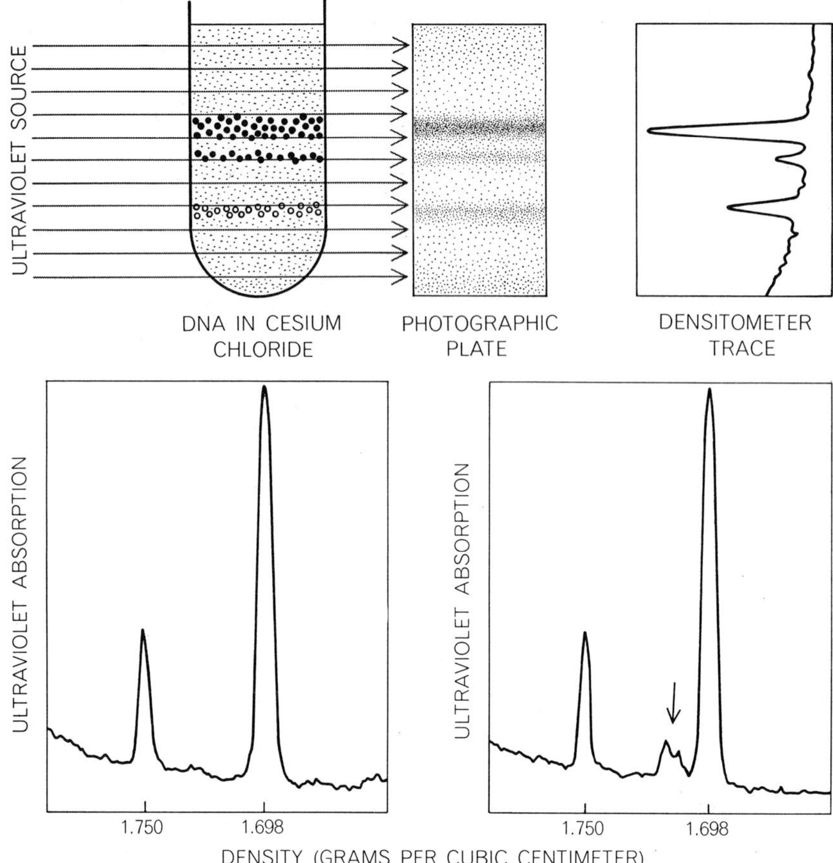


**R-FACTOR DNA is isolated by density-gradient centrifugation. DNA from *Proteus* cells is suspended in a cesium chloride solution and spun in a high-speed centrifuge. The cesium chloride establishes a density gradient and the DNA forms bands in the solution according to its density. The DNA pattern is photographed in ultraviolet, which is absorbed by DNA, and the photograph is scanned by a densitometer (*top*). The densitometer trace derived from *Proteus* without *R* factor (*bottom left*) shows a band at 1.698 grams per cubic centimeter that is characteristic of *Proteus* DNA and a reference band at 1.750. The trace from *Proteus* with *R* factor (*right*) has extra bands (*arrow*) at 1.710 and 1.716, representing *R*-factor DNA.**

each determinant had its own mode of action. Permeability may be involved in the case of some drugs, but it is now clear that other processes are at work. S. Okamoto and Y. Suzuki of the National Institute of Health in Japan and Mrs. Datta and P. Kontomichalou in Britain have shown that bacteria bearing various *R* factors synthesize particular enzymes that inactivate specific drugs, thereby rendering them harmless to the bacteria.

The public health threat posed by infectious drug resistance is measured by the range of bacterial hosts it affects, the number of drugs to which it imparts resistance and the prevalence of certain practices in medicine, agriculture and food processing that tend to favor its spread. *R* factors can be transferred not only to shigellae but also to *Salmonella,* one species of which causes typhoid fever; to *Vibrio cholerae,* the agent of cholera; to the plague bacillus *Pasteurella pestis* and to *Pseudomonas aeruginosa,* which causes chronic purulent infections. In addition, more than 90 percent of the agents of urinary tract infections, including *E. coli, Klebsiella, Citrobacter* and *Proteus,* now carry *R* factors.

(These organisms are all gram-negative bacteria; *R* factors seem not to be transferable to the gram-positive bacteria, which include streptococci and staphylococci. A somewhat similar form of transmissible resistance has been discovered in staphylococci, however. There are cytoplasmic genes, or plasmids, in some staphylococci that determine the production of penicillinase, an enzyme that inactivates penicillin. Richard P. Novick of the Public Health Research Institute in New York and Stephen I. Morse of Rockefeller University recently showed that these plasmids can be transduced to drug-sensitive staphylococci both in the test tube and in laboratory animals. The actual clinical importance of this process remains to be determined.)

The *R* factors seem to be acquiring resistance genes for an increasing number of antibiotics. The original factors, it will be remembered, imparted resistance to sulfanilamide, streptomycin, chloramphenicol and tetracycline. In 1963 G. Lebek of West Germany discovered a factor that causes resistance to these four drugs and also to the neomycin-kanamycin group of antibiotics. In 1965 Mrs. Datta and Kontomichalou reported a new determinant of resistance to aminobenzyl penicillin (ampicillin). In 1966 H. W. Smith and Sheila Halls of the Animal Health Trust in Britain found factors imparting resistance to the synthetic antibacterial drug furazolidone. This year David H. Smith of the Harvard Medical School reported *R*-factor-controlled resistance to gentamycin and spectinomycin. We must assume that additional drug-resistance determinants will appear and proliferate as new antibiotics come into use.

This is implicit in the mechanism of infectious resistance. *R* factors are common in *E. coli,* which are often present in the intestinal tracts of human beings and animals. When a person or an animal becomes infected with a susceptible disease organism, the *R* factor is readily transferred to the new population. Although the frequency of transfer of *R* factors is not high even in the laboratory, and is reduced by the presence of bile salts and fatty acids in the intestine, recipient bacteria bearing the *R* factor are given a selective advantage as soon as drug therapy begins, and they soon predominate.

In addition to being ineffective and helping to spread resistance, "shotgun" treatment of an infection with drugs to which it is resistant causes undesirable side effects. It is therefore important to culture the causative agent, determine its drug-resistance pattern and institute treatment with a drug to which it is not resistant; that is the only way to combat the multiple-resistance strains. As more is learned about the *R* factor, new forms of therapy may be developed—possibly utilizing the acridine dyes, which attack drug-resistant as well as sensitive cells and can also eliminate *R* factors from cells.

In many parts of the world antibiotics are routinely incorporated in livestock feeds to promote fattening and are also used to control animal diseases. Anderson and Mrs. Datta have shown clearly that the presence of antibiotics in livestock exerts a strong selective pressure in favor of organisms—particularly salmonellae—with *R* factors and plays an important role in the spread of infectious resistance. Meat and other foodstuffs are also treated with antibiotics and synthetic drugs as preservatives in many countries, and this too may help to spread *R* factors and carry them to man. Unless we put a halt to the prodigal use of antibiotics and synthetic drugs we may soon be forced back into the preantibiotic era of medicine.

One final note. Typhoid, cholera and plague bacilli are obviously much more difficult to combat if they are resistant to drug therapy. There are grounds for believing that the military in some countries are investigating the potentialities of *R* factors as weapons of bacteriological warfare.

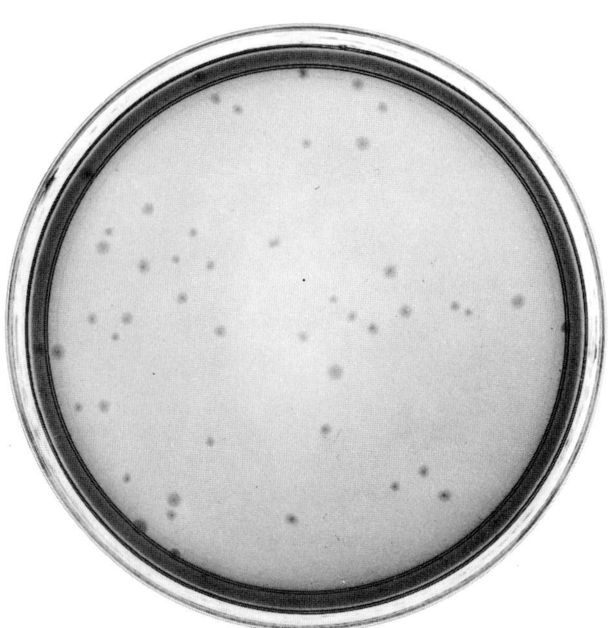

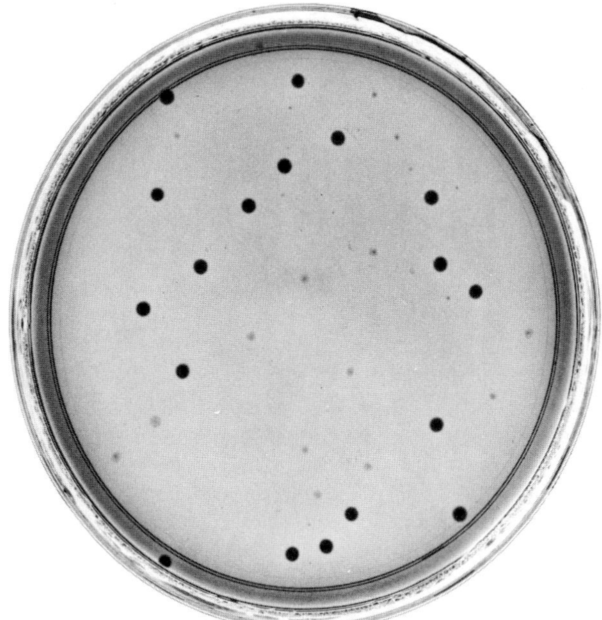

**SENSITIVITY TEST conducted in Smith's laboratory at Harvard demonstrates infectious drug resistance. A culture of *Salmonella typhimurium* with an *R* factor controlling resistance to four drugs is mixed with drug-sensitive *E. coli*. A portion of the mixed culture is immediately plated on a medium containing the drugs (*left*). Only *Salmonella* colonies (*gray*) appear. After the mixed culture has incubated, the plating procedure is repeated, and now *E. coli* colonies (*black*) grow as well (*right*): the *R* factor was transferred.**

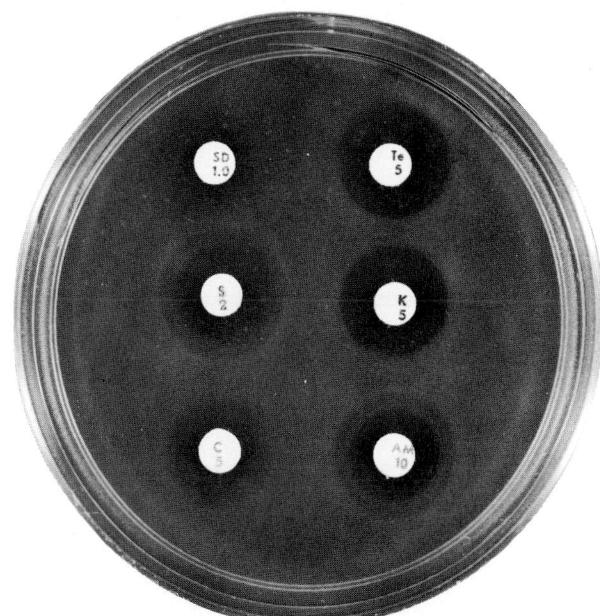

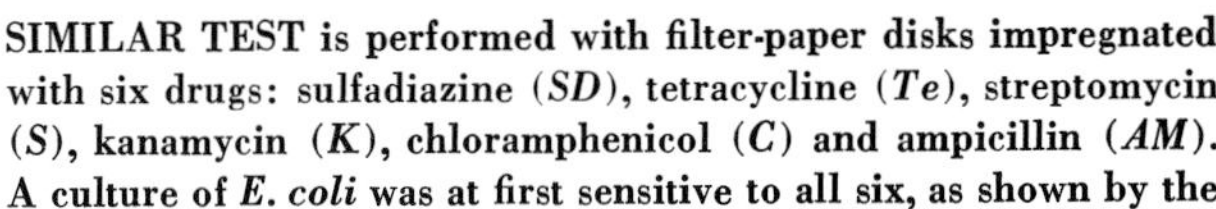

**SIMILAR TEST is performed with filter-paper disks impregnated with six drugs: sulfadiazine (*SD*), tetracycline (*Te*), streptomycin (*S*), kanamycin (*K*), chloramphenicol (*C*) and ampicillin (*AM*). A culture of *E. coli* was at first sensitive to all six, as shown by the dark zones around each disk where the bacteria have been killed (*left*). After the culture was incubated with a strain of *Klebsiella*, taken from a patient, that was resistant to all the drugs but ampicillin, the *E. coli* too were resistant to all but ampicillin (*right*).**

# 8

# *Genes Outside the Chromosomes*

RUTH SAGER
*January 1965*

The science of genetics encompasses any and all systems involved in the heredity of living organisms, but in practice it has been largely restricted to the study of chromosomal genes: the molecular carriers of heredity that are arranged in linear order in the threadlike chromosomes in the nucleus of the living cell. The fact is, however, that more information is involved in the heredity of organisms than is contained in the chromosomes. Evidence of additional genetic systems has been accumulating slowly for some time. Within the past few years studies in my laboratory at Columbia University have demonstrated the existence of a second genetic system that is quite amenable to analysis. This system consists of a class of genetic determinants located elsewhere than in the chromosomes. These determinants are particulate elements, stably replicated and transmitted from one generation to the next; they can exist in either normal or mutant form and they influence a wide variety of cell traits. They are probably composed, like the chromosomal genes, of nucleic acid. No one has yet isolated one of these determinants, and they have still to be located in a specific region of the cell, but on the basis of genetic analysis they fully deserve to be called nonchromosomal genes.

The first example of a nonchromosomal gene was described in 1908 by the German botanist Carl Correns, one of the three biologists who had rediscovered the laws of inheritance enunciated in 1865 by Gregor Mendel. Correns thought it likely that all organisms had more than one genetic system, and he and his students investigated the problem in a number of plant species. Interest in nonchromosomal inheritance was soon overshadowed, however, by the explosive development of chromosomal genetics. The convincing experiments that correlated the distribution of Mendel's "factors" of heredity with the observable details of chromosome behavior focused attention on the chromosomal genes to the exclusion of other hereditary phenomena. Biologists tended to forget that nothing in the orderly methodology of chromosomal genetics precluded the existence of additional genetic systems. Although a few hundred well-authenticated examples of nonchromosomal traits have been described in organisms as diverse as insects, flowering plants, algae, yeasts and fungi, this evidence has largely been ignored in textbooks and in the mainstream of genetic theory. This is not to say, of course, that it was not best to concentrate on the chromosomal system first. It is the most accessible system, and it must not only be understood but also brought under control in order to conduct experiments dealing with nonchromosomal inheritance.

In principle the difference in the patterns of inheritance of the two systems is elementary. The salient fact of Mendelian, or chromosomal, heredity is that genes from the male and female parents contribute equally to the genetic constitution of the progeny. This is true in diploid organisms, whose body cells have two sets of chromosomes, and in the more primitive haploid organisms, which have only one set [*see top illustration on page 68*]. A nonchromosomal gene is identified by its failure to follow the Mendelian pattern. Instead the nonchromosomal genes from the female par-

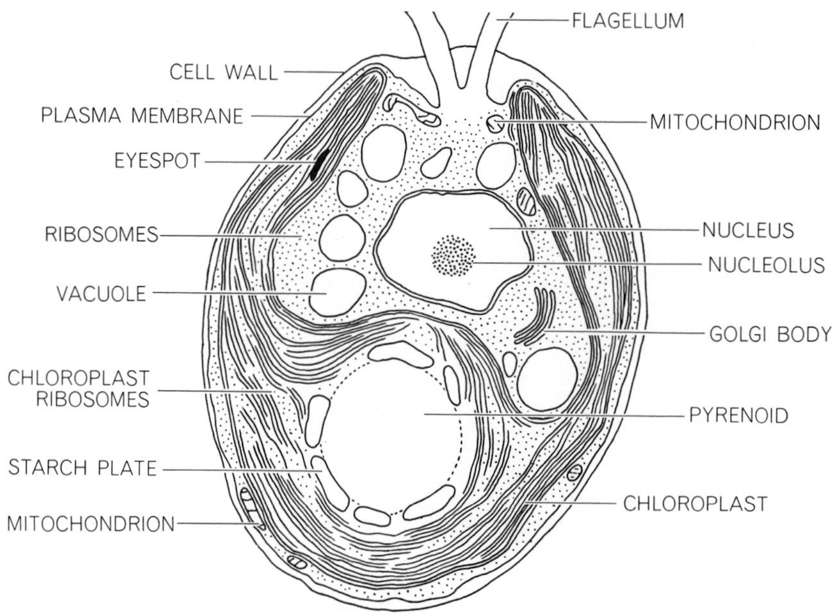


**ORGANELLES of *Chlamydomonas* cell include the chloroplast for photosynthesis, ribosomes for protein synthesis and mitochondria for energy conversion. The chromosomes are in the nucleus. The flagella provide motility and the eyespot is a rudimentary eye.**

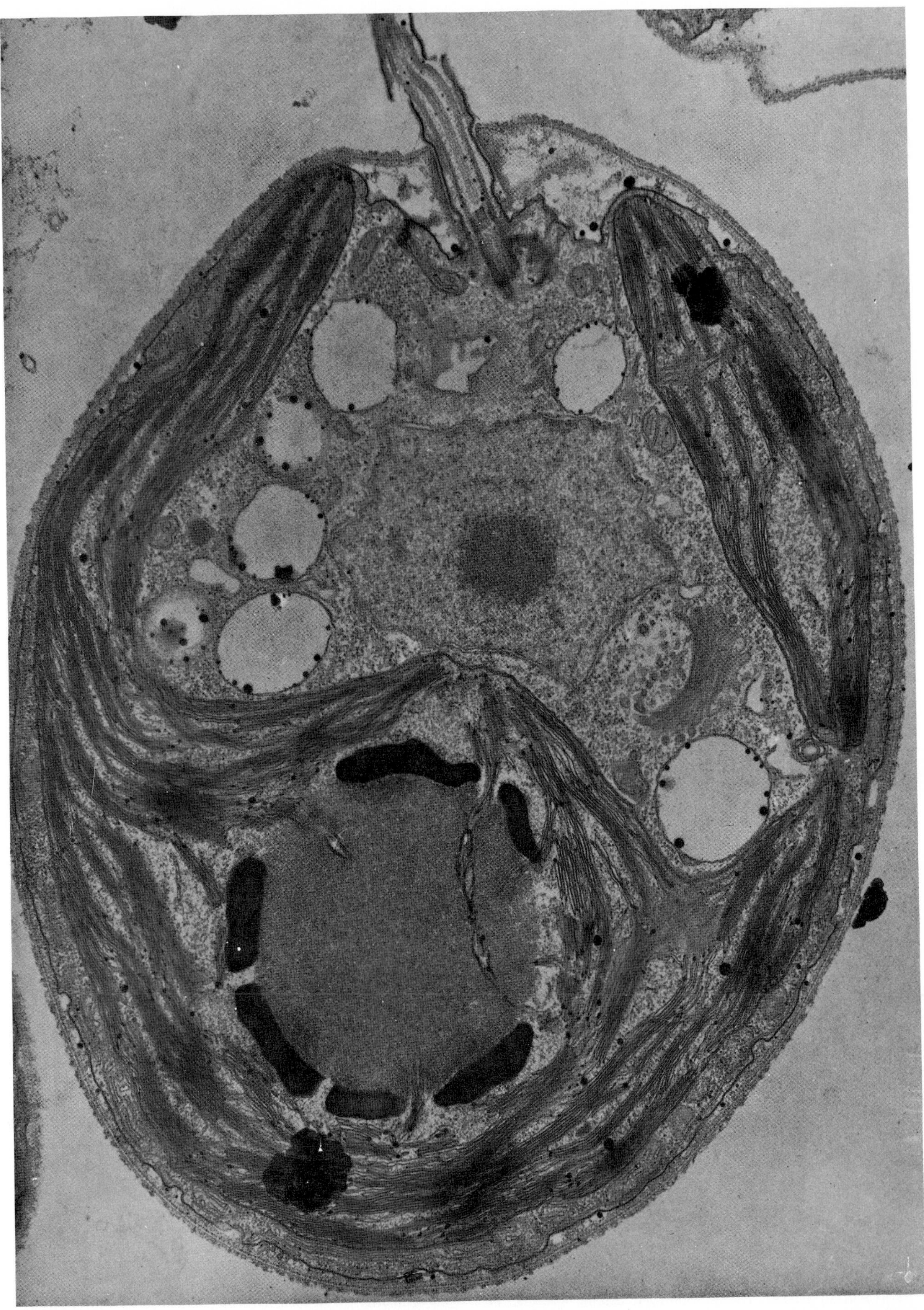

CHLAMYDOMONAS, the one-celled alga that is the subject of the author's genetic experiments, is enlarged about 26,000 diameters in an electron micrograph made by George Palade of the Rockefeller Institute. Major organelles are identified on opposite page.

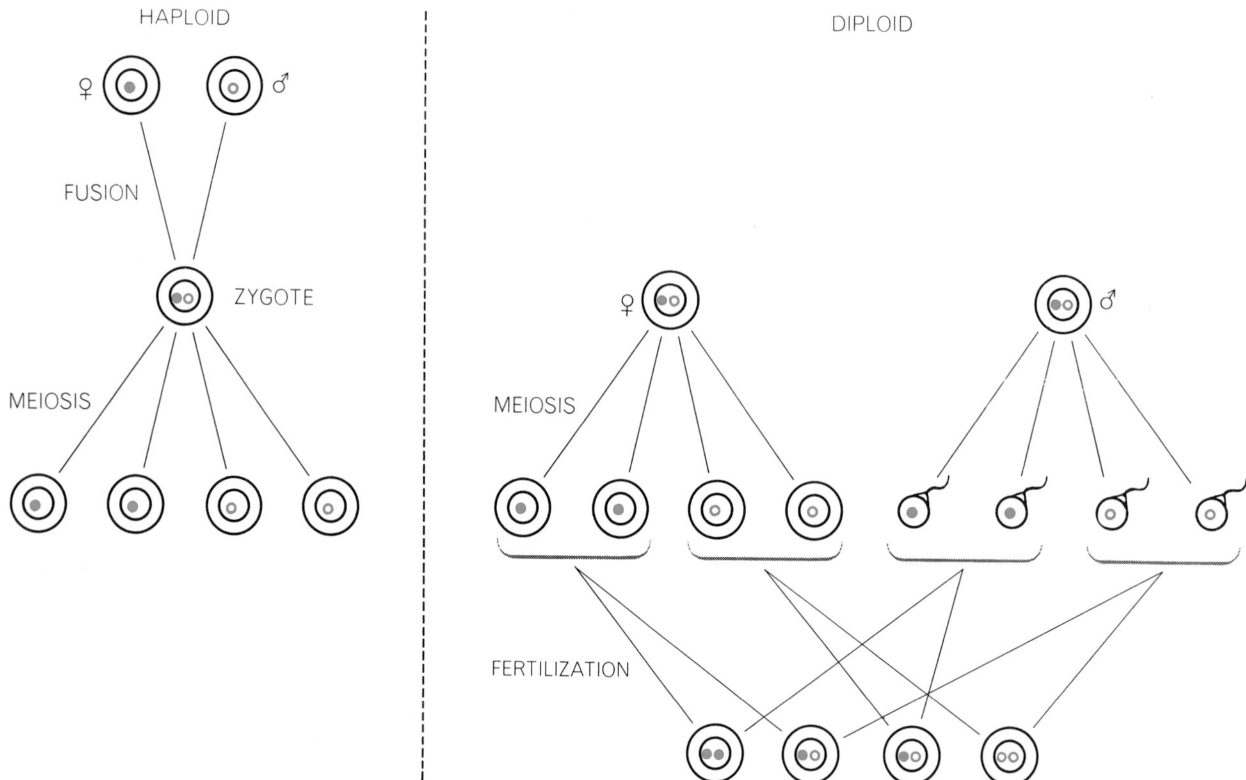


**CHROMOSOMAL GENES** are transmitted as shown here. In haploid organisms such as *Chlamydomonas* (*left*) two cells fuse; the resulting zygote undergoes meiosis to produce four progeny. Higher plants and animals are diploid (*right*). Precursors of female and male germ cells undergo meiosis to form germ cells (eggs and sperm or pollen) and fertilization follows. In haploid and diploid organisms chromosomal genes (*colored dots and circles*) from male and female parent contribute equally to progeny.

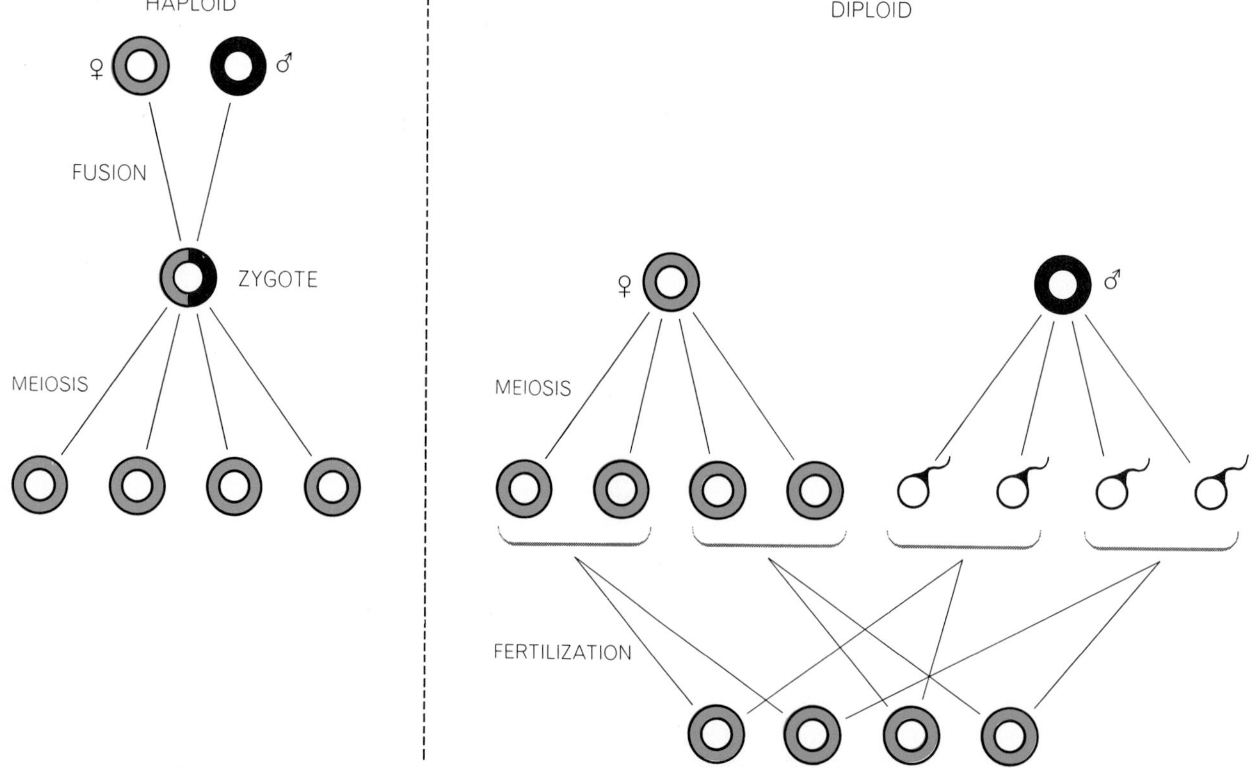


**NONCHROMOSOMAL GENES**, however, exhibit maternal inheritance. In haploid (*left*) and diploid (*right*) organisms all progeny ordinarily get all their nonchromosomal genes (*colored rings*) from the female parent; genes from the male (*black*) are lost.

ent are usually transmitted, during the sexual cell-division process called meiosis, to all the progeny and those from the male to none [*see bottom illustration on opposite page*]. The most obvious explanation for this maternal inheritance pattern in most organisms is that the nonchromosomal genes are outside the nucleus in the cytoplasm of the cell; in higher plants and animals it is the female germ cell, not the male pollen or sperm, that supplies most of the cytoplasm for the fertilized egg. This explanation does not account for maternal inheritance in some haploid organisms, however; there are microorganisms in which the male and female contribute equal amounts of cytoplasm in fusing to form new cells. The mechanism of maternal inheritance must therefore involve more than the mere presence or absence of certain cell constituents.

Whatever the mechanism, maternal inheritance is the primary criterion for the recognition of nonchromosomal genes. To classify an inheritable trait as being chromosomal or nonchromosomal one needs only to study its pattern of transmission in test crosses. Two major difficulties account for the fact that in spite of this apparent ease of analysis the investigation of nonchromosomal genes has made little progress until recently.

The first difficulty has been in finding mutant genes. Genetics is based on comparisons between "wild type," or normal, genes and mutant, or altered, forms. An unusual trait—say white eyes in a normally red-eyed fruit fly—signals the presence of a mutant gene that can be followed through generations and studied in detail. Spontaneous mutations of nonchromosomal genes in natural populations are rare in comparison with those of the chromosomal genes. Moreover, the radiations and chemical compounds with which geneticists induce chromosomal mutations have not been effective in producing nonchromosomal mutations. The result is that material with which to study the nonchromosomal system—and even to decide just how important it is in any organism's total heredity—has been hard to come by.

The second difficulty in the study of nonchromosomal heredity lies in the maternal pattern of inheritance that most clearly distinguishes it. As long as all of the female parent's nonchromosomal genes, and none of the male parent's, are transmitted to all the progeny, one cannot bring to bear the standard techniques of genetic analysis, which depend on the manner in which the two parents' genes are distributed and expressed among the progeny.

We have been able to overcome both of these technical difficulties in my laboratory by working with a very cooperative microorganism, the green alga *Chlamydomonas*. This organism belongs to the class of the phytoflagellates, haploid organisms on the main line of evolution to both the higher plants and the animals. Although it can be handled in the laboratory as easily as a simple bacterium, *Chlamydomonas* contains within its single cell the primitive counterparts of such specialized tissues as eye, kidney and muscle. The organelles, or subcellular organs, of *Chlamydomonas* are similar to those of the cells of higher forms. There are mitochondria for generating energy, ribosomes for synthesizing proteins, a large chloroplast for photosynthesis and other such typical structures. In addition the organism reproduces sexually, having two "mating types," or sexes, determined by a single chromosomal gene. Other genes have been located in all of its eight chromosomes.

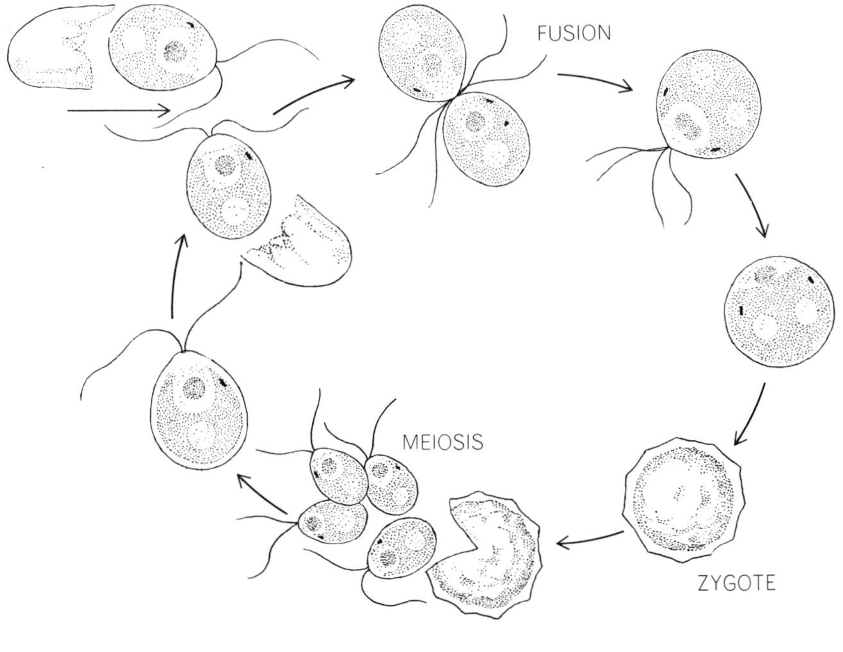

**LIFE CYCLE of *Chlamydomonas* includes a sexual phase. Two cells of opposite sex shed their cell walls and form a zygote that undergoes meiosis and produces four progeny cells. These may then divide asexually (*not shown here*) to form clones, or colonies of cells.**

Some years ago we began to search for a compound that would induce nonchromosomal genes of *Chlamydomonas* to mutate. We found the first such mutant by accident. We had placed cells susceptible to the antibiotic streptomycin on a culture medium that contained the antibiotic, and about one in a million cells survived and multiplied to form streptomycin-resistant colonies. Each colony resulted from a mutation—from streptomycin-sensitive (*ss*) to streptomycin-resistant (*sr*). Crossing tests showed that most of these mutations were chromosomal but that about 10 percent of them had a maternal-inheritance pattern: they were nonchromosomal. We found that the chromosomal mutations were spontaneous, arising at random before the cells were placed on a streptomycin-containing medium. The nonchromosomal-gene mutations, however, arose only after sensitive cells had been grown in the presence of toxic but sublethal concentrations of streptomycin. They were induced mutations.

The question then arose whether the drug was inducing specific mutations to drug-resistance or acting as a general mutagen. Subsequent studies made it clear that the streptomycin was acting nonspecifically, and that it could induce mutations in many different nonchromosomal genes. Under special inducing conditions each cell treated with streptomycin multiplies to form a colony, and we find mutant cells of various kinds in almost every colony. This remarkably high level of mutagenic efficiency is quite different from that of chromosomal-gene mutagens, which affect only a small fraction of a population.

Do nonchromosomal genes control a class of cell traits different from the

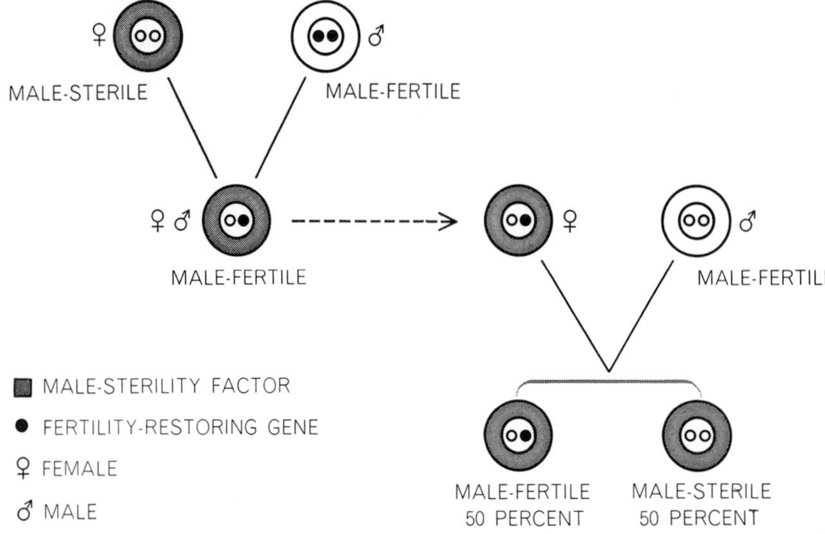


**INTERACTION** of chromosomal and nonchromosomal genes in corn is shown. The chromosomal fertility-restoring gene (*black dot*) counteracts the nonchromosomal male-sterility factor (*gray ring*); progeny (*middle left*) are fertile. When female plants of this line are crossed with plants lacking the fertility-restorer, however, half of the progeny are male-sterile, because the fertility-restoring gene is distributed to only half of the progeny.

traits determined by chromosomal genes? Probably not. We have so far studied about 30 different nonchromosomal-gene mutations, and most of them resemble chromosomal mutations in their effects; in many cases both kinds of gene affect the identical traits. Even when they influence different traits, the two genetic systems are apparently closely integrated in their action. For example, in *Chlamydomonas* there is a chromosomal gene *A* (for "amplifier"); this gene has no apparent effect in streptomycin-sensitive cells, but in the resistant mutants it greatly amplifies the cell's resistance to the antibiotic. It is equally effective in amplifying resistance brought about by either chromosomal or nonchromosomal genes. This interaction is purely physiological; gene *A* has no influence on the inheritance of either kind of resistant mutation.

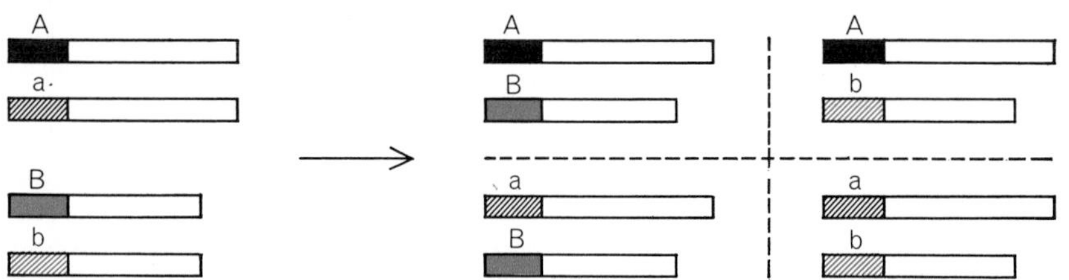


**UNLINKED RECOMBINATION** is diagrammed for two genes on different chromosomes (*left*). In the course of meiosis the four alleles may be assorted in four different ways.

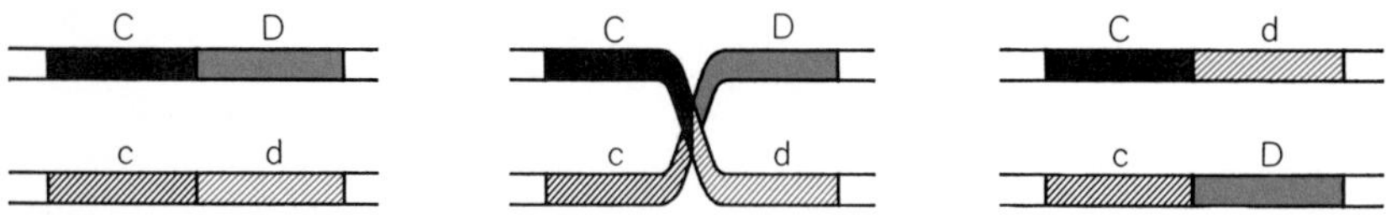


**LINKED RECOMBINATION** can result from "crossing over." Alleles that were together on a chromosome (*left*) are separated and appear in new combinations in progeny (*right*).

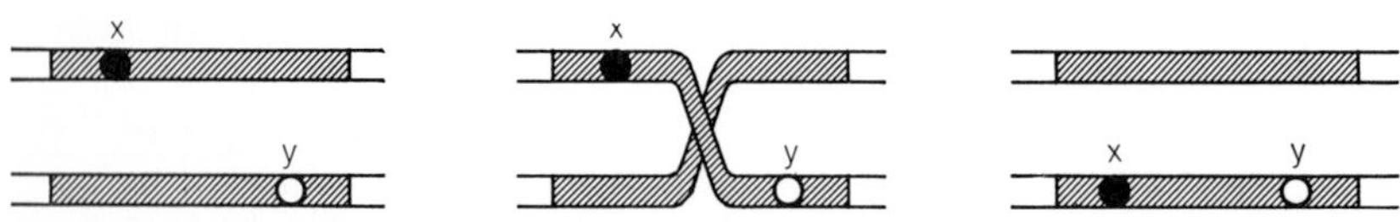


**INTRAGENIC RECOMBINATION** involves a crossing-over of two portions of one gene. By this process two alleles with mutations (*x, y*) might give rise to a normal gene (*right*).

Similar interactions of chromosomal and nonchromosomal genes have been described in yeast and corn; in the latter the system is of considerable economic importance. Corn plants are hermaphrodites with both male and female flower parts, respectively the pollen-bearing tassel and the ear. Several different nonchromosomal-gene mutations have been found that interfere with normal pollen formation, causing a corn plant to be "male-sterile" although it still produces normal ears. Pollen sterility is valuable in hybrid-seed production because the sterile plants cannot fertilize themselves and cross-fertilization is ensured. Sometimes, however, this male-sterility is suppressed by one of several chromosomal genes called "fertility-restorers." In the presence of the proper restorer gene the nonchromosomal trait of sterility cannot express itself and fertile pollen is produced. Then, in plants of the next generation that have lost the restorer gene, the nonchromosomal gene is again operative [*see top illustration on this page*]. Pollen formation is thus controlled by sets of interacting chromosomal and nonchromosomal genes in a manner analogous to the streptomycin-resistance system in *Chlamydomonas*.

Equipped with a variety of mutant nonchromosomal genes produced by streptomycin treatment, we were ready to face the hard questions: What are these genes and how do they function? We began to approach these questions by applying the classical methods of genetic analysis. The genetic studies I shall report here are interesting not only for the information they revealed but also as a demonstration of the power of genetic analysis.

The two basic processes observed in such analysis are "segregation" and "recombination." Segregation refers to the manner in which a pair of alleles, or alternative forms of the same gene, that come from the two parents are distributed to the progeny. As can be seen in the top illustration on the preceding page, chromosomal genes from both parents are distributed in a strict one-to-one ratio because half of the products of meiosis get the maternal chromosome and half get the paternal one. In the case of nonchromosomal genes [*bottom figure on page 68*] there is no segregation: all the progeny get the maternal genes.

Recombination is the process by which genetic material from the two parents is mixed and redistributed among the progeny. If, for example, two genes are on different chromosomes, or "unlinked," they segregate independently in meiosis. A progeny cell may therefore get either allele of each gene [*see third illustration from the bottom on page 70*]. If two genes are on the same chromosome, or "linked," their chances of passing together to the same progeny cell depend on their distance apart along the chromosome. Recombination between linked genes results from an exchange process that occurs more or less at random along the chromosome during meiosis [*see second illustration from bottom on opposite page*]. Recent genetic studies have shown that recombination can occur not only between genes but also within a gene: a progeny cell can receive a gene that is a recombinant of portions of two alleles [*see bottom illustration on page 70*].

When we set out to look for segregation and recombination of nonchromosomal genes, we of course came up against the problem to which I have alluded: the difficulty of working with a maternal-inheritance system. Nonchromosomal genes in *Chlamydomonas* follow the typical one-parent transmission pattern. The two mating types are designated $mt^-$ and $mt^+$; only the $mt^+$ cell regularly transmits its nonchromosomal genes to the progeny. In such a system segregation and recombination would seem to be precluded. We knew, however, that exceptions to the maternal-inheritance rule had been reported in several organisms. As we worked with the streptomycin-resistance and streptomycin-sensitivity genes we observed that once in a great while the nonchromosomal genes from both parents, rather than from just one, came through to the progeny. We still do not know precisely how this happens, but we have learned to capitalize on the event. We plate the zygotes, the cells that result from fusion, on culture media that favor the survival of the exceptional zygotes in which both parents' nonchromosomal genes survive [*see illustration below*].

By means of this technique Zenta Ramanis and I select exceptional zygotes resulting from multifactor crosses and trace the subsequent segregation and recombination of the nonchromosomal genes. To illustrate the kind of results we have obtained I shall describe a cross involving two pairs of mutant nonchromosomal genes and three unlinked pairs of chromosomal genes that served as markers. In this cross the female parent was designated, on the basis of the genes it carried, $ac_1$ *sd act-r ms-r* $mt^+$; the male parent was $ac_2$ *sr act-s ms-s* $mt^-$. The nonchromosomal genes

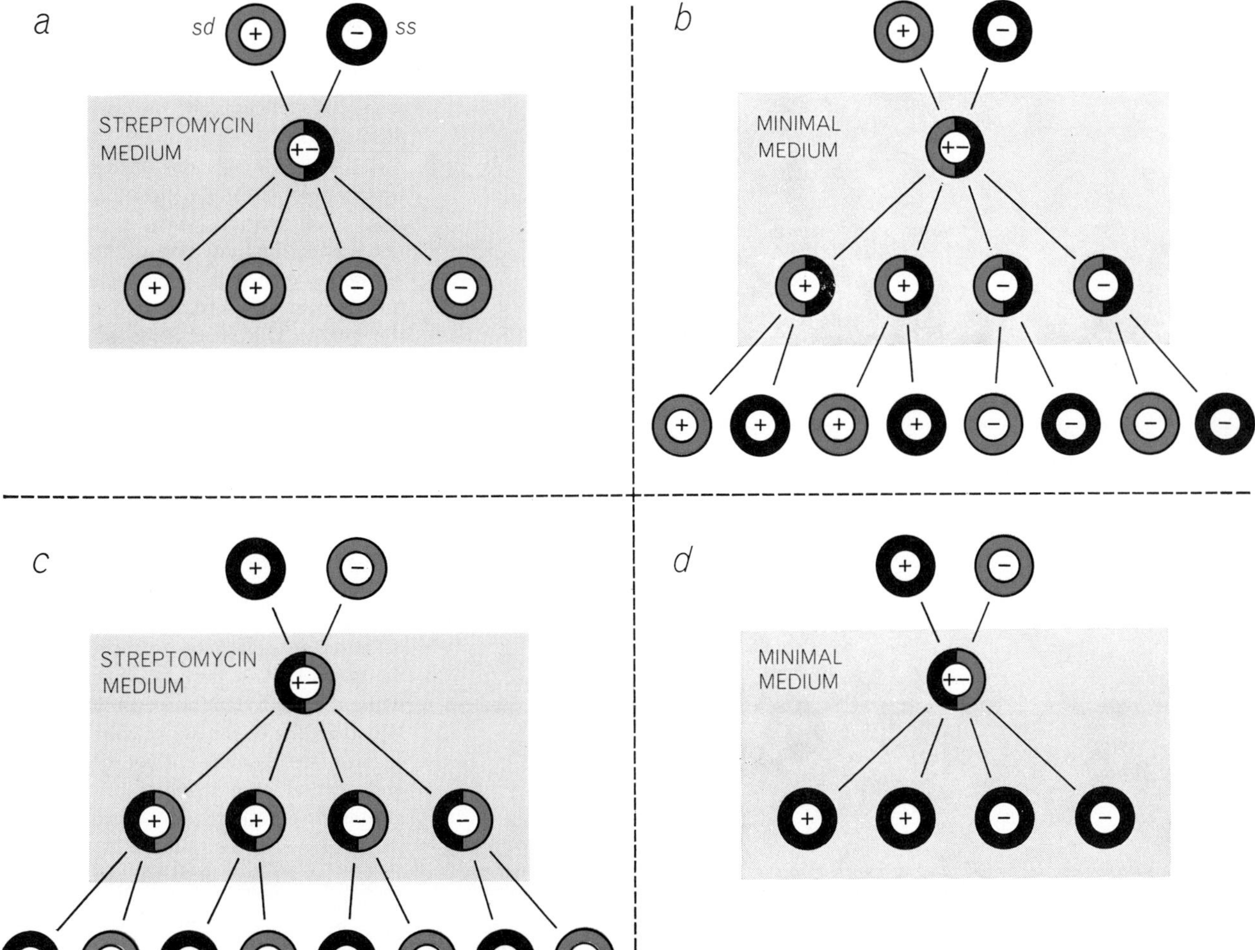


**SELECTION of exceptional zygotes that transmit the male parent's nonchromosomal genes is explained. Ordinarily a cross (*a*) between streptomycin-dependent ("*sd*," *colored ring*) female (+) and streptomycin-sensitive ("*ss*," *black ring*) male (—) cells yields only *sd* cells like the female parent, and the zygotes must be plated on streptomycin medium if live progeny are to be recovered. Similarly, in the reciprocal cross (*d*) most zygotes have only *ss* progeny like the female parent. To select exceptional zygotes that transmit nonchromosomal genes from the male parent, one plates the upper cross on minimal medium (*b*) and the lower cross on streptomycin medium (*c*). In both cases exceptional zygotes are selected that transmit both *sd* and *ss* genes to all progeny, which grow equally well with or without streptomycin. Each progeny cell divides asexually to form a clone containing pure *sd* and pure *ss* cells.**

are $ac_1$ and $ac_2$ and also *sd* and *sr*; $ac_1$ and $ac_2$ are mutants that block photosynthesis and thereby create a requirement for acetate in the medium; *sd* stands for streptomycin-dependence and *sr* for streptomycin-resistance. Among the chromosomal genes *act-r* and *act-s* determine resistance to the antibiotic actidione, *ms-r* and *ms-s* affect resistance to the substance methionine sulfoximine, and $mt^+$ and $mt^-$ of course determine the mating type.

More than 99 percent of the zygotes resulting from this cross give rise to progeny that segregate normally for the three pairs of chromosomal genes but that like the female parent are all $ac_1$ *sd*. The 1 percent of exceptional zygotes, which we selected out by plating the cells on a medium containing no streptomycin, give quite different results: all the nonchromosomal genes are preserved and transmitted to the progeny, each of which therefore contains one set of chromosomal genes and two sets of nonchromosomal genes. The three chromosomal pairs, segregating normally, serve to identify the four products of meiosis [*see illustration at left*].

Each of the four kinds of progeny cell begins to divide in the asexual process called mitosis to form clones of cells that remain genetically identical with respect to their set of chromosomal genes. Now, however, the nonchromosomal genes begin to segregate! At each division some daughter cells arise that have only one of each pair of nonchromosomal genes—cells that are pure *sd* or *sr*, pure $ac_1$ or $ac_2$. By classifying samples of progeny after different numbers of cell divisions and determining the frequencies with which each type appears, we have begun to learn how nonchromosomal genes segregate and recombine. These are our preliminary findings:

1. Segregation begins during the first few doublings after meiosis. After four or five cell divisions between 50 and 60 percent of the progeny are still mixed in nonchromosomal-gene constitution, but the rest are pure for one or both of the genes. The occurrence of segregation just after meiosis tells us that at this stage each gene must be present in one copy or a very few copies.

2. The $ac_1/ac_2$ pair segregates in an average ratio of one to one, indicating that equal numbers of copies of the two versions are present in the mixed cells. Individual clones vary greatly, however, in their $ac_1/ac_2$ ratio, suggesting that segregation may not be equational; that is, a mixed cell may give rise to one daughter that is pure $ac_1$ or pure $ac_2$ while the other daughter remains mixed. Whether or not the *sd*/*sr* pair also segregates one to one is not yet established, and several other aspects of the nonchromosomal segregation process are still under investigation.

3. The $ac_1/ac_2$ and the *sd*/*sr* pairs segregate independently of each other in both time and space. After several doublings we recover, on the average, equal numbers of the parental combinations $ac_1$ *sd* and $ac_2$ *sr* and of the recombinants $ac_1$ *sr* and $ac_2$ *sd*. In other words, the two pairs of nonchromosomal genes behave as if they are carried on separate particles.

4. Each daughter cell always receives one or another form of each nonchromo-

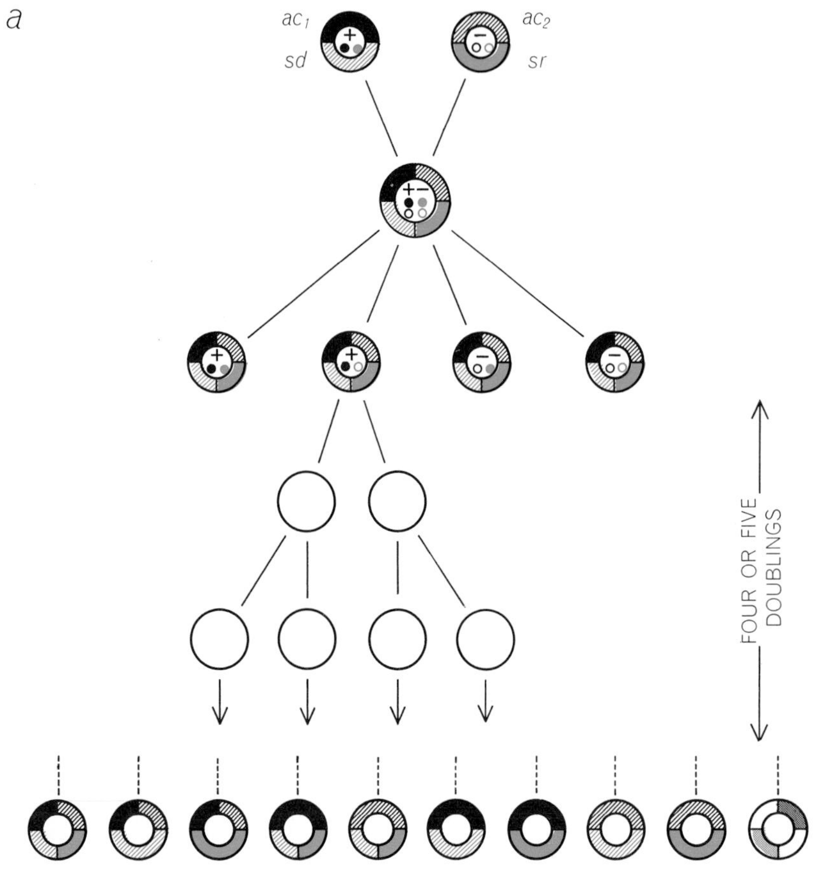


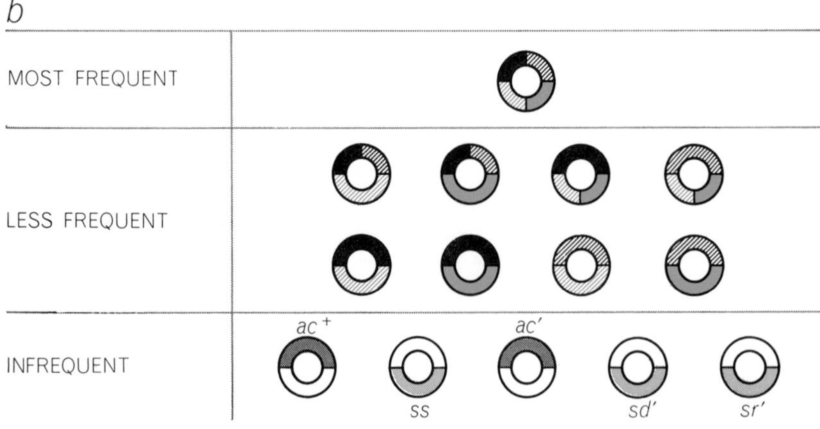


**EXPERIMENTAL CROSS is diagrammed (*a*) involving cells with three pairs of chromosomal genes and two pairs of nonchromosomal genes (*see text*). The exceptional zygotes transmit to the progeny both sets of nonchromosomal genes. Progeny cells divide and the nonchromosomal genes begin to segregate. The bottom row of the diagram shows, schematically, the kinds of cell that result. Some are still mixed, others pure for one or both of the genes; still others (*far right*) have new mutant or wild-type recombinants. The lower diagram (*b*) indicates the approximate proportions in which various combinations appear. Cells carrying a new recombinant may carry any form (*white segments*) of the other gene.**

somal gene. In this sense $ac_1$ and $ac_2$ are alleles, as are *sd* and *sr*.

5. In addition to the parental nonchromosomal alleles, several alleles not present in the parents arise from this cross. There are clones that are *ss* like the wild-type strain from which *sd* and *sr* arose by mutation, or $ac^+$ like the wild-type strain of which $ac_1$ and $ac_2$ are mutants, and also new mutant *sr′*, *sd′* and *ac′* clones with levels of resistance, dependence or acetate requirement different from those of either parent. Genetic analysis of these strains by backcrossing experiments shows that the new types are genetically stable; they can be considered true intragenic recombinants analogous to the intragenic chromosomal recombinants shown in the bottom illustration on page 6. We think that *ss* arose from an exchange between *sr* and *sd* in which both mutated areas of the gene were replaced by normal sections, and that $ac^+$ arose by a similar exchange between $ac_1$ and $ac_2$.

Data of this kind make it possible to state that in *Chlamydomonas* the nonchromosomal genes we have been studying are particles carrying primary genetic information. Each gene is present in one copy or a few copies per cell, but the number may be different for different genes. The nonchromosomal genetic system, which constitutes a sizable fraction of the organism's heredity, usually shows the characteristic maternal inheritance in crosses. Occasionally, however, nonchromosomal genes from both parents are passed to the progeny, where they segregate in subsequent cell divisions.

The results show, moreover, that the nonchromosomal genes are not distributed at random during cell division but by means of a highly oriented mechanism. This is demonstrated first of all by the fact that each nonchromosomal gene, although present in only one copy or a few copies, is transmitted regularly at each cell division. Second, the mutated and normal forms of nonchromosomal genes behave like alleles when they are present together in the same cells, segregating so that each daughter cell gets at least one of them. Finally, intragenic recombination between nonchromosomal alleles requires some kind of consistent close pairing of two structures in order for a precise exchange of precise portions of a gene to take place.

The occurrence of close pairing also says something about the chemical composition of nonchromosomal genes. At the present time the nucleic acids are the only large molecules known to have the physical and chemical potentialities for the close pairing and intramolecular exchange implied by the recombination process. We therefore consider our genetic data the first evidence for the nucleic acid composition of nonchromosomal genes. There is supporting evidence from biochemical studies.

In most organisms the primary genetic material is deoxyribonucleic acid (DNA). Most of it is in the cell nucleus, but apparently not quite all of it. In the *Chlamydomonas* chloroplast we have found a DNA fraction with distinctive chemical and physical properties. We can distinguish it from nuclear DNA because various DNA's differ in what is called their base composition. All DNA contains the same four bases in two linked pairs: the base adenine paired with thymine and guanine paired with cytosine. The two pairs occur in different proportions, however, in DNA from different organisms or even from different parts of the same cell. The base composition is reflected in the density of a DNA sample, so it is possible to separate two DNA's by spinning them in a high-speed centrifuge. When DNA is centrifuged and its fractions are tested for ultraviolet absorption at 260 millimicrons (the wavelength most readily absorbed by DNA), the optical density of each fraction shows its DNA content. When we centrifuge an extract of DNA from whole *Chlamydomonas* cells, we find two DNA's, one of them a light fraction that accounts for only about 5 percent of the total. When we test an extract of DNA from intact chloroplasts, this light fraction accounts for about 40 percent of the total [*see illustration below*]. It is clearly a special chloroplast DNA.

Chloroplast DNA has now been found in a number of other algae and in higher plants. Similar "satellite bands" of nonnuclear DNA are observed in extracts from some animal cells. Recently David J. L. Luck and Edward Reich of the Rockefeller Institute have for the first time been able to demonstrate directly that DNA is localized in mitochondria, which are common to all plant and animal cells.

The existence of both extrachromosomal genes and extranuclear DNA calls for a new and sharply revised picture of cell heredity. Primary genetic

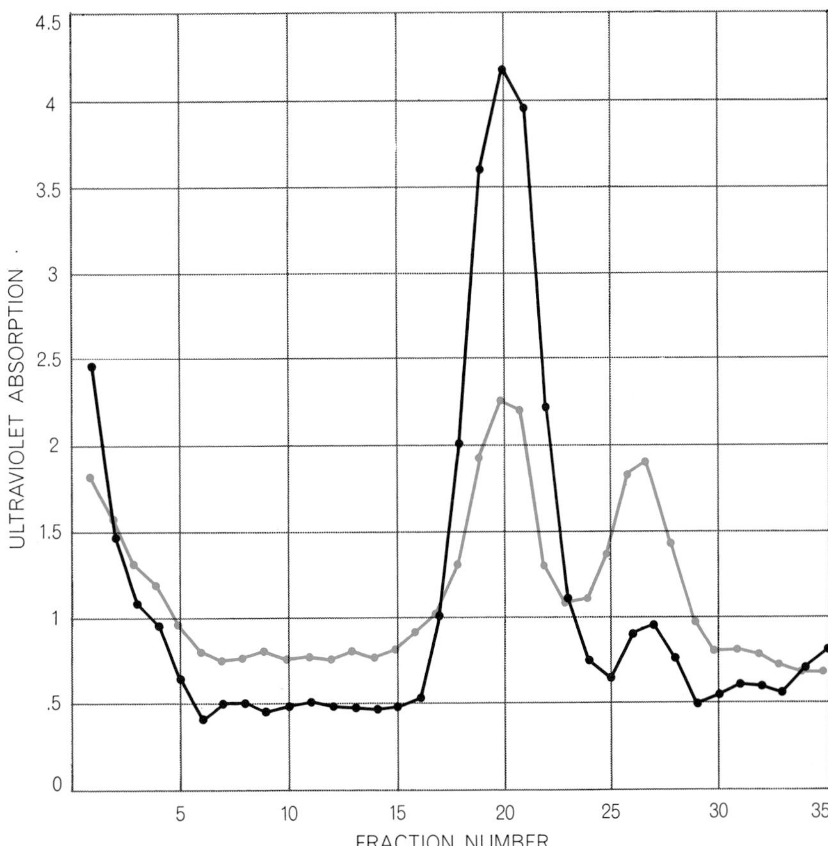


**CHLOROPLAST DNA, or deoxyribonucleic acid, is demonstrated by measuring the optical density in the ultraviolet of fractions of two DNA's: an extract from whole cells (*black curve*) and one from chloroplasts (*color*). A light fraction, about 5 percent of the whole-cell DNA, accounts for much of the chloroplast DNA: it must be localized in that organelle.**

information is apparently distributed at many sites within cells in addition to sites on the nuclear chromosomes. How many sites there are and how many genetic systems and what their chemical identity is remain open questions. Because a substantial fraction of the nonchromosomal genes studied so far influence the functioning of chloroplasts and mitochondria, it seems a fair possibility that organelle DNA may be the carrier of some nonchromosomal genes. No experimental evidence has yet demonstrated, however, that organelle DNA has a genetic role. Nonchromosomal genes carried elsewhere in the cell could influence the activity of an organelle, so the mere fact that a nonchromosomal gene affects the development of a chloroplast, for example, proves nothing about the location of that gene. Moreover, a number of chromosomal genes are known to influence the function of chloroplasts and mitochondria.

Some investigators have reasoned from the available data that the chloroplasts and mitochondria must be self-replicating bodies, but I do not think the significant question is whether or not organelles are autonomous. The smallest self-replicating unit is the cell; subcellular organs are only components of a highly integrated system. The problem is rather: How do the chromosomal and nonchromosomal genetic systems interact and what is the particular role of each gene in each system?

It seems likely that nonchromosomal genes affect other subcellular systems as well as the chloroplasts and mitochondria. The *sd* and *sr* genes in *Chlamydomonas*, for example, appear not to affect either of these organelles. We have been trying to find out if it is the ribosomes that are affected by the mutations. Our evidence indicates that ribosomes of the *sd* and *sr* strains do not differ from those of the wild-type *ss* strain in their protein-synthesizing ability in the presence of streptomycin, so apparently these genes are not acting at the level of ribosome formation. It seems likely, therefore, that there are sites of nonchromosomal gene action other than the chloroplast, the mitochondrion and the ribosome. I should also point out that none of our results precludes the possibility that cells contain still other genetic systems not based on nucleic acids and that hereditary mechanisms may involve principles as yet unknown.

The further analysis of nonchromosomal heredity will require the application of methods similar to those developed for the chromosomal system. The inherent power and high resolution of genetic analysis ensure its continuing usefulness; provided that genes can be made to mutate, they can be identified even if they exist at concentrations far below the resolving power of physical or chemical methods. At the same time the identification of extranuclear DNA has opened the way to a correlated line of investigation, and future studies of nonchromosomal genes will require the most intimate interlocking of evidence from these complementary disciplines.

What is the significance, finally, of the existence of more than one genetic system in cells, systems operating with some local autonomy? Are the cytoplasmic systems vestigial remnants of primitive forms of life or even, as has also been suggested, the remains of what was initially some kind of parasitic invasion of cells? I think the nonchromosomal systems must, on the contrary, be playing an essential role in the life of modern cells, or they would long since have disappeared. I think, moreover, that they played a role from the very beginning. Life began, I would speculate, with the emergence of a stabilized tripartite system: nucleic acids for replication, a photosynthetic or chemosynthetic system for energy conversion, and protein enzymes to catalyze the two processes. Such a tripartite system could have been the ancestor of chloroplasts and mitochondria and perhaps of the cell itself. In the course of evolution these primitive systems might have coalesced into the larger framework of the cell, in which most of the primary genetic information was shifted to specialized structures: the chromosomes.

Why was all the genetic information not shifted to the chromosomes? I think the answer may lie in the cell's requirements for flexibility in growth. Chromosome replication is closely geared to the cellular division cycle, but organelles seem not to be governed entirely by that cycle; to some extent chloroplasts and mitochondria grow independently of cell division in direct response to environmental stresses. The same may be true of other cytoplasmic systems. Under these circumstances it may be desirable that some of the genetic information involved in organelle development be free to replicate at times different from those at which the chromosomal DNA does. The nonchromosomal genetic systems, then, may be of continuing importance in providing flexibility for organelle growth in response to a changing environment.

# II
# *The Nature of the Gene*

# II

# *The Nature of the Gene*

## INTRODUCTION

There is no direct resemblance between the architectural plans for a building and the building itself. The type of problem faced by the geneticist concerned with how the genetic blueprint determines the nature of the organism is comparable to determining how architectural plans are transformed into a constructed building. The genetic studies of the first half of the century showed that this rather formidable problem could be greatly simplified. The discreteness of genes and their relatively specific effects on the phenotype of the organism suggested that the blueprint was, at least to a first approximation, simply an enormous collection of genes, each controlling a specific vital step. The question then became: in what way does the gene exert its controlling influence in the life of the organism?

Usually, mere observation of the consequences of a gene defect provides no simple answer to this question. The loss of function of a single gene usually has manifold effects on the organism. The geneticist then becomes a detective seeking clues to the primary causes of an often-complex hereditary aberration.

It was the pioneering work of Garrod early in this century that led eventually to the realization that the primary functional property of the genes was to control the presence and activities of specific proteins—mostly enzymes, which carry out the principal metabolic and synthetic activities of cells. This primary functional property was first clearly and decisively demonstrated by George W. Beadle and Edward L. Tatum. Their beautiful work is described by Beadle in the first article in this section, "The Genes of Men and Molds."

That genes control the presence of specific enzymes in cells clarifies but does not answer the question of how genes work. What differentiates one gene from another? Wherein lies the specificity of a gene? In chemical terms, how do genes exert their controlling influences over the vast array of proteins in the cell?

The knowledge that genes are composed of DNA helped formulate the hypotheses that led to solutions to these questions. It appeared inevitable that the gene must consist of a stretch of DNA and thus a large number of similar elements. The work of Seymour Benzer, described in his article "The Fine Structure of the Gene," showed that the gene was in fact a linear array of recombinable elements that act together in determining a specific gene function. This work was truly a *tour de force*, for Benzer was able by purely genetic techniques to analyze in great detail the structure of a pair of genes for which, to this day, the protein products remain unidentified.

This genetic system, the rII genes of the bacterial virus T4, provided material for another triumph of formal genetic analysis, which is described by F. H. C. Crick in the article "The Genetic Code." The uniqueness of the gene resides in the specific sequence of nucleotides within the DNA, just as the uniqueness of a word resides in the sequence of letters of which it is composed. The attempt to determine the way in which this unique sequence of nucleotides in DNA specifies the sequence of amino acids in proteins was termed "The Coding Problem." Ingenious exploitation of a special class of mutations in one of the genes analyzed by Benzer enabled Crick and his colleagues to deduce a number of the coding properties of genes.

These properties and the mechanisms that account for them are described in more detail in the article by Crick, "The Genetic Code: III." The nucleotide sequence in the segment of DNA that is defined as the gene is specified by a three-letter code that determines the specific sequence in which the component parts of the protein—the amino acids—are assembled. The most compelling proof for this concept of the relationship between the structure of the gene and the structure of the protein that it controls is described in the article by Charles Yanofsky. This work, which proves the colinearity of gene and protein, illustrates the extent to which the molecular geneticist can probe with genetic and biochemical skill into the detailed business of gene function.

These articles demonstrate that our understanding of what genes do and how they do it has progressed at a prodigious rate since the work of Beadle and Tatum. We started this section by asking how the genetic blueprint of an organism controls its nature and then concerned ourselves largely with how individual genes work. The last article in this section, "The Genetics of a Bacterial Virus" by R. S. Edgar and R. H. Epstein, describes their attempts to analyze the functions of the ensemble of genes of the bacterial virus T4. In a sense, then, their work is an attempt to pull all the pieces together and see to what extent the structure and function of an organism (albeit a tiny and incomplete one) can be accounted for in terms of the functions of its genes.

# 9
# *The Genes of Men and Molds*

GEORGE W. BEADLE
*September 1948*

EIGHTY-FIVE years ago, in the garden of a monastery near the village of Brünn in what is now Czechoslovakia, Gregor Johann Mendel was spending his spare moments studying hybrids between varieties of the edible garden pea. Out of his penetrating analysis of the results of his studies there grew the modern theory of the gene. But like many a pioneer in science, Mendel was a generation ahead of his time; the full significance of his findings was not appreciated until 1900.

In the period following the "rediscovery" of Mendel's work biologists have developed and extended the gene theory to the point where it now seems clear that genes are the basic units of all living things. They are the master molecules that guide the development and direct the vital activities of men and amoebas.

Today the specific functions of genes in plants and animals are being isolated and studied in detail. One of the most useful genetic guinea pigs is the red bread mold *Neurospora crassa*. Its genes can conveniently be changed artificially and the part that they play in the chemical alteration and metabolism of cells can be analyzed with considerable precision. We are learning what sort of material the genes are made of, how they affect living organisms and how the genes themselves, and thereby heredity, are affected by forces in their environment. Indeed, in their study of genes biologists are coming closer to an understanding of the ultimate basis of life itself.

It seems likely that life first appeared on earth in the form of units much like the genes of present-day organisms. Through the processes of mutation in such primitive genes, and through Darwinian natural selection, higher forms of life evolved—first as simple systems with a few genes, then as single-celled forms with

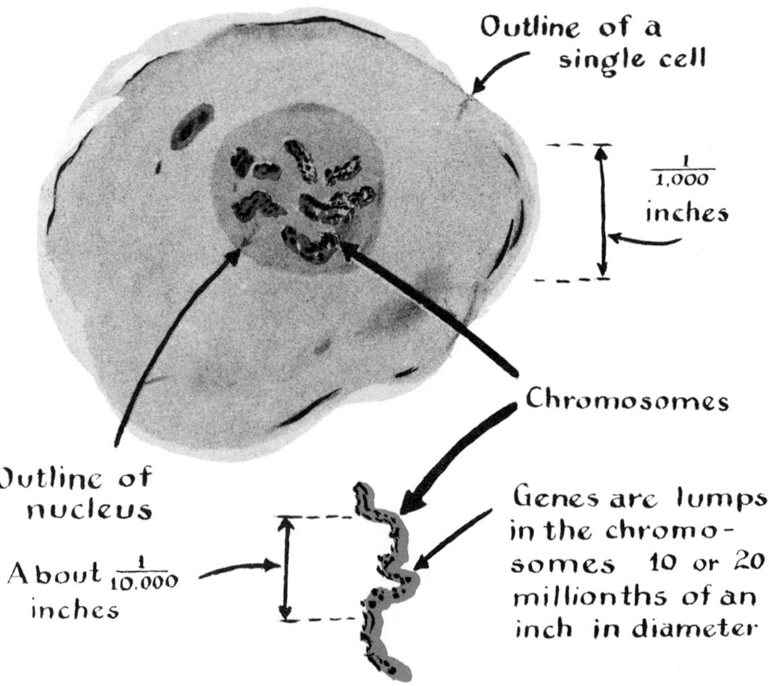


THE CELL is the site of nearly all the interactions between the gene and its environment. The genes themselves are located in the chromosomes, shown above in the stage before cell divides, duplicating each gene in the process.

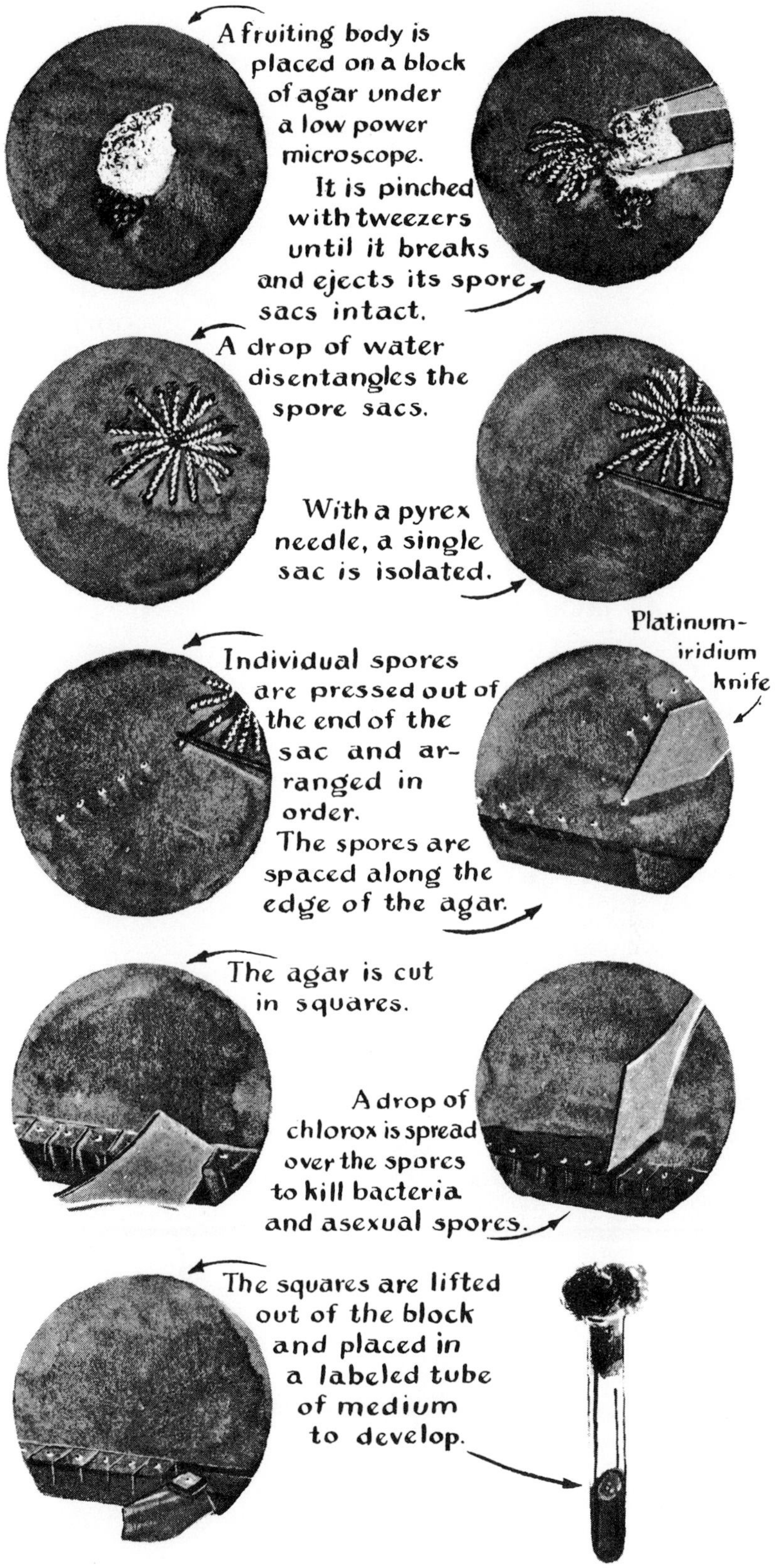

**THE MOLD *Neurospora* is an admirable organism for the study of genes, mainly because of its unusually simple reproductive apparatus. This may be neatly dissected to isolate a single complete set of genes. The sequence of steps in the drawing at the right shows how the tiny fruiting body of the mold is taken apart in the laboratory. With the aid of a microscope, the laboratory worker is able to spread out a set of spore sacs, each containing eight spores. One spore sac may then be separated from the others, and its spores carefully removed. The individual spores are lined up on a block of agar and finally planted in a test tube which contains all the substances that are normally required for the mold to grow.**

A fruiting body is placed on a block of agar under a low power microscope.
It is pinched with tweezers until it breaks and ejects its spore sacs intact.
A drop of water disentangles the spore sacs.
With a pyrex needle, a single sac is isolated.
Platinum-iridium knife
Individual spores are pressed out of the end of the sac and arranged in order.
The spores are spaced along the edge of the agar.
The agar is cut in squares.
A drop of chlorox is spread over the spores to kill bacteria and asexual spores.
The squares are lifted out of the block and placed in a labeled tube of medium to develop.

many genes, and finally as multicellular plants and animals.

What do we know about these genes that are so all-important in the process of evolution, in the development of complex organisms, and in the direction of those vital processes which distinguish the living from the non-living worlds?

In the first place, genes are characterized by students of heredity as the units of inheritance. What is meant by this may be illustrated by examples of some inherited traits in man.

Blue-eyed people may differ by a single gene from those with brown eyes. This eye-color gene exists in two forms, which for convenience may be designated *B* and *b*.

Every person begins as a single cell a few thousandths of an inch in diameter—a cell that comes into being through the fusion of an egg cell from the mother and a sperm cell from the father. This fertilized egg carries two representatives of the eye-color gene, one from each parent. Depending on the parents, there are therefore three types of individuals possible so far as this particular gene is concerned. They start from fertilized eggs represented by the genetic formulas *BB*, *Bb* and *bb*. The first two types, *BB* and *Bb*, will develop into individuals with brown eyes. The third one, *bb*, will have blue eyes. You will note that when both forms of the gene are present the individual is brown-eyed. This is because the form of the gene for brown eyes is *dominant* over its alternative form for blue eyes. Conversely, the form for blue eyes is said to be *recessive*.

During the division of the fertilized egg cell into many daughter cells, which through growth, division and specialization give rise to a fully developed person, the genes multiply regularly with each cell division. As a result each of the millions of cells of a fully developed individual carries exact copies of the two representatives of the eye-color gene

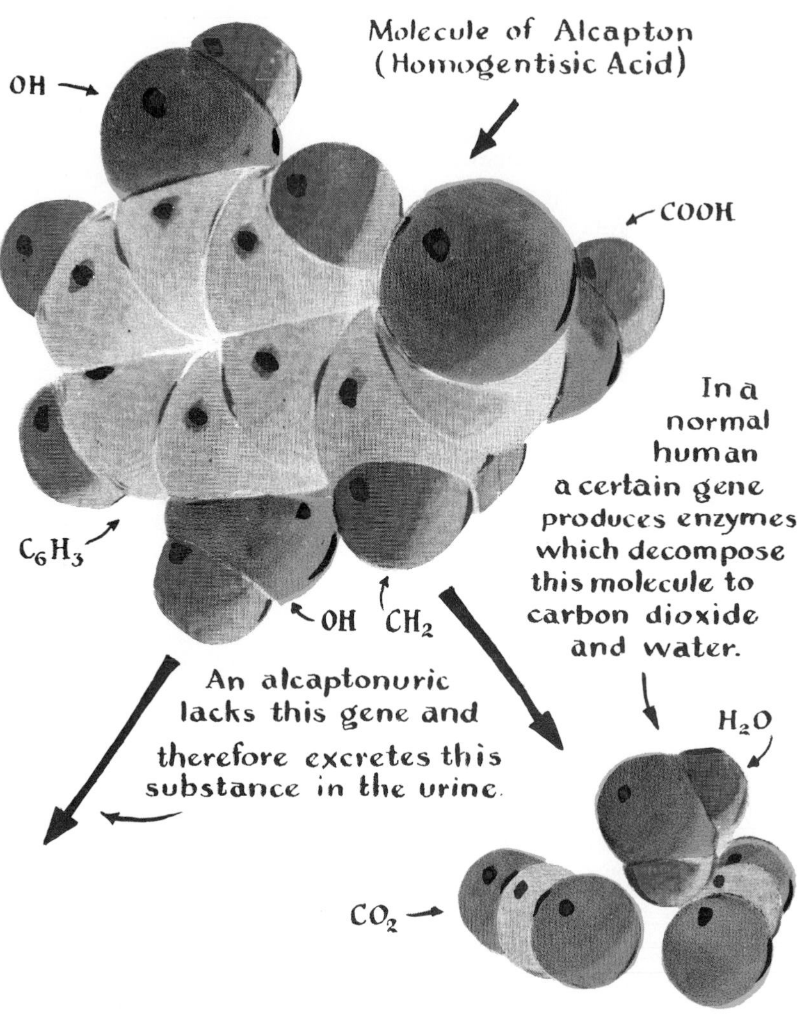
Molecule of Alcapton
(Homogentisic Acid)
OH
COOH
$C_6H_3$
OH
$CH_2$
In a normal human a certain gene produces enzymes which decompose this molecule to carbon dioxide and water.
An alcaptonuric lacks this gene and therefore excretes this substance in the urine.
$H_2O$
$CO_2$

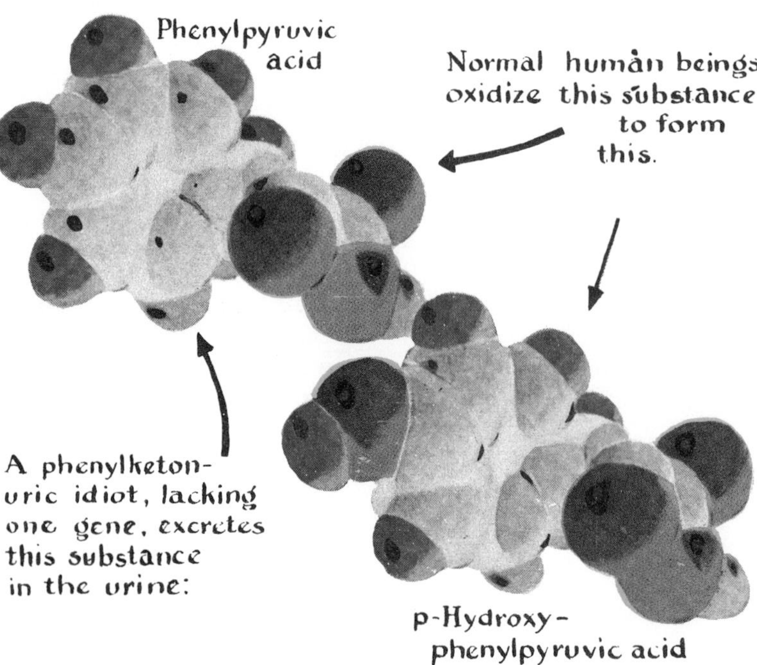
Phenylpyruvic acid
Normal human beings oxidize this substance to form this.
A phenylketonuric idiot, lacking one gene, excretes this substance in the urine:
p-Hydroxy-phenylpyruvic acid

**DEFECTIVE GENES** in man can cause serious hereditary disorders. The chemical basis of two such disorders is shown in the drawings on this page. The large molecule in the drawing at the left is homogentisic acid, or alcapton. In most human beings a single gene produces an enzyme which is capable of breaking alcapton down to carbon dioxide and water. When the gene that produces the enzyme is faulty, however, alcapton is not decomposed. It must be eliminated in the urine, to which it gives a dark color. This excretion of alcapton in the urine is called alcaptonuria. The drawing at the bottom of this page shows the basis of a much more serious genetic disorder. The biochemical apparatus of most human beings, again, is able to transform phenylpyruvic acid into p-hydroxy phenylpyruvic acid. Those who cannot transform it are called phenylketonurics. Phenylketonuria is characterized by extreme feeble-mindedness. Most phenylketonurics are imbeciles or idiots; a few are low-grade morons. The faulty genes that are responsible for both are recessive. This means that they are expressed only when two such genes are paired in the union of an egg and sperm cell. Thus most of the genes responsible for these disorders are carried by normal people without being expressed.

which has been contributed by the parents.

In the formation of egg and sperm cells, the genes are again reduced from two to one per cell. Therefore a mother of the type *BB* forms egg cells carrying only the *B* form of the gene. A type *bb* mother produces only *b* egg cells. A *Bb* mother, on the other hand, produces both *B* and *b* egg cells, in equal numbers on the average. Exactly corresponding relations hold for the formation of sperm cells.

With these facts in mind it is a simple matter to determine the types of children expected to result from various unions. Some of these are indicated in the following list:

| *Mother* | *Father* | *Children* |
|---|---|---|
| *BB* (brown) | *BB* (brown) | All *BB* (brown) |
| *Bb* (brown) | *Bb* (brown) | ¼ *BB* (brown)<br>½ *Bb* (brown)<br>¼ *bb* (blue) |
| *BB* (brown) | *bb* (blue) | All *Bb* (brown) |
| *Bb* (brown) | *bb* (blue) | ½ *Bb* (brown)<br>½ *bb* (blue) |
| *bb* (blue) | *bb* (blue) | All *bb* (blue) |

This table shows that while it is expected that some families in which both parents have brown eyes will include blue-eyed children, parents who are both blue-eyed are not expected to have brown-eyed children.

**LIFE CYCLE of the mold *Neurospora* is illustrated in the drawing at the right. The hyphal fusion of Sex A and Sex a at the bottom of the page is taken as a starting point. *Neurospora* enters a sexual stage rather similar to the union of sperm and egg cells in higher organisms. The union produces a fertile egg, in which two complete sets of genes are paired. The fertile egg cell then divides (*center of drawing*), and divides again. This produces four nuclei, each of which has only a single set of genes. Lined up in a spore sac, the four nuclei divide once more to produce four pairs of nuclei that are genetically identical. A group of spore sacs is gathered in a fruiting body. The sacs and the spores may then be dissected by the technique outlined on page 79. Following this, the germinating spores (*top of page*) may be planted in test tubes containing the necessary nutrients. It is at this point that genetic defects can be exposed by changing the constitution of the medium. Here also *Neurospora* may be allowed to multiply by asexual means. This makes it possible to grow large quantities of the mold without genetic change for convenient chemical analysis. The entire life cycle of the mold takes only 10 days, another reason why *Neurospora* is an exceptionally useful experimental organism.**

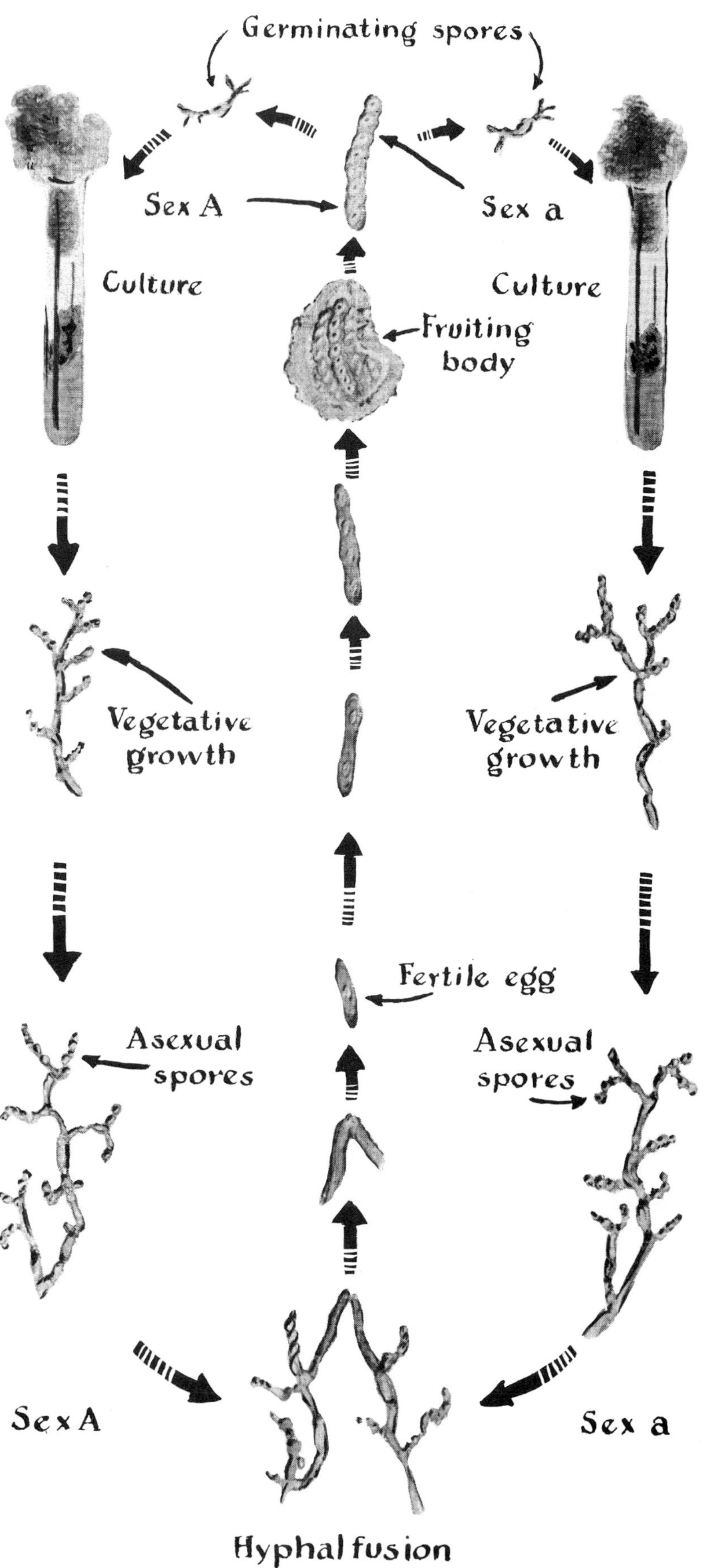


It is important to emphasize conditions that may account for apparent exceptions to the last rule. The first is that eye-color inheritance in man is not completely worked out genetically. Probably other genes besides the one used as an example here are concerned with eye color. It may therefore be possible, when these other genes are taken into account, for parents with true blue eyes to have brown-eyed children. A second factor which accounts for some apparent exceptions is that brown-eyed persons of the *Bb* type may have eyes so light brown that an inexperienced observer may classify them as blue. Two parents of this type may, of course, have a *BB* child with dark brown eyes.

Another example of an inherited trait in man is curly hair. Ordinary curly hair, such as is found frequent'y in people of European descent, is dominant to straight hair. Therefore parents with curly hair may have straight-haired children but straight-haired parents do not often have children with curly hair. Again there are other genes concerned, and the simple rules based on a one-gene interpretation do not always hold.

## Defective Genes

Eye-color and hair-form genes have relatively trivial effects in human beings. Other known genes are concerned with traits of deeper significance. One of these involves a rare hereditary disease in which the principal symptom is urine that turns black on exposure to air. This "inborn error of metabolism," as the English physician and biochemist Sir Archibald Garrod referred to it, has been known to medical men for probably 300 years. Its biochemical basis was established in 1859 by the German biochemist C. Bödeker, who showed that darkening of urine is due to a specific chemical substance called alcapton, later identified chemically as 2,5-dihydroxyphenylacetic acid. The disease is known as alcaptonuria, meaning "alcapton in the urine."

Alcaptonuria is known to result from a gene defect. It shows typical Mendelian inheritance, like blue eyes, but the defective form of the gene is much less frequent in the population than is the recessive form of the eye-color gene.

The excretion of alcapton is a result of the body's inability to break it down by oxidation. Normal individuals possess an enzyme (a protein-containing catalyst, often called a biocatalyst) which makes possible a reaction by which alcapton is further oxidized. This enzyme is absent in alcaptonurics. As a result alcaptonurics cannot degrade alcapton to carbon dioxide and water as normal individuals do.

Alcaptonuria is of special interest genetically and biochemically because it gives us a clue as to what genes do and how they do it. It is clear that the normal kind of gene is essential for the production of the enzyme necessary for the breakdown of alcapton. If the cells of an individual contain only the recessive or inactive form of the gene, no enzyme is formed, alcapton accumulates and is excreted in the urine. The relations between gene and chemical reaction are shown in the diagram at the top of page 80.

A hereditary error of metabolism related biochemically to alcaptonuria is phenylketonuria, a rare disease in which phenylpyruvic acid is excreted in the urine. Like alcaptonuria, this metabolic defect is inherited as a simple Mendelian recessive. It is more serious in its consequences, however, because it is invariably associated with feeble-mindedness of an extreme kind. Most phenylketonurics are imbeciles or idiots; a few are low-grade morons. It should be made clear, however, that only a small fraction of feeble-minded persons are of this particular genetic type.

Phenylketonurics excrete phenylpyruvic acid because they cannot oxidize it, as normal individuals can, to a closely related derivative differing from phenylpyruvic acid by having one more oxygen atom per molecule (*see diagram at the bottom of page 80*). Again it is evident that the normal form of a gene is essential for the carrying out of a specific chemical reaction.

Man, however, is far from an ideal organism in which to study genes. His life cycle is too long, his offspring are too few, his choice of a mate is not often based on a desire to contribute to the knowledge of heredity, and it is inconvenient to subject him to a complete chemical analysis. As a result, most of what we have learned about genes has come from studies of such organisms as garden peas, Indian corn plants and the fruit fly *Drosophila.*

In these and other plants and animals there are many instances in which genes seem to be responsible for specific chemical reactions. It is believed that in most or all of these cases they act as pattern molecules from which enzymes are copied.

Many enzymes have been isolated in a pure crystalline state. All of them have proved to be proteins or to contain proteins as essential parts. Gene-enzyme relations such as those considered above suggest that the primary function of genes may be to serve as models from which specific kinds of enzyme proteins are copied. This hypothesis is strengthened by evidence that some genes control the presence of proteins that are not parts of enzymes.

For example, normal persons have a specific blood protein that is important in blood clotting. Bleeders, known as hemophiliacs, differ from non-bleeders by a single gene. Its normal form is presumed to be essential for the synthesis of the specific blood-clotting protein. Hemophilia, incidentally, is almost completely limited to the male because it is sex-linked; that is, it is carried in the so-called X chromosome, which is concerned with the determination of sex. As is well known, this hereditary disorder has been carried for generations by some of the royal families of Europe.

The genes that determine blood types in man and other animals direct the production of so-called antigens. These are giant molecules which apparently derive their specificity from gene models, and which are capable of inducing the formation of specific antibodies.

### Neurospora

The hypothesis that genes are concerned with the elaboration of giant protein molecules has been tested by experiments with the red mold *Neurospora.* This fungus has many advantages in the study of what genes do. It has a short life cycle—only 10 days from one sexual spore generation to the next. It multiplies profusely by asexual spores. The result is that any strain can be multiplied a millionfold in a few days without any genetic change. Each of the cell nuclei that carry the genes of the bread mold has only a single set of genes instead of the two sets found in the cells of man and other higher organisms. This means that recessive genes are not hidden by their dominant counterparts.

During the sexual stage, in which molds of opposite sex reactions come together, there is a fusion comparable to that between egg and sperm in man. The fusion nucleus then immediately undergoes two divisions in which genes are reduced again to one per cell. The four products formed from a single fusion nucleus by these divisions are lined up in a spore sac. Each divides again so as to produce pairs of nuclei that are genetically identical. The eight resulting nuclei are included in eight sexual spores, each one-thousandth of an inch long. This life cycle of *Neurospora* is shown in the illustration on page 81.

Using a microscope, a skilled laboratory worker can dissect the sexual spores from the spore sac in orderly sequence. Each of them can be planted separately in a culture tube (*see illustration on page 79*). If the two parental strains differ by a single gene, four spores always carry descendants of one form of the gene and four carry descendants of the other. Thus if a yellow and a white strain are crossed, there occur in each spore sac four spores that will give white molds and four that will give yellow.

The red bread mold is almost ideally suited for chemical studies. It can be grown in pure culture on a chemically known medium containing only nitrate, sulfate, phosphate, various other inorganic substances, sugar and biotin, a vitamin of the B group. From these relatively

**EXPERIMENT to determine the role of a single *Neurospora* gene essentially consists in disabling a gene and tracking down its missing biochemical function. Spores of the mold are first exposed to radiation that will cause mutation, *i.e.*, change in a gene. This culture is then crossed with another. The spores resulting from this union are then planted in a medium that contains all the substances that normal *Neurospora* needs for growth, plus a few that the mold normally manufactures for itself. All the spores, including those which may carry a defective gene, germinate on this medium. Spores from these same cultures are then planted in a medium that contains only the bare minimum of substances required by *Neurospora.* Four of the cultures fail to grow, indicating that they have lost the power to manufacture one substance that *Neurospora* normally synthesizes. In test tubes at the bottom of opposite page, the detailed identification of exactly what synthetic power has been lost is begun by planting the defective culture in media that contain (1) all substances required by the normal mold plus vitamins, and (2) all substances plus amino acids. When mold grows on first medium, it appears it has lost the power to synthesize vitamin.**

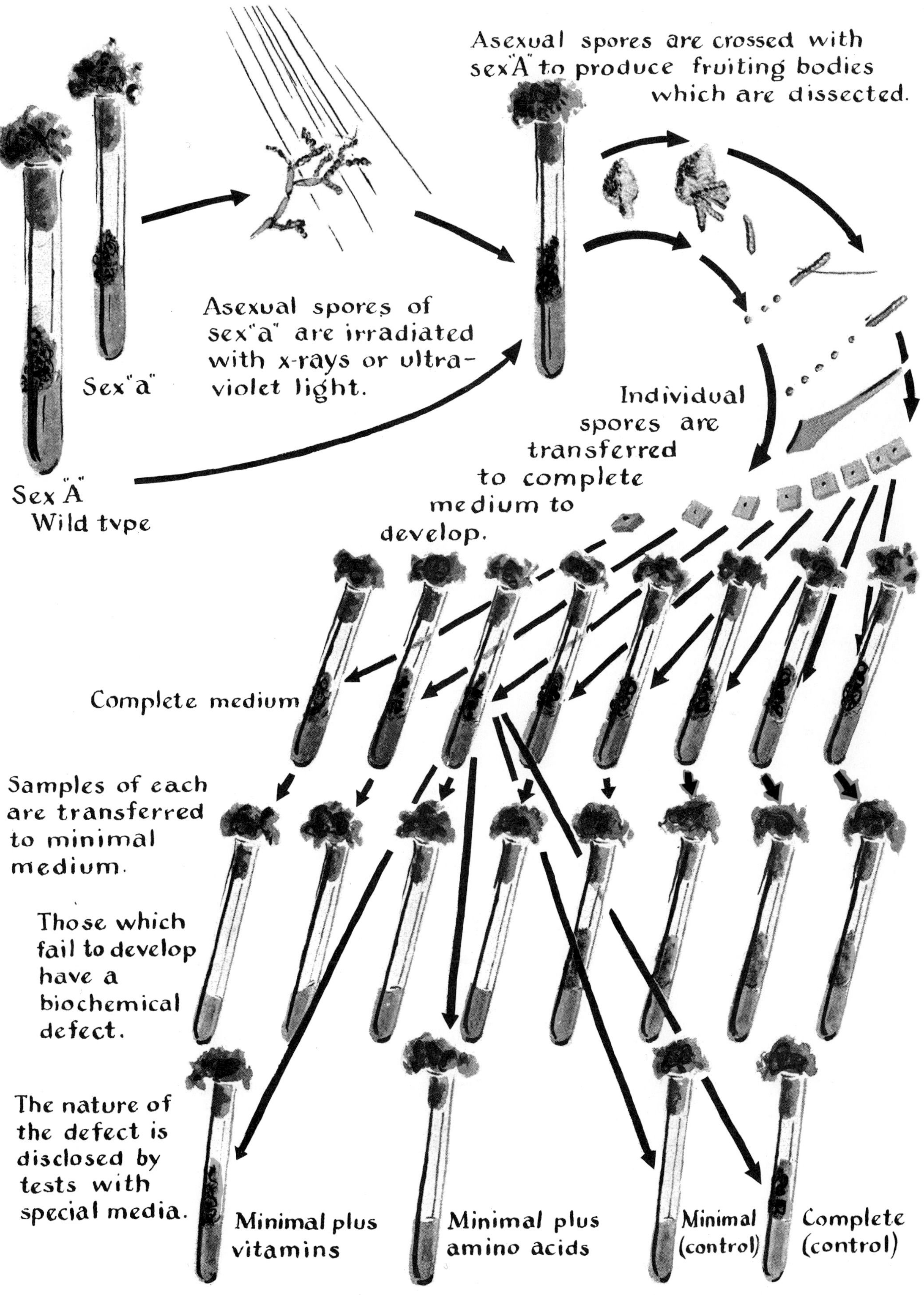
Asexual spores are crossed with sex "A" to produce fruiting bodies which are dissected.
Sex "a"
Sex "A" Wild type
Asexual spores of sex "a" are irradiated with x-rays or ultra-violet light.
Individual spores are transferred to complete medium to develop.
Complete medium
Samples of each are transferred to minimal medium.
Those which fail to develop have a biochemical defect.
The nature of the defect is disclosed by tests with special media.
Minimal plus vitamins
Minimal plus amino acids
Minimal (control)
Complete (control)

simple starting materials, the mold produces all the constituent parts of its protoplasm. These include some 20 amino acid building blocks of proteins, nine water-soluble vitamins of the B group, and many other organic molecules of vital biological significance.

To one interested in what genes do in a human being, it might at first thought seem a very large jump from a man to a mold. Actually it is not. For in its basic metabolic processes, protoplasm—Thomas Huxley's physical stuff of life—is very much the same wherever it is found.

If the many chemical reactions by which a bread mold builds its protoplasm out of the raw materials at its disposal are catalyzed by enzymes, and if the proteins of these enzymes are copied from genes, it should be possible to produce molds with specific metabolic errors by causing genes to mutate. Or to state the problem somewhat differently, one ought to be able to discover what genes do by making them defective.

The simplicity of this approach can be illustrated by an analogy. The manufacture of an automobile in a factory is in some respects like the development of an organism. The workmen in the factory are like genes—each has a specific job to do. If one observed the factory only from the outside and in terms of the cars that come out, it would not be easy to determine what each worker does. But if one could replace able workers with defective ones, and then observe what happened to the product, it would be a simple matter to conclude that Jones puts on the radiator grill, Smith adds the carburetor, and so forth. Deducing what genes do by making them defective is analogous and equally simple in principle.

It is known that changes in genes—mutations—occur spontaneously with a low frequency. The probability that a given gene will mutate to a defective form can be increased a hundredfold or more by so-called mutagenic (mutation producing) agents. These include X-radiation, neutrons and other ionizing radiations, ultraviolet radiation, and mustard gas. Radiations are believed to cause mutations by literally "hitting" genes in a way to cause ionization within them or by otherwise causing internal rearrangements of the chemical bonds.

A bread-mold experiment to test the hypothesis that genes control enzymes and metabolism can be set up in the manner shown in the diagrams on pages 83 and 85. Asexual spores are X-rayed or otherwise treated with mutagenic agents. Following a sexual phase of the life cycle, descendants of mutated genes are recovered in sexual spores. These are grown separately, and the molds that grow from them are tested for ability to produce the molecules out of which they are built.

If a gene essential for the production of vitamin B-1 by the mold is made defective, then B-1 must be supplied in the medium if a mold is to develop from a spore carrying the defective gene. But in the present state of our knowledge it is not possible to produce mutations in specific genes at will. By X-raying, for example, any one or more of several thousand genes may be mutated, or in many cases none at all will be changed. There is no known method of predicting which of the genes, if any, will be hit. It is therefore necessary to grow presumptive mutant spores on a medium supplemented with protoplasmic building blocks of which the formation could be blocked if defective genes were present.

**PHOTOMICROGRAPH of *Neurospora* shows the structure of its fine red tendrils. This photograph, supplied through the courtesy of Life Magazine, was made by Herbert Gehr in the genetics laboratory of E. L. Tatum at Yale.**

**IN CONTINUATION of the experiment begun on page 83, the strain of *Neurospora* that carries a defective gene is put through another series of steps. On page 83 it had been determined that the strain in question did not grow in the absence of vitamins. This indicated that the defective gene was involved in the synthesis of a vitamin. Now the question is: exactly what vitamin? This may be found by planting the strain carrying the defective gene on a group of minimal media, each of which is supplemented by a single vitamin. The mold will then grow on the medium which contains the vitamin that it has lost the power to synthesize. In the experiment outlined on the opposite page, the missing vitamin turns out to be pantothenic acid, a vitamin of the B group. When this has been established, further experiments must be run to determine whether the deficiency of the strain involves a single gene. This is done by crossing the strain bearing the defective gene with a normal strain. All the spores from the union flourish in a medium supplemented with pantothenic acid. When they are planted in a medium that does not contain pantothenic acid, however, only four cultures grow. This is proof that one gene is involved.**

Molds grown on such supplemented medium may grow normally either (1) by making a particular essential part themselves or (2) by taking it ready-made from the culture medium, as they must do if the gene involved in making it is defective. The two possibilities can be distinguished by trying to grow the mold on an unsupplemented medium and on media to which single supplements are added.

Following heavy ultraviolet treatment, about two sexual spores out of every hundred tested carry defective forms of those genes which are necessary for the production of essential substances supplied in the supplemented medium. For example, strain number 5531 of the mold cannot manufacture the B-vitamin pantothenic acid. For normal growth it requires an external supply of this vitamin just as human beings do.

How do we know that the inability of the mold to produce its own pantothenic acid involves a gene defect? The only way this question can be answered at present is by seeing if inability to make pantothenic acid behaves in crosses as a single unit of inheritance.

The answer is that it does. If the mold that cannot make pantothenic acid is crossed with a normal strain of the other sex, the resulting spore sacs invariably contain four spores that produce molds like one parent and four that produce

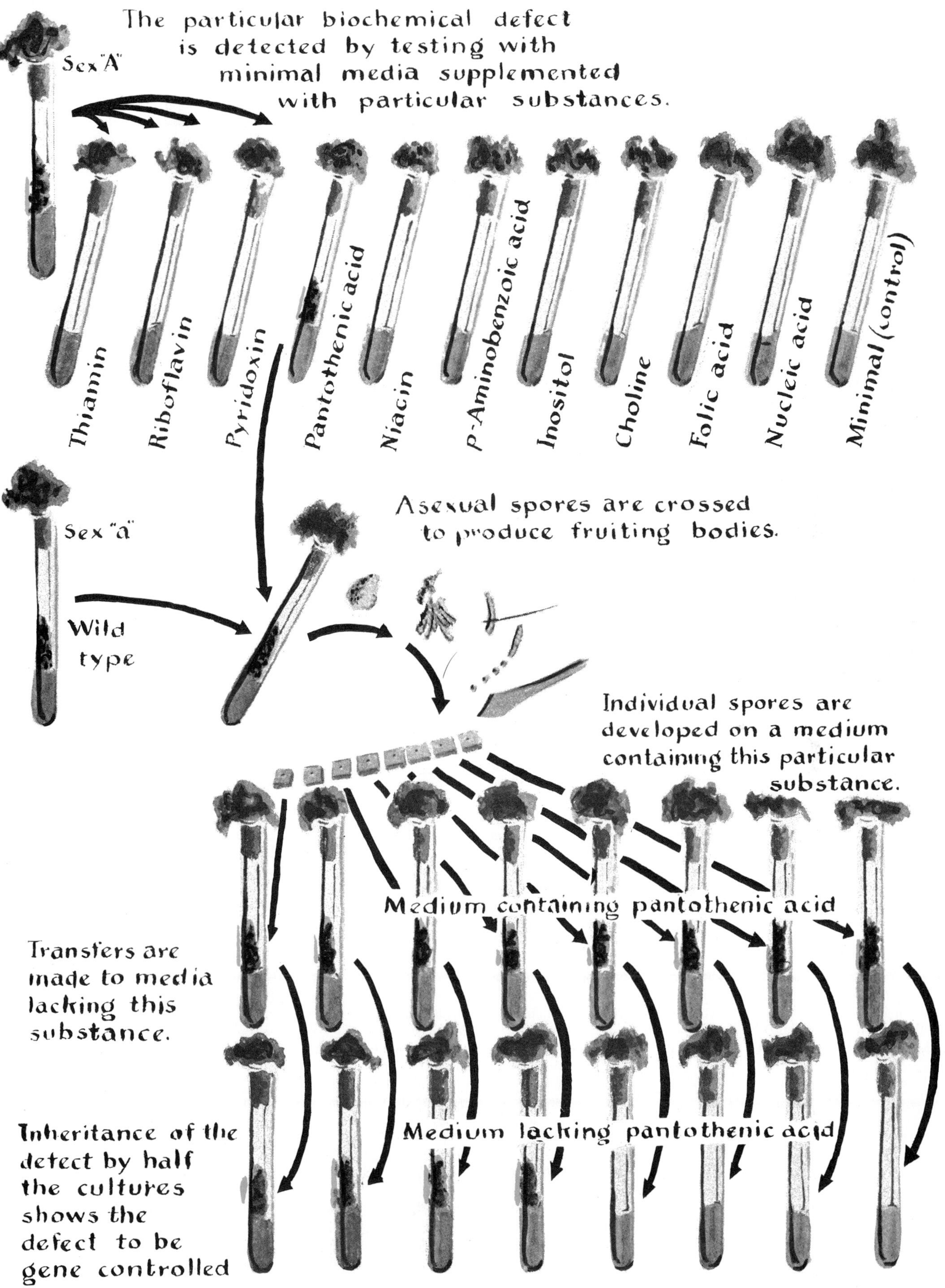
The particular biochemical defect is detected by testing with minimal media supplemented with particular substances.
Sex "A"
Thiamin
Riboflavin
Pyridoxin
Pantothenic acid
Niacin
p-Aminobenzoic acid
Inositol
Choline
Folic acid
Nucleic acid
Minimal (control)
Sex "a"
Wild type
Asexual spores are crossed to produce fruiting bodies.
Individual spores are developed on a medium containing this particular substance.
Medium containing pantothenic acid
Transfers are made to media lacking this substance.
Medium lacking pantothenic acid
Inheritance of the defect by half the cultures shows the defect to be gene controlled

strains like the other parent. Four daughter molds out of each set of eight from a spore sac are able to make pantothenic acid, and four are not (*see page 85*).

In a similar way, genes concerned with many other specific bread-mold chemical reactions have been mutated. In each case that has been studied in sufficient detail to be sure of the relation, it has been found that single genes are directly concerned with single chemical reactions.

An example that illustrates not only that genes are concerned with specific chemical reactions but also how mutant types can be used as tools for the study of metabolic processes involves the production of the amino acid tryptophane and the vitamin niacin (also known as nicotinic acid) by bread mold. Several steps in the synthesis of tryptophane, an indispensable component of the protoplasm of all organisms, have been shown to be gene-controlled. These have been used to show that bread mold forms this component by combining indole and the amino acid serine.

It has been found that indole, in turn, is made from anthranilic acid. If the second gene in the series in the accompanying diagram is made defective, anthranilic acid cannot be converted to indole, and if the mold carrying this gene in defective form is grown on a small amount of tryptophane it accumulates anthranilic acid in much the same way as an alcaptonuric accumulates alcapton. The accumulated anthranilic acid has been chemically identified in the culture medium of such a defective strain.

A recent report that rats fed on diets rich in tryptophane did not need niacin suggested to animal biochemists that possibly niacin is made from tryptophane. Following this lead, studies were made of the strains of bread mold which require ready-made tryptophane and niacin. They gave clear evidence that the bread mold does indeed derive its niacin from tryptophane. Intermediates in the chain of reactions by which the conversion is made were then identified (*see drawing on the opposite page*).

## Men and Molds

The tryptophane-niacin relation so clearly disclosed by bread mold mutants has an interesting relation to the dietary deficiency disease pellagra in man. In the past this disease has been variously attributed to poor quality of dietary proteins, to a toxic factor in Indian corn, and to lack of a vitamin. When, in 1937, C. A. Elvehjem of Wisconsin demonstrated that niacin would bring about spectacular cures of black tongue, a disease of dogs like pellagra in man, the problem seemed to be solved. It was very soon found that pellagra in man, too, is cured by small amounts of niacin in the diet. The alternative hypotheses were promptly forgotten, even though the facts that led to them were not explained by niacin alone.

The tryptophane-niacin relation now makes it clear that the protein quality theory also is correct. Good quality proteins contain plenty of tryptophane. If this is present in sufficient amounts in the diet, niacin appears not to be needed. The corn toxin theory also has a reasonable basis. There appear to be chemical substances in this grain that interfere with the body's utilization of tryptophane and niacin in such a way as to increase the requirements of those two materials.

Another point of interest in connection with the tryptophane-niacin story is that it illustrates again that, in terms of basic protoplasmic reactions, pretty much the same things go on in men and molds. It is supposed that in much the same way as a single gene is in control of the enzyme by which alcaptonuria is broken down in man, genes of the bread mold guide chemical reactions indirectly through their control of enzyme proteins. In most instances the enzymes involved have not yet been studied directly.

Bread-mold studies have contributed strong support to the hypothesis that each gene controls a single protein. But they have not proved it to the satisfaction of all biologists. There remains a possibility that some genes possess several distinct functions and that such genes were automatically excluded by the experimental procedure followed.

What is the process by which genes direct the formation of specific proteins? This is a question to which the answer is not yet known. There is evidence that genes themselves contain proteins combined with nucleic acids to form giant nucleoprotein molecules hundreds of times larger than the relatively simple molecules pictured on the opposite page. And it has been suggested that genes direct the building of non-genic proteins in essentially the same way in which they form copies of themselves.

The general question of how proteins are synthesized by living organisms is one of the great unsolved problems of biology. Until we have made headway toward its solution, it will not be possible to understand growth, normal or abnormal, in anything but superficial terms.

Do all organisms have genes? All sexually reproducing organisms that have been investigated by geneticists demonstrably possess them. Until recently there was no simple way of determining whether bacteria and viruses also have them. As a result of very recent investigations it has been found that some bacteria and some bacterial viruses perform a kind of sexual reproduction in which hereditary units like genes can be quite clearly demonstrated.

By treatment of bacteria with mutagenic agents, mutant types can be produced that parallel in a striking manner those found in the bread mold. These make it almost certain that bacterial genes are functionally like the genes of molds.

So we can sum up by asserting that genes are irreducible units of inheritance in viruses, single-celled organisms and in many-celled plants and animals. They are organized in threadlike chromosomes which in higher plants and animals are carried in organized nuclei. Genes are probably nucleoproteins that serve as patterns in a model-copy process by which new genes are copied from old ones and by which non-genic proteins are produced with configurations that correspond to those of the gene templates.

Through their control of enzyme proteins many genes show a simple one-to-one relation with chemical reactions. Other genes appear to be concerned primarily with the elaboration of antigens—giant molecules which have the property of inducing antibody formation in rabbits or other animals.

It is likely that life first arose on earth as a genelike unit capable of multiplication and mutation. Through natural selection of the fittest of these units and combinations of them, more complex forms of life gradually evolved.

GENES DIRECT a sequence of vital chemical reactions in *Neurospora.* Each of the molecules shown in the models on the opposite page is made up of the atoms hydrogen (*white spheres*), oxygen (*light color*), carbon (*black*) and nitrogen (*dark color*). Reactions involving the genes switch these atoms around to manufacture one molecule out of another. Beginning at the upper left, a single gene is known to be involved in the synthesis of anthranilic acid. Two genes are then involved in making anthranilic acid into indole, with an unknown intermediate indicated by a question mark. Indole is combined with serine to make the amino acid tryptophane, with water left over. Tryptophane is made into kynurenine. Two genes transform kynurenine into 3-hydroxy-anthranilic acid, again with an unknown intermediate molecule. Two genes finally synthesize the last product of the chain: niacin, the B vitamin that is an essential of both plant and animal life. This sequence of events is also involved in the human nutritional disease pellagra. A diet poor in the amino acid tryptophane obviously will lead to a deficiency of niacin, which causes the symptoms of pellagra. Therefore supplying either tryptophane or niacin to patient will alleviate disease.

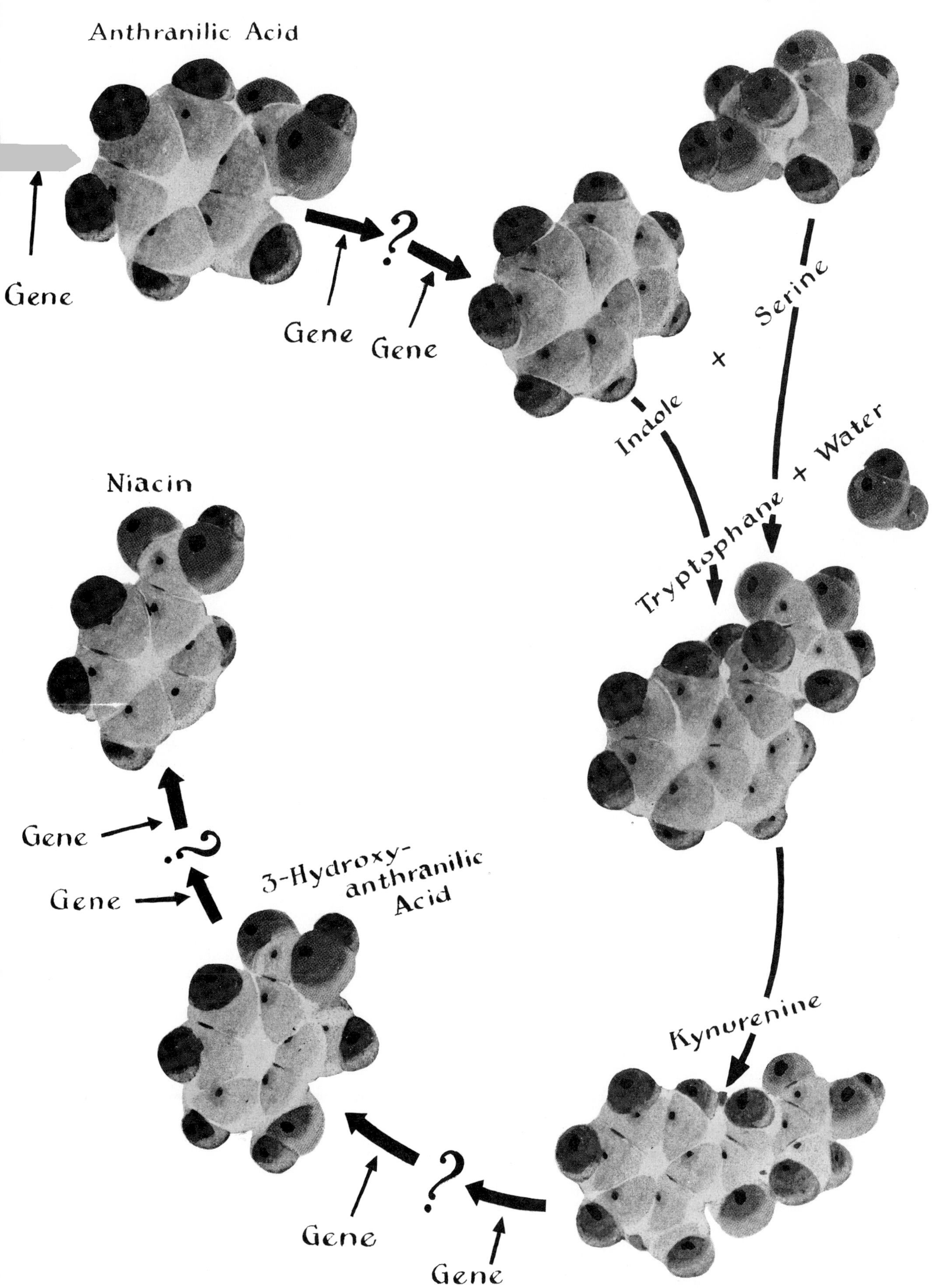
Anthranilic Acid
Gene
Gene
Gene
Serine
+
Indole
+ Water
Tryptophane
Niacin
Gene
Gene
3-Hydroxy-
anthranilic
Acid
Kynurenine
Gene
Gene

# 10

# *The Fine Structure of the Gene*

SEYMOUR BENZER

*January 1962*

Much of the work of science takes the form of map making. As the tools of exploration become sharper, they reveal finer and finer details of the region under observation. In the December, 1961 issue of *Scientific American* John C. Kendrew of the University of Cambridge described the mapping of the molecule of the protein myoglobin, revealing a fantastically detailed architecture. A living organism manufactures thousands of different proteins, each to precise specifications. The "blueprints" for all this detail are stored in coded form within the genes. In this article we shall see how it is possible to map the internal structure of a single gene, with the revelation of detail comparable to that in a protein.

It has been known since about 1913 that the individual active units of heredity—the genes—are strung together in one-dimensional array along the chromosomes, the threadlike bodies in the nucleus of the cell. By crossing such organisms as the fruit fly *Drosophila,* geneticists were able to draw maps showing the linear order of various genes that had been marked by the occurrence of mutations in the organism. Most geneticists regarded the gene as a more or less indivisible unit. There seemed to be no way to attack the questions "Exactly what is a gene? Does it have an internal structure?"

In recent years it has become apparent that the information-containing part of the chromosomal chain is in most cases a giant molecule of deoxyribonucleic acid, or DNA. (In some viruses the hereditary material is ribonucleic acid, or RNA.) Indeed, the threadlike molecule of DNA can be seen in the electron microscope [*see bottom illustration on opposite page*]. For obtaining information about the fine structure of DNA, however, modern methods of genetic analysis are a more powerful tool than even the electron microscope.

It is important to understand why this fine structure is not revealed by conventional genetic mapping, as is done with fruit flies. Genetic mapping is possible because the chromosomes sometimes undergo a recombination of parts called crossing over. By this process, for example, two mutations that are on different chromosomes in a parent will sometimes emerge on the same chromosome in the progeny. In other cases the progeny will inherit a "standard" chromosome lacking the mutations seen in the parent. It is as if two chromosomes lying side by side could break apart at any point and recombine to form two new chromosomes, each made up of parts derived from the original two. As a matter of chance two points far apart will recombine frequently; two points close together will recombine rarely. By carrying out many crosses in a large population of fruit flies one can measure the frequency—meaning the ease—with which different genes will recombine, and from this one can draw a map showing the parts in correct linear sequence. This technique has been used to map the chromosomes of many organisms. Why not, then, use the technique to map mutations inside the gene? The answer is that points within the same gene are so close together that the chance of detecting recombination between them would be exceedingly small.

In the study of genetics, however, everything hinges on the choice of a suitable organism. When one works with fruit flies, one deals with at most a few thousand individuals, and each generation takes roughly 20 days. If one works with a microorganism, such as a bacterium or, better still, a bacterial virus (bacteriophage), one can deal with billions of individuals, and a generation takes only minutes. One can therefore perform in a test tube in 20 minutes an experiment yielding a quantity of genetic data that would require, if humans were used, the entire population of the earth. Moreover, with microorganisms special tricks enable one to select just those individuals of interest from a population of a billion. By exploiting these advantages it becomes possible not only to split the gene but also to map it in the utmost detail, down to the molecular limits of its structure.

## Replication of a Virus

An extremely useful organism for this fine-structure mapping is the T4 bacteriophage, which infects the colon bacillus. T4 is one of a family of viruses that has been most fruitfully exploited by an entire school of molecular biologists founded by Max Delbrück of the California Institute of Technology. The T4 virus and its relatives each consist of a head, which looks hexagonal in electron micrographs, and a complex tail by which the virus attaches itself to the bacillus wall [*see top illustration on opposite page*]. Crammed within the head of the virus is a single long-chain molecule of DNA having a weight about 100 million times that of the hydrogen atom. After a T4 virus has attached itself to a bacillus, the DNA molecule enters the cell and dictates a reorganization of the cell machinery to manufacture 100 or so copies of the complete virus. Each copy consists of the DNA and at least six distinct protein components. To make these components the invading DNA specifies the formation of a series of special enzymes, which themselves are proteins. The entire process is controlled by the battery of genes that constitutes the DNA molecule.

According to the model for DNA de-

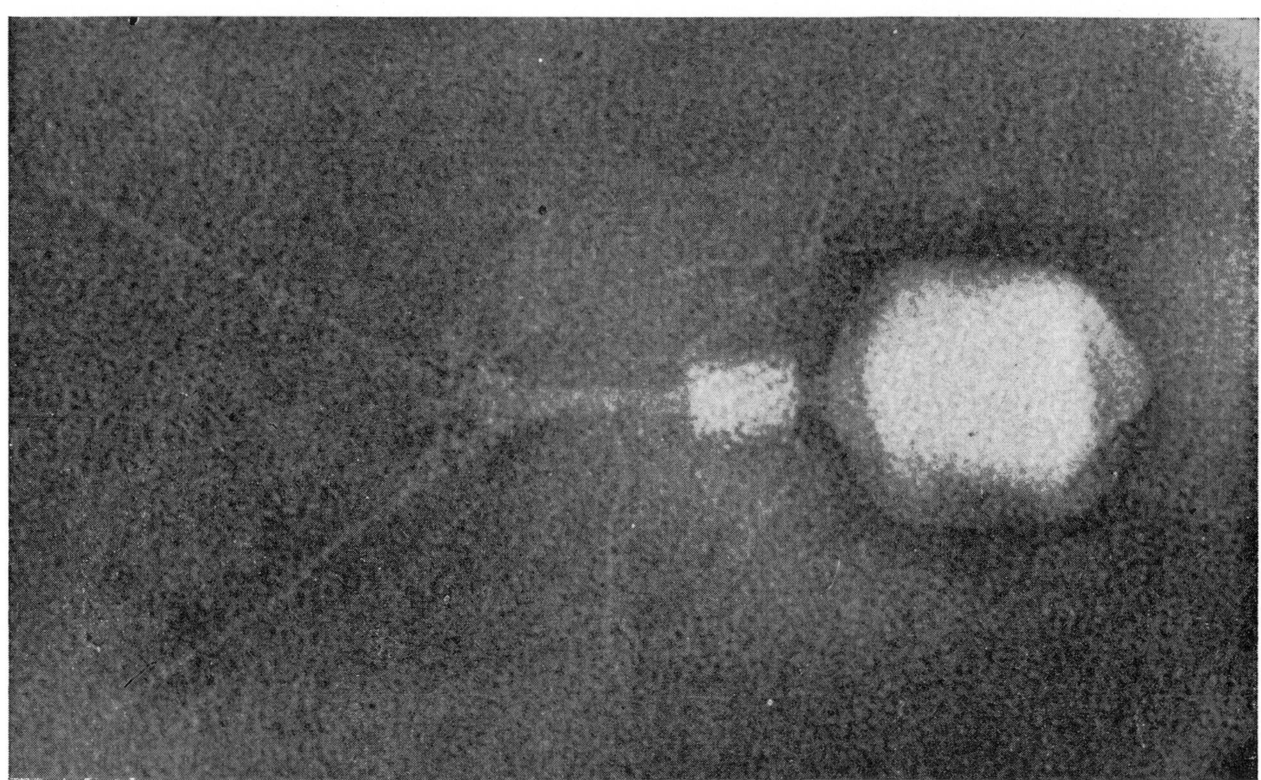

T2 BACTERIOPHAGE, magnified 500,000 diameters, is a virus that contains in its head complete instructions for it own replication. To replicate, however, it must find a cell of the colon bacillus into which it can inject a giant molecule of deoxyribonucleic acid (DNA). This molecule, comprising the genes of the phage, subverts the machinery of the cell to make about 100 copies of the complete phage. The mutations that occasionally arise in the DNA molecule during replication enable the geneticist to map the detailed structure of individual genes. The electron micrograph was made by S. Brenner and R. W. Horne at the University of Cambridge.

MOLECULE OF DNA is the fundamental carrier of genetic information. This electron micrograph shows a short section of DNA from calf thymus; its length is roughly that of the *r*II region in the DNA of T4 phage studied by the author. The DNA molecule in the phage would be about 30 feet long at this magnification of 150,000 diameters. The white sphere, a polystyrene "measuring stick," is 880 angstrom units in diameter. The electron micrograph was made by Cecil E. Hall of the Massachusetts Institute of Technology.

vised by James D. Watson and F. H. C. Crick, the DNA molecule resembles a ladder that has been twisted into a helix. The sides of the ladder are formed by alternating units of deoxyribose sugar groups and phosphate groups. The rungs, which join two sugar units, are composed of pairs of nitrogenous bases: either adenine paired with thymine or guanine paired with cytosine. The particular sequence of bases provides the genetic code of the DNA in a given organism. The DNA in the T4 virus contains some 200,000 base pairs, which, in amount of information, corresponds to much more than that contained in this article. Each base pair can be regarded as a letter in a word. One word (of the DNA code) may specify which of 20-odd amino acids is to be linked into a polypeptide chain. An entire paragraph might be needed to specify the sequence of amino acids for a polypeptide chain that has functional activity. Several polypeptide units may be needed to form a complex protein.

One can imagine that "typographical" errors may occur when DNA molecules are being replicated. Letters, words or sentences may be transposed, deleted or even inverted. When this occurs in a daily newspaper, the result is often humorous. In the DNA of living organisms typographical errors are never funny and are often fatal. We shall see how these errors, or mutations, can be used to analyze a small portion of the genetic information carried by the T4 bacteriophage.

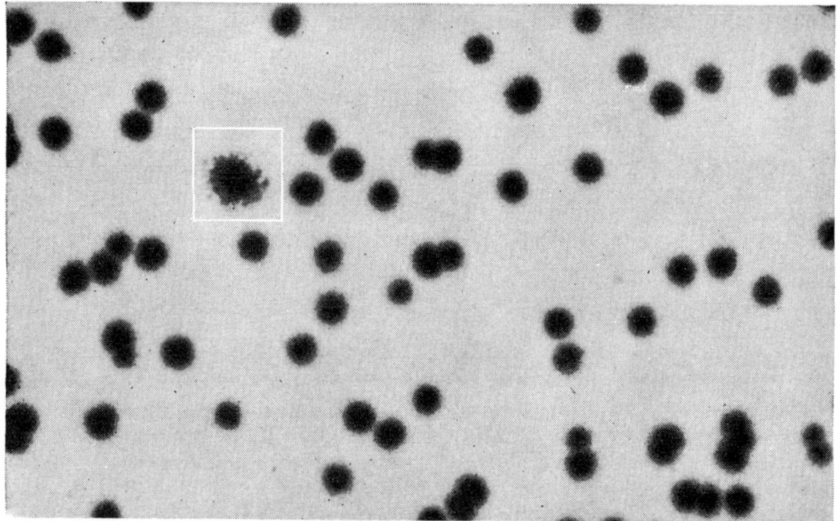

**SPONTANEOUS MUTATIONAL EVENT is disclosed by the one mottled plaque (*square*) among dozens of normal plaques produced when standard T4 phage is "plated" on a layer of colon bacilli of strain B. Each plaque contains some 10 million progeny descended from a single phage particle. The plaque itself represents a region in which cells have been destroyed. Mutants found in abnormal plaques provide the raw material for genetic mapping.**

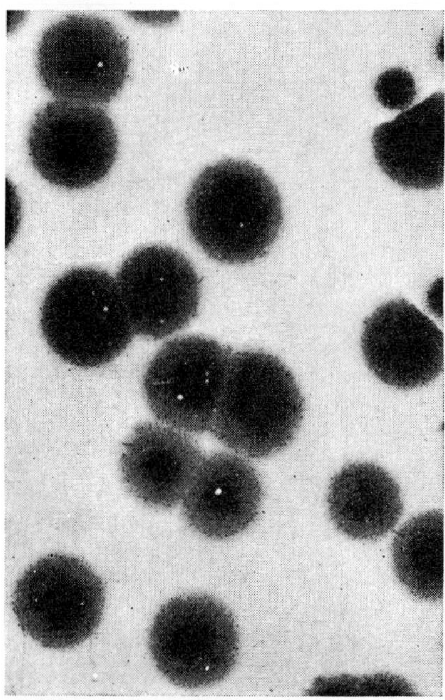

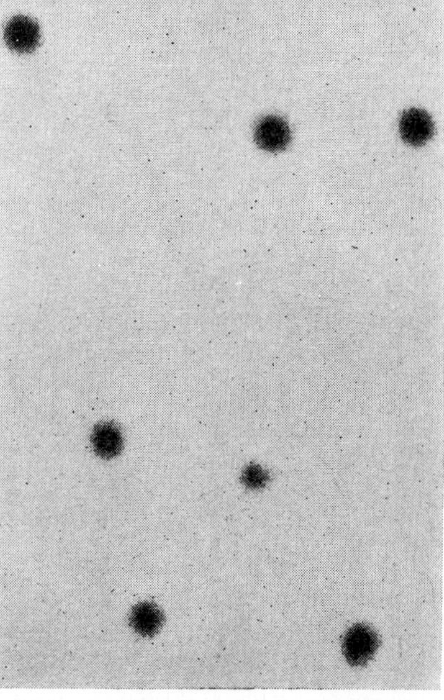

**DUPLICATE REPLATINGS of mixed phage population obtained from a mottled plaque, like that shown at top of page, give contrasting results, depending on the host. Replated on colon bacilli of strain B (*left*), *r*II mutants produce large plaques. If the same mixed population is plated on strain K (*right*), only standard type of phage produce plaques.**

## Genetic Mapping with Phage

Before examining the interior of a gene let us see how genetic experiments are performed with bacteriophage. One starts with a single phage particle. This provides an important advantage over higher organisms, where two different individuals are required and the male and female may differ in any number of respects besides their sex. Another simplification is that phage is haploid, meaning that it contains only a single copy of its hereditary information, so that none of its genes are hidden by dominance effects. When a population is grown from a single phage particle, using a culture of sensitive bacteria as fodder, almost all the descendants are identical, but an occasional mutant form arises through some error in copying the genetic information. It is precisely these errors in reproduction that provide the key to the genetic analysis of the structure [*see upper illustration on pages 92 and 93*].

Suppose that two recognizably different kinds of mutant have been picked up; the next step is to grow a large population of each. This can be done in two test tubes in a couple of hours. It is now easy to perform a recombination experiment. A liquid sample of each phage population is added to a culture of bacterial cells in a test tube. It is arranged that the phage particles outnumber the bacterial cells at least three to one, so that each cell stands a good chance of being infected by both mutant forms of phage DNA. Within 20 minutes about 100 new phage particles are formed within each cell and are released when the cell bursts. Most of the progeny will resemble one or the other parent. In a few of them, however, the genetic information from the two parents may have been recombined to form a DNA molecule that is not an exact copy of the molecule possessed by either parent but a combination of the two. This new recombinant phage particle can carry both mutations or neither of them [*see lower illustration on next two pages*].

When this experiment is done with various kinds of mutant, some of the mutant genes tend to recombine almost independently, whereas others tend to be tightly linked to each other. From such experiments Alfred D. Hershey and Raquel Rotman, working at Washington University in St. Louis, were able to construct a genetic map for phage showing an ordered relationship among the various kinds of mutation, as had been done earlier with the fruit fly *Drosophila* and other higher organisms. It thus appears that the phage has a kind of chromosome —a string of genes that controls its hereditary characteristics.

One would like to do more, however, than just "drosophilize" phage. One would like to study the internal structure of a single gene in the phage chromosome. This too can be done by recombination experiments, but instead of choosing mutants of different kinds one chooses mutants that look alike (that is, have modifications of what is apparently the same characteristic), so that they are likely to contain errors in one or another part of the same gene.

Again the problem is to find an experimental method. When looking for mutations in fruit flies, say a white eye or a bent wing, one has to examine visually every fruit fly produced in the experiment. When working with phage, which reproduce by the billions and are invisible except by electron microscopy, the trick is to find a macroscopic method for identifying just those individuals in which recombination has occurred.

Fortunately in the T4 phage there is a class of mutants called *r*II mutants that can be identified rather easily by the appearance of the plaques they form on a given bacterial culture. A plaque is a clear region produced on the surface of a culture in a glass dish where phage particles have multiplied and destroyed the bacterial cells. This makes it possible to count individual phage particles without ever seeing them. Moreover, the shape and size of the plaques are hereditary characteristics of the phage that can be easily scored. A plaque produced in several hours will contain about 10 million phage particles representing the progeny of a single particle. T4 phage of the standard type can produce plaques on either of two bacterial host strains, B or K. The standard form of T4 occasionally gives rise to *r*II mutants that are easily noticed because they produce a distinctive plaque on B cultures. The key to the whole mapping technique is that

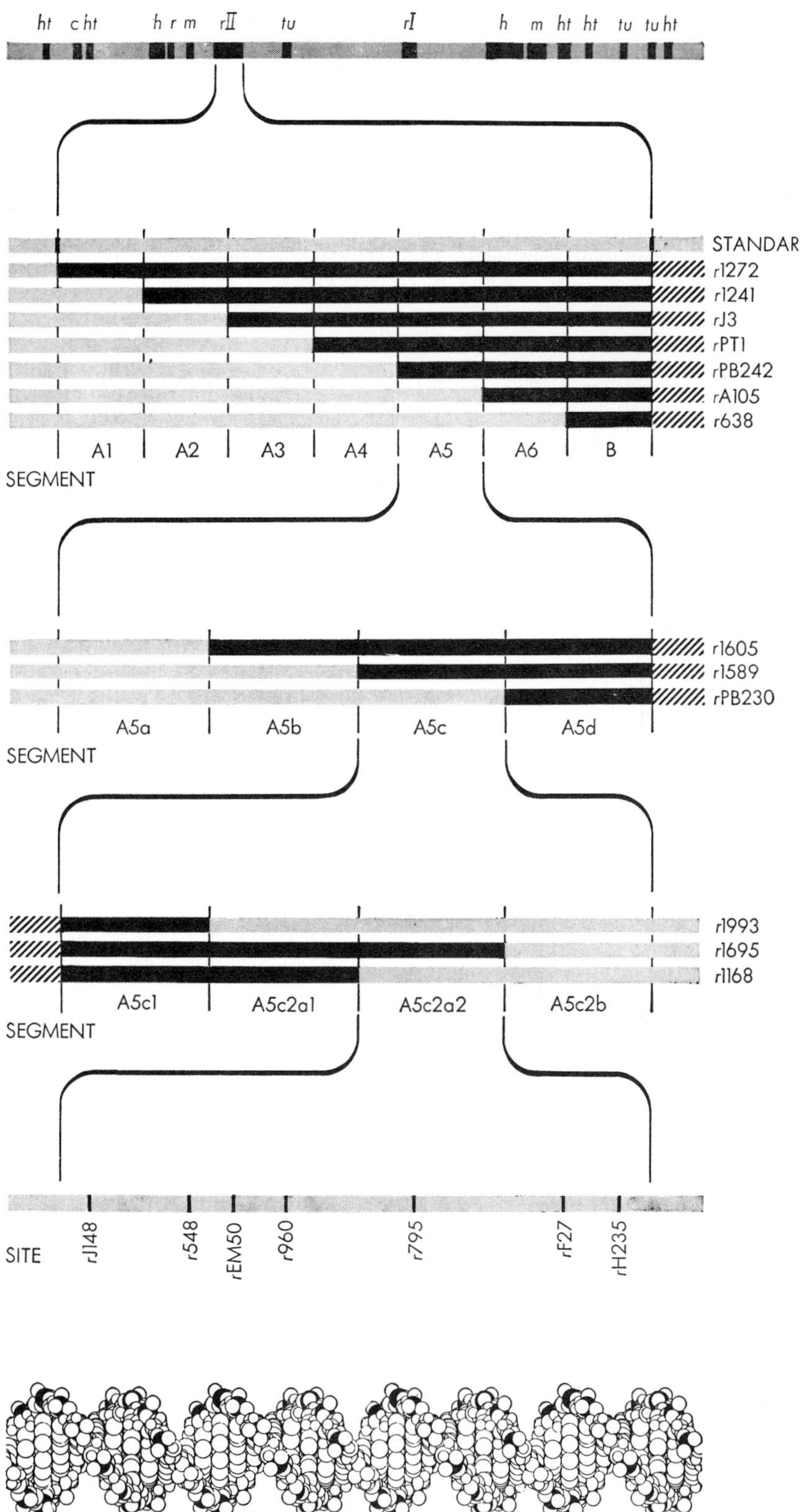


**MAPPING TECHNIQUE localizes the position of a given mutation in progressively smaller segments of the DNA molecule contained in the T4 phage. The *r*II region represents to start with only a few per cent of the entire molecule. The mapping is done by crossing an unknown mutant with reference mutants having deletions (*dark gray tone*) of known extent in the *r*II region (*see illustration of method on page 8*). The order and spacing of the seven mutational sites in the bottom row are still tentative. Each site probably represents the smallest mutable unit in the DNA molecule, a single base pair. The molecular segment (*extreme bottom*), estimated to be roughly in proper scale, contains a total of about 40 base pairs.**

these mutants do not produce plaques on K cultures.

Nevertheless, an *r*II mutant can grow normally on bacterial strain K if the cell is simultaneously infected with a particle of standard type. Evidently the standard DNA molecule can perform some function required in K that the mutants cannot. This functional structure has been traced to a small portion of the DNA molecule, which in genetic maps of the T4 phage is designated the *r*II region.

To map this region one isolates a number of independently arising *r*II mutants (by removing them from mutant plaques visible on B) and crosses them against one another. To perform a cross, the two mutants are added to a liquid culture of B cells, thereby providing an opportunity for the progeny to recombine portions of genetic information from either parent. If the two mutant versions are due to typographical errors in different parts of the DNA molecule, some individuals of standard type may be regenerated. The standards will produce plaques on the K culture, whereas the mutants cannot. In this way one can easily detect a single recombinant among a billion progeny. As a consequence one can "resolve" two *r*II mutations that are extremely close together. This resolving power is enough to distinguish two mutations that are only one base pair apart in the DNA molecular chain.

What actually happens in the recombination of phage DNA is still a matter of conjecture. Two defective DNA molecules may actually break apart and rejoin to form one nondefective molecule, which is then replicated. Some recent evidence strongly favors this hypothesis. Another possibility is that in the course of replication a new DNA molecule arises from a process that happens to copy only the good portions of the two mutant molecules. The second process is called copy choice. An analogy for the two different processes can be found in the methods available for making a good tape recording of a musical performance from two tapes having defects in different places. One method is to cut the defects out of the two tapes and splice the good sections together. The second method (copy choice) is to play the two tapes and record the good sections on a third tape.

## Mapping the *r*II Mutants

A further analogy with tape recording will help to explain how it has been established that the *r*II region is a simple linear structure. Given three tapes, each with a blemish or deletion in a different place, labeled *A*, *B* and *C*, one can imagine the deletions so located that deletion *B* overlaps deletion *A* and deletion *C*, but that *A* and *C* do not overlap each other. In such a case a good performance can be re-created only by recombining *A* and *C*. In mutant forms of phage DNA containing comparable deletions the existence of overlapping can be established by recombination experiments of just the same sort.

To obtain such deletions in phage one looks for mutants that show no tendency to revert to the standard type when they reproduce. The class of nonreverting mutants automatically includes those in which large alterations or deletions have occurred. (By contrast, *r*II mutants that revert spontaneously behave as if their alterations were localized at single points). The result of an exhaustive study covering hundreds of nonreverting *r*II mutants shows that all can be represented as containing deletions of one size or another in a single linear structure. If the structure were more complex, containing, for example, loops or branches, some mutations would have been expected to overlap in such a way as to make it impossible to represent them in a linear map. Although greater complexity cannot be absolutely excluded, all observations to date are satisfied by the postulate of simple linearity.

Now let us consider the *r*II mutants that do, on occasion, revert spontaneously when they reproduce. Conceivably they arise when the DNA molecule of the phage undergoes an alteration of a single base pair. Such "point" mutants are those that must be mapped if one is to probe the fine details of genetic structure. However, to test thousands of point mutants against one another for recombination in all possible pairs would

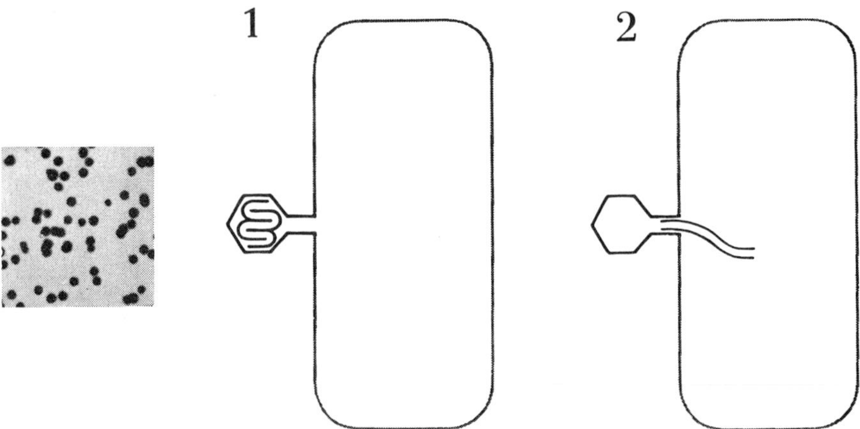


**REPLICATION AND MUTATION occur when a phage particle infects a bacillus cell. The experiment begins by isolating a few standard particles from a normal plaque (*photograph at far left*) and growing billions of progeny in a broth culture of strain B colon bacilli. A sample of the broth is then spread on a Petri dish containing the same strain, on which the**

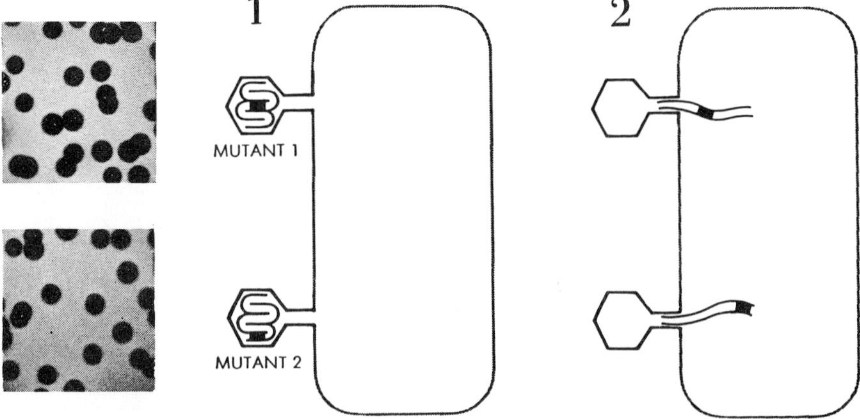


**PROCESS OF RECOMBINATION permits parts of the DNA of two different phage mutants to be reassembled in a new DNA molecule that may contain both mutations or neither of them. Mutants obtained from two different cultures (*photographs at far left*) are introduced into a broth of strain B colon bacilli. Crossing occurs (*1*) when DNA from each mutant type**

require millions of crosses. Mapping of point mutations by such a procedure would be totally impracticable.

The way out of this difficulty is to make use of mutants of the nonreverting type, whose deletions divide up the *r*II region into segments. Each point mutant is tested against these reference deletions. The recombination test gives a negative result if the deletion overlaps the point mutation and a positive result (over and above the "noise" level due to spontaneous reversion of the point mutant) if it does not overlap. In this way a mutation is quickly located within a particular segment of the map. The point mutation is then tested against a second group of reference mutants that divide this segment into smaller segments, and so on [*see illustration on pages 96 and 97*]. A point mutation can be assigned by this method to any of 80-odd ordered segments.

The final step in mapping is to test against one another only the group of mutants having mutations within each segment. Those that show recombination are concluded to be at different sites, and each site is then named after the mutant indicating it. (The mutants themselves have been assigned numbers according to their origin and order of discovery.) Finally, the order of the sites within a segment can be established by making quantitative measurements of recombination frequencies with respect to one another and neighbors outside the segment.

## The Functional Unit

Thus we have found that the hereditary structure needed by the phage to multiply in colon bacilli of strain K consists of many parts distinguishable by mutation and recombination. Is this region to be thought of as one gene (because it controls one characteristic) or as hundreds of genes? Although mutation at any one of the sites leads to the same observed physiological defect, it does not necessarily follow that the entire structure is a single functional unit. For instance, growth in strain K could require a series of biochemical reactions, each controlled by a different portion of the region, and the absence of any one of the steps would suffice to block the final result. It is therefore of interest to see whether or not the *r*II region can be subdivided into parts that function independently.

This can be done by an experiment known as the *cis-trans* comparison. It will be recalled that the needed function can be supplied to a mutant by simultaneous infection of the cell with standard phage; the standard type supplies an intact copy of the genetic structure, so that

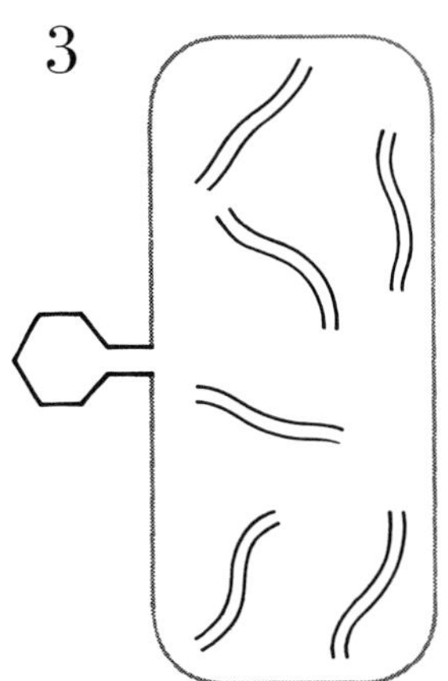


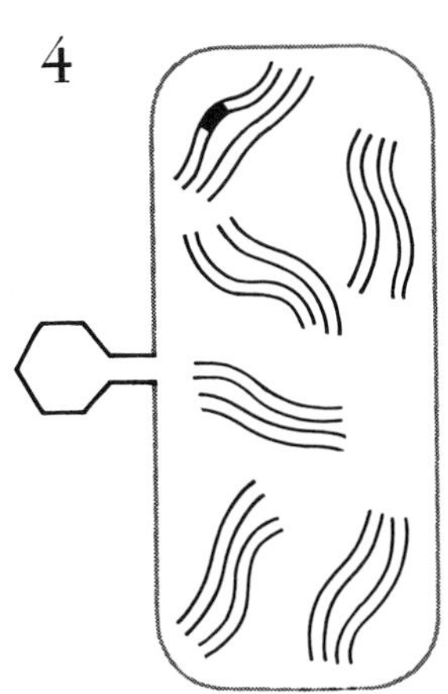


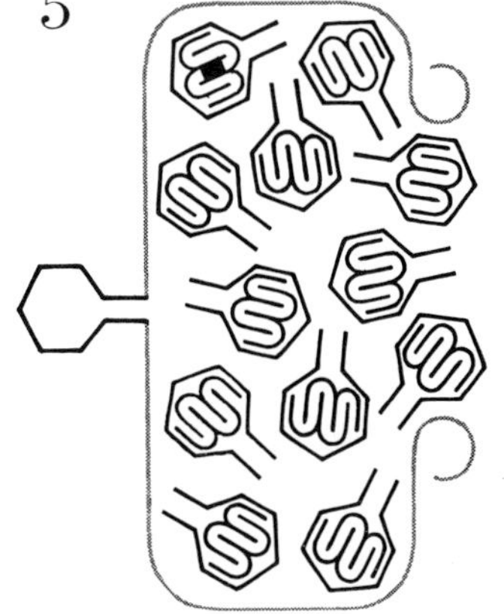


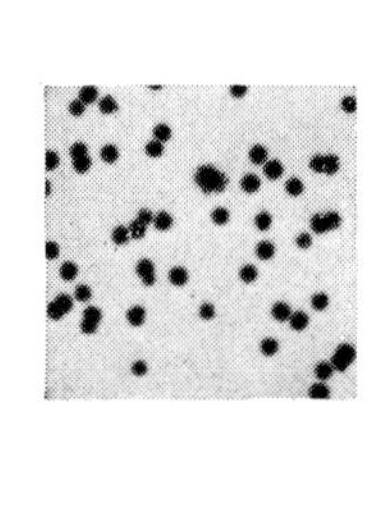

**mutants and standard phage produce different plaque types. The diagrams show a bacillus infected by a single standard phage. The DNA molecule from the phage enters the cell (*2*) and is replicated (*3 and 4*). Among scores of perfect replicas, one may contain a mutation (*dark patch*). Encased in protein jackets, the phage particles finally burst out of the cell (*5*). When a mutant arises during development of a plaque, the mixture of its mutant progeny and standard types makes plaque look mottled (*photograph at right*).**

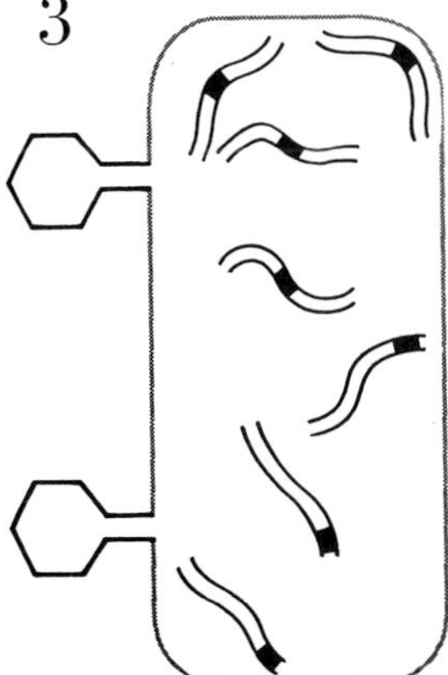


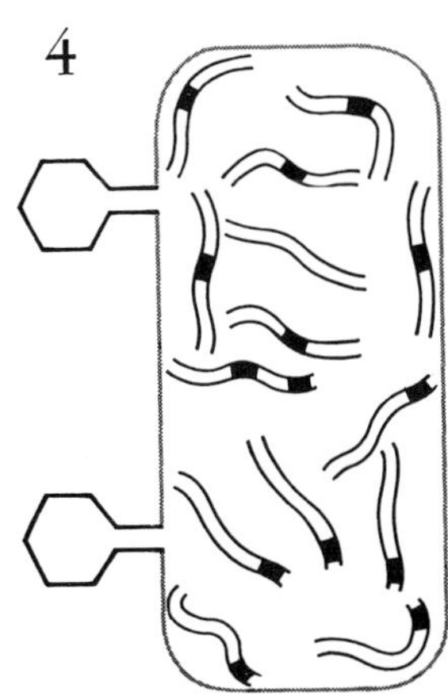


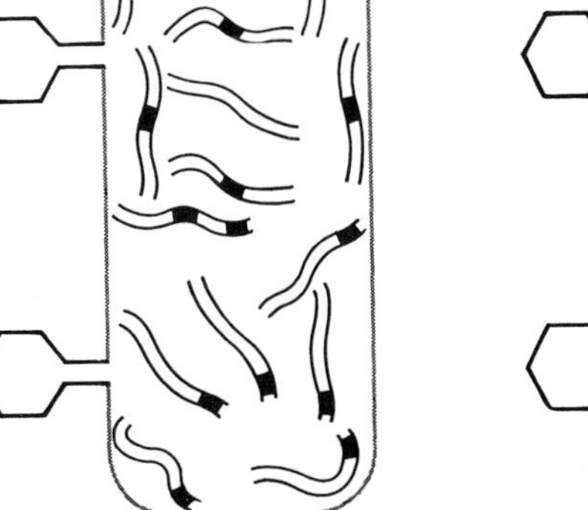

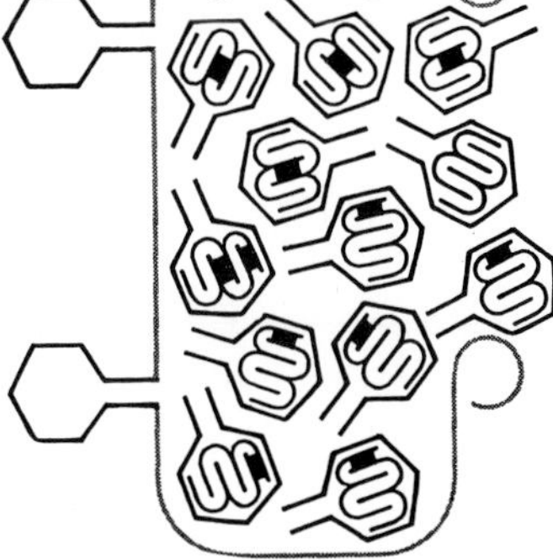


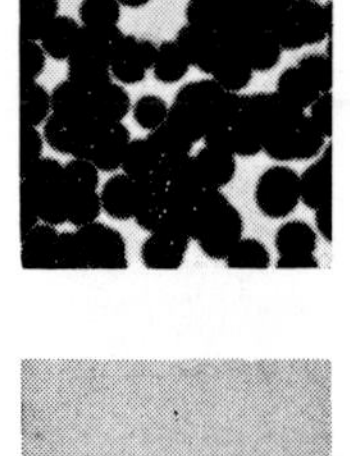

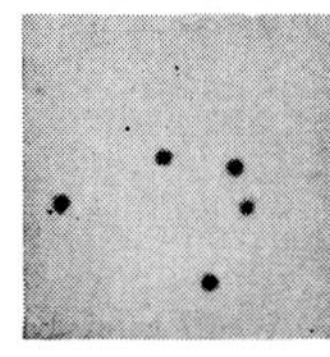

**infects a single bacillus. Most of the DNA replicas are of one type or the other, but occasionally recombination will produce either a double mutant or a standard recombinant containing neither mutation. When the progeny of the cross are plated on strain B (*top photograph at far right*), all grow successfully, producing many plaques. Plated on strain K, only the standard recombinants are able to grow (*bottom photograph at right*). A single standard recombinant can be detected among as many as 100 million progeny.**

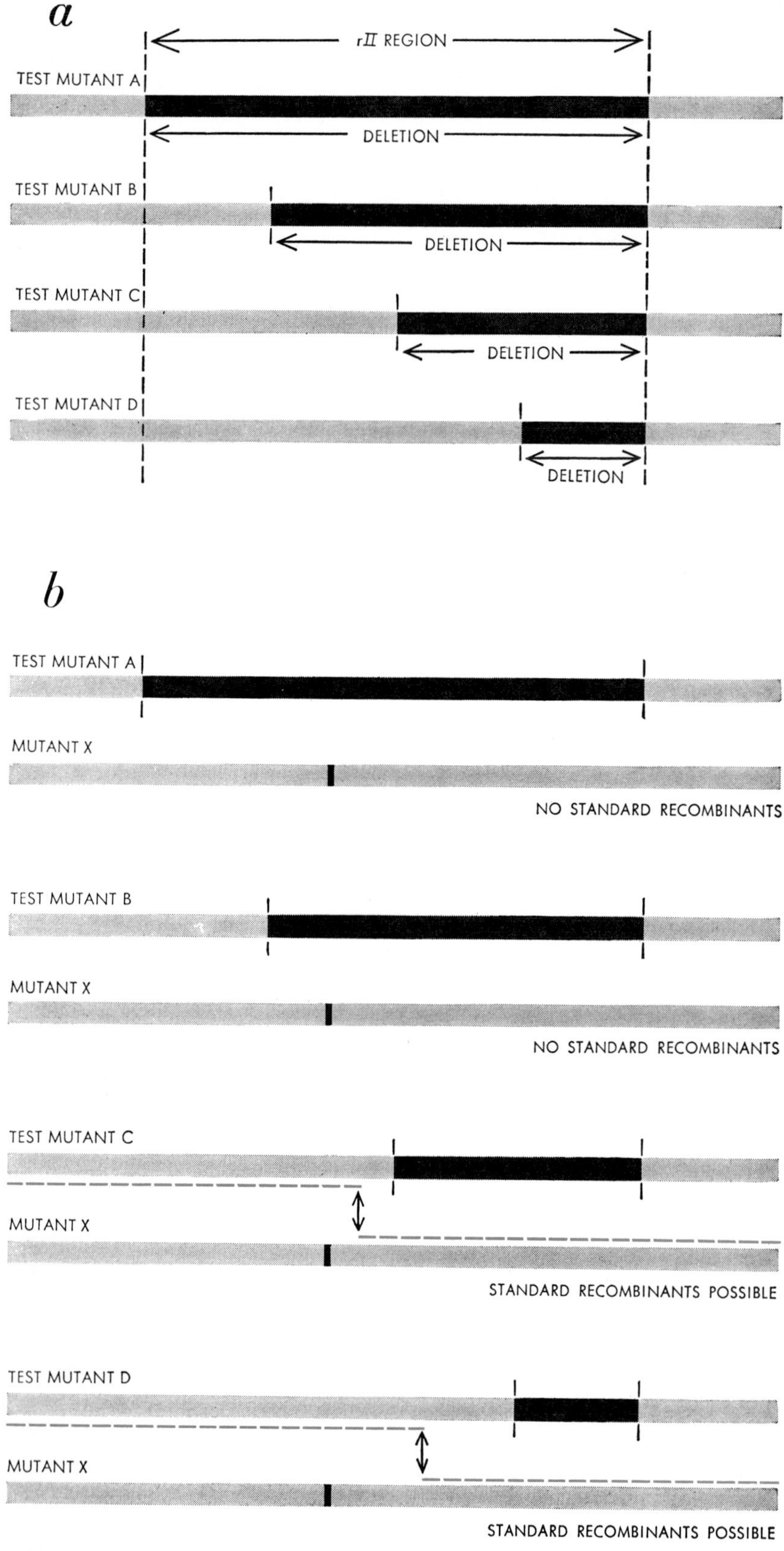


**DELETION MAPPING is done by crossing an unknown mutant with a selected group of reference mutants (*four at top*) whose DNA molecules contain deletions—or what appear to be deletions—of known length in the *r*II region. Thus when mutant *X* is crossed with test mutants *A* and *B*, no standard recombinants are observed because both copies of the DNA molecule are defective at the same place. When *X* is crossed with *C* and *D*, however, standard recombinants can be formed, as indicated by broken lines and arrows. By using other reference mutants with appropriate deletions the location of *X* can be further narrowed.**

it does not matter what defect the *r*II mutant has and both types are enabled to reproduce. Now suppose the intact structure of the standard type could be split into two parts. If this were to destroy the activity, the two parts could be regarded as belonging to a single functional unit. Although the experiment as such is not feasible, one can do the next best thing. That is to supply piece *A* intact by means of a mutant having a defect in piece *B*, and to use a mutant with a defect in piece *A* to supply an intact piece *B*. If the two pieces *A* and *B* can function independently, the system should be active, since each mutant supplies the function lacking in the other. If, however, both pieces must be together to be functional, the split combination should be inactive.

The actual experimental procedure is as follows. Let us imagine that one has identified two mutational sites in the *r*II region, *X* and *Y*, and that one wishes to know if they lie within the same functional unit. The first step is to infect cells of strain K with the two different mutants, *X* and *Y;* this is called the *trans* test because the mutations are borne by different DNA molecules. Now in K the decision as to whether or not the phage will function occurs very soon after infection and *before* there is any opportunity for recombination to take place. To carry out a control experiment one needs a double mutant (obtainable by recombination) that contains both *X* and *Y* within a single phage particle. When cells of strain K are infected with the double mutant and the standard phage, the experiment is called the *cis* test since one of the infecting particles contains both mutations in a single DNA molecule. In this case, because of the presence of the standard phage, normal replication is expected and provides the control against which to measure the activity observed in the *trans* test. If, in the *trans* test, the phage fails to function or shows only slight activity, one can conclude that *X* and *Y* fall within the same functional unit. If, on the other hand, the phage develops actively, it is probable (but not certain) that the sites lie in different functional units. (Certainty in this experiment is elusive because the products of two defective versions of the same functional unit, tested in a *trans* experiment, will sometimes produce a partial activity, which may be indistinguishable from that produced by a *cis* experiment.)

As applied to *r*II mutants, the test divides the structure into two clear-cut parts, each of which can function independently of the other. The functional units have been called cistrons, and we say that the *r*II region is composed of an A cistron and a B cistron.

We have, then, genetic units of various sizes: the small units of mutation and recombination, much larger cistrons and finally the *r*II region, which includes both cistrons. Which one of these shall we call the gene? It is not surprising to find geneticists in disagreement, since in classical genetics the term "gene" could apply to any one of these. The term "gene" is perfectly acceptable so long as one is working at a higher level of integration, at which it makes no difference which unit is being referred to. In describing data on the fine level, however, it becomes essential to state unambiguously which operationally defined unit one is talking about. Thus in describing experiments with *r*II mutants one can speak of the *r*II "region," two *r*II "cistrons" and many *r*II "sites."

Some workers have proposed using the word "gene" to refer to the genetic unit that provides the information for one enzyme. But this would imply that one should not use the word "gene" at all, short of demonstrating that a specific enzyme is involved. One would be quite hard pressed to provide this evidence in the great majority of cases in which the term has been used, as, for example, in almost all the mutations in *Drosophila*. A genetic unit should be defined by a genetic experiment. The absurdity of doing otherwise can be seen by imagining a biochemist describing an enzyme as that which is made by a gene.

We have seen that the topology of the *r*II region is simple and linear. What can be said about its topography? Are there local differences in the properties of the various parts? Specifically, are all the subelements equally mutable? If so, mutations should occur at random throughout the structure and the topography would then be trivial. On the other hand, sites or regions of unusually high or low mutability would be interesting topographic features. To answer this question one isolates many independently arising *r*II mutants and maps each one to see if mutations tend to occur more frequently at certain points than at others. Each mutation is first localized into a main segment, then into a smaller segment, and finally mutants of the same small segment are tested against each other. Any that show recombination are said to define different sites. If two or more reverting mutants are found to show no detectable recombination with each other, they are considered to be

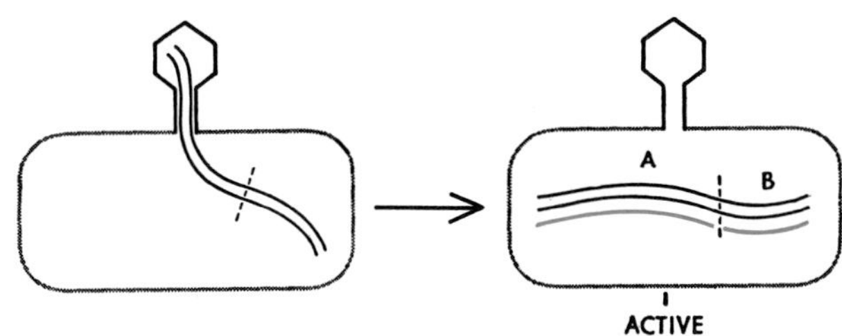


**PHAGE ACTIVITY requires that the coded information inside functional units of the DNA molecule be available intact. The *r*II region consists of two functional units called A cistron and B cistron. When both are present intact (*right*), the phage actively replicates inside colon bacillus of strain K. Colored lines indicate effective removal of coded information.**

*a*

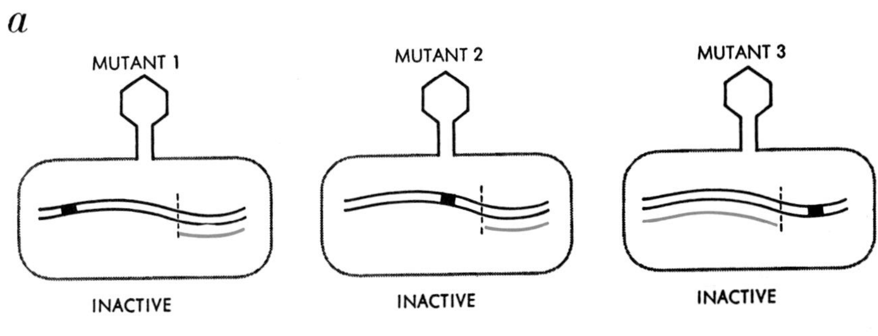


*b*

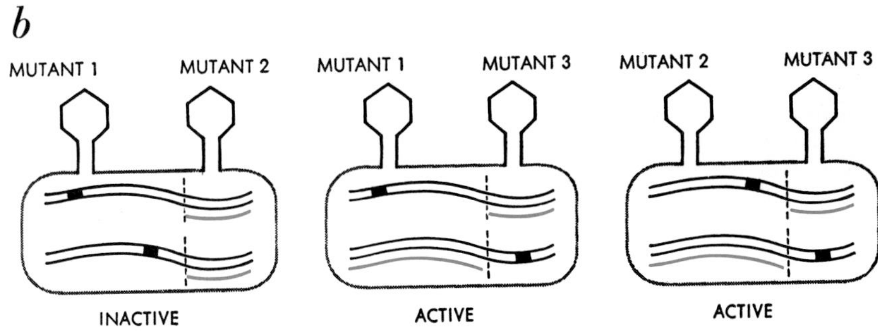


*c*

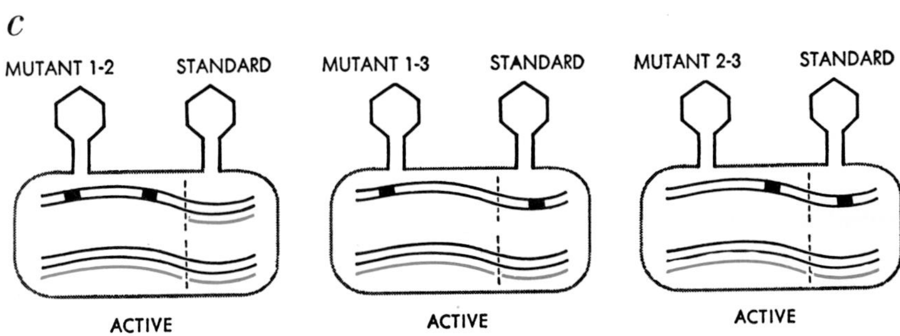


***CIS-TRANS* TEST determines the size of functional units. In bacillus of strain K, T4 phage is active only if both A and B cistrons are provided intact; hence mutants *1*, *2* and *3* are inactive. (The sites of mutation have been previously established.) Tests with the three mutants taken two at a time (*b*) show that sites *1* and *2* must be in the same cistron. A test of each mutant with standard phage (*c*) provides a control; in this case all are active.**

repeats, and one of them is chosen to represent the site in further tests. A set of distinct sites is thereby obtained, each with its own group of repeats. The designation of a mutant as a repeat is, of course, tentative, since in principle it remains possible that a more sensitive test could show some recombination.

The illustration on the next two pages shows a map of the *r*II region with each occurrence of a spontaneous mutation indicated by a square. These mutations, as well as other data from induced mutations, subdivide the map into more than 300 distinct sites, and the distribution of repeats is indeed far from random. The topography for spontaneous mutation is evidently quite complex, the structure consisting of elements with widely different mutation rates.

Spontaneous mutation is a chronic disease; a spontaneous mutant is simply one for which the cause is unknown. By using chemical mutagens such as nitrous acid or hydroxylamine, or physical agents such as ultraviolet light, one can alter the DNA in a more controlled manner and induce mutations specifically. A method of inducing specific mutations has long been the philosophers' stone of genetics. What the genetic alchemist desired, however, was an effect that could be directed at the gene controlling a particular characteristic. Chemical mutagenesis is highly specific but not in this way. When Rose Litman and Arthur B. Pardee at the University of California discovered the mutagenic effect of 5-bromouracil on phage, they regarded it as a nonspecific mutagen because mutations were induced that affected a wide assortment of different phage characteristics. This nonspecificity resulted because each functional gene is a structure with many parts and is bound to contain a number of sites that are responsive to any particular mutagen. Therefore the rate at which mutation is

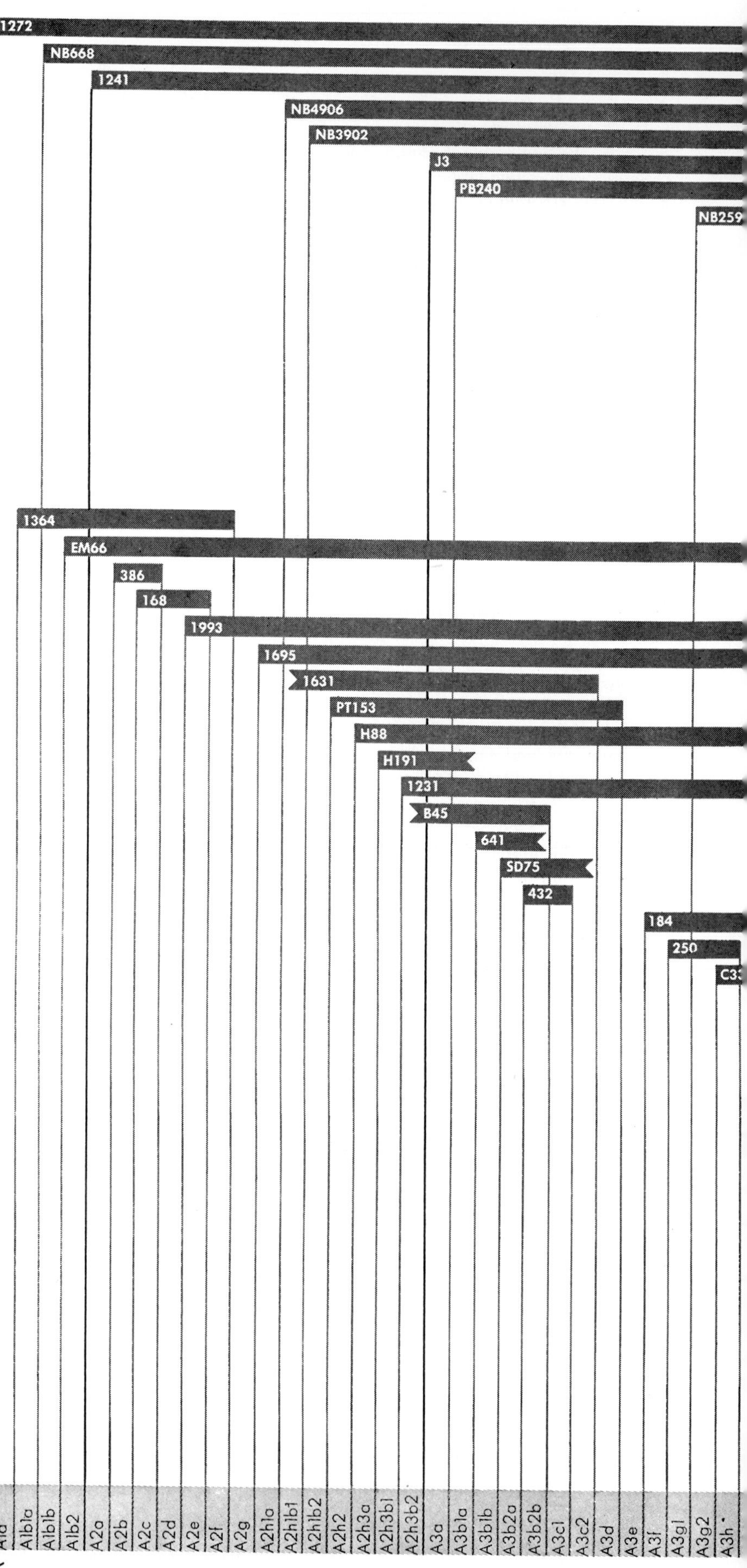


**DELETION MAP shows the reference mutants that divided the *r*II region into 80 segments. These mutants behave as if various sections of the DNA molecule had been deleted or inactivated, and as a class they do not revert, or back-mutate, spontaneously to produce standard phage. Mutants that do revert usually act as if the mutation is localized at a single point on the DNA molecule. Where this point falls in the *r*II region is determined by systematically crossing the revertible mutant with these reference deletion mutants, as illustrated on page 94. The net result is to assign the point mutation to smaller and smaller segments of the map.**

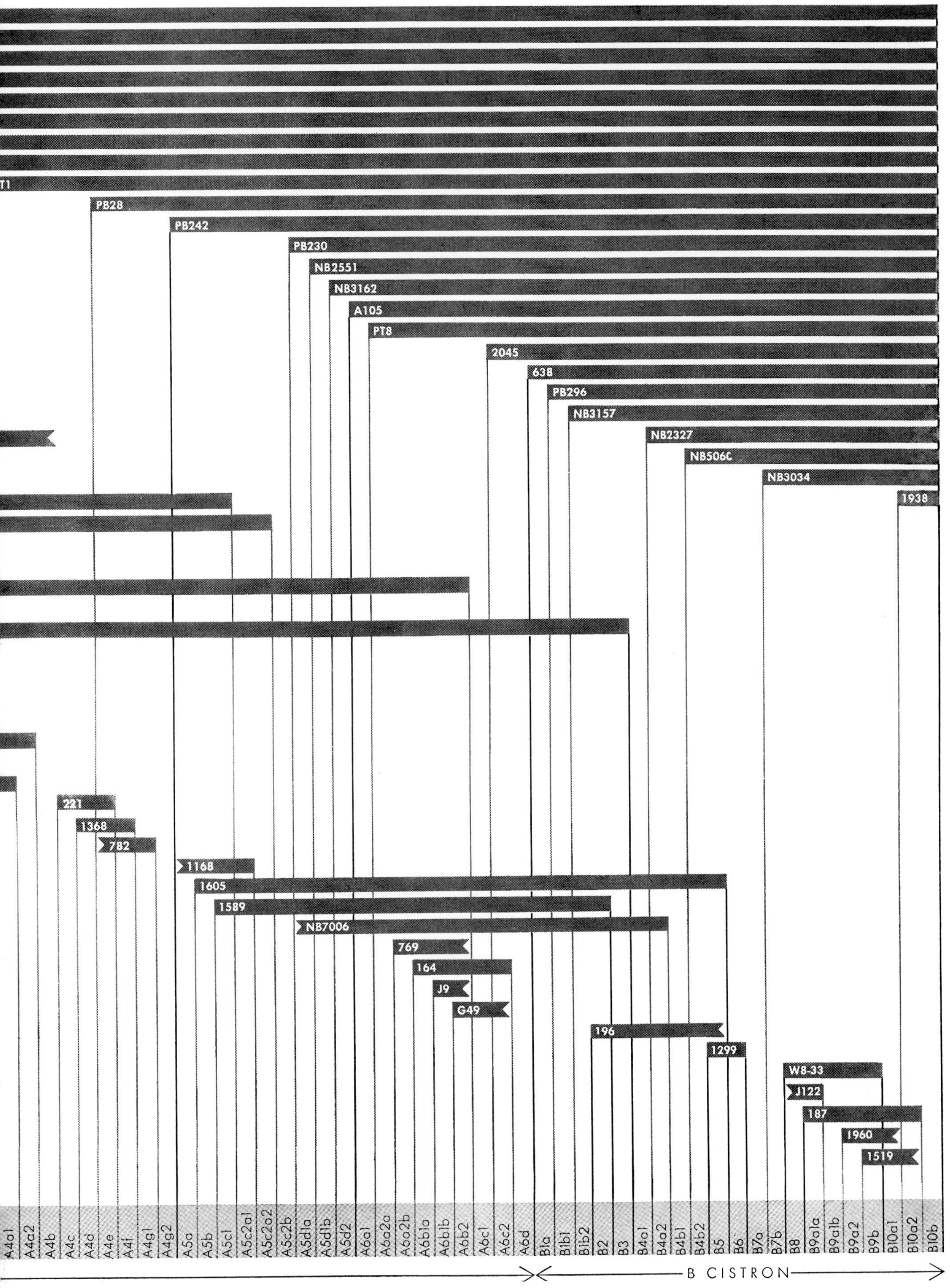

T1
PB28
PB242
PB230
NB2551
NB3162
A105
PT8
2045
638
PB296
NB3157
NB2327
NB506C
NB3034
1938
221
1368
782
1168
1605
1589
NB7006
769
164
J9
G49
196
1299
W8-33
J122
187
1960
1519
A4a1
A4a2
A4b
A4c
A4d
A4e
A4f
A4g1
A4g2
A5a
A5b
A5c1
A5c2a1
A5c2a2
A5c2b
A5d1a
A5d1b
A5d2
A6a1
A6a2a
A6a2b
A6b1a
A6b1b
A6b2
A6c1
A6c2
A6d
B1a
B1b1
B1b2
B2
B3
B4a1
B4a2
B4b1
B4b2
B5
B6
B7a
B7b
B8
B9a1a
B9a1b
B9a2
B9b
B10a1
B10a2
B10b
B CISTRON

induced in various genes is more or less the same. By fine-structure genetic analysis, however, Ernst Freese and I, working in our laboratory at Purdue University, have found that 5-bromouracil increases the mutation rate at certain sites by a factor of 10,000 or more over the spontaneous rate, while producing no noticeable change at some other sites. This indicates a high degree of specificity indeed, but at the level within the cistron. Furthermore, other mutagens specifically alter other sites. The response of part of the B cistron to a variety of mutagens is shown in the illustration on the following two pages.

Each site in the genetic map can, then, be characterized by its spontaneous mutability and by its response to various mutagens. By this means many different kinds of site have been found. Some response patterns are represented at only a single site in the entire structure; for example, the prominent spontaneous hot spot in segment B4. This is at first surprising, because according to the Watson-Crick model for DNA the structure should consist of only two types of element, adenine-thymine (AT) pairs and guanine-hydroxymethylcytosine (GC) pairs. One possible explanation for the uneven reactivity among various sites is that the response may depend not

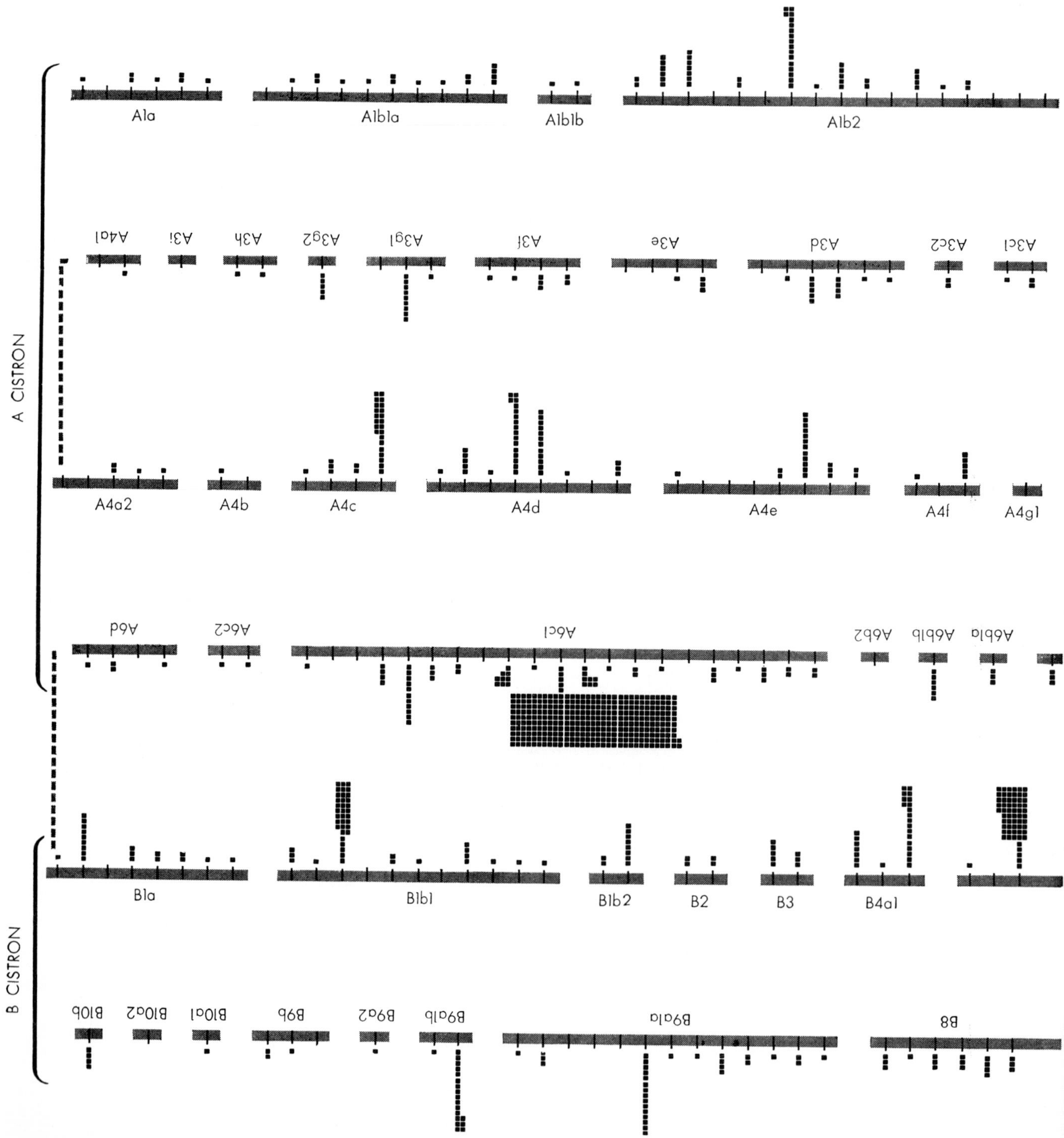


**FREQUENCY OF SPONTANEOUS MUTATIONS at various sites is shown in this complete map of the *r*II region. Alternate rows have been deliberately inverted to indicate that the region is a continuous molecular thread. Each spontaneous mutation at a site**

only on the particular base pair at a site but also very much on the type and arrangement of neighboring base pairs.

Once a site is identified it can be further characterized by the ease with which a particular mutagen makes reverse mutations produce phage of standard type. Combining such studies with studies of the chemical mechanism of mutagenesis, it may be possible eventually to translate the genetic map, bit by bit, into the actual base sequence.

## Saturation of the Map

How far is the map from being run into the ground? Since many of the sites are represented by only one occurrence of a mutation, it is clear that there must still exist some sites with zero occurrences, so that the map as it stands is not saturated. From the statistics of the distribution it can be estimated that there must exist, in addition to some 350 sites now known, at least 100 sites not yet discovered. The calculation provides only a minimum estimate; the true number is probably larger. Therefore the map at the present time cannot be more than 78 per cent saturated.

Everything that we have learned about the genetic fine structure of T4 phage is compatible with the Watson-

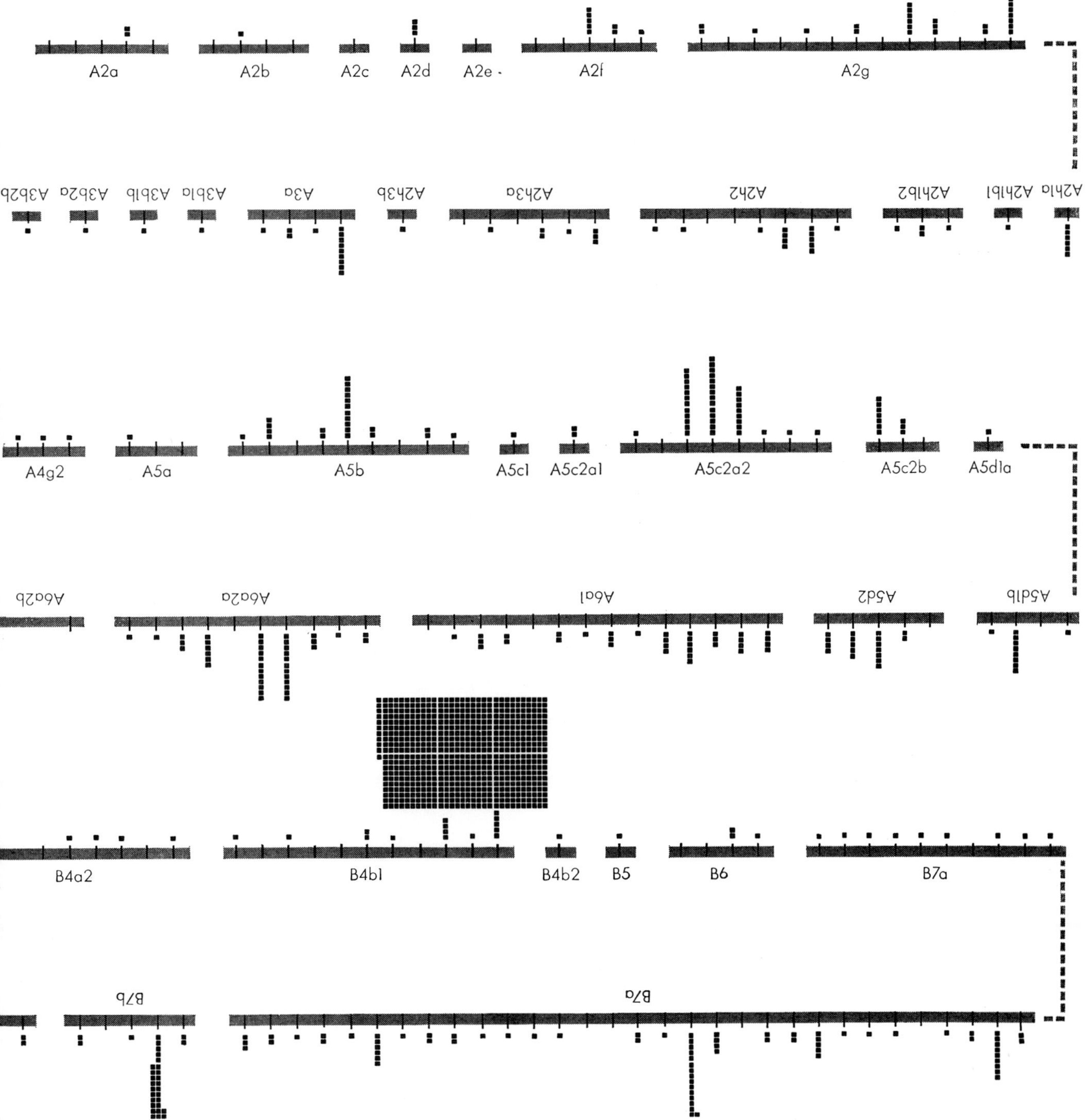


**is represented by a small black square. Sites without squares are known to exist because they can be induced to mutate by use of chemical mutagens or ultraviolet light (*see illustration on next two pages*), but they have not been observed to mutate spontaneously.**

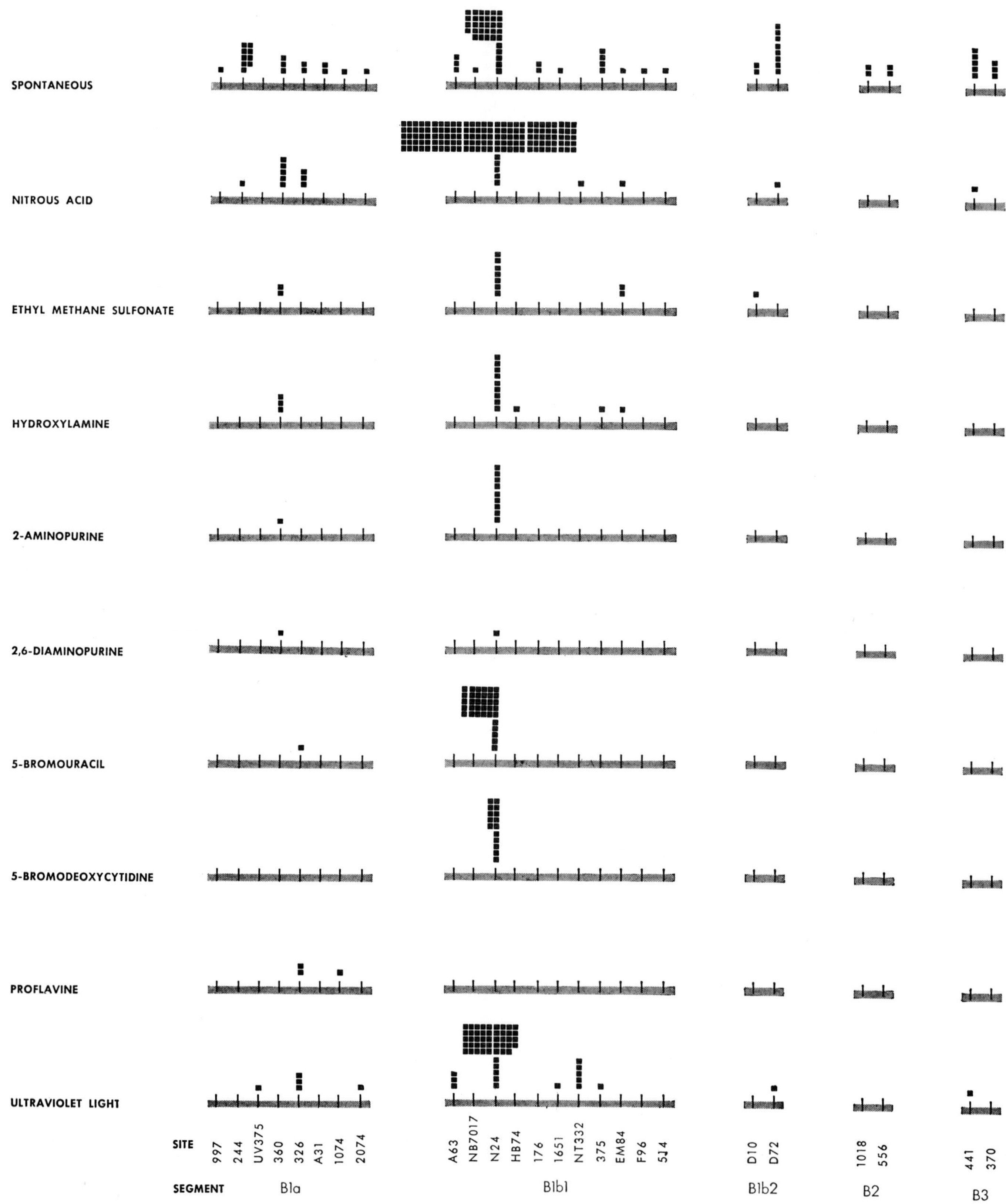


RESPONSE OF PHAGE TO MUTAGENS is shown for a portion of the B cistron. The total number of mutations studied is not the same for each mutagen. It is clear, nevertheless, that mutagenic action is highly specific at certain sites. For example, site EM26,

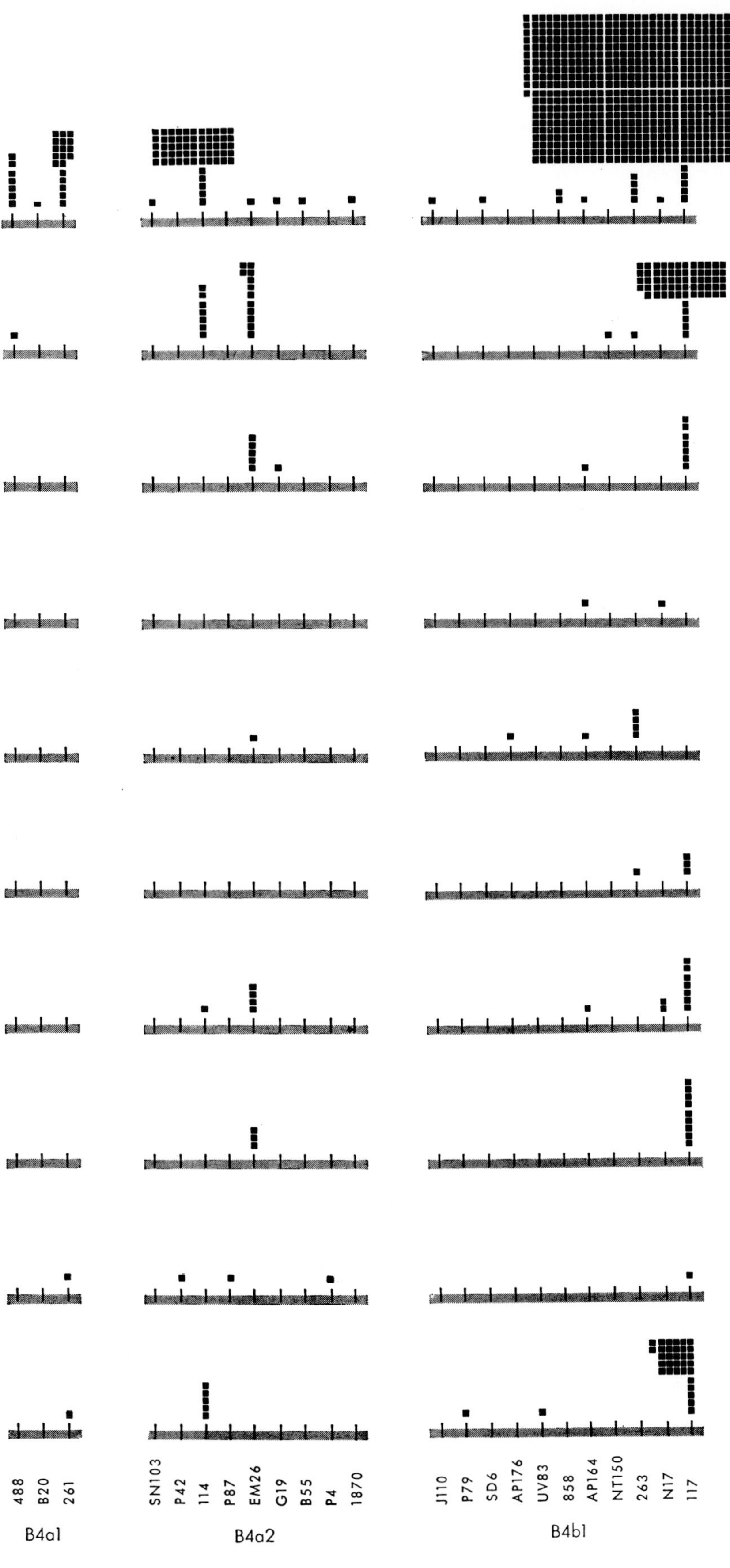


which resists spontaneous mutation, responds readily to certain mutagens. However, site 117 in segment B4b1 is more apt to mutate spontaneously than in response to a mutagen.

Crick model of the DNA molecule. In this model the genetic information is contained in the specific order of bases arranged in a linear sequence. The four-letter language of the bases must somehow be translated into the 20-letter language of the amino acids, so that at least several base pairs must be required to specify one amino acid, and an entire polypeptide chain should be defined by a longer segment of DNA. Since the activity of the resulting enzyme, or other protein, depends on its precise structure, this activity should be impaired by any of a large number of changes in the DNA base sequence leading to amino acid substitutions.

One can also imagine that certain changes in base sequence can lead to a "nonsense" sequence that does not specify any amino acid, and that as a result the polypeptide chain cannot be completed. Thus the genetic unit of function should be vulnerable at many different points within a segment of the DNA structure. Considering the monotonous structure of the molecule, there is no obvious reason why recombination should not be possible at every link in the molecular chain, although not necessarily with the same probability. In short, the Watson-Crick model leads one to expect that the functional units—the genes of traditional genetics—should consist of linear segments that can be finely dissected by mutation and recombination.

## Mapping Other Genes

The genetic results fully confirm these expectations. All mutations can in fact be represented in a strictly linear map, the functional units correspond to sharply defined segments, and each functional unit is divisible by mutation and recombination into hundreds of sites. Mutations are induced specifically at certain sites by agents that interact with the DNA bases. Although the data on mutation rates are complex, it is quite probable that they can be explained by interactions between groups of base pairs.

In confining this investigation to *r*II mutants of T4, attention has been focused on a tiny bit of hereditary material constituting only a few per cent of the genetic structure of a virus, enabling the exploration to be carried almost to the limits of the molecular structure. Similar results are being obtained in many other microorganisms and even in higher organisms such as corn. Given techniques for handling cells in culture in the appropriate way, man too may soon be a subject for genetic fine-structure analysis.

# 11

# *The Genetic Code*

F. H. C. CRICK
*October 1962*

Within the past year important progress has been made in solving the "coding problem." To the biologist this is the problem of how the information carried in the genes of an organism determines the structure of proteins.

Proteins are made from 20 different kinds of small molecule—the amino acids—strung together into long polypeptide chains. Proteins often contain several hundred amino acid units linked together, and in each protein the links are arranged in a specific order that is genetically determined. A protein is therefore like a long sentence in a written language that has 20 letters.

Genes are made of quite different long-chain molecules: the nucleic acids DNA (deoxyribonucleic acid) and, in some small viruses, the closely related RNA (ribonucleic acid). It has recently been found that a special form of RNA, called messenger RNA, carries the genetic message from the gene, which is located in the nucleus of the cell, to the surrounding cytoplasm, where many of the proteins are synthesized [see "Messenger RNA," by Jerard Hurwitz and J. J. Furth; SCIENTIFIC AMERICAN, February, 1962].

The nucleic acids are made by joining up four kinds of nucleotide to form a polynucleotide chain. The chain provides a backbone from which four kinds of side group, known as bases, jut at regular intervals. The order of the bases, however, is not regular, and it is their precise sequence that is believed to carry the genetic message. The coding problem can thus be stated more explicitly as the problem of how the sequence of the four bases in the nucleic acid determines the sequence of the 20 amino acids in the protein.

The problem has two major aspects, one general and one specific. Specifically one would like to know just what sequence of bases codes for each amino acid. Remarkable progress toward this goal was reported early this year by Marshall W. Nirenberg and J. Heinrich Matthaei of the National Institutes of Health and by Severo Ochoa and his colleagues at the New York University School of Medicine. [Editor's note: Brief accounts of this work appeared in "Science and the Citizen" for February and March. This article was planned as a companion to one by Nirenberg, now in preparation, which will deal with the biochemical aspects of the genetic code.]

The more general aspect of the coding problem, which will be my subject, has to do with the length of the genetic coding units, the way they are arranged in the DNA molecule and the way in which the message is read out. The experiments I shall report were performed at the Medical Research Council Laboratory of Molecular Biology in Cambridge, England. My colleagues were Mrs. Leslie Barnett, Sydney Brenner, Richard J. Watts-Tobin and, more recently, Robert Shulman.

The organism used in our work is the bacteriophage T4, a virus that infects the colon bacillus and subverts the biochemical machinery of the bacillus to make multiple copies of itself. The infective process starts when T4 injects its genetic core, consisting of a long strand of DNA, into the bacillus. In less than 20 minutes the virus DNA causes the manufacture of 100 or so copies of the complete virus particle, consisting of a DNA core and a shell containing at least six distinct protein components. In the process the bacillus is killed and the virus particles spill out. The great value of the T4 virus for genetic experiments is that many generations and billions of individuals can be produced in a short time. Colonies containing mutant individuals can be detected by the appearance of the small circular "plaques" they form on culture plates. Moreover, by the use of suitable cultures it is possible to select a single individual of interest from a population of a billion.

Using the same general technique, Seymour Benzer of Purdue University was able to explore the fine structure of the A and B genes (or cistrons, as he prefers to call them) found at the "*r*II" locus of the DNA molecule of T4 [see "The Fine Structure of the Gene," by Seymour Benzer, which begins on page 88 in this book]. He showed that the A and B genes, which are next to each other on the virus chromosome, each consist of some hundreds of distinct sites arranged in linear order. This is exactly what one would expect if each gene is a segment, say 500 or 1,000 bases long, of the very long DNA molecule that forms the virus chromosome [*see illustration on following page*]. The entire DNA molecule in T4 contains about 200,000 base pairs.

### The Usefulness of Mutations

From the work of Benzer and others we know that certain mutations in the A and B region made one or both genes inactive, whereas other mutations were only partially inactivating. It had also been observed that certain mutations were able to suppress the effect of harmful mutations, thereby restoring the function of one or both genes. We suspected that the various—and often puzzling—consequences of different kinds of mutation might provide a key to the nature of the genetic code.

We therefore set out to re-examine the effects of crossing T4 viruses bearing mutations at various sites. By growing two different viruses together in a common culture one can obtain "recombinants" that have some of the properties

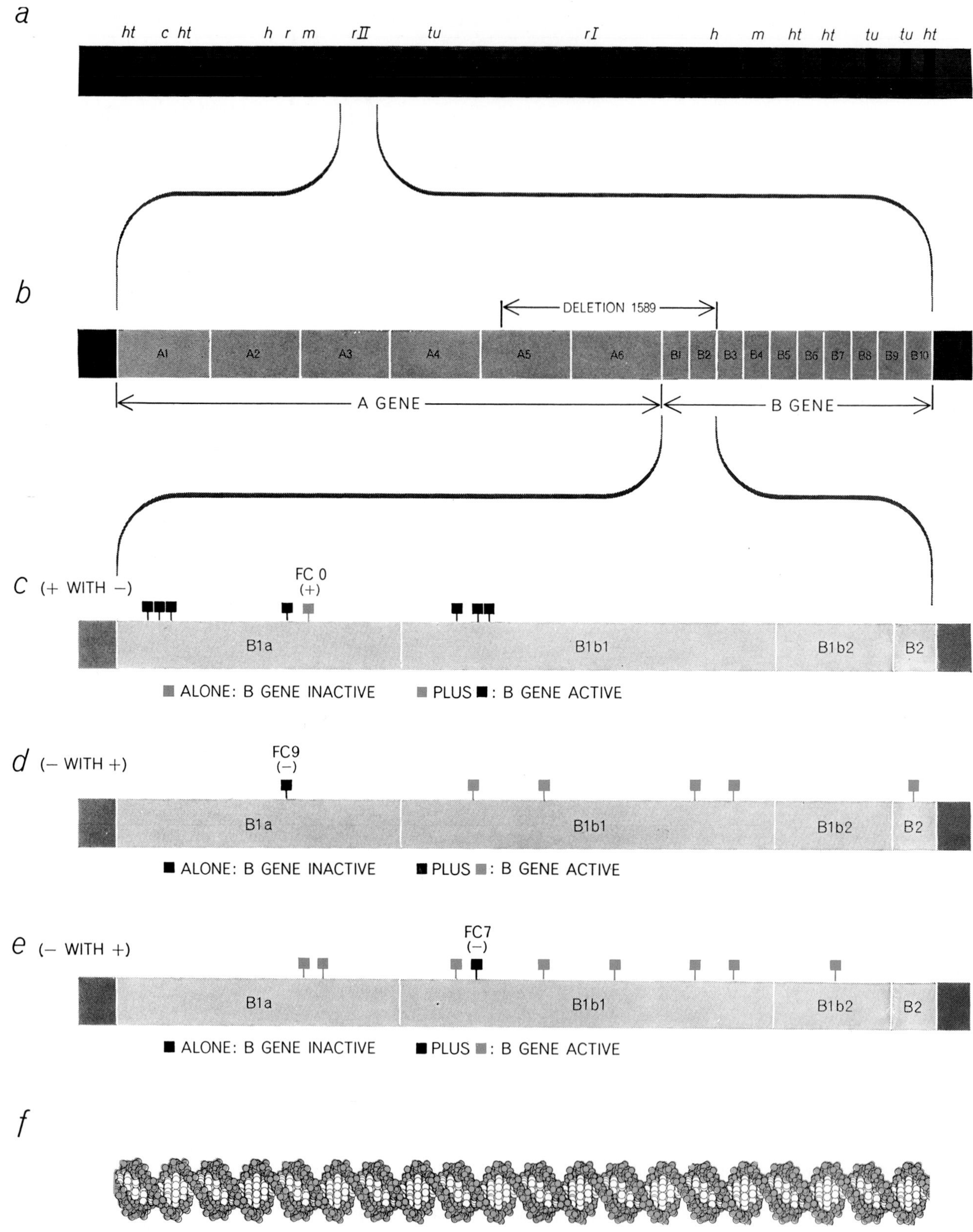


*r*II REGION OF THE T4 VIRUS represents only a few per cent of the DNA (deoxyribonucleic acid) molecule that carries full instructions for creating the virus. The region consists of two genes, here called A and B. The A gene has been mapped into six major segments, the B gene into 10 (*b*). The experiments reported in this article involve mutations in the first and second segments of the B gene. The B gene is inactivated by any mutation that adds a molecular subunit called a base (*colored square*) or removes one (*black square*). But activity is restored by simultaneous addition and removal of a base, as shown in *c*, *d* and *e*. An explanation for this recovery of activity is illustrated on page 107. The molecular representation of DNA (*f*) is estimated to be approximately in scale with the length of the B1 and B2 segments of the B gene. The two segments contain about 100 base pairs.

of one parent and some of the other. Thus one defect, such as the alteration of a base at a particular point, can be combined with a defect at another point to produce a phage with both defects [*see upper illustration below*]. Alternatively, if a phage has several defects, they can be separated by being crossed with the "wild" type, which by definition has none. In short, by genetic methods one can either combine or separate different mutations, provided that they do not overlap.

Most of the defects we shall be considering are evidently the result of adding or deleting one base or a small group of bases in the DNA molecule and not merely the result of altering one of the bases [*see lower illustration on this page*]. Such additions and deletions can be produced in a random manner with the compounds called acridines, by a process that is not clearly understood. We think they are very small additions or deletions, because the altered gene seems to have lost its function completely; mutations produced by reagents capable of changing one base into another are often partly functional. Moreover, the acridine mutations cannot be reversed by such reagents (and vice versa). But our strongest reason for believing they are additions or deletions is that they can be combined in a way that suggests they have this character.

To understand this we shall have to go back to the genetic code. The simplest sort of code would be one in which a small group of bases stands for one particular acid. This group can scarcely be a pair, since this would yield only 4 × 4, or 16, possibilities, and at least 20 are needed. More likely the shortest code group is a triplet, which would provide 4 × 4 × 4, or 64, possibilities. A small group of bases that codes one amino acid has recently been named a codon.

The first definite coding scheme to be proposed was put forward eight years ago by the physicist George Gamow, now at the University of Colorado. In this code adjacent codons overlap as illustrated on the following page. One consequence of such a code is that only certain amino acids can follow others. Another consequence is that a change in a single base leads to a change in three adjacent amino acids. Evidence gathered since Gamow advanced his ideas makes an overlapping code appear unlikely. In the first place there seems to be no restriction of amino acid sequence in any of the proteins so far examined. It has also been shown that typical mutations change only a single amino acid in the polypeptide chain of a protein. Although it is theoretically possible that the genetic code may be partly overlapping, it is more likely that adjacent codons do not overlap at all.

Since the backbone of the DNA molecule is completely regular, there is nothing to mark the code off into groups of three bases, or into groups of any other size. To solve this difficulty various ingenious solutions have been proposed. It was thought, for example, that the code might be designed in such a way that if the wrong set of triplets were chosen, the message would always be complete nonsense and no protein would

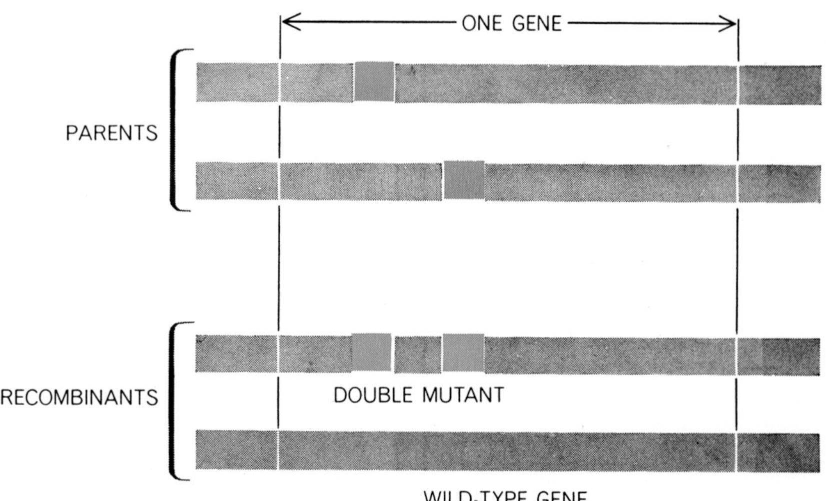


**GENETIC RECOMBINATION provides the means for studying mutations. Colored squares represent mutations in the chromosome (DNA molecule) of the T4 virus. Through genetic recombination, the progeny can inherit the defects of both parents or of neither.**

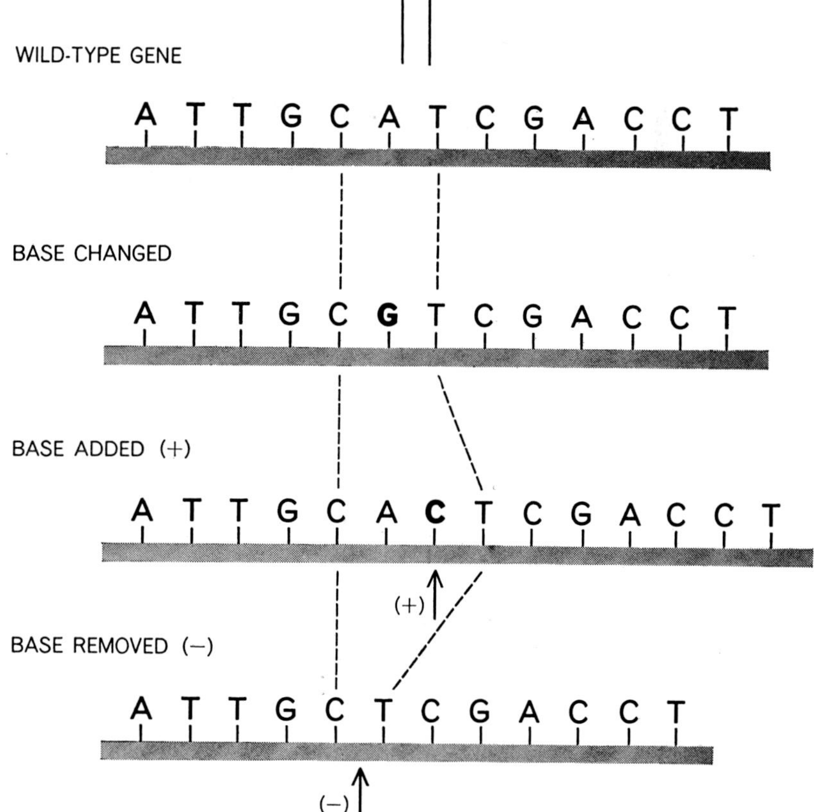


**TWO CLASSES OF MUTATION result from introducing defects in the sequence of bases (A, T, G, C) that are attached to the backbone of the DNA molecule. In one class a base is simply changed from one into another, as A into G. In the second class a base is added or removed. Four bases are adenine (A), thymine (T), guanine (G) and cytosine (C).**

be produced. But it now looks as if the most obvious solution is the correct one. That is, the message begins at a fixed starting point, probably one end of the gene, and is simply read three bases at a time. Notice that if the reading started at the wrong point, the message would fall into the wrong sets of three and would then be hopelessly incorrect. In fact, it is easy to see that while there is only one correct reading for a triplet code, there are two incorrect ones.

If this idea were right, it would immediately explain why the addition or the deletion of a base in most parts of the gene would make the gene completely nonfunctional, since the reading of the genetic message from that point onward would be totally wrong. Now, although our single mutations were always without function, we found that if we put certain pairs of them together, the gene would work. (In point of fact we picked up many of our functioning double mutations by starting with a nonfunctioning mutation and selecting for the rare second mutation that restored gene activity, but this does not affect our argument.) This enabled us to classify all our mutations as being either plus or minus. We found that by using the following rules we could always predict the behavior of any pair we put together in the same gene. First, if plus is combined with plus, the combination is nonfunctional. Second, if minus is combined with minus, the result is nonfunctional. Third, if plus is combined with minus, the combination is nonfunctional if the pair is too widely separated and functional if the pair is close together.

The interesting case is the last one. We could produce a gene that functioned, at least to some extent, if we combined a plus mutation with a minus mutation, provided that they were not too far apart.

To make it easier to follow, let us assume that the mutations we called plus really had an extra base at some point and that those we called minus had lost a base. (Proving this to be the case is rather difficult.) One can see that, starting from one end, the message would be read correctly until the extra base was reached; then the reading would get out of phase and the message would be wrong until the missing base was reached, after which the message would come back into phase again. Thus the genetic message would not be wrong over a long stretch but only over the short distance between the plus and the minus. By the same sort of argument one can see that for a triplet code the combination plus with plus or minus with minus should never work [*see illustration on opposite page*]

We were fortunate to do most of our work with mutations at the left-hand end of the B gene of the *r*II region. It appears that the function of this part of the gene may not be too important, so that it may not matter if part of the genetic message in the region is incorrect. Even so, if the plus and minus are too far apart, the combination will not work.

## Nonsense Triplets

To understand this we must go back once again to the code. There are 64 possible triplets but only 20 amino acids to be coded. Conceivably two or more triplets may stand for each amino acid. On the other hand, it is reasonable to expect that at least one or two triplets may not represent an amino acid at all but have some other meaning, such as "Begin here" or "End here." Although such hypothetical triplets may have a meaning of some sort, they have been named nonsense triplets. We surmised that sometimes the misreading produced in the region lying between a plus and a minus mutation might by chance give rise to a nonsense triplet, in which case the gene might not work.

We investigated a number of plus-with-minus combinations in which the distance between plus and minus was relatively short and found that certain combinations were indeed inactive when we might have expected them to function. Presumably an intervening nonsense triplet was to blame. We also found cases in which a plus followed by a minus worked but a minus followed by a plus did not, even though the two mutations appeared to be at the same sites, although in reverse sequence. As I have indicated, there are two wrong ways to read a message; one arises if the plus is to the left of the minus, the other if the plus is to the right of the minus. In cases where plus with minus gave rise to an active gene but minus with plus did not, even when the mutations evidently occupied the same pairs of sites, we concluded that the intervening misreading produced a nonsense triplet in one case but not in the other. In confirmation of this hypothesis we have been able to modify such nonsense triplets by mutagens that turn one base into another, and we have thereby restored the gene's activity. At the same time we have been able to locate the position of the nonsense triplet.

Recently we have undertaken one

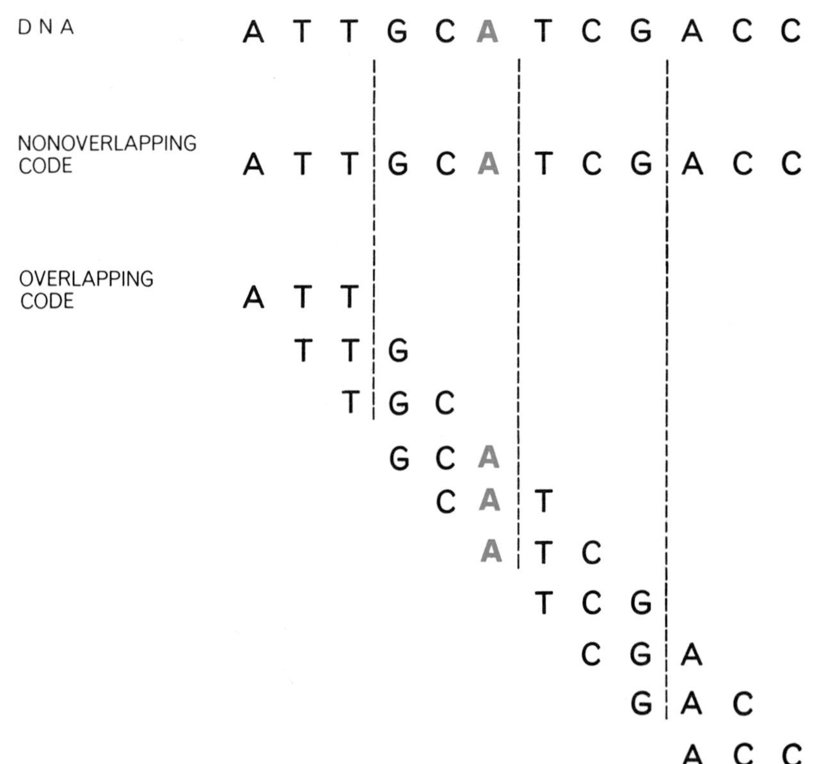


**PROPOSED CODING SCHEMES show how the sequence of bases in DNA can be read. In a nonoverlapping code, which is favored by the author, code groups are read in simple sequence. In one type of overlapping code each base appears in three successive groups.**

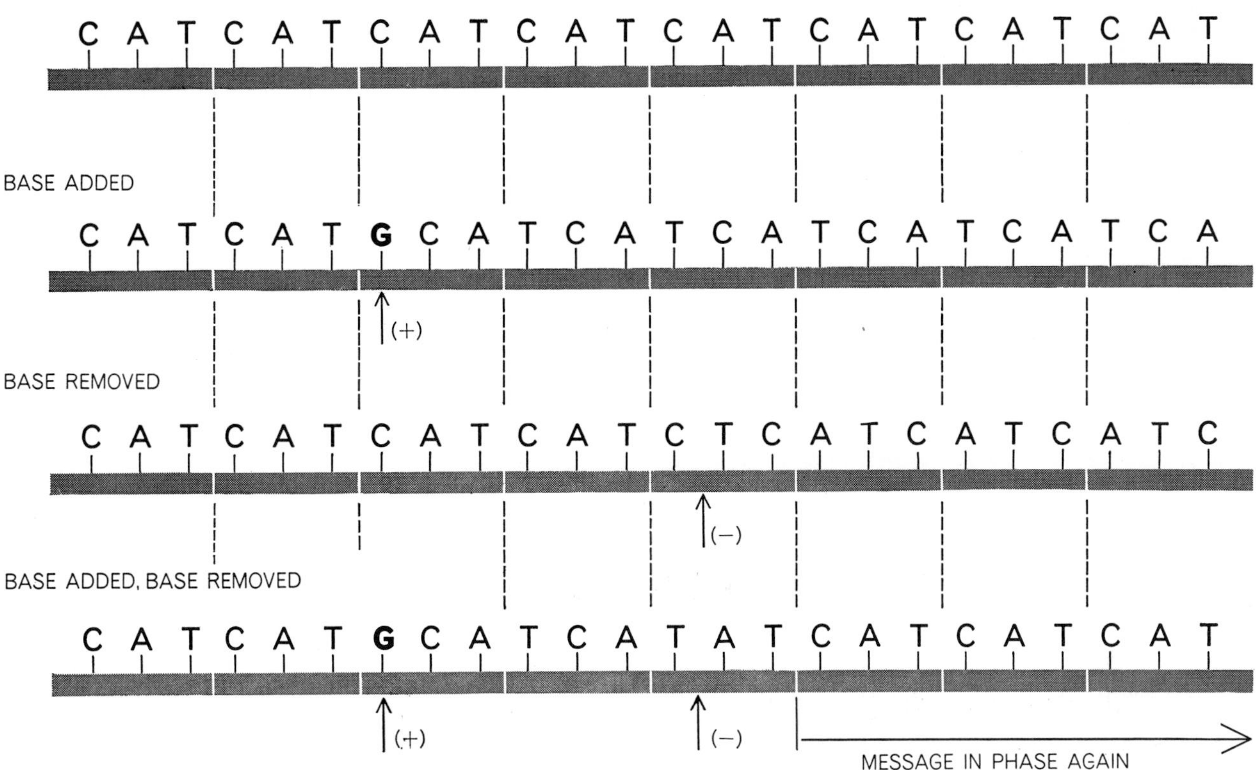


**EFFECT OF MUTATIONS that add or remove a base is to shift the reading of the genetic message, assuming that the reading begins at the left-hand end of the gene. The hypothetical message in the wild-type gene is CAT, CAT... Adding a base shifts the reading to TCA, TCA... Removing a base makes it ATC, ATC... Addition and removal of a base puts the message in phase again.**

other rather amusing experiment. If a single base were changed in the left-hand end of the B gene, we would expect the gene to remain active, both because this end of the gene seems to be unessential and because the reading of the rest of the message is not shifted. In fact, if the B gene remained active, we would have no way of knowing that a base had been changed. In a few cases, however, we have been able to destroy the activity of the B gene by a base change traceable to the left-hand end of the gene. Presumably the change creates a nonsense triplet. We reasoned that if we could shift the reading so that the message was read in different groups of three, the new reading might not yield a nonsense triplet. We therefore selected a minus and a plus that together allowed the B gene to function, and that were on each side of the presumed nonsense mutation. Sure enough, this combination of three mutants allowed the gene to function [*see top illustration on page 109*]. In other words, we could abolish the effect of a nonsense triplet by shifting its reading.

All this suggests that the message is read from a fixed point, probably from one end. Here the question arises of how one gene ends and another begins, since in our picture there is nothing on the backbone of the long DNA molecule to separate them. Yet the two genes A and B are quite distinct. It is possible to measure their function separately, and Benzer has shown that no matter what mutation is put into the A gene, the B function is not affected, provided that the mutation is wholly within the A gene. In the same way changes in the B gene do not affect the function of the A gene.

## The Space between the Genes

It therefore seems reasonable to imagine that there is something about the DNA between the two genes that isolates them from each other. This idea can be tested by experiments with a mutant T4 in which part of the *r*II region is deleted. The mutant, known as T4 1589, has lost a large part of the right end of the A gene and a smaller part of the left end of the B gene. Surprisingly the B gene still shows some function; in fact this is why we believe this part of the B gene is not too important.

Although we describe this mutation as a deletion, since genetic mapping shows that a large piece of the genetic information in the region is missing, it does not mean that physically there is a gap. It seems more likely that DNA is all one piece but that a stretch of it has been left out. It is only by comparing it with the complete version—the wild type—that one can see a piece of the message is missing.

We have argued that there must be a small region between the genes that separates them. Consequently one would predict that if this segment of the DNA were missing, the two genes would necessarily be joined. It turns out that it is quite easy to test this prediction, since by genetic methods one can construct double mutants. We therefore combined one of our acridine mutations, which in this case was near the beginning of the A gene, with the deletion 1589. Without the deletion present the acridine mutation had no effect on the B function, which showed that the genes were indeed separate. But when 1589 was there as well, the B function was completely destroyed [*see top illustration on page 108*]. When the genes were joined, a change far away in the A gene knocked out the B gene completely. This strongly suggests that the reading proceeds from one end.

We tried other mutations in the A

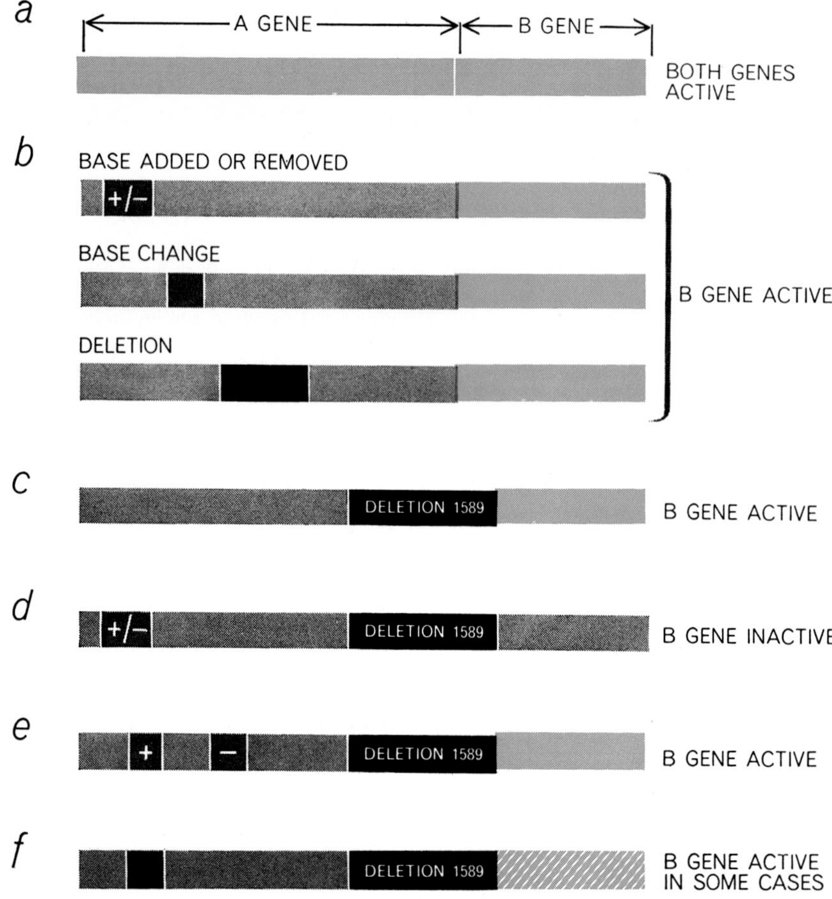


**DELETION JOINING TWO GENES makes the B gene vulnerable to mutations in the A gene. The messages in two wild-type genes (*a*) are read independently, beginning at the left end of each gene. Regardless of the kind of mutation in A, the B gene remains active (*b*). The deletion known as 1589 inactivates the A gene but leaves the B gene active (*c*). But now alterations in the A gene will often inactivate the B gene, showing that the two genes have been joined in some way and are read as if they were a single gene (*d, e, f*).**

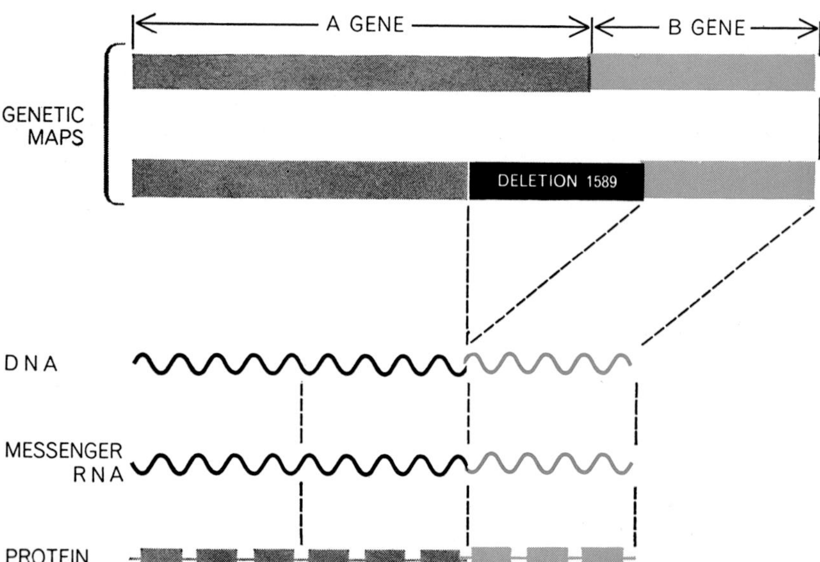


**PROBABLE EFFECT OF DELETION 1589 is to produce a mixed protein with little or no A-gene activity but substantial B activity. Although the conventional genetic map shows the deletion as a gap, the DNA molecule itself is presumably continuous but shortened. In virus replication the genetic message in DNA is transcribed into a molecule of ribonucleic acid, called messenger RNA. This molecule carries the message to cellular particles known as ribosomes, where protein is synthesized, following instructions coded in the DNA.**

gene combined with 1589. All the acridine mutations we tried knocked out the B function, whether they were plus or minus, but a pair of them (plus with minus) still allowed the B gene to work. On the other hand, in the case of the other type of mutation (which we believe is due to the change of a base and not to one being added or subtracted) about half of the mutations allowed the B gene to work and the other half did not. We surmise that the latter are nonsense mutations, and in fact Benzer has recently been using this test as a definition of nonsense.

Of course, we do not know exactly what is happening in biochemical terms. What we suspect is that the two genes, instead of producing two separate pieces of messenger RNA, produce a single piece, and that this in turn produces a protein with a long polypeptide chain, one end of which has the amino acid sequence of part of the presumed A protein and the other end of which has most of the B protein sequence—enough to give some B function to the combined molecule although the A function has been lost. The concept is illustrated schematically at the bottom of this page. Eventually it should be possible to check the prediction experimentally.

## How the Message Is Read

So far all the evidence has fitted very well into the general idea that the message is read off in groups of three, starting at one end. We should have got the same results, however, if the message had been read off in groups of four, or indeed in groups of any larger size. To test this we put not just two of our acridine mutations into one gene but three of them. In particular we put in three with the same sign, such as plus with plus with plus, and we put them fairly close together. Taken either singly or in pairs, these mutations will destroy the function of the B gene. But when all three are placed in the same gene, the B function reappears. This is clearly a remarkable result: two blacks will not make a white but three will. Moreover, we have obtained the same result with several different combinations of this type and with several of the type minus with minus with minus.

The explanation, in terms of the ideas described here, is obvious. One plus will put the reading out of phase. A second plus will give the other wrong reading. But if the code is a triplet code, a third plus will bring the message back into phase again, and from then on to the end it will be read correctly. Only between the pluses will the message be wrong [*see illustration below*].

Notice that it does not matter if plus is really one extra base and minus is one fewer; the conclusions would be the same if they were the other way around. In fact, even if some of the plus mutations were indeed a single extra base, others might be two fewer bases; in other words, a plus might really be minus minus. Similarly, some of the minus mutations might actually be plus plus. Even so they would still fit into our scheme.

Although the most likely explanation is that the message is read three bases at a time, this is not completely certain. The reading could be in multiples of three. Suppose, for example, that the message is actually read six bases at a time. In that case the only change needed in our interpretation of the facts is to assume that all our mutants have been changed by an even number of bases. We have some weak experimental evidence that this is unlikely. For instance, we can combine the mutant 1589 (which joins the genes) with medium-sized deletions in the A cistron. Now, if deletions were random in length, we should expect about a third of them to allow the B function to be expressed if the message is indeed read three bases at a time, since those deletions that had lost an exact multiple of three bases should allow the B gene to function. By the same reasoning only a sixth of them should work (when combined with 1589) if the reading proceeds six at a time. Actually we find that the B gene is active in a little more than a third. Taking all the evidence together, however, we find that although three is the most likely coding unit, we cannot completely rule out multiples of three.

There is one other general conclusion we can draw about the genetic code. If we make a rough guess as to the actual size of the B gene (by comparing it with another gene whose size is known approximately), we can estimate how many bases can lie between a plus with minus combination and still allow the B gene to function. Knowing also the frequency with which nonsense triplets are created in the misread region between the plus and minus, we can get some idea whether there are many such triplets or only a few. Our calculation suggests that nonsense triplets are not too common. It seems, in other words, that most of the 64 possible triplets, or codons, are not nonsense, and therefore they stand for amino acids. This implies that probably more than one codon can stand for one amino acid. In the jargon of the trade, a code in which this is true is "degenerate."

In summary, then, we have arrived at three general conclusions about the genetic code:

1. The message is read in nonover-

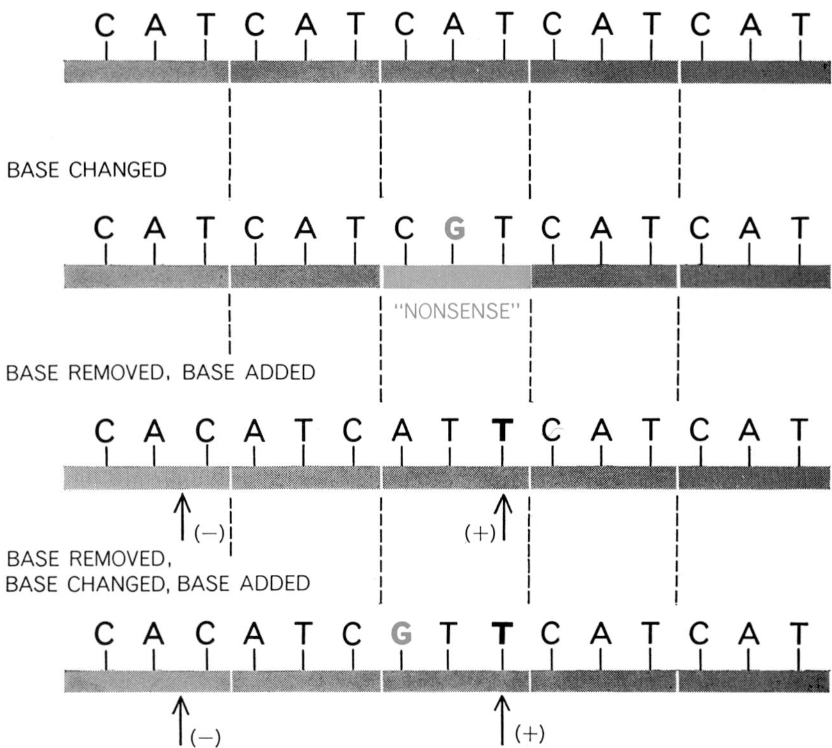


**NONSENSE MUTATION is one creating a code group that evidently does not represent any of the 20 amino acids found in proteins. Thus it makes the gene inactive. In this hypothetical case a nonsense triplet, CGT, results when an A in the wild-type gene is changed to G. The nonsense triplet can be eliminated if the reading is shifted to put the G in a different triplet. This is done by recombining the inactive gene with one containing a minus-with-plus combination. In spite of three mutations, the resulting gene is active.**

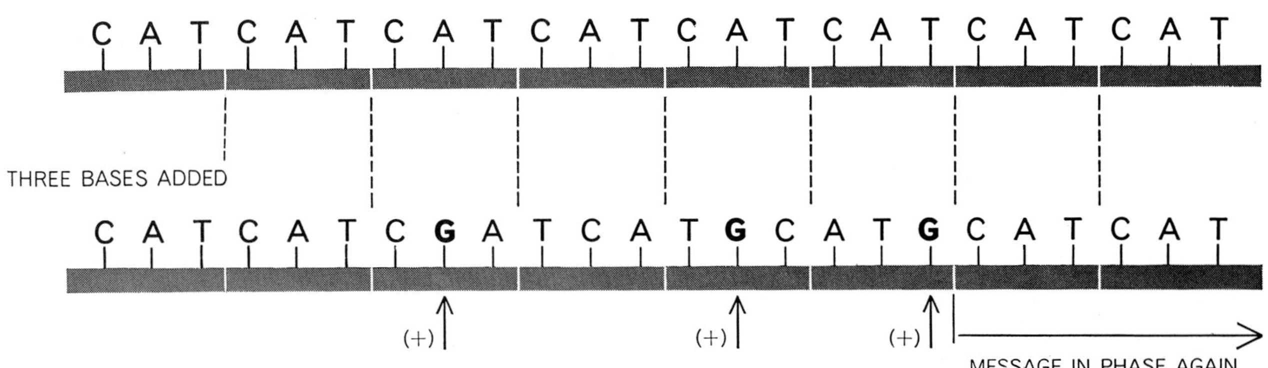


**TRIPLE MUTATION in which three bases are added fairly close together spoils the genetic message over a short stretch of the gene but leaves the rest of the message unaffected. The same result can be achieved by the deletion of three neighboring bases.**

lapping groups from a fixed point, probably from one end. The starting point determines that the message is read correctly into groups.

2. The message is read in groups of a fixed size that is probably three, although multiples of three are not completely ruled out.

3. There is very little nonsense in the code. Most triplets appear to allow the gene to function and therefore probably represent an amino acid. Thus in general more than one triplet will stand for each amino acid.

It is difficult to see how to get around our first conclusion, provided that the B gene really does code a polypeptide chain, as we have assumed. The second conclusion is also difficult to avoid. The third conclusion, however, is much more indirect and could be wrong.

Finally, we must ask what further evidence would really clinch the theory we have presented here. We are continuing to collect genetic data, but I doubt that this will make the story much more convincing. What we need is to obtain a protein, for example one produced by a double mutation of the form plus with minus, and then examine its amino acid sequence. According to conventional theory, because the gene is altered in only two places the amino acid sequences also should differ only in the two corresponding places. According to our theory it should be altered not only at these two places but also at all places in between. In other words, a whole string of amino acids should be changed. There is one protein, the lysozyme of the T4 phage, that is favorable for such an approach, and we hope that before long workers in the U.S. who have been studying phage lysozyme will confirm our theory in this way.

The same experiment should also be useful for checking the particular code schemes worked out by Nirenberg and Matthaei and by Ochoa and his colleagues. The phage lysozyme made by the wild-type gene should differ over only a short stretch from that made by the plus-with-minus mutant. Over this stretch the amino acid sequence of the two lysozyme variants should correspond to the same sequence of bases on the DNA but should be read in different groups of three.

If this part of the amino acid sequence of both the wild-type and the altered lysozyme could be established, one could check whether or not the codons assigned to the various amino acids did indeed predict similar sequences for that part of the DNA between the base added and the base removed.

# 12
# *The Genetic Code: III*

F. H. C. CRICK
*October 1966*

The hypothesis that the genes of the living cell contain all the information needed for the cell to reproduce itself is now more than 50 years old. Implicit in the hypothesis is the idea that the genes bear in coded form the detailed specifications for the thousands of kinds of protein molecules the cell requires for its moment-to-moment existence: for extracting energy from molecules assimilated as food and for repairing itself as well as for replication. It is only within the past 15 years, however, that insight has been gained into the chemical nature of the genetic material and how its molecular structure can embody coded instructions that can be "read" by the machinery in the cell responsible for synthesizing protein molecules. As the result of intensive work by many investigators the story

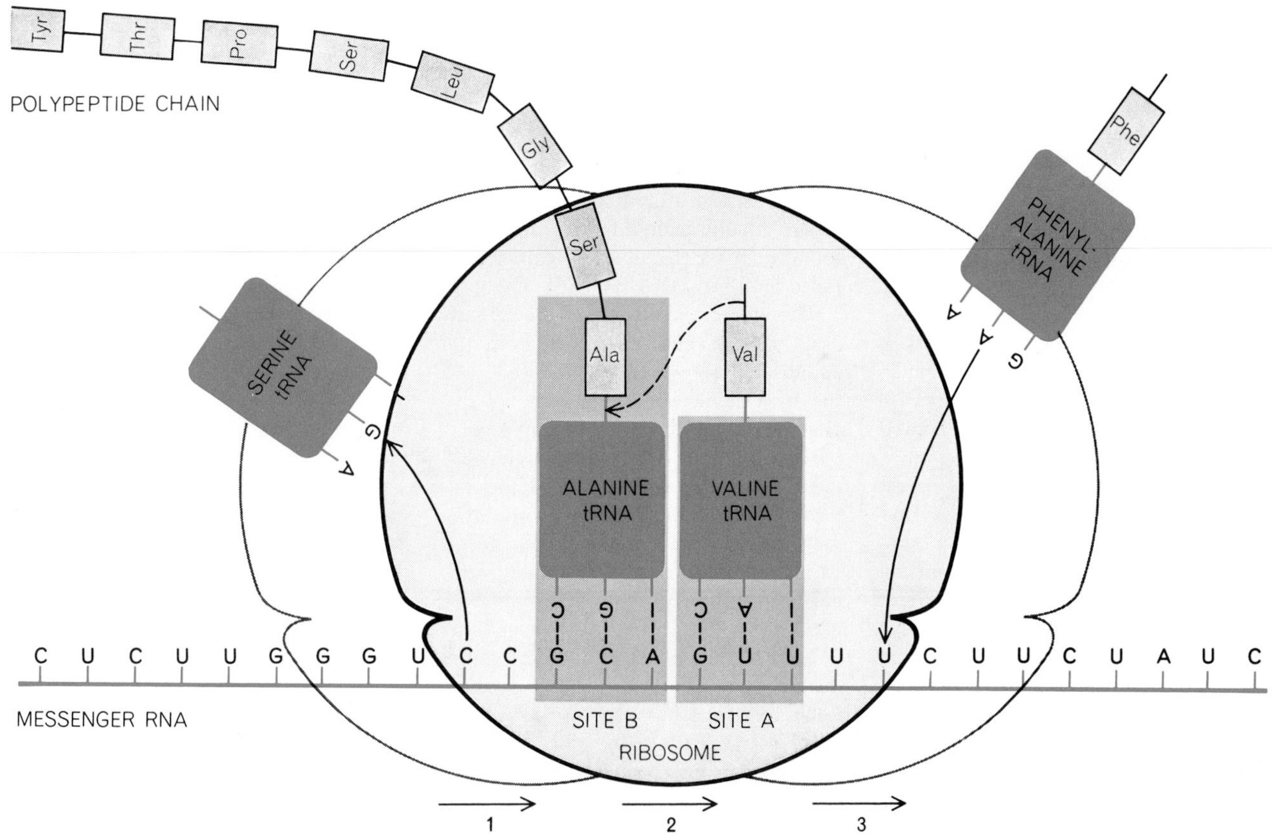


**SYNTHESIS OF PROTEIN MOLECULES is accomplished by the intracellular particles called ribosomes. The coded instructions for making the protein molecule are carried to the ribosome by a form of ribonucleic acid (RNA) known as "messenger" RNA. The RNA code "letters" are four bases: uracil (U), cytosine (C), adenine (A) and guanine (G). A sequence of three bases, called a codon, is required to specify each of the 20 kinds of amino acid, identified here by their abbreviations. (A list of the 20 amino acids and their abbreviations appears on the next page.) When linked end to end, these amino acids form the polypeptide chains of which proteins are composed. Each type of amino acid is transported to the ribosome by a particular form of "transfer" RNA (tRNA), which carries an anticodon that can form a temporary bond with one of the codons in messenger RNA. Here the ribosome is shown moving along the chain of messenger RNA, "reading off" the codons in sequence. It appears that the ribosome has two binding sites for molecules of tRNA: one site (*A*) for positioning a newly arrived tRNA molecule and another (*B*) for holding the growing polypeptide chain.**

| AMINO ACID | ABBREVIATION |
|---|---|
| ALANINE | Ala |
| ARGININE | Arg |
| ASPARAGINE | AspN |
| ASPARTIC ACID | Asp |
| CYSTEINE | Cys |
| GLUTAMIC ACID | Glu |
| GLUTAMINE | GluN |
| GLYCINE | Gly |
| HISTIDINE | His |
| ISOLEUCINE | Ileu |
| LEUCINE | Leu |
| LYSINE | Lys |
| METHIONINE | Met |
| PHENYLALANINE | Phe |
| PROLINE | Pro |
| SERINE | Ser |
| THREONINE | Thr |
| TRYPTOPHAN | Tryp |
| TYROSINE | Tyr |
| VALINE | Val |

**TWENTY AMINO ACIDS constitute the standard set found in all proteins. A few other amino acids occur infrequently in proteins but it is suspected in each case that they originate as one of the standard set and become chemically modified after they have been incorporated into a polypeptide chain.**

of the genetic code is now essentially complete. One can trace the transmission of the coded message from its original site in the genetic material to the finished protein molecule.

The genetic material of the living cell is the chainlike molecule of deoxyribonucleic acid (DNA). The cells of many bacteria have only a single chain; the cells of mammals have dozens clustered together in chromosomes. The DNA molecules have a very long backbone made up of repeating groups of phosphate and a five-carbon sugar. To this backbone the side groups called bases are attached at regular intervals. There are four standard bases: adenine (A), guanine (G), thymine (T) and cytosine (C). They are the four "letters" used to spell out the genetic message. The exact sequence of bases along a length of the DNA molecule determines the structure of a particular protein molecule.

Proteins are synthesized from a standard set of 20 amino acids, uniform throughout nature, that are joined end to end to form the long polypeptide chains of protein molecules [*see illustration at left*]. Each protein has its own characteristic sequence of amino acids. The number of amino acids in a polypeptide chain ranges typically from 100 to 300 or more.

The genetic code is not the message itself but the "dictionary" used by the cell to translate from the four-letter language of nucleic acid to the 20-letter language of protein. The machinery of the cell can translate in one direction only: from nucleic acid to protein but not from protein to nucleic acid. In making this translation the cell employs a variety of accessory molecules and mechanisms. The message contained in DNA is first transcribed into the similar molecule called "messenger" ribonucleic acid—messenger RNA. (In many viruses—the tobacco mosaic virus, for example—the genetic material is simply RNA.) RNA too has four kinds of bases as side groups; three are identical with those found in DNA (adenine, guanine and cytosine) but the fourth is uracil (U) instead of thymine. In this first transcription of the genetic message the code letters A, G, T and C in DNA give rise respectively to U, C, A and G. In other words, wherever A appears in DNA, U appears in the RNA transcription; wherever G appears in DNA, C appears in the transcription, and so on. As it is usually presented the dictionary of the genetic code employs the letters found in RNA (U, C, A, G) rather than those found in DNA (A, G, T, C).

The genetic code could be broken easily if one could determine both the amino acid sequence of a protein and the base sequence of the piece of nucleic acid that codes it. A simple comparison of the two sequences would yield the code. Unfortunately the determination of the base sequence of a long nucleic acid molecule is, for a variety of reasons, still extremely difficult. More indirect approaches must be used.

Most of the genetic code first became known early in 1965. Since then additional evidence has proved that almost all of it is correct, although a few features remain uncertain. This article describes how the code was discovered and some of the work that supports it.

*Scientific American* has already presented a number of articles on the genetic code. In one of them [see "The Genetic Code" on page 103] I explained that the experimental evidence (mainly indirect) suggested that the code was a triplet code: that the bases on the messenger RNA were read three at a time and that each group corresponded to a particular amino acid. Such a group is called a codon. Using four symbols in groups of three, one can form 64 distinct triplets. The evidence indicated that most of these stood for one amino acid or another, implying that an amino acid was usually represented by several codons. Adjacent amino acids were coded by adjacent codons, which did not overlap.

In a sequel to that article ["The Genetic Code: II," Offprint 153] Marshall W. Nirenberg of the National Institutes of Health explained how the composition of many of the 64 triplets had been determined by actual experiment. The technique was to synthesize polypeptide chains in a cell-free system, which was made by breaking open cells of the colon bacillus (*Escherichia coli*) and extracting from them the machinery for protein synthesis. Then the system was provided with an energy supply, 20 amino acids and one or another of several types of synthetic RNA. Although the exact sequence of bases in each type was random, the proportion of bases was known. It was found that each type of synthetic messenger RNA directed the incorporation of certain amino acids only.

By means of this method, used in a quantitative way, the *composition* of many of the codons was obtained, but the *order* of bases in any triplet could not be determined. Codons rich in G were difficult to study, and in addition a few mistakes crept in. Of the 40 codon compositions listed by Nirenberg in his article we now know that 35 were correct.

## The Triplet Code

The main outlines of the genetic code were elucidated by another technique invented by Nirenberg and Philip Leder. In this method no protein synthesis occurs. Instead one triplet at a time is used to bind together parts of the machinery of protein synthesis.

Protein synthesis takes place on the comparatively large intracellular structures known as ribosomes. These bodies travel along the chain of messenger RNA, reading off its triplets one after another and synthesizing the polypeptide chain of the protein, starting at the amino end ($NH_2$). The amino acids do not diffuse to the ribosomes by themselves. Each amino acid is joined chemically by a special enzyme to one of the codon-recognizing molecules known both as soluble RNA (sRNA) and transfer RNA (tRNA). (I prefer the latter designation.) Each tRNA molecule has its own triplet of bases, called an anticodon, that recognizes the relevant codon on the messenger RNA by pairing bases with it [*see illustration on page 111*].

Leder and Nirenberg studied which amino acid, joined to its tRNA molecules, was bound to the ribosomes in the presence of a particular triplet, that is, by a "message" with just three letters. They did so by the neat trick of passing the mixture over a nitrocellulose filter that retained the ribosomes. All the tRNA molecules passed through the filter except the ones specifically bound to the ribosomes by the triplet. Which they were could easily be decided by using mixtures of amino acids in which one kind of amino acid had been made artificially radioactive, and determining the amount of radioactivity absorbed by the filter.

For example, the triplet GUU retained the tRNA for the amino acid valine, whereas the triplets UGU and UUG did not. (Here GUU actually stands for the trinucleoside diphosphate GpUpU.) Further experiments showed that UGU coded for cysteine and UUG for leucine.

Nirenberg and his colleagues synthesized all 64 triplets and tested them for their coding properties. Similar results have been obtained by H. Gobind Khorana and his co-workers at the University of Wisconsin. Various other groups have checked a smaller number of codon assignments.

Close to 50 of the 64 triplets give a clearly unambiguous answer in the binding test. Of the remainder some evince only weak binding and some bind more than one kind of amino acid. Other results I shall describe later suggest that the multiple binding is often an artifact of the binding method. In short, the binding test gives the meaning of the majority of the triplets but it does not firmly establish all of them.

The genetic code obtained in this way, with a few additions secured by other methods, is shown in the table below. The 64 possible triplets are set out in a regular array, following a plan

| FIRST LETTER | SECOND LETTER: U | | C | | A | | G | | THIRD LETTER |
|---|---|---|---|---|---|---|---|---|---|
| U | UUU | Phe | UCU | Ser | UAU | Tyr | UGU | Cys | U |
| | UUC | | UCC | | UAC | | UGC | | C |
| | UUA | Leu | UCA | | UAA | OCHRE | UGA | ? | A |
| | UUG | | UCG | | UAG | AMBER | UGG | Tryp | G |
| C | CUU | Leu | CCU | Pro | CAU | His | CGU | Arg | U |
| | CUC | | CCC | | CAC | | CGC | | C |
| | CUA | | CCA | | CAA | GluN | CGA | | A |
| | CUG | | CCG | | CAG | | CGG | | G |
| A | AUU | Ileu | ACU | Thr | AAU | AspN | AGU | Ser | U |
| | AUC | | ACC | | AAC | | AGC | | C |
| | AUA | | ACA | | AAA | Lys | AGA | Arg | A |
| | AUG | Met | ACG | | AAG | | AGG | | G |
| G | GUU | Val | GCU | Ala | GAU | Asp | GGU | Gly | U |
| | GUC | | GCC | | GAC | | GGC | | C |
| | GUA | | GCA | | GAA | Glu | GGA | | A |
| | GUG | | GCG | | GAG | | GGG | | G |

**GENETIC CODE, consisting of 64 triplet combinations and their corresponding amino acids, is shown in its most likely version. The importance of the first two letters in each triplet is readily apparent. Some of the allocations are still not completely certain, particularly for organisms other than the colon bacillus (*Escherichia coli*). "Amber" and "ochre" are terms that referred originally to certain mutant strains of bacteria. They designate two triplets, UAA and UAG, that may act as signals for terminating polypeptide chains.**

that clarifies the relations between them.

Inspection of the table will show that the triplets coding for the same amino acid are often rather similar. For example, all four of the triplets starting with the doublet AC code for threonine. This pattern also holds for seven of the other amino acids. In every case the triplets *XY*U and *XY*C code for the same amino acid, and in many cases *XY*A and *XY*G are the same (methionine and tryptophan may be exceptions). Thus an amino acid is largely selected by the first two bases of the triplet. Given that a triplet codes for, say, valine, we know that the first two bases are GU, whatever the third may be. This pattern is true for all but three of the amino acids. Leucine can start with UU or CU, serine with UC or AG and arginine with CG or AG. In all other cases the amino acid is uniquely related to the first two bases of the triplet. Of course, the converse is often not true. Given that a triplet starts with, say, CA, it may code for either histidine or glutamine.

## Synthetic Messenger RNA's

Probably the most direct way to confirm the genetic code is to synthesize a messenger RNA molecule with a strictly defined base sequence and then find the amino acid sequence of the polypeptide produced under its influence. The most extensive work of this nature has been done by Khorana and his colleagues. By a brilliant combination of ordinary chemical synthesis and synthesis catalyzed by enzymes, they have made long RNA molecules with various repeating sequences of bases. As an example, one RNA molecule they have synthesized has the sequence UGUGUGUGUGUG.... When the biochemical machinery reads this as triplets the message is UGU–GUG–UGU–GUG.... Thus we expect that a polypeptide will be produced with an alternating sequence of two amino acids. In fact, it was found that the product is Cys–Val–Cys–Val.... This evidence alone would not tell us which triplet goes with which amino acid, but given the results of the binding test one has no hesitation in concluding that UGU codes for cysteine and GUG for valine.

In the same way Khorana has made chains with repeating sequences of the type *XYZ*... and also *XXYZ*.... The type *XYZ*...would be expected to give a "homopolypeptide" containing one amino acid corresponding to the triplet *XYZ*. Because the starting point is not clearly defined, however, the homopolypeptides corresponding to *YZX*... and *ZXY*... will also be produced. Thus poly-AUC makes polyisoleucine, polyserine and polyhistidine. This confirms that AUC codes for isoleucine, UCA for serine and CAU for histidine. A repeating sequence of four bases will yield a single type of polypeptide with a repeating sequence of four amino acids. The general patterns to be expected in each case are set forth in the table on this page. The results to date have amply demonstrated by a direct biochemical method that the code is indeed a triplet code.

Khorana and his colleagues have so far confirmed about 25 triplets by this method, including several that were quite doubtful on the basis of the binding test. They plan to synthesize other sequences, so that eventually most of the triplets will be checked in this way.

## The Use of Mutations

The two methods described so far are open to the objection that since they do not involve intact cells there may be some danger of false results. This objection can be met by two other methods of checking the code in which the act of protein synthesis takes place inside the cell. Both involve the effects of genetic mutations on the amino acid sequence of a protein.

It is now known that small mutations are normally of two types: "base substitution" mutants and "phase shift" mutants. In the first type one base is changed into another base but the total number of bases remains the same. In the second, one or a small number of bases are added to the message or subtracted from it.

There are now extensive data on base-substitution mutants, mainly from studies of three rather convenient proteins: human hemoglobin, the protein of tobacco mosaic virus and the *A* protein of the enzyme tryptophan synthetase obtained from the colon bacillus. At least 36 abnormal types of human hemoglobin have now been investigated by many different workers. More than 40 mutant forms of the protein of the tobacco mosaic virus have been examined by Hans Wittmann of the Max Planck Institute for Molecular Genetics in Tübingen and by Akita Tsugita and Heinz Fraenkel-Conrat of the University of California at Berkeley [see "The Genetic Code of a Virus," by Heinz Fraenkel-Conrat; Scientific American Offprint No. 193]. Charles Yanofsky and his group at Stanford University have characterized about 25 different mutations of the *A* protein of tryptophan synthetase.

| RNA BASE SEQUENCE | READ AS | AMINO ACID SEQUENCE EXPECTED |
|---|---|---|
| $(XY)_n$ | . . . X Y X \| Y X Y \| X Y X \| Y X Y . . . | αβαβ |
| $(XYZ)_n$ | . . . X Y Z \| X Y Z \| X Y Z . . . | ααα |
| | . . . Y Z X \| Y Z X \| Y Z X . . . | βββ |
| | . . . Z X Y \| Z X Y \| Z X Y . . . | γγγ |
| $(XXYZ)_n$ | . . . X X Y \| Z X X \| Y Z X \| X Y Z . . . | αβγδαβγδ |
| $(XYXZ)_n$ | . . . X Y X \| Z X Y \| X Z X \| Y X Z . . . | αβγδαβγδ |

**VARIETY OF SYNTHETIC RNA's with repeating sequences of bases have been produced by H. Gobind Khorana and his colleagues at the University of Wisconsin. They contain two or three different bases (*X, Y, Z*) in groups of two, three or four. When introduced into cell-free systems containing the machinery for protein synthesis, the base sequences are read off as triplets (*middle*) and yield the amino acid sequences indicated at the right.**

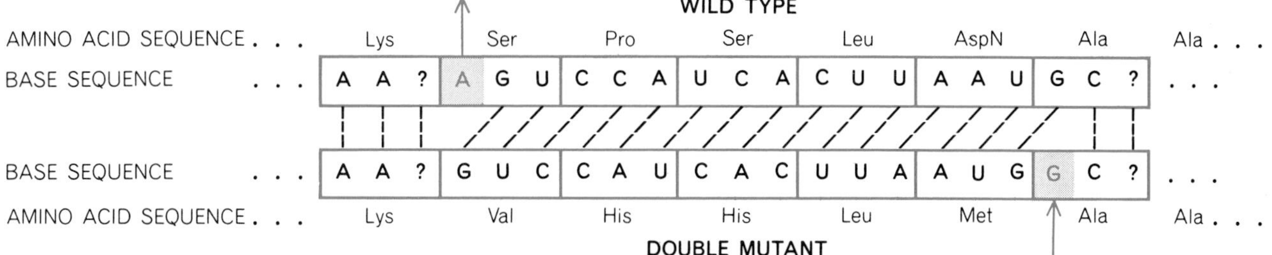


**"PHASE SHIFT" MUTATIONS help to establish the actual codons used by organisms in the synthesis of protein. The two partial amino acid sequences shown here were determined by George Streisinger and his colleagues at the University of Oregon. The sequences are from a protein, a type of lysozyme, produced by the bacterial virus T4. A pair of phase-shift mutations evidently removed one base, A, and inserted another, G, about 15 bases farther on. The base sequence was deduced theoretically from the genetic code.**

The remarkable fact has emerged that in every case but one the genetic code shows that the change of an amino acid in a polypeptide chain could have been caused by the alteration of a single base in the relevant nucleic acid. For example, the first observed change of an amino acid by mutation (in the hemoglobin of a person suffering from sickle-cell anemia) was from glutamic acid to valine. From the genetic code dictionary on page 113 we see that this could have resulted from a mutation that changed either GAA to GUA or GAG to GUG. In either case the change involved a single base in the several hundred needed to code for one of the two kinds of chain in hemoglobin.

The one exception so far to the rule that all amino acid changes could be caused by single base changes has been found by Yanofsky. In this one case glutamic acid was replaced by methionine. It can be seen from the genetic code dictionary that this can be accomplished only by a change of *two* bases, since glutamic acid is encoded by either GAA or GAG and methionine is encoded only by AUG. This mutation has occurred only once, however, and of all the mutations studied by Yanofsky it is the only one not to back-mutate, or revert to "wild type." It is thus almost certainly the rare case of a double change. All the other cases fit the hypothesis that base-substitution mutations are normally caused by a single base change. Examination of the code shows that only about 40 percent of all the possible amino acid interchanges can be brought about by single base substitutions, and it is only these changes that are found in experiments. Therefore the study of actual mutations has provided strong confirmation of many features of the genetic code.

Because in general several codons stand for one amino acid it is not possible, knowing the amino acid sequence, to write down the exact RNA base sequence that encoded it. This is unfortunate. If we know which amino acid is changed into another by mutation, however, we can often, given the code, work out what that base change must have been. As an example, glutamic acid can be encoded by GAA or GAG and valine by GUU, GUC, GUA or GUG. If a mutation substitutes valine for glutamic acid, one can assume that only a single base change was involved. The only such change that could lead to the desired result would be a change from A to U in the middle position, and this would be true whether GAA became GUA or GAG became GUG.

It is thus possible in many cases (not in all) to compare the nature of the base change with the chemical mutagen used to produce the change. If RNA is treated with nitrous acid, C is changed to U and A is effectively changed to G. On the other hand, if double-strand DNA is treated under the right conditions with hydroxylamine, the mutagen acts only on C. As a result some C's are changed to T's (the DNA equivalent of U's), and thus G's, which are normally paired with C's in double-strand DNA, are replaced by A's.

If 2-aminopurine, a "base analogue" mutagen, is added when double-strand DNA is undergoing replication, it produces only "transitions." These are the same changes as those produced by hydroxylamine—plus the reverse changes. In almost all these different cases (the exceptions are unimportant) the changes observed are those expected from our knowledge of the genetic code.

Note the remarkable fact that, although the code was deduced mainly from studies of the colon bacillus, it appears to apply equally to human beings and tobacco plants. This, together with more fragmentary evidence, suggests that the genetic code is either the same or very similar in most organisms.

The second method of checking the code using intact cells depends on phase-shift mutations such as the addition of a single base to the message. Phase-shift mutations probably result from errors produced during genetic recombination or when the DNA molecule is being duplicated. Such errors have the effect of putting out of phase the reading of the message from that point on. This hypothesis leads to the prediction that the phase can be corrected if at some subsequent point a nucleotide is deleted. The pair of alterations would be expected not only to change two amino acids but also to alter all those encoded by bases lying between the two affected sites. The reason is that the intervening bases would be read out of phase and therefore grouped into triplets different from those contained in the normal message.

This expectation has recently been confirmed by George Streisinger and his colleagues at the University of Oregon. They have studied mutations in the protein lysozyme that were produced by the T4 virus, which infects the colon bacillus. One phase-shift mutation involved the amino acid sequence ...Lys—Ser—Pro—Ser—Leu—AspN—Ala—Ala—Lys.... They were then able to construct by genetic methods a double phase-shift mutant in which the corresponding sequence was ...Lys–Val–His–His–Leu–Met–Ala–Ala–Lys....

Given these two sequences, the reader should be able, using the genetic code dictionary on page 113, to decipher uniquely a short length of the nucleic acid message for both the original protein and the double mutant and thus deduce the changes produced by each of the phase-shift mutations. The correct result is presented in the illustration above. The result not only confirms several rather doubtful codons, such as UUA for leucine and AGU for serine, but also shows which codons are actually involved in a genetic message. Since the technique is difficult, however, it may not find wide application.

Streisinger's work also demonstrates what has so far been only tacitly as-

| ANTICODON | CODON |
|---|---|
| U | A<br>G |
| C | G |
| A | U |
| G | U<br>C |
| I | U<br>C<br>A |

**"WOBBLE" HYPOTHESIS has been proposed by the author to provide rules for the pairing of codon and anticodon at the *third* position of the codon. There is evidence, for example, that the anticodon base I, which stands for inosine, may pair with as many as three different bases: U, C and A. Inosine closely resembles the base guanine (G) and so would ordinarily be expected to pair with cytosine (C). Structural diagrams for standard base pairings and wobble base pairings are illustrated at the bottom of this page.**

sumed: that the two languages, both of which are written down in a certain direction according to convention, are in fact translated by the cell in the same direction and not in opposite directions. This fact had previously been established, with more direct chemical methods, by Severo Ochoa and his colleagues at the New York University School of Medicine. In the convention, which was adopted by chance, proteins are written with the amino ($NH_2$) end on the left. Nucleic acids are written with the end of the molecule containing a "5 prime" carbon atom at the left. (The "5 prime" refers to a particular carbon atom in the 5-carbon ring of ribose sugar or deoxyribose sugar.)

### Finding the Anticodons

Still another method of checking the genetic code is to discover the three bases making up the anticodon in some particular variety of transfer RNA. The first tRNA to have its entire sequence worked out was alanine tRNA, a job done by Robert W. Holley and his collaborators at Cornell University [see "The Nucleotide Sequence of a Nucleic Acid," by Robert W. Holley; SCIENTIFIC AMERICAN, February, 1966]. Alanine tRNA, obtained from yeast, contains 77 bases. A possible anticodon found near the middle of the molecule has the sequence IGC, where I stands for inosine, a base closely resembling guanine. Since then Hans Zachau and his colleagues at the University of Cologne have established the sequences of two closely related serine tRNA's from yeast, and James Madison and his group at the U.S. Plant, Soil and Nutrition Laboratory at Ithaca, N.Y., have worked out the sequence of a tyrosine tRNA, also from yeast.

A detailed comparison of these three sequences makes it almost certain that the anticodons are alanine–IGC, serine–IGA and tyrosine–GΨA. (Ψ stands for pseudo-uridylic acid, which can form the same base pairs as the base uracil.) In addition there is preliminary evidence from other workers that an anticodon for valine is IAC and an anticodon for phenylalanine is GAA.

All these results would fit the rule that the codon and anticodon pair in an antiparallel manner, and that the pairing in the first two positions of the codon is of the standard type, that is, A pairs with U and G pairs with C. The pairing in the third position of the codon is more complicated. There is now good experimental evidence from both Nirenberg and Khorana and their co-workers that one tRNA can recognize several codons, provided that they differ only in the last place in the codon. Thus Holley's alanine tRNA appears to recognize GCU, GCC and GCA. If it recognizes GCG, it does so only very weakly.

### The "Wobble" Hypothesis

I have suggested that this is because of a "wobble" in the pairing in the third place and have shown that a reasonable theoretical model will explain many of the observed results. The suggested rules for the pairing in the third position of the anticodon are presented in the table at the top of this page, but this theory is still speculative. The rules for the first two places of the codon seem reasonably secure, however, and can be used as partial confirmation of the genetic code. The likely codon-anticodon pairings for valine, serine, tyrosine, alanine and phenylalanine satisfy the standard base pairings in the first two places and the wobble hypothesis in the third place [*see illustration on page 117*].

Several points about the genetic code remain to be cleared up. For example, the triplet UGA has still to be allocated.

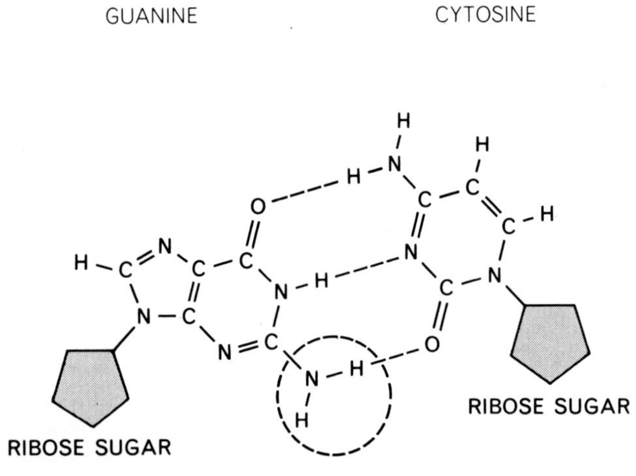

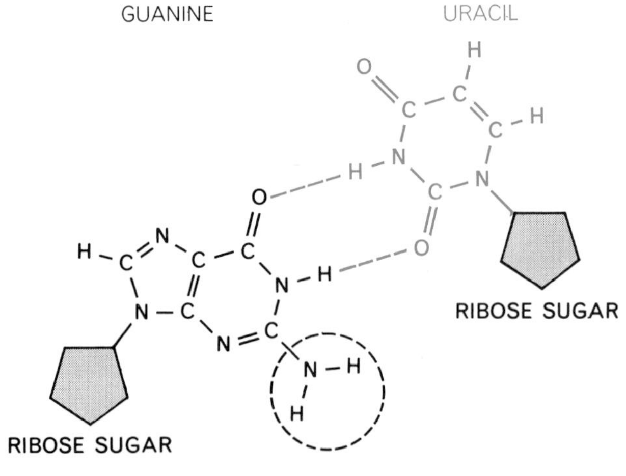

GUANINE    CYTOSINE    GUANINE    URACIL

RIBOSE SUGAR    RIBOSE SUGAR    RIBOSE SUGAR    RIBOSE SUGAR

**STANDARD AND WOBBLE BASE PAIRINGS both involve the formation of hydrogen bonds when certain bases are brought into close proximity. In the standard guanine-cytosine pairing (*left*) it is believed three hydrogen bonds are formed. The bases are shown as they exist in the RNA molecule, where they are attached to 5-carbon rings of ribose sugar. In the proposed wobble pairing (*right*) guanine is linked to uracil by only two hydrogen bonds. The base inosine (I) has a single hydrogen atom where guanine has an amino ($NH_2$) group (*broken circle*). In the author's wobble hypothesis inosine can pair with U as well as with C and A (*not shown*).**

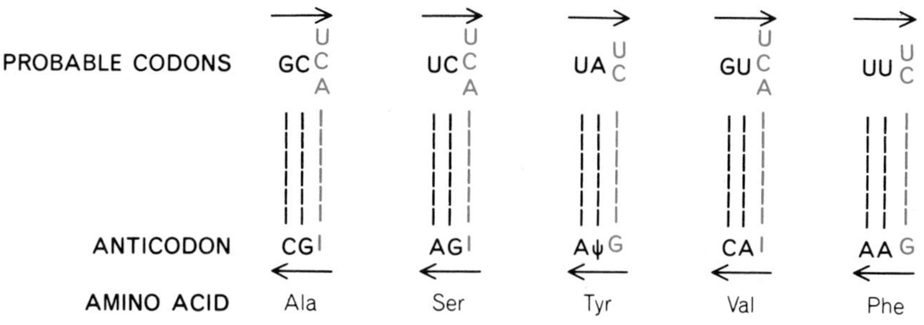

The punctuation marks—the signals for "begin chain" and "end chain"—are only partly understood. It seems likely that both the triplet UAA (called "ochre") and UAG (called "amber") can terminate the polypeptide chain, but which triplet is normally found at the end of a gene is still uncertain.

The picturesque terms for these two triplets originated when it was discovered in studies of the colon bacillus some years ago that mutations in other genes (mutations that in fact cause errors in chain termination) could "suppress" the action of certain mutant codons, now identified as either UAA or UAG. The terms "ochre" and "amber" are simply invented designations and have no reference to color.

A mechanism for chain initiation was discovered fairly recently. In the colon bacillus it seems certain that formylmethionine, carried by a special tRNA, can initiate chains, although it is not clear if all chains have to start in this way, or what the mechanism is in mammals and other species. The formyl group (CHO) is not normally found on finished proteins, suggesting that it is probably removed by a special enzyme. It seems likely that sometimes the methionine is removed as well.

It is unfortunately possible that a few codons may be ambiguous, that is, may code for more than one amino acid. This is certainly not true of most codons. The present evidence for a small amount of ambiguity is suggestive but not conclusive. It will make the code more difficult to establish correctly if ambiguity can occur.

| | | | | | |
|---|---|---|---|---|---|
| PROBABLE CODONS → | GC U/C/A | UC U/C/A | UA U/C | GU U/C/A | UU U/C |
| ANTICODON ← | CGI | AGI | AψG | CAI | AAG |
| AMINO ACID | Ala | Ser | Tyr | Val | Phe |

**CODON-ANTICODON PAIRINGS take place in an antiparallel direction. Thus the anticodons are shown here written backward, as opposed to the way they appear in the text. The five anticodons are those tentatively identified in the transfer RNA's for alanine, serine, tyrosine, valine and phenylalanine. Color indicates where wobble pairings may occur.**

## Problems for the Future

From what has been said it is clear that, although the entire genetic code is not known with complete certainty, it is highly likely that most of it is correct. Further work will surely clear up the doubtful codons, clarify the punctuation marks, delimit ambiguity and extend the code to many other species. Although the code lists the codons that *may* be used, we still have to determine if alternative codons are used equally. Some preliminary work suggests they may not be. There is also still much to be discovered about the machinery of protein synthesis. How many types of tRNA are there? What is the structure of the ribosome? How does it work, and why is it in two parts? In addition there are many questions concerning the control of the rate of protein synthesis that we are still a long way from answering.

When such questions have been answered, the major unsolved problem will be the structure of the genetic code. Is the present code merely the result of a series of evolutionary accidents, so that the allocations of triplets to amino acids is to some extent arbitrary? Or are there profound structural reasons why phenylalanine has to be coded by UUU and UUC and by no other triplets? Such questions will be difficult to decide, since the genetic code originated at least three billion years ago, and it may be impossible to reconstruct the sequence of events that took place at such a remote period. The origin of the code is very close to the origin of life. Unless we are lucky it is likely that much of the evidence we should like to have has long since disappeared.

Nevertheless, the genetic code is a major milestone on the long road of molecular biology. In showing in detail how the four-letter language of nucleic acid controls the 20-letter language of protein it confirms the central theme of molecular biology that genetic information can be stored as a one-dimensional message on nucleic acid and be expressed as the one-dimensional amino acid sequence of a protein. Many problems remain, but this knowledge is now secure.

# 13
# *Gene Structure and Protein Structure*

CHARLES YANOFSKY
*May 1967*

The present molecular theory of genetics, known irreverently as "the central dogma," is now 14 years old. Implicit in the theory from the outset was the notion that genetic information is coded in linear sequence in molecules of deoxyribonucleic acid (DNA) and that the sequence directly determines the linear sequence of amino acid units in molecules of protein. In other words, one expected the two molecules to be colinear. The problem was to prove that they were.

Over the same 14 years, as a consequence of an international effort, most of the predictions of the central dogma have been verified one by one. The results were recently summarized in these pages by F. H. C. Crick, who together with James D. Watson proposed the helical, two-strand structure for DNA on

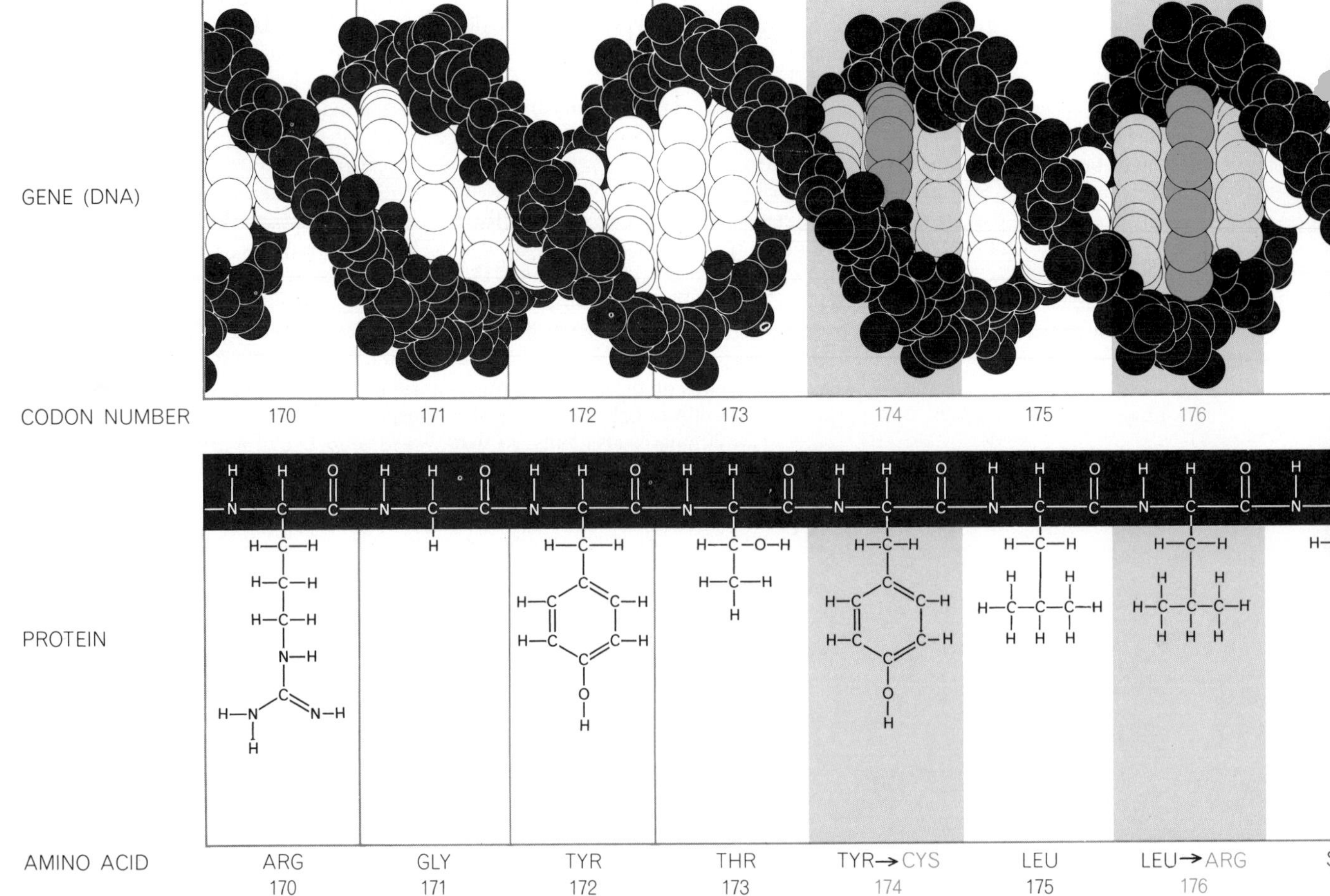


**STRUCTURES OF GENE AND PROTEIN have been shown to bear a direct linear correspondence by the author and his colleagues at Stanford University. They demonstrated that a particular sequence of coding units (codons) in the genetic molecule deoxyribonucleic acid, or DNA (*top*), specifies a corresponding sequence of amino acid units in the structure of a protein molecule (*bottom*). In the DNA molecule depicted here the black spheres represent repeating units of deoxyribose sugar and phosphate, which form the helical backbones of the two-strand molecule. The white spheres connecting the two strands represent complementary pairs of the four kinds of base that provide the "letters" in which the genetic message is written. A sequence of three bases attached to**

which the central dogma is based [see "The Genetic Code: III," by F. H. C. Crick, which begins on page 111 in this book]. Here I shall describe in somewhat more detail how our studies at Stanford University demonstrated the colinearity of genetic structure (as embodied in DNA) and protein structure.

Let me begin with a brief review. The molecular subunits that provide the "letters" of the code alphabet in DNA are the four nitrogenous bases adenine (A), guanine (G), cytosine (C) and thymine (T). If the four letters were taken in pairs, they would provide only 16 different code words—too few to specify the 20 different amino acids commonly found in protein molecules. If they are taken in triplets, however, the four letters can provide 64 different code words, which would seem too many for the efficient specification of the 20 amino acids. Accordingly it was conceivable that the cell might employ fewer than the 64 possible triplets. We now know that nature not only has selected the triplet code but also makes use of most (if not all) of the 64 triplets, which are called codons. Each amino acid but two (tryptophan and methionine) are specified by at least two different codons, and a few amino acids are specified by as many as six codons. It is becoming clear that the living cell exploits this redundancy in subtle ways. Of the 64 codons, 61 have been shown to specify one or another of the 20 amino acids. The remaining three can act as "chain terminators," which signal the end of a genetic message.

A genetic message is defined as the amount of information in one gene; it is the information needed to specify the complete amino acid sequence in one polypeptide chain. This relation, which underlies the central dogma, is sometimes expressed as the one-gene-one-enzyme hypothesis. It was first clearly enunciated by George W. Beadle and Edward L. Tatum, as a result of their studies with the red bread mold *Neurospora crassa* around 1940. In some cases a single polypeptide chain constitutes a complete protein molecule, which often acts as an enzyme, or biological catalyst. Frequently, however, two or more polypeptide chains must join together in order to form an active protein. For example, tryptophan synthetase, the enzyme we used in our colinearity studies, consists of four polypeptide chains: two alpha chains and two beta chains.

How might one establish the colinearity of codons in DNA and amino acid units in a polypeptide chain? The most direct approach would be to separate the two strands of DNA obtained from some organism and determine the base sequence of that portion of a strand which is presumed to be colinear with the amino acid sequence of a particular protein. If the amino acid sequence of the protein were not already known, it too would have to be established. One could then write the two sequences in adjacent columns and see if the same codon (or its synonym) always appeared adjacent to a particular amino acid. If it

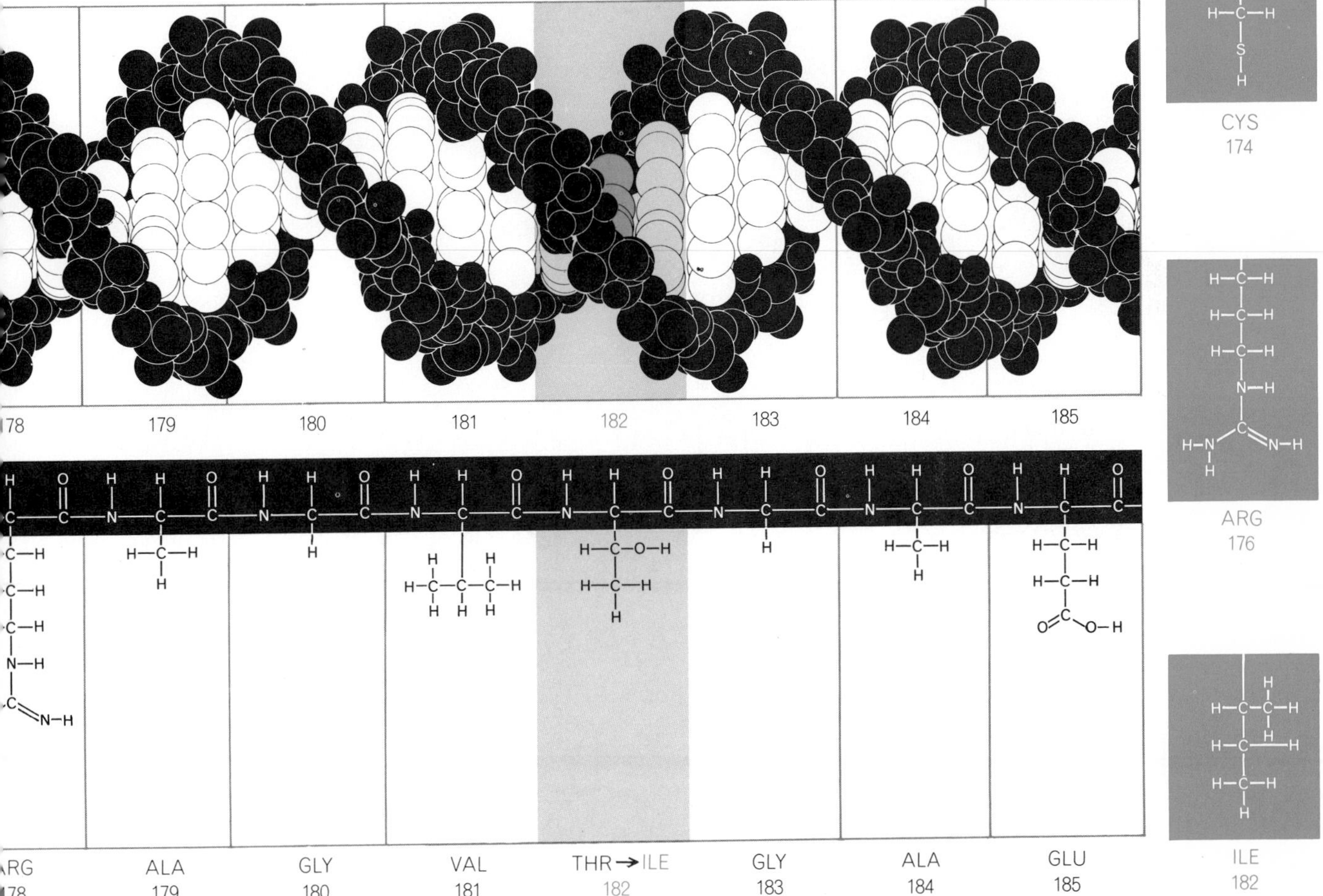


**one strand of DNA is a codon and specifies one amino acid. The amino acid sequence illustrated here is the region from position 170 through 185 in the *A* protein of the enzyme tryptophan synthetase produced by the bacterium *Escherichia coli.* It was found that mutations in the *A* gene of *E. coli* altered the amino acids at three places (*174, 176 and 182*) in this region of the *A* protein. (A key to the amino acid abbreviations can be found on page 5.) The three amino acids that replace the three normal ones as a result of mutation are shown at the extreme right. Each replacement is produced by a mutation at one site (*dark color*) in the DNA of the *A* gene. In all, the author and his associates correlated mutations at eight sites in the *A* gene with alterations in the *A* protein.**

did, a colinear relation would be established. Unfortunately this direct approach cannot be taken because so far it has not been possible to isolate and identify individual genes. Even if one could isolate a single gene that specified a polypeptide made up of 150 amino acids (and not many polypeptides are that small), one would have to determine the sequence of units in a DNA strand consisting of some 450 bases.

It was necessary, therefore, to consider a more feasible way of attacking the problem. An approach that immediately suggests itself to a geneticist is to construct a genetic map, which is a representation of the information contained in the gene, and see if the map can be related to protein structure. A genetic map is constructed solely on the basis of information obtained by crossing individual organisms that differ in two or more hereditary respects (a refinement of the technique originally

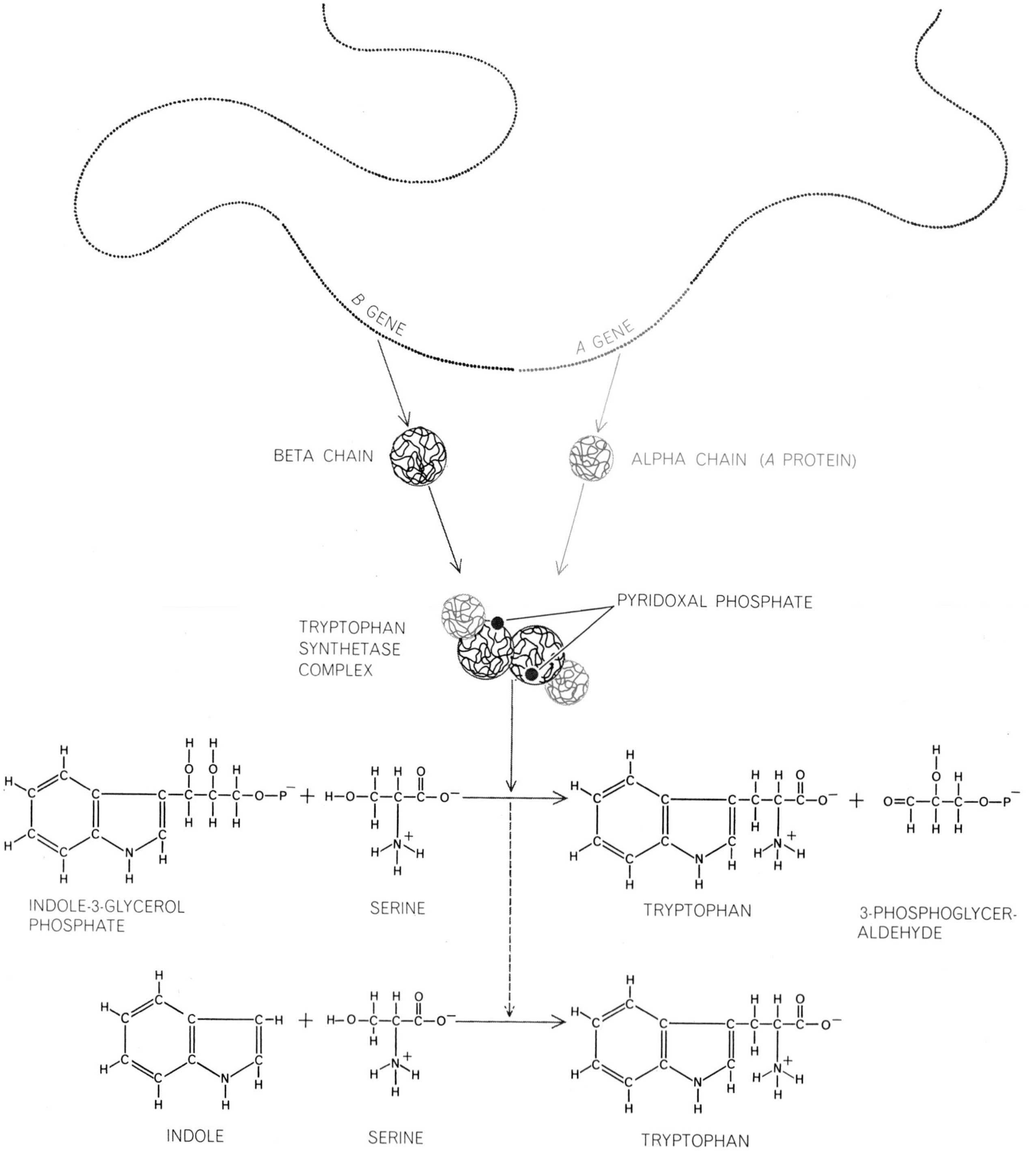

**GENETIC CONTROL OF CELL'S CHEMISTRY is exemplified by the two genes in *E. coli* that carry the instructions for making the enzyme tryptophan synthetase. The enzyme is actually a complex of four polypeptide chains: two alpha chains and two beta chains. The alpha chain is the *A* protein in which changes produced by mutations in the *A* gene have provided the evidence for gene-protein colinearity. One class of *A*-protein mutants retains the ability to associate with beta chains but the complex is no longer able to catalyze the normal biochemical reaction: the conversion of indole-3-glycerol phosphate and serine to tryptophan and 3-phosphoglyceraldehyde. But the complex can still catalyze a simpler nonphysiological reaction: the conversion of indole and serine to tryptophan.**

used by Gregor Mendel to demonstrate how characteristics are inherited).

By using bacteria and bacterial viruses in such studies one can catalogue the results of crosses involving millions of individual organisms and thereby deduce the actual distances separating the sites of mutational changes in a single gene. The distances are inferred from the frequency with which parent organisms, each with at least one mutation in the same gene, give rise to offspring in which neither mutation is present. As a result of the recombination of genetic material the offspring can inherit a gene that is assembled from the mutation-free portions of each parental gene. If the mutational markers lie far apart on the parental genes, recombination will frequently produce mutation-free progeny. If the markers are close together, mutation-free progeny will be rare [*see bottom illustration on next page*].

In his elegant studies with the "*r*II" region of the chromosome of the bacterial virus designated T4, Seymour Benzer, then at Purdue University, showed that the number of genetically distinguishable mutation sites on the map of the gene approaches the estimated number of base pairs in the DNA molecule corresponding to that gene. (Mutations involve pairs of bases because the bases in each of the two entwined strands of the DNA molecule are paired with and are complementary to the bases in the other strand. If a mutation alters one base in the DNA molecule, its partner is eventually changed too during DNA replication.) Benzer also showed that the only type of genetic map consistent with his data is a map on which the sites altered by mutation are arranged linearly. Subsequently A. D. Kaiser and David Hogness of Stanford University demonstrated with another bacterial virus that there is a linear correspondence between the sites on a genetic map and the altered regions of a DNA molecule isolated from the virus. Thus there is direct experimental evidence indicating that the genetic map is a valid representation of DNA structure and that the map can be employed as a substitute for information about base sequence.

This, then, provided the basis of our approach. We would pick a suitable organism and isolate a large number of mutant individuals with mutations in the same gene. From recombination studies we would make a fine-structure genetic map relating the sites of the mutations. In addition we would have to be able to isolate the protein specified by that gene and determine its amino acid sequence. Finally we would have to analyze the protein produced by each mutant (assuming a protein were still produced) in order to find the position of the amino acid change brought about in its amino acid sequence by the mutation. If gene structure and protein structure were colinear, the positions at which amino acid changes occur in the protein should be in the same order as the positions of the corresponding mutationally altered sites on the genetic map. Although this approach to the question of colinearity would require a great deal of work and much luck, it was logical and experimentally feasible. Several research groups besides our own set out to find a suitable system for a study of this kind.

The essential requirement of a suitable system was that a genetically

| | | | | | |
|---|---|---|---|---|---|
| ALA | ALANINE | GLY | GLYCINE | PRO | PROLINE |
| ARG | ARGININE | HIS | HISTIDINE | SER | SERINE |
| ASN | ASPARAGINE | ILE | ISOLEUCINE | THR | THREONINE |
| ASP | ASPARTIC ACID | LEU | LEUCINE | TRP | TRYPTOPHAN |
| CYS | CYSTEINE | LYS | LYSINE | TYR | TYROSINE |
| GLN | GLUTAMINE | MET | METHIONINE | VAL | VALINE |
| GLU | GLUTAMIC ACID | PHE | PHENYLALANINE | | |

**AMINO ACID ABBREVIATIONS identify the 20 amino acids commonly found in all proteins. Each amino acid is specified by a triplet codon in the DNA molecule (*see below*).**

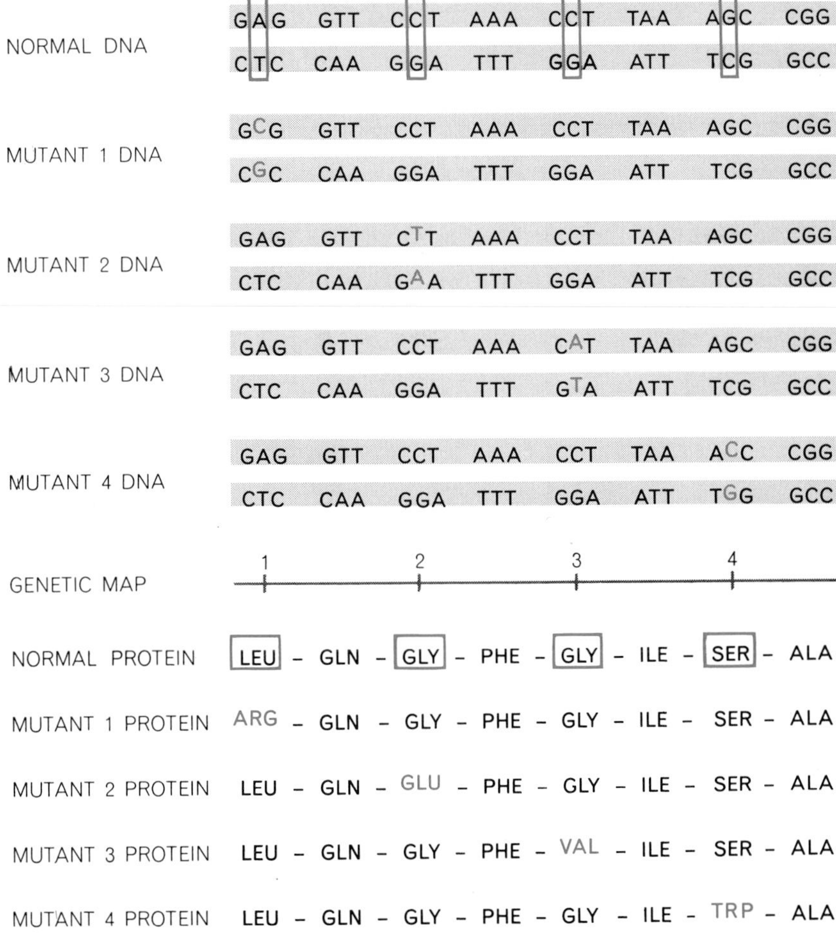


**GENETIC MUTATIONS can result from the alteration of a single base in a DNA codon. The letters stand for the four bases: adenine (A), thymine (T), guanine (G) and cytosine (C). Since the DNA molecule consists of two complementary strands, a base change in one strand involves a complementary change in the second strand. In the four mutant DNA sequences shown here (*top*) a pair of bases (*color*) is different from that in the normal sequence. By genetic studies one can map the sequence and approximate spacing of the four mutations (*middle*). By chemical studies of the proteins produced by the normal and mutant DNA sequences (*bottom*) one can establish the corresponding amino acid changes.**

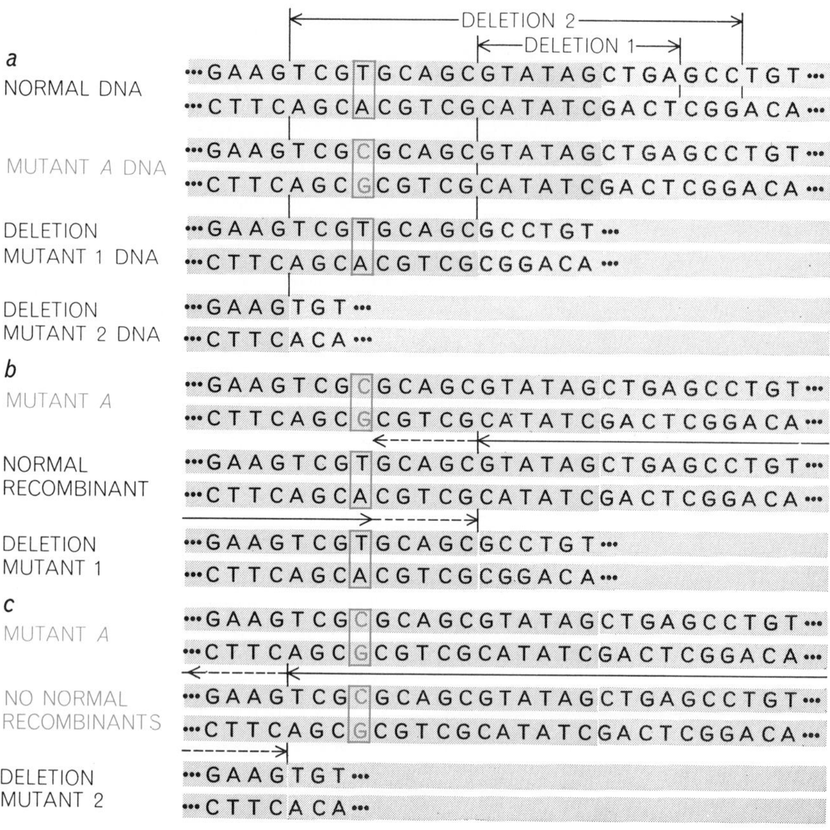


**"DELETION" MUTANTS provide one approach to making a genetic map. Here (*a*) normal DNA and mutant *A* differ by only one base pair (*C–G* has replaced *T–A*) in a certain portion of the *A* gene (*colored area*). In deletion mutant 1 a sequence of 10 base pairs, including six pairs from the *A* gene, has been spontaneously deleted. In deletion mutant 2, 22 base pairs, including 15 pairs from the *A* gene, have been deleted. By crossing mutant *A* with the two different deletion mutants in separate experiments (*b, c*), one can tell whether the mutated site (*C–G*) in the *A* gene falls inside or outside the deleted regions. A normal-type recombinant will appear (*b*) only if the altered base pair falls outside the deleted region.**

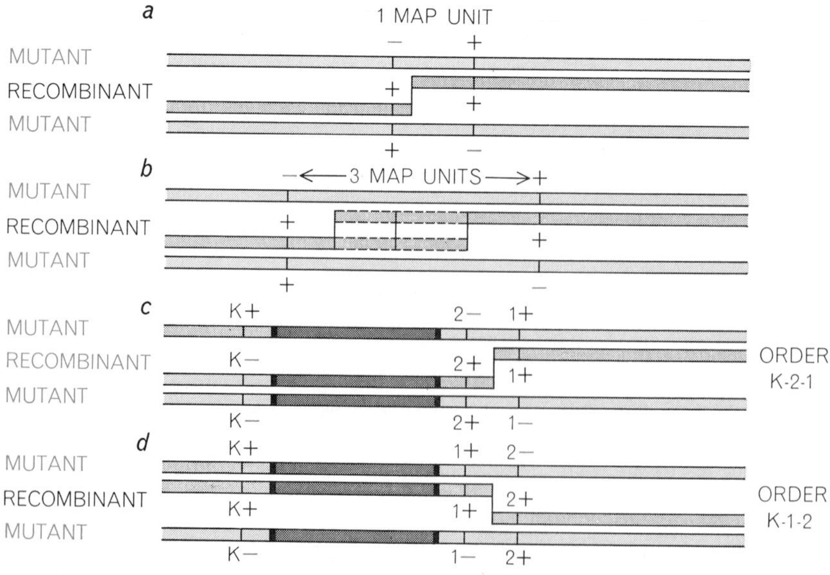


**OTHER MAPPING METHODS involve determination of recombination frequency (*a, b*) and the distribution of outside markers (*c, d*). The site of a mutational alteration is indicated by "–," the corresponding unaltered site by "+." If the altered sites are widely spaced (*b*), normal recombinants will appear more often than if the altered sites are close together (*a*). In the second method the mutants are linked to another gene that is either normal ($K^+$) or mutated ($K^-$). Recombinant strains that contain $1^+$ and $2^+$ will carry the $K^-$ gene if the correct order is *K*–2–1. They will carry the $K^+$ gene if the order is *K*–1–2.**

mappable gene should specify a protein whose amino acid sequence could be determined. Since no such system was known we had to gamble on a choice of our own. Fortunately we were studying at the time how the bacterium *Escherichia coli* synthesizes the amino acid tryptophan. Irving Crawford and I observed that the enzyme that catalyzed the last step in tryptophan synthesis could be readily separated into two different protein species, or subunits, one of which could be clearly isolated from the thousands of other proteins synthesized by *E. coli.* This protein, called the tryptophan synthetase *A* protein, had a molecular weight indicating that it had slightly fewer than 300 amino acid units. Furthermore, we already knew how to force *E. coli* to produce comparatively large amounts of the protein—up to 2 percent of the total cell protein—and we also had a collection of mutants in which the activity of the tryptophan synthetase *A* protein was lacking. Finally, the bacterial strain we were using was one for which genetic procedures for preparing fine-structure maps had already been developed. Thus we could hope to map the *A* gene that presumably controlled the structure of the *A* protein.

To accomplish the mapping we needed a set of bacterial mutants with mutational alterations at many different sites on the *A* gene. If we could determine the amino acid change in the *A* protein of each of these mutants, and discover its position in the linear sequence of amino acids in the protein, we could test the concept of colinearity. Here again we were fortunate in the nature of the complex of subunits represented by tryptophan synthetase.

The normal complex consists of two *A*-protein subunits (the alpha chains) and one subunit consisting of two beta chains. Within the bacterial cell the complex acts as an enzyme to catalyze the reaction of indole-3-glycerol phosphate and serine to produce tryptophan and 3-phosphoglyceraldehyde [*see illustration on page 120*]. If the *A* protein undergoes certain kinds of mutations, it is still able to form a complex with the beta chains, but the complex loses the ability to catalyze the reaction. It retains the ability, however, to catalyze a simpler reaction when it is tested outside the cell: it will convert indole and serine to tryptophan. There are still other kinds of *A*-gene mutants that evidently lack the ability to form an *A* protein that can combine with beta chains; thus these strains are not able to catalyze even the simpler reaction. The first class of mutants—those that produce an *A* protein that is still able to combine with beta chains and exhibit catalytic activity when they are tested outside the cell—proved to be the most important for our study.

A fine-structure map of the *A* gene was constructed on the basis of genetic crosses performed by the process called transduction. This employs a particular bacterial virus known as transducing phage *P1kc*. When this virus multiplies in a bacterium, it occasionally incorporates a segment of the bacterial DNA within its own coat of protein. When the virus progeny infect other bacteria, genetic material of the donor bacteria is introduced into some of the recipient cells. A fraction of the recipients survive the infection. In these survivors segments of the bacterium's own genetic material pair with like segments of the "foreign" genetic material and recombination between the two takes place. As a result the offspring of an infected bacterium can contain characteristics inherited from its remote parent as well as from its immediate one.

In order to establish the order of mutationally altered sites in the *A* gene we have relied partly on a set of mutant bacteria in which one end of a deleted segment of DNA lies within the *A* gene. In each of these "deletion" mutants a segment of the genetic material of the bacterium was deleted spontaneously. Thus each deletion mutant in the set retains a different segment of the *A* gene. This set of mutants can now be crossed with any other mutant in which the *A* gene is altered at only a single site. Recombination can give rise to a normal gene only if the altered site does not fall within the region of the *A* gene that is missing in the deletion mutant [*see top illustration on opposite page*]. By crossing many *A*-protein mutants with the set of deletion mutants one can establish the linear order of many of the mutated sites in the *A* gene. The ordering is limited only by the number of deletion mutants at one's disposal.

A second method, which more closely

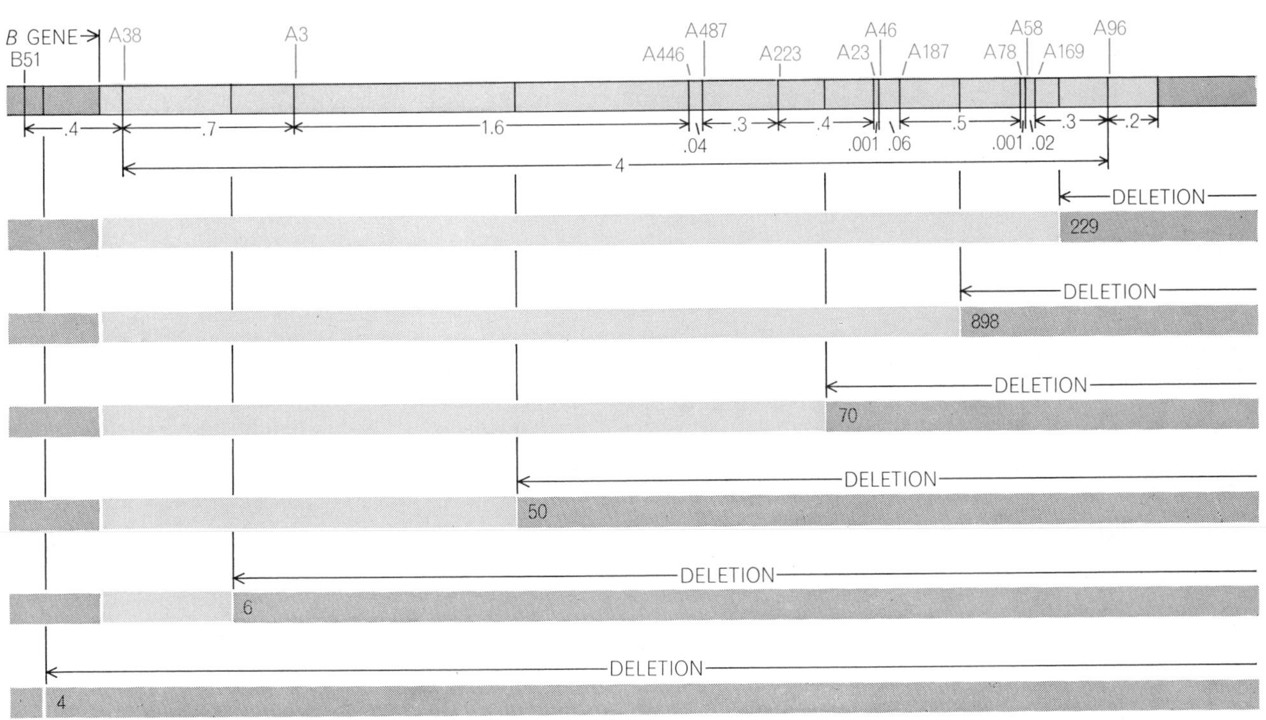


**MAP OF *A* GENE shows the location of mutationally altered sites, drawn to scale, as determined by the three genetic-mapping methods illustrated on the opposite page. The total length of the *A* gene is slightly over four map units (probably 4.2). Below map are six deletion mutants that made it possible to assign each of the 12 *A*-gene mutants to one of six regions within the gene. The more sensitive mapping methods were employed to establish the order of mutations and the distance between mutation sites within each region.**

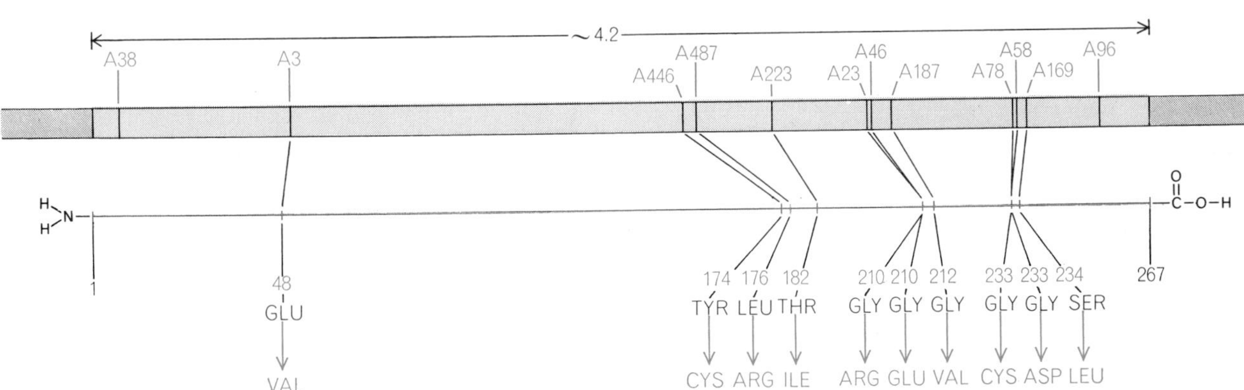


**COLINEARITY OF GENE AND PROTEIN can be inferred by comparing the *A*-gene map (*top*) with the various amino acid changes in the *A* protein (*bottom*), both drawn to scale. The amino acid changes associated with 10 of the 12 mutations are also shown.**

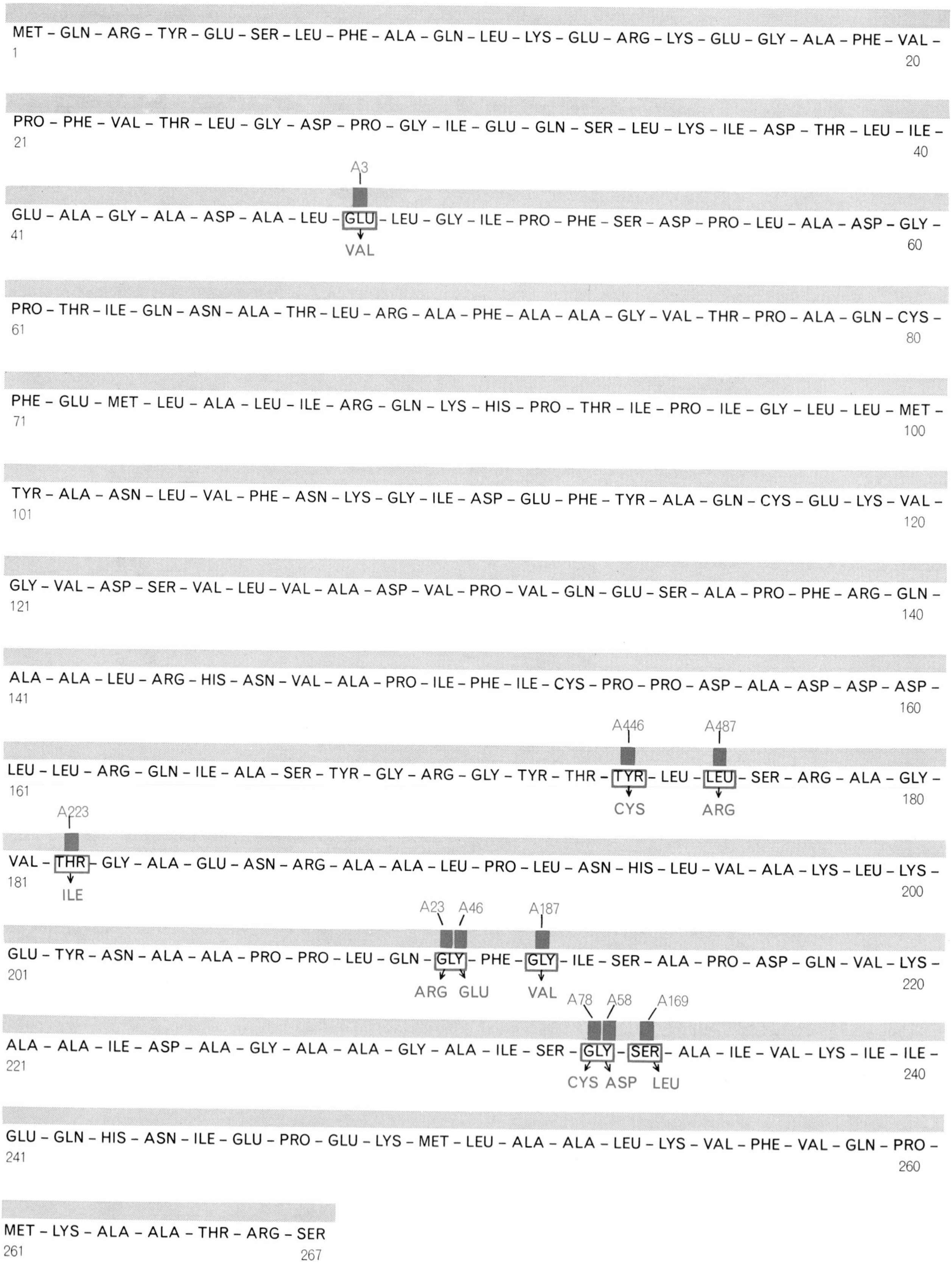

MET – GLN – ARG – TYR – GLU – SER – LEU – PHE – ALA – GLN – LEU – LYS – GLU – ARG – LYS – GLU – GLY – ALA – PHE – VAL –
1 ... 20

PRO – PHE – VAL – THR – LEU – GLY – ASP – PRO – GLY – ILE – GLU – GLN – SER – LEU – LYS – ILE – ASP – THR – LEU – ILE –
21 ... 40

A3

GLU – ALA – GLY – ALA – ASP – ALA – LEU – [GLU] – LEU – GLY – ILE – PRO – PHE – SER – ASP – PRO – LEU – ALA – ASP – GLY –
41 ... 60

VAL

PRO – THR – ILE – GLN – ASN – ALA – THR – LEU – ARG – ALA – PHE – ALA – ALA – GLY – VAL – THR – PRO – ALA – GLN – CYS –
61 ... 80

PHE – GLU – MET – LEU – ALA – LEU – ILE – ARG – GLN – LYS – HIS – PRO – THR – ILE – PRO – ILE – GLY – LEU – LEU – MET –
71 ... 100

TYR – ALA – ASN – LEU – VAL – PHE – ASN – LYS – GLY – ILE – ASP – GLU – PHE – TYR – ALA – GLN – CYS – GLU – LYS – VAL –
101 ... 120

GLY – VAL – ASP – SER – VAL – LEU – VAL – ALA – ASP – VAL – PRO – VAL – GLN – GLU – SER – ALA – PRO – PHE – ARG – GLN –
121 ... 140

ALA – ALA – LEU – ARG – HIS – ASN – VAL – ALA – PRO – ILE – PHE – ILE – CYS – PRO – PRO – ASP – ALA – ASP – ASP – ASP –
141 ... 160

A446 A487

LEU – LEU – ARG – GLN – ILE – ALA – SER – TYR – GLY – ARG – GLY – TYR – THR – [TYR] – LEU – [LEU] – SER – ARG – ALA – GLY –
161 ... 180

CYS ARG

A223

VAL – [THR] – GLY – ALA – GLU – ASN – ARG – ALA – ALA – LEU – PRO – LEU – ASN – HIS – LEU – VAL – ALA – LYS – LEU – LYS –
181 ... 200

ILE

A23 A46 A187

GLU – TYR – ASN – ALA – ALA – PRO – PRO – LEU – GLN – [GLY] – PHE – [GLY] – ILE – SER – ALA – PRO – ASP – GLN – VAL – LYS –
201 ... 220

ARG GLU VAL

A78 A58 A169

ALA – ALA – ILE – ASP – ALA – GLY – ALA – ALA – GLY – ALA – ILE – SER – [GLY] – [SER] – ALA – ILE – VAL – LYS – ILE – ILE –
221 ... 240

CYS ASP LEU

GLU – GLN – HIS – ASN – ILE – GLU – PRO – GLU – LYS – MET – LEU – ALA – ALA – LEU – LYS – VAL – PHE – VAL – GLN – PRO –
241 ... 260

MET – LYS – ALA – ALA – THR – ARG – SER
261 ... 267

**AMINO ACID SEQUENCE OF *A* PROTEIN** is shown side by side with a ribbon representing the DNA of the *A* gene. It can be seen that 10 different mutations in the gene produced alterations in the amino acids at only eight different places in the *A* protein. The explanation is that at two of them, 210 and 233, there were a total of four alterations. Thus at No. 210 the mutation designated A23 changed glycine to arginine, whereas mutation A46 changed glycine to glutamic acid. At No. 233 glycine was changed to cysteine by one mutation (A78) and to aspartic acid by another mutation (A58). On the genetic map A23 and A46, like A78 and A58, are very close.

resembles traditional genetic procedures, relies on recombination frequencies to establish the order of the mutationally altered sites in the *A* gene with respect to one another. By this method one can assign relative distances—map distances—to the regions between altered sites. The method is often of little use, however, when the distances are very close.

In such cases we have used a third method that involves a mutationally altered gene, or genetic marker, close to the *A* gene. This marker produces a recognizable genetic trait unrelated to the *A* protein. What this does, in effect, is provide a reading direction so that one can tell whether two closely spaced mutants, say No. 58 and No. 78, lie in the order 58–78, reading from the left on the map, or vice versa [*see bottom illustration on page 122*].

With these procedures we were able to construct a genetic map relating the altered sites in a group of mutants responsible for altered *A* proteins that could themselves be isolated for study. Some of the sites were very close together, whereas others were far apart [*see upper illustration on page 123*]. The next step was to determine the nature of the amino acid changes in each of the mutationally altered proteins.

| SEGMENT OF PROTEIN | MUTANT H11 | C140 | B17 | B272 | H32 | B278 | C137 | H36 | A489 | C208 | NORMAL |
|---|---|---|---|---|---|---|---|---|---|---|---|
| I | + | + | + | + | + | + | + | + | + | + | + |
| II | – | + | + | + | + | + | + | + | + | + | + |
| III | – | – | + | + | + | + | + | + | + | + | + |
| IV | – | – | – | + | + | + | + | + | + | + | + |
| V | – | – | – | – | + | + | + | + | + | + | + |
| VI | – | – | – | – | – | + | + | + | + | + | + |
| VII | – | – | – | – | – | – | + | + | + | + | + |
| VIII | – | – | – | – | – | – | – | + | + | + | + |
| IX | – | – | – | – | – | – | – | – | + | + | + |
| X | – | – | – | – | – | – | – | – | – | + | + |
| XI | – | – | – | – | – | – | – | – | – | – | + |

GENETIC MAP: H11 C140 B17 B272 H32 B278 C137 H36 A489 C208

**INDEPENDENT EVIDENCE FOR COLINEARITY of gene and protein structure has been obtained from studies of the protein that forms the head of the bacterial virus T4D. Sydney Brenner and his co-workers at the University of Cambridge have found that mutations in the gene for the head protein alter the length of head-protein fragments. In the table "+" indicates that a given segment of the head protein is produced by a particular mutant; "—" indicates that the segment is not produced. When the genetic map was plotted, it was found that the farther to the right a mutation appears, the longer the fragment of head protein.**

It was expected that each mutant of the *A* protein would have a localized change, probably involving only one amino acid. Before we could hope to identify such a specific change we would have to know the sequence of amino acids in the unmutated *A* protein. This was determined by John R. Guest, Gabriel R. Drapeau, Bruce C. Carlton and me, by means of a well-established procedure. The procedure involves breaking the protein molecule into many short fragments by digesting it with a suitable enzyme. Since any particular protein rarely has repeating sequences of amino acids, each digested fragment is likely to be unique. Moreover, the fragments are short enough—typically between two and two dozen amino acids in length—so that careful further treatments can release one amino acid at a time for analysis. In this way one can identify all the amino acids in all the fragments, but the sequential order of the fragments is still unknown. This can be established by digesting the complete protein molecule with a different enzyme that cleaves it into a uniquely different set of fragments. These are again analyzed in detail. With two fully analyzed sets of fragments in hand, it is not difficult to find short sequences of amino acids that are grouped together in the fragment of one set but that are divided between two fragments in the other. This provides the clue for putting the two sets of fragments in order. In this way we ultimately determined the identity and location of each of the 267 amino acids in the unmutated *A* protein of tryptophan synthetase.

Simultaneously my colleagues and I were examining the mutants of the *A* protein to identify the specific sites of mutational changes. For this work we used a procedure first developed by Vernon M. Ingram, now at the Massachusetts Institute of Technology, in his studies of naturally occurring abnormal forms of human hemoglobin. This procedure also uses an enzyme (trypsin) to break the protein chain into peptides, or polypeptide fragments. If the peptides are placed on filter paper wetted with certain solvents, they will migrate across the paper at different rates; if an electric potential is applied across the paper, the peptides will be dispersed even more, depending on whether they are negatively charged, positively charged or uncharged under controlled conditions of acidity. The former separation process is chromatography; the latter, electrophoresis. When they are employed in combination, they produce a unique "fingerprint" for each set of peptides obtained by digesting the *A* protein from a particular mutant bacterium. The positions of the peptides are located by spraying the filter paper with a solution of ninhydrin and heating it for a few minutes at about 70 degrees centigrade. Each peptide reacts to yield a characteristic shade of yellow, gray or blue.

When the fingerprints of mutationally altered *A* proteins were compared with the fingerprint of the unmutated protein, they were found to be remarkably similar. In each case, however, there was a difference. The mutant fingerprint usually lacked one peptide spot that appears in the nonmutant fingerprint and exhibited a spot that the nonmutant fingerprint lacks. The two peptides would presumably be related to each other with the exception of the change resulting from the mutational event. One can isolate each of the peptides and compare their amino acid composition. Guest, Drapeau, Carlton and I, together with D. R. Helinski and U. Henning, identified the amino acid substitutions in each of a variety of altered *A* proteins.

The final step was to compare the locations of these changes in the *A* protein with the genetic map of the mutationally altered sites. There could be no doubt that the amino acid sequence of the *A* protein and the map of the *A* gene are in fact colinear [*see lower illustration on page 123*].

One can also see that the distances between mutational sites on the map of the *A* gene correspond quite closely to the distances separating the corresponding amino acid changes in the *A* protein. In two instances two separate mutational changes, so close as to be almost at the same point on the genetic map, led to changes of the same amino acid in the unmutated protein. This is to be expected if a codon of three bases in DNA is required to specify a single amino acid in a protein. Evidently the most closely spaced mutational sites in our genetic map represent alterations in two bases within a single codon.

Thus our studies have shown that each

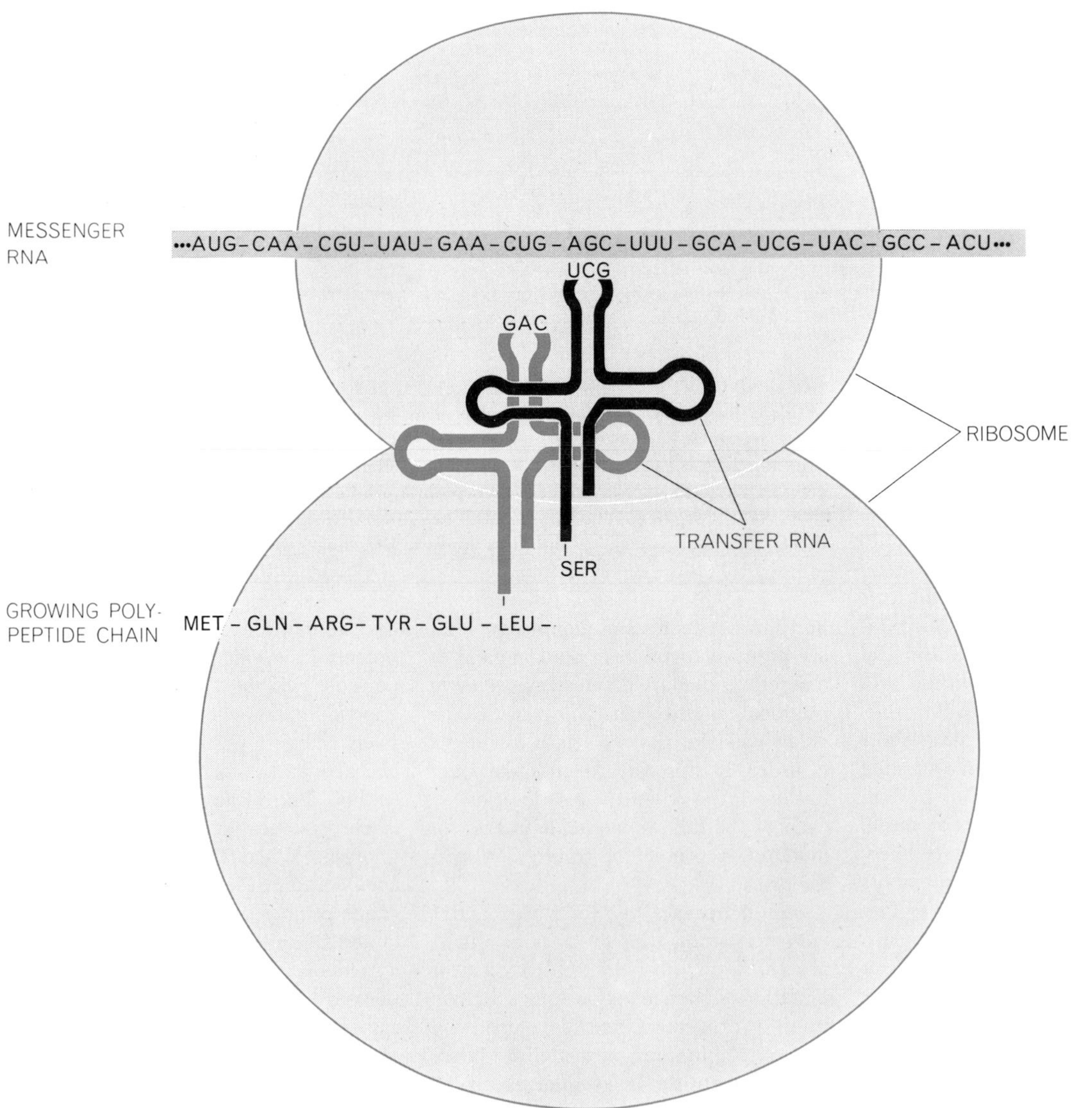


**SCHEME OF PROTEIN SYNTHESIS,** according to the current view, involves the following steps. Genetic information is transcribed from double-strand DNA into single-strand messenger ribonucleic acid (RNA), which becomes associated with a ribosome. Amino acids are delivered to the ribosome by molecules of transfer RNA, which embody codons complementary to the codons in messenger RNA. The next to the last molecule of transfer RNA to arrive (*color*) holds the growing polypeptide chain while the arriving molecule of transfer RNA (*black*) delivers the amino acid that is to be added to the chain next (serine in this example). The completed polypeptide chain, either alone or in association with other chains, is the protein whose specification was originally embodied in DNA.

AGY AGX AGY AGY
CGZ GGZ UAX ACZ UAX XUZ CUZ UCZ CGZ GCZ GGZ GUZ ACW GGZ GCZ GAY AAX CGZ GCZ GCZ XUZ
– ARG – GLY – TYR – THR – TYR – LEU – LEU – SER – ARG – ALA – GLY – VAL – THR – GLY – ALA – GLU – ASN – ARG – ALA – ALA – LEU –
170 190

CCZ XUZ AAX CAX XUZ GUZ GCZ AAY XUZ AAY GAY UAX AAX GCZ GCZ CCZ CCZ XUZ CAY GGA
PRO – LEU – ASN – HIS – LEU – VAL – ALA – LYS – LEU – LYS – GLU – TYR – ASN – ALA – ALA – PRO – PRO – LEU – GLN – GLY –
191 210

AGX
UUX GGZ AUW UCZ GCZ CCZ GAX CAY GUZ AAY GCZ GCZ AUW GAX GCZ GGZ GCZ GCZ GGZ GCZ
PHE – GLY – ILE – SER – ALA – PRO – ASP – GLN – VAL – LYS – ALA – ALA – ILE – ASP – ALA – GLY – ALA – ALA – GLY – ALA –
211 230

AGX
AUW UCZ GGX UCZ GCZ AUW GUZ AAY AUW AUW GAY CAY CAX AAX AUW GAY CCZ GAY AAY AUG
ILE – SER – GLY – SER – ALA – ILE – VAL – LYS – ILE – ILE – GLU – GLN – HIS – ASN – ILE – GLU – PRO – GLU – LYS – MET –
231 250

W = U, C or A X = U or C Y = A or G Z = U, C, A or G

**PROBABLE CODONS IN MESSENGER RNA** that determines the sequence of amino acids in the *A* protein are shown for 81 of the protein's 267 amino acid units. The region includes seven of the eight mutationally altered positions (*colored boxes*) in the *A* protein. The codons were selected from those assigned to the amino acids by Marshall Nirenberg and his associates at the National Institutes of Health and by H. Gobind Khorana and his associates at the University of Wisconsin. Codons for the remaining 186 amino acids in the *A* protein can be supplied similarly. In most cases the last base in the codon cannot be specified because there are usually several synonymous codons for each amino acid. With a few exceptions the synonyms differ from each other only in the third position.

unique sequence of bases in DNA—a sequence constituting a gene—is ultimately translated into a corresponding unique linear sequence of amino acids—a sequence constituting a polypeptide chain. Such chains, either by themselves or in conjunction with other chains, fold into the three-dimensional structures we recognize as protein molecules. In the great majority of cases these proteins act as biological catalysts and are therefore classed as enzymes.

The colinear relation between a genetic map and the corresponding protein has also been convincingly demonstrated by Sydney Brenner and his co-workers at the University of Cambridge. The protein they studied was not an enzyme but a protein that forms the head of the bacterial virus T4. One class of mutants of this virus produces fragments of the head protein that are related to one another in a curious way: much of their amino acid sequence appears to be identical, but the fragments are of various lengths. Brenner and his group found that when the chemically similar regions in fragments produced by many mutants were matched, the fragments could be arranged in order of increasing length. When they made a genetic map of the mutants that produced these fragments, they found that the mutationally altered sites on the genetic map were in the same order as the termination points in the protein fragments. Thus the length of the fragment of the head protein produced by a mutant increased as the site of mutation was displaced farther from one end of the genetic map [*see illustration on page 125*].

The details of how the living cell translates information coded in gene structure into protein structure are now reasonably well known. The base sequence of one strand of DNA is transcribed into a single-strand molecule of messenger ribonucleic acid (RNA), in which each base is complementary to one in DNA. Each strand of messenger RNA corresponds to relatively few genes; hence there are a great many different messenger molecules in each cell. These messengers become associated with the small cellular bodies called ribosomes, which are the actual site of protein synthesis [*see illustration on page 126*]. In the ribosome the bases on the messenger RNA are read in groups of three and translated into the appropriate amino acid, which is attached to the growing polypeptide chain. The messenger also contains in code a precise starting point and stopping point for each polypeptide.

From the studies of Marshall Nirenberg and his colleagues at the National Institutes of Health and of H. Gobind Khorana and his group at the University of Wisconsin the RNA codons corresponding to each of the amino acids are known. By using their genetic code dictionary we can indicate approximately two-thirds of the bases in the messenger RNA that specifies the structure of the A-protein molecule. The remaining third cannot be filled in because synonyms in the code make it impossible, in most cases, to know which of two or more bases is the actual base in the third position of a given codon [*see illustration above*]. This ambiguity is removed, however, in two cases where the amino acid change directed by a mutation narrows down the assignment of probable codons. Thus at amino acid position 48 in the *A*-protein molecule, where a mutation changes the amino acid glutamic acid to valine, one can deduce from the many known changes at this position that of the two possible codons for glutamic acid, GAA and GAG, GAG is the correct one. In other words, GAG (specifying glutamic acid) is changed to GUG (specifying valine). The other position for which the codon assignment can be made definite in this way is No. 210. This position is affected by two different mutations: the amino acid glycine is replaced by arginine in one case and by glutamic acid in the other. Here one can infer from the observed amino acid changes that of the four possible codons for glycine, only one—GGA—can yield by a single base change either arginine (AGA) or glutamic acid (GAA).

Knowledge of the bases in the messenger RNA for the *A* protein can be translated, of course, into knowledge of the base pairs in the *A* gene, since each base pair in DNA corresponds to one of the bases in the RNA messenger. When the ambiguity in the third position of most of the codons is resolved, and when we can distinguish between two quite different sets of codons for arginine, leucine and serine, we shall be able to write down the complete base sequence of the *A* gene—the base sequence that specifies the sequence of the 267 amino acids in the *A* protein of the enzyme tryptophan synthetase.

# 14

# *The Genetics of a Bacterial Virus*

R. S. EDGAR and
R. H. EPSTEIN
*February 1965*

Viruses, the simplest living things known to man, have two fundamental attributes in common with higher forms of life: a definite architecture and the ability to replicate that architecture according to the genetic instructions encoded in molecules of nucleic acid. Yet in viruses life is trimmed to its bare essentials. A virus particle consists of one large molecule of nucleic acid wrapped in a protective coat of protein. The virus particle can do nothing for itself; it is able to reproduce only by parasitizing, or infecting, a living host cell that can supply the machinery and materials for translating the viral genetic message into the substance and structure of new virus particles. Since a virus is an isolated packet of genetic information unencumbered by the complex supporting systems characteristic of living cells, it is a peculiarly suitable subject for genetic investigation. One can study the molecular basis of life by identifying the individual genes in viral nucleic acid and learning what part each plays in the formation of virus progeny. That is what we have been doing for the past four years, working with the T4 bacteriophage, a virus that infects the colon bacterium *Escherichia coli.*

The T4 virus is one of the most complex viral structures. About .0002 millimeter long, the T4 particle consists of a head in the shape of a bipyramidal hexagonal prism and a tail assembly with several components. The head is a protein membrane stuffed with a long, tightly coiled molecule of deoxyribonucleic acid (DNA). The protein tail plays a role in attaching the virus to the host bacterial cell and injecting the viral DNA through the cell wall. Six tail fibers resembling tentacles bring the virus to the surface of the cell; a flat end plate fitted with prongs anchors the virus there as the muscle-like sheath of the tail contracts to extrude the viral DNA through a hollow core into the cell.

Within a few minutes after the DNA enters the bacterium the metabolism of the infected bacterial cell undergoes a profound change. The cell's own DNA is degraded and its normal business —the synthesis of bacterial protein— ceases; synthetic activity has come under the control of the viral DNA, which takes over the synthesizing apparatus of the cell to direct the synthesis of new types of protein required for the production of new virus particles. The first proteins to appear include enzymes needed for the replication of the viral DNA, which has components not present in bacterial DNA and for the synthesis of which there are therefore no bacterial enzymes. Once these "early enzymes" are available the replication of viral DNA begins. Soon thereafter a new class of proteins appears in the cell: the proteins that will be required for the head membrane and tail parts. About 15 minutes after the viral DNA was first injected new viral DNA begins to condense in the form of heads; protein components assemble around these condensates and soon whole virus particles are completed. For perhaps 10 minutes the synthesis and assembly of DNA and protein components continue and mature virus particles accumulate. The lysis, or dissolution, of the infected cell brings this process to an abrupt halt. Some 200 new virus particles are liberated to find new host cells to infect and so repeat the cycle of reproduction.

The remarkable sequence of synthesis, assembly and lysis is directed by the message borne by the genes of the viral DNA. Each gene is a segment of the DNA molecule, a twisted molecular ladder in which the rungs are pairs of nitrogenous bases: either adenine paired with thymine or guanine paired with cytosine. (In T4 DNA the cytosine is hydroxymethyl cytosine.) The sequence of base pairs in the DNA molecule, like

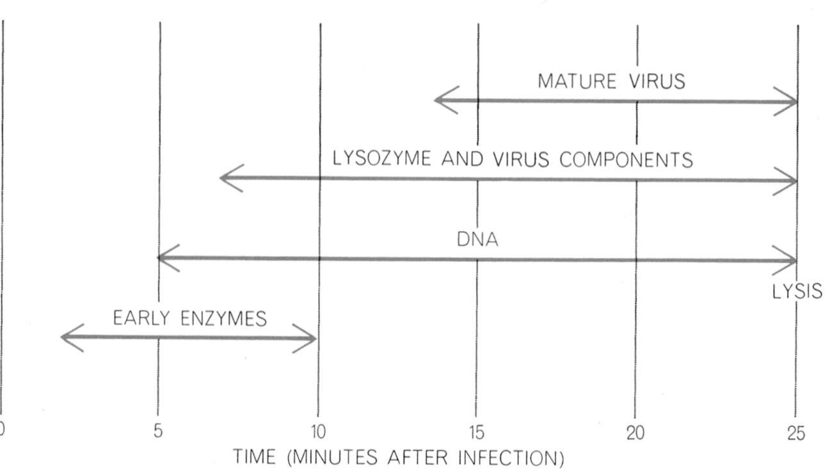


**VIRUS INFECTION of a colon bacterium (at 37 degrees centigrade) proceeds on schedule, with the sequence of syntheses leading up to the lysis, or dissolution, of the host cell.**

the sequence of letters in a word, spells out the information for the assembly of amino acids into protein molecules; a gene is defined as a segment of DNA sufficient to encode a single protein molecule. Since the average protein molecule consists of about 200 amino acid units and the code of DNA requires three base pairs per amino acid, the average length of a gene should be about 600 base pairs. Since there are about 200,000 base pairs in a molecule of T4 DNA, we began by assuming that the molecule contains several hundred genes and initiates the production of several hundred proteins in the host cell. Our task was first to map the location in the T4 DNA molecule of as many genes as possible and then to associate these genes with specific functions.

In order to identify a gene, map its location and learn its function one must find a gene that has undergone mutation: a molecular mistake that occurs like a typographical error in the sequence of base pairs and results either in genetic nonsense, meaning the inability to form protein, or in "missense," meaning the formation of faulty protein. Once a mutation occurs it is copied in successive replications of the DNA and reveals itself by its malfunction in protein synthesis. A mutation therefore serves as a marker for a gene. Moreover, by comparing the growth of a mutant strain of an organism with the growth of a "wild type," or normal, strain one can often infer the normal function of the gene under examination.

The trouble is that most mutations important enough to be recognized and studied are lethal; that is, they result in offspring that cannot survive, or at least cannot reproduce. How, then, can one study lethal mutations? In advanced plants and animals there are two copies of every gene, and it is possible to study "recessive" mutations that are lethal only when they happen to occur in both copies. Less advanced forms of life such as molds, bacteria and viruses, however, have only one copy of each gene, so some other method of studying lethal mutations must be found.

One such method was developed by George W. Beadle and Edward L. Tatum for the study of mutations in the genes of molds and bacteria. The genes that can be investigated by this method are those that direct the synthesis of enzymes required for the formation of nutrients, such as amino acids and vitamins, that are essential to the mold or bacterial cell. In these cases a mutation, although inherently lethal, will not pre-

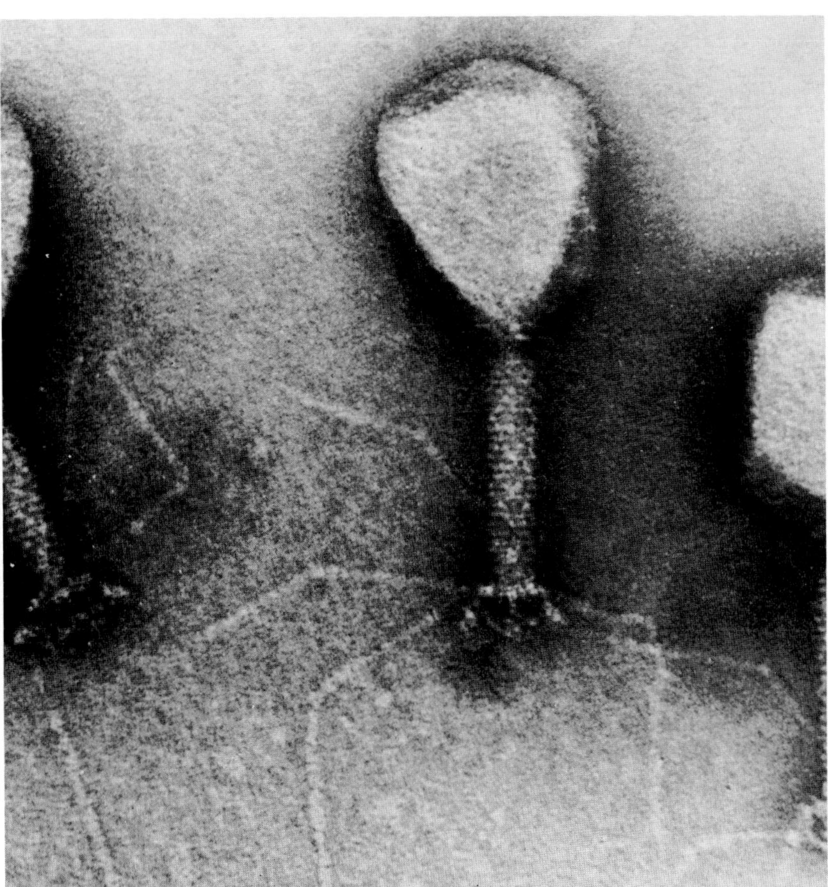

**T4 BACTERIOPHAGE** is enlarged about 300,000 diameters in an electron micrograph made by Michael Moody of the California Institute of Technology. The preparation was negatively stained with electron-dense uranyl acetate, which makes the background dark.

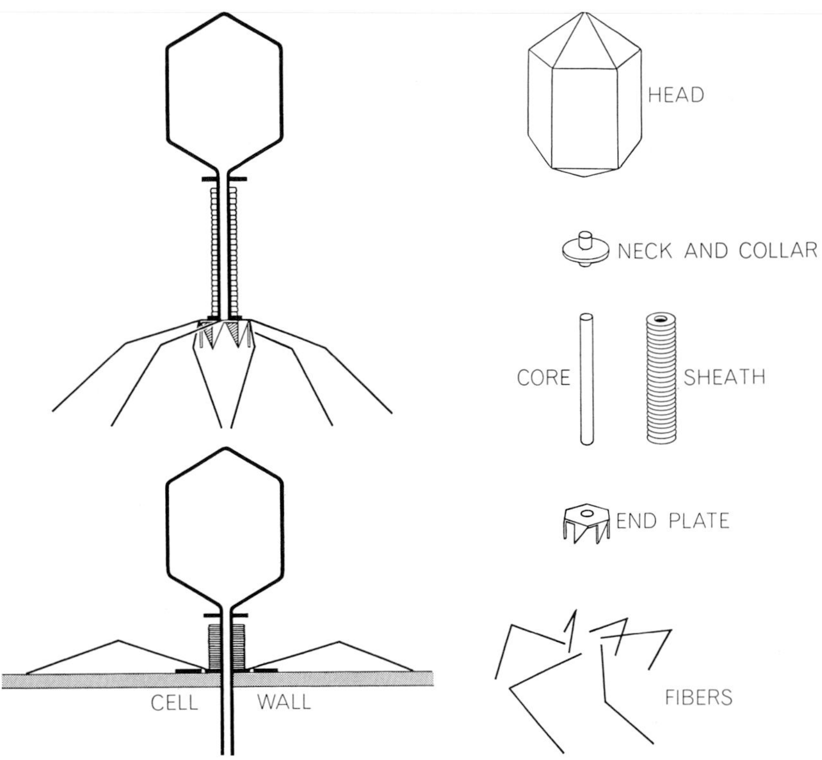


**T4 COMPONENTS** are diagrammed. A complete virus particle is shown at top left. Below it is a particle attached to a bacterial cell wall, with its sheath contracted and its hollow core penetrating the cell wall. The various components are shown separately at the right.

ROLE OF INDIVIDUAL GENES of the T4 bacterial virus was investigated by the authors. This electron micrograph made by E. Boy de la Tour of the University of Geneva shows a complete virus particle with its hexagonal head and springlike tail assembly (*upper center*) and a number of "polyheads": cylindrical tubes of hexagonally arranged protein subunits that were not assembled into virus heads. The failure in assembly is due to a mutation in gene No. 20. The enlargement is about 270,000 diameters.

vent cell growth if the missing nutrient is supplied by the experimenter: it is a "conditional" lethal mutation. Such mutations are restricted to genes whose function can be supplanted by the experimenter. Our aim is to study mutations that affect the synthesis and assembly of virus components, and we had no way of supplying proteins or pieces of virus to infected cells. We needed other kinds of conditional lethal mutations.

One of us (Edgar), working at the California Institute of Technology, has dealt primarily with a class of mutations that are temperature-sensitive: they render the gene inactive at one temperature but not at temperatures a few degrees lower. An example of such a gene in a higher animal is the gene that controls the hair pigment in Siamese cats. The gene is inactive at body temperature, with the result that most of the cat's coat is white. On the cooler parts of the body—the paws, the tip of the tail, the nose and the ears—the gene becomes functional and the hair is pigmented. Of course, this defect is not lethal to the cat, but similar mutations in genes with functions essential to an organism are conditional lethal mutations if one can control the temperature. A strain of T4 bacteriophage with a temperature-sensitive lethal mutation, for example, grows perfectly well if it is incubated on bacteria at 25 degrees centigrade but not if it is incubated at 42 degrees. Temperature-sensitive mutations can occur in many different genes, since what they do is simply render a protein—regardless of its particular function—more readily inactivated by heat. They apparently do so by substituting one amino acid for another at some sensitive point in the structure of the protein molecule; in other words, they are "missense" mutations.

Epstein has worked with another class of conditional lethal mutations: the "amber" mutations, which he developed at Cal Tech and has studied primarily in the laboratory of Edouard Kellenberger at the University of Geneva. (We call them the amber mutations because they were discovered with the help of a graduate student named Bernstein, and *bernstein* is the German word for "amber"; it is often safer to give a new discovery a silly name than a speculatively descriptive one!) In these mutations the conditional property is not temperature-sensitivity but the ability of a virus to grow in certain host cells. Whereas the wild-type T4 virus grows equally well in colon bacteria of strains *B* and *CR*, amber mutants grow only in *CR*. Apparently only *CR* bacteria are able to translate the mutant message into protein properly; in strain *B* the mutant gene is translated into protein only up to the point of mutation and the resulting protein fragment is inactive. In other words, amber mutations are translated as "nonsense" in strain *B* but as "sense," or at worst as "missense," in strain *CR*. Again we could expect the amber mutations to occur in many different genes, since these mutations affect the overall translatability of any affected gene rather than the ability of specific genes to direct the synthesis of specific proteins.

Mutations arise at random in the normal course of virus infection and reproduction; we amplify the process by treating virus particles with one of a variety of chemical mutagens. We then plate the virus on cultures of colon bacteria. Any amber or temperature-sensitive mutant reveals itself by its failure to grow under "restrictive" conditions, that is, on strain *B* in the case of an amber mutation or at 42 degrees in the case of a temperature-sensitive mutation. In this manner we have isolated more than 1,000 amber and temperature-sensitive mutant strains. The mutations, however, occur at random at various sites in the many genes of the viral DNA. Since we are trying to identify genes, not merely mutations, we need to determine which mutant strains contain mutations affecting the same gene.

We do this by performing complementation tests [*see illustration on next two pages*]. The test consists in infecting bacteria simultaneously with two mutant viruses under restrictive conditions in which each mutant alone would be unable to grow in the bacterial cells.

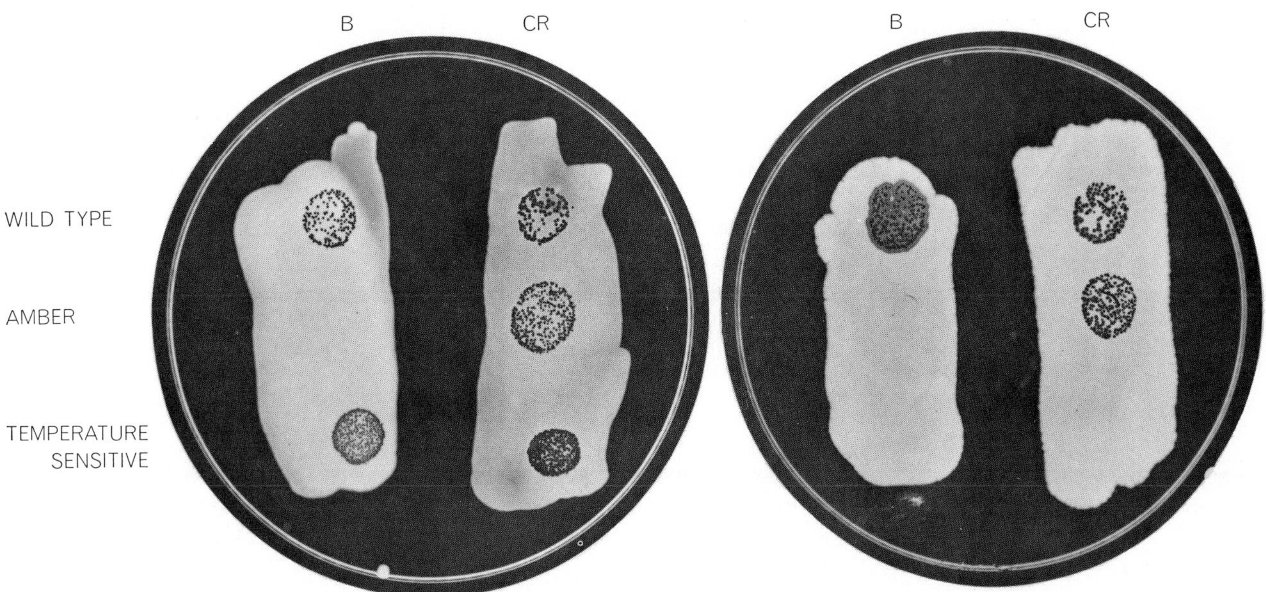


**GROWTH CHARACTERISTICS** of "wild type" virus and "amber" and temperature-sensitive mutants are compared. The photographic prints were made by exposing actual Petri dishes in an enlarger. On each dish bacterial strains *B* and *CR* had been streaked, with drops of virus suspensions placed on each streak, and the plates had been incubated at two temperatures, as shown. The amber mutants grew only on strain *CR*, the temperature-sensitive mutants grew only at 25 degrees C. and the wild-type virus grew under all conditions.

Infection of strain *B* bacteria at high temperature is restrictive for both amber and temperature-sensitive mutants. If, under these restrictive conditions, a yield of progeny virus is produced from cells infected by two mutants, the mutations must be complementary defects. Each mutant can perform the function the other mutant is unable to perform, and we can conclude that the two mutations are in different genes. If, on the other hand, the doubly infected bacteria produce no progeny virus, the two mutant strains must be unable to complement each other. Their mutations must affect a common function, and we conclude that they are in the same gene.

When complementation tests are applied to amber mutants, the results are clear-cut. These mutants, when tested against one another, fall into mutually exclusive classes: mutations in different genes result in full complementation no matter how they are paired, whereas mutations within the same gene fail to complement each other no matter how they are paired. In the case of temperature-sensitive mutants, however, the results are equivocal: some of the mutants display "intragenic" complementation and yield virus progeny even under restrictive conditions. Apparently two different "missense" mutations can give rise to "hybrid" proteins that, although altered, are nevertheless complete and functional. The amber mutants, as we have mentioned, involve "nonsense" mutations and therefore would not be expected to show intragenic complementation. Since both amber and temperature-sensitive mutations occur in many genes, the ambers provide a check on the equivocal temperature-sensitive results.

By means of complementation tests we subdivided our many hundreds of amber and temperature-sensitive mutants into separate groups, each of which identifies one gene of the virus; our mutations turned out to be located in 56 different genes. The next step was

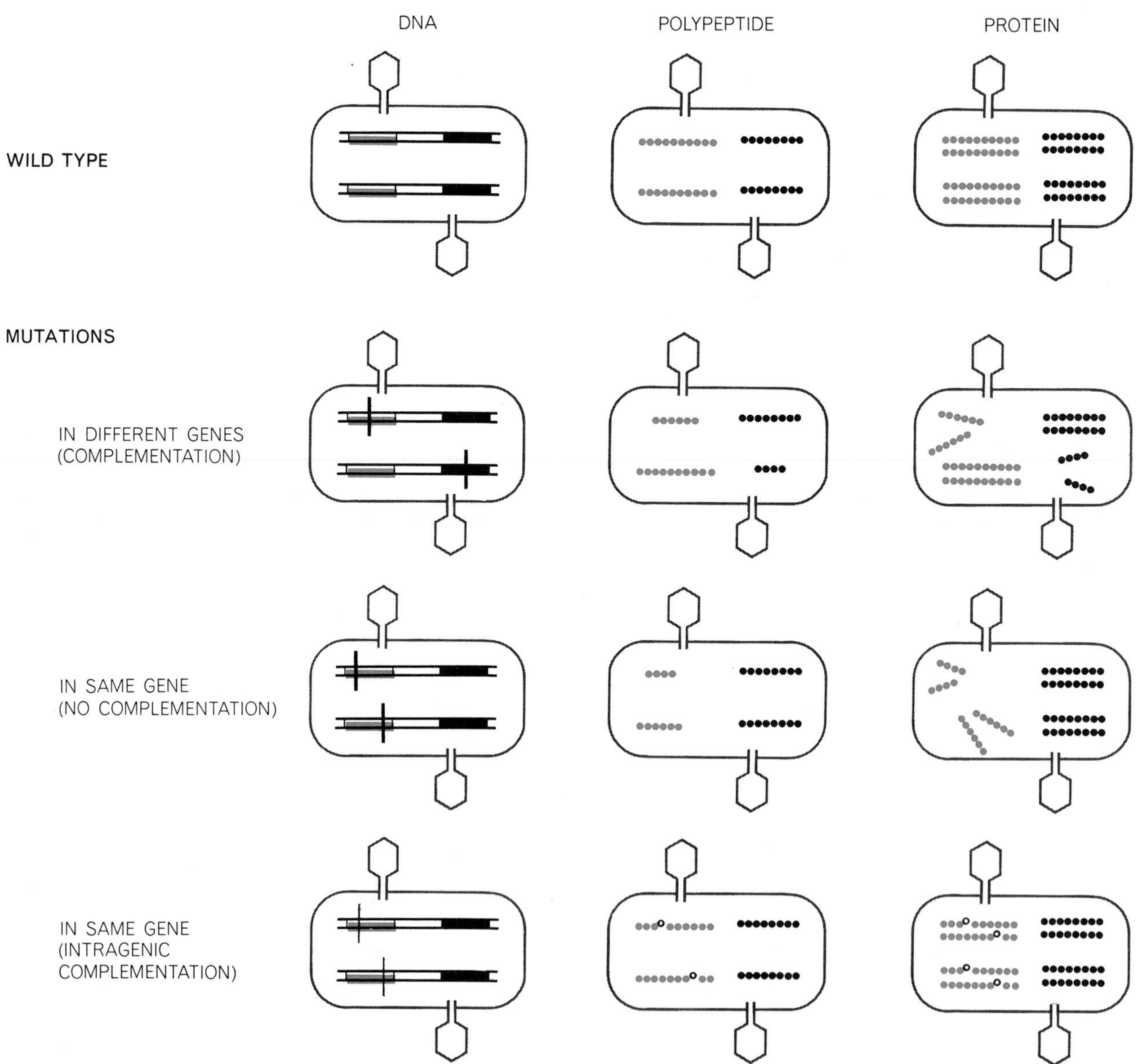


**COMPLEMENTATION TEST identifies individual genes. The top row shows how, in wild-type virus, two genes of the deoxyribonucleic acid (DNA) molecule (*color and black*) might direct the synthesis of two polypeptide chains that form proteins and end up as virus components. An infection with wild-type virus results in a large number of plaques on a bacterial culture (*right*). If two mutations being tested occur in different genes, one gene makes the protein the other cannot make; they complement each other and virus particles are produced (*second row*). Two mutations in the same gene will ordinarily not complement each other, as**

to locate those genes, and four that had been identified earlier by other investigators, on a genetic map—a representation of the position of the genes in relation to one another.

Such a map is constructed on the basis of recombination, the process by which the genetic material from two parents is mixed in the progeny. In viruses recombination can occur when viruses of two different strains infect the same cell. The mechanism of recombination is still poorly understood, but it probably involves the breakage of DNA molecules and the reassociation of pieces derived from both strains to form a new "hybrid" DNA molecule. Recombination between two different mutants can result in some virus progeny that carry both mutations and in some wild-type viruses with no mutations. The wild-type recombinations can be recognized by their ability to multiply under restrictive conditions. The closer together two genes are on the DNA molecule, the less likely it is that breaks and reunions will occur between them, so the frequency of recombination is a measure of the distance between the two genes. We infect a bacterial culture with two strains that are mutant, say, in genes *a* and *x* respectively, and incubate it under "permissive" conditions in which both mutants can grow. Among millions of virus progeny of such a cross there will be some wild-type recombinants. By plating measured amounts of the progeny under permissive and under restrictive conditions we can determine what fraction of the progeny are wild-type. From this we calculate the frequency of recombination between genes *a* and *x* and thus the distance between them.

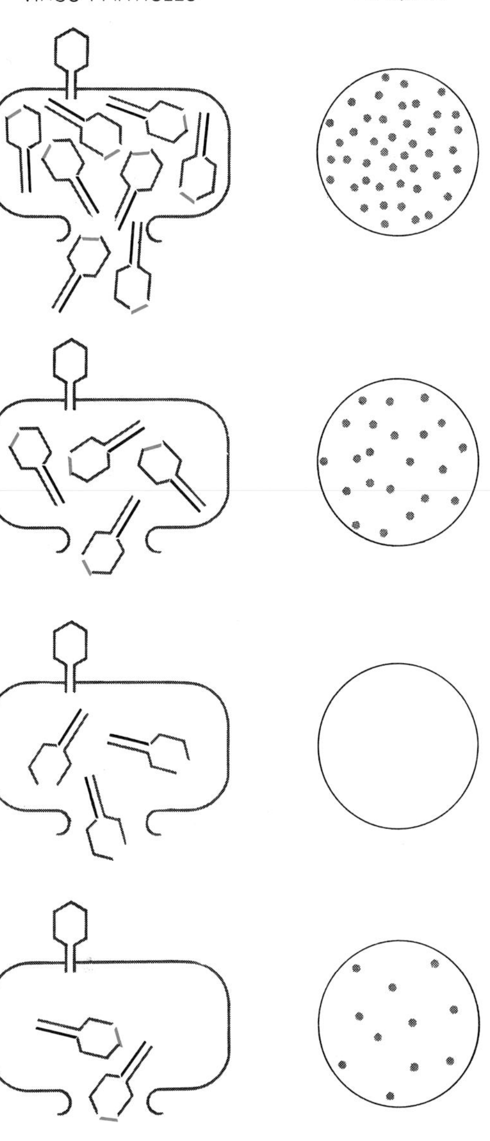


**seen in the third row. In some cases involving the temperature-sensitive mutants, however, "intragenic complementation" occurs: some virus is produced in spite of errors in polypeptide synthesis (*bottom row*).**

By plotting the results of hundreds of crosses we constructed a genetic map of the T4 DNA molecule [*see illustration on next page*]. A remarkable feature of the map is that it has no "ends" and must be drawn as a linear array that closes on itself—a circle. This is rather surprising, since it has been established by electron microscopy and other means that the actual form of the T4 DNA molecule is that of a strand with two ends. (Just to confuse matters, some other viruses do have circular molecules!) Why the map should be circular is not yet known with certainty. It is probably because different viral DNA molecules have different sequences of genes, all of them circular permutations of the same basic sequence. In alphabetical terms, it is as if one DNA were *a, b . . . y, z* while another were *n, o . . . z, a . . . l, m*. In the second case *z* and *a* would be "closely linked" and would map close together.

Recombination occurs between mutation sites within genes as well as between genes, so we have been able to make a number of "intragenic" maps. These show that the genes are not uniform in size. Although most of them are quite small, each accounting for about half of 1 percent of the length of the map, gene No. 34 is about 20 times larger, and genes No. 35 and No. 43 are also outsized. Average gene size is therefore not a precise indicator of the number of genes in the virus. It looks as if the mutations discovered to date cover about half of the map, so we conclude that roughly half of the genes remain to be discovered. Unfortunately a kind of law of diminishing returns seems to be taking effect: for every 100 new mutants we isolate and test we are lucky to discover one new gene. Apparently amber and temperature-sensitive mutations are rare in the genes that are as yet undiscovered. We are devising new techniques with which to seek them out, but there will probably be a number of genes that are simply not susceptible to the conditional-lethal procedure. This could be because neither amber nor temperature-sensitive mutations occur in them or because, if they do occur, the loss of gene function is not lethal and the mutation therefore goes unnoticed.

While attempting to uncover the remaining genes, we have begun to determine the functions performed by the genes already identified. The mutants were originally detected because of their inability to produce progeny virus under restrictive conditions. In order to investigate the nature of the abortive infections more closely in an attempt to find out just what step in the growth cycle goes awry, we have employed a large number of mutants involving several different defects in each of the 60 genes. We chose just a few aspects of bacteriophage growth to examine, largely because they are easy to observe or measure and because they provide information on the major events of the cycle.

1. Can the infecting mutant virus accomplish the disruption of the bacterial DNA molecule? With the phase microscope one can observe whether or not the bacterial nucleoid, or DNA-containing body, disintegrates. So far every mutant we have tested has been able to disrupt the host DNA, so it is clear that in every case the infective process is at least initiated.

2. Does DNA synthesis occur in the infected cell? After the disruption of the bacterial nucleoid all host functions cease. Any new DNA that is revealed in chemical tests is viral DNA and an indication that the genes responsible for DNA synthesis are operative.

3. Do the infected cells lyse at the normal time? During the last half of the growth cycle an enzyme, lysozyme, is synthesized that is responsible for disrupting the cell wall. Normal lysis indicates that this enzyme is synthesized and does its work.

4. Are complete virus particles or components such as heads and tails produced in the infected cells? Electron

microscopy tells us the extent to which protein virus components have been synthesized and assembled in an infected cell.

Our data indicate that the various genes can be assigned to two groups. There are genes that appear to govern early steps in the infective process, as indicated by the fact that they affect DNA synthesis, and genes that appear to govern later steps, as indicated by their role in the maturation of new phage particles.

The major class of "early" genes includes those that are essential if any DNA synthesis is to occur. Mutations in these genes must cause the loss of some enzyme function necessary for DNA synthesis. Seven genes of this type have been identified, the precise function of one of which has been determined: John M. Buchanan and his co-workers at the Massachusetts Institute of Technology have found that gene No. 42 controls the synthesis of an enzyme necessary for the manufacture of hydroxymethyl cytosine, one of the four bases in the T4 DNA molecule.

The "no DNA" mutants reveal an interesting regulatory feature of gene action. Not only is there no DNA synthesis in cells infected by these mutants, but also the cells do not lyse and no virus components are made. It appears that the decoding of the late-functioning genes depends somehow on the prior synthesis of viral DNA. Buchanan's group has found, moreover, that in these cells any of the early enzymes that are

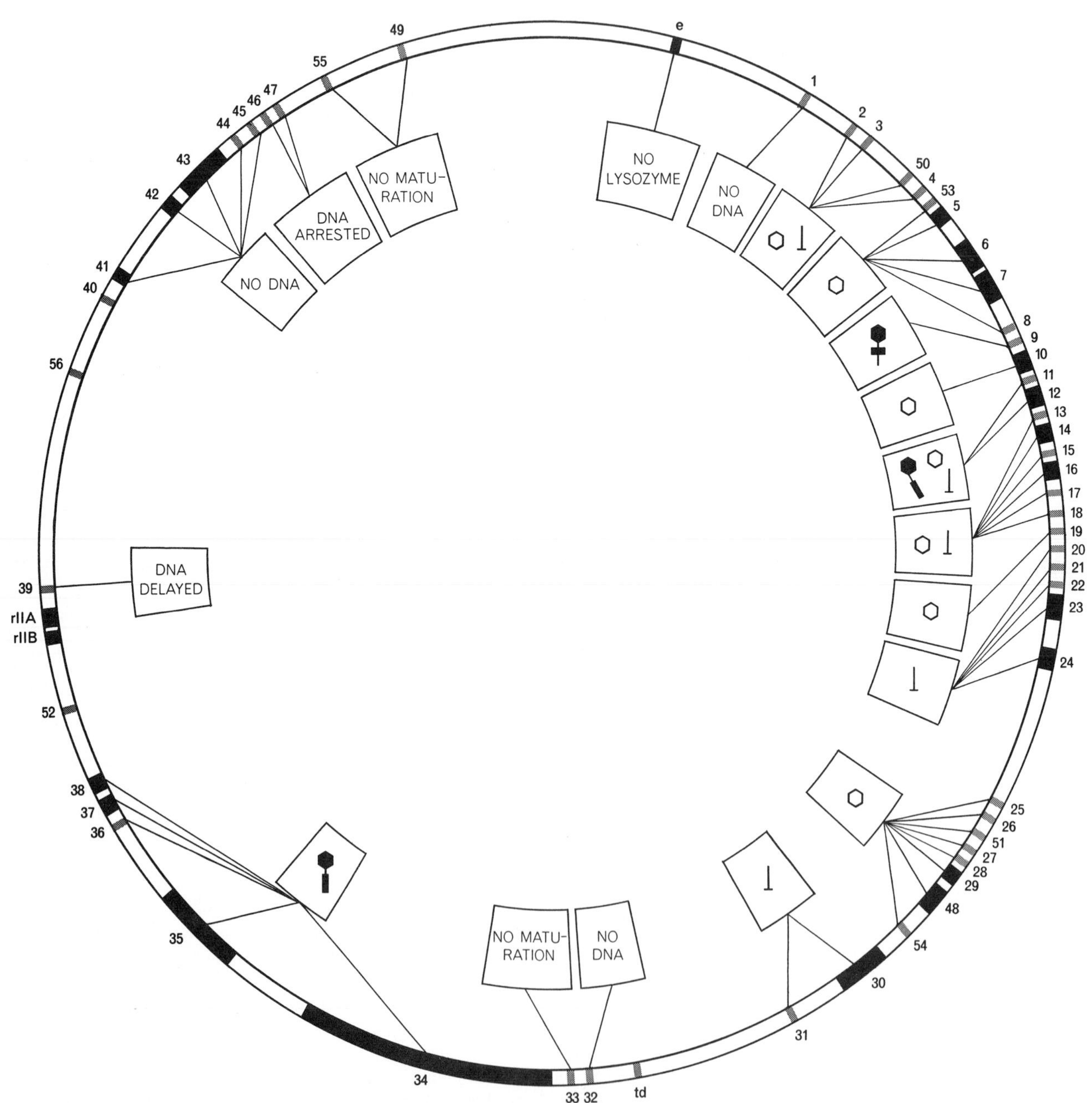


**GENE MAP OF T4 shows the relative positions of the 60 genes identified to date and the major physiological properties of mutants defective in various genes. Minimum length is shown for some of the genes (*black segments*) but is not yet known for others (*gray*). The boxes indicate deficiencies in synthesis associated with mutations in some genes or, in the case of other genes, the components that are present in defective lysates of corresponding mutants. There may be no DNA synthesis or it may be delayed or arrested. There may be no virus maturation at all. Synthesis and lysis may proceed normally but, as shown by the symbols, incomplete viruses may be produced, ranging from heads or tails only to complete particles lacking tail fibers (*genes No. 34 through No. 38*).**

not eliminated by the particular mutation continue to be synthesized well beyond the normal shutoff time of 10 minutes. It appears, then, that in the absence of normal DNA synthesis some timing mechanism for switching early genes off and turning late ones on fails to function.

Among the early genes some others have been found that appear to delay or modify DNA synthesis or to block the activity of late genes without disturbing DNA synthesis, but the manner in which they function is still obscure.

Most of the genes—about 40 of those we have identified so far—clearly play roles in forming and assembling the virus components. Mutations in these morphogenetic genes seem not to affect the synthesis of DNA or the lysis of the cell. What happens is that no infective progeny virus particles are produced, only bits and pieces of virus. For example, mutations in genes No. 20 through No. 24 result in the production of normal numbers of virus tails but no heads; mutations in the segment from gene No. 25 through No. 54 produce heads but no tails; mutations in genes No. 34 through No. 38 produce particles that are complete except for the tail fibers. Presumably the defective gene in each case is concerned with synthesis or assembly of the missing component.

A glance at the map [*opposite page*] shows that the arrangement of the genes in the DNA molecule is far from random: genes with like functions tend to fall into clusters. Similar clusters of certain genes in bacteria are called "operons," and all the genes within an operon function together as a unit under the control of separate regulatory genes. There is no indication that the clusters in viral DNA act as operons; the available evidence suggests, indeed, that each gene acts independently. Still, it is difficult to believe the clustering does not reflect in some meaningful way a high degree of coordination in the activities of the genes.

The large number of genes associated with morphogenesis is of particular interest. What do all these genes do? There is evidence that only a few of them are concerned with the actual synthesis of protein components. For example, the head of the virus particle is made up of about 300 identical protein

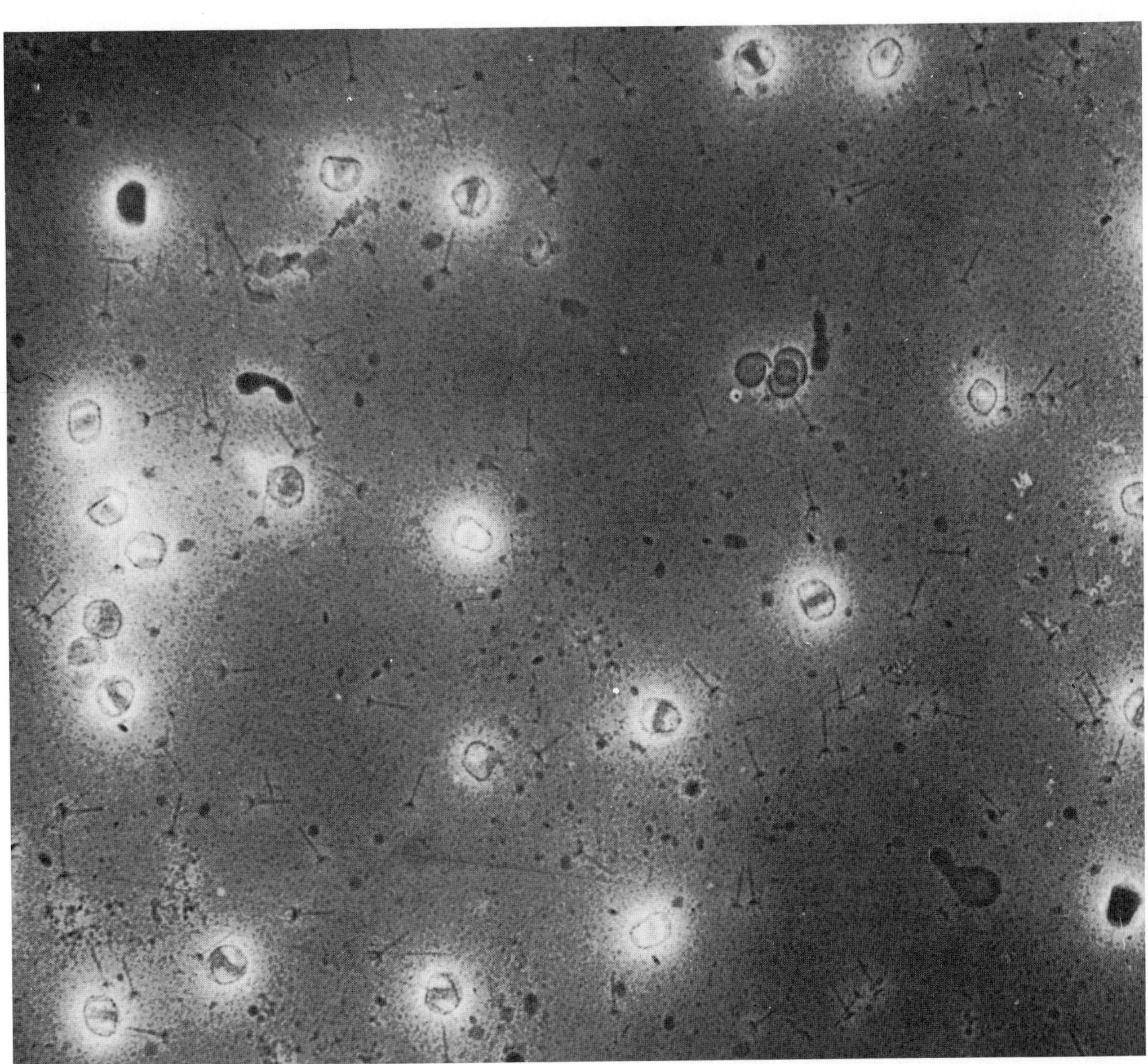

DEFECTIVE LYSATE of a temperature-sensitive strain mutant in gene No. 18 is enlarged about 60,000 diameters in this electron micrograph made by Edgar. Heads and tails have been formed but not assembled, and most of the heads are empty of DNA. The lysate was negatively stained with phosphotungstic acid, which filled the empty virus heads, and the exposed plate was printed as a negative.

subunits aggregated in a precise pattern; if there are any other protein molecules in the head membrane, they must be present in very small amounts. Yet at least seven genes and probably more are involved in the production of virus heads. Sydney Brenner and his associates at the University of Cambridge have found that just one of these genes, No. 23, is responsible for the actual synthesis of the protein subunits; cells infected with mutants defective in any other genes contain normal numbers of the subunits. The other genes must therefore be concerned with the assembly of the units rather than with their synthesis. When gene No. 20 is defective, for instance, the subunits assemble in the form of long cylindrical tubes instead of forming hexagonal heads [*see illustration on page 130*].

At this time we can only speculate as to the precise roles of the many morphogenetic genes. One possibility is that the proteins made by all of them are incorporated into the virus but in minor amounts that have escaped detection. Such minor components might be necessary to serve as the hinges, joints, nuts and bolts of the virus. Another possibility is that the proteins made by some of these late genes do not appear in the completed virus at all but instead play accessory roles in the assembly process —perhaps "gluing" subunits together in the specific configurations necessary for the proper construction of the virus. This notion of accessory morphogenetic genes is somewhat novel to many students of virus structure, who have generally believed that the assembly of viruses comes about through a spontaneous "crystallization" of subunits. In other words, it has been assumed that the form of a virus is inherent in its structural components. Although this may be true of viruses with simple spherical or cylindrical forms, it may not be true of viruses with more complex forms. The study of the effects of mutations on the assembly of viruses should serve as a powerful tool with which to explore this problem.

The relation between genes and form should be of general interest. Life is characterized by the complexity of its architecture. This complexity is manifested at all levels of organization, from molecules to the assemblages of specialized cells that make up higher animals and plants. The building blocks of all living things are, like virus particles, intricate molecular aggregations. Knowing how a bacteriophage such as T4 is put together may help us to understand the origins of form in all living systems.

# III
# *From Gene to Organism*

# III

# *From Gene to Organism*

## INTRODUCTION

We turn now to an old riddle—old but ever fresh. The successive mitotic divisions through which the egg becomes a multicellular organism seem designed to guarantee that all of the somatic cells deriving from the egg will be identical in their genic composition. The genes are the ultimate seats of specific synthesis; they determine what biochemical jobs the cell can do—what it can become. The cells of the organism become very different from each other: one may specialize in producing hemoglobin, and another may become a neuron, an islet cell in the pancreas, a spermatogonial cell, or any of thousands of other distinctive cell-types. How does this come about if the cells all have the same genic composition? In the nicely regulated and interrelated processes of normal development, how is the blueprint for the organism represented and realized in the germinal material and its products?

The articles in this section bridge a five-year period, from 1964 to 1969. They bridge, too, the levels of increasing complexity with which our subject must deal—from a turning-on (or off) of sets of gene activities in a bacterial cell to the mechanisms underlying organ formation in a mammal. We begin with Jean-Pierre Changeux's "The Control of Biochemical Reactions." The frame of reference here is largely that of a bacterial cell as an organism—an "automatic chemical factory designed to make the most economical use of the energy available to it." We do not ordinarily think of bacteria within a genetically uniform culture as "differentiating" in the same sense as do the cells of a complex multicellular organism. Nevertheless, genetically identical bacteria may become biochemically distinct when they are placed in different environments, and the story of how they do so, as told by Changeux, is the story of one of the most fruitful sets of investigation in modern biology.

Then on to the evidences of localized activity in the chromosomes of insect cells. In the article "Chromosome Puffs," Wolfgang Beermann and Ulrich Clever paint the visual picture of differential gene activation—the patterns in place and time of the sequences from DNA to RNA to product. Their study carries us a large step further, too, as it relates the action of the hormone ecdysone to the activity of single chromosomal sites in the extensive differentiation that occurs during the molting period.

The scope of our understanding of the chemical integrators of organism development—the hormones as activators of sets of genes—is further broadened in the article by Eric H. Davidson, "Hormones and Genes." Many hormones exert their effects by activating numbers of separate but functionally coordinated genes. But the effects of a particular hormone, such as estrogen, are different for a liver cell, for example, than for a uterine cell. You will ponder, with the author, the question: "How are the sets of genes that are activated by a given hormone selected?"

The problems of morphogenesis are viewed in a quite different context in William B. Wood and R. S. Edgar's account of "Building a Bacterial Virus." The work is satisfying in two rather different connections. First, the bacterial virus is itself an organism, and the authors use genetic techniques to establish the separate assembly lines for tail, head, and tail-fibers, which combine to form active virus particles. Second, their consideration of virus morphogenesis encourages us to think about the assembly of other supermolecular aggregates within cells—the cell organelles that play so prominent a part in determining what the cell is, what it becomes, and what it does.

This section began with a premise that mitosis seems designed to guarantee genetic identity among the cellular progeny of the fertilized egg. You may have objected that this premise has its exceptions. For example, units of cytoplasmic DNA need not be regularly distributed to daughter cells. Do they segregate into diverse cell lines, and do they therefore provide a basis for irreversible cell differentiation as some of them are lost from particular cell lines? Other possibilities may have come to the reader's attention; for instance, you may have read about the systematic chromosome-elimination that takes place during the development of gall midges (see "How Cells Specialize," by Michail Fischberg and Antonie Blackler, SCIENTIFIC AMERICAN Offprint 94), or the hypothesis that some kind of somatic scrambling of nucleic information may regulate part of the structure of antibodies (see "The Structure of Antibodies," by R. R. Porter, SCIENTIFIC AMERICAN Offprint 1083). In the fifth article of this section, "Transplanted Nuclei and Cell Differentiation," J. B. Gurdon tells how, through the ingenious technique of nuclear transplantation, one can establish that the nuclei of fully differentiated amphibian cells retain all of the potentialities for the development of whole normal animals. Specialization of cells, therefore, does not require the elimination of chromosomes or of genes in either the nucleus or the cytoplasm; it involves mechanisms by which cytoplasmic components bring about selected changes in gene activity in cells containing all of the egg's original genetic information.

Ernst Hadorn's article on "Transdetermination in Cells" points strongly to the same kind of conclusion. It begins with an account of transplanting imaginal disks of *Drosophila* larvae into the body cavities of other larvae. The individual cells of the disks prove to be already programmed for specific differentiation. Appropriate cells of different types move about until they recognize one another, and come together to form determined structures. Long before these cells show any visible differences, their future course of differentiation and interaction seems to have been established. They can be maintained as cultures in the abdomens of adult flies, multiplying indefinitely without revealing their apparently fixed potentialities; but when they are again transplanted back into the larva, and come under the influences of the larval hormones of metamorphosis, they take their usual, set paths. The major part of Hadorn's article, however, deals with switches to new paths of cell heredity, in which cells display the coordinated actions of quite different sets of genes. These switches may take place even in cell populations that originate from a single ancestral cell. How they do so is a key current question of developmental biology. The *Drosophila* transplantation technique offers a promising set of tools for investigating this question.

We end this section with Norman K. Wessells and William J. Rutter's "Phases in Cell Differentiation." Here we are at home among the animals most like ourselves; we deal with the differentiation of the mammalian pancreas. The evidence suggests a series of "regulatory transitions," each initiating a new step on the stairway of differentiation. These developmental phases involve coordinated switch-ons of hundreds of genes, as others are switched off. Tissue culture techniques are shown to offer substantial advantages over studies of these processes in intact organisms. By this point the reader will find the now-familiar vocabulary of molecular and cellular biology applied to the processes of mammalian organ formation. Descriptive embryology has entered a dynamic new phase.

The articles in this section seem to tell a coherent story. Much has been learned in recent years, and the complex problems of development and differentiation now seem relatively approachable. This does not mean, however, that we now understand how the fertilized egg turns into an integrated organism composed of many very different kinds of cells, working healthfully together. We are only beginning to emerge from the Dark Ages in this challenging part of the field of Biology. The next decade may be our renaissance.

# 15

# *The Control of Biochemical Reactions*

JEAN-PIERRE CHANGEUX
*April 1965*

The analogy between a living organism and a machine holds true to a remarkable extent at all levels at which it is investigated. To be sure, living things are machines with exceptional powers, set apart from other machines by their ability to adapt to the environment and to reproduce themselves. Yet in all their functions they seem to obey mechanistic laws. An organism can be compared to an automatic factory. Its various structures work in unison, not independently; they respond quantitatively to given commands or stimuli; the system regulates itself by means of automatic controls consisting of specific feedback circuits.

These principles have long been recognized in the behavior of living organisms at the physiological level. In response to the tissues' need for more oxygen during exercise the heart speeds up its pumping of blood; in response to a rise in the blood-sugar level the pancreas increases its secretion of insulin. Now analogous systems have been discovered at work within the living cell. The new findings of molecular biology show that the cell is a mechanical microcosm: a chemical machine in which the various structures are interdependent and controlled by feedback systems quite similar to the systems devised by engineers who specialize in control theory. In this article we shall survey the experimental findings and hypotheses that have developed from the viewpoint that the cell is a self-regulating machine.

We can think of the cell as a completely automatic chemical factory designed to make the most economical use of the energy available to it. It manufactures certain products—for example proteins—by means of series of reactions that constitute its production lines, and most of the energy goes to power these processes. Regulating the production lines are control circuits that themselves require very little energy. Typically they consist of small, mobile molecules that act as "signals" and large molecules that act as "receptors" and translate the signals into biological activity.

The elementary machines of the cellular factory are the biological catalysts known as enzymes. The synthesis of any product (for example a specific protein) entails a series of steps, each of which calls for a specific enzyme. Obviously there are two possible ways in which the cell can control its output of a given product: (1) it may change the number of machines (enzyme molecules) available for some step in the chain or (2) it may change their rate of operation. Therefore in order to reduce the output of the product in question the cell may cut down the number of enzyme molecules or inhibit some of them or do both.

An excellent demonstration of such control has been obtained in experiments with the common bacterium *Escherichia coli*. The experiments involved the bacterial cell's production of the amino acid L-isoleucine, which it uses, along with other amino acids, to make proteins. Would the cell go on synthesizing this amino acid if it already had more than it needed for building proteins? L-isoleucine labeled with radioactive atoms was added to the medium in which the bacteria were growing; the experiments showed that when the substance was present in excess, the bacteria ceased to produce it. The amount of the amino acid in the cell in this case serves as the signal controlling its synthesis: if the amount is below a certain level, the cell produces more L-isoleucine; if it rises above that level, the cell stops producing L-isoleucine. Like the temperature level in a house with a thermostatically regulated heating system, the level of L-isoleucine in the cell exerts negative-feedback control on its own production.

How is the control carried out? H. Edwin Umbarger and his colleagues, working in the laboratory of the Long Island Biological Association, found that the presence of an excess of L-isoleucine has two effects on the cell: it inhibits the activity of the enzyme (L-threonine deaminase) needed for the first step in the chain of synthesizing reactions, and it stops production by the cell of all the enzymes (including L-threonine deaminase) required for L-isoleucine synthesis. Curiously it turned out that the two control mechanisms are independent of each other. By experiments with mutant strains of *E. coli* it was found that one mutation deprived the cell of the ability represented by the inhibition of L-threonine deaminase by L-isoleucine; another mutation deprived it of the ability to halt production of the entire set of enzymes. The two mutations were located at different places on the bacterial chromosome. Therefore it is clear that the two control mechanisms are completely separate.

Let us first examine the type of mechanism that controls the manufacture of enzymes. It was Jacques Monod and Germaine Cohen-Bazire of the Pasteur Institute in Paris who discovered the phenomenon of repression: the inhibition of enzyme synthesis by the presence of the product, the product serving as a signal that the enzymes are not needed. The signal substance

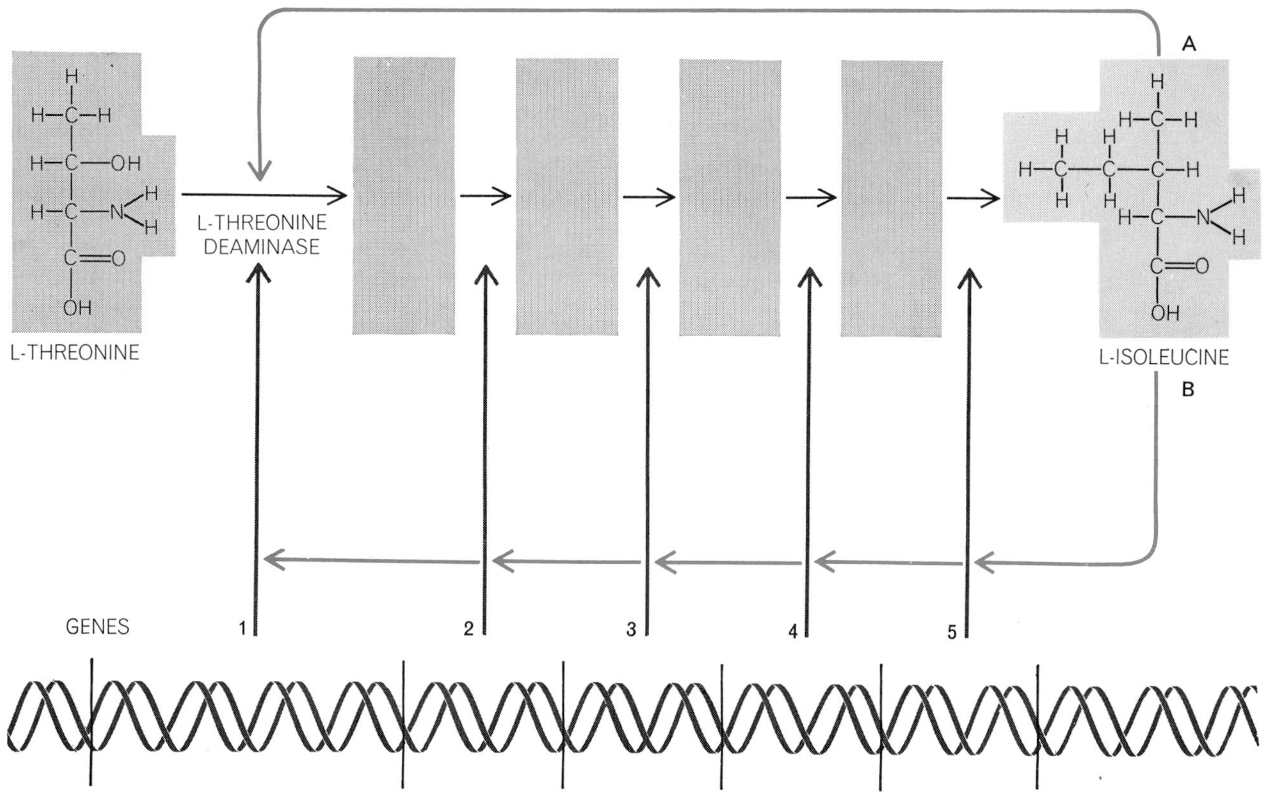


**TWO FEEDBACK SYSTEMS** control the biosynthesis of cell products, as shown here for the synthesis of the amino acid L-isoleucine in the bacterium *Escherichia coli.* The end product of the synthesizing chain acts as a regulatory signal that inhibits the activity of the first enzyme in the chain, L-threonine deaminase (*A*), and also represses the synthesis of all the enzymes (*B*).

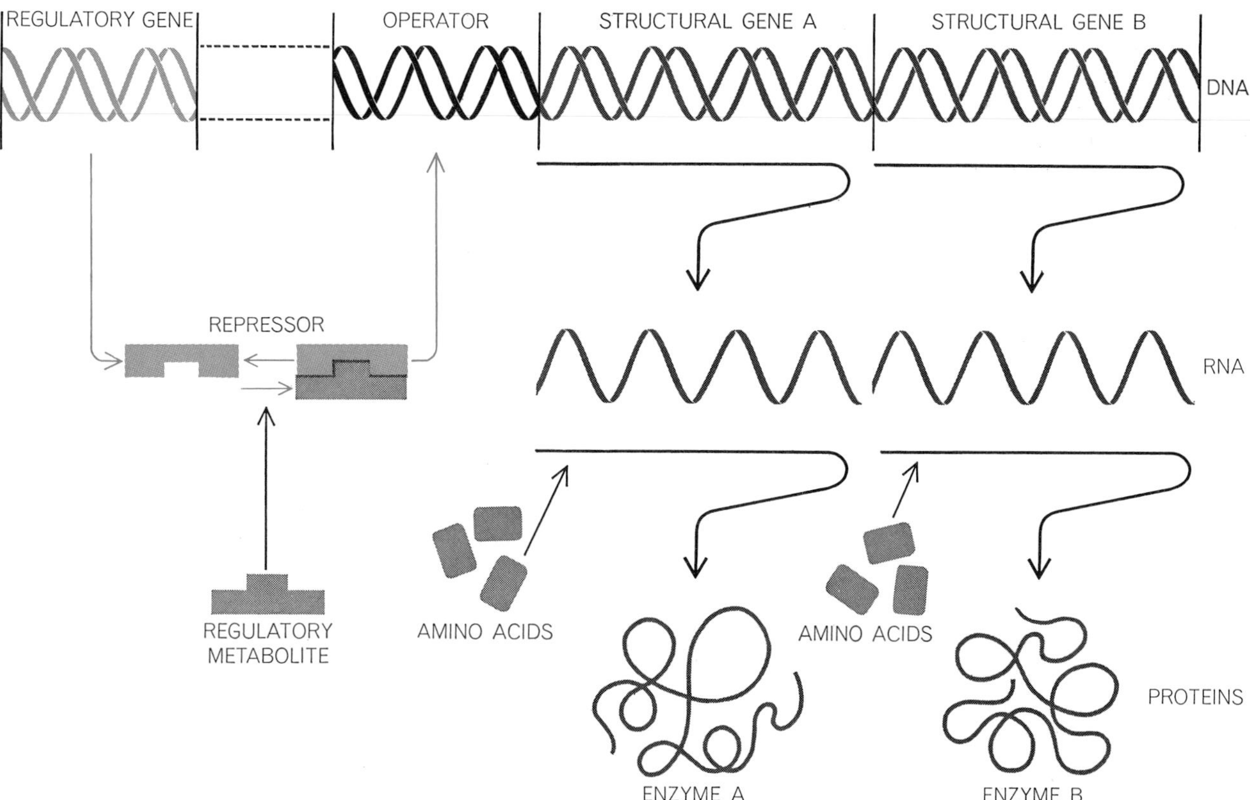


**CONTROL OF PROTEIN SYNTHESIS** by a genetic "repressor" was proposed by François Jacob and Jacques Monod. A regulatory gene directs the synthesis of a molecule, the repressor, that binds a metabolite acting as a regulatory signal. This binding either activates or inactivates the repressor, depending on whether the system is "repressible" or "inducible." In its active state the repressor binds the genetic "operator," thereby causing it to switch off the structural genes that direct the synthesis of the enzymes.

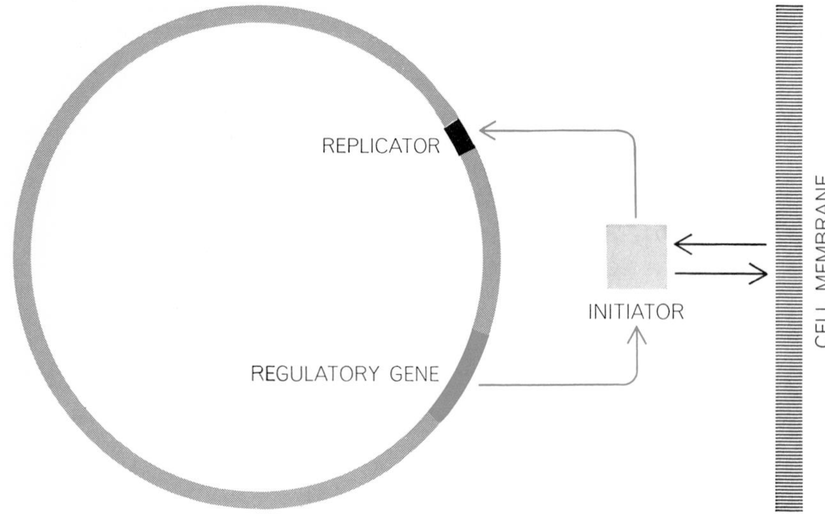


**REPLICATION OF DNA of a bacterial chromosome may be under a control like that of protein synthesis. A regulatory gene directs the synthesis of an "initiator," which receives a signal (perhaps from the cell membrane) that makes it act on the "replicator."**

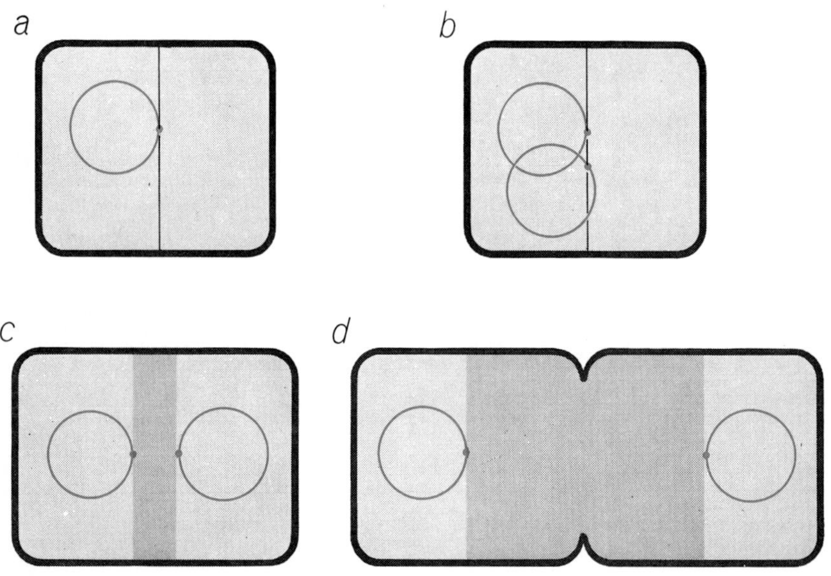


**ROLE OF CELL MEMBRANE in replication is suggested by the fact that a bacterial chromosome is attached to a point on the membrane (*a*). It could be a signal from the membrane that initiates the formation of daughter chromosomes (*b*). Then the membrane begins to grow, separating the points of attachment (*c*) until the cell is ready to divide (*d*).**

in their experiments was the amino acid tryptophan. They found that when the medium in which *E. coli* cells were growing contained an abundance of tryptophan, the cells stopped producing tryptophan synthetase, the enzyme required for the synthesis of the amino acid. This efficient behavior has since been demonstrated in many cells, not only bacteria but also the cells of higher organisms. The addition of an essential product to the cells' growth medium results in a negative-feedback signal that causes them to stop synthesizing enzymes they do not need.

In other systems the response of the cell is not negative but positive. We have been considering signals that repress the synthesis of enzymes; the cell can also respond to signals calling on it to produce enzymes. An example of such a situation is that the cell is confronted with a compound it must break down into substances it requires for growth.

The "induction" of enzyme synthesis in cells was discovered at the turn of the century by Frédéric Dienert of the Agronomical Institute in France. He was studying the effect of a yeast (*Saccharomyces ludwigii*) in fermenting the milk sugar lactose. He found that strains of the yeast that had been grown for several generations in a medium containing lactose would begin to work on the sugar immediately, causing it to start fermenting within an hour. These cells had a high level of lactase, an enzyme that specifically breaks down lactose. Yeast cells that had not been grown in lactose lacked this enzyme, and not surprisingly they failed to ferment lactose on being introduced to the sugar. After 14 hours, however, fermentation of the sugar did get under way; it developed that the presence of the lactose had induced the yeast to produce the enzyme lactase. The adaptation was quite specific: only lactose caused the yeast to synthesize this enzyme; other sugars failed to do so.

In recent years Monod and François Jacob of the Pasteur Institute have worked out some of the basic mechanisms of enzymatic adaptation by the cell, in both the repression and induction aspects. First they discovered that a single mutation in *E. coli* could eliminate the control of lactase synthesis by lactose: the mutant cells produced lactase just as well in the absence of lactose as in its presence. In these cells only the triggering effect was changed; the enzyme they produced was exactly the same as that synthesized by nonmutant strains. In other words, it appeared that the rate of production of the enzyme was controlled by one gene and that the structure of the enzyme was determined by quite another gene. This was confirmed by genetic experiments that showed that the "regulatory gene" and the "structural gene" were indeed in separate positions on the bacterial chromosome.

How does the regulatory gene work? Arthur B. Pardee, Jacob and Monod found that it causes the cell to produce a "repressor" molecule that controls the functioning of the structural gene. In the absence of lactose the repressor molecule prevents the structural gene from directing the synthesis of lactase molecules. The repressor does not act on the structural gene directly; it binds itself to a special structure that is closely linked on the chromosome with the structural gene for the enzyme and with several other genes involved in lactose metabolism. This special genetic structure is called an "operator." The binding of the repressor to the operator causes the latter to switch off the activity of the adjacent structural genes, and in this way it blocks the complex series of events that would lead to synthesis of the enzyme.

Jacob and Monod have shown that this scheme of control applies to any category of "adaptive" enzymes [*see bottom illustration on page 141*]. The repression and induction of enzymes can be regarded as opposite sides of the same coin. In a repressible system the binding of the regulatory signal on the repressor activates the repressor so that it blocks the synthesis of the enzyme. In an inducible system, on the other hand, the binding of the inducing signal on the repressor *inactivates* the repressor, thus releasing the cell machinery to synthesize the enzyme. Mutant cells that lose the repressive machinery need no inducer: they synthesize the enzyme almost limitlessly without requiring any induction signal.

In brief, the various repressors in the cell are specialized receptors, each capable of recognizing a specific signal. And within its chromosomes a cell possesses instructions for synthesizing a wide variety of enzymes, each of which can be evoked simply by the presentation of the appropriate signal to the appropriate repressor.

The cell's selection of chromosomal records for transcription is so efficient as to seem almost "conscious." Actually, however, the responses of the cell are automatic, and like any other automatic mechanism they can be "tricked." It is as though a vending machine were made to work by a false coin: certain artificial compounds closely resembling lactose are excellent inducers of lactase but cannot be broken down by the enzyme. This means that the cell is tricked into spending energy to make an enzyme it cannot use. The signal works, but it is a false alarm. Trickery in the opposite direction is also possible. There is an analogue of tryptophan, called 5-methyl tryptophan, that acts as a repressive signal, causing the cell to stop its production of tryptophan. But 5-methyl tryptophan cannot be incorporated into protein in place of the genuine amino acid. Without that essential amino acid the cell stops growing and dies of starvation. Thus the false signal in effect acts as an antibiotic.

If chemical signals control the production of enzymes, may they not also control the more generalized activities of the cell, notably its self-replication? Jacob, Sydney Brenner and François Cuzin, working cooperatively at the Pasteur Institute and at the Laboratory of Molecular Biology at the University of Cambridge, recently discovered evidence of such a chemical control. They investigated the replication of the unique circular chromosome of *E. coli*. The synthesis of the deoxyribonucleic acid (DNA) of the chromosome, they found, is initiated by a signaling molecule that corresponds to the repressor of enzyme synthesis. The "initiator" has a positive effect rather than a repressive one. Like the repressor of enzyme synthesis, it is synthesized under the direction of a regulatory gene for replication. As the cell prepares for division, the initiator receives orders from the cell membrane and triggers the replication of its DNA by activating a genetic structure called the replicator (analogous to the "operator" of enzyme synthesis). Not much information has been gathered so far about the signal that prompts the initiator or about the

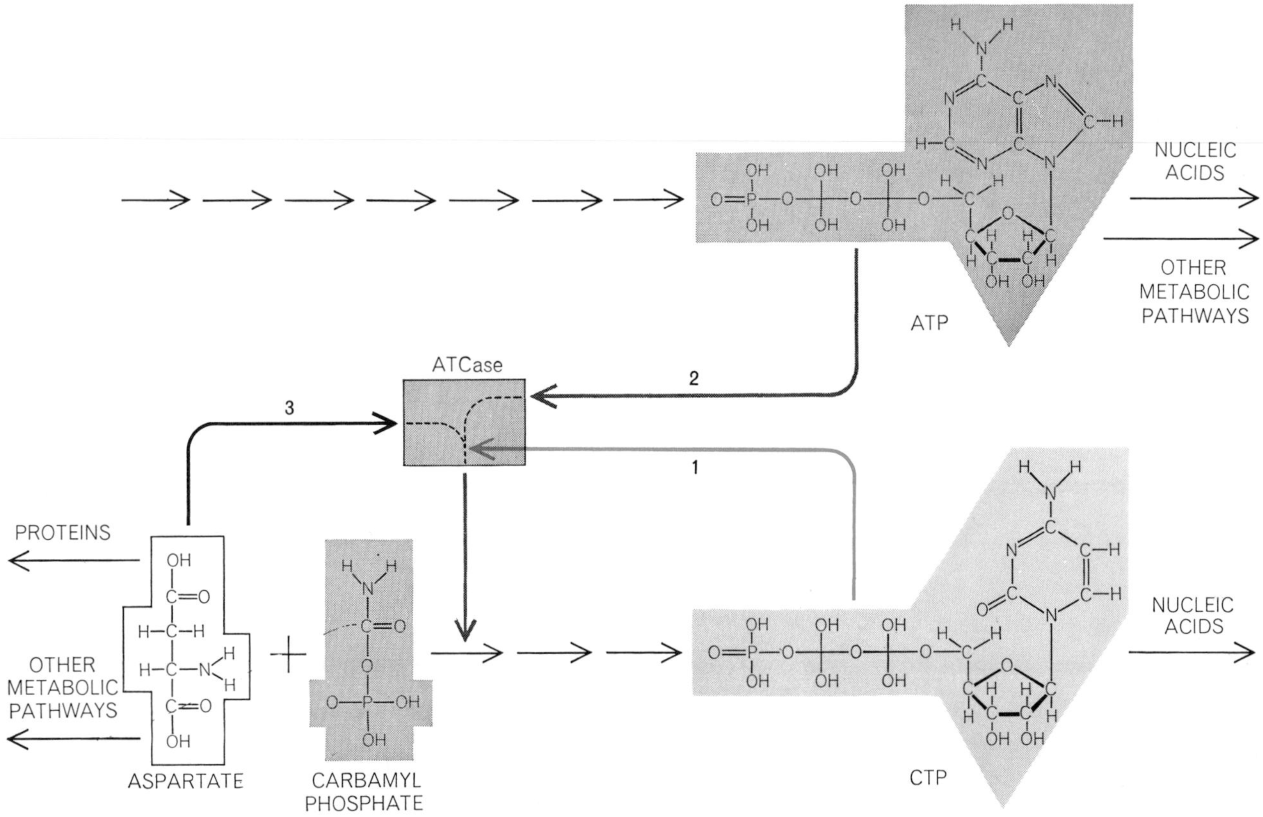

**TWO NUCLEOTIDES, adenosine triphosphate (ATP) and cytidine triphosphate (CTP), are required by the cell in fixed proportions, so their production is regulated by interconnected feedback mechanisms operating on the first enzymes in the synthetic chains. In the case of CTP the enzyme is aspartate transcarbamylase (ATCase). It is inhibited by an excess of CTP (*1*), activated by an excess of ATP (*2*) and must also recognize and respond to the "cooperative" effects of aspartate, its substrate (*3*), which also plays a role in protein synthesis. Notice that ATP, CTP and aspartate have different shapes. How, then, can they all "fit" ATCase chemically?**

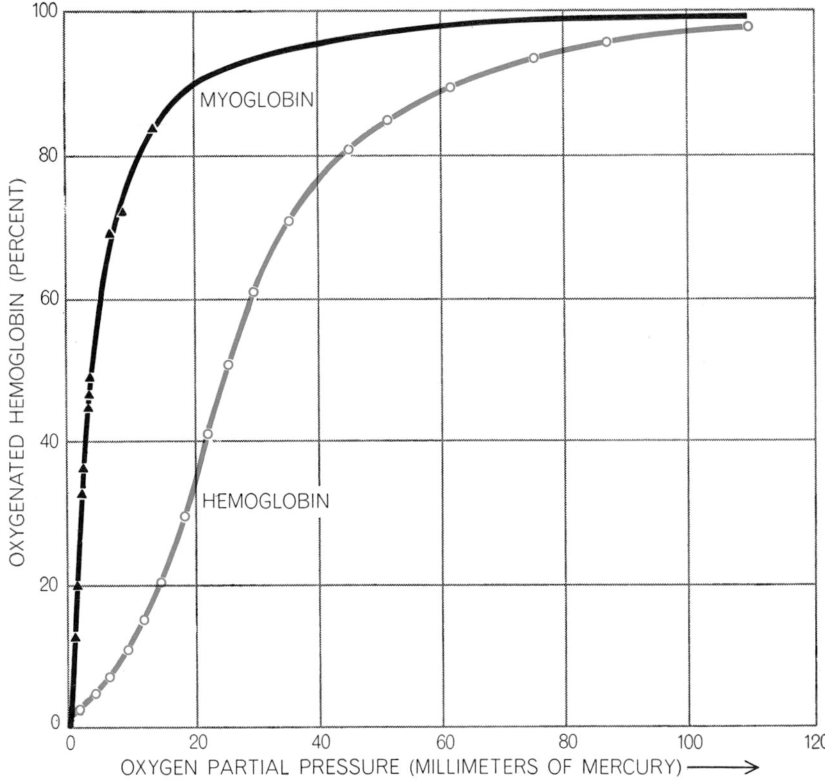


**HEMOGLOBIN, like an enzyme, is a large molecule that binds a small one (oxygen) at specific sites. The curves show the rate of oxygen-binding by hemoglobin (*color*) and myoglobin (*black*), a related oxygen-carrier in muscle. The myoglobin curve is a hyperbola but the hemoglobin curve is *S*-shaped. Hemoglobin binds best at higher oxygen concentrations (in the lungs); the binding of a few oxygen molecules favors the binding of more.**

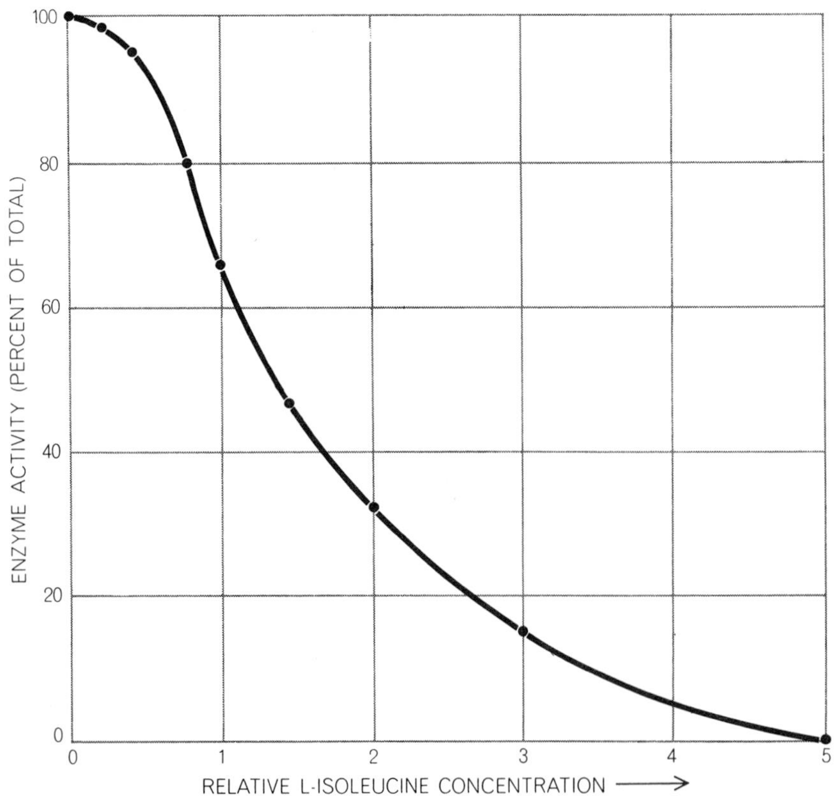


**"COOPERATIVE EFFECT" occurs in regulatory enzymes as in hemoglobin. This curve shows the inhibition of L-threonine deaminase by L-isoleucine. The curve's *S* shape indicates that the effect of the regulatory signal is significant only above a threshold value.**

details of the machinery it sets in motion, but it seems clear that cell division has its own system of chemical control and that it can adjust itself to the composition of the growth medium.

We have been considering the control of the synthesis of enzymes; now let us turn to the control of their activity. As I have mentioned, Umbarger and his colleagues found that the presence of L-isoleucine would not only cause *E. coli* to stop synthesizing the enzymes needed for its production but also inhibit the activity of the first enzyme in the chain leading to the formation of the amino acid. The phenomenon of control of enzyme activity had already been noted earlier in the 1950's by Aaron Novick and Leo Szilard of the University of Chicago. They had shown that an excess of tryptophan in the *E. coli* cell halted the cell's production of tryptophan immediately, which means that the signal inhibited the activity of enzymes already present in the cell. Umbarger went on to investigate the direct effect of L-isoleucine on the enzymes that synthesize it; these had been extracted from the cell. He demonstrated that L-isoleucine inhibited the first enzyme in the chain (L-threonine deaminase), and only the first. This action was extremely specific; no other amino acid—not even D-isoleucine, the mirror image of L-isoleucine—had any effect on the enzyme's activity.

One must pause to remark on the extraordinary economy and efficiency of this control system. As soon as the supply of L-isoleucine reaches an adequate level, the cell stops making it at once. The signal acts simply by turning off the activity of the first enzyme; that is enough to stop the whole production line. Most remarkable of all, once this first enzyme has been synthesized the control costs the cell no expenditure of energy whatever; this is shown by the fact that the amino acid will act to inhibit the enzyme outside the cell without any energy being supplied. A factory with control relays that require no energy for their operation would be the ultimate in industrial efficiency!

The L-isoleucine control system of *E. coli* is only one example of this type of regulation in the living cell. It has now been demonstrated that similar circuits control the cell's production of the other amino acids, vitamins and other major substances, including the purine and pyrimidine bases that are the precursors of DNA.

In all these cases the control is nega-

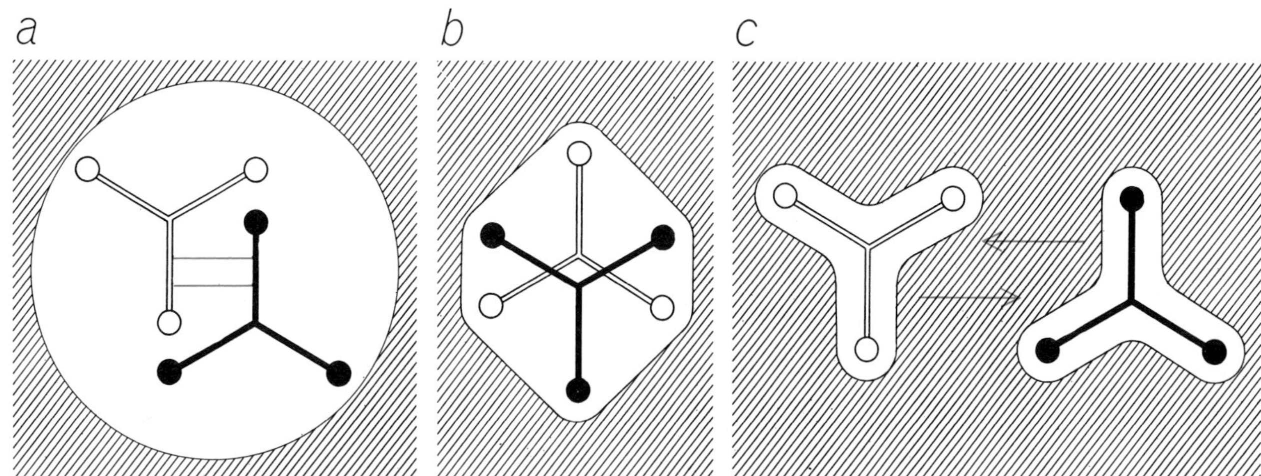


**REGULATORY PROPERTY** of an enzyme might be explained in three different ways. A regulatory signal (*open shape*) might combine with the substrate (*black shape*), participating directly in the chemical reaction it is controlling (*a*). But no such compounds have been found. A signal could simply get in the way of the substrate, excluding it from the enzyme's active site by "steric hindrance" (*b*). The different shapes of substrates and signals preclude this, and in any case steric hindrance could only account for enzyme inhibition, not activation. The only plausible hypothesis, confirmed by experiments with several enzymes, is that the signals and the substrate fit different sites on the enzyme and that the regulatory interactions of these sites are "allosteric," or indirect (*c*).

tive; that is, it involves the inhibition of enzymes. There are opposite situations, of course, in which the control system *activates* an enzyme when the circumstances call for it. An excellent example of such a positive control has to do with the cell's storage and use of energy.

Animal cells store reserve energy in the form of glycogen, or animal starch. Glycogen is synthesized from a precursor—glucose-6-phosphate—in three enzymatic steps. First glucose-6-phosphate is made into glucose-1-phosphate; then glucose-1-phosphate is made into uridine diphosphate D-glucose. Finally uridine diphosphate D-glucose is made into glycogen. When the cell has a good supply of energy, it produces considerable amounts of glucose-6-phosphate. This serves as a signal for stimulating the synthesis of glycogen. The signal works at the third step: the presence of a high level of glucose-6-phosphate strongly activates the enzyme that brings about the conversion of uridine diphosphate D-glucose into glycogen. On the other hand, when the supply of working energy in the cell falls to a low level, so that it must draw on the reserve stored in glycogen, it becomes necessary to activate an enzyme that splits the glycogen (the enzyme known as glycogen phosphorylase). One chemical signal known to be capable of activating this enzyme is adenosine monophosphate (AMP). AMP is a product of the splitting of adenosine triphosphate (ATP), the principal source of the cell's working energy, and an accumulation of AMP therefore indicates that the cell has used up its energy. The AMP signal activates the glycogen-splitting enzyme; the enzyme splits the glycogen molecule; the splitting releases energy, and the energy then is used to regenerate ATP.

The cell thus possesses mechanisms for two types of control of enzyme activity: negative (inhibited enzymes) and positive (activated enzymes). There are

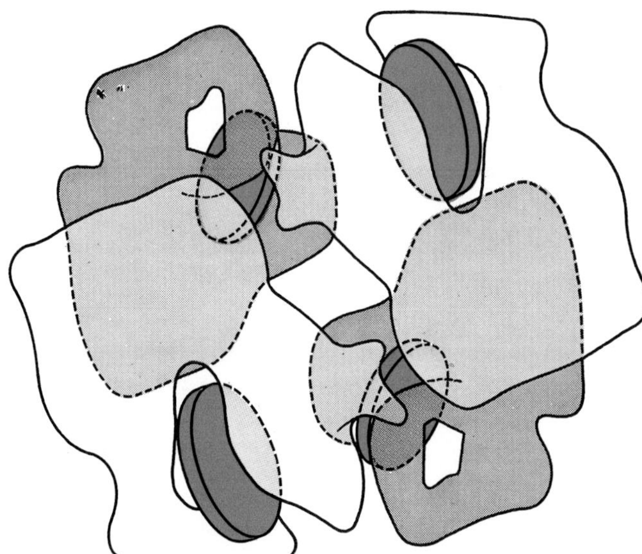

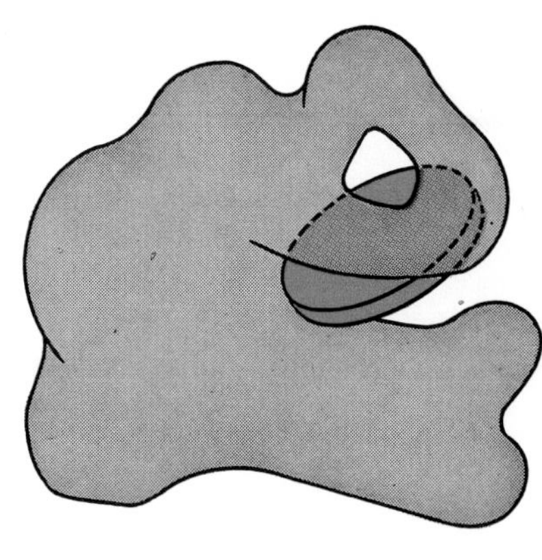

**MOLECULE OF HEMOGLOBIN,** shown (*left*) in very simplified form, has four heme groups (*color*), each of which is borne on a subunit, or chain, that is very similar to a myoglobin molecule (*right*). The heme groups of hemoglobin, each of which is a binding site for an oxygen molecule, are relatively far apart. Cooperative interactions among them must therefore be "allosteric."

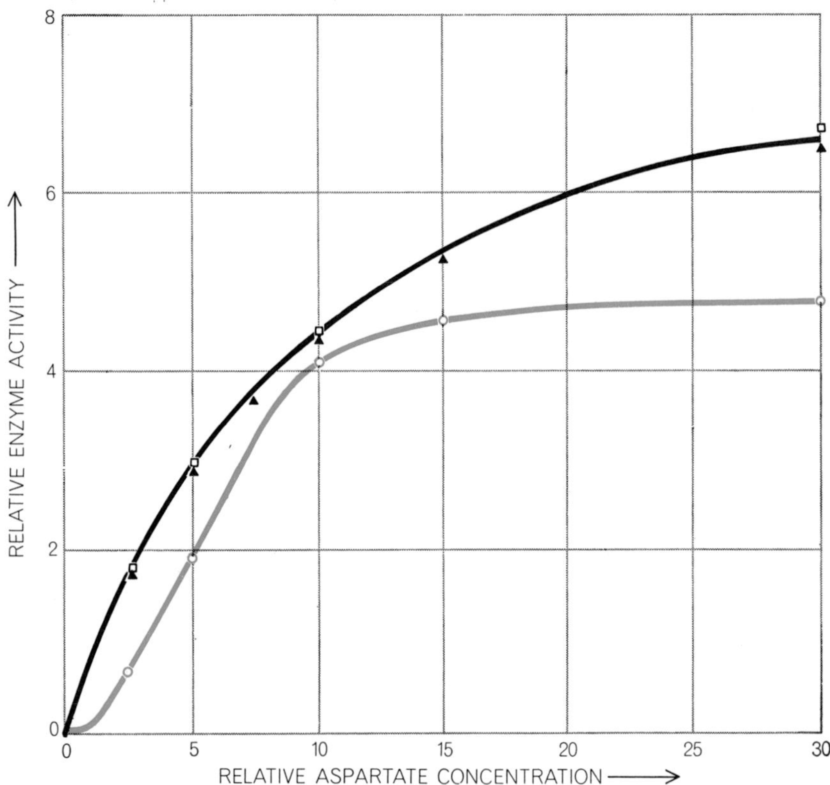


**DESENSITIZATION** of an enzyme affects all its regulatory properties. The substrate saturation curve of natural ATCase (*color*) is S-shaped as a result of the cooperative effect. If the enzyme is denatured by heating, the cooperative effect is lost (*black curve*). So is the effect of feedback inhibition by CTP, as shown by the fact that the curve is the same whether the enzyme is assayed without CTP (*triangles*) or with CTP added (*squares*).

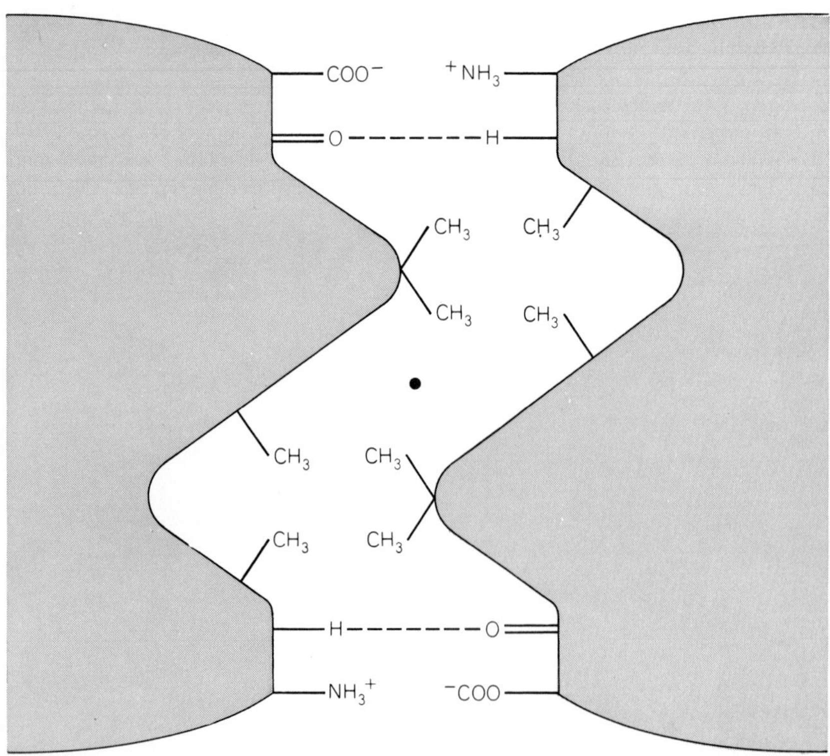


**ALLOSTERIC PROTEINS** are assumed by Monod, Jeffries Wyman and the author to be polymers, molecules composed of identical subunits, that have a definite axis of symmetry (*black dot*). A cross section through such a molecule (made up in this case of two subunits) shows how the symmetry results from the chemical bonds by which the units are associated.

situations in which both methods operate simultaneously. Consider, for example, the synthesis of a nucleic acid. It is assembled from purine and pyrimidine bases, combined in certain definite proportions. The purines and pyrimidines are synthesized on parallel production lines. For the sake of economy they should be produced roughly in the proportions in which they will be used.

This implies that the rate of production by each production line should feed back to control the output by the other. Such a system of mutual regulation must employ both negative and positive controls. Exactly this kind of system has been demonstrated in experiments with *E. coli* conducted by John C. Gerhart and Pardee at the University of California at Berkeley and at Princeton University. They showed that the output of the pyrimidine production line is controlled not only by its own end product (which inhibits the first enzyme in the synthetic sequence) but also by the end product of the purine production line, which counteracts the inhibition by the pyrimidine end product in vitro. Indeed, the purine end product can activate the pyrimidine production directly when no pyrimidine product is present! In short, the enzyme involved here is inhibited by one signal and activated by another.

Several enzymes involved in regulation have also been found to respond in this way to different signals. Moreover, this is not the only exceptional property of these enzymes. Let us now consider another property that will clarify the mechanism by which they are controlled.

A clue to this property seems to lie in the shape of the curve describing the rate at which the enzymes react with their substrates: the substances whose changes they catalyze. Ordinarily the rate of reaction of an enzyme increases as the concentration of substrate is increased. The increase is described by an experimental curve that fits a hyperbola. This kind of curve expresses the fact that the first step in the transformation of the substrate by the enzyme is the binding of the substrate to a specific attachment site on the enzyme.

When the concentration of substrate is increased, molecules of substrate tend to occupy more and more binding sites. Since the number of enzyme molecules is limited, at high concentrations of substrate nearly all the binding sites are occupied. At this point the rate of reaction levels off, hence the hyperbolic shape of the curve. The regulatory enzymes, surprisingly, do not exactly follow this pattern: their reaction rate increases with the concentration of substrate but often the curve is sigmoid (S-shaped) rather than hyperbolic.

When one reflects on the saturation curve of the regulatory enzymes, one notes that it is strikingly like the curve describing the saturation of the hemoglobin of the blood with oxygen. There too the reaction rate traces a sigmoid curve; this remarkable property is related to hemoglobin's physiological function of carrying oxygen from the lungs to other tissues. In the lungs, where the oxygen pressure is high, the hemoglobin is readily charged with the gas; in the tissues, where the oxygen pressure is low, the hemoglobin readily discharges its oxygen. Consider now, however, the myoglobin of muscle tissue. It takes on oxygen, but its oxygenation follows a hyperbolic curve like the classical one for enzymes. A comparative chart shows that when the pressure of oxygen is increased, the amount of oxygen bound by hemoglobin increases faster than the amount bound by myoglobin [*see top illustration on page 144*]. It looks as if the first oxygen molecules picked up by the hemoglobin favor the binding of others—as if there is cooperation among the oxygen molecules in binding themselves to the carrier. Oxygen thus plays the role of a regulatory signal for its own binding.

Similarly, cooperation may be the key to the sigmoid pattern of binding activity in many of the regulatory enzymes. An example of such an enzyme is threonine deaminase. Here again physiological function is evident. The substrate of threonine deaminase is the amino acid threonine. If the amount of this amino acid falls to a very low level in the cell, the cell cannot synthesize proteins. In the absence of threonine, it would be a waste of energy to make isoleucine, the end product of the chain of which threonine deaminase is the first step; hence the economy-geared control system of the cell calls off the production of the second amino acid. In other words, threonine deaminase will not be active and isoleucine will not be produced unless at least threshold concentrations of threonine are present in the cell. In this situation threonine plays the role of regulatory signal for the reaction of which it is the specific substrate; it is an activator of its own transformation.

The most remarkable part of the story is that such cooperative effects are not restricted to the binding of substrate but also operate in the binding of more familiar regulatory signals: specific inhibitors or activators. Regulatory enzymes appear to be built in such a way that they not only recognize the configuration of specific substrates as signals but also gauge their response to whether or not the substrates and regulatory signals are present in certain threshold concentrations. (This is strongly reminiscent, of course, of electric relays—and, one may add, of nerve cells—which react only if the signal has a certain threshold strength.) The regulatory enzymes are thus capable of integrating several signals—both positive and negative—that modulate their activity.

We come now to the question: How do the regulatory relays work? The signals (either activators or inhibitors) are usually small molecules, and the receptor is a regulatory enzyme. In chemical terms, how does the enzyme translate and integrate the signals it receives? The answer to this question applies not only to regulatory enzymes but also to any other molecule that mediates a regulatory interaction. Since little is known about many of these molecules, the model I shall now describe is based on the experimental results obtained from regulatory enzymes. It seems legitimate, however, to extend the model to any category of regulatory molecule.

The question presents a biochemist with a difficult paradox. A molecule can "recognize" a message only in terms of geometry, that is, the shape or configuration of the molecule bearing the message. In this case the message is supposed to cause the enzyme to carry out (or refrain from carrying out) a certain reaction: conversion of a specific substrate into a specific product. Yet the molecule bearing the message often has no structural likeness to either the substrate or the product! How, then, can it promote or interfere with the enzyme's performance of its specific catalytic action on this substrate?

Considering several possible explana-

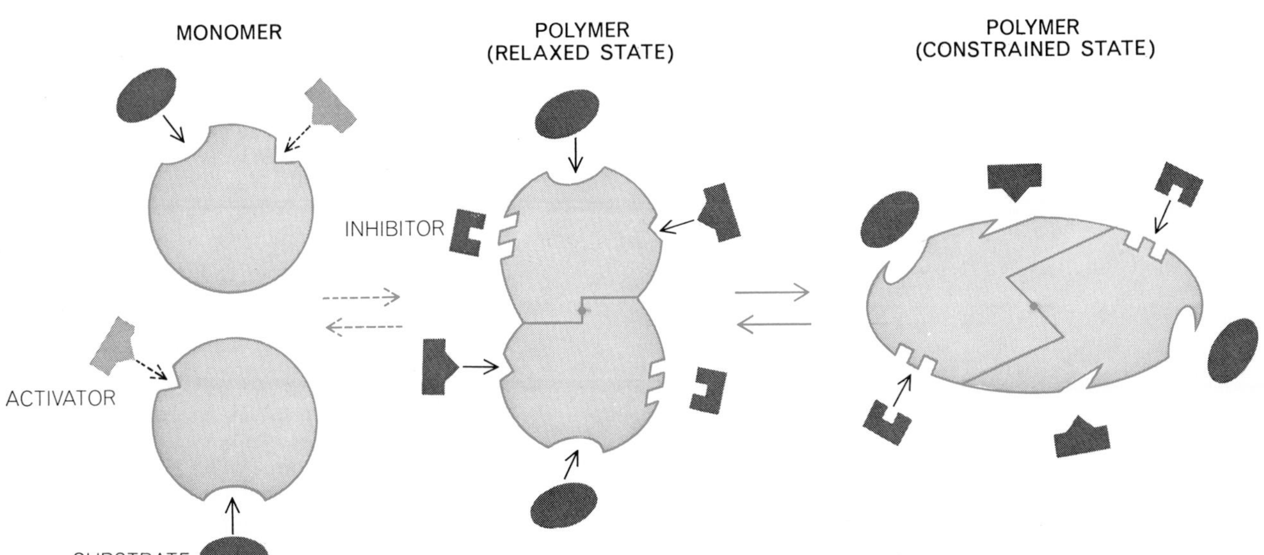


**REGULATORY CHANGES in an allosteric molecule are conceived of as arising from its shifting back and forth between two states. The polymeric molecule is made up of several monomers (two in this case), as shown at left. The polymer can exist in a "relaxed" state (*middle*) or a "constrained" state (*right*). In one condition it binds substrate and activators; in the other state it binds inhibitors. The binding of a signal tilts the balance toward one or the other state but the molecule's symmetry is preserved.**

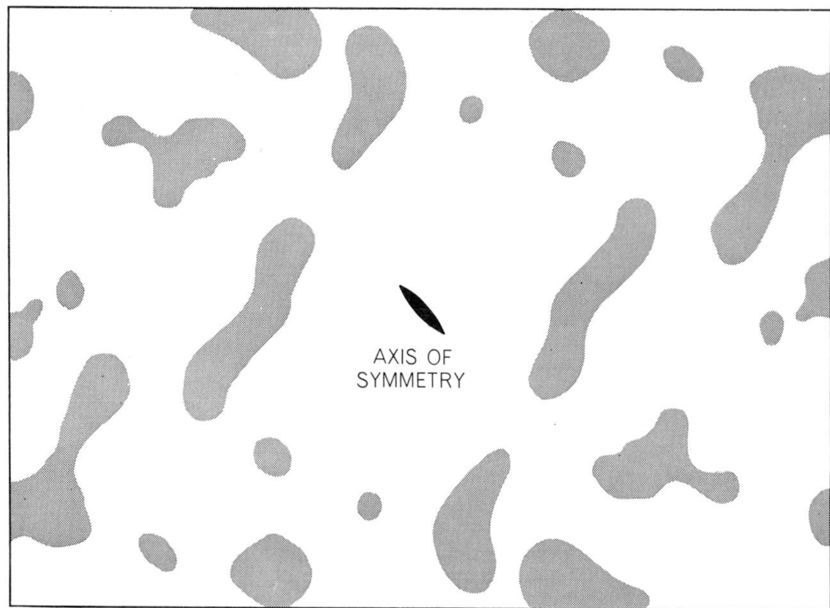


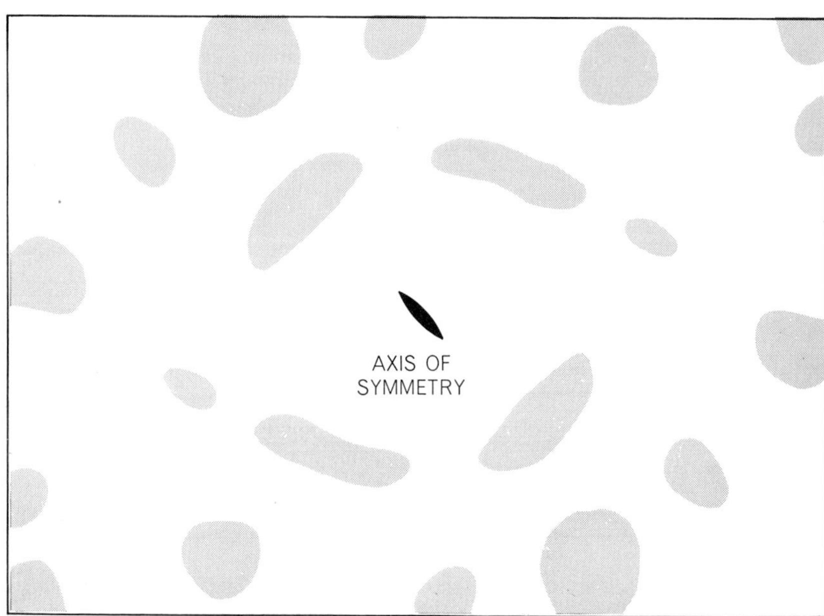


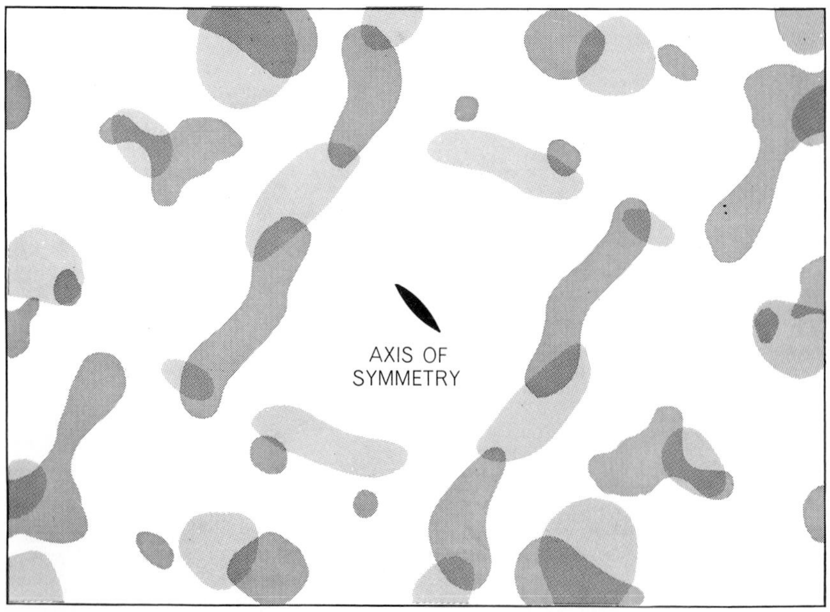


**EXPERIMENTAL DATA supporting the allosteric model come from X-ray diffraction maps of hemoglobin made by M. F. Perutz and his colleagues at the University of Cambridge. The contour lines based on electron densities suggest the shapes of the subunit chains of oxygenated hemoglobin (*top*), reduced hemoglobin (*middle*) and the two superposed (*bottom*). A conformational change of the kind proposed in the model on the preceding page is evident, as is preservation of the molecule's axis of symmetry.**

tions, Monod, Jacob and I have concluded that the only plausible one is that the signal and the substrate fit into separate binding sites on the enzyme and that the signal takes effect by an interaction between these sites [*see top illustration on page 145*]. There is strong experimental evidence in favor of this model. One of the most convincing lines of evidence is the recent discovery by Gerhart that the regulatory enzyme aspartate transcarbamylase has a binding site for its substrate on one subunit of the molecule and a site for an inhibitor of its activity on another subunit. When the subunits are split apart, one retains the ability to recognize the substrate, the other the ability to recognize the inhibitor.

We must now inquire into the nature of the interaction of these two categories of sites on the enzyme. How does the binding of a molecule at one site affect the binding of another molecule at the other site? The best clue to an understanding of the mechanism of the interaction seems to lie in a property of regulatory enzymes that I have already mentioned: the sigmoid curve describing their binding of substrate or of signal molecules, which indicates a cooperative effect among those molecules. Again it is instructive to consider the analogy of the binding of oxygen molecules by hemoglobin.

The hemoglobin molecule has four hemes that are well separated from one another; each is a binding site for an oxygen molecule. In view of the separation between the sites, their cooperation in binding oxygen must be "allosteric," or indirect. Myoglobin, which has only one binding site, binds oxygen hyperbolically (that is, without any control); hemoglobin, with its four sites, binds oxygen in a sigmoid pattern. It seems, therefore, that the key to hemoglobin's cooperative, controlled binding of oxygen lies in the molecule's four-part structure.

Now consider a regulatory enzyme. The binding of any particular molecule (substrate, inhibitor or activator) is sigmoid and therefore a cooperative affair; this implies that there is a set of reception sites for each specific molecule. There also appears to be interaction among the binding sites for different molecules, such as substrate and activator or substrate and inhibitor. Surprisingly the experimental evidence suggests that both types of allosteric interaction—that among the sites binding a particular molecule and that among the sites binding different molecules—may depend on one and the same mechanism, embodied in the structure of the enzyme molecule.

The most striking evidence comes from experiments in the alteration of the structure of regulatory enzyme molecules. Gerhart and Pardee at Berkeley and Princeton and I at the Pasteur Institute, working independently, have found that by changing the molecular structure of aspartate transcarbamylase or L-threonine deaminase (by means of heat, bacterial mutation or certain other procedures) it is possible to "desensitize" these regulatory enzymes so that they are no longer affected by a feedback inhibitor. They are still capable, however, of reacting with their respective substrates. The interesting point is that a change in the enzyme's structure eliminates, along with the negative interaction of the feedback inhibitor and the substrate, all the cooperative interactions in the enzyme molecule. This applies particularly to the binding of the substrate, which changes from a sigmoid to a hyperbolic pattern.

What, then, is the crucial structural feature that accounts for the allosteric interactions within the enzyme molecule? Again hemoglobin offers a clue.

We have noted that the hemoglobin molecule is a four-part structure. It comprises four heme units, each of which is attached to a distinct chain of amino acid units. This molecule is thus made up of four subunits, each of which is so similar to a myoglobin molecule that hemoglobin can be considered essentially a combination of four myoglobin molecules. Hemoglobin displays cooperative interaction, whereas myoglobin does not; hence this property evidently is associated with its four-part structure. Now, experiments show that the binding of oxygen by hemoglobin is connected in some way with an adjustment in the bonding between the subunits making up the molecule [see "The Hemoglobin Molecule," by M. F. Perutz; SCIENTIFIC AMERICAN Offprint No. 196]. The same turns out to be true of many of the regulatory enzymes; their binding of smaller molecules also depends on the adjustment of the bonds holding together their subunits.

On the strength of the experimental findings, Monod, Jeffries Wyman and I have proposed a model picturing the working of the regulatory enzyme system [*see illustration on page 147*]. It suggests that the enzyme molecule consists of a set of identical subunits, each subunit containing just one specific site for each of the molecules it may bind to itself, either substrate molecules or regulatory signals. Now, if a molecule is made up of a definite and limited number of subunits, the implication is that it has an axis of symmetry. Let us say that the enzyme molecule can switch back and forth between two states, and that in each state its symmetry is preserved. The two symmetrical states differ in the energy of bonding between the subunits: in the more relaxed state the enzyme molecule will preferentially bind activator and substrate; in the more constrained state it will bind inhibitor. Whichever compound it binds (substrate, inhibitor or activator) will tip the balance so that it then favors the binding of that category of small molecule. A change in the relative concentrations of substrate and signals may, depending on their molecular structure, tip the balance one way or the other. Thus the model indicates how the enzyme molecule's binding sites may interact, either cooperatively or antagonistically. It suggests that the enzyme may integrate different messages simply by adopting a characteristic state of spontaneous equilibrium between two states.

The major conclusion from the study of the regulatory enzymes is that their powers of control and regulation depend entirely on the form of their molecular structure. Built into that structure, as into a computer, is the capacity to recognize and integrate various signals. The enzyme molecule responds to the signals automatically with structural modifications that will determine the rate of production of the product in question. How did these biological "computers" come into being? Obviously they must owe their remarkable properties to nature's game of genetic mutation and selection, which in eons of time has refined their construction to a peak of exquisite efficiency.

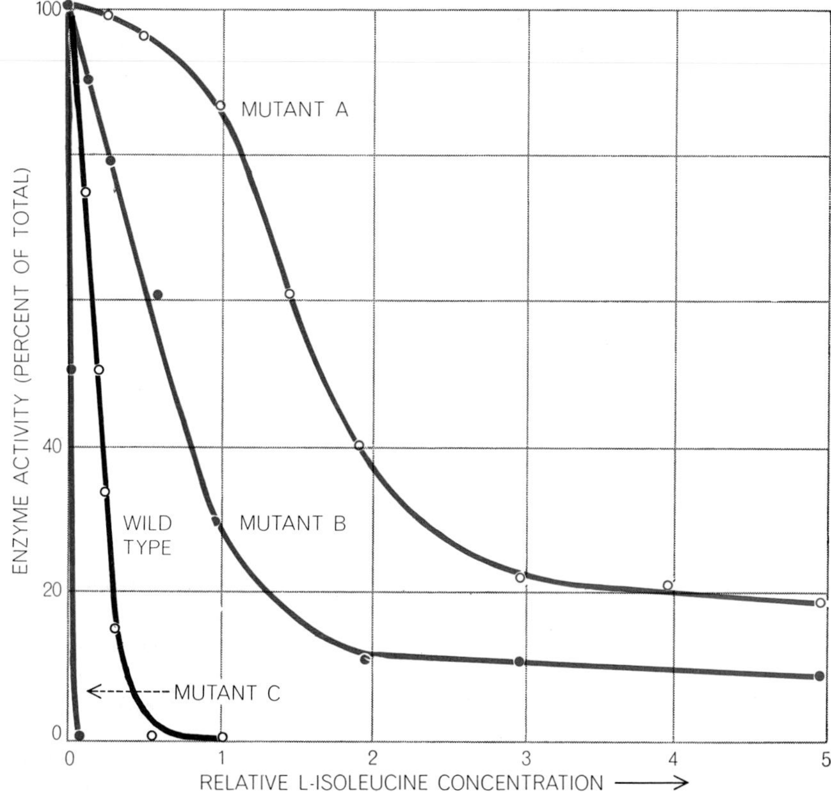


**MUTATIONS in the structural gene for L-threonine deaminase in *E. coli* affect the regulatory properties of the enzyme. Mutant enzymes respond differently to feedback inhibition.**

# 16

# *Chromosome Puffs*

WOLFGANG BEERMANN and
ULRICH CLEVER

*April 1964*

The genetic material performs two functions that are basic to life: it replicates itself and it ultimately directs all the manifold chemical activities of every living cell. The first function is expressed at the time of cell division in the manufacture of more of the genetic material: deoxyribonucleic acid (DNA). The second is accomplished during the "interphase" between cell divisions; DNA directs the synthesis of ribonucleic acid (RNA), which in turn directs the synthesis of proteins, which as enzymes in turn catalyze the other reactions of the cell. In this way RNA translates the genetic information of DNA into the language of physiology and growth, into the everyday processes of synthesis and metabolism.

As readers of SCIENTIFIC AMERICAN are aware, the work of elucidating the genetic code is now being carried out by investigators in laboratories throughout the world, largely by the breeding and statistical study of certain bacteria and the viruses that infect them. In recent years our laboratory at the Max Planck Institute for Biology in Tübingen and several other laboratories have adopted somewhat different techniques for investigating the relation between DNA and RNA in the genetic material of higher organisms—those belonging to the insect order Diptera, such as the fruit fly *Drosophila* and the midge *Chironomus*. In these insects, as in all higher organisms, the DNA resides in the structures called chromosomes. In certain exceptionally large cells of *Drosophila* and *Chironomus* we have found that we can actually see the ultimate units of heredity—the genes—at work. These active genes take the form of "puffs" scattered here and there along the giant chromosomes of the giant cells. We have found that the puffs produce RNA and that the RNA made in one puff differs from the RNA made in another. Observations of the puffs have also enabled us to trace the time patterns of gene activity in several tissues of developing insect larvae. Furthermore, by administering hormones and other substances we can start, stop and prevent some of these activities.

The giant chromosomes were first observed late in the last century, but it was not until 1933 that Emil Heitz and Hans Bauer of the University of Hamburg recognized them as chromosomes. By 1933 breeding studies of the fruit fly had resulted in detailed "maps" on which genes were placed in relation to each other along the chromosomes. The genes, however, were still conceptions rather than physical entities, and the chromosomes had been recognized only during cell division, when they are coiled like a spring and present a condensed, rodlike appearance. During interphase, when they are directing cellular activity, the chromosomes in typical cells are virtually invisible because, although they are long, they are so thin that they can be seen only at the extremely high magnifications provided by the electron microscope, a comparatively recent invention.

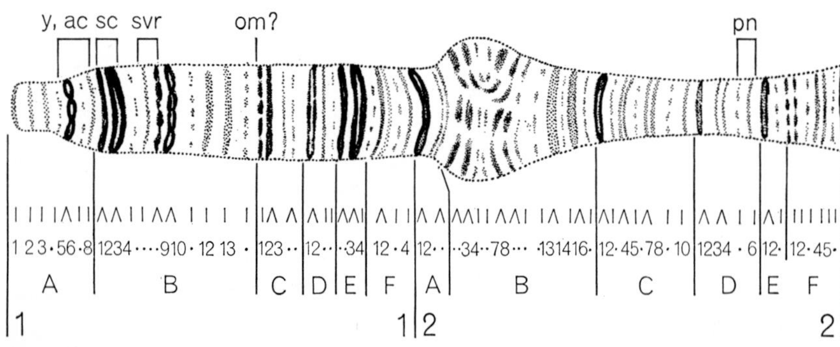


**TIP OF A GIANT CHROMOSOME from the salivary gland of the fruit fly *Drosophila melanogaster* is shown in this diagram. The reference system below it was devised by Calvin B. Bridges of the California Institute of Technology. The letters and brackets above it mark certain sites known to be associated with specific bodily characteristics. For example, the "y" at left denotes the band or gene responsible for yellow body color.**

Heitz and Bauer realized that giant chromosomes, which are clearly visible in the light microscope, are the equivalent of the interphase chromosomes of typical cells. In the words of T. S. Painter of the University of Texas, the giant salivary-gland chromosomes of fruit fly larvae were "the material of which every geneticist had been dreaming. The way led to the lair of the gene." Intensive work by Painter and others in the U.S., including H. J. Muller, Calvin B. Bridges and Milislav Demerec, soon identified specific characteristics of flies with particular loci, or bands, on the giant chromosomes. Since then the bands have been considered the material equivalent of the conceptual Mendelian genes.

The giant chromosomes are found primarily in well-differentiated organs that are engaged in vigorous metabolic activity, such as salivary glands, intestines and the Malpighian tubules (excretory

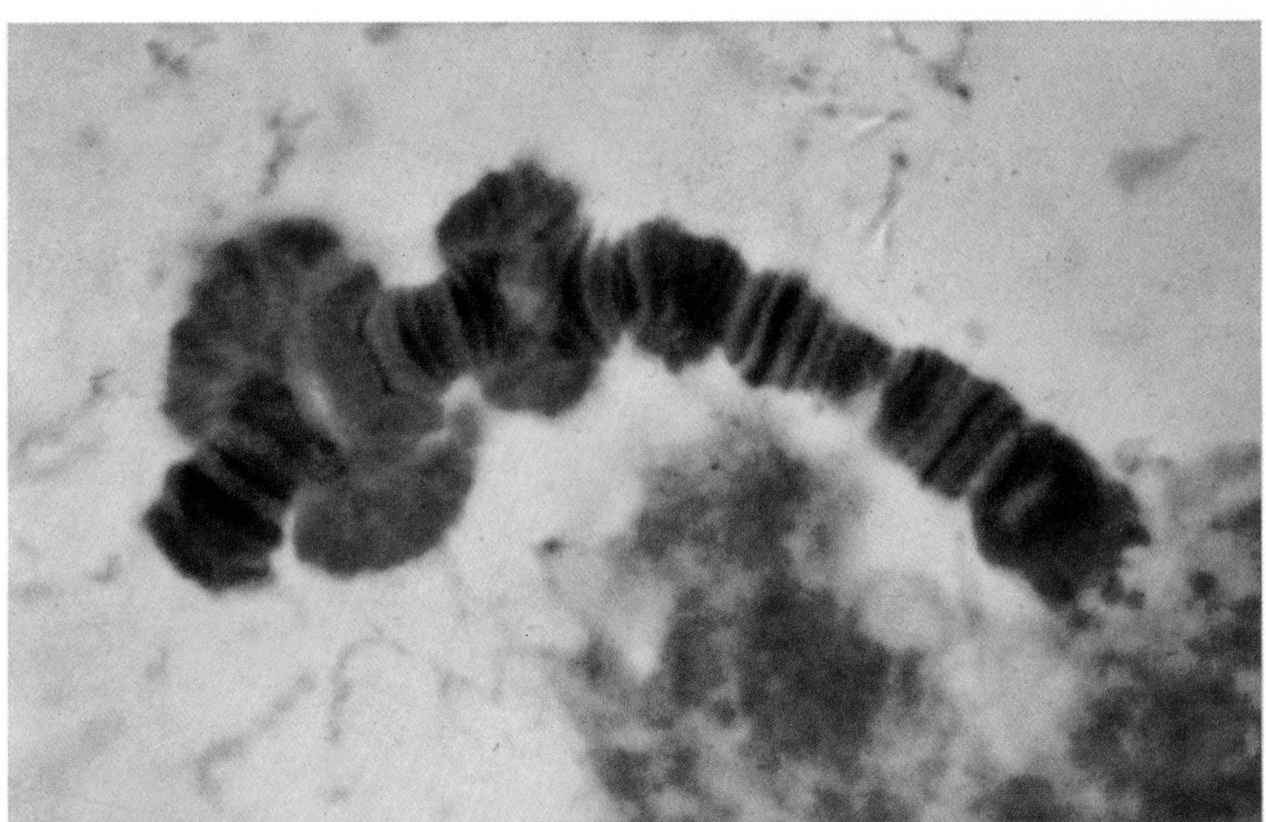

**CHROMOSOME PUFFS are the protuberances on the left-hand portion of the giant chromosome in this photomicrograph. Very large puffs, of which two are seen, are called Balbiani rings. Protein has been stained green, deoxyribonucleic acid (DNA) brown.**

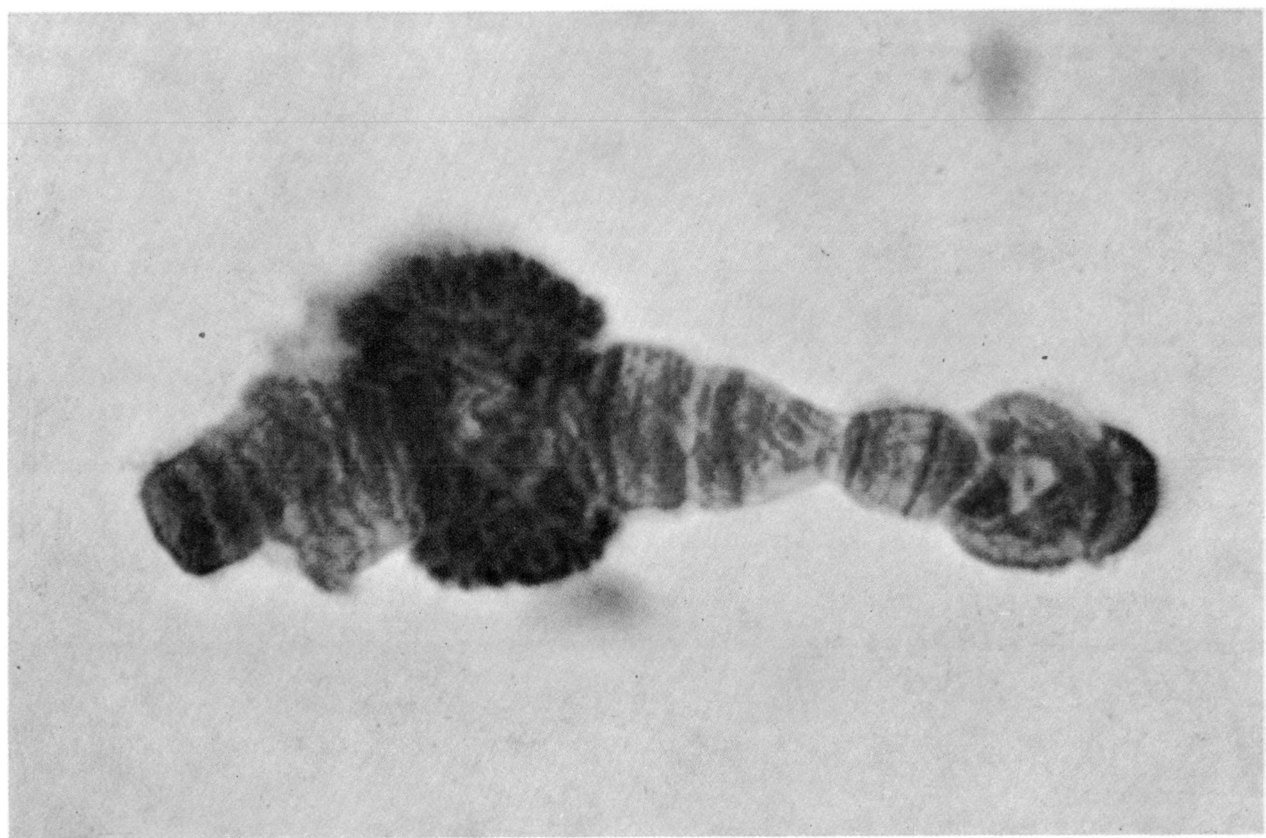

**PRODUCT OF PUFFS, ribonucleic acid (RNA), is reddish-violet when dyed with toluidine blue. Here the DNA is blue. The photomicrographs on this page show two different specimens of the giant chromosome IV from the salivary gland of the midge *Chironomus tentans*. Both were made at the Max Planck Institute for Biology in Tübingen. The magnification in each is some 2,500 diameters.**

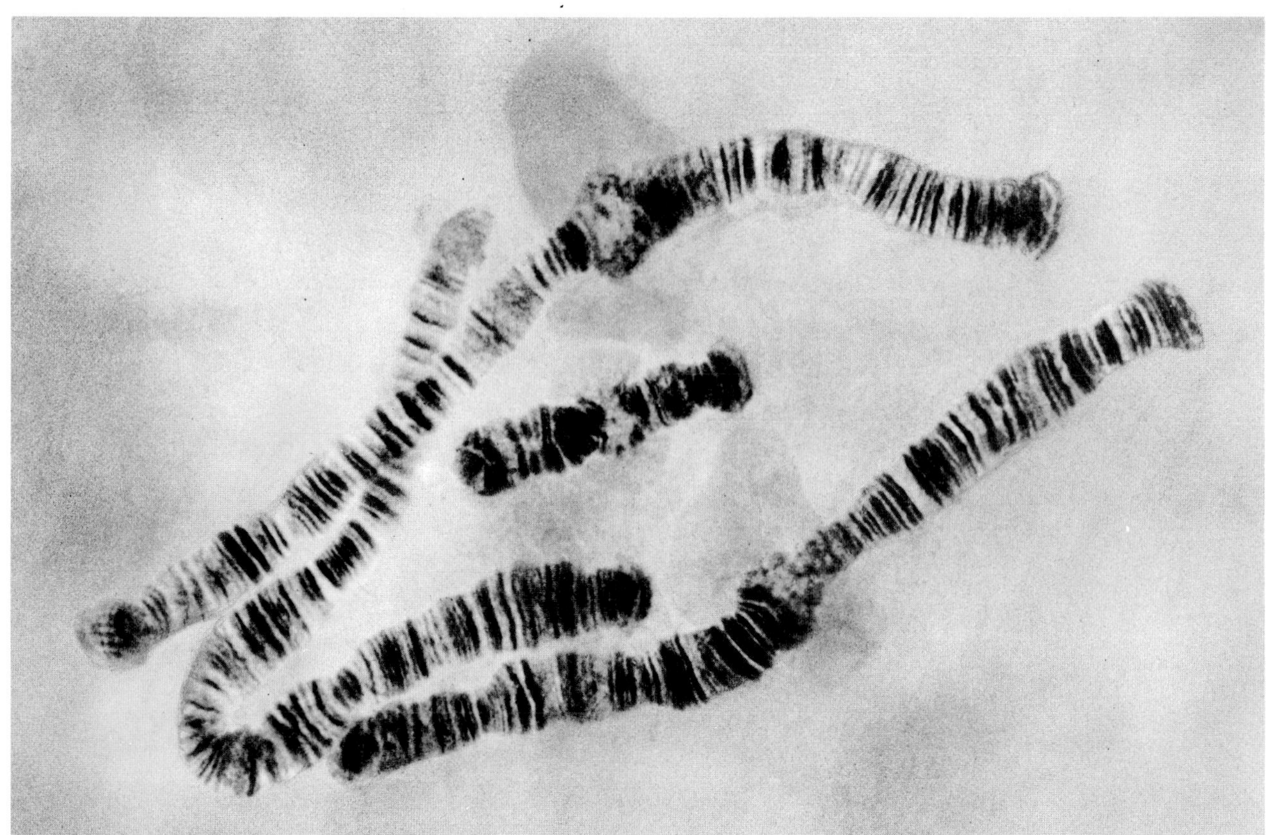

SET OF FOUR GIANT CHROMOSOMES from a cell in the salivary gland of *Ch. tentans* is here magnified some 700 diameters. The enlarged regions on two of the long chromosomes are nucleoli. Chromosome IV is the shortest of the four; it has a Balbiani ring. The banding pattern on each of the four chromosomes is visible in corresponding giant chromosomes from entirely different tissues.

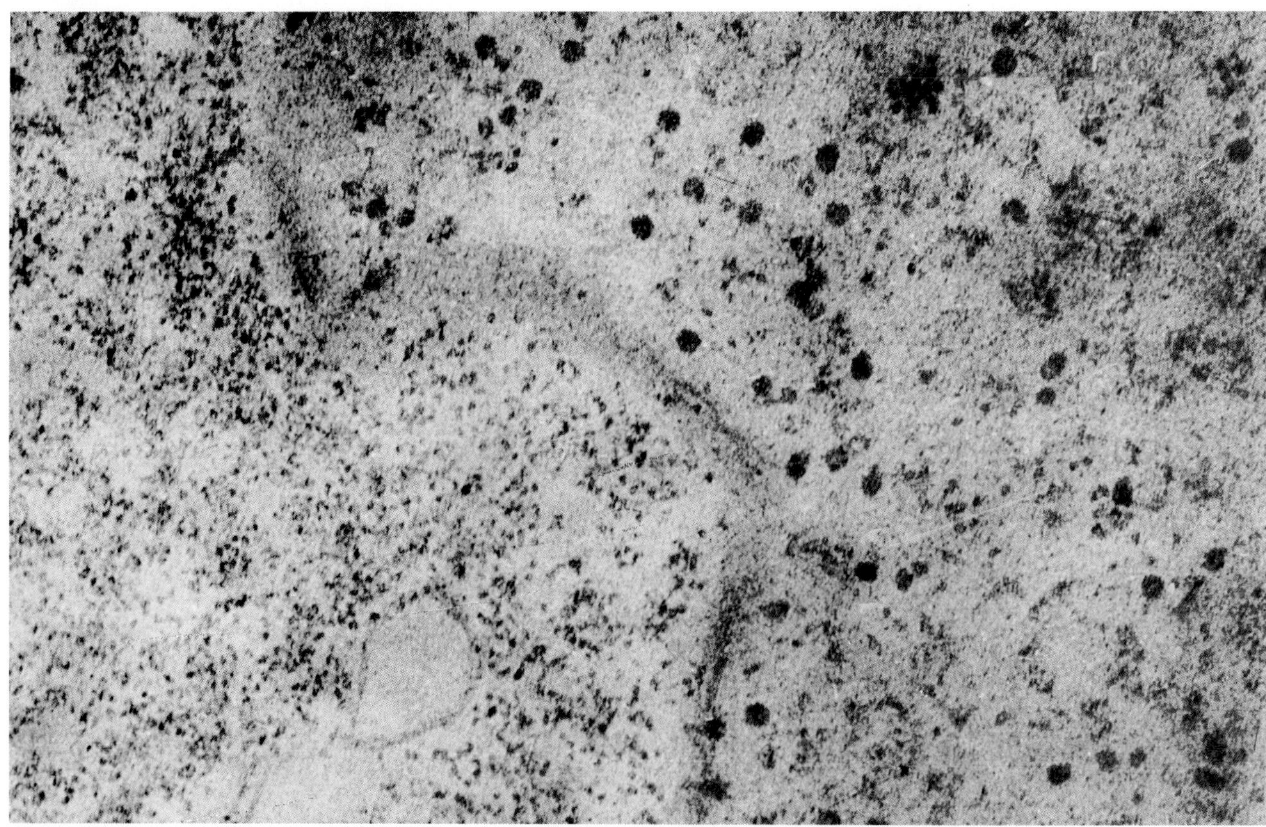

MIGRATING GRANULES, consisting of ribonucleic acid and protein (*right*), are penetrating pores in the nuclear membrane (*bottom center*) in this electron micrograph. Cytoplasm of cell is to left of membrane. The small particles in it are ribosomes, the sites of protein synthesis. The RNA in the large particles may be on its way to the ribosomes to act as a template for proteins.

organs). These tissues grow by an increase in cell size rather than in cell number. Apparently the giant cells require more genetic material than typical cells do; as they expand, the chromosomes replicate again and again and also increase in length. Along individual chromosome fibers there are numerous dense spots where presumably the structure is drawn into tight folds. These locations are called chromomeres.

As the chromosome filaments in giant cells increase in number, those of a particular chromosome remain tightly bound together; each chromomere is fastened to the homologous, or matching, chromomere of the neighboring filaments. Such locations become the bands, which are also known as chromomeres. The chromosome that results from this growth process is said to be "polytene": it has a multistrand structure resembling a rope. At full size the giant chromosomes are almost 100 times thicker and more than 10 times longer than the chromosomes of typical cells at cell division.

The bands, which vary in thickness, contain a high concentration of DNA and histone, a protein associated with DNA. The spaces between the bands, known as interbands, contain a very low concentration of these substances. It was discovered in 1933 that each giant chromosome in a set within a cell has its own characteristic sequence, or pattern, of banding and that, even more striking, every detail of the pattern recurs with the utmost precision in the homologous giant chromosome of every individual of the species.

In the past most cell geneticists were so occupied with localizing the genes in the salivary-gland chromosomes that they did not investigate the giant chromosomes in other tissues. Yet the presence of such chromosomes in cells with quite different functions poses an obvious challenge to the biologist interested in development and differentiation. It had long been held that every cell of an individual possesses exactly the same set of chromosomes and the same pattern of genes. Giant chromosomes in a variety of tissues provided an opportunity for testing this idea, that is, for determining if the special metabolic condition or function of a cell influences in any way the state of its chromosomes and genes. For example, in spite of the constancy of the banding pattern found in salivary-gland chromosomes, the same chromosomes in other organs of the same species might present a different banding pattern. If this were true, the localization of genes in specific bands would lose all general meaning.

Assertions that different tissues have different banding patterns were actually made 15 years ago by Curt Kosswig and Atif Şengün of the University of Istanbul. One of us (Beermann, then working in the laboratory of Hans Bauer at the Max Planck Institute for Marine Biology in Wilhelmshaven) checked these claims by a detailed comparative study of the banding of giant chromosomes from four different tissues of the midge *Chironomus tentans*. Independently Clodowaldo Pavan and Martha E. Breuer of the University of São Paulo carried out similar investigations on the fly *Rhynchosciara angelae*. We could not find any detectable variation in the arrangement and sequence of bands along the chromosomes in different tissues. The uniformity of chromosome banding lends strong support to the basic concept that the linear arrangement of the genes as mapped in breeding experiments corresponds to the pattern of the bands on giant chromosomes.

At the same time, however, we found that chromosomal differentiation of a very interesting kind does exist. The fine structure of individual bands can differ with respect to puffs that are in one location on a chromosome in one tissue and in another location on the same chromosome at another time or in another tissue. These localized modifications in chromosome structure of various Diptera had been noted many years earlier, but their possible significance was overlooked.

The coherence of the chromosome filaments is loosened at the puffed regions. The loosening always starts at a single band. In small puffs a particular band simply loses its sharp contour and presents a diffuse, out-of-focus appearance in the microscope. At other loci or at other times a band may look as though it had "exploded" into a large ring of loops around the chromosome [*see top illustrations on next two pages*]. Such doughnut-like structures are called Balbiani rings, after E. G. Balbiani of the Collège de France, who first described them in 1881. Puffing is thought to be due to the unfolding or uncoiling of individual chromomeres in a band. On observing that specific tissues and stages of development are characterized by

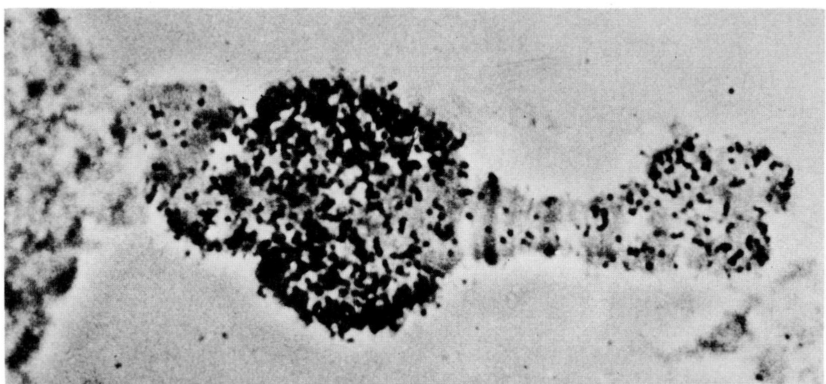

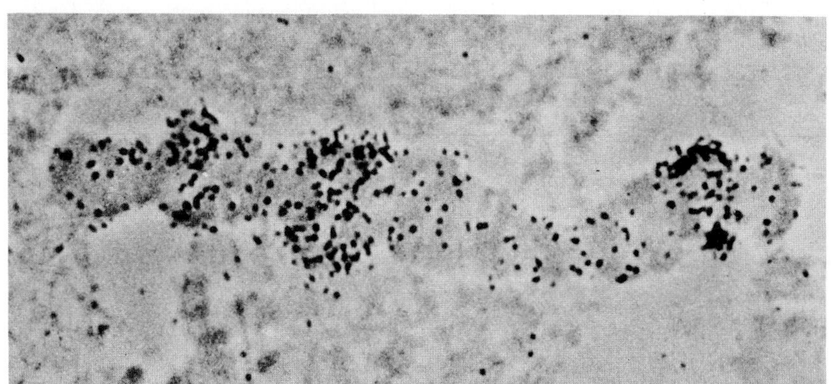

**INHIBITION OF PUFFING and of RNA synthesis is accomplished by treatment with the antibiotic actinomycin D. At top an autoradiogram of a chromosome IV of *Ch. tentans* shows the incorporation of much radioactive uridine (*black spots*), which takes place during the production of RNA, as explained in the text. Another chromosome IV (*bottom*) that had been puffing shows puff regression and little radioactivity after half an hour of treatment with minute amounts of actinomycin D, which inhibits RNA synthesis by DNA.**

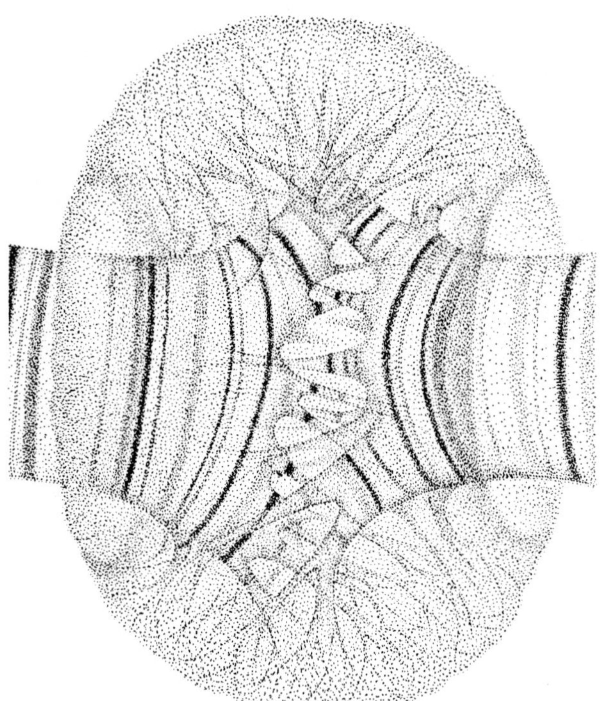

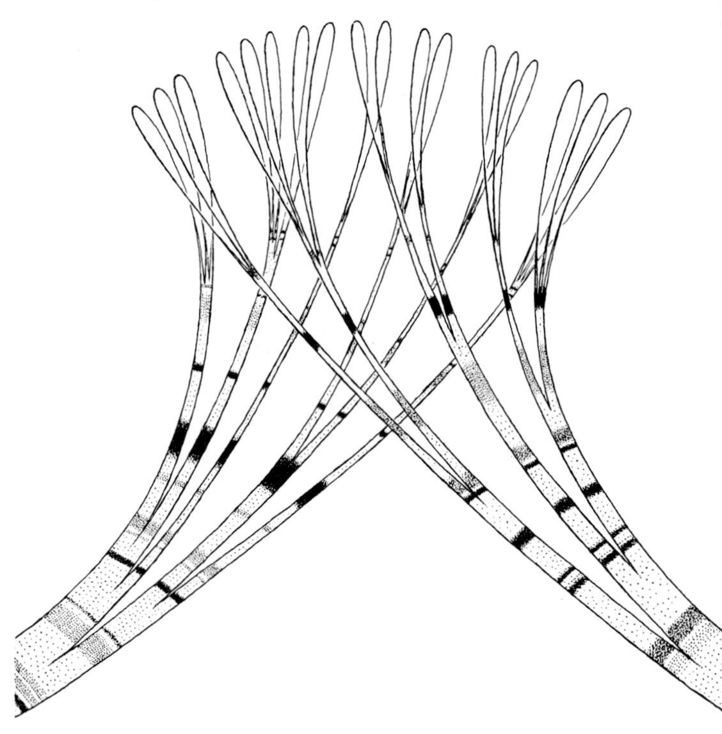

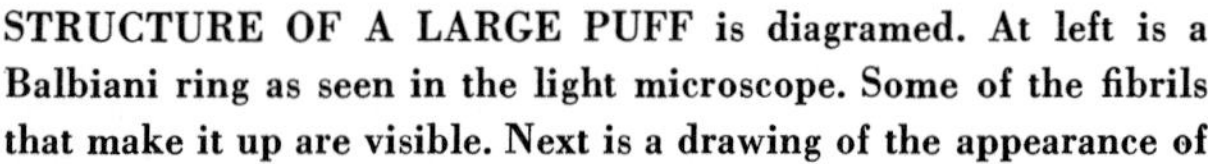

**STRUCTURE OF A LARGE PUFF is diagramed. At left is a Balbiani ring as seen in the light microscope. Some of the fibrils that make it up are visible. Next is a drawing of the appearance of a few of the fibrils at very high magnification in the light microscope. The much greater magnification provided by the electron microscope (*third from left*) shows two puff fibrils with granules**

definite puff patterns, one of us (Beermann) postulated in 1952 that a particular sequence of puffs represents a corresponding pattern of gene activity. At about the same time, Pavan and Breuer arrived at a comparable conclusion based on their experiments with *Rhynchosciara*.

If differential gene activation does in fact occur, one would predict that genes in a specific type of cell will regularly puff whereas the same gene in another type of cell will not. A gene of exactly this kind has been discovered in *Chironomus*. A group of four cells near the duct of the salivary gland of the species *Chironomus pallidivittatus* produces a granular secretion. The same cells in the closely related species *Ch. tentans* give off a clear, nongranular fluid. In hybrids of the two species this characteristic follows simple Mendelian laws of heredity. We have been able to localize the difference in a group of fewer than 10 bands in one of *Chironomus'* four chromosomes; the chromosome is designated IV. The granule-producing cells of *Ch. pallidivittatus* have a puff associated with this group of bands, a puff that is entirely absent at the corresponding loci of chromosome IV in *Ch. tentans*. In hybrids the puff appears only on the chromosome coming from the *Ch. pallidivittatus* parent; the hybrid produces a far smaller number of granules than that parent. Moreover, the size of the puff is positively correlated with the number of granules. This reveals quite clearly the association between a puff and a specific cellular product.

Such analysis can demonstrate only that a specific relation exists between certain puffed genes and certain cell functions. We therefore sought to find a biochemical method for showing that puffing patterns along chromosomes are in fact patterns of gene activity. According to the current hypothesis the sequence of the four bases that characterize DNA—guanine, adenine, thymine and cytosine—represents a code for the sequence of the 20 kinds of amino acid unit that make up a protein. Most, if not all, protein synthesis takes place not in the nucleus of the cell but in the surrounding cytoplasm. The DNA always remains in the nucleus. As a result the instructions supplied by DNA must be carried to the cytoplasm, where the translation is made. The carrier and translator of the DNA information is thought to be the special form of RNA called messenger RNA. Each DNA molecule serves as a template for a specific messenger RNA molecule, which then acts as a template in the synthesis of a particular protein. Hence what we have termed gene activity becomes equivalent to the rate of production of messenger RNA at each gene.

It has been known for some time that chromosome puffs contain significant amounts of RNA. As we have noted, the normal, unpuffed bands chiefly contain DNA and histone. In general the amount of these compounds remains unchanged in the transition from a band to a puff, whereas the amount of RNA increases considerably. The presence of RNA is beautifully demonstrated by metachromatic dyes such as toluidine blue, which simultaneously stains RNA red-violet and DNA a shade of blue. A great increase in the amount of RNA, however, is not sufficient to demonstrate that RNA synthesis is the main function of puffs. For one thing, some dyes show that a protein other than histone accumulates in the puffs along with RNA. Perhaps it too is made there.

In order to find out if RNA is the major puff product, Claus Pelling of our laboratory employed the technique of autoradiography. His "tracer" was uridine, a substance the cell tends to use to make RNA rather than DNA, that had been labeled with the radioactive isotope hydrogen 3 (tritium). He injected the uridine into *Chironomus* larvae, which he later killed. When giant cells from the larvae were placed in contact with a photographic emulsion, the radioactive loci in their chromosomes darkened the emulsion [*see illustration*

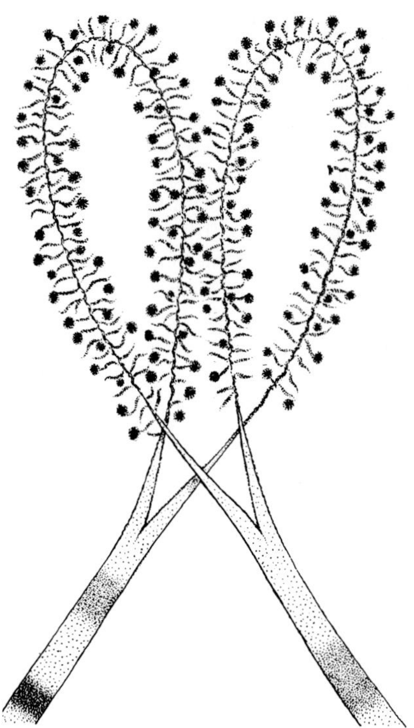

**that are believed to be messenger RNA produced by the genes. In particularly small puffs the loops cannot actually be observed.**

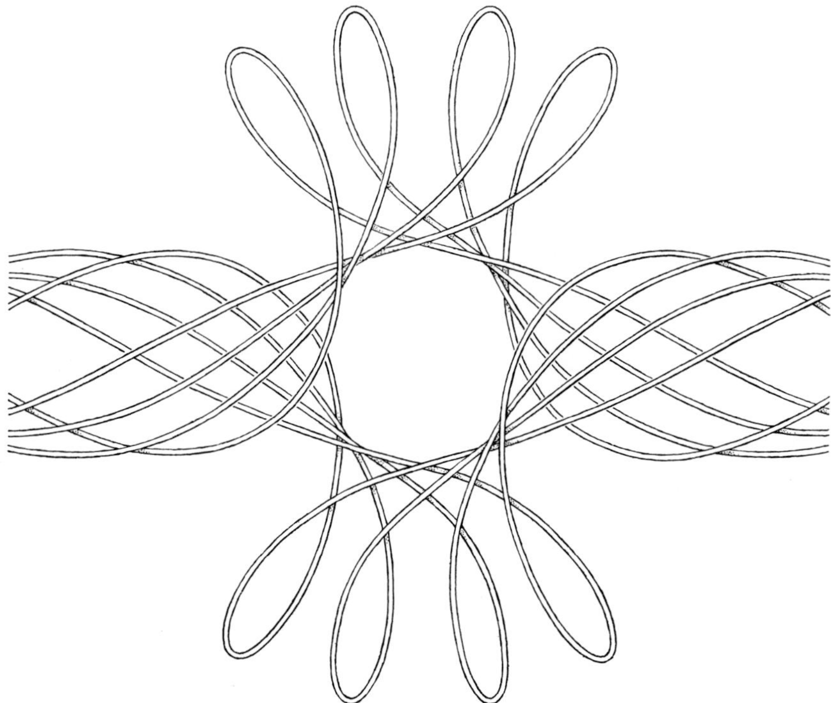

**SCHEMATIC REPRESENTATION of how a large puff is formed shows fibrils untwisted and "popped out" of the cable-like structure. A giant chromosome in reality contains thousands of fibrils. Those untwisted here are tightly coiled when in the form of bands.**

*on page 153*]. In every case in which Pelling killed the larvae soon after injection, sometimes as quickly as two minutes afterward, only the puffs, the Balbiani rings and the nucleoli were labeled. (Nucleoli are large deposits of RNA and protein that are formed in all types of cells by chromosomal regions known as nucleolar organizers. Presumably they are involved in the formation of ribosomes, which are the sites of protein synthesis in the cytoplasm.) The rest of the chromosomal material and the cytoplasm showed very little radioactive label until long after the injection.

When the preparations were treated with an enzyme that decomposes RNA before placing them in contact with the emulsion, the label was absent. Pelling demonstrated further that the rate of RNA synthesis is closely correlated with the relative size of the puffs. The administration of the antibiotic actinomycin D, a specific inhibitor of any RNA synthesis that depends directly on DNA, stopped the formation of RNA.

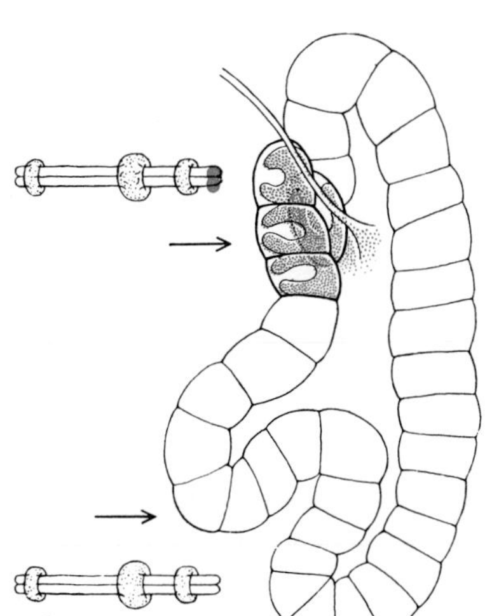

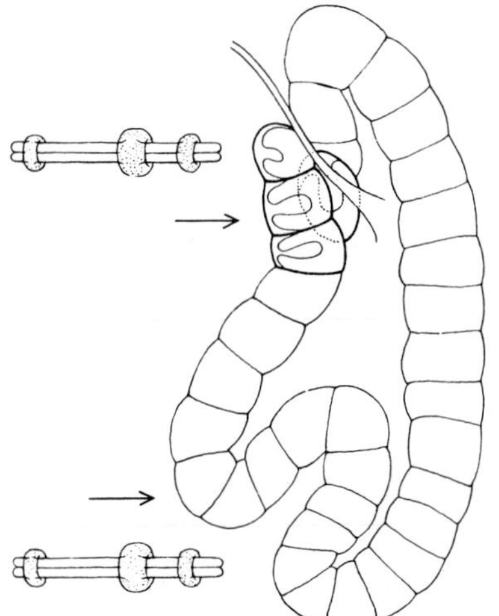

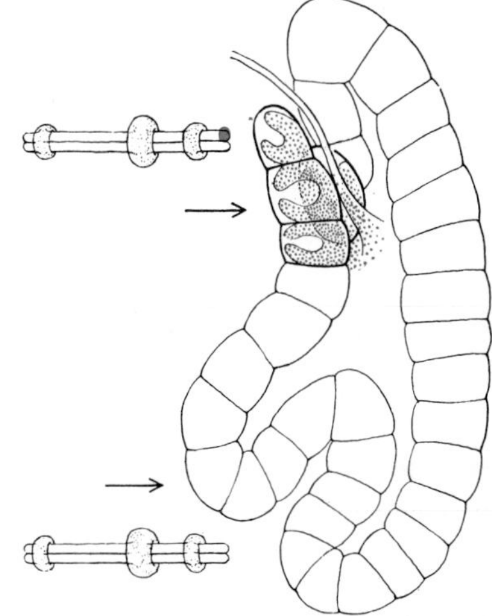

**DIFFERENTIAL GENE ACTIVATION occurs on homologous, or matching, chromosomes of *Chironomus*. Four salivary-gland cells in the species *Ch. pallidivittatus* (*left*) produce granules (*colored stippling*). The species *Ch. tentans* (*center*) makes no granules. Chromosome IV from the four granule-producing cells (*at left of cells*) has a puff at one end (*color*), whereas the same chromosome from other cells (*lower left*) of the same gland and from all salivary-gland cells of *Ch. tentans* have no puff there. (In each case the chromosome inherited from both parents is shown.) Hybrids of the two species (*right*) have a puff only on the chromosome from the *Ch. pallidivittatus* parent in the four granule-producing cells. As indicated in the drawing, they make far fewer granules.**

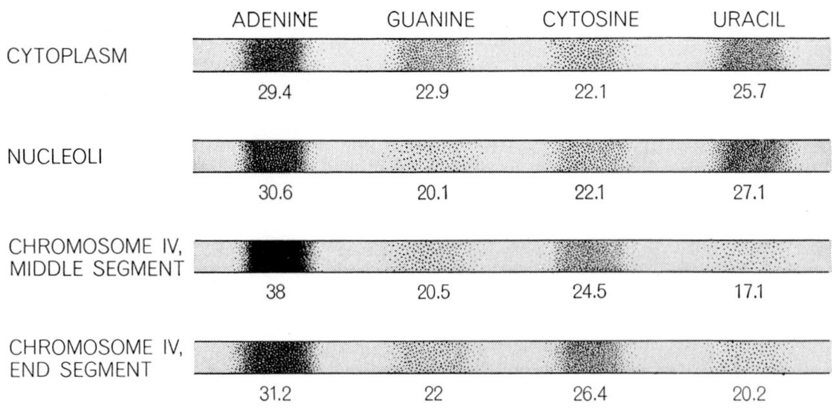


**RNA'S FROM VARIOUS REGIONS of salivary-gland cells of *Chironomus* differ from one another in percentages of the four RNA bases. The RNA from each place was decomposed by an enzyme and the bases were then separated by electrophoresis on rayon threads. Samples from chromosome IV differ widely from nucleolar and cytoplasmic RNA's.**

This proved that the synthesis was taking place at the site of the DNA in the chromosome. All these results agree with the assumption that the pattern of puffing along chromosomes is a quantitative reflection of the pattern of synthetic activities from gene to gene.

The protein in the puffs, in contrast to RNA, takes up little or no radioactive material if we inject the larvae with radioactively labeled leucine or another labeled amino acid. The labeled protein always appears first in the cytoplasm and does not reach the chromosomes for at least an hour. One of us (Clever) obtained the same result when he injected labeled leucine together with the hormone ecdysone, which elicits puffing at several sites in a short time. We have concluded, therefore, that the puff protein is made elsewhere than in the chromosome. Probably some of this protein is the enzyme RNA polymerase, which presides at the synthesis of RNA.

In order to learn if the RNA made in the puffs is messenger RNA, we collaborated with Jan-Erik Edström of the University of Göteborg in Sweden. He has developed an elegant microelectrophoretic technique that makes it possible to determine the base composition of very small amounts of RNA. He applies RNA that has been decomposed by an enzyme to moist rayon threads, which are then laid between two electric poles. Thereafter the different bases move different distances along the threads in a given time. Their quantities can be determined by photometry and their relative proportions established. We made separate analyses of the proportions of bases in RNA's from various parts of the salivary-gland cells of *Chironomus*, including the cytoplasm, the nucleoli, the entire chromosome I and the three large Balbiani rings of chromosome IV. This involved, among other things, cutting several hundred IV chromosomes into three pieces. The base compositions of all these RNA's differ from one another. The RNA's of the cytoplasm and the nucleoli appear to be nearly identical, but both differ from the RNA of the entire chromosome I and particularly from the RNA of puffs. In addition, there are slight but significant differences among the RNA's of the three Balbiani rings. One conclusion is that puff RNA certainly represents a special type of RNA. Is it therefore messenger RNA? An unusual feature of its base composition suggests that it is.

The RNA of salivary-gland chromosome puffs consistently contains more adenine than uracil—twice as much in the case of one Balbiani ring. (RNA contains uracil in place of the thymine in DNA.) Deviations from a one-to-one ratio are also found with respect to guanine and cytosine. In typical DNA the ratios of adenine to thymine and of guanine to cytosine invariably equal one because the bases are paired in the double-strand helix of the DNA molecule. RNA, being single-stranded, is not subject to this rule. In the case of messenger RNA, however, if one assumes that both strands of DNA make complementary copies of RNA, the ratios of adenine to uracil and guanine to cytosine should also be one. Most investigators confirm this expectation. Our data, on the other hand, strongly suggest that puff RNA is a copy of only one DNA strand. This appears to us to be a more reasonable way to make messenger RNA, since in protein synthesis only one of the two putative RNA copies of double-strand DNA could serve as a template. Messenger RNA fractions similar in composition to ours have now been discovered in other organisms.

Evidence for the physical movement of our messenger RNA has been found recently in electron micrographs of sections through the Balbiani rings that reveal the presence of ribonucleoprotein (RNA and protein) particles. In other electron micrographs such particles are seen floating freely in the nuclear sap and through pores in the nuclear membrane [*see bottom illustration on page 152*]. They break up in the cytoplasm. We believe these particles carry the messenger RNA to the ribosomes, where it would serve as the template for the synthesis of proteins.

In the hope of delineating at least some of the forces that control the behavior of genes, one of us (Clever) set out to learn about the conditions under which puffs are produced or changed. Since insect metamorphosis has been studied rather fully, a good starting point seemed to be the changes of the puff pattern in the course of metamorphosis.

Insect metamorphosis is the transformation from the larva to the adult. In the higher insects, to which the Diptera belong, it begins with the molting of the larva into the pupa and ends with the molting of the pupa into the imago, or adult. The moltings are caused by the hormone ecdysone, which is produced by the prothorax gland located in the thorax. So far this is the only insect hormone that has been purified. Because ecdysone affects single cells directly, injection of it induces changes related to molting in all cells of the insect body.

First we examined the time relation between the changes in puffing of individual loci and the metamorphic processes in the larvae. In the great majority of the puffed loci in the salivary glands, phases in which a puff is produced alternate with phases in which a puff is absent [*see illustration on opposite page*]. Some of the phases of puff formation have no recognizable connection with the molting process. Other puffs, however, appear regularly only after the molting of the larva has begun; some at the start of molting, others later. Apparently these chromosomal sites participate in metabolic processes that take place in the cell only during the molting stage. Finally, a third group of puffs, which are found in larvae of all ages, always become particularly large during metamorphosis. This indicates that some components of the metabolic process not specific to molting are intensified at that time.

Further experiments and observations have given some indication of how ecdysone regulates the activity of single sites during molting. In the first place, the hormone not only initiates the process; it must also be present continuously in the hemolymph, or blood, of the insect if molting is to continue. The secretion of ecdysone may stop for a time in *Chironomus* larvae that had begun to molt. In such larvae all the puffs characteristic of molting are absent, which shows that the hormone controls the pattern of gene activity specific to molting. Hans-Joachim Becker of the University of Marburg confirmed this by knotting a thread around *Drosophila* larvae at the start of metamorphosis so that the prothorax gland and part of the salivary gland were in front of the knot and another part of the salivary gland was behind it, cut off from the prothorax secretions. After a time he killed the larvae and found that the puff pattern of metamorphosis was absent in the salivary-gland cells behind the knot but present in cells in front of it.

In detail ecdysone affects the puffs in a variety of ways. If we inject the hormone into *Chironomus* larvae, most of the puffs do not react until long afterward. For some the interval is a few hours, for others one or more days, and this is independent of the quantity of ecdysone. Two puffs, on the other hand, appear quite soon after the injection of ecdysone into larvae that have not begun to molt. One puff arises in 15 to 30 minutes at locus 18-C of chromosome I, the other in 30 to 60 minutes at locus 2-B of chromosome IV. These are the earliest observable gene activations produced so far by the administration of ecdysone. At both loci the higher the dosage of hormone, the longer the puffs last. The injection of more hormone slows the regression of the puffs at both loci, and if ecdysone is injected after the puffs have regressed, they swell up again. From this we conclude that the cause of puff regression is the elimination of the hormone.

The two loci exhibit different reaction thresholds. At locus 18-C on chromosome I a minimum ecdysone concentration of about $10^{-7}$ microgram (one ten-trillionth of a gram) per milligram of larval weight is required to induce puffing. The locus 2-B on chromosome IV reacts only to about $10^{-6}$ microgram per milligram of larval weight. In these concentrations there can be no more than 100 ecdysone molecules at each of the chromosome strands in a puff, assuming that each giant chromosome

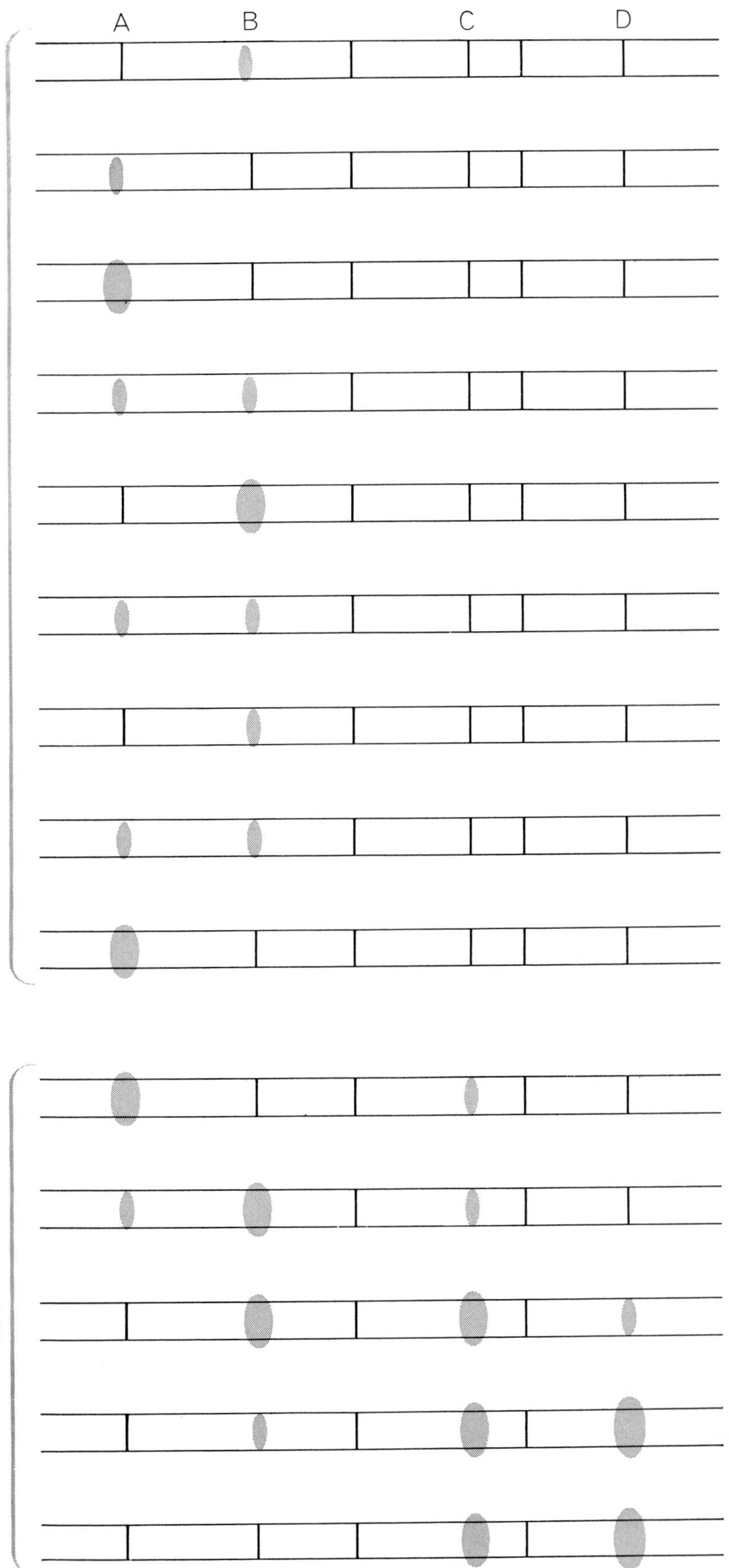


**SEQUENCE OF PUFFING** at four sites (*A, B, C, D*) of one chromosome in the salivary gland of *Ch. tentans* is diagramed. Some bands that do not puff are also shown. Starting from the top, the changes occur before and during the molt that begins pupation.

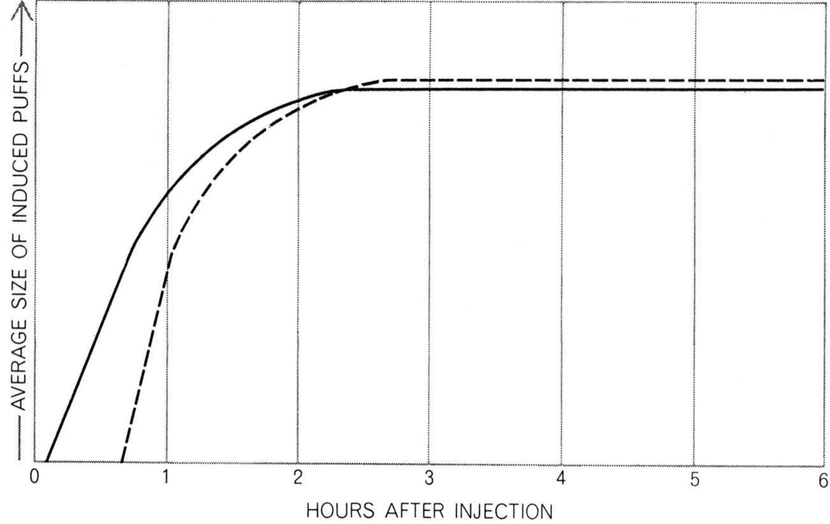


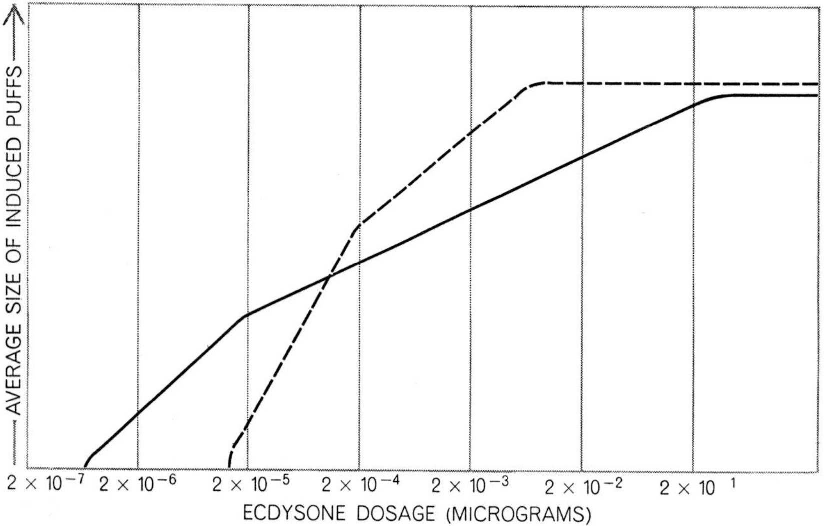


**INDUCED PUFFING follows injection of the hormone ecdysone at locus 18-C of chromosome I in *Ch. tentans* (*solid curves*) and locus 2-B of chromosome IV (*broken curves*). Upper diagram shows time schedule of puffs, lower diagram the relation of puff size to quantity of hormone. Dosage is in micrograms per milligram of total weight of larva.**

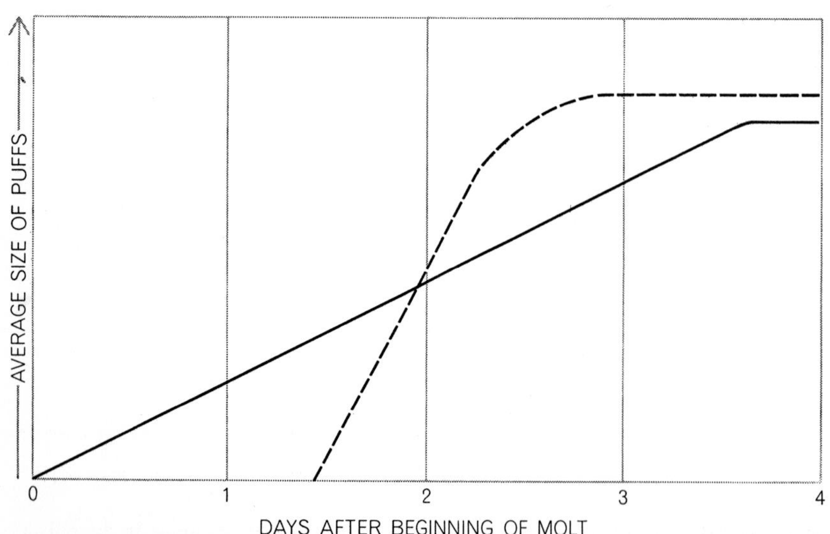


**NORMAL PUFFING SEQUENCE at loci 18-C of chromosome I (*solid curve*) and 2-B of chromosome IV (*broken curve*) follows a schedule different from that of induced puffing.**

consists of 10,000 to 20,000 single strands and that the hormone is distributed evenly throughout the larva. The puff at locus 2-B attains maximum size at a lower concentration than that at locus 18-C.

By applying these findings to a record of the growth and change of the two sites during normal molting, we find that the hormone level apparently increases gradually during metamorphosis. Thus the different activity patterns of these two loci can be explained as responses to the same factor: the changing hormone concentration.

In the case of locus 2-B, however, ecdysone is not the only active agent. The puff at this locus begins to regress during the second half of the prepupal phase (the last larval stage before pupation) even though the puff at locus 18-C persists to the end of the phase. In larvae that are ready to pupate, the puff at locus 2-B has usually regressed altogether. Yet when we inject hemolymph from these prepupae into young larvae, puffing is induced both at locus 18-C and locus 2-B. The former puff is quite large, which indicates that the hemolymph from the older larvae still contains ecdysone in high concentration. Evidently in these larvae, although not in their hemolymph, there is an antagonistic factor that actively represses puffing at locus 2-B in spite of the presence of ecdysone. This demonstrates that in higher organisms the activity of genes falls under the control of more than one factor.

Whereas gene activity at the loci 18-C and 2-B is subject to stringent regulation by very specific factors, other tests show that later puffs elsewhere are not the result of changes in the concentration of ecdysone. Rather they behave in an all-or-nothing manner that appears to depend on the duration and size of puffing at the two sites of earliest reaction.

We know nothing as yet about the mechanism by which the hormone regulates the genes. We are not even certain of the exact point of action of the hormone, although we would like to believe that it is the gene itself. The induction of puffing can be prevented with inhibitors of nucleic acid metabolism such as actinomycin or mitomycin, but inhibitors of protein synthesis, such as chloramphenicol and puromycin, have no apparent effect on the puffing. Thus ecdysone does not seem to act through the stimulation of protein synthesis in the cytoplasm, or to depend for its action on this synthesis. Only further investigation will solve such problems.

# 17

# *Hormones and Genes*

ERIC H. DAVIDSON

*June 1965*

In the living cell the activities of life proceed under the direction of the genes. In a many-celled organism the cells are marshaled in tissues, and in order for each tissue to perform its role its cells must function in a cooperative manner. For more than a century biologists have studied the ways in which tissue functions are controlled, providing the organism with the flexibility it needs to adapt to a changing environment. Gradually it has become clear that among the primary controllers are the hormones. Thus whereas the genes control the activities of individual cells, these same cells constitute the tissues that respond to the influence of hormones.

New experimental evidence is now making it possible to complete this syllogism: it is being found that hormones can affect the activity of genes. Hormones of the most diverse sources, molecular structure and physiological influence appear able to rapidly alter the pattern of genetic activity in the cells responsive to them. The establishment of a link between hormones and gene action completes a conceptual bridge stretching from the molecular level to ecology and animal behavior.

In order to understand the nature of the link between hormones and genes it will be useful to review briefly what is known of how genes function in differentiated, or specialized, cells. One of the most striking examples of cell specialization in animals is the red blood cell, the protein content of which can be more than 90 percent hemoglobin. It has been shown that in man the ability to manufacture a given type of hemoglobin is inherited; this provides a clear case of a differentiated-cell function under genetic control. Hemoglobin also furnishes an example of another principle that is fundamental to the study of differentiation: the specialized character of a cell depends on the type and quantity of proteins in it, and therefore the process of differentiation is basically the process of developing a specific pattern of protein synthesis. Some cells, such as red blood cells and the cells of the pancreas that produce digestive enzymes, specialize in synthesizing one kind of protein; other cells specialize in synthesizing an entire set of protein enzymes to manufacture nonprotein end products, for example glycogen, or animal starch (which is made by liver cells), and steroid hormones (which are made by cells of the adrenal cortex).

If one understood the means by which the type and quantity of protein made by cells was controlled, one would have taken a long step toward understanding the nature of the differentiated cell. Part of this objective has been attained: we now know something of how genes act and how proteins are synthesized. A protein owes its properties to the sequence of amino acid subunits in its chainlike molecule. The genes of most organisms consist of deoxyribonucleic acid (DNA), the chainlike molecules of which are made up of nucleotide subunits. The sequence of nucleotides in a single gene determines the sequence of amino acids in a single protein.

The protein is not assembled directly on the gene; instead the cell copies the sequence of nucleotides in the gene by synthesizing a molecule of ribonucleic acid (RNA). This "messenger" RNA moves away from the gene to the small bodies called ribosomes. On the ribosomes, which contain their own unique kind of RNA, the amino acids are assembled into protein. In the assembly process each molecule of amino acid is identified and moved into position through its attachment to a specific molecule of a third kind of RNA: "transfer" RNA. It can therefore be said that the characteristics of the cell are determined at the level of "gene transcription"—the synthesis of messenger and ribosomal RNA.

Each differentiated cell in a many-celled organism contains a complete set of the organism's genes. It is obvious, however, that in such a cell only a small fraction of the genes are actually functioning; the gene for hemoglobin is not active in a skin cell and the assortment of genes active in a liver cell is not the same as the assortment active in an adrenal cell. The active genes release their information in the form of messenger RNA and the inactive genes do not. Exactly how the inactive genes are repressed is not clearly understood, but the repression seems to involve a chemical combination between DNA and the proteins called histones; it has been shown that histones inhibit the synthesis of messenger RNA in the isolated nuclei of calf-thymus cells, and similar results have been obtained with the nuclei of other kinds of cell. In any case it is clear that the characteristics of the cell are the result of variable gene activity. The prime question becomes: How are the genes selectively turned on or selectively repressed during the life of the cell?

Gene action is often closely linked to cell function in terms of time. It has been demonstrated that genes can exercise immediate control over the activities of differentiated cells—particularly very active or growing cells—and over cells that are going through some change of state. In many specialized cells at least part of the messenger RNA

| HORMONE | SOURCE | CHEMICAL NATURE | FUNCTION |
|---|---|---|---|
| ECDYSONE | INSECT PROTHORACIC GLAND | STEROID | Causes molting, initiation of adult development and puparium formation. |
| GLUCOCORTICOIDS (CORTISONE) | ADRENAL CORTEX | STEROID | Causes glycogen synthesis in liver.<br>Causes redistribution of fat throughout organism.<br>Alters nitrogen balance.<br>Causes complete revision of white blood cell type frequencies.<br>Is required for muscle function.<br>Alters central nervous system excitation threshold.<br>Affects connective tissue differentiation.<br>Promotes healing.<br>Induces appearance of new enzymes in liver.<br>Affects almost all tissues. |
| INSULIN | PANCREAS (ISLETS OF LANGERHANS) | POLYPEPTIDE | Affects entry rate of carbohydrates, amino acids, cations and fatty acids into cells.<br>Promotes protein synthesis.<br>Affects glycogen synthetic activity.<br>Stimulates fat synthesis.<br>Stimulates acid mucopolysaccharide synthesis.<br>Affects almost all tissues. |
| ESTROGEN | OVARY | STEROID | Promotes appearance of secondary sexual characteristics.<br>Increases synthesis of contractile and other proteins in uterus.<br>Increases synthesis of yolk proteins in fowl liver.<br>Increases synthesis of polysaccharides.<br>Affects rates of glycolysis, respiration and substrate uptake into cells.<br>Probably affects almost all tissues. |
| ALDOSTERONE | ADRENAL CORTEX | STEROID | Controls sodium and potassium excretion and cation flux across many internal body membranes. |
| PITUITARY ACTH | ANTERIOR PITUITARY | POLYPEPTIDE | Stimulates glucocorticoid synthesis by adrenal cortex.<br>Stimulates adrenal protein synthesis and glucose uptake.<br>Inhibits protein synthesis in adipose tissue.<br>Stimulates fat breakdown. |
| PITUITARY GH | ANTERIOR PITUITARY | PROTEIN | Stimulates all anabolic processes.<br>Affects nitrogen balance, water balance, growth rate and all aspects of protein metabolism.<br>Stimulates amino acid uptake and acid mucopolysaccharide synthesis.<br>Affects fat metabolism.<br>Probably affects all tissues. |
| THYROXIN | THYROID | THYRONINE DERIVATIVE | Affects metabolic rate, growth, water and ion excretion.<br>Promotes protein synthesis.<br>Is required for normal muscle function.<br>Affects carbohydrate levels, transport and synthesis.<br>Probably affects all tissues. |

**HORMONES DISCUSSED IN THIS ARTICLE** are listed according to their source, their chemical nature and their effects, which are usually quite diverse. Pituitary GH is the pituitary growth hormone. The steroid hormones share a basic molecular skeleton consisting of adjoining four-ring structures. The polypeptide hormones and the protein hormones consist of chains of amino acid subunits.

produced by the active genes decays in a matter of hours, and therefore the genes must be continuously active for protein synthesis to continue normally. Other differentiated cells display the opposite characteristic, in that gene activity occurs at a time relatively remote from the time at which the messenger RNA acts. The very existence of this time element in gene control of cell function indicates how extensive that control is. Furthermore, certain genes can be alternately active and inactive over a short period; for example, if a leaf is bleached by being kept in the dark and is then exposed to light, it immediately begins to manufacture messenger RNA for the synthesis of chlorophyll.

The sum of such observations is that the patterns of gene activity in the living cell are in a state of continuous flux. For a cell in a many-celled organism, however, it is essential that the genetic apparatus be responsive to external conditions. The cell must be able to meet changing situations with altered metabolism, and if all the cells in a tissue are to alter their metabolism in a coordinated way, some kind of organized external control is needed. Evidence obtained from experiments with a number of biological systems suggests that such control is obtained by externally modulating the highly variable activity of the cellular genetic apparatus. The studies that will be reviewed here are cases of this general proposition; in these cases the external agents that alter the pattern of gene activity are hormones.

Many efforts have been made to explain the basis of hormone action. It has been suggested that hormones are coenzymes (that is, cofactors in enzymatic reactions), that they activate key enzymes, that they modify the outer membrane of cells and that they directly affect the physical state of structures within the cell. For each hypothesis there is evidence from studies of one or several hormones. As an example, experiments with the pituitary hormone vasopressin, which causes blood vessels to constrict and decreases the excretion of urine by the kidney, strongly support the conclusion that the hormone attaches itself to the outer membrane of the cells on which it acts.

To these hypotheses has been added the new one that hormones act by regulating the genetic apparatus, and many investigators have undertaken to study the effects of hormones on gene activity. It turns out that the gene-regulation hypothesis is more successful than the others in explaining some of the most puzzling features of hormone activity, such as the time lag between the administration of some hormones and the initial appearance of their effects, and also the astonishing variety of these effects [*see illustration on opposite page*]. There can be no doubt that some hormone action is independent of gene activity, but it has now been shown that a wide variety of hormones can affect such activity. This conclusion is strongly supported by the fact that each of these same hormones is powerless to exert some or all of its characteristic effects when the genes of the cells on which it acts are prevented from functioning.

The genes can be blocked by the remarkably specific action of the antibiotic actinomycin D. The antibiotic penetrates the cell and forms a complex with the cell's DNA; once this has happened the DNA cannot participate in the synthesis of messenger RNA. The specificity of actinomycin is indicated by the fact that it does not affect other activities of the cell: protein synthesis, respiration and so on. These activities continue until the cellular machinery stops because it is starved for messenger RNA. In high concentrations actinomycin totally suppresses the synthesis of messenger RNA; in lower concentrations it depresses this synthesis and appears to prevent it from developing at new sites.

So far the greatest number of studies of the effects of hormones on genes have been concerned with the steroid hormones, particularly the estrogens produced by the ovaries. This work has been carried forward by many investigators in many laboratories. It has been found that when the ovaries are removed from an experimental animal and then estrogen is administered to the animal at a later date, the synthesis of protein by cells in the uterus of the animal increases by as much as 300 percent. The increase is detected by measuring the incorporation of radioactively labeled amino acids into uterine protein, or by testing the capacity for protein synthesis of homogenized uterine tissue removed from the animal at various times after the administration of estrogen. Added proof that these observations have to do with the synthesis of protein is provided by the fact that the stimulating effects of estrogen are blocked by the antibiotic puromycin, which specifically inhibits protein synthesis.

In these experiments the principal rise in protein synthesis is first observed between two and four hours after estrogen treatment. Less than 30 minutes after the treatment, however, there is a dramatic increase in the rate of RNA synthesis. When actinomycin is used to block the rise in RNA synthesis, the administration of estrogen has no effect on protein synthesis! What this means is that since the diverse metabolic changes brought about in uterine cells by estrogen are all mediated by protein enzymes, none of the changes can occur unless the estrogen has induced gene action. Among the changes are the increased synthesis of amino acids from glucose, the increased evolution of carbon dioxide and the increased synthesis of the fatty lipids and phospholipids. It is not surprising to find that none of these metabolic changes in uterine cells can be detected when estrogen is administered to an animal that has first been treated with actinomycin.

The effect of estrogen on the synthesis of RNA is not limited to messenger RNA. There is also an increase in the manufacture of the other two kinds of RNA: transfer RNA and ribosomal RNA. The administration of estrogen first stimulates the production of messenger RNA and transfer RNA. The genes responsible for the synthesis of ribosomal RNA become active somewhat later, and the number of ribosomes per cell increases. One of the earliest changes brought about by estrogen, however, is an increase in the activity of the enzyme RNA-DNA polymerase. This enzyme appears to be responsible for all RNA synthesis in such cells.

Two main conclusions can be drawn from these various observations. First, there can be no reasonable doubt that treatment with estrogenic hormones results in activation at the gene level, and that many of the well-known effects of estrogen on uterine cells result from this gene activation. Second, it is clear that a considerable number of genes must be activated in order to account for the many different responses of the cells to estrogen. Consider only the fact that estrogen stimulates the production of three different kinds of RNA. At least two different genes are known to be associated with the synthesis of ribosomal RNA, and each cell needs to manufacture perhaps as many as 60 species of transfer RNA. As for messenger RNA, the variety of the changes induced by estrogen implies that under such influences it too must be produced

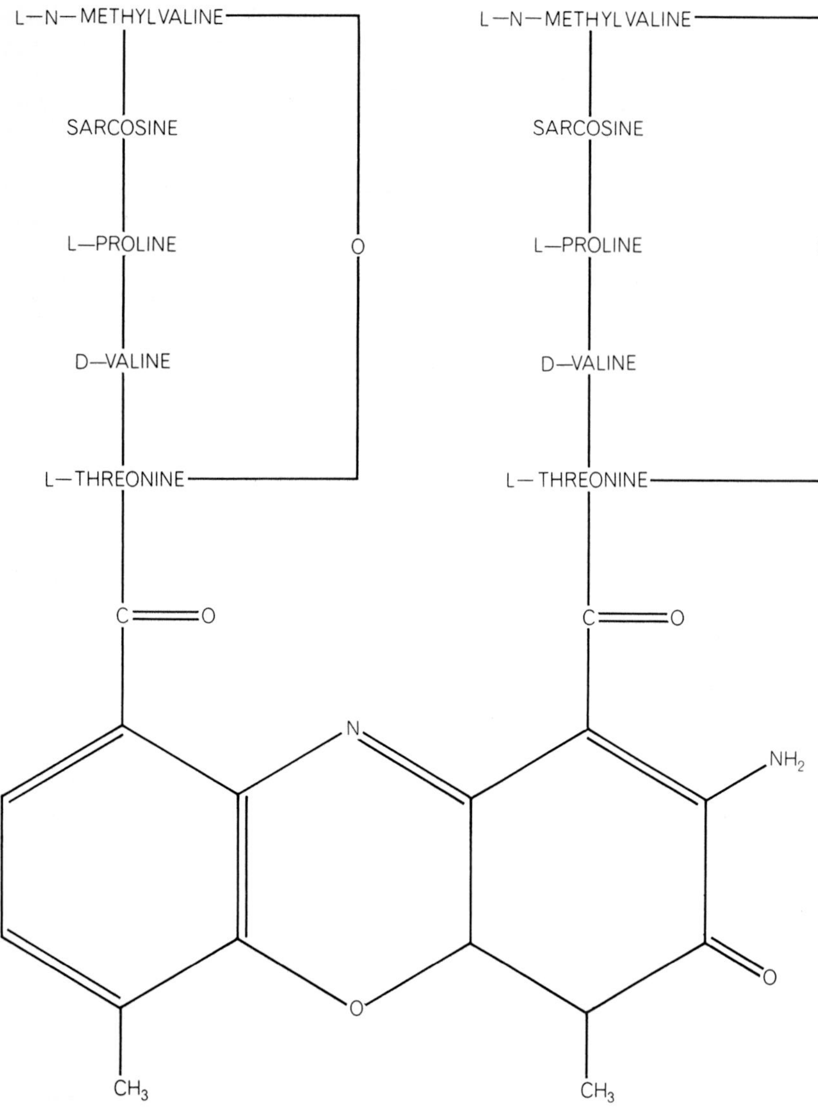


**ANTIBIOTIC ACTINOMYCIN D has a complex chemical structure. The antibiotic blocks the participation of the genetic material in the synthesis of ribonucleic acid (RNA); thus it can be used in studies to determine whether or not a given hormone stimulates gene activity.**

in a number of molecular species. We are therefore confronted with a major mystery of gene regulation: How can a single hormone activate an entire set of functionally related but otherwise quite separate genes, and activate them in a specific sequence and to a specific degree?

The question can be sharpened somewhat by considering the effect of estrogen not on uterine cells but on the cells of the liver. When an egg is being formed in a hen, the estrogen produced by the hen's ovaries stimulates its liver to produce the yolk proteins lipovitellin and phosvitin. Obviously a rooster does not need to synthesize these proteins, but if it is treated with estrogen, its liver will make them in large amounts! A more unequivocal example of the selective activation of repressed genes by a hormone could scarcely be imagined. What is more, experiments by E. N. Carlsen and his co-workers at the University of California School of Medicine (Los Angeles) have demonstrated that this gene-activating effect of estrogen is remarkably specific. Phosvitin is an unusual protein in that nearly half of its subunits are of one kind: they are residues of the amino acid serine. Carlsen and his colleagues found that estrogen most strongly stimulates liver cells to produce the particular species of transfer RNA that is associated with the incorporation of this amino acid into protein.

The effect of estrogen on liver cells is thus quite different from its effect on uterine cells. Indeed, it has long been recognized that hormonal specificity resides less in the hormone than in the "target" cell. We are now, however, able to ask new questions: How are the sets of genes that are activated by a given hormone selected? Are these genes somehow preset for hormonal activation? How does the hormone interact not only with the gene itself but also with the cell's entire system of genetic regulation?

The male hormone testosterone has also been shown to operate by gene activation. Like the estrogens, the male sex hormones can give rise to dramatic increases of RNA synthesis in various cells. In experiments on male and female rats it has been found that the effect of testosterone on the liver cells of a female is somewhat different from that on the liver cells of a castrated male. In both cases the hormone causes an increase in the *amount* of messenger RNA produced, but in the female it also brings about the synthesis of a new *variety* of messenger RNA. This effect, like the ability of estrogen to stimulate a rooster's liver cells to produce egg-yolk proteins, provides a new approach for examining the whole question of sexual differentiation.

Apart from the sex hormones, the principal steroids in mammals are those secreted by the adrenal cortex. One group of adrenocortical hormones is typified by cortisone; this hormone and its relatives are known for their quite different effects in different tissues. Only a fraction of these effects have been studied from the standpoint of gene activation, and there is much evidence to indicate that some of them are not mediated by the genes. Some responses to cortisone, however, do appear to be the consequence of gene activation.

If the adrenal glands are removed from an experimental animal and cortisone is administered later, the hormone induces in the liver cells of the animal the production of a number of new proteins. Among these proteins are enzymes required for the synthesis of glucose (but not the breakdown of glucose) and enzymes involved in the metabolism of amino acids. Moreover, cortisone steps up the total production of protein by the liver cells. The effect of cortisone on the synthesis of messenger RNA is apparent as soon as five minutes after the hormone has been administered; within 30 minutes the amount of RNA produced has increased two to three times and probably includes not

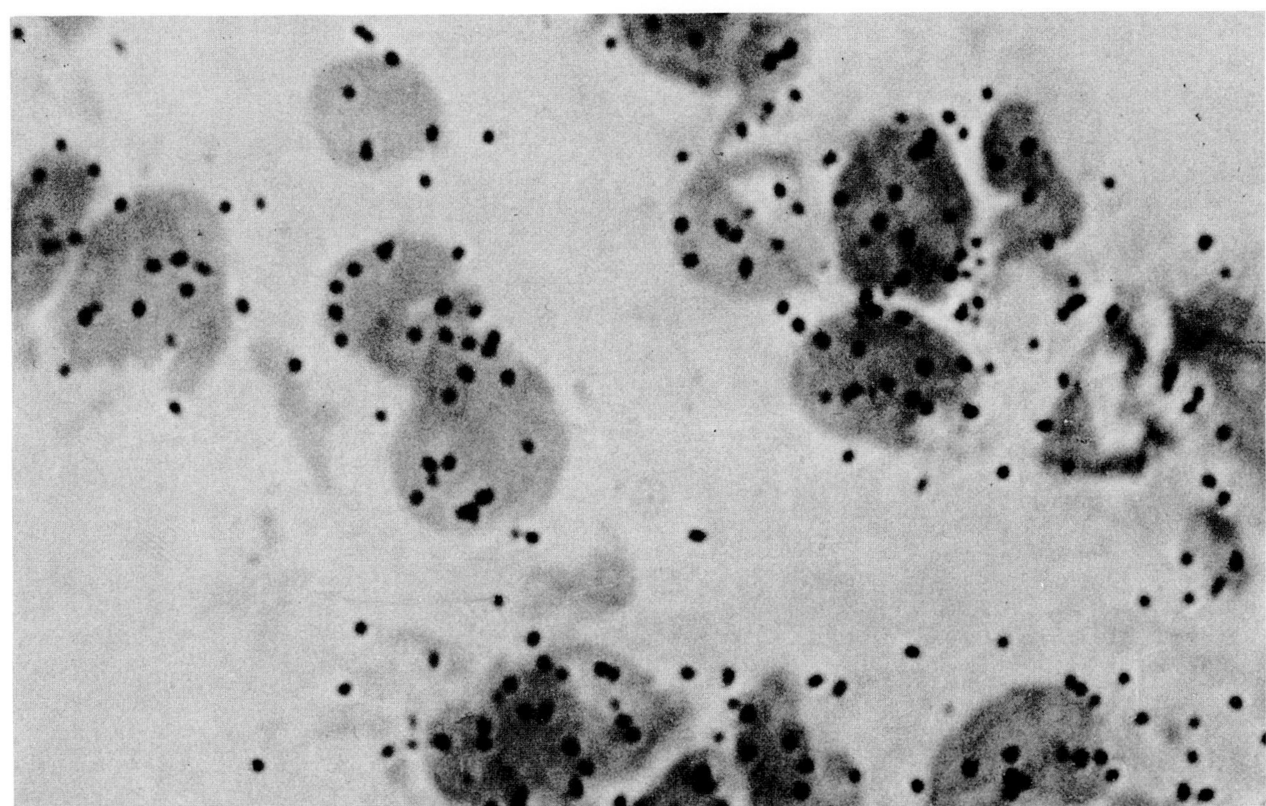

HORMONE IS LOCALIZED IN NUCLEI of cells in this radioautograph made by George A. Porter, Rita Bogoroch and Isidore S. Edelman of the University of California School of Medicine (San Francisco). The hormone aldosterone was radioactively labeled and administered to a preparation of toad bladder tissue. When the tissue was radioautographed, the hormone revealed its presence by black dots. The dots appear predominantly in the nuclei (*dark gray areas*) of the cells rather than in the cytoplasm (*light gray areas*).

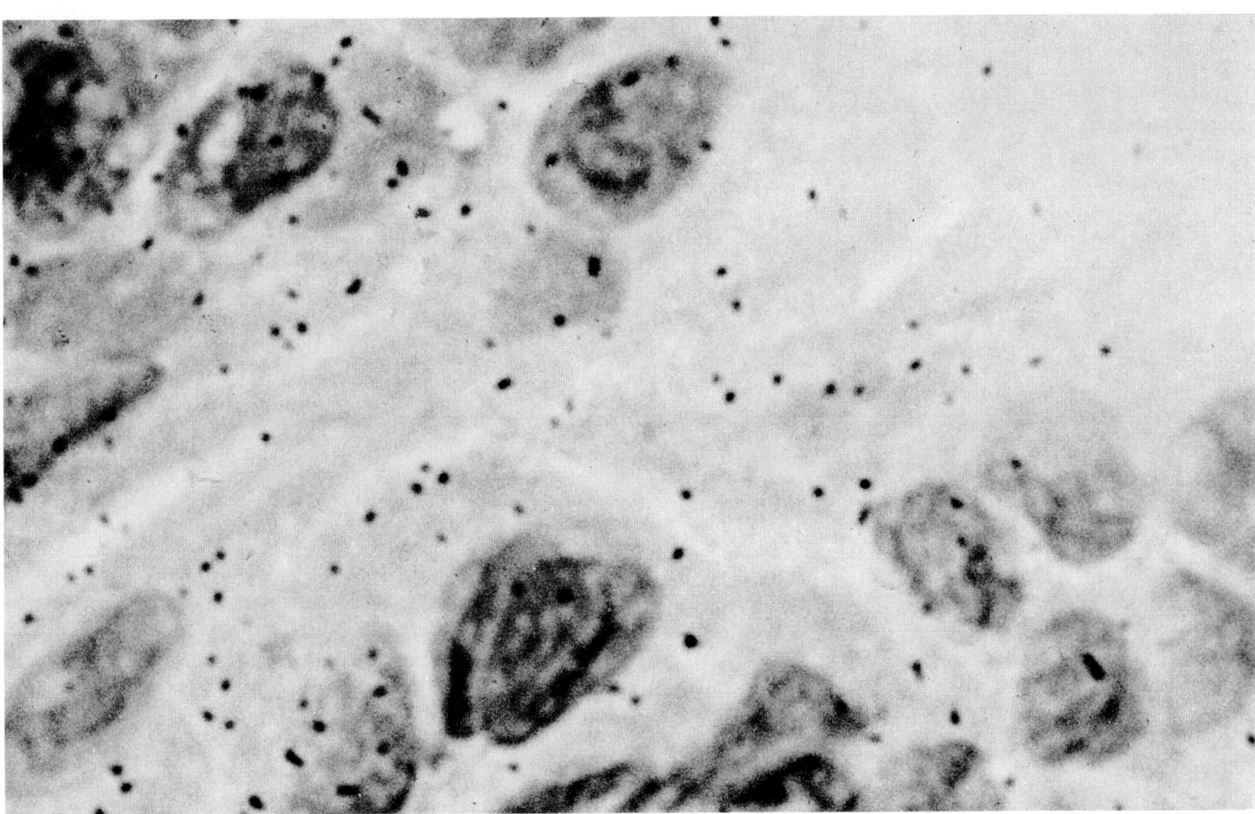

ANOTHER HORMONE IS NOT LOCALIZED in the nuclei in this radioautograph made by the same investigators. Here the hormone was progesterone, and it too was labeled and administered to toad bladder tissue. The dots are distributed more or less at random.

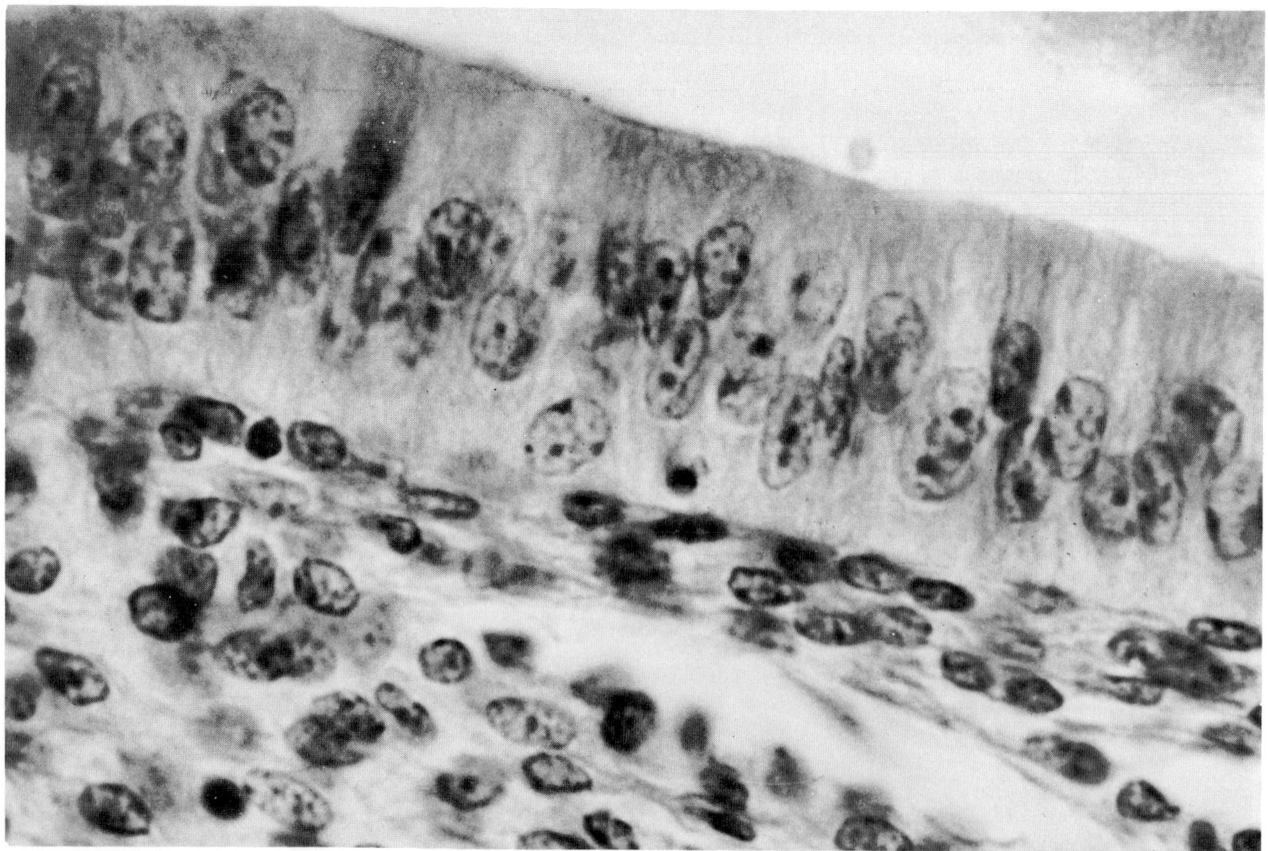

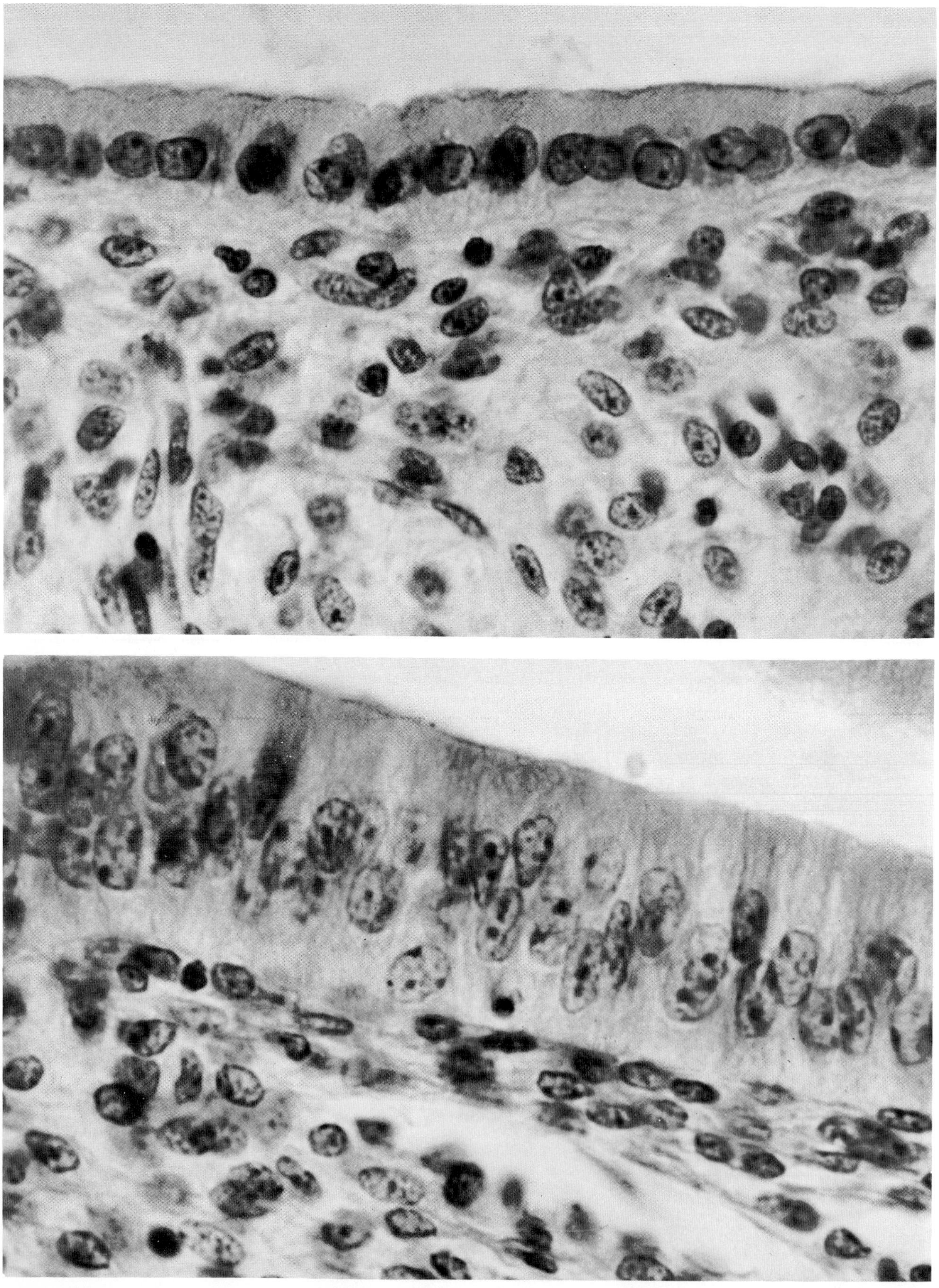

EFFECT OF ESTROGEN ON CELLS in the uterus of rats is demonstrated in these photomicrographs made by Sheldon J. Segal and G. P. Talwar of the Rockefeller Institute. The photomicrograph at top shows uterine cells from a rat that had not been treated with estrogen; the layer of cells at the surface of the tissue is relatively thin. The photomicrograph at bottom shows uterine cells from a rat that had been treated with the hormone; the layer of cells is much thicker. The effect involves enhanced synthesis of protein.

only messenger RNA but also ribosomal RNA. These events are followed by the increase in enzyme activity. Olga Greengard and George Acs of the Institute for Muscle Disease in New York have shown that if the animal is treated with actinomycin before cortisone is administered, the new enzymes fail to appear in its liver cells.

Another clear case of the activation of genes by an adrenocortical hormone has been demonstrated by Isidore S. Edelman, Rita Bogoroch and George A. Porter of the University of California School of Medicine (San Francisco). They employed the hormone aldosterone, which regulates the passage through the cell membrane of sodium and potassium ions. Tracer studies with radioactively labeled aldosterone showed that when the bladder cells of a toad were exposed to the hormone, the molecules of hormone penetrated all the way into the nuclei of the cells [*see illustrations on page 163*]. About an hour and a half after the aldosterone has reached its peak concentration within the cells the movement of sodium ions across the cell membrane increases. It appears that this facilitation of sodium transport is brought about by proteins the cell is induced to make, because it will not occur if the cells have been treated beforehand with puromycin, the drug that blocks the synthesis of protein. Moreover, treatment of the cells with actinomycin will block the aldosterone-induced increase in sodium transport through the membrane. Thus the experiments indicate that aldosterone activates genes in the nucleus and gives rise to proteins—that is, enzymes—that speed up the passage of sodium ions across the membrane.

Ecdysone, a steroid hormone of insects, is also believed to be a gene activator. The evidence for this conclusion has been provided by Wolfgang Beermann and his colleagues at the Max Planck Institute for Biology in Tübingen [see "Chromosome Puffs," by Wolfgang Beermann and Ulrich Clever, which begins on page 150 in this book]. If the larva of an insect lacks ecdysone, the development of the larva is indefinitely arrested at a stage preceding its metamorphosis into a pupa. Only when, in the course of normal development, the concentration of ecdysone in the tissues of the larva begins to rise does further differentiation take place; the larva then advances to metamorphosis. Ecdysone has been of especial interest to cell biologists because it has been observed

**ROOSTER TREATED WITH ESTROGEN** (*bottom*) **is compared with a normal rooster** (*top*)**. The signs of femaleness induced by estrogen include changes in comb and plumage.**

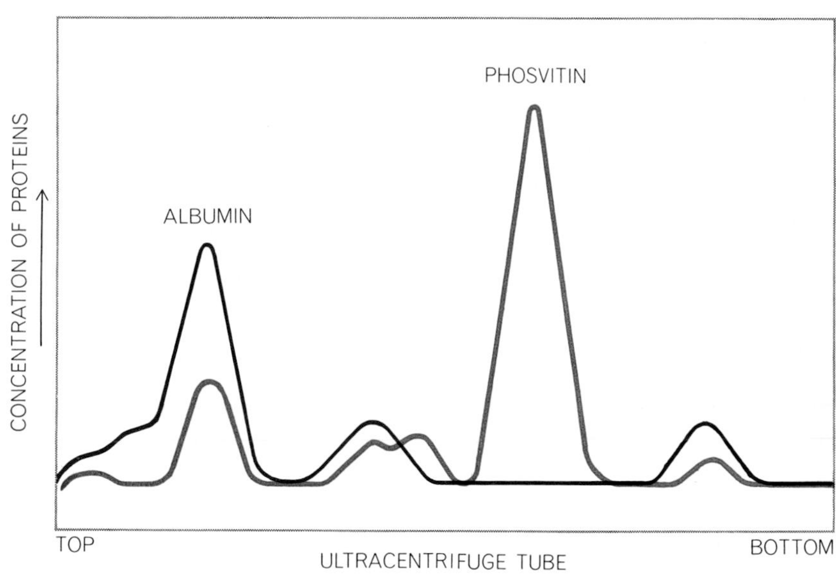


**ULTRACENTRIFUGE PATTERNS show that phosvitin, a yolk protein found only in hens, is present in serum extracted from a bird that had been injected with estrogen** (*colored curve*) **but not in serum from a bird used as a control** (*black curve*)**. Each curve gives the concentration of proteins as they are separated out of a mixture by an ultracentrifuge.**

to cause startling changes in the chromosomes within the nuclei of the cells affected by it. Studies of this kind are possible in insects because the cells of certain insect tissues have giant chromosomes that can easily be examined in the microscope. These "polytene" chromosomes develop in many kinds of differentiated cell by means of a process in which the chromosomes repeatedly replicate but do not separate.

In some polytene chromosomes genetic loci, or specific regions, have a distended, diffuse appearance [*see illustration below*]. Biologists regard these regions, which have been named "puffs," as sites of intense gene activity. Evidence for this conclusion is provided by radioautograph studies, which show that the puffs are localized sites of intense RNA synthesis. In such studies a molecular precursor of RNA is radioactively labeled and after it has been incorporated into RNA reveals its presence as a black dot in the emulsion of the radioautograph. According to the view of differentiation presented in this article, different genes should be active in different types of cell, and this appears to be the case in insect cells with polytene chromosomes. In many different kinds of cell—salivary-gland cells, rectal-gland cells and excretory-tubule cells—the giant chromosomes have a different constellation of puffs; this suggests that different sets of genes are active, a given gene being active in one cell and quiescent in another.

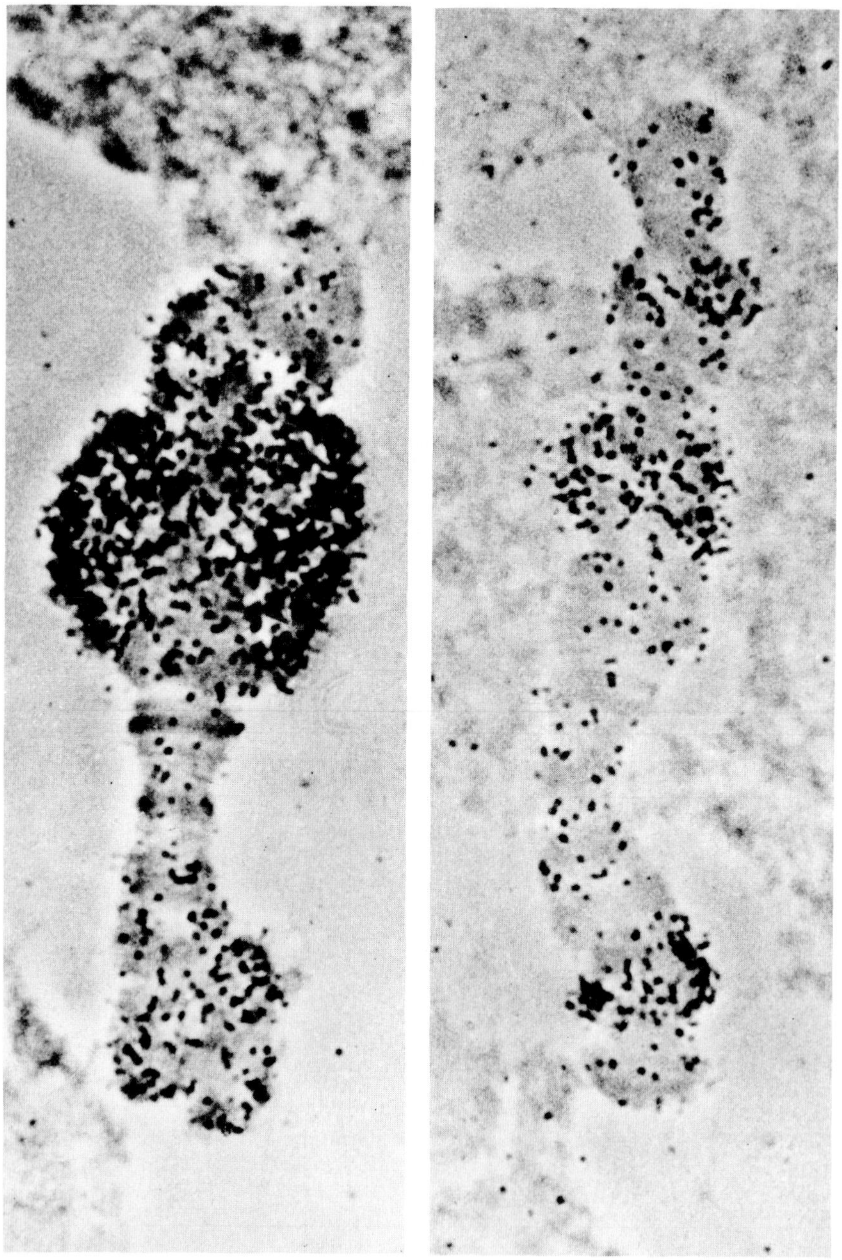

On the polytene chromosomes of insect salivary-gland cells new puffs develop as metamorphosis begins. This is where ecdysone comes into the picture: the hormone seems to be capable of inducing the appearance of specific new puffs. When a minute amount of ecdysone is injected into an insect larva, a specific puff appears on one of its salivary-gland chromosomes; when a slightly larger amount of ecdysone is injected, a second puff materializes at a different chromosomal location. In the normal course of events the concentration of ecdysone increases as the larva nears metamorphosis; therefore there exists a mechanism whereby the more sensitive genetic locus can be aroused first. This example of hormone action at the gene level, which is directly visible to the investigator, seems to have provided some of the strongest evidence for the regulation of gene action by hormones. The effect of ecdysone, which is clearly needed for differentiation, appears to be to arouse quiescent genes to visible states of activity. In this way the specific patterns of gene activity required for differentiation are provided.

**"PUFF" ON A GIANT CHROMOSOME from the salivary gland of the midge *Chironomus tentans* appears after administration of the insect hormone ecdysone. In the radioautograph at left the round area at top center is a puff. The black dots result from the fact that the midge was given radioactively labeled uridine, which is a precursor of RNA. The concentration of dots in the puff indicates that it is actively synthesizing RNA. In the radioautograph at right is a chromosome from a fly that had been treated with actinomycin before receiving ecdysone. No puff has occurred and RNA synthesis appears to be muted. The radioautographs were made by Claus Pelling of the Max Planck Institute for Biology in Tübingen.**

What about nonsteroid hormones? Here the overall picture is not as clearcut. The effects of some hormones are quite evidently due to gene activation, and yet other effects of the same hormones are not blocked by the administration of actinomycin; a small sample of these effects is listed in the illustration on the opposite page. As for the hormonal effects that are quite definitely not genetic, they fall into one of the following categories.

(1) Some hormones act on specific enzymes; for example, the thyroid hormone thyroxin promotes the dissociation of the enzyme glutamic dehydrogenase. (2) Other hormones, for instance insulin and vasopressin, act on systems that transport things through cell mem-

| HORMONE | EVIDENCE FOR HORMONAL ACTION BY GENE ACTIVATION. | EVIDENCE THAT HORMONAL ACTION IS CLEARLY INDEPENDENT OF IMMEDIATE GENE ACTIVATION. |
|---|---|---|
| PITUITARY GROWTH HORMONE | General stimulation of protein synthesis.<br>Stimulation of rates of synthesis of ribosomal RNA, transfer RNA and messenger RNA within 90 minutes in liver.<br>Effect blocked with actinomycin. | |
| PITUITARY ACTH | Stimulates adrenal protein synthesis.<br>Messenger RNA and total RNA synthesis stimulated. | Steroid synthesis in isolated adrenal sections is independent of RNA synthesis and is insensitive to actinomycin D. |
| THYROXIN | Promotes new messenger RNA synthesis within 10 to 15 minutes of administration, promotes stimulation of all classes of RNA by 60 minutes.<br>Promotes increase in RNA–DNA polymerase at 10 hours, later promotes general increase in protein synthesis. | Causes isolated, purified glutamic dehydrogenase to dissociate to the inactive form.<br>Affects isolated mitochondria in vitro. |
| INSULIN | Promotes 100 percent increase in rate of RNA synthesis.<br>Causes striking change in messenger RNA profile within 15 minutes of administration to rat diaphragm; effect blocked with actinomycin.<br>Actinomycin-sensitive induction of glucokinase activity. | Actinomycin-insensitive increase in ATP synthesis and in glucose transport into cells; mechanism appears to involve insulin binding to cell membrane, occurs at 0 degrees C. |
| VASOPRESSIN | | Actinomycin-insensitive promotion of water transport in isolated bladder preparation under same conditions in which aldosterone action is blocked by actinomycin. |

**SUMMARY OF EXPERIMENTAL EVIDENCE is given in table. Facts indicating that hormones activate the genes (*middle column*) are compared with facts suggesting that hormonal action does not entail the immediate activation of the genes (*column at right*).**

branes; indeed, it is believed that both of these hormones attach themselves directly to the membranes whose function they affect. (3) Still other hormones rapidly activate a particular enzyme; phosphorylase, a key enzyme in determining the overall rate at which glycogen is broken down, is converted from an inactive form by several hormones, including epinephrine, glucagon and ACTH.

This does not alter the fact that many nonsteroid hormones operate at the gene level. Some of the best evidence for this statement is provided by studies of several hormones made by Chev Kidson and K. S. Kirby of the Chester Beatty Research Institute of the Royal Cancer Hospital in London. They separately injected rats with thyroxin, testosterone, cortisone and insulin and then measured the synthesis of messenger RNA by the rats' liver cells [*see illustration on page 168*]. The most striking aspect of their measurements is the extremely short time lag between the administration of the hormone and the change in the pattern of gene activity. The activation of genes in the nuclei of the affected cells occurs so quickly that one is tempted to assume that it is an initial effect of the hormone.

Here, however, we come face to face with a basic problem that must be solved in any attempt to explain the exact molecular mechanism of hormone action. The problem is simply that of identifying the initial site of reaction in a cell exposed to a hormone. Does a hormone move directly to the chromosome and exert its effect, so to speak, "in person"? As we have seen, aldosterone does appear to enter the nucleus, but there is little real evidence that other hormones do so.

For many years biologists have been looking for the "receptor" substance of various hormones. The discovery that hormones ultimately act on genes makes this search all the more interesting. The evidence presented here only goes as far as to prove that an early stage in the operation of many hormones is the selective stimulation of genetic activity in the target cell. The molecules of the hormones range in size and structure from the tiny molecule of thyroxin to the unique multi-ring molecule of a steroid and the giant molecule of a protein; how these various molecules similarly affect the genetic apparatus of their target cells remains an intriguing mystery.

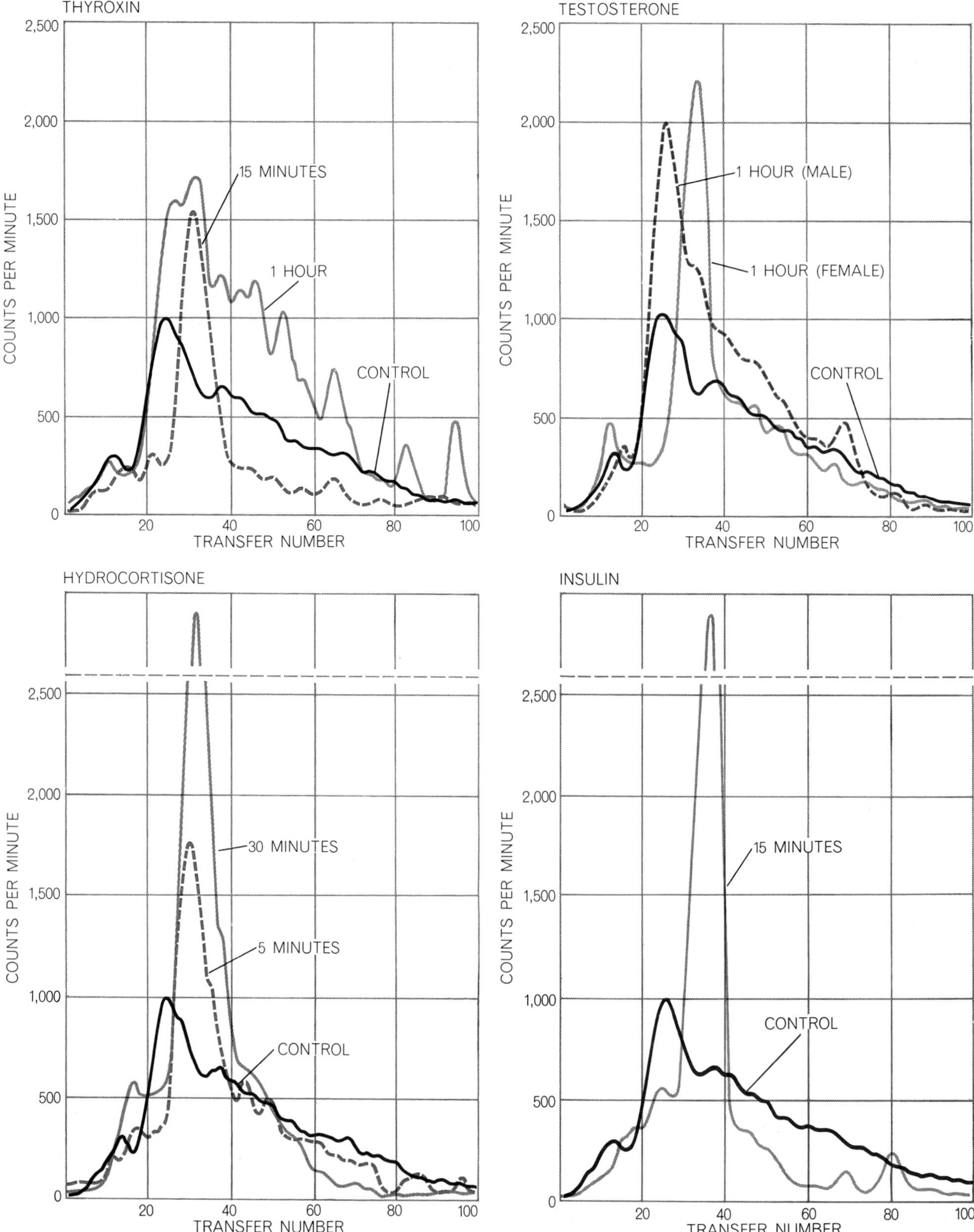


**GENETIC ACTIVITY OF SEVERAL HORMONES** is indicated by measurements made by Chev Kidson and K. S. Kirby of the Chester Beatty Research Institute in London. Their basic technique was first to administer to rats radioactively labeled orotic acid, which is a precursor of RNA. The tissues of the rat then incorporated the radioactive label into new RNA. Next liver tissue was removed from the rat and the species of RNA called "messenger" RNA was extracted from its cells. When the messenger RNA was analyzed by the method of countercurrent distribution, it gave rise to a characteristic curve (*black "Control" curve in each graph*); "Transfer number" refers to a stage of transfer in the countercurrent-distribution process and "Counts per minute" to the radioactivity of the solution at that point. Then, in separate measurements, rats were first given one of a number of hormones (*top left of each graph*) and shortly thereafter radioactively labeled orotic acid. The curves (*color*) of the messenger RNA obtained from such rats were entirely different, depending on the time that had elapsed before the administration of the orotic acid or on the sex of the animal (*top right*).

# 18
# *Building a Bacterial Virus*

WILLIAM B. WOOD and
R. S. EDGAR
*July 1967*

Slice an orange in half, squeeze the juice into a pitcher and then drop in the rind. It comes as no surprise that the orange does not reconstitute itself. If, on the other hand, the components of the virus that causes the mosaic disease of tobacco are gently dissociated and then brought together under the proper conditions, they do reassociate, forming complete, infectious virus particles. The tobacco mosaic virus consists of a single strand of ribonucleic acid with several thousand identical protein subunits assembled around it in a tubular casing. The orange, of course, is a large and complex structure composed of a variety of cell types incorporating many different kinds of proteins and other materials. Yet both orange and virus are examples of biological architecture that must arise as a consequence of the action of genes.

Molecular biologists have now provided a fairly complete picture of how

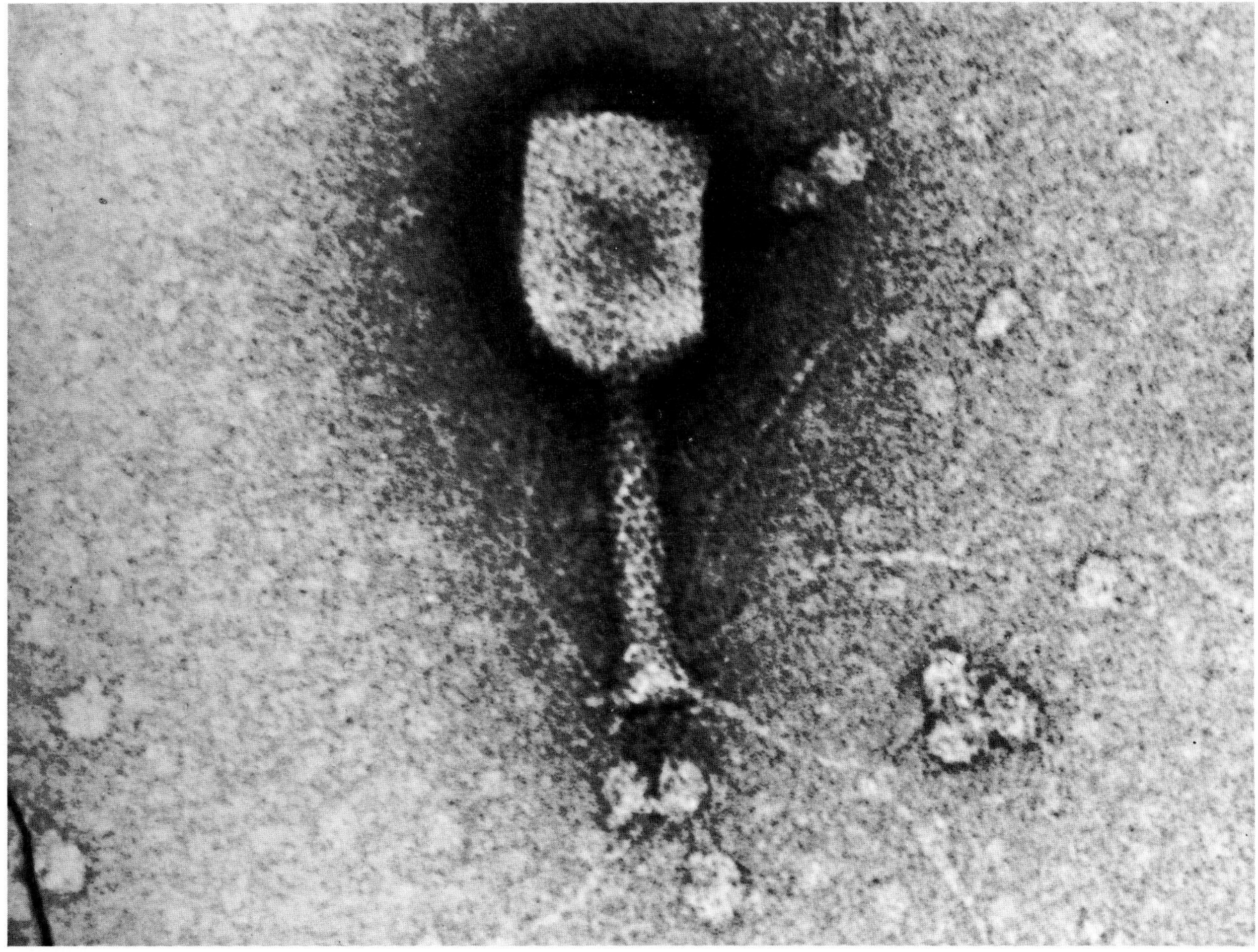

**COMPLETE T4 PARTICLE was built by assembling component parts in the test tube. The virus is enlarged about 300,000 diameters in this electron micrograph made, like the ones on the next page, by Jonathan King of the California Institute of Technology.**

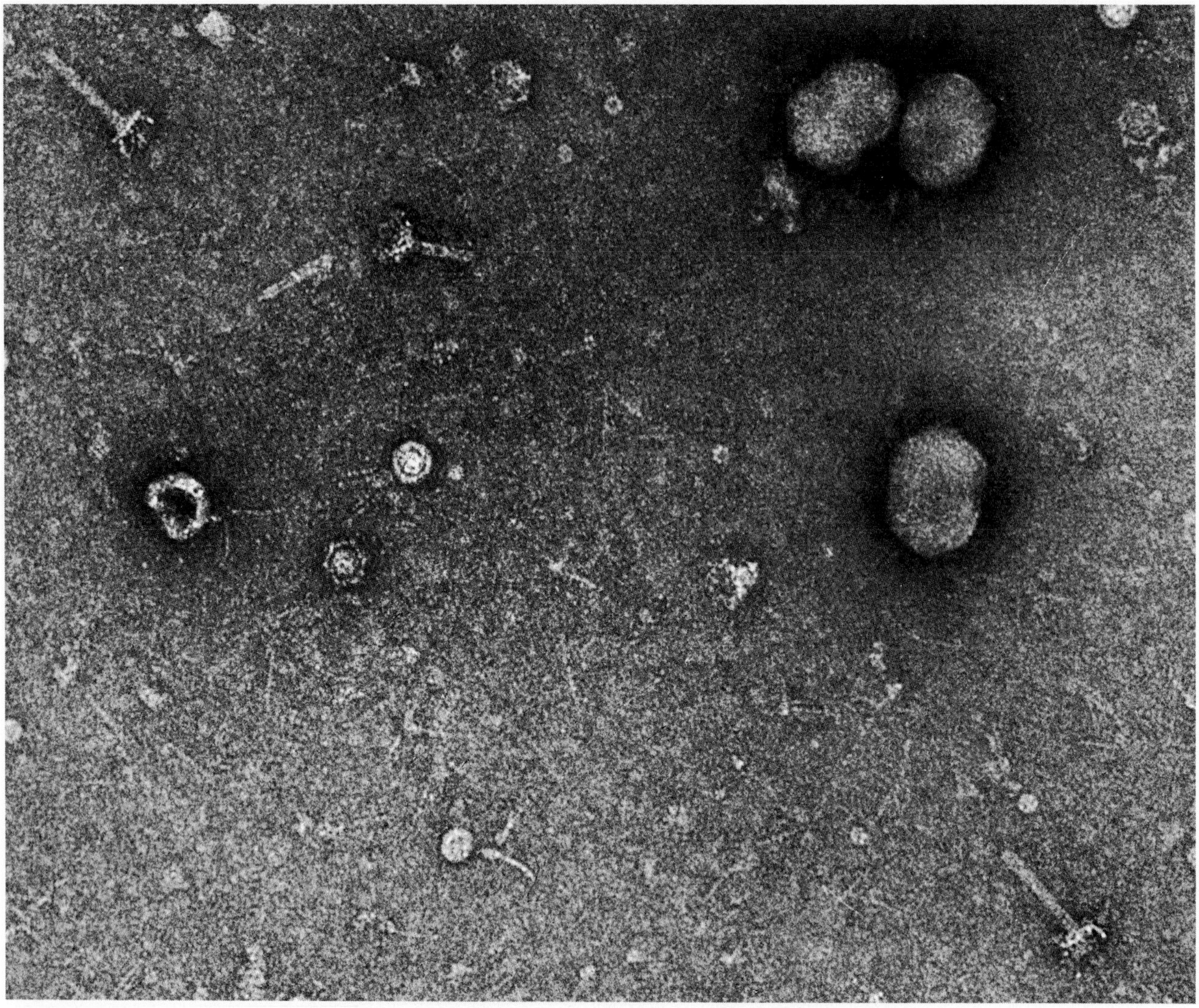

UNASSEMBLED PARTS of the T4 virus are present in this extract. It was prepared by infecting colon bacilli with a mutant virus defective in gene No. 18, which specifies the synthesis of the sheath (*see upper illustration on page 5*). The result is the accumulation of all major components except the sheath: heads, free tail fibers and "naked" tails consisting of cores and end plates.

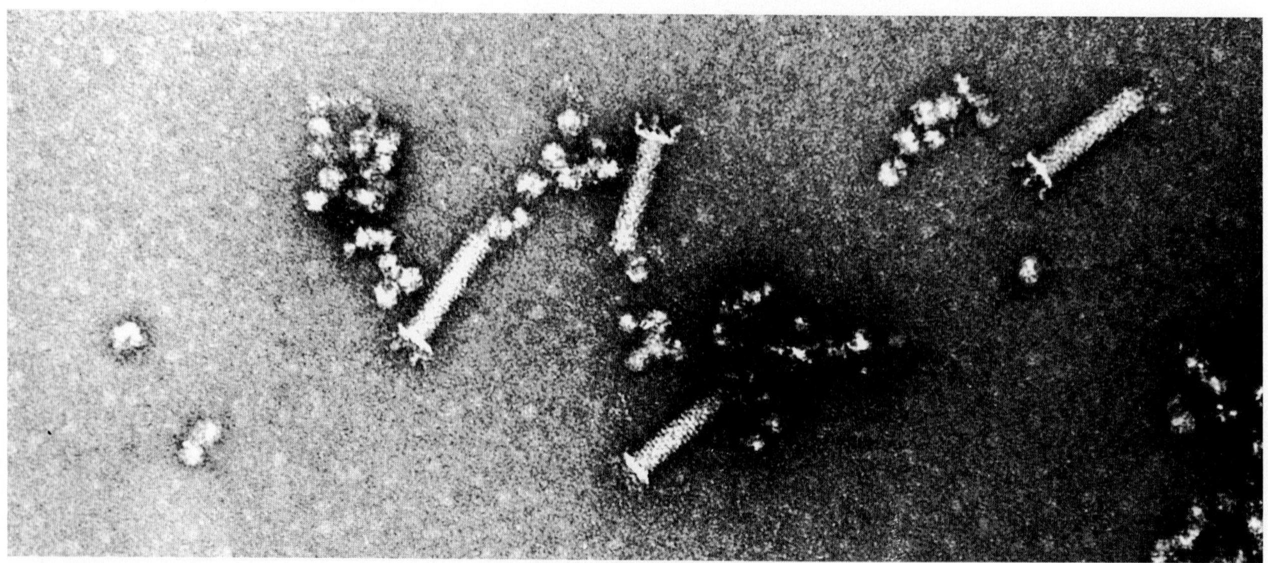

COMPLETE TAILS, enclosed in sheaths, were produced by a different mutant, defective in a gene involved in head formation. The tails were separated from the resulting extract (along with some spherical bacterial ribosomes) by being spun in a centrifuge. If the tails are added to the extract (*top photograph*), they combine with the heads and free fibers in it to form infectious virus.

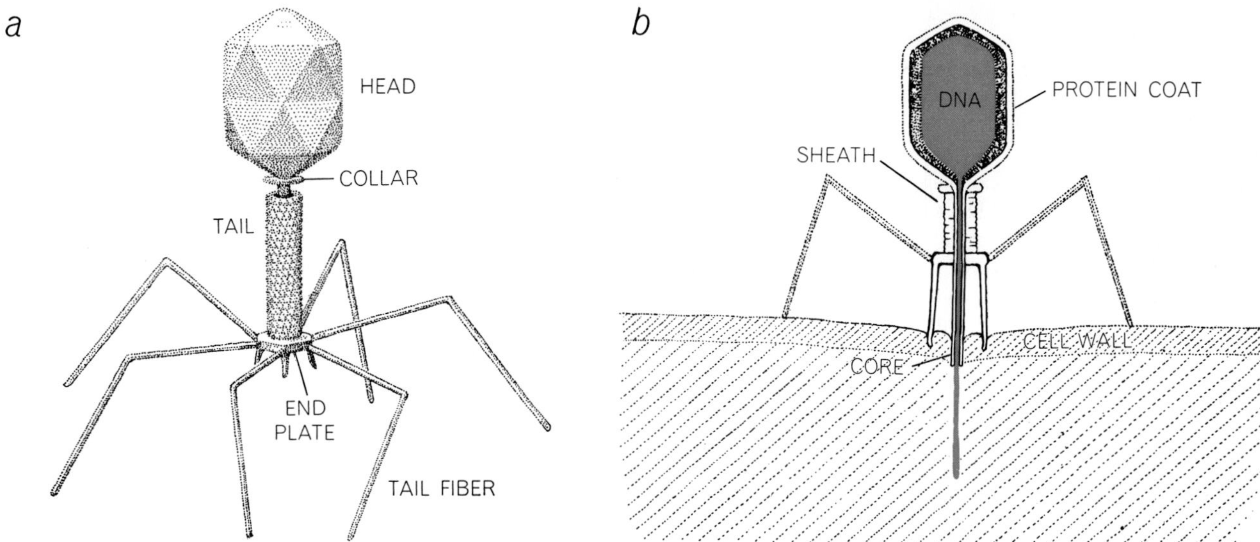


**T4 BACTERIAL VIRUS is an assembly of protein components (*a*). The head is a protein membrane, shaped like a kind of prolate icosahedron with 30 facets and filled with deoxyribonucleic acid (DNA). It is attached by a neck to a tail consisting of a hollow core surrounded by a contractile sheath and based on a spiked end plate to which six fibers are attached. The spikes and fibers affix the virus to a bacterial cell wall (*b*). The sheath contracts, driving the core through the wall, and viral DNA enters the cell.**

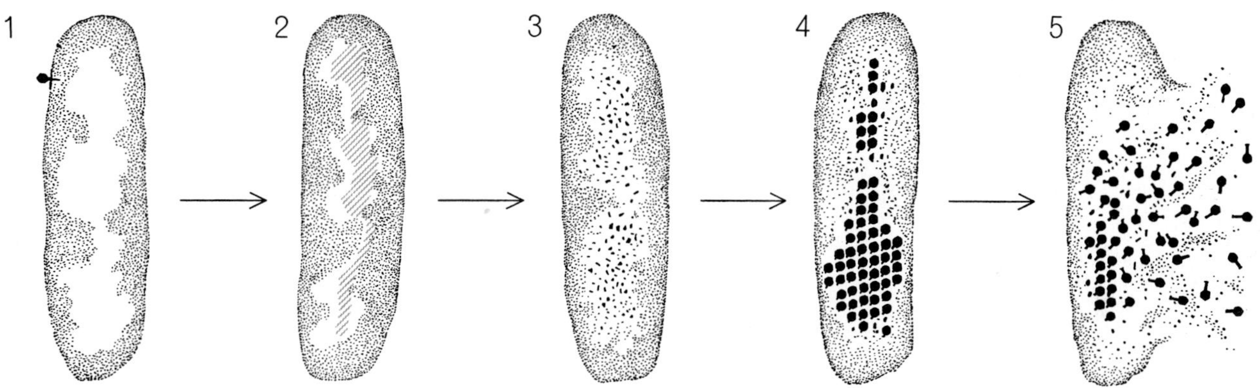


**VIRAL INFECTION begins when viral DNA (*color*) enters a bacterium (*1*). Bacterial DNA is disrupted and viral DNA replicated (*2*). Synthesis of viral structural proteins (*3*) and their assembly into virus (*4*) continues until the cell bursts, releasing particles (*5*).**

genes carry out their primary function: the specification of protein structure. The segment of nucleic acid (DNA or RNA) that constitutes a single gene specifies the chain of amino acids that comprises a protein molecule. Interactions among the amino acids cause the chain to fold into a unique configuration appropriate to the enzymatic or structural role for which it is destined. In this way the information in one gene determines the three-dimensional structure of a single protein molecule.

Where does the information come from to direct the next step: the assembly of many kinds of protein molecules into more complex structures? To build the relatively simple tobacco mosaic virus no further information is required; the inherent properties of the strand of RNA and the protein subunits cause them to interact in a unique way that results in the formation of virus particles. Clearly such a self-assembly process cannot explain the morphogenesis of an orange. At some intermediate stage on the scale of biological complexity there must be a point at which self-assembly becomes inadequate to the task of directing the building process. Working with a virus that may be just beyond that point, the T4 virus that infects the colon bacillus, we have been trying to learn how genes supply the required additional information.

Although the T4 virus is only a few rungs up the biological ladder from the tobacco mosaic virus, it is considerably more complex. Its DNA, which comprises more than 100 genes (compared with five or six in the tobacco mosaic virus), is coiled tightly inside a protein membrane to form a polyhedral head. Connected to the head by a short neck is a springlike tail consisting of a contractile sheath surrounding a central core and attached to an end plate, or base, from which protrude six short spikes and six long, slender fibers.

The life cycle of the T4 virus begins with its attachment to the surface of a colon bacillus by the tail fibers and spikes on its end plate. The sheath then contracts, driving the tubular core of the tail through the wall of the bacterial cell and providing an entry through which the DNA in the head of the virus can pass into the bacterium. Once inside, the genetic material of the virus quickly takes over the machinery of the cell. The bacterial DNA is broken down, production of bacterial protein stops and within less than a minute the cell has begun to manufacture viral proteins under the control of the injected virus genes. Among the first proteins to be made are the enzymes needed for viral DNA replication, which begins five minutes after infection. Three minutes later a second set of genes starts to direct the synthesis of the structural proteins that will form the head components and the tail components, and the process of viral morphogenesis begins. The first completed virus particle materializes 13 minutes after infection. Synthesis of both the DNA and the protein components continues for 12 more minutes until about 200 virus particles have accumulated within the cell. At this point a viral enzyme, lysozyme, attacks the cell wall from the inside to break open the bacterium and liberate the new viruses for a subsequent round of infection.

Additional insight into this process has come from studying strains of T4 carrying mutations—molecular defects that arise randomly and infrequently in the viral DNA during the course of its replication [see "The Genetics of a Bacterial Virus," by R. S. Edgar and R. H. Epstein, which begins on page 128 in this book]. When a mutation is present, the protein specified by the mutant gene is synthesized in an altered form. This new protein is often nonfunctional, in which case the development of the virus stops at the point where the protein is required. Normally such a mutation has little experimental use, since the virus in

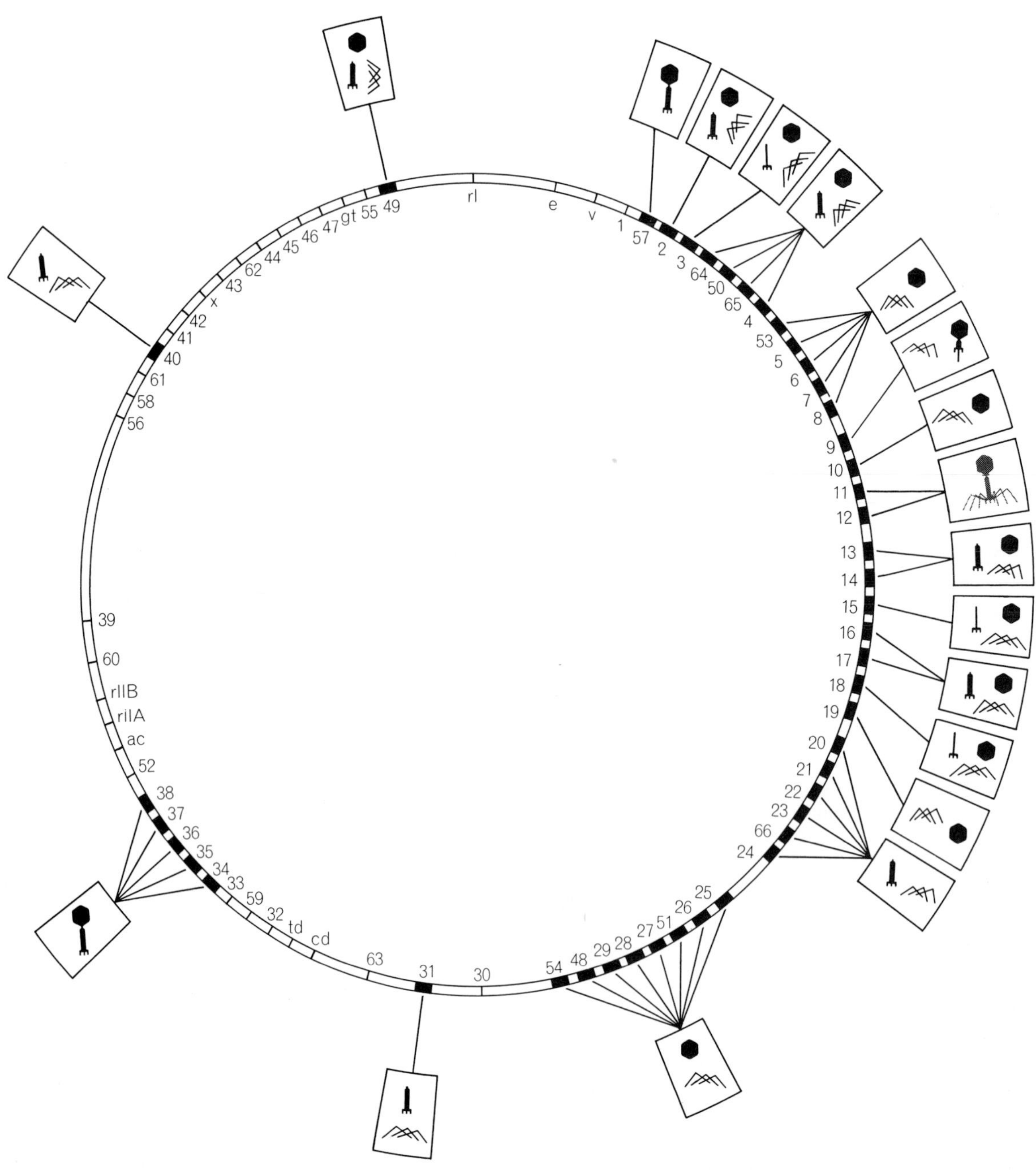


**GENETIC MAP of the T4 virus shows the relative positions of more than 75 genes so far identified on the basis of mutations. The solid black segments of the circle indicate genes with morphogenetic functions. The boxed diagrams show which viral components are seen in micrographs of extracts of cells infected by mutants defective in each morphogenetic gene. A defect in gene No. 11 or 12 produces a complete but fragile particle. Heads, all tail parts, sheaths or fibers are the missing components in other extracts.**

which it arises is dead and hence cannot be recovered for study. Edgar and Epstein, however, found mutations that are only "conditionally lethal": the mutant protein is produced in either a functional or a nonfunctional form, depending on the conditions of growth chosen by the experimenter. Under "permissive" conditions reproduction is normal, so that the mutants can be cultured and crossed for genetic studies. Under "restrictive" conditions, however, viral development comes to a halt at the step where the protein is needed, and by determining the point at which development is blocked the investigator can infer the normal function of the mutated gene. In this way a number of conditionally lethal mutations have been assigned to different genes, have been genetically mapped and have been tested for their effects on viral development under restrictive conditions [*see illustration on page 172*].

In the case of genes that control the later stages of the life cycle, involving the assembly of virus particles, mutations lead to the accumulation of unassembled viral components. These can be identified with the electron microscope. By noting which structures are absent as a result of mutation in a particular gene, we learn about that morphogenetic gene's normal function. For example, genes designated No. 23, No. 27 and No. 34 respectively appear to control steps in the formation of the head, the tail and the tail fibers; these are the structures that are missing from the corresponding mutant-infected cells.

A blockage in the formation of one of these components does not seem to affect the assembly of the other two, which accumulate in the cell as seemingly normal and complete structures. This information alone provides some insight into the assembly process. The virus is apparently not built up the way a sock is knitted—by a process starting at one end and adding subunits sequentially until the other end is reached. Instead, construction seems to follow an assembly-line process, with three major branches that lead independently to the formation of heads, tails and tail fibers. The finished components are combined in subsequent steps to form the virus particle.

A second striking aspect of the genetic map is the large number of genes controlling the morphogenetic process. More than 40 have already been discovered, and a number probably remain to be identified. If all these genes specify proteins that are component parts of the virus, then the virus is considerably more complex than it appears to be. Alternatively, however, some gene products

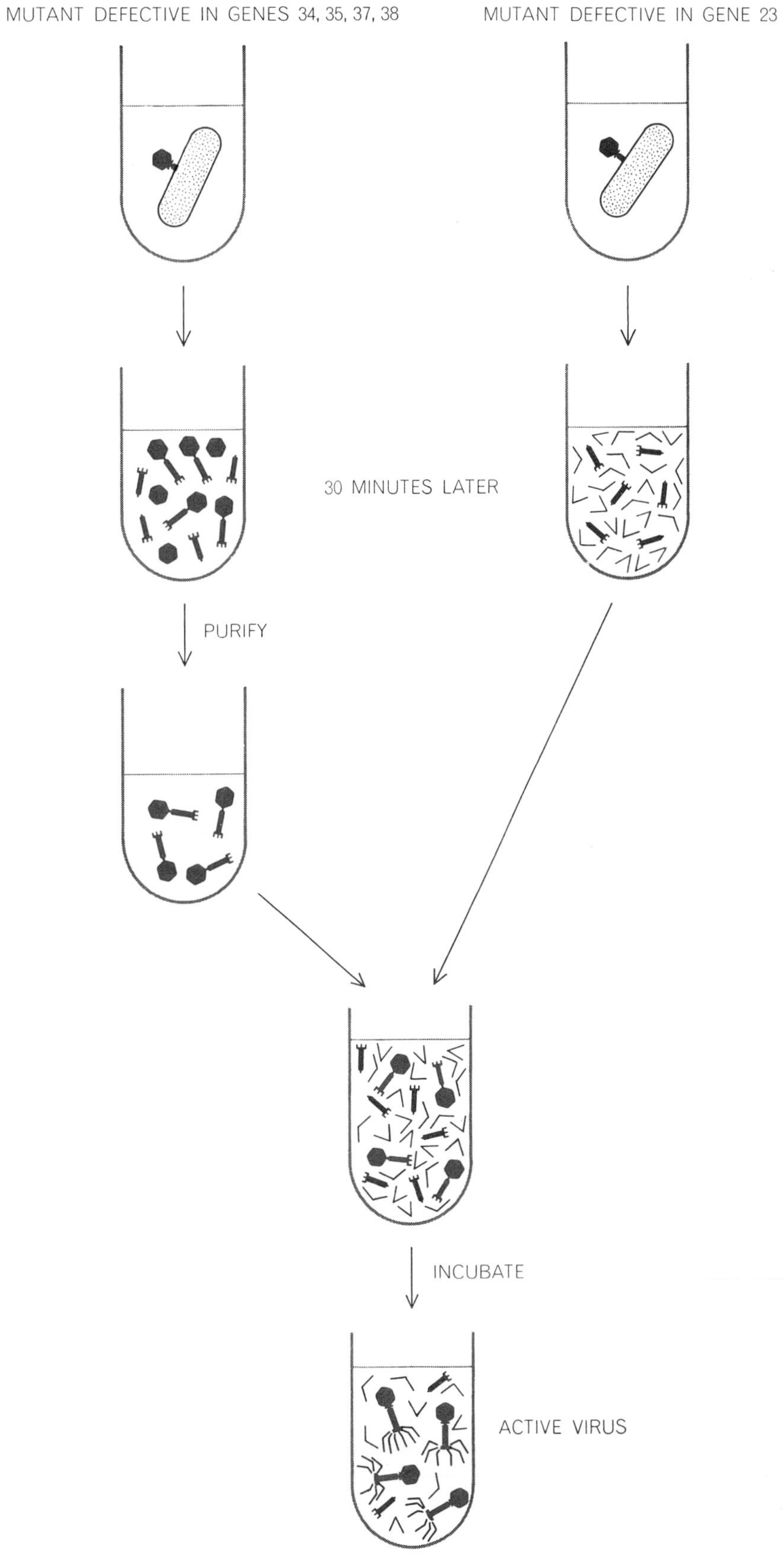

**TAIL FIBERS** are attached to fiberless particles in the experiment diagrammed here. Cells are infected with a virus (*color*) bearing defective tail-fiber genes. The progeny particles, lacking fibers, are isolated with a centrifuge. A virus with a head-gene mutation (*black*) infects a second bacterial culture, providing an extract containing free tails and fibers. When the two preparations are mixed and incubated at 30 degrees centigrade, the fiberless particles are converted to infectious virus particles by the attachment of the free fibers.

may play directive roles in the assembly process without contributing materially to the virus itself. Studies of seven genes controlling formation of the virus's head support this possibility [see "The Genetic Control of the Shape of a Virus," by Edouard Kellenberger; SCIENTIFIC AMERICAN Offprint 1058].

In order to determine the specific functions of the many gene products involved in morphogenesis, it seemed necessary to seek a way to study individual assembly steps under controlled conditions outside the cell. One of us (Edgar) is a geneticist by training, the other (Wood) a biochemist. The geneticist is inclined to let reproductive processes take their normal course and then, by analyzing the progeny, to deduce the molecular events that must have occurred within the organism. The biochemist is eager to break the organism open and search among the remains for more direct clues to what is going on inside. For our current task a synthesis of these two approaches has proved to be most fruitful. Since it seemed inconceivable that the T4 virus could be built from scratch like the tobacco mosaic virus, starting with nucleic acid and individual protein molecules, we decided to let cells infected with mutants serve as sources of preformed viral components. Then we would break open the cells and, by determining how the free parts could be assembled into complete infectious virus, learn the sequence of steps in assembly, the role of each gene product and perhaps its precise mode of action.

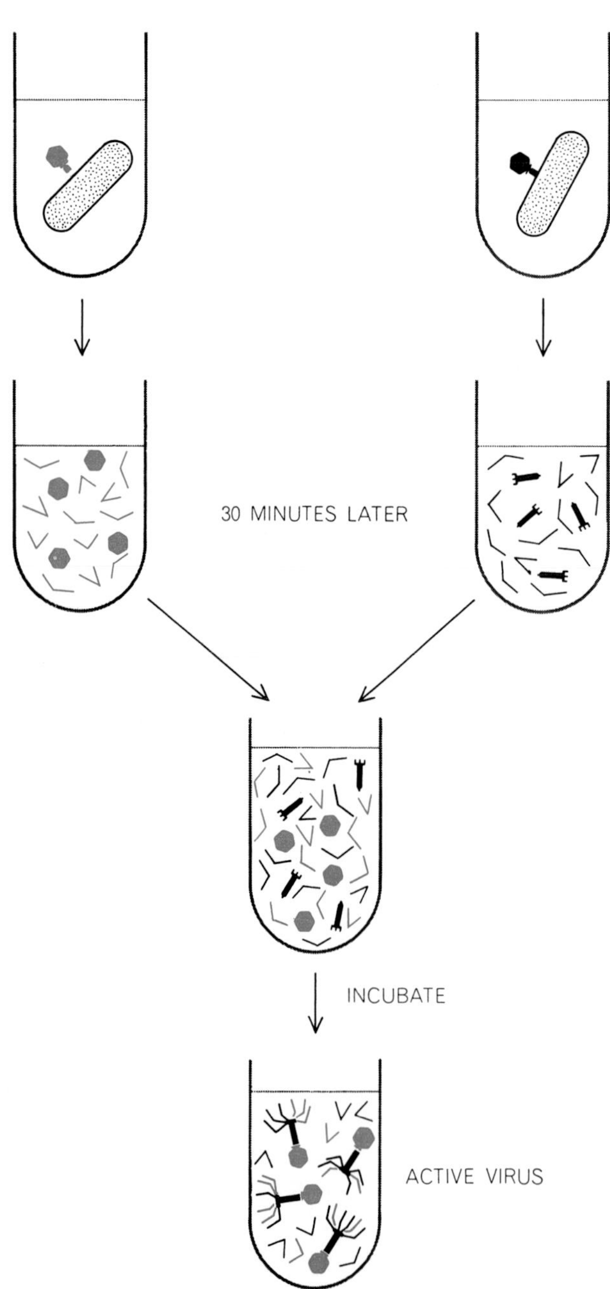


**TWO ASSEMBLY REACTIONS occur in this experiment: union of heads and tails and attachment of fibers. One virus (*color*), with a defective tail gene, produces heads and fibers. Another (*black*), with a mutation in a head gene, produces tails and fibers. When the two extracts are mixed and incubated, the parts assemble to produce infectious virus.**

Our first experiment was an attempt to attach tail fibers to the otherwise complete virus particle—a reaction we suspected was the terminal step in morphogenesis. Cells infected with a virus bearing mutations in several tail fiber genes (No. 34, 35, 37 and 38) were broken open, and the resulting particles —complete except for fibers and noninfectious—were isolated by being spun in a high-speed centrifuge. Other cells, infected with a gene No. 23 mutant that was defective in head formation, were similarly disrupted to make an extract containing free fibers and tails but no heads. When a sample of the particles was incubated with the extract, the level of infectious virus in the mixture increased rapidly to 1,000 times its initial value. Electron micrographs of samples taken from the mixture at various times showed that the particles were indeed acquiring tail fibers as the reaction proceeded.

In that first experiment the production of infectious virus required only one kind of assembly reaction—the attachment of completed fibers to completed particles. We went on to test more demanding mixtures of defective cell extracts. For example, with a mutant blocked in head formation and another one blocked in tail formation we prepared two extracts, one containing tails and free tail fibers but no heads and another containing heads and free tail fibers but no tails. When a mixture of these two extracts also gave rise to a large number of infectious viruses, we concluded that at least two reactions must have occurred: the attachment of heads to tails and the attachment of fibers to the resulting particles.

By infecting bacilli with mutants bearing defects in different genes concerned with assembly, we prepared 40 different extracts containing viral components but no infectious virus. When we tested the extracts by mixing pairs of them in many of the appropriate combinations, some mixtures produced active virus and others showed no detectable activity. The production of infective virus implied that the two extracts were complementing each other in the test tube, that each was supplying a component that was missing or defective in the other and that could be assembled into complete, active virus under our experimental conditions. Lack of activity, on the other hand, suggested that both extracts were deficient in the same viral component—a component being defined as a subassembly unit that functions in our experimental system. By analyzing the pattern of positive and negative results we could find out how many functional components we were dealing with.

It developed that there are at least 13 such components. That is, analysis of our pair combinations produced 13 complementation groups, the members of which did not complement one another but did complement any member of any other group. Two of these groups were quite large [*see illustration below*]. Since one gene produces one protein and since each extract has a different defective gene product, a mixture of any two extracts should include all the proteins required for building the virus. The fact that members of these large groups do not complement one another must mean that our experimental system is not as efficient as an infected cell; whatever the gene products that are missing in each of these extracts do, they cannot do it in the test tube.

The idea that a complementation group consisted of extracts deficient in

| EXTRACT GROUP | MUTANT GENES | COMPONENTS PRESENT | INFERRED DEFECT |
|---|---|---|---|
| I | 5, 6, 7, 8, 10, 25, 26, 27, 28, 29, 48, 51, 53 | | TAIL |
| II | 20, 21, 22, 23, 24, 31 | 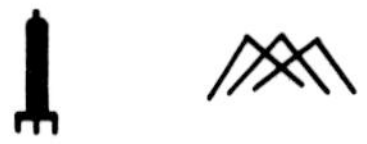 | HEAD (FORMATION) |
| | 2, 4, 16, 17, 49, 50, 64, 65 |  | HEAD (COMPLETION) |
| III | 54 |  | TAIL CORE |
| IV | 13, 14 |  | ? |
| V | 15 | | |
| VI | 18 | | |
| VII | 9 |  | ? |
| VIII | 11 |  | ? |
| IX | 12 | | |
| X | 37, 38 |  | TAIL FIBERS |
| XI | 36 | | |
| XII | 35 | | |
| XIII | 34 | | |

**COMPLEMENTATION TESTS defined 13 groups of defective extracts, as described in the text. Mixing any two extracts in a single group fails to produce infectious virus in the test tube, but mixing any two members of different groups yields infectious virus. Apparently each group represents the genes concerned with the synthesis of a component that is functional under experimental conditions. The precise nature of the defect in some extracts, and hence the function of the missing gene product, could not be identified on the basis of the structures recognized in electron micrographs and remained to be determined by additional experiments.**

the same functional component could be checked against the earlier electron micrograph results. Micrographs of the 12 defective extracts of Group I, for example, all show virus heads and tail fibers but no tails. Each of these extracts must therefore be deficient in a gene product that has to do with a stage of tail formation that cannot be carried out in our extracts. The second large complementation group appeared at first to be anomalous in terms of electron micrography: some extracts contained only tails and tail fibers, whereas others contained heads as well. Tests against extracts known to contain active tails revealed, however, that these heads—although they looked whole—could not combine to produce active virus in the test tube. In other words, heads, like tails, must be nearly completed within an infected cell before they become active for complementation. The early stages of head formation are still inaccessible to study in mixed extracts.

The remaining defective extracts gave rise to active virus in almost all possible pair combinations, segregating into another 11 complementation groups. With a total of 13 groups, there must be at least 12 assembly steps that can occur in mixtures of extracts. The defects recognizable in micrographs suggest what some of these steps must be: the completion and union of heads and tails, the assembly of tail fibers and the attachment of fibers to head-tail particles. These, then, are the steps that can be studied further in our present experimental system. We have in effect a virus-building kit, some of whose more intricate parts have been preassembled at the cellular factory.

Our next experiments were designed to determine the normal sequence of assembly reactions and further characterize those whose nature remained ambiguous. Examples of the latter were the steps controlled by genes No. 13, 14, 15 and 18. Defects in the corresponding gene products resulted in the accumulation of free heads and tails, suggesting that they are somehow involved in head-tail union. It was unclear, however, whether these gene products are required for the attachment process itself or for completion of the head or the tail before attachment. We could distinguish the alternatives by complementation tests using complete heads and tails. These we isolated from the appropriate extracts in the centrifuge, taking advantage of their large size in relation to the other materials present. On the basis of the evidence for the independent assembly of heads and tails, we assumed that

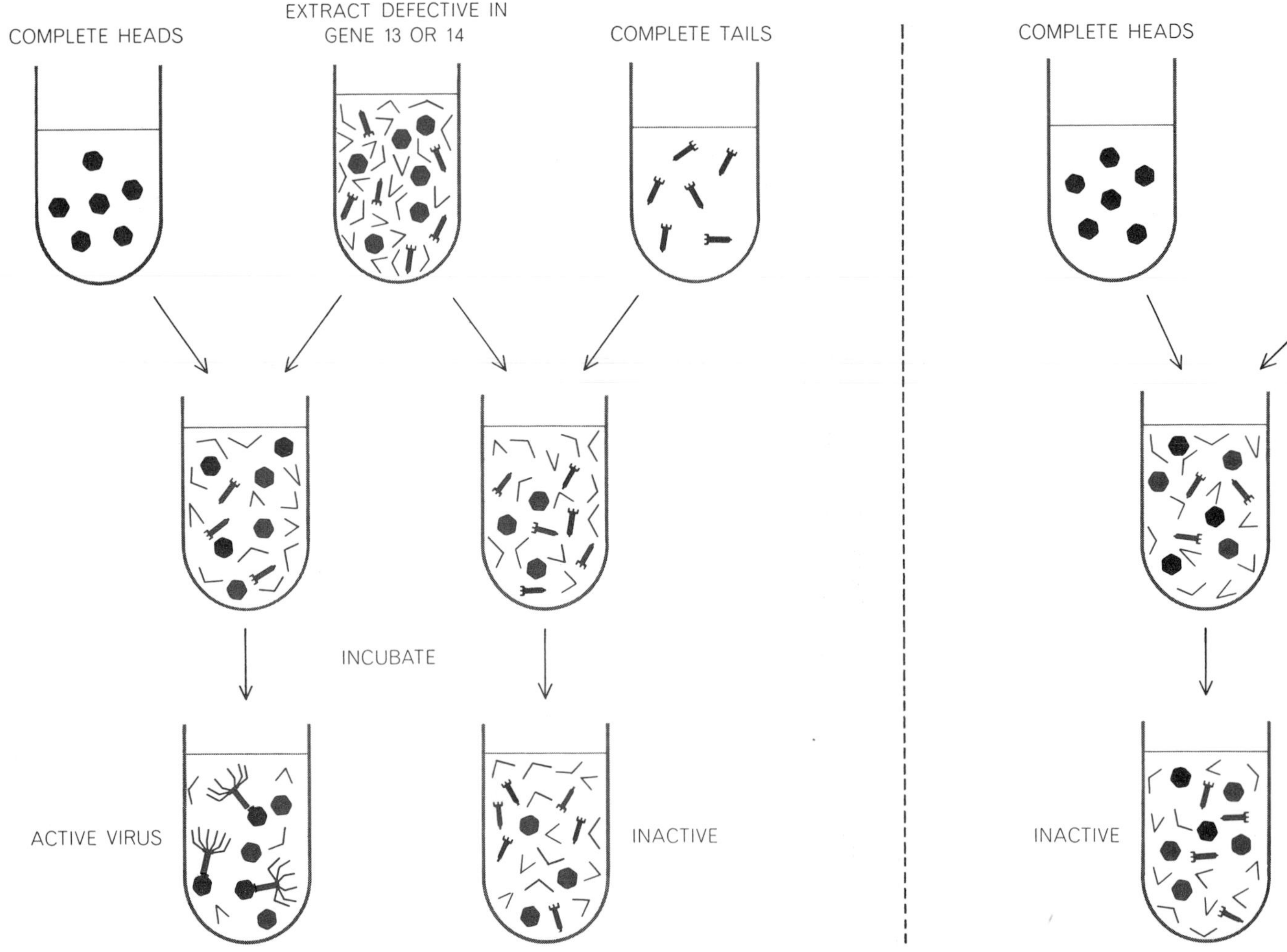


**ASSEMBLY DEFECTS of mutants (*color*) that seem to produce complete heads, tails and fibers are identified, using isolated complete heads and tails (*black*) as test reagents. When complete heads are added to some extracts to be tested (*left*), infectious virus is produced, but the addition of complete tails is ineffective. This indicates that the tails made by these mutants must be functional,**

the heads we isolated from a tail-defective extract would be complete, as would tails isolated from a head-defective extract.

The results of the tests were unambiguous. The addition of isolated heads to extracts lacking the products of gene No. 13 or 14 resulted in virus production, whereas the addition of tails did not. We could therefore conclude that the components missing from these extracts normally affect the head structure, and that genes No. 13 and 14 control head completion rather than tail completion or head-tail union. The remaining two of the four extracts gave the opposite result; these were active with added tails but not with added heads, indicating that genes No. 15 and 18 are involved in the completion of the tail. All four of these steps must precede the attachment of heads to tails, since defects in any of

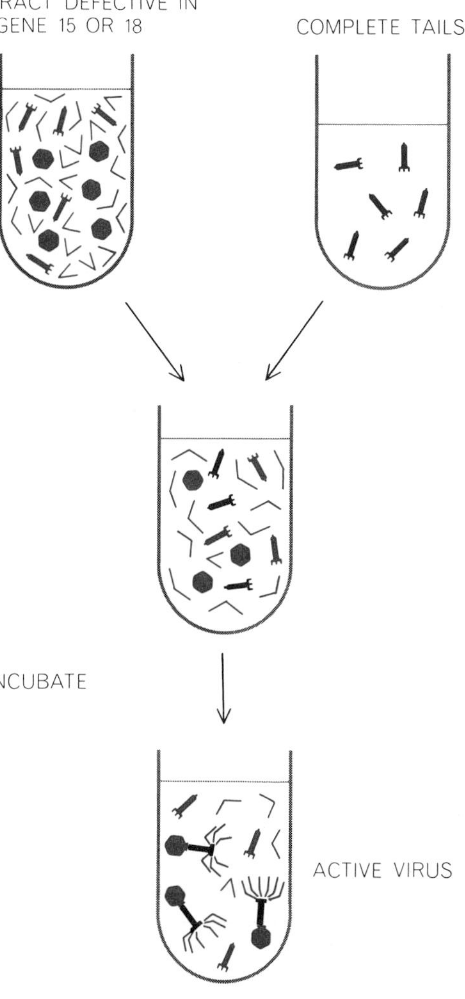


**implying that the heads must be defective. In the case of other mutants (*right*), such tests indicate that the tails must be defective.**

the corresponding genes block head-tail union.

By manipulating extracts blocked at other stages we worked out the remaining steps in the assembly process with the help of Jonathan King and Jeffrey Flatgaard. The various reactions were characterized and their sequence determined by many experiments similar to those described above. In addition, more detailed electron micrographs of defective components helped to clarify the nature of some individual steps. For example, knowing that genes No. 15 and 18 were concerned with tail completion, we went on to find just what each one did. Electron micrographs showed that in the absence of the No. 18 product no contractile sheaths were made. If No. 18 was functional but No. 15 was defective, the sheath units were assembled on the core but were unstable and could fall away. The addition of the product of gene No. 15 (and of No. 3 also, as it turned out) supplied a kind of "button" at the upper end of the core and thus apparently stabilized the sheath.

The results to date of this line of investigation can be summarized in the form of a morphogenetic pathway [*see illustration on page 178*]. As we had thought, it consists of three principal independent branches that lead respectively to the formation of the head, the tail and the tail fibers.

The earliest stages of head morphogenesis are controlled by six genes. These genes direct the formation of a precursor that is identifiable as a head in electron micrographs but is not yet functional in extract-complementation experiments. Eight more gene products must act on this precursor to produce a head structure that is active in complementation experiments. This active structure undergoes the terminal step in head formation (the only one so far demonstrated in the test tube): conversion to the complete head that is able to unite with the tail. The nature of this conversion, which is controlled by genes No. 13 and 14, remains unclear. A likely possibility would be that these genes control the formation of the upper neck and collar, but evidence on this point is lacking. The attachment of head structures to tails has never been observed in extracts prepared with mutants defective in gene No. 13 or 14, or with any of the preceding class of eight genes. It therefore appears that completion of the head is a prerequisite for the union of heads and tails.

The earliest structure so far identified in the morphogenesis of the tail is the end plate. It is apparently an intricate bit of machinery, since 15 different gene products participate in its formation. All the subsequent steps in tail formation can be demonstrated in the test tube. The core is assembled on the end plate under the control of the products of gene No. 54 and probably No. 19; the resulting structure appears as a tail without a sheath. The product of gene No. 18 is the principal structural component of the sheath, which is somehow stabilized by the products of genes No. 3 and 15. Tails without sheaths do not attach themselves to head structures, indicating that the tail as well as the head must be completed before head-tail union can occur. Moreover, unattached tail structures are never fitted with fibers, suggesting that these can be added only at a later stage of assembly.

Completed heads and tails unite spontaneously, in the absence of any additional factors, to produce a precursor particle that interacts in a still undetermined manner with the product of gene No. 9, resulting in the complete head-plus-tail particle. It is only at this point that tail fibers can become attached to the end plate.

At least five gene products participate in the formation of the tail fiber. In the first step, which has not yet been demonstrated in extracts, the products of genes No. 37 and 38 combine to form a precursor corresponding in dimensions to one segment of the finished fiber. This precursor then interacts sequentially with the products of genes No. 36, 35 and 34 to produce the complete structure. Again the completion of a major component—in this case the tail fiber—appears to be a prerequisite for its attachment, since we have never seen the short segments linked to particles.

The final step in building the virus is the attachment of completed tail fibers to the otherwise finished particle. We have studied this process in reaction mixtures consisting of purified particles and a defective extract containing complete tail fibers but no heads or tails. When we divided the extract into various fractions, we found that it supplies two components, both of which are necessary for the production of active virus. One of these of course is the tail fiber. The other is a factor whose properties suggest that

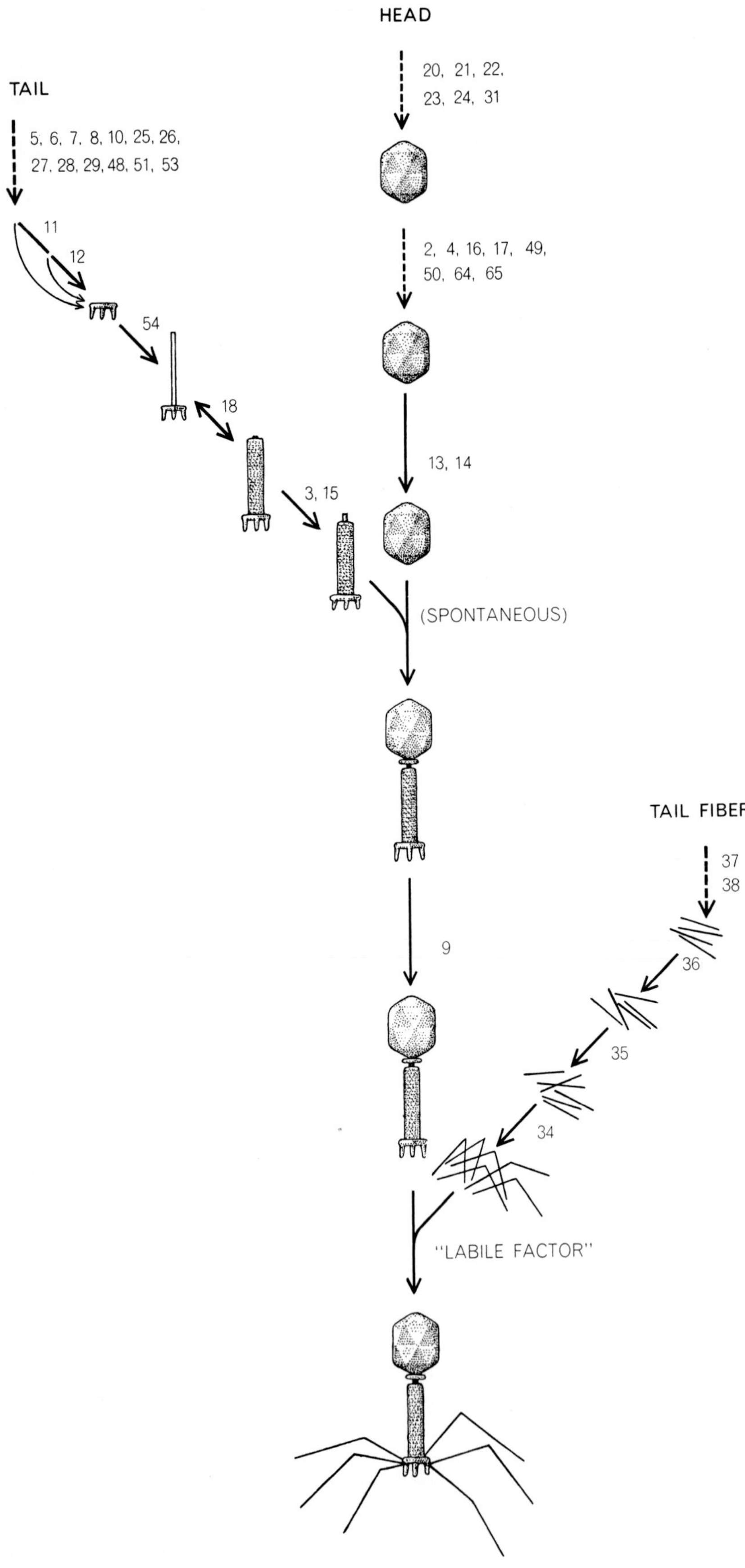


**MORPHOGENETIC PATHWAY** has three principal branches leading independently to the formation of heads, tails and tail fibers, which then combine to form complete virus particles. The numbers refer to the gene product or products involved at each step. The solid portions of the arrows indicate the steps that have been shown to occur in extracts.

it might be an enzyme. For one thing, the rate at which fibers are attached depends on the level of this factor present in the reaction mixture, and yet the factor does not appear to be used up in the process. Moreover, the rate of attachment depends on the temperature of incubation—increasing by a factor of about two with every rise in temperature of 10 degrees centigrade. These characteristics suggest that the factor could be catalyzing the formation of bonds between the fibers and the tail end plate. At the moment we can only speculate on its possible mechanism of action, since the chemical nature of these bonds is not yet known; we call it simply a "labile factor," not an enzyme. Although no gene controlling the factor has yet been discovered, we assume that its synthesis must be directed by the virus, since it is not found in extracts of uninfected bacteria.

The T4 assembly steps so far accomplished and studied in the test tube represent only a fraction of the total number. Already, however, it is apparent that there is a high degree of sequential order in the assembly process; restrictions are somehow imposed at each step that prevent its occurrence until the preceding step has been completed. Only two exceptions to this rule have been discovered. The steps controlled by genes No. 11 and 12, which normally occur early in the tail pathway, can be bypassed when these gene products are lacking. In that case the tail is completed, attaches itself to a head and acquires tail fibers, but the result is a fragile, defective particle. The particle can, however, be converted to a normal active virus by exposure to an extract containing the missing gene products. These are the only components whose point of action in the pathway appears to be unimportant.

The problem has now reached a tantalizing stage. A partial sequence of gene-controlled assembly steps can be written, but the manner in which the corresponding gene products contribute to the process remains unclear, and the questions posed at the beginning of this article cannot yet be answered definitively. There is the suggestion that the attachment of tail fibers is catalyzed by a virus-induced enzyme. If this finding is substantiated, it would overthrow the notion that T4 morphogenesis is entirely a self-assembly process. Continued investigation of this reaction and the assembly steps that precede it can be expected to provide further insight into how genes control the building of biological structures.

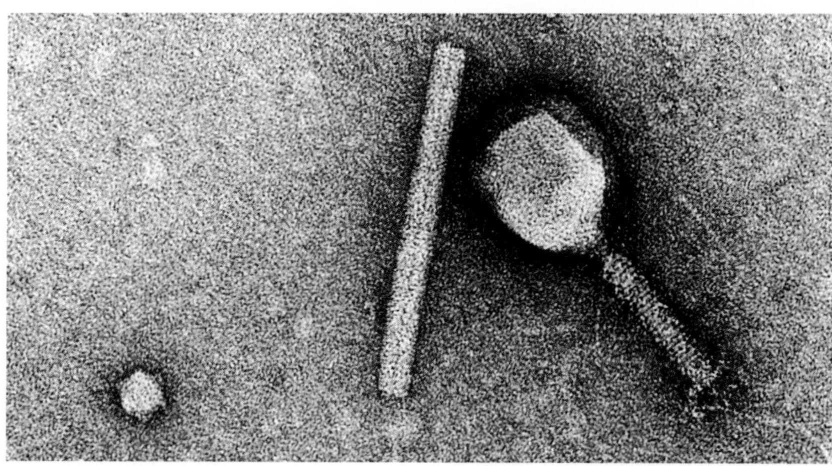

TWO SIMPLER VIRUSES are shown with the T4 in an electron micrograph made by Fred Eiserling of the University of California at Los Angeles. The icosahedral ΦX174 virus (*left*) infects the colon bacillus, as does the T4. The rod-shaped tobacco mosaic virus reassembles itself in the test tube after dissociation. The enlargement is 200,000 diameters.

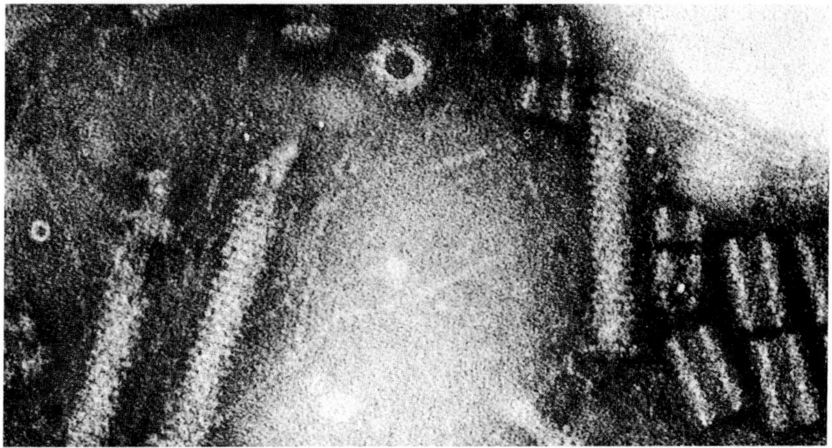

COMPLEX STRUCTURE of the T4 tail is shown in an electron micrograph made by E. Boy de la Tour of the University of Geneva. The parts were obtained by breaking down virus particles, not by synthesis, which is why fibers are attached to tails. The hollow interiors of the free core (*top right*) and pieces of sheath are delineated by dark stain that has flowed into them. There are end-on views of pieces of core (*left*) and sheath (*top center*).

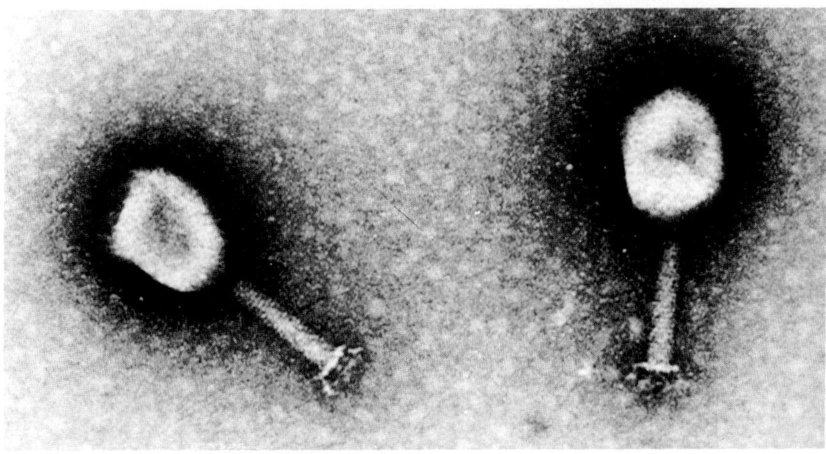

FIBERLESS PARTICLES, otherwise complete, are the products of infection by a mutant defective in one of the fiber-forming genes. Heads, tails and fibers are each formed by a subassembly line (*see illustration on page 178*). The electron micrograph was made by King.

# 19

# *Transplanted Nuclei and Cell Differentiation*

J. B. GURDON

*December 1968*

The means by which cells first come to differ from one another during animal development has interested humans for nearly 2,000 years, and it still constitutes one of the major unsolved problems of biology. Much of the experimental work designed to investigate the problem has been done with amphibians such as frogs and salamanders because their eggs and embryos are comparatively large and are remarkably resistant to microsurgery. As with most animal eggs, the early events of amphibian development are largely independent of the environment, and the processes leading to cell differentiation must involve a redistribution and interaction of constituents already present in the fertilized egg.

Several different kinds of experiment have revealed the dependence of cell differentiation on the activity of the genes in the cell's nucleus. This is clearly shown by the nonsurvival of hybrid embryos produced by fertilizing the egg of one species (after removal of the egg's nucleus) with the sperm of another species. Such hybrids typically die before they reach the gastrula stage, the point in embryonic development at which major cell differences first become obvious. Yet the hybrids differ from nonhybrid embryos only by the substitution of some of the nuclear genes. If gene activity were not required for gastrulation and further development, the hybrids should survive as well as nonhybrids. The importance of the egg's non-nuclear material—the cytoplasm—in early development is apparent in the consistent relation that is seen to exist between certain regions in the cytoplasm of a fertilized egg and certain kinds or directions of cell differentiation. It is also evident in the effect of egg cytoplasm on the behavior of chromosomes [see "How Cells Specialize," by Michail Fischberg and Antonie W. Blackler; SCIENTIFIC AMERICAN Offprint No. 94]. Such facts have justified the belief that the early events in cell differentiation depend on an interaction between the nucleus and the cytoplasm.

Nuclear transplantation is a technique that has enormously facilitated the analysis of these interactions between nucleus and cytoplasm. It allows the nucleus from one of several different cell types to be combined with egg cytoplasm in such a way that normal embryonic development can take place. Until this technique was developed the only kind of nucleus that could be made to penetrate an egg was the nucleus of a sperm cell, and this was obviously of limited use for an analysis of those interactions between nucleus and cytoplasm that lead to the majority of cell differences in an individual.

The technique was first applied to the question primarily responsible for its development. The question is whether or not the progressive specialization of cells during development is accompanied by the loss of genes no longer required in each cell type. For example, does an intestine-cell nucleus retain the genes needed for the synthesis of hemoglobin, the protein characteristic of red blood cells, and a nerve-cell nucleus the genes needed for making myosin, a protein characteristic of muscle cells? If unwanted genes are lost, the possibility exists that it is the progressive loss of different genes that itself determines the specialization of cells, as August Weismann originally proposed in 1892. The clearest alternative is that all genes are retained in all cells and that the genes are inactive in those cells in which they are not required. Before describing the nuclear-transplant experiments that distinguish between these two possibilities, we must outline the methods used to transplant living cell nuclei into eggs.

The aim of a nuclear-transplant experiment is to insert the nucleus of a specialized cell into an unfertilized egg whose nucleus has been removed. Ingenious attempts in this direction were made many years ago by constricting an egg just after fertilization and then letting one of the early-division nuclei that appeared in the nucleated half of the egg enter the non-nucleated half. This method, however, is applicable only to the nuclei of early embryos whose cells are not normally regarded as being specialized. The first real success in transplanting living cell nuclei into animal eggs was achieved in 1952 by Robert W. Briggs and Thomas J. King, both of whom were working at the Institute for Cancer Research in Philadelphia. Their method, which has been generally adopted in subsequent work, involves three steps [*see illustration on page 183*]. Owing to the fortunate circumstance that the unfertilized egg of an amphibian has its nucleus (in the form of chromosomes) located just under the surface of the egg at a point visible through the microscope, it is not difficult to obtain an egg with no nucleus. This can be done by removing the region of the egg that contains chromosomes with a needle or by killing the nuclear material with ultraviolet radiation. The second step is to dissociate a tissue into separate cells,

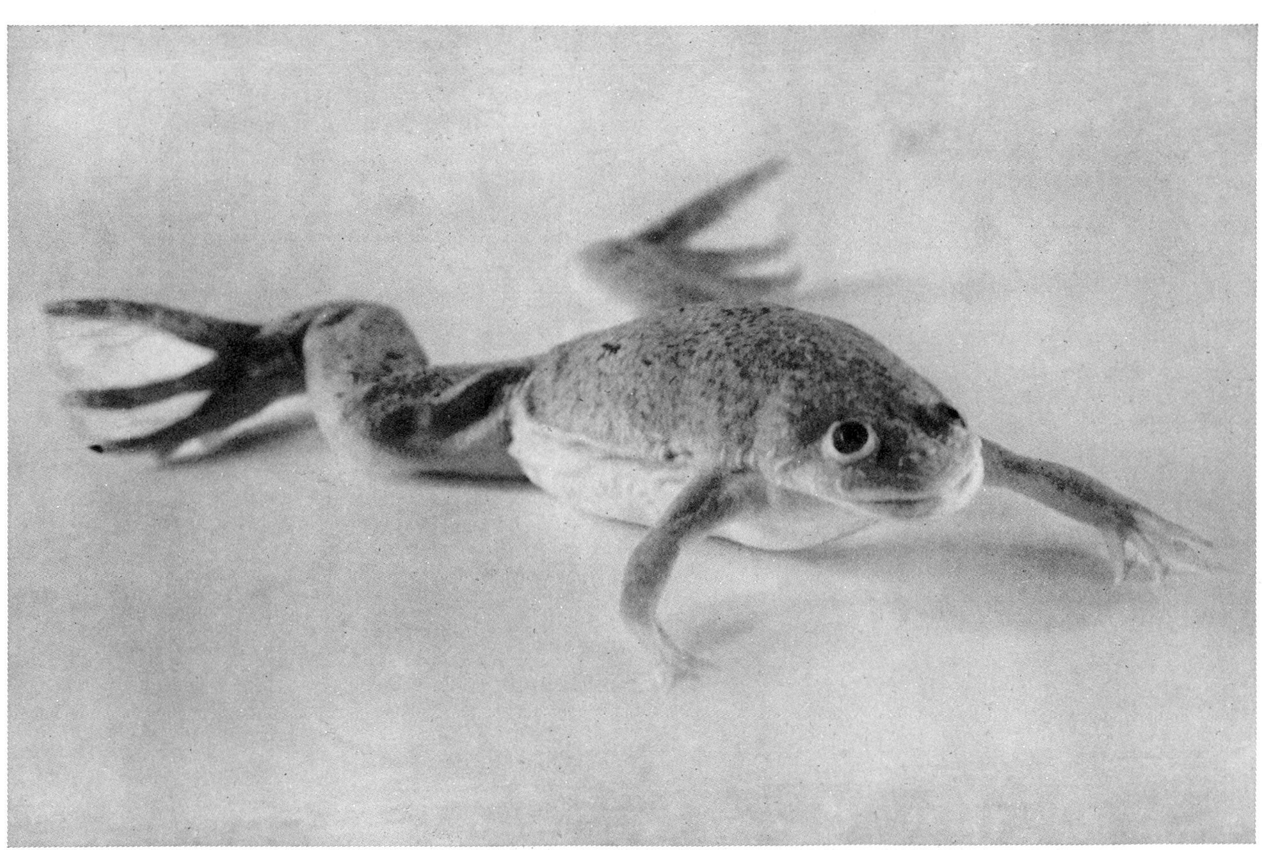

**NORMAL FROG was raised in the author's laboratory at the University of Oxford from an egg that had been fertilized in the usual way by a sperm. The frogs used in the experiments described in this article were the South African clawed species *Xenopus laevis*.**

**TRANSPLANT FROG was raised from an egg cell from which the nucleus had been removed and into which the nucleus of an intestine cell had been transplanted. The frog is normal in all respects, indicating that intestine-cell nucleus has full range of genes.**

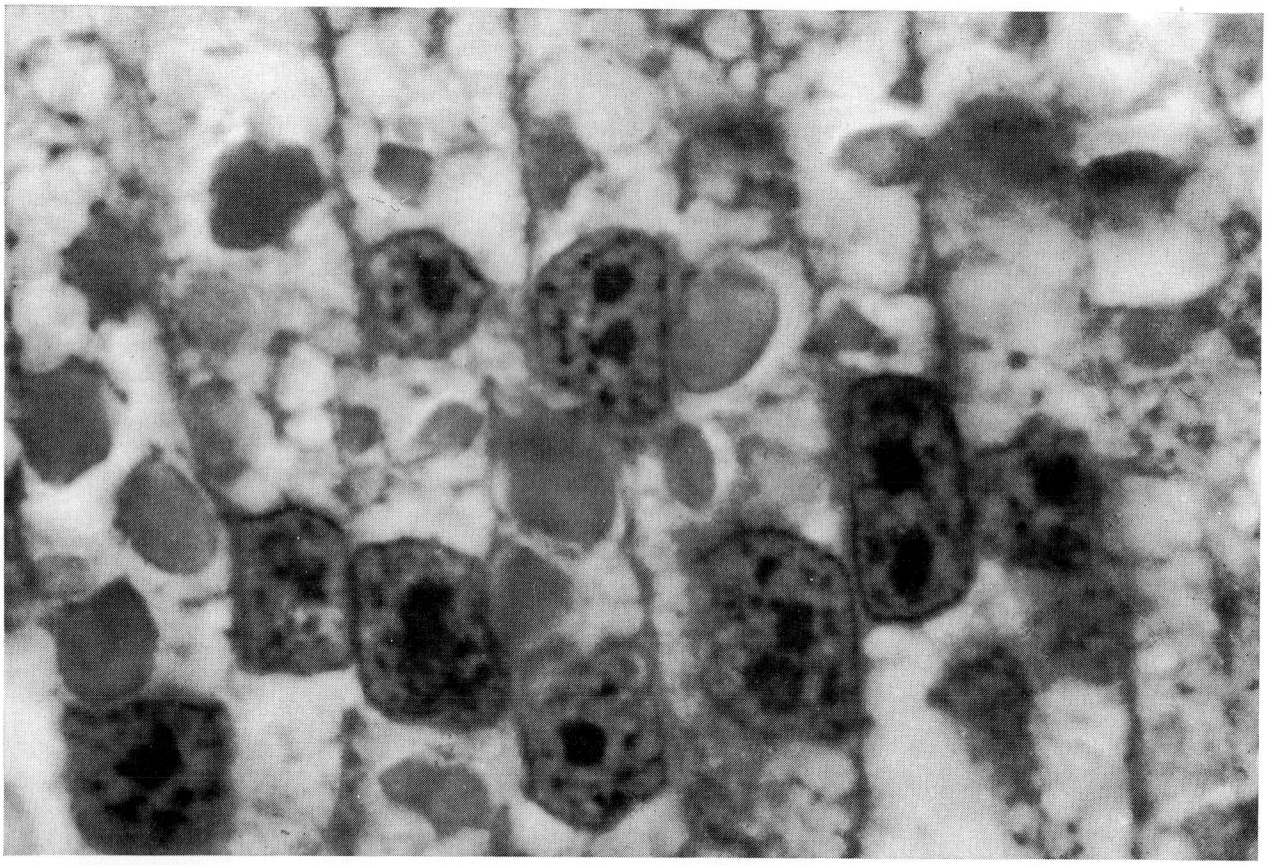

**DONOR CELLS** used in the author's transplant experiments were taken from the epithelial layer of intestine. The micrograph shows the cells' characteristic columnar shape, central nucleus and yolk contents. Cells will be dissociated in preparation for transplant.

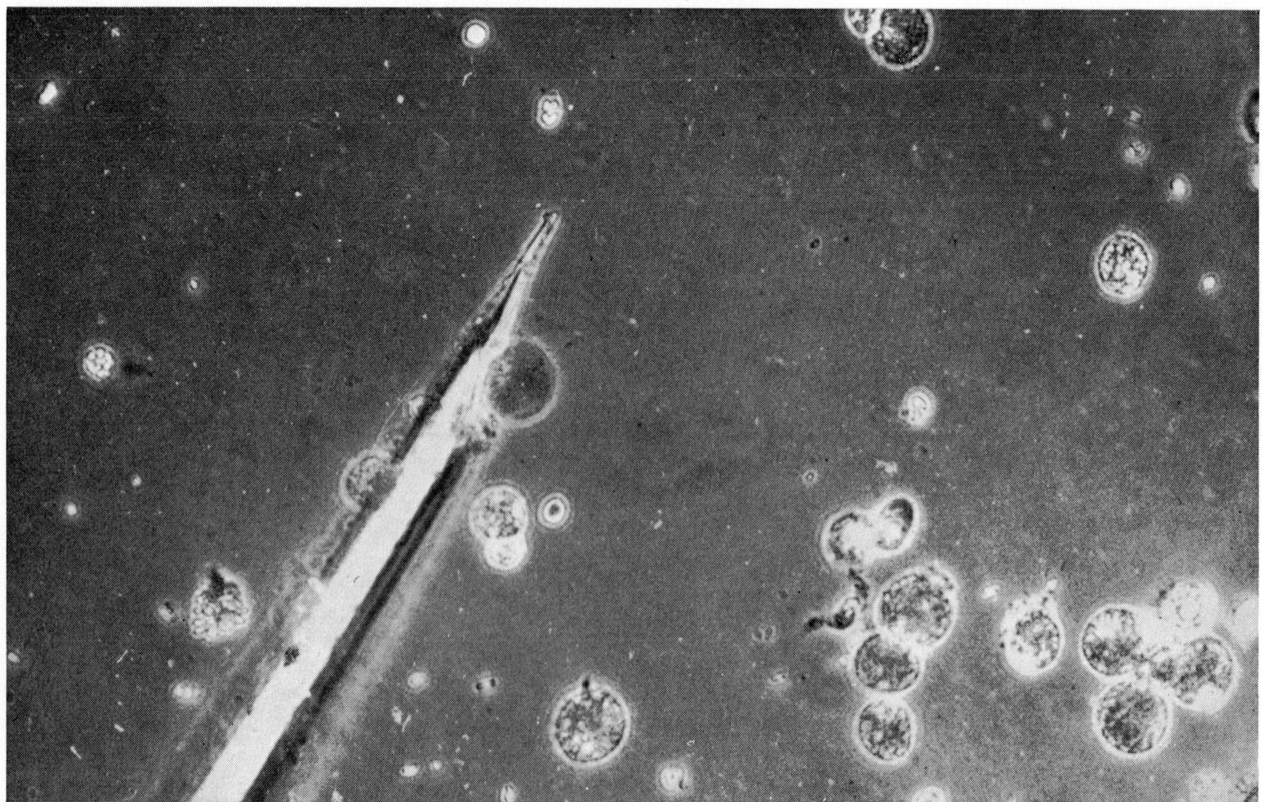

**SINGLE NUCLEI** from intestine epithelium cells are obtained for transplanting by sucking the whole cell into a micropipette (*left*). Smaller in diameter than the cell, the pipette breaks the cell wall; only the nucleus and a coating of cytoplasm enter the host egg.

each of which can be used to provide a donor nucleus for transplantation. The cells separate from one another in a medium lacking calcium and magnesium ions, which are removed from the embryo more quickly by adding to the medium a chelating substance such as Versene.

The third and most difficult stage in the procedure involves the insertion of the donor-cell nucleus into the enucleated egg. Briggs and King found that this can be done by sucking an isolated cell into a micropipette that is small enough to break the cell wall but large enough to leave the nucleus still surrounded by cytoplasm. This compromise is required because the nucleus in an unbroken cell does not make the necessary response to egg cytoplasm, and conversely a bare nucleus without surrounding cytoplasm is readily damaged by exposure to any artificial medium. The broken cell with its cytoplasm-protected nucleus is injected into the recipient egg. The amount of donor-cell cytoplasm injected is very small and does not have any effect.

A useful extension of the basic nuclear-transplant technique is called serial nuclear transplantation. It involves the same procedure as the one just described except that instead of the donor nuclei being taken from the cells of an embryo or larva reared from a fertilized egg, they are taken from a young embryo that is itself the result of a nuclear-transplant experiment. The effect is the same as in the vegetative propagation of plants, namely the production of a clone: a population consisting of many individuals all having an identical set of genes in their nuclei.

One other feature of nuclear-transplant experiments is of the greatest importance for their interpretation. It is the use of a nuclear marker whereby the division products of a transplanted nucleus can be distinguished from those of the host egg nucleus. A nuclear marker is virtually indispensable where attention is to be paid to the development of a very small percentage of eggs that have received transplanted nuclei, since one cannot otherwise be sure that an occasional error in enucleation by hand or by ultraviolet irradiation has not occurred. Only by the presence of a marker in the nuclei of a transplant embryo does one have proof of its origin.

A nuclear marker must be replicated and therefore be genetic. One of the most useful for nuclear transplantation is found in a mutant line of the South African clawed frog *Xenopus laevis*, discovered at the University of Oxford by Mi-

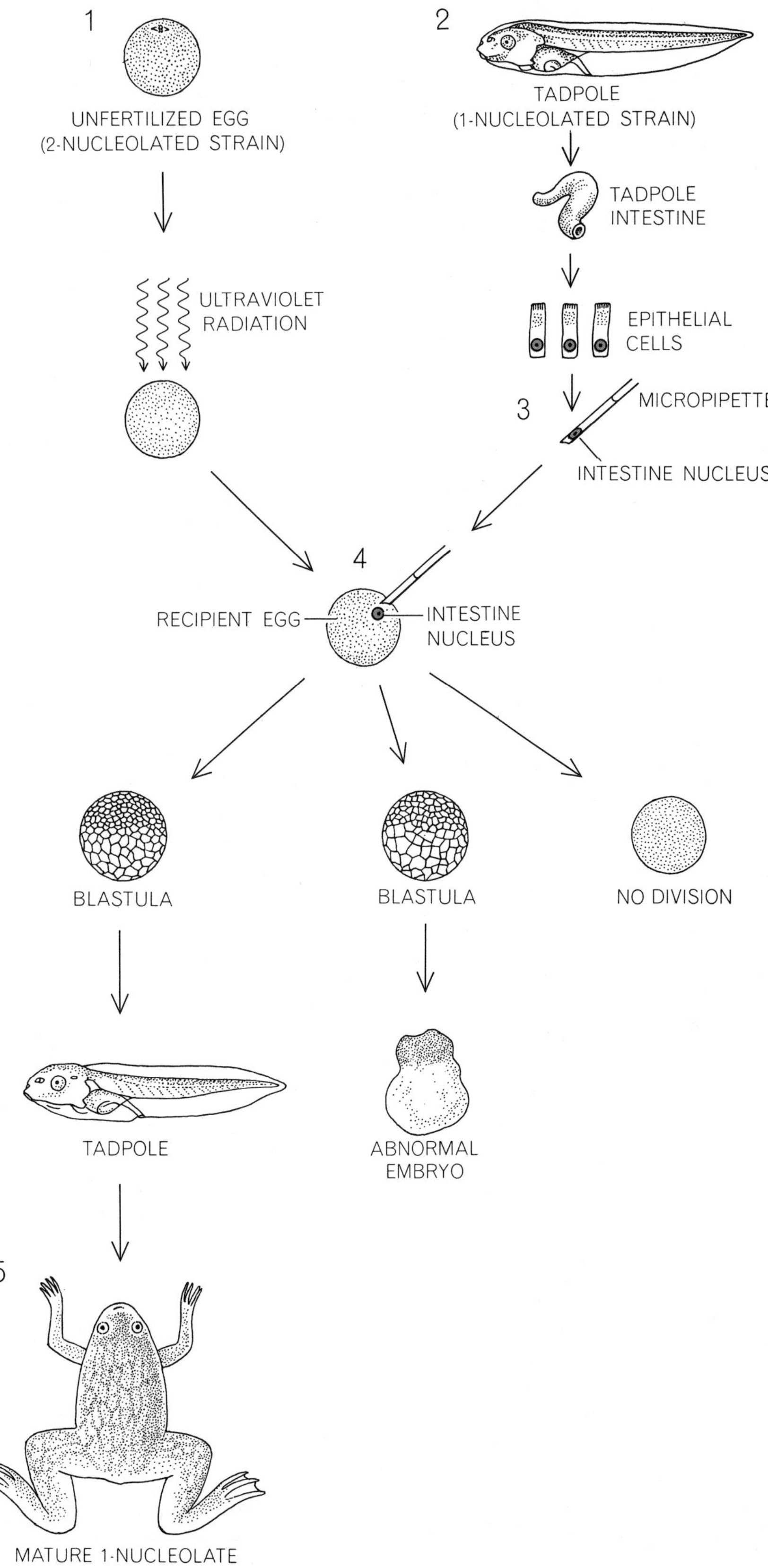


**TRANSPLANT PROCEDURE starts with preparation of a frog's unfertilized egg (*1*) for receipt of a cell nucleus by destroying its own nucleus through exposure to ultraviolet radiation. Next, intestine is taken from a tadpole that has begun to feed (*2*) and cells are taken from its epithelial layer. A single epithelial cell is then drawn into a micropipette; the cell walls break (*3*), leaving the nucleus free. The intestine-cell nucleus is transplanted into the prepared egg (*4*), which is allowed to develop. In some 1 percent of transplants the egg develops into a frog that has one nucleolus in its nucleus instead of the usual two (*5*).**

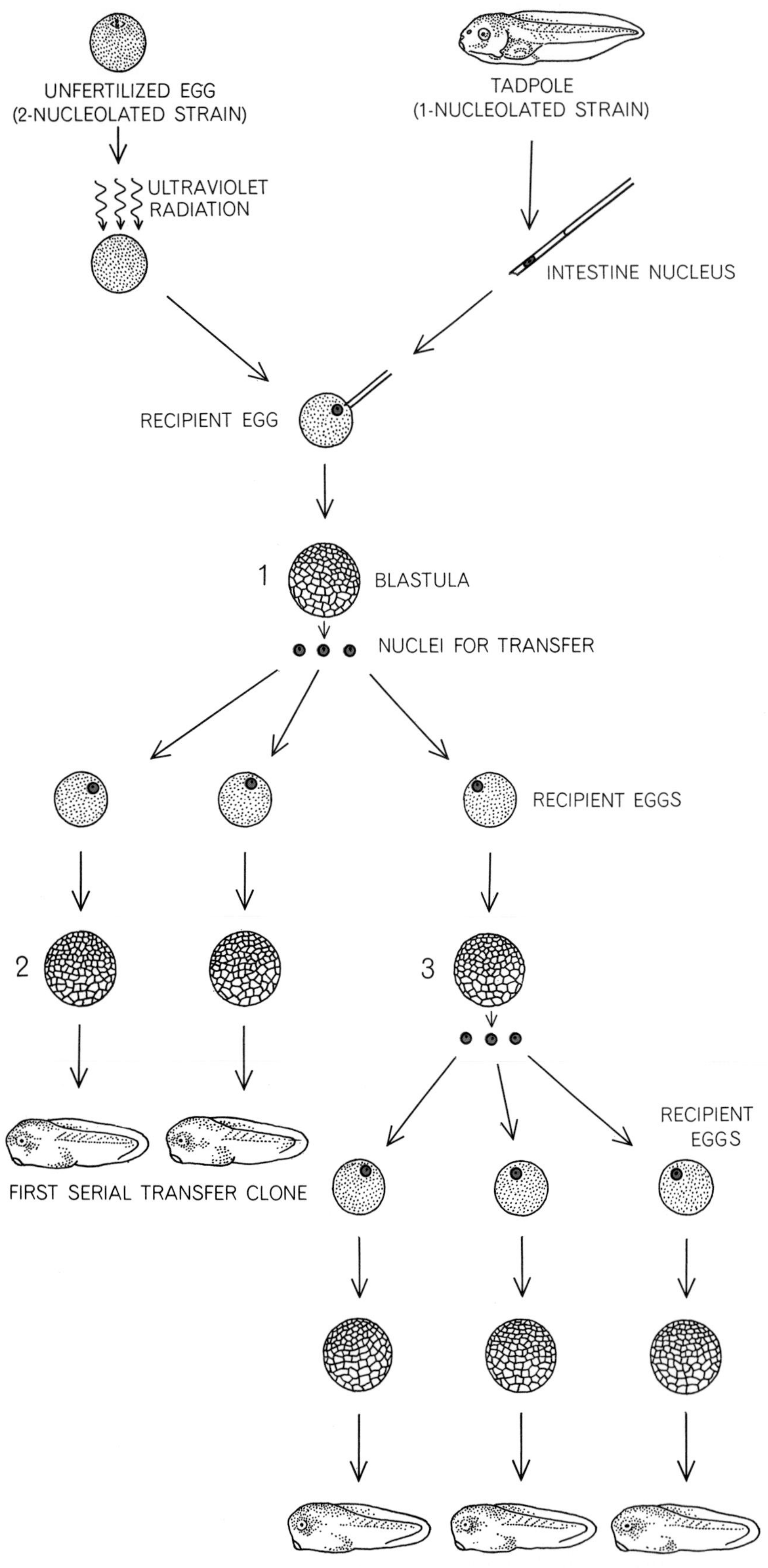


**SERIAL TRANSPLANTS involve the same first steps as the transplantation procedure illustrated on the preceding page. At the blastula stage (*1*) the cells of a transplant-embryo are dissociated. The genetically identical nuclei from these cells are then transplanted into enucleate eggs, giving rise to a clone: a population comprised of genetically identical individuals (*2*). The procedure can be continued indefinitely (*3*).**

chail Fischberg. The nuclei of most normal frog cells contain two of the bodies called nucleoli; the nuclei of cells carrying the mutation never have more than one. This mutation is almost ideal as a nuclear marker because a sample of cells taken from any tissue at any developmental stage beyond the blastula (the hollow sphere from which the gastrula arises) can be readily classified as being mutant or not.

We can now return to the question of whether or not genes are lost in the course of normal cell differentiation. Nuclear-transfer experiments are performed to answer this question on the assumption that if the combination of egg cytoplasm with a transplanted nucleus can develop into a normal embryo possessing all cell types, then the transplanted nucleus cannot have lost the genes essential for pathways of cell differentiation other than its own. For example, if a normal embryo containing a specialized cell type such as blood cells can be obtained by transplanting an intestine-cell nucleus into an enucleated egg, then the genes responsible for the synthesis of hemoglobin cannot have been lost from the intestine-cell nucleus in the course of cell differentiation. The only assumption here is that a gene, once lost, cannot be regained in the course of a few cell generations. It happens that the best evidence for the retention of genes in fully differentiated cells comes from two series of experiments carried out at Oxford on eggs of the frog *Xenopus*.

The fully differentiated cells used for these experiments were taken from the epithelial layer of the intestine of mutant tadpoles that had begun to feed. Intestine epithelium cells have a "brush border," a structure that is present only in cells specialized for absorption and that is assumed to have arisen as a result of the activity of certain intestine-cell genes. Not all the cells of the intestine are epithelial, but when the epithelial cells are dissociated, they can be distinguished from the other cell types by their large content of yolk, by the ease with which they dissociate in a medium that contains Versene and sometimes by their retention of the brush border.

The first experiments with intestine-cell nuclei were designed to show that at least *some* of these nuclei possess *all* the genes necessary for the differentiation of all cell types, and therefore that some of the transplant embryos derived from intestine nuclei could be reared into normal adult frogs. Both male and female adult frogs, fertile and normal in every respect, have in fact been obtained from transplanted intestine nuclei [*see bottom illustration on page 181*]. Although only about 1.5 percent of the eggs with transplanted intestine nuclei developed into adult frogs, all of these frogs carried the mutant nuclear marker in their cells; their existence therefore proves that at least some intestine cells possess as many different kinds of nuclear genes as are present in a fertilized egg.

Subsequent experiments with intestine nuclei were designed to show that *many* of these nuclei have retained genes required for the differentiation of at least *some* quite different cell types. In these experiments the criterion for gene retention was the differentiation of functional muscle and nerve cells by nuclei whose mitotic ancestors had already promoted the differentiation of intestine cells. Functional muscle and nerve cells are present in any nuclear-transplant embryo that shows the small twitching movements, or muscular responses, characteristic of developing tadpoles just before they swim. Out of several hundred intestine nuclear transfers, about 2.5 percent of the injected eggs developed as far as the muscular-response stage or further. The reason why the remainder did not reach this stage is not necessarily because that proportion of intestine nuclei lack the necessary genes. In some cases it is known to be the inability of certain recipient eggs to withstand injection; in others it is the incomplete replication of some of the transplanted nuclei or their daughter nuclei during cleavage. In either case a nuclear-transplant embryo should contain some cells with normal nuclei as well as some abnormal cells responsible for the early death of the embryo.

Serial nuclear transplantation offered a means of overcoming both difficulties. A sample of nuclear-transplant embryos whose development was so abnormal they would have died before reaching the muscular-response stage provided nuclei for serial transplant clones. Many of the serial transplant clones included embryos which developed as far as the muscular-response stage or beyond it. By adding the proportion of nuclei shown by first transplants to be able to support muscular-response differentiation to the proportion shown by serial transplantation to possess this capacity, we can conclude that at least 20 percent of the intestine epithelium cells must have retained the genes necessary for muscle-cell and nerve-cell differentiation [*see illustration below*].

There is no reason to believe that muscle-cell or nerve-cell genes have been lost or permanently inactivated in the remaining 80 percent of transplanted intestine nuclei. There are many reasons why it might not have been possible to demonstrate their presence. For example, about 50 percent of all the eggs that received intestine nuclei failed to divide even once. When a sample of these eggs was sectioned, they were found to contain either no nucleus at all or else a nucleus that was still inside an intact intestine cell. In the first instance the nucleus presumably stuck to the injection pipette and was never deposited in the egg; in

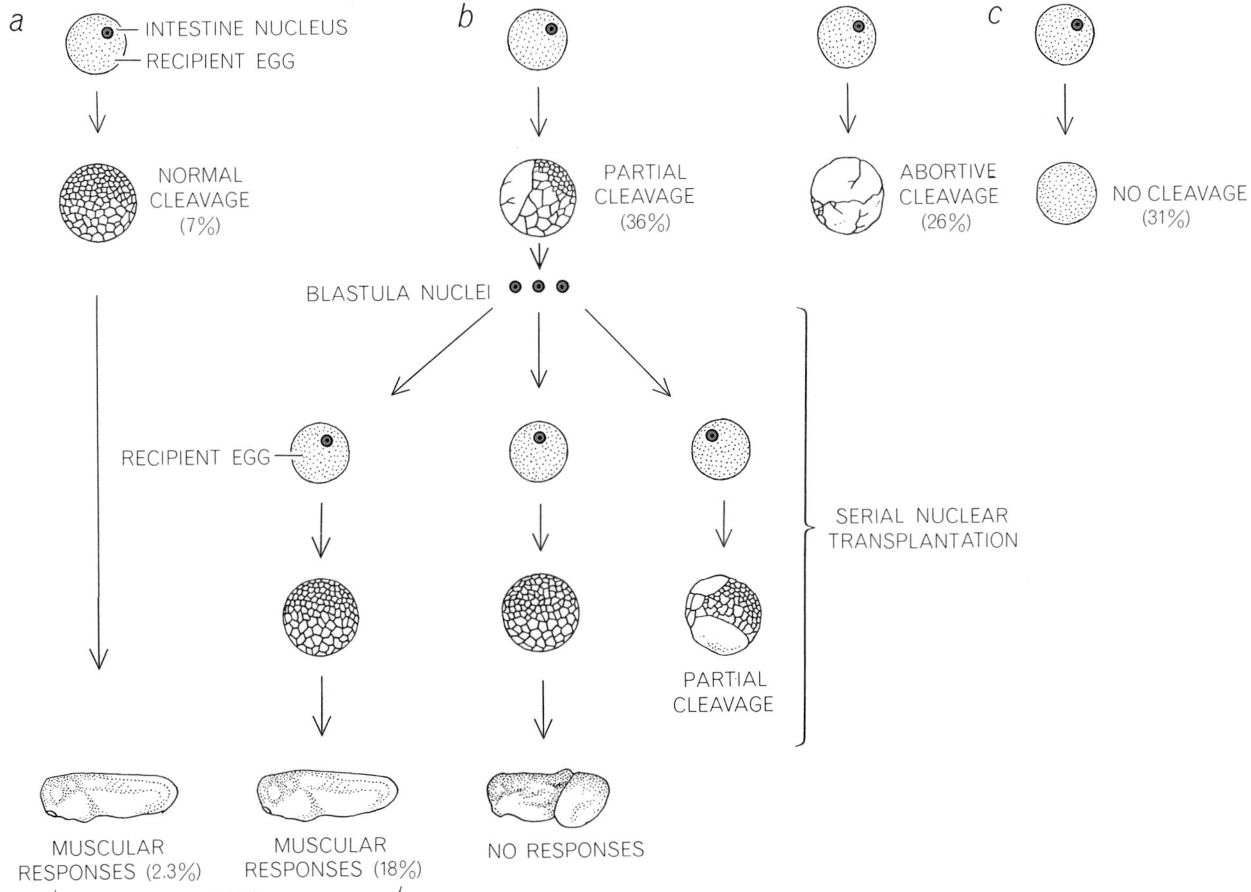


**NERVE AND MUSCLE TISSUE in nuclear-transplant embryos shows that 20 percent of the nuclei of differentiated cells still possess the genes necessary for other kinds of function. The criterion of success is development of an embryo to the point at which small twitching motions are observed; this was demonstrated for more than 2 percent of all intestine nuclei using first transfers (*a*), and in 18 percent using serial transfers (*b*). Many eggs that failed to develop (*c*) contained either no transplant nuclei or still unruptured cells.**

the second, the donor cell was never broken so as to liberate its nucleus, a technical error that is easy to make with very small cells. In both cases the developmental capacity of the intestine nuclei was not tested, and the recipient eggs that failed to divide should not be counted in the results.

It is clear from these experiments that the loss or permanent inactivation of genes does not necessarily accompany the normal differentiation of animal cells. This conclusion is not inconsistent with a recent finding: the "amplification" of the genes responsible for synthesizing the ribonucleic acid (RNA) of the subcellular particles called ribosomes. This phenomenon was demonstrated in amphibian oöcytes, the cells that give rise to mature eggs. The nuclear-transfer experiments just described do not exclude such amplification, which simply alters the number of copies of one kind of gene in a nucleus. Instead they show that specialized cells always have at least one copy of every different gene.

The inability of some transplanted nuclei to support normal development has attracted considerable interest because it is always found that the proportion of nuclei showing a restricted developmental capacity increases as the cells from which they are taken become differentiated. Furthermore, serial nuclear-transplant experiments conducted by Briggs and King (and subsequently by others) have shown that all the embryos in a clone derived from one original nuclear transplant often suffer from the same abnormality, whereas the embryos in a clone derived from another original transplant may suffer from a different abnormality. Some of the abnormalities of nuclear-transplant embryos can therefore be attributed to nuclear changes that can be inherited.

The discovery that these changes arise as a result of nuclear transplantation, and not in the course of normal cell differentiation, was an important one. This was first established by Marie A. DiBerardino of the Institute for Cancer Research, who made a detailed analysis of the number and shape of chromosomes in nuclear-transplant embryos. Abnormal embryos were usually found to suffer from chromosome abnormalities that were not present in the donor embryos, a finding that at once explains why the factors causing many of the developmental abnormalities of nuclear-transplant embryos are inherited. The fact that chromosome abnormalities arise after nuclear transplantation does not necessarily mean that they are of no interest; there could be a connection between the kind of chromosome abnormality encountered and the cell type of the donor nucleus concerned. In spite of an intensive search, however, no such relationship has yet been found.

The origin of these chromosome abnormalities is probably to be understood as an incompatibility between the very slow rate of division of differentiating cells—only one division every two days or more—and the rapid rate of division in an egg, which starts to divide (and causes any injected nucleus to try to divide) about an hour after injection. Unless an injected nucleus can complete the replication of its chromosomes within this brief period, they will be torn apart and broken at division. This concept is supported by the observation, made at Oxford in collaboration with my colleagues C. F. Graham and K. Arms, that many transplanted nuclei continue to synthesize the genetic material DNA right up to the time of the first nuclear division, whereas sperm and egg nuclei always complete this synthesis well be-

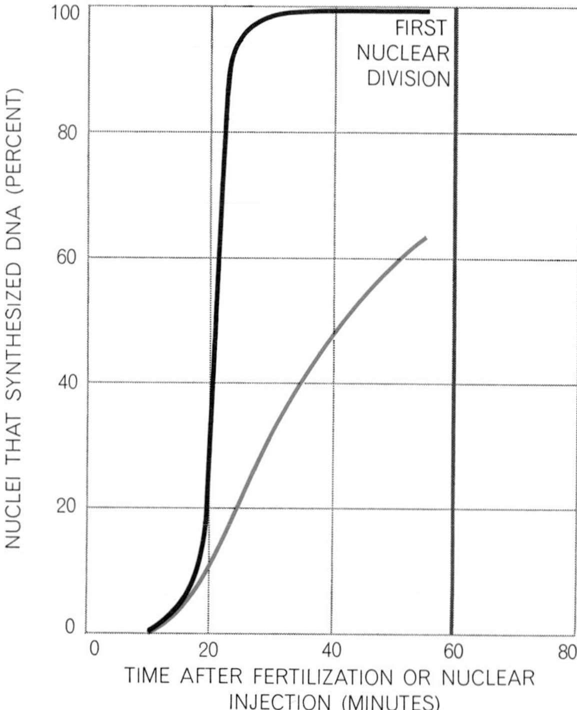


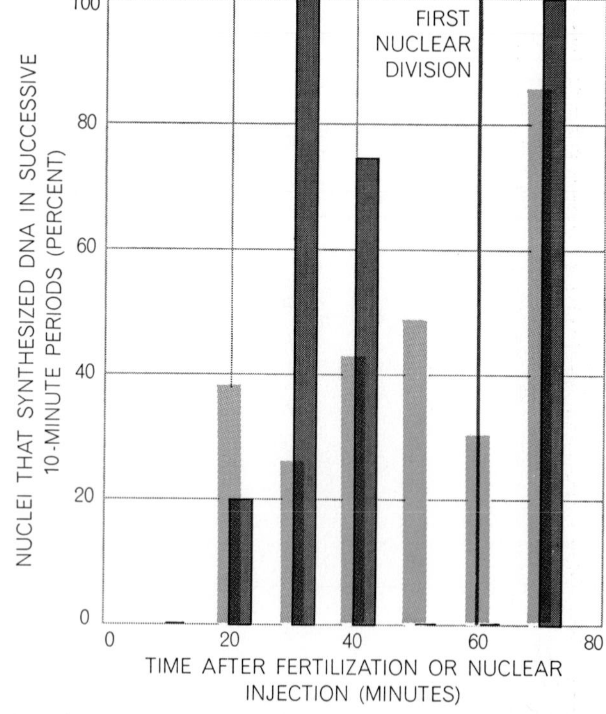


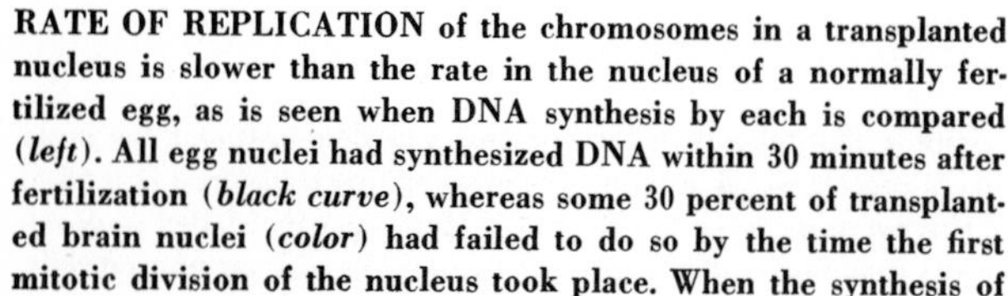
**RATE OF REPLICATION** of the chromosomes in a transplanted nucleus is slower than the rate in the nucleus of a normally fertilized egg, as is seen when DNA synthesis by each is compared (*left*). All egg nuclei had synthesized DNA within 30 minutes after fertilization (*black curve*), whereas some 30 percent of transplanted brain nuclei (*color*) had failed to do so by the time the first mitotic division of the nucleus took place. When the synthesis of DNA was assessed at 10-minute intervals (*right*), the nuclei of fertilized eggs (*black*) were found to have ceased synthesis, an indication of completed chromosome replication, before time for mitotic division. Some 30 percent of transplanted gastrula nuclei, however, were still making DNA (*color*). Failure of transplant nuclei to complete replication before division results in damage to the chromosomes as they are torn apart, producing abnormalities.

fore division. Presumably molecules associated with the DNA of specialized cells prevent the chromosomes of such cells from undergoing replication as rapidly as those of sperm nuclei, thereby leading to the chromosome abnormalities commonly observed in nuclear-transplant embryos.

Having concluded that the specialization of cells involves the differential activity of genes present in all cells, rather than the selective elimination of unwanted genes, we can now consider how genes are activated or repressed during early embryonic development. Nuclear transplantation has been used to demonstrate that the signals to which genes or chromosomes respond are normal constituents of cell cytoplasm. This information has come from experiments in which the nucleus of a cell carrying out one kind of activity is combined with the enucleated cytoplasm of a cell whose nucleus would normally be active in quite another way. One of two results is to be expected: either the transplanted nucleus should continue its previous activity or it should change function so as to conform to that of the host cell to whose cytoplasm it has been exposed. For the purposes of these experiments changes in nuclear activity have to be recognized by the appearance of direct gene products and not by the much less direct criterion of the normality of nuclear-transplant embryo development. Many of these experiments have been carried out in collaboration with Donald D. Brown of the Carnegie Institution of Washington or with another of my Oxford colleagues, H. R. Woodland.

The first experiments were designed to find out if the different functions performed by any one gene—the synthesis of DNA, the synthesis of RNA and chromosome condensation in preparation for cell division—are determined by cytoplasmic constituents. Three kinds of host cell were used: unfertilized but activated eggs whose nucleus would normally synthesize DNA but no RNA; growing oöcytes in which the nucleus synthesized RNA but not DNA, and oöcytes maturing into eggs, in which situation the nucleus consists of condensed chromosomes arranged in the "spindle" of cell division, and synthesizes neither RNA nor DNA. Two kinds of test nuclei were used: nuclei from adult brain tissue, which synthesize RNA but almost never synthesize DNA or divide, and nuclei from embryonic tissue at the mid-blastula stage of development; mid-blastula nuclei do not synthesize RNA but

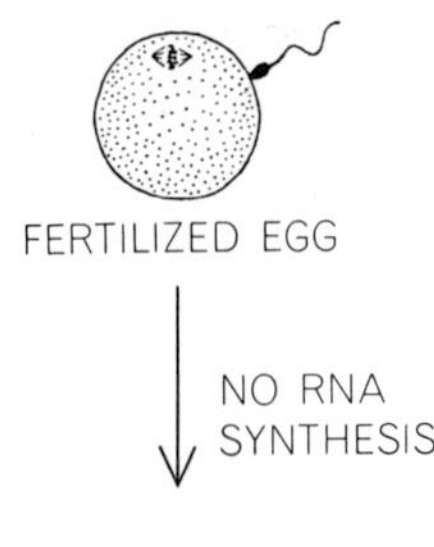


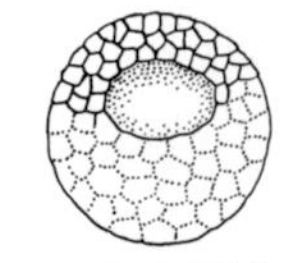


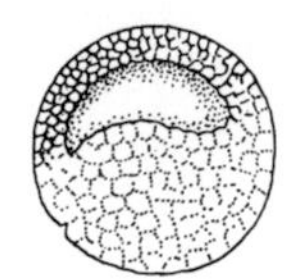


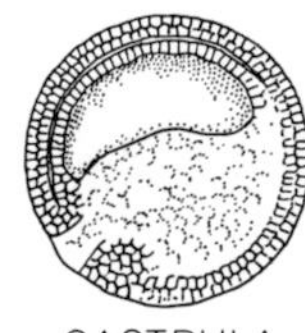


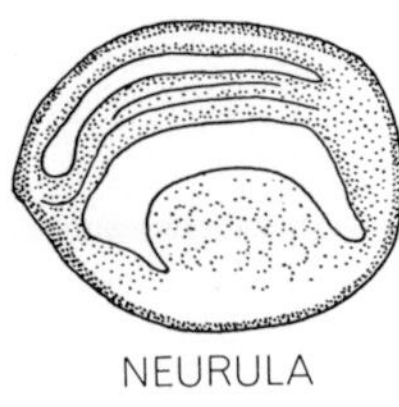


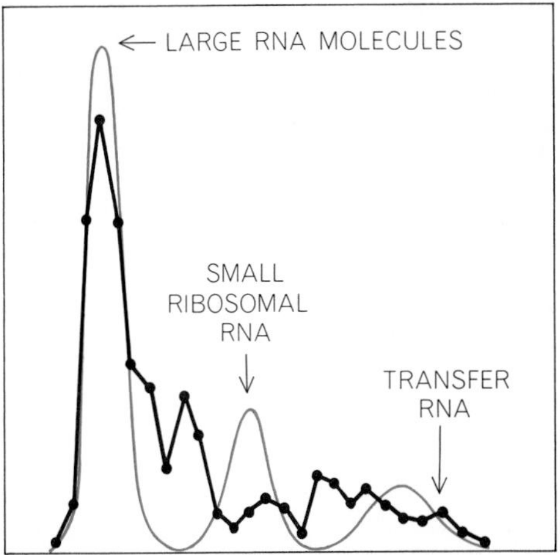


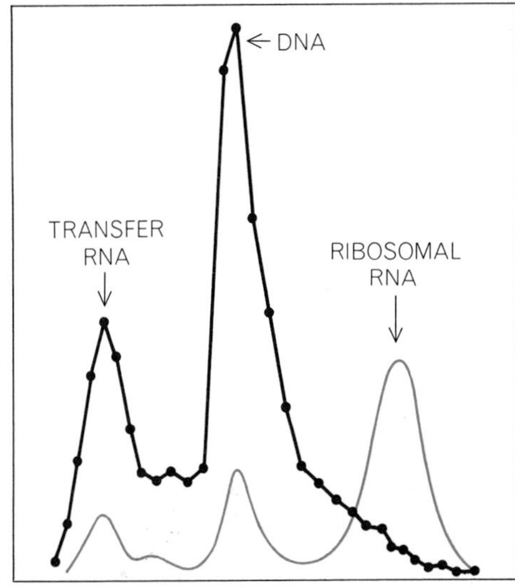


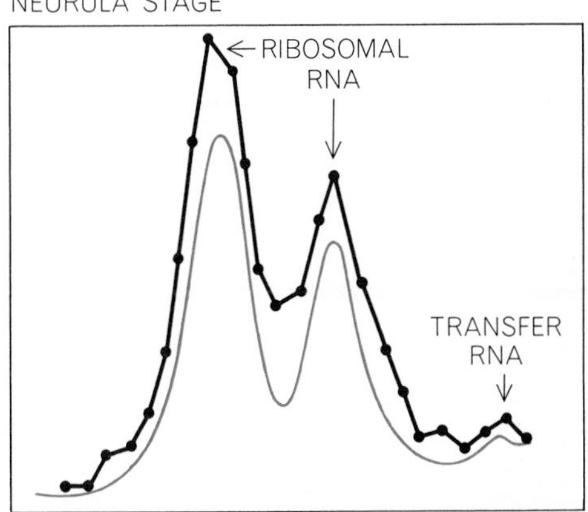


**CYTOPLASMIC CONTROL of the activities of genes is demonstrated in a nuclear-transplant experiment. As a normal embryo develops (*left*), early growth stages are marked by the nuclear synthesis of different varieties of RNA. The nucleus of a cell from an embryo at the neurula stage, which synthesizes all varieties of RNA, is transplanted into an egg. At first the transplanted nucleus halts RNA synthesis. On reaching the blastula stage (*top right*) the nucleus starts to synthesize large RNA molecules. In this graph and the other ones the black curve indicates the extent to which radioactive precursors of various nucleic acids are incorporated. The quantity of preexisting nucleic acids is in red. At the late-blastula stage (*middle right*) transfer RNA, but not ribosomal RNA, is synthesized. By the neurula stage (*bottom right*) nucleic acid synthesis has gone on to include the making of ribosomal RNA.**

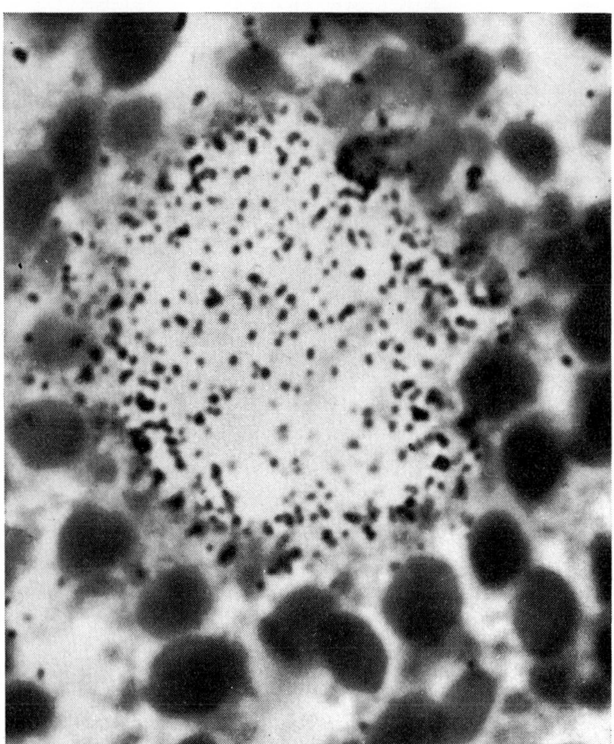

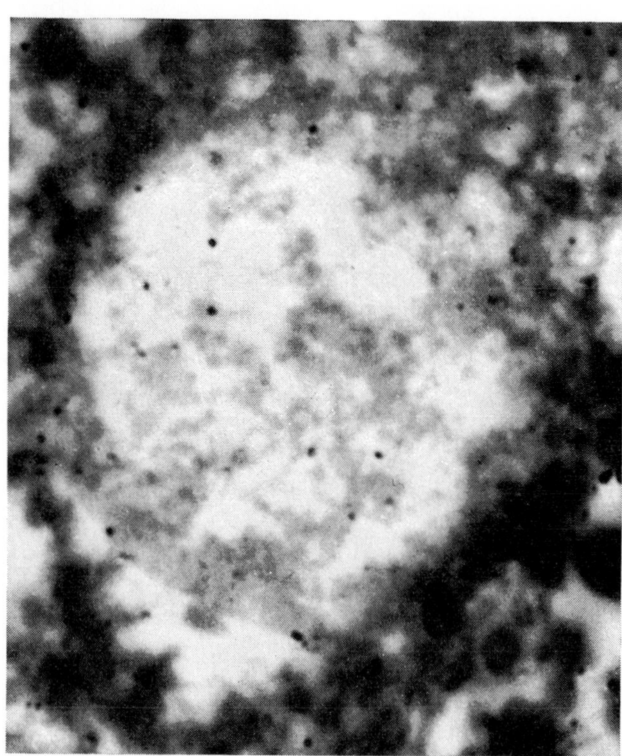

CHANGE IN ACTIVITY of a transplanted nucleus to conform to the normal activity of the host cell's missing nucleus is shown in two photomicrographs. Nuclei from embryos at the mid-blastula stage of development, which synthesize DNA but not RNA, have been injected into an oöcyte (*left*) and an egg (*right*). Oöcytes, the cells that grow into eggs, synthesize RNA but not DNA; eggs do the opposite. A substance that is a precursor of RNA, labeled with a radioactive isotope, was injected simultaneously. The many black dots, formed by the radioactive molecules, show that the substance was incorporated in newly synthesized RNA in the oöcyte, a factor in the host cell's cytoplasm having altered the injected nuclei's activity. The nuclei in the egg, however, make no RNA.

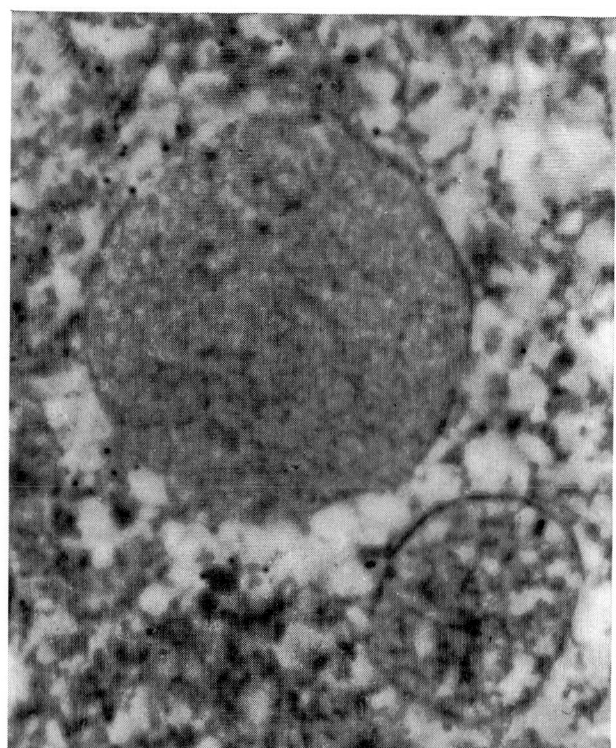

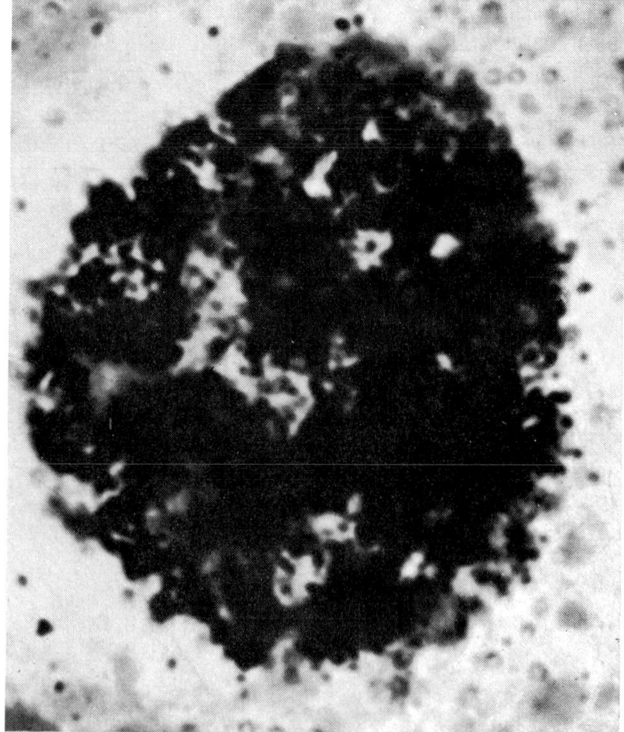

REVERSE RESULTS are shown in two other photomicrographs. The nuclei to be injected are from frog brain cells; they synthesize RNA but almost never DNA. The radioactively labeled precursor is one that is taken up only in DNA synthesis. In the oöcyte (*left*), where the synthesis of RNA is progressing, the nucleus contains no radioactive DNA. The intense radioactivity in the egg (*right*) shows that the injected brain nuclei have switched from RNA to DNA synthesis in response to a factor in the host cell's cytoplasm.

synthesize DNA and divide about every 20 minutes. For technical reasons it was desirable to inject each host cell with many nuclei, even though this can prevent the subsequent division of the injected cell. The results were clear: In all respects tested the transplanted nuclei changed their function within one or two hours so as to conform to the function characteristic of the normal host-cell nucleus [*see illustrations on opposite page*]. Mid-blastula nuclei injected into growing oöcytes stopped synthesizing DNA and dividing and entered a continuous phase of RNA synthesis that lasted for as long as the injected oöcytes survived in culture (about three days). Adult brain nuclei injected into eggs stopped RNA synthesis and began DNA synthesis. When the same nuclei were injected into maturing oöcytes, they synthesized neither RNA nor DNA but were rapidly converted into groups of chromosomes on spindles.

The next set of experiments was designed to find out if cytoplasmic components can repress or activate genes, that is, if they can select which genes in a nucleus will be active at any one time. Advantage was taken of the natural dissociation that exists in the time of synthesis of different classes of RNA during the early embryonic development of *Xenopus*. The work of several investigators has established the following sequence of events in *Xenopus* embryos. For the first 10 divisions after fertilization no nuclear RNA synthesis can be detected. Just after this—at the mid-late-blastula stage—the cells synthesize large RNA molecules, which are believed not to include ribosomal RNA but which are likely to include "messenger" RNA. Toward the end of the blastula stage "transfer" RNA synthesis is first detected; this is followed a few hours later, during the formation of the gastrula, by the synthesis of ribosomal RNA.

The extent to which these events are under cytoplasmic control has been investigated by transplanting into enucleated eggs single nuclei from embryonic tissue at the neurula stage of development, the one that follows the gastrula stage. As the nuclear-transplant embryos develop, RNA precursor substances that have been labeled with radioactive atoms (for example uridine triphosphate labeled with tritium, the radioactive form of hydrogen) are used to determine the classes of RNA being synthesized at each stage. Autoradiography has shown that a neurula nucleus, which synthesizes each main kind of RNA, stops all detectable RNA synthesis, that is, it no longer incorporates labeled RNA precursors, within an hour of transplantation into egg cytoplasm. Furthermore, chromatography and other kinds of analysis show that, when the transplant embryos are reared through the blastula and gastrula stages, they synthesize heterogeneous RNA, transfer RNA and ribosomal RNA in turn and in the same sequence as do embryos reared from fertilized eggs.

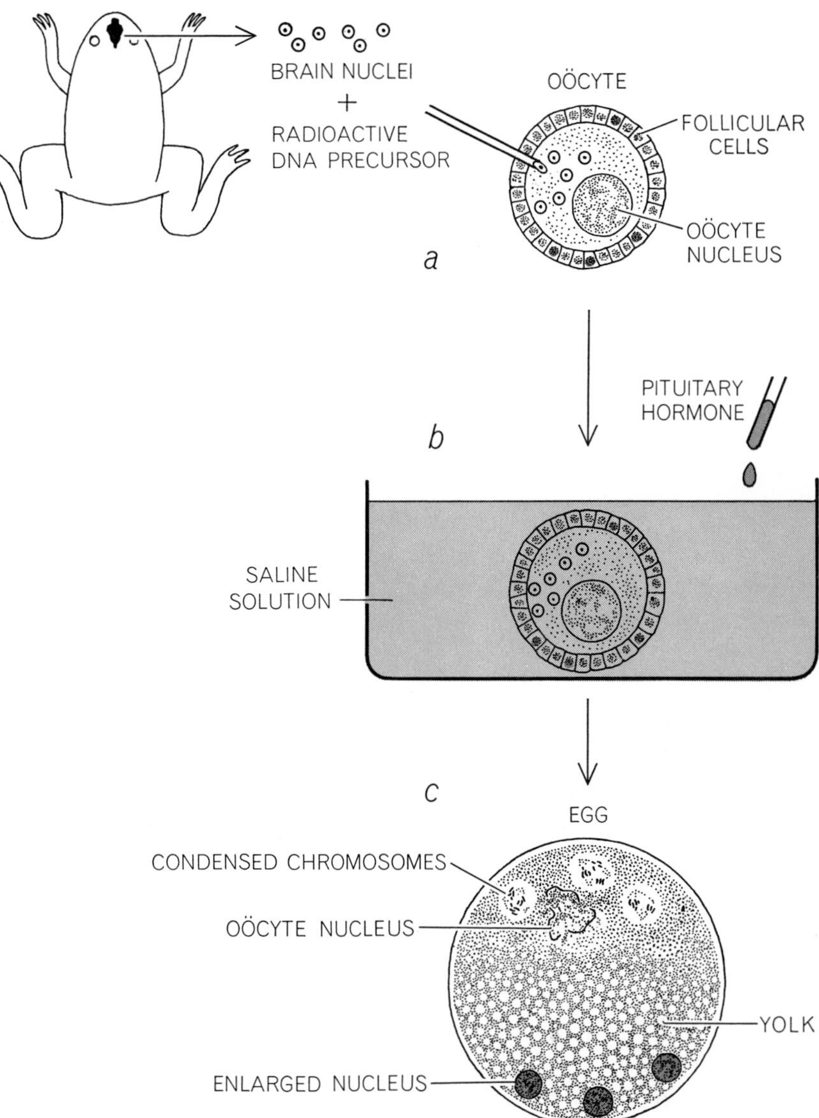


**INDUCING FACTOR in egg cytoplasm that alters the activity of the nucleus was found to be absent at the oöcyte level of egg-cell development; when brain nuclei and labeled DNA precursor were injected into oöcytes (*a*), the nuclei did not synthesize DNA. After the oöcyte was brought to maturity (*b*), however, the inducing factor made its appearance (*c*). Brain nuclei near the oöcyte's ruptured nucleus underwent chromosomal condensation like that of the oöcyte nucleus. Brain nuclei in the yolky region began to synthesize DNA.**

Taken together, these experiments have shown that changes in the type of gene product (for example the synthesis of RNA or DNA), as well as changes in the selection of genes that are active (for example the synthesis of different types of RNA), can be experimentally induced. Since a high proportion of transplanted neurula nuclei support entirely normal development, the results show that egg cytoplasm must contain constituents responsible for independently controlling the activity of different classes of genes in normal living nuclei.

We can now consider what is perhaps the most interesting question of all: What is the mechanism by which cytoplasmic components bring about changes in gene activity? Of the various changes in chromosome and gene activity that can be experimentally induced in transplanted nuclei, special attention has been devoted to the induction of DNA synthesis by egg cytoplasm. It is

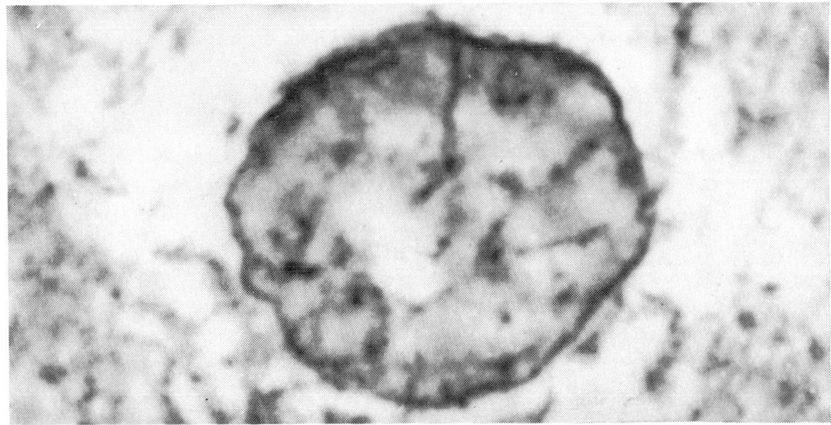

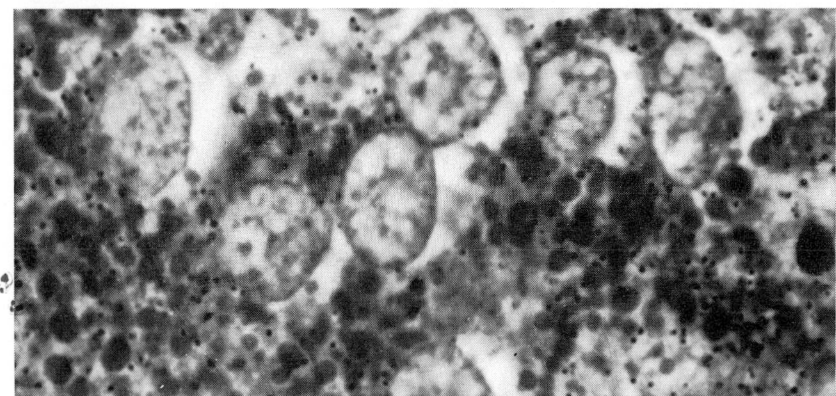

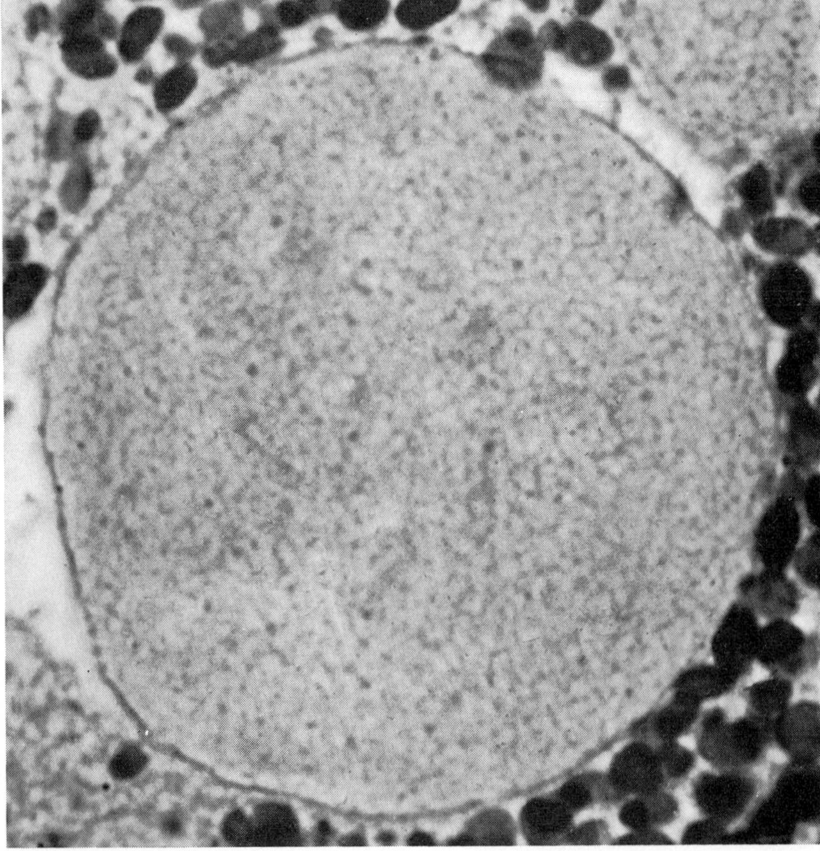

**BLASTULA NUCLEI swell up following their injection into an egg (*top*) or into an oöcyte (*bottom*). Within the oöcyte the enlargement may be as much as 200 times. The center micrograph shows blastula nuclei of normal dimensions; the bottom micrograph, a nucleus 48 hours after injection. Genetic material in the nucleus is dispersed during swelling.**

easier to analyze than other changes, and it seems likely to exemplify certain general principles of cytoplasmic regulation in early embryonic development.

The origin of the cytoplasmic condition that induces DNA synthesis has been investigated by injecting adult brain nuclei, together with a radioactive labeling substance, into growing and maturing oöcytes. The inducing factor appears just after an increase in the level of pituitary hormone has caused an oöcyte to mature into an egg, an event that is accompanied by intensive RNA and protein synthesis [*see illustration on preceding page*].

Concerning the identity of the inducing factor, the first candidate to be considered was simply the presence of an adequate supply of DNA precursor substances. Woodland, however, has injected growing oöcytes with 10 times the amount of all four common DNA precursors believed to be present in the mature egg. One of the precursors, thymidine triphosphate, had been labeled with tritium. In spite of the availability of these precursors, the brain nuclei did not incorporate the labeled thymidine into DNA. Although this experiment requires further analysis before DNA precursors can be excluded as inducers of DNA synthesis, it encourages a search in other directions.

The next candidate to be considered was DNA polymerase, an enzyme that promotes the incorporation of precursor substances into new DNA in a way that is specified by the composition of the preexisting "template" DNA. DNA polymerase activity in living cells has been tested by introducing purified DNA and tritium-labeled thymidine into eggs. In collaboration with Max Birnstiel of the University of Edinburgh we have established that the injected DNA serves as a template for synthesis of the same kind of DNA. When DNA and labeled thymidine are introduced into oöcytes, no DNA replication can be detected. This means that the cytoplasmic factor inducing DNA synthesis in eggs includes DNA polymerase or something that activates this enzyme. It is doubtful, however, that this is the *only* constituent of the inducer. If it were, the injection of egg cytoplasm (which contains DNA polymerase) into oöcytes might be expected to induce DNA synthesis, a result that is not in fact obtained. This experiment, in which purified DNA is replicated in the cytoplasm of unfertilized eggs, also serves to demonstrate that constituents of injected brain nuclei other than their DNA are not required in order to initiate the particular reaction being discussed here.

The last aspect of this reaction on which some information is available concerns the mechanism by which the inducing factors in the cytoplasm interact with the DNA in the nucleus. It was noticed several years ago by Stephen Subtelny, now at Rice University (and subsequently by others), that transplanted nuclei increase in volume soon after they have been injected into eggs. A pronounced swelling is also observed in nuclei injected into oöcytes; the swelling is therefore not directly related to a particular type of nuclear response. During this nuclear enlargement chromatin (which contains the genetic material in the nucleus) becomes dispersed and, as Arms has demonstrated, cytoplasmic protein also enters the swelling nuclei. While working at Oxford, Robert W. Merriam of the State University of New York at Stony Brook found a close temporal relation between the passage of cytoplasmic protein into enlarging nuclei and the initiation of DNA synthesis. The interpretation of these events currently favored by those of us involved in the experiments is that the nuclear swelling and chromatin dispersion facilitate the association of cytoplasmic regulatory molecules with chromosomal genes, thereby leading to a change in gene activity of a kind determined by the nature of the molecules that enter the nucleus.

The experiments described here have established two general conclusions. First, nuclear genes are not necessarily lost or permanently inactivated in the course of cell differentiation. Second, major changes in chromosome function as well as in different kinds of gene activity can be experimentally induced by normal constituents of living cell cytoplasm. The same type of experiment is now proving useful in attempts to determine the identity of the cytoplasmic components and their mode of action.

We have had to restrict our attention to what can be described as sequential changes in gene activity, that is, differences between one developmental stage and the next. These may be compared with regional variations in nuclear activity, that is, differences between one part of an embryo and another at the same developmental stage. The latter are hard to study biochemically because of the difficulty in obtaining enough material. There is no obvious reason, however, why the processes leading to the two types of differentiation should be fundamentally different.

Experiments analogous to those described here have been conducted with bacteria infected with viruses, with nuclear transplantation in protozoans and with fusion in mammalian cells. Each kind of material is well suited for certain problems; nuclear transplantation utilizing amphibian eggs and cell nuclei is especially suited to the analysis of processes that lead to the first major differences between cells. Only after these differences have been established by constituents of egg cytoplasm are cells able to respond differentially to other important agents that guide development, such as inducer substances and hormones. Finally, the technique of nuclear transplantation may be used to introduce cell components other than the nucleus into the cytoplasm of different living cells; this is likely to be of great value for the more detailed analysis of early development and cell differentiation.

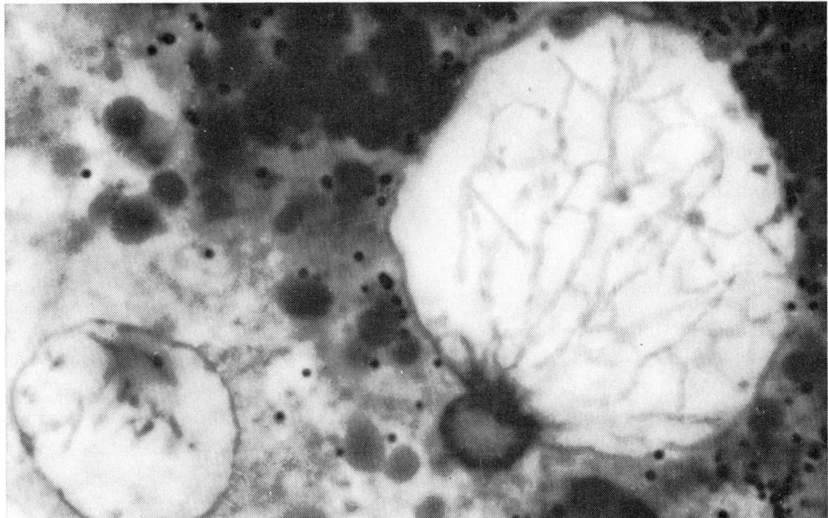

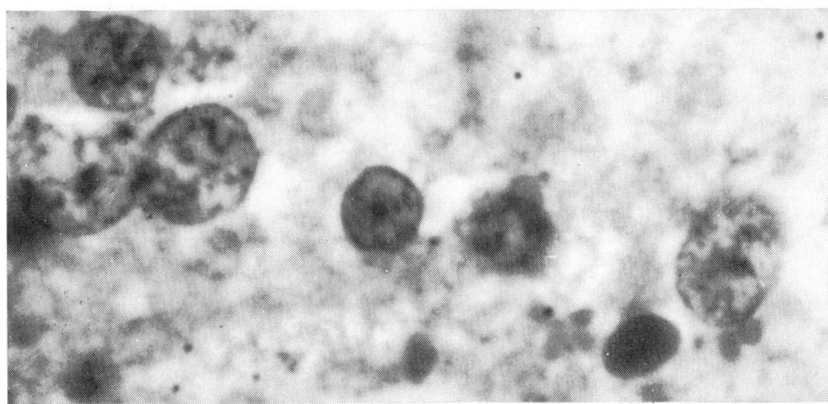

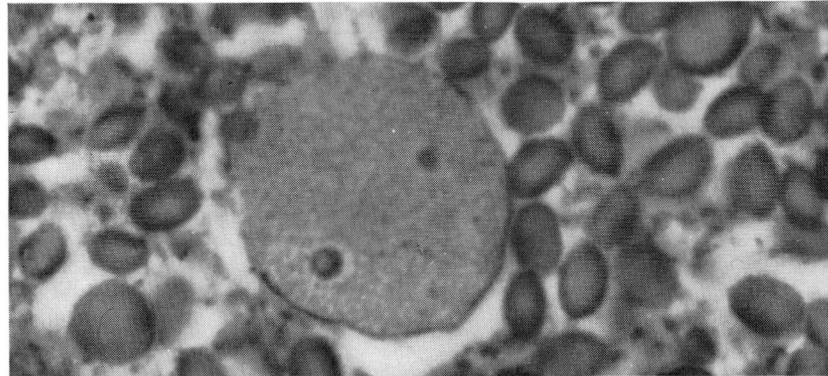

**BRAIN NUCLEI also swell up when injected into an egg (*top*) or an oöcyte (*bottom*), but they do not enlarge as much as blastula nuclei. The center micrograph shows brain nuclei of normal dimensions. During enlargement, dispersal of genetic material and entry of cytoplasmic protein into the nucleus facilitate contact of cytoplasmic regulators with the genes.**

# 20
# *Transdetermination in Cells*

ERNST HADORN
*November 1968*

The investigation of how the genetic material guides the development of cells and organisms is surely one of the most exciting enterprises of modern biology. It has already yielded many illuminating discoveries, but it is still full of surprising phenomena. We have encountered such a phenomenon in experiments conducted in our laboratory at the University of Zurich.

Let me first explain what prompted our experiments. We wanted to find out when and under what circumstances the future of embryonic cells is determined, so that the cells will proceed to specialize and form the various differentiated structures of the adult organism. It is well established that all the embryonic cells start out with an identical full set of genes, carrying instructions for the specific development and functioning of every kind of cell in the organism. It is generally assumed that a distinct pathway of differentiation is determined for cells when specific groups of genes are activated, some groups directing the differentiation of muscle cells; some, nerve cells; some, gut cells, and so on. We say that such cells are in a certain determined state. Thus determination is an important event in which cells of identical genotype begin to differ in their characteristics, or phenotype. At what stage in the embryo's development is this determination established? How stable is such a state? Will it be inherited by all cells descended from a determined ancestor, or can the cells perhaps be switched into new pathways?

The organism we chose for our experiments is the fruit fly *Drosophila melanogaster*. This insect has rendered prodigious service to genetics, but at first it seems much less suitable for studies of development than the embryos of amphibians and sea urchins. For our special purpose, however, *Drosophila* offered a most convenient feature.

In the early embryonic development of higher insects such as flies, bees and beetles a remarkable separation of cells occurs. In one class of embryonic cells final differentiation proceeds without delay, that is, it is initiated as soon as the cells become established by the process of cleavage and rudiment formation. These cells give rise to the body of the insect larva, with all its functioning organs. Another class of cells is set apart, however, in what are called imaginal disks. Although the cells of the imaginal disks are in contact with their differentiating neighbors, they remain in an embryonic state throughout the larval period. In this first period of insect life the cells of the disks divide. Therefore a few thousand of them are present in each disk when the larva is ready to become a pupa and proceed with metamorphosis. The location of a few of the imaginal disks in the larva of *Drosophila* is shown at bottom right in the illustration on the opposite page.

During metamorphosis most of the larval organs break down within the case of the pupa. Their cells eventually disintegrate. At the same time, and only at this time, the cells of the disks lose their embryonic character and differentiate into the specific body tissues of the imago, or adult fly. Each of the disks furnishes its part of the metamorphosed insect, as the illustration shows. One disk, the genital disk, contains cells that will develop into the structures of the sex organs, the back parts of the abdomen and the hind gut. There is a disk for each of the six future legs, and three pairs of disks combine to form the head. (Of these latter disks only the eye-antenna disk appears in the illustration.)

Each disk forms a set of different structures. A leg disk, for instance, contains cells that respectively develop into claws, tarsal parts, tibia, femur, trochanter, coxa and adjoining parts of the thorax. Moreover, in each section of leg many different functions must be conducted. Accordingly bristles and hairs of various sizes and shapes are formed by different cells and are distributed in definite patterns.

Imaginal disks can be dissected out of an insect larva. In our experiments with *Drosophila* we cut the disks with a fine tungsten needle and implant the pieces in the body cavity of other larvae. Then, when the host larvae metamorphose into

**CELLS FROM A LARVA of the fruit fly *Drosophila* were kept alive and undifferentiated by transplanting them into the abdomens of 100 successive generations of adult fly hosts; alterations in the genetic control of the cells are plotted in the illustration on the opposite page. The cells were taken from the cell cluster, or imaginal disk, destined to form the adult fly's sex organs and part of the gut and abdomen (*blue area in the larva and fly*). At each transfer some of the cells were transplanted into larvae that underwent metamorphosis, as did the transplanted cells. By the eighth transfer (*B*) the cells' initial genetic controls had begun to change; metamorphosis produced head parts (*red*) and leg parts (*yellow*) and soon thereafter wing parts (*green*) and thorax parts (*brown*). By the 56th generation none of the cells in any subculture any longer matured into genital components. The letters *A* through *G* in the illustration refer to the structures shown in photomicrographs on page 198. The subcultures connected by broken lines (*below D*) showed mutant abnormalities.**

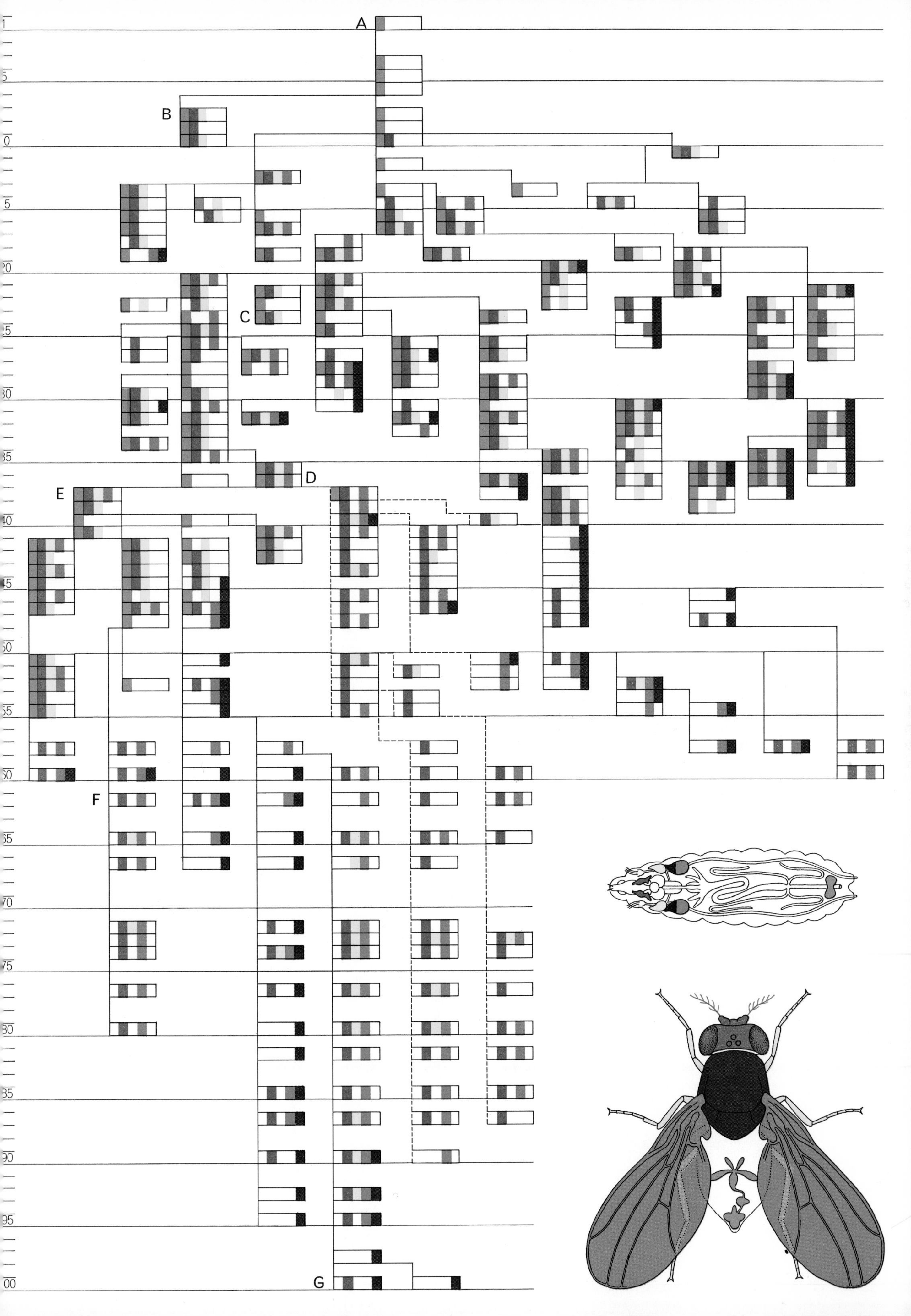
A
B
C
D
E
F
G

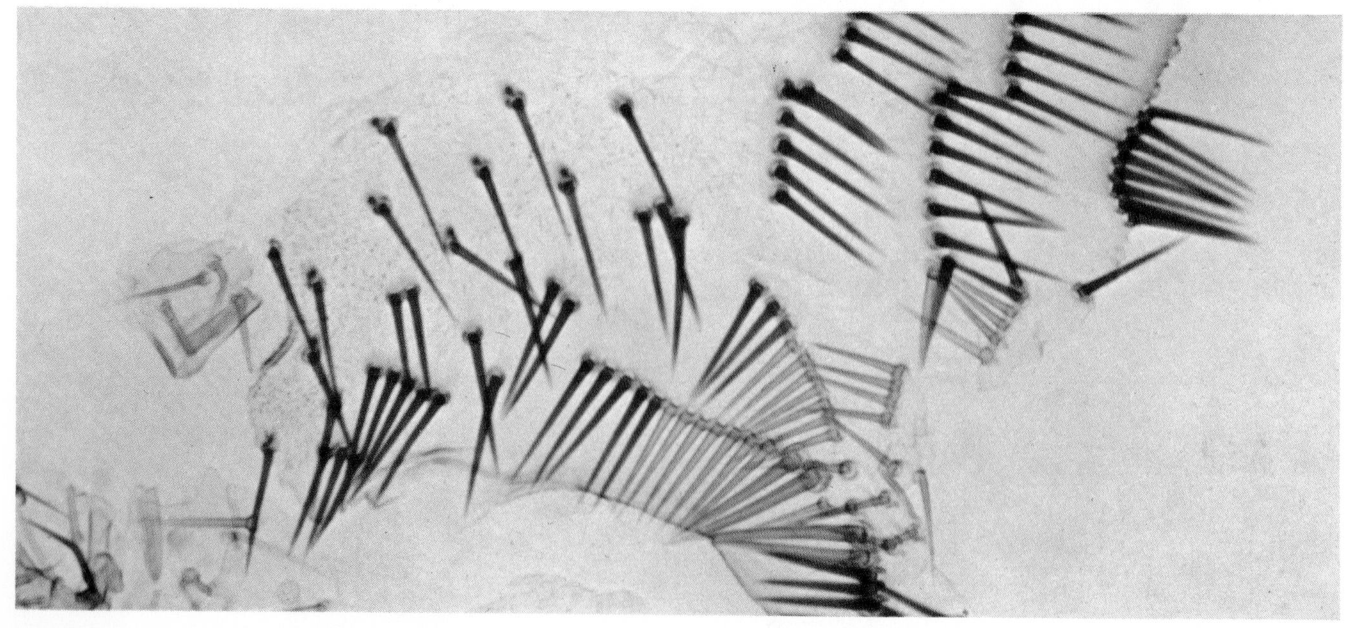

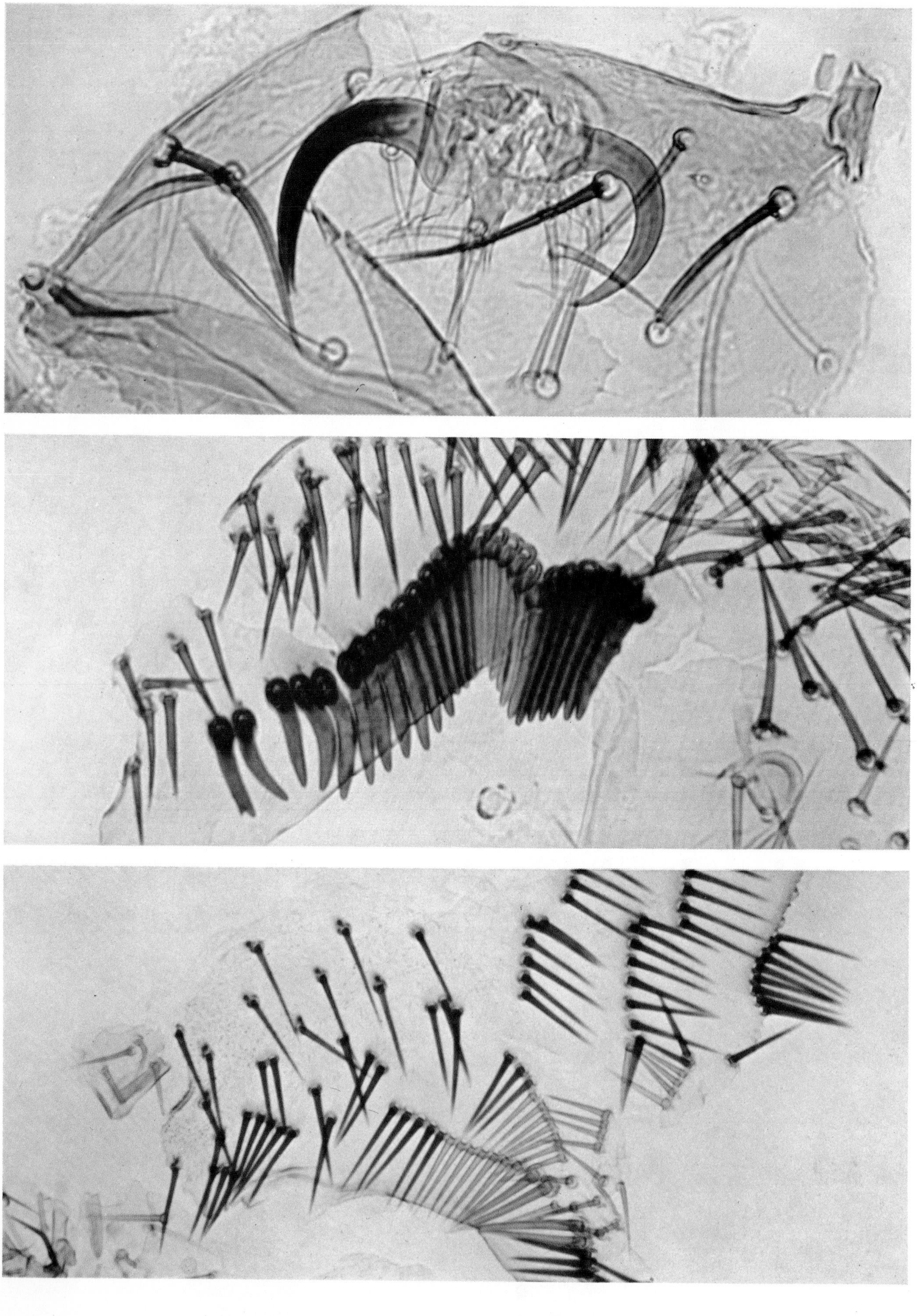

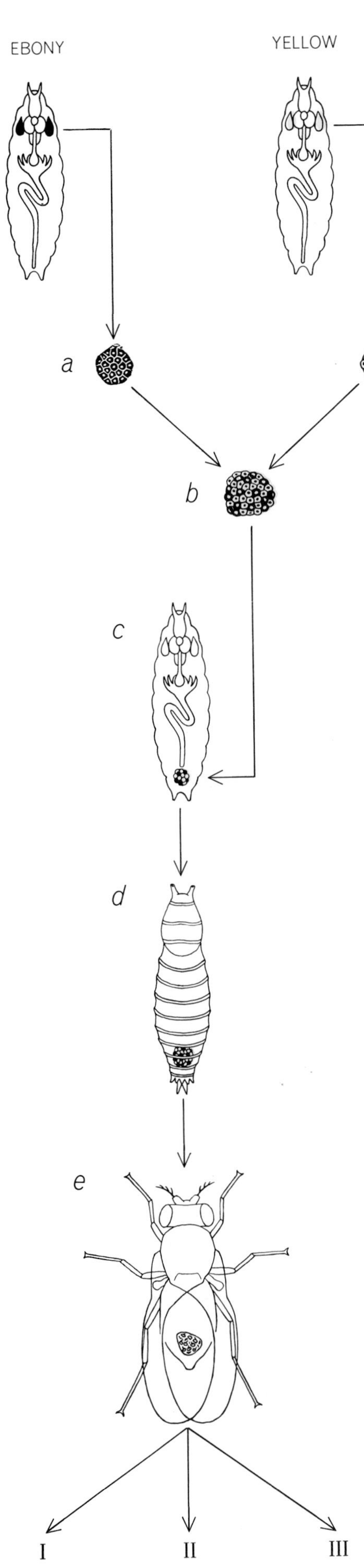


pupae, the implants also differentiate into adult structures. For example, when a piece of implanted disk contains cells determined for eyes, a fully developed eye is found in the host's abdomen. By much experimentation we found that each imaginal disk contains a mosaic of different cell populations; these rudiments are determined for particular structures within a body region. Some mosaic regions of a male genital disk will develop into a sperm pump, other regions will differentiate into various elements of the penis, still others furnish anal plates and hind gut. In short, the future differentiation of the cells is already determined in the larval stage.

We need more detailed information, however. What about the individual cells? Could it be shown that each cell in a mosaic region of an imaginal disk is primed for a particular differentiation, regardless of the company in which it might find itself later?

We investigated this question by breaking disks down into individual cells and then implanting a mixture of cells from different disks into a host larva [*see drawing at left*]. The fruit fly is particularly convenient for this kind of experiment because of the availability of readily identifiable mutant forms, which differ from one another in visible features such as body color and bristle shape. Therefore we could mix disk cells from, say, a yellow mutant and a dark ebony mutant, allow the cells to clump together and then see where each cell ended up when the implant formed adult structures in the host larva.

It turned out that the individual cells of disks are indeed already programmed for specific differentiation. They collaborate in forming normal structures. For example, yellow leg cells and ebony leg cells combine to produce leg parts that are an ebony and yellow mosaic [*see photomicrographs on opposite page*]. My collaborators Rolf Nöthiger, Heinz Tobler and Antonio Garcia-Bellido performed many experiments, testing mixtures of imaginal-disk cells in various combinations. The results were consistent: Somehow during the development of the host fly those cells in the implanted mixture that carry a particular differentiation program recognize one another and move about until they come together to form precisely the indicated structure. In contrast, cells from different disk regions never combine but separate from one another. A cell from a leg disk never joins one from a wing disk, and wing-base cells do not collaborate with wing-tip cells to form a base or a tip.

We were able to conclude, therefore, that every cell in an imaginal disk of a fruit-fly larva is individually determined for a specific differentiation. This determination is fixed long before any morphological differences of the cells can be detected. We worked with such determined but not yet differentiated genital-disk cells in the experiments I shall now describe.

As we started this work we wondered what would happen if we implanted parts of larval disks directly into adult fruit flies. With such a procedure the pupal stage would be bypassed. When we tried the experiment, there were three quite unexpected results.

First, when larval cells were transplanted into the abdomen of an adult fly, they divided and grew without limit. In the normal course of events the same cells would have stopped dividing as soon as pupal metamorphosis set in. Under the influence of the insect hormone ecdysone they would have begun to differentiate into adult structures such as bristles, hairs, claws and gland cells. The unlimited growth of these cultures has now continued for six years. Since a host fly lives for about a month, we have had to transfer the proliferating cultures about every two weeks. The oldest cultures are now in their 160th transfer generation. For the permanent maintenance of a culture we need only take a small part of an implant to found the next transfer generation. It is thus easy to start many subcultures from a single culture.

The second result was even more surprising. Although the cultured cells have

**MIXTURE OF LEG CELLS** from the larvae of fruit flies that belong to separate color strains is formed by removing part of a leg imaginal disk from each larva (*a, top of drawing at left*). The bits of disk are then separated into their individual cells, mixed together (*b*) and implanted in the body cavity of a host larva (*c*). The host pupates (*d*) and the implanted cells mature. When removed from the adult fly (*e*), the leg organs are partly yellow and partly ebony in color (*see photomicrographs on opposite page*). The organs include (*I, top photograph*) a leg tip with one ebony and one yellow claw, (*II, middle*) an ebony-and-yellow sex comb and (*III, bottom*) ebony-and-yellow tibia bristles. The mixed colors show that cells with the same differentiation potential will find one another and join to form normal organs.

now lived for years in adult flies, they retain their original embryonic character. No differentiation has been observed. Apparently the pupal phase is indispensable for the initiation of differentiation. From this observation we conclude that ecdysone, which is dominant during pupal metamorphosis, is no longer present or active in the imago. The adult body fluid in which the cultured cells float seems to be much like the body fluid of the larva.

The third result is encountered when parts of cultures are transplanted back into larvae. Such implants will now metamorphose along with their new host. With some exceptions, to which I shall refer later, the test pieces taken from the permanent cultures differentiate normally into those adult structures for which they were determined. The capacity for normal development is therefore propagated in our cultures; during all the transfer generations the genetic endowment needed for adult differentiation remains intact. Such stability is unusual for cells in permanent cultures. In all the dividing cells of the cultures we find only the normal set of chromosomes.

What I have said so far is prologue to the main chapter of our story. Let us now start a permanent culture with half of a male genital disk. One half, the stem piece, is implanted in the abdomen of an adult fly. The other half, the test piece, we implant in a larva within which it will differentiate. The stem piece grows, and after two weeks it is removed from its first adult host. It is again divided into a stem piece, which will maintain the culture, and a test piece, which shows us what kind of adult differentiation it is capable of. After a few transfers the cultures usually grow faster, so that it becomes possible to pass more than one test piece through metamorphosis and to implant several stem pieces and found branched sublines [*see illustration below*].

In the course of 160 transfer generations over a period of six years, several thousand test implants were studied. This experimental situation clearly offers a unique experimental opportunity. At any time we can find out what capacities for differentiation are present in the proliferating cells.

In the first few transfer generations

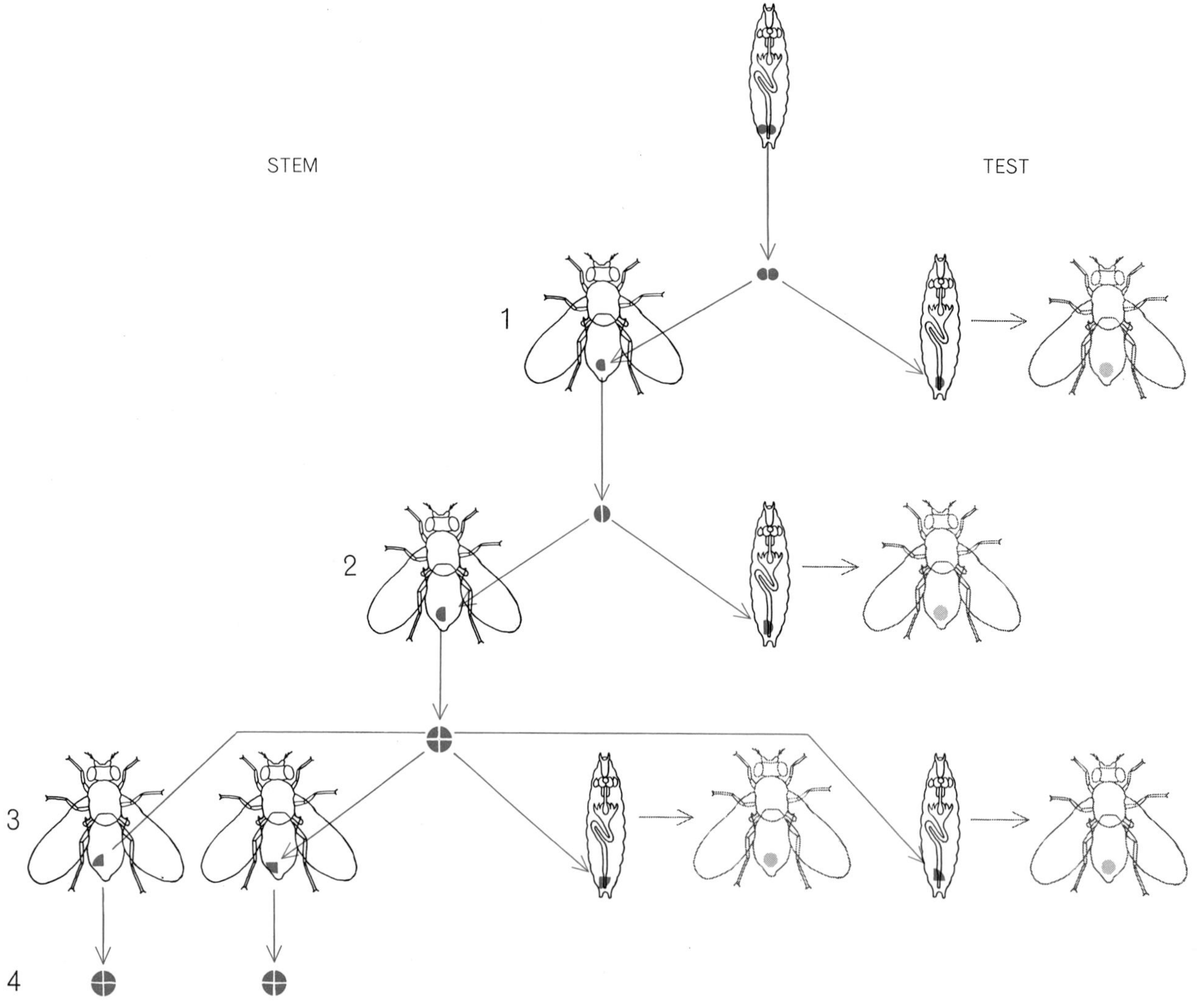


**SUCCESSIVE TRANSFERS** of cells grown from a single bit of imaginal disk is diagrammed over four generations. A bit of genital disk from a larva is divided between an adult host (*1, left*) and a larva (*right*), which is allowed to mature so that the condition of the transferred cells can be checked. The process is repeated (*2*) by retransplanting the disk cells of the "stem" host. By the next transplant (*3*) the stem disk cells have increased sufficiently to provide two transplants for testing and two transplants into stem hosts. In the next generation (*4*) the number quadruples. Disk cells have survived and multiplied for more than 150 transfer generations.

the test pieces from the genital disk developed in accordance with their original determination, that is, they differentiated "autotypically" into genital and anal organs. Thereafter came the surprise. Parts of the test implants developed instead into head organs or leg parts. In other words, populations of cells no longer differentiated as their ancestors had. They now produced "allotypic" organs that in normal development arise only from the cells of a head imaginal disk or a leg disk. For the change from one developmental pathway to another we introduced the term transdetermination.

The entire story of one typical culture can be followed in the illustration on page 193. If the reader will study the illustration, he will see that in the eighth transfer generation (after about four months) head and leg structures appear. A further transdetermination leads in Transfer Generation 13 to wings. In Transfer Generation 19 thorax cells become established for the first time. After many culture transfers we were able to make some generalizations about determination and transdetermination. First, we found that a state of determination can be replicated without any change when the cells divide; the characteristics of a determined cell population are transferred by "cell heredity." This holds true for the initial autotypic state as well as for newly acquired allotypic states after they become established by transdetermination. For example, one can see in the illustration that genital and anal organs appear in the test implants without interruption from Transfer Generation 1 to Transfer Generation 55. By the same token, however, newly initiated allotypic cell populations are also multiplied by cell heredity. In certain sublines head structures, legs, wings and thorax parts appear continuously for years.

For each given state of determination there exists a distinct probability or frequency of transdetermination in a specific direction. In the cells of genital disks a transdetermination of the first order leads with about the same frequency to head or leg cells. From these allotypic states a transdetermination of the second order results in wing cells. We have never observed a direct switch from genital to wing cells. Finally rudiments of thorax arise by a further transdetermination from wing cells.

These sequences are summarized in the bottom illustration at the right. Some of the changes are indicated as

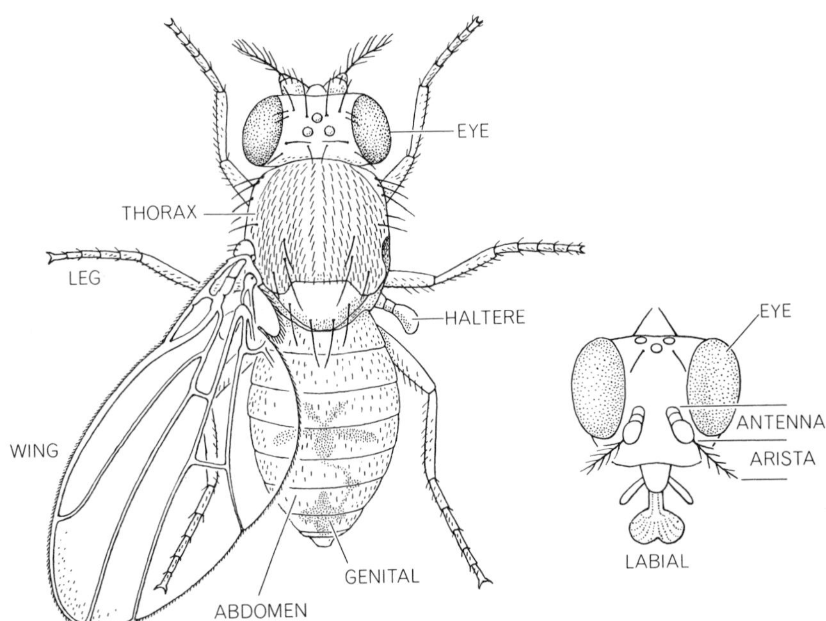


**ADULT FLY'S ORGANS** are formed by the maturation of a number of imaginal disks that are present in the body of the larva. The sex organs and portions of the gut and abdomen are formed from the cells of the genital disk and the head develops from three pairs of disks; each of the fly's six legs develops from a separate disk. Disk cells that are kept in an undifferentiated state by successive transplants into adult hosts for years will continue to mature into the appropriate organs when allowed to undergo metamorphosis. After a number of transfer generations, however, a change takes place and the disk cells turn into adult tissue of some other kind. The "transdetermination" generally follows the same sequence (*see illustration below*) and the change is preserved over many more generations.

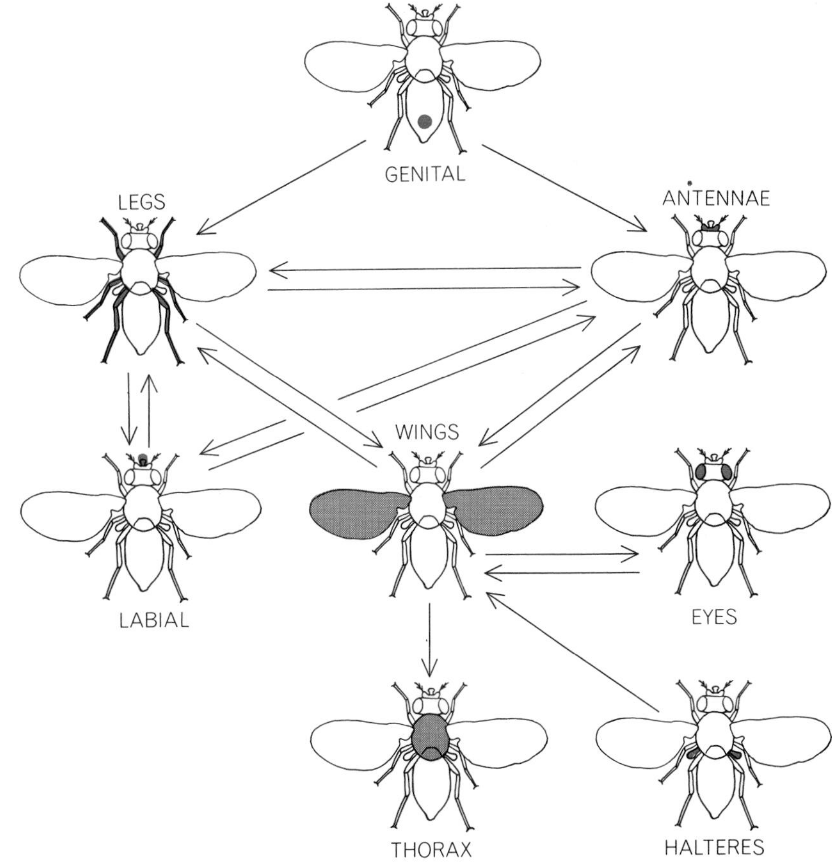


**TRANSDETERMINATION SEQUENCE** undergone by seven kinds of imaginal-disk cells is shown by arrows. Genital cells, for example, may change into leg or antenna cells, whereas leg and antenna cells may become labial or wing cells. In most instances the final transdetermination is from wing cell to thorax cell; the change to thorax appears to be irreversible.

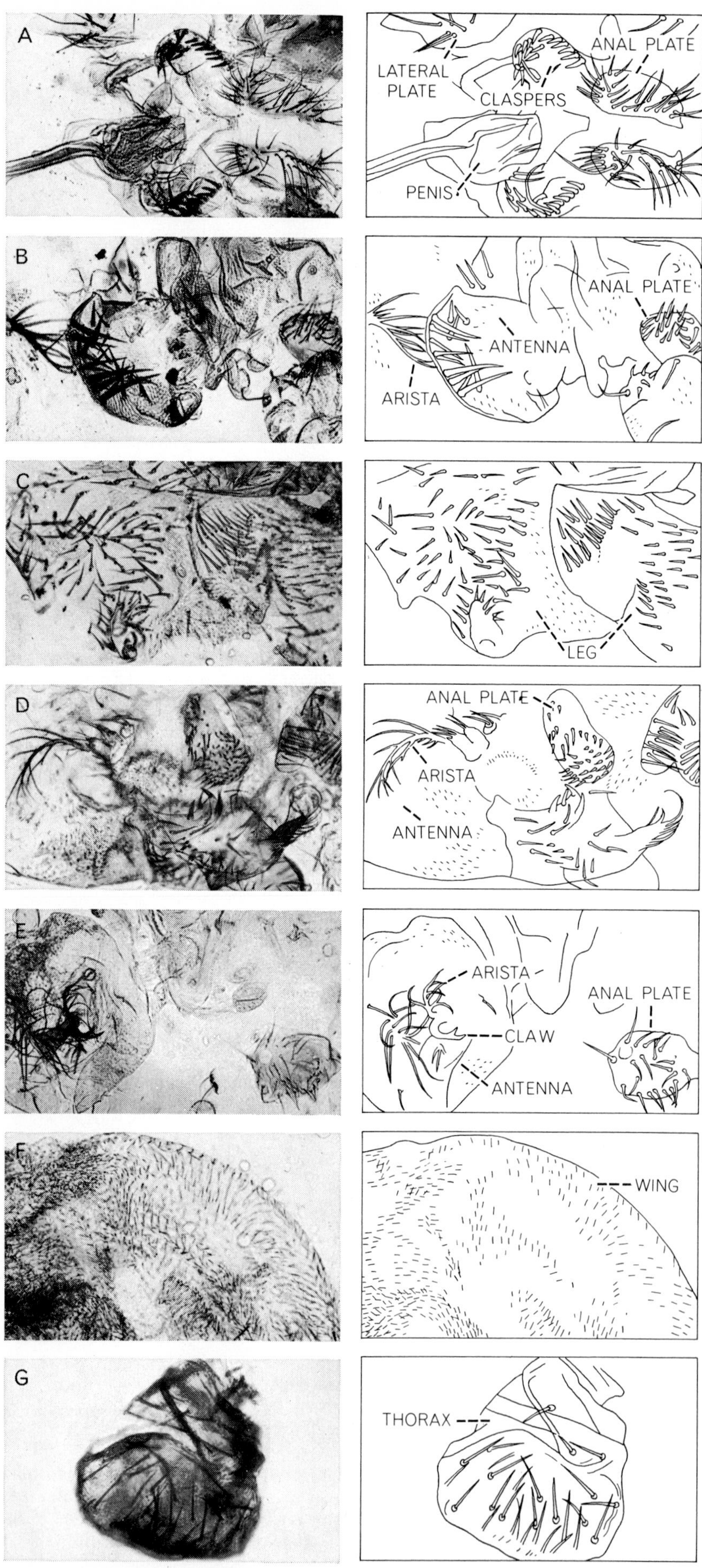


being reversible, others as being irreversible. For instance, it seems impossible for dividing thoracic cells to revert to genital cells. The dynamics of transdetermination lead to a shift in the frequency of different kinds of differentiation in the test implants. The initial states represented by the autotypic structures decrease or even disappear, whereas the "end stations," such as thorax, become increasingly frequent.

I have given the experiments with genital disks as an example; the same behavior is observed for other imaginal disks. The cells of these disks, which are initially determined for head and mouth parts, for wing parts or for thorax parts, also undergo transdetermination in the course of prolonged culturing. Each series of changes follows its own sequence.

Transdetermination leads, as we have seen, to a change in cell heredity. What factors cause this switching and on what mechanisms might it be based? A satisfactory answer to these complex questions cannot yet be given. We can only offer a few tentative speculations. One fact, established by my co-workers Tobler and Hansruedi Wildermuth, is that the frequency of transdetermination is correlated with the rate of proliferation of the cells: the faster they multiply, the higher the probability of a change in the cell's heredity. We therefore interpret transdetermination as a phenomenon somehow brought about by changes in the dynamics of proliferation on which the functional interaction of different cellular components might depend.

Stability of determination is ensured only when the same group of genes remains active in subsequent cell generations. Transdetermination leading to a

**TRANSDETERMINATIONS** of cells from a genital disk during 100 successive transfer generations are shown in a series of photomicrographs (*left; organs formed before and after transdetermination are identified in the accompanying drawings*). The generations are those illustrated on page 193. Mature structures from the first transfer generation (*A*) are all normal products of genital cells. By the eighth generation (*B*) some genital cells have matured into head-cell components. A specimen from the 24th generation (*C*) shows leg structures. In the 37th generation (*D*) and the 38th (*E*), specimens still form genital as well as head structures. The wing parts (*F*) are from the 60th generation and thorax structures (*G*), the final products, are from the 100th generation.

long-lasting new type of cell heredity, however, requires the action of new sets of genes. From the findings of molecular genetics we know that genes are activated or repressed by certain effector molecules. When our cells divide and grow, the effectors will be diluted and new ones must be synthesized. Differences and changes in the rate of cell division could in different populations of a culture perhaps affect the inventory of cellular components that control the activity of genes, and a new equilibrium between interacting factors might result. Members of a formerly active gene team that established determination for legs, say, could become inactivated. Their role would then be taken over by wing-determining genes.

Only further experiments can bring a better understanding of all the things that happen in transdetermination. I should like, however, to call attention to one additional fact. We have seen that cells determined for genital organs can give rise to descendants that develop alternatively into an entire set of other body parts. As my collaborator Walter Gehring has shown, sequences of diverse differentiations arise even in cell populations (clones) derived from a single cell. Therefore normal determination is certainly not based on an irreversible state of cellular units. Competence for entering many different pathways of development is a persistent characteristic of cells.

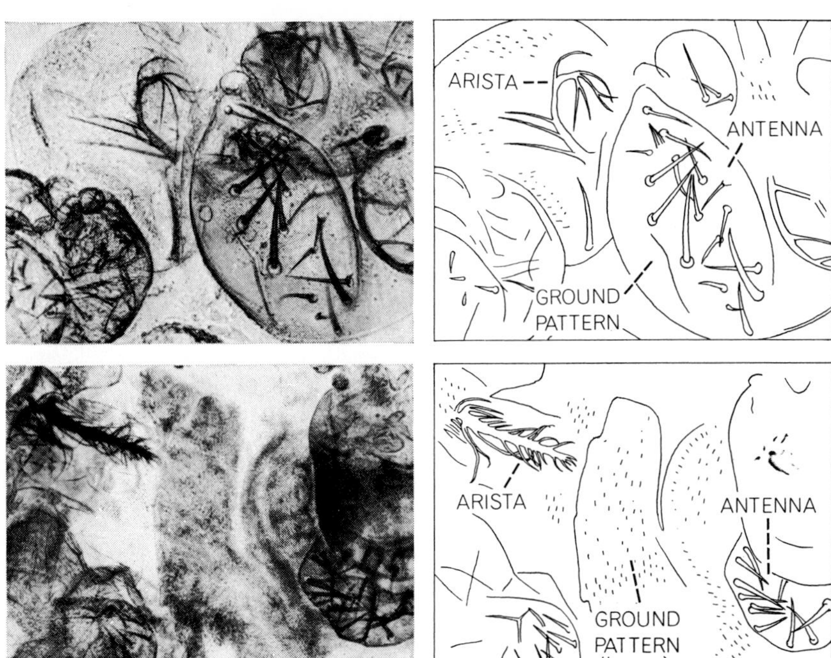


**NORMAL AND ABNORMAL STRUCTURES are shown in two photomicrographs and identified in the accompanying drawings. At top is mature tissue from the 31st transfer generation of genital-disk cells; it has changed into normal head-cell components. At bottom are the same components but in abnormal forms, as found in tissue from the 41st transfer generation. The abnormalities, probably caused by mutations, have proved to be hereditary.**

Thus far we have dealt with changes in the cultures that lead from one normal type of differentiation to another normal type. Indeed, allotypic thorax implants cannot, even after six years of culturing, be distinguished from autotypic test pieces taken directly from a fresh larval control disk. In addition to such "normotypic" differentiation, however, we encountered in a few sublines an "anormotypic" type of development [*see illustration at right*]. In a subline that could be followed from Transfer Generation 38 to Transfer Generation 90 the small hairs and protuberances (trichomes) that cover the cuticle between the bristles look coarse and somewhat distorted. Similarly, the arista, a structure protruding from the antenna, is abnormal in shape. On the other hand, the bristles are all normal. In another subline the ground pattern is fully normal, but here the capacity for forming bristles is lost. Such anormotypic lines might be due to classical mutations. Reversion to normality was never observed. Once established, the abnormal characteristics breed true as long as we have observed the descendants. A further developmental type became established in other sublines. These cultures have an extremely high rate of growth, but the test pieces either disappear during metamorphosis or pass the pupal stage without any visible change; in any case, the ability for adult differentiation is lost. Perhaps such types of cell heredity are due to the kind of mutations that in whole animals would be called developmental lethals.

All the events revealed in the cultures of cells from imaginal disks should now be compared with processes going on in other biological systems. Here I shall mention only a few obvious similarities. Stability in cell heredity—the kind of stability observed in our proliferating cultures—is encountered in, among other tissues, the epidermis of human skin and in human bone marrow. There the stem cells remain undifferentiated, but they breed true throughout life by replicating a specific state of determination.

Transdetermination might also have biological counterparts. Consider certain types of regeneration. Biologists working on leg regeneration in amphibians have demonstrated that formerly specialized and differentiated cells have offspring that become determined for new tasks. In insects and crustaceans one observes occasionally that from an antenna stump a leg or an eye can regenerate. Such allotypic performance is probably based on the same cellular events as transdetermination. Moreover, in *Drosophila* quite a few mutants of the type called homoiotic are known that lead to allotypic organs. One such mutation is termed aristapedia; it steers the developmental fate of the cells in the antenna disks so that leg parts instead of antennal structures become differentiated. The end effect is the same as in our transdetermining cells, but one should not overlook the differences in principle. The effects of homoiotic mutations are due to changes in the genetic material, whereas for transdetermination we postulate a change in the control systems that only regulate the activity of unmutated genes.

Since we observe in cultures so many changes in the proliferating behavior of cells and in their heredity, one is tempted to compare certain of our results with cancerous growths and aberrant differentiation in tumors. I think it is too early for any detailed comparison. All one can say at present is that cancer cells, like transdetermined cells, depart from the pathways of their ancestors and enter on new pathways of development.

# 21

# *Phases in Cell Differentiation*

NORMAN K. WESSELLS and
WILLIAM J. RUTTER
*March 1969*

How does a single cell—the fertilized egg—give rise to the many different cell types of a multicellular organism? After fertilization of an egg by a spermatozoon the number of cells in the developing embryo increases dramatically. Soon three classes of cells can be distinguished: ectodermal and endodermal cells respectively make up an outer and an inner embryonic layer, and mesodermal cells constitute an intermediate layer. These cells come to be arranged in groups that develop into recognizable tissues and organs. Ultimately several hundred kinds of cells can be distinguished in an adult mammal. The process of functional and structural specialization of cells is called differentiation, and the mechanisms controlling differentiation remain major mysteries of biology.

Experimental work done by embryologists since 1900 has demonstrated that interactions between cells play an important role in differentiation. The development of some organs, including the pancreas, the liver and the lungs, depends on discrete sets of cells derived from the embryonic gut endoderm and also on adjacent mesodermal cells; the brain, the mammary glands and the limbs involve combinations of certain ectodermal and mesodermal cells. Experiments with intact embryos and with laboratory cultures of combinations of tissues indicate that for normal development the two interacting tissues must be adjacent to each other. A fundamental and still unanswered question is: How does one cell influence the development of another?

What happens within cells during development to make one cell type different from another? Since nearly all cells have some physiological activities in common, all must contain some of the same enzymes and structures, and mere quantitative differences among cells in this common metabolic machinery would not in themselves confer unique characteristics on a given cell. For that a cell has a group of specific proteins responsible for its specialized functions: muscle cells contain contractile proteins, plasma cells make antibodies, red blood cells form hemoglobin and so on. What controls these qualitative differences? This is a second fundamental and still unanswered question.

In this article we shall approach these two major questions by discussing the differentiation of the mammalian pancreas. First we shall outline the development of the pancreas and then describe a detailed investigation of interactions between cells and the regulation of specific protein synthesis in this organ. The view of differentiation gleaned from this work, carried out in our laboratories at Stanford University and the University of Washington, may be generally applicable to other organ systems.

The mammalian pancreas manufactures enzymes that digest foodstuffs and also secretes two hormones that regulate the metabolism of carbohydrates. Different populations of cells are involved in the two functions. The exocrine cells make about a dozen specific proteins that are involved in the breakdown of proteins, carbohydrates, fats and nucleic acids. Some are synthesized as active enzymes and others as inactive precursors called zymogens; all of them are stored within the cells in granules. The exocrine cells are arranged in clusters called acini and secrete the contents of their granules into a duct system that leads into the small intestine. There the zymogens are converted to active enzymes, and digestion takes place.

There are two kinds of endocrine cell. The *B* cells synthesize insulin, the hormone that regulates the uptake of blood glucose and its storage as glycogen in muscle and fat. Insulin is stored in beta granules. Donald F. Steiner of the University of Chicago has demonstrated that the synthesis of insulin (like the synthesis of the protein-digesting enzymes) has two steps: first the formation of a precursor called proinsulin and then its conversion to insulin by an enzyme. The site of this conversion and the specific enzymes involved are not known. The *A* cells produce the hormone glucagon, a protein that controls the degradation of glycogen in the liver. Glucagon is stored in alpha granules. Both *B* cells and *A* cells are in the islets of Langerhans, spherical aggregates of cells that lie near blood capillaries, so that the hormones pass easily into the blood for transport to their sites of action elsewhere in the body.

In the embryo of a mouse or a rat the formation of the pancreas begins about midway through the gestation period. A group of cells in the upper wall of the primitive gut begins to bulge upward on the ninth day of gestation in the mouse, and about 36 hours later in the rat. The base of this evagination constricts, giving rise to the future pancreatic duct; the upper part, the pancreatic epithelium, expands into the surrounding mesodermal tissue. (Note the proximity of endoderm and mesoderm.) In the next three or four days rapid cell division continues and the cells become arranged in typical acini and primitive islets. During the later stages of gestation cell division tends to be restricted to the peripheral regions of the pancreas; near the center one can detect the signs of advanced differentiation: zymogen granules in the acinar cells and alpha

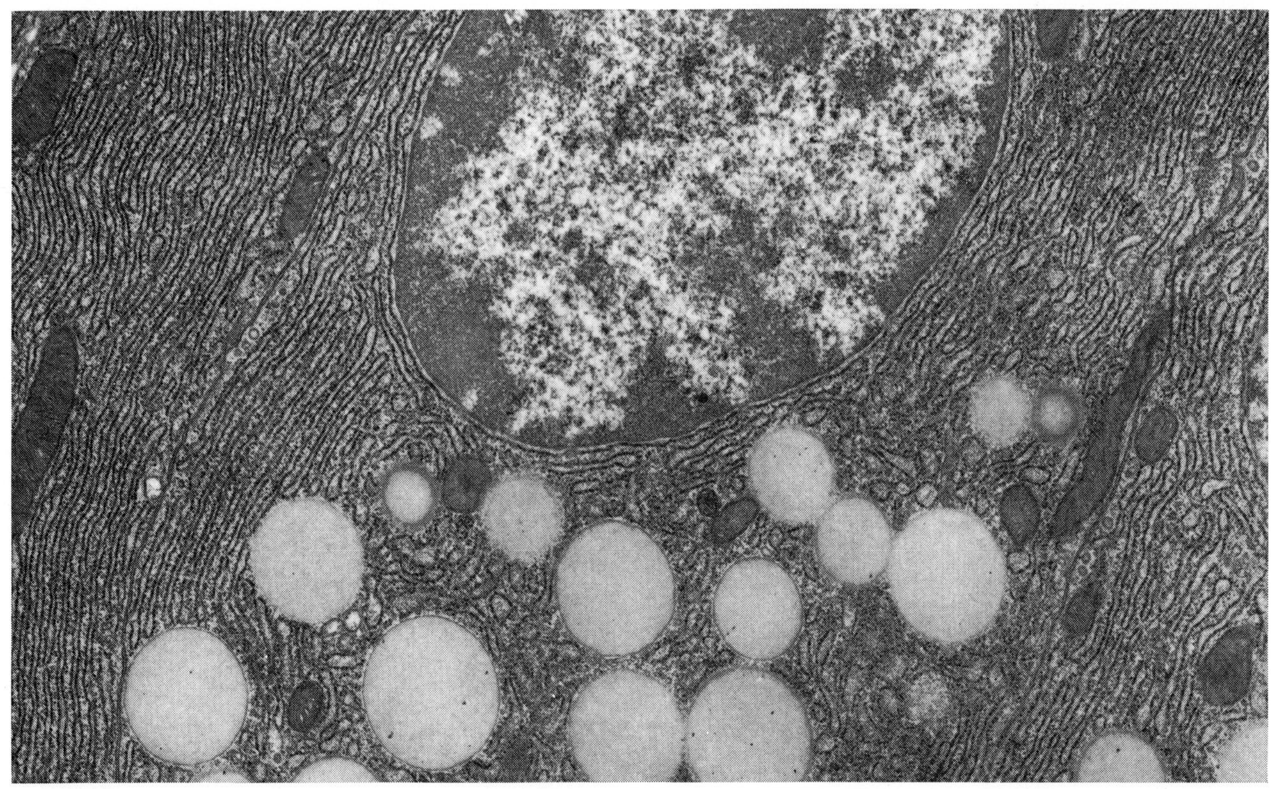

EXOCRINE CELLS of the pancreas secrete zymogens, proteins that are precursors of digestive enzymes. In this electron micrograph made in the laboratory of one of the authors (Wessells) a thin section of exocrine cells from the adult mouse pancreas is enlarged 14,000 diameters. The extensive endoplasmic reticulum, a folded membrane studded with ribosomes (*small black dots*), is the site of protein synthesis. Zymogens are stored in zymogen granules (*spheroidal vesicles at bottom*) until they are secreted into the intestine.

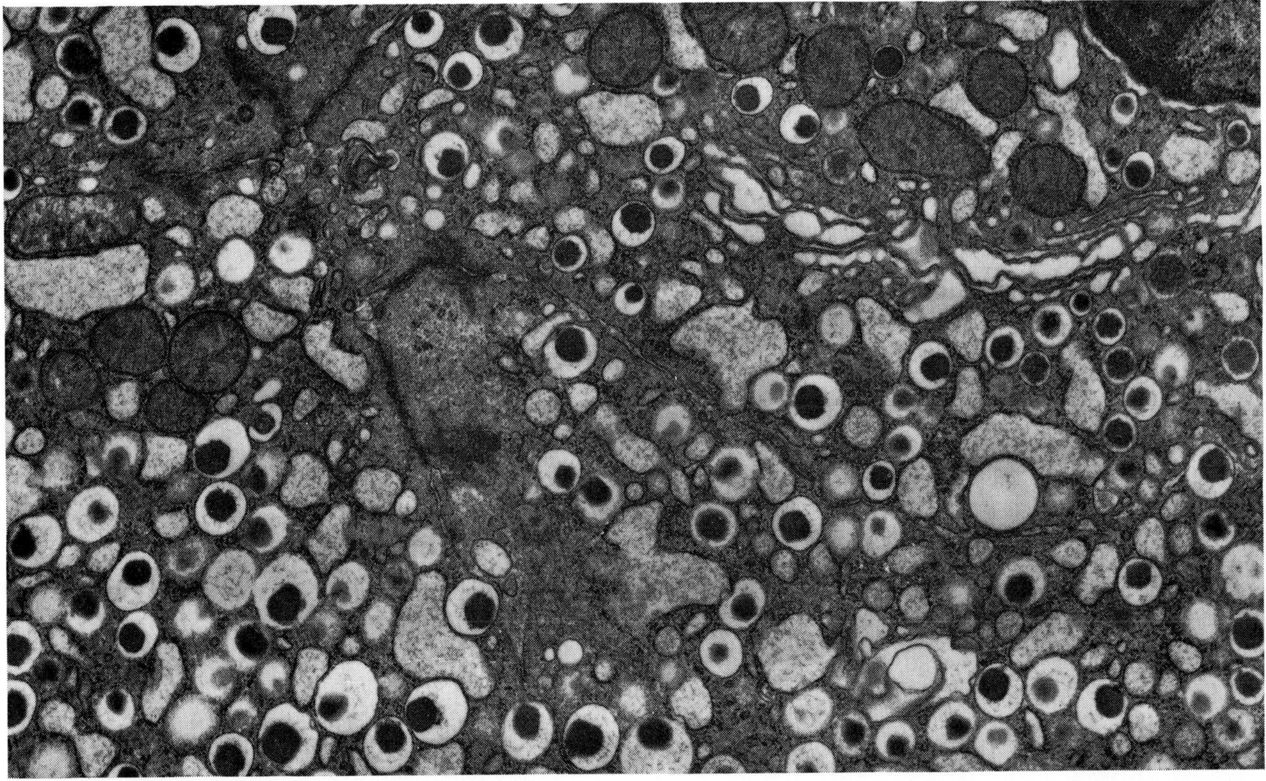

ENDOCRINE CELLS of the pancreas secrete the hormones glucagon and insulin. In this electron micrograph several adult mouse *B* cells, which secrete insulin, are enlarged 17,000 diameters. The light spheroidal vesicles present in large numbers are beta granules and the dense material in many of the granules is thought to be insulin or proinsulin, the precursor protein of the hormone.

and beta granules in the endocrine cells. As birth approaches, cell division stops in the peripheral acini and they too differentiate.

To establish the course of differentiation at the molecular level one can monitor the specific functional attributes of the various cell types. For the exocrine cells the digestive enzymes produced in the largest quantities are the best indexes of differentiation; for the endocrine cells insulin and glucagon serve the same purpose. The problem is to measure these proteins specifically and sensitively in the mixture of several hundred different proteins that are present in the cells. We developed microassays based on specific catalytic activity (for the individual digestive enzymes), the ability of specific antibodies to recognize single species of proteins (for insulin and glucagon) and such distinctive physical characteristics of the protein molecules as their size and their mobility in an electric field.

With these assay procedures we discovered that both the exocrine enzymes and glucagon and insulin are present at concentrations many thousands of times higher in functional pancreatic cells than in other cells of the adult organism; these proteins are indeed cell-specific. This finding gave us a considerable experimental advantage, since it meant we could attribute the proteins to pancreas cells even when we measured them in extracts of several cell types, such as pancreas plus gut. We set out to determine the pattern of accumulation of the specific products and relate it to the development of acini and islets and the secretory granules.

By measuring the activity of the major exocrine enzymes present in extracts of tissues we established their developmental profiles during gestation [*see top illustration on page 6*]. Two significant and rather unexpected features of their accumulation patterns became apparent. First, the concentration of proteins (expressed as enzyme molecules per cell) does not change in a simple way, rising from zero to finite high levels in one step. Instead there are three discrete stages in the developmental process. In what we call State I the enzymes are present at a relatively low concentration; then the concentration is 1,000 times or more higher (State II), and finally the concentrations of individual enzymes are adjusted a little with respect to one another (State III). Even the low State I level was much higher than the level in other cell types, and the relative proportions of the individual enzymes approximate those found in State II, indicating that State I is a stage in development unique to the pancreas rather than simply a reflection of lack of development.

The second unexpected feature of the developmental profiles is that the concentrations of the exocrine enzymes do not change as a coordinated set during

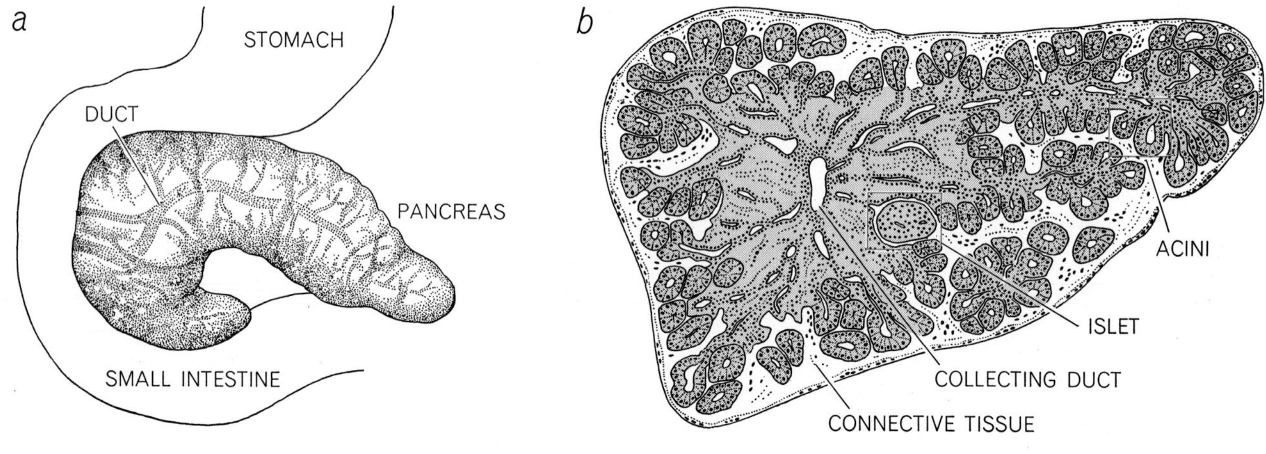


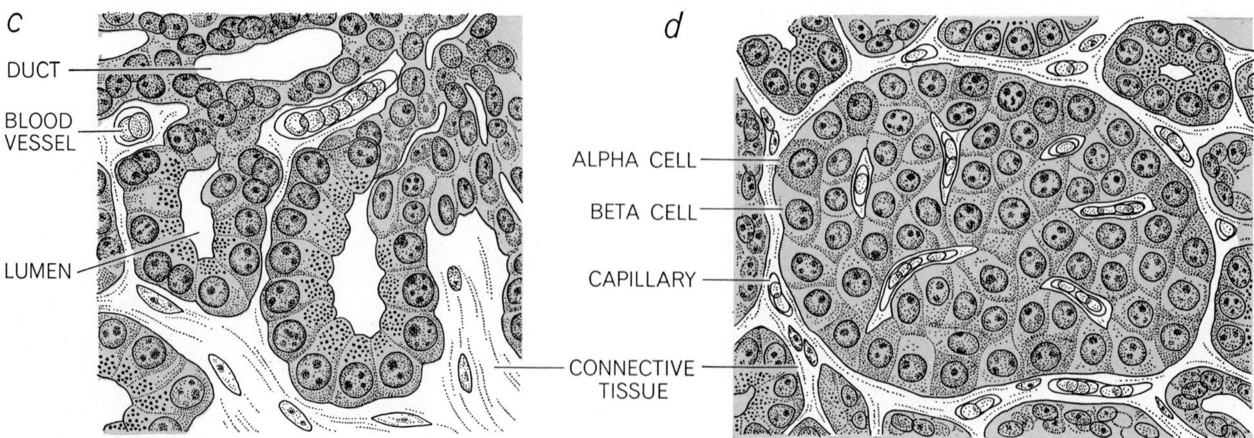


**PANCREAS is a secretory organ that lies behind the stomach, as indicated in the schematic drawing of mouse viscera (*a*). A cross section of a small portion of the pancreas (*b*) contains cell clusters called acini, ranged along a duct system, and distinctive cell groups called islets of Langerhans. The acini are composed of exocrine cells with zymogen granules (*c*); islets contain endocrine cells (*d*).**

the transition from State I to State II; the pattern of accumulation between 15 and 19 days differs significantly for the various proteins. For example, trypsin and carboxypeptidase *B* lag significantly behind the other proteins; lipase *B* activity is first detected in the rat at 18 days, when the other enzymes have already reached high levels. The exocrine proteins, then, are not all regulated as a unit. On the other hand, the relative concentrations of certain groups of enzymes—for instance lipase *A* and carboxypeptidase *A*, trypsin and carboxypeptidase *B*—appear to change in concert; the curves for the former pair are similar, for example, and their midway points between low and high levels occur at the same time. Perhaps the simplest way to explain the developmental pattern of these proteins is to assume that small sets of them are regulated together and that these sets are synchronized to rise together over a certain period of embryonic development.

With the aid of the electron microscope we correlated these changes in protein content with the appearance of intracellular structures. During the period of great amplification of enzyme content between State I and State II we observed a dramatic formation of the rough endoplasmic reticulum, the site of enzyme synthesis, and saw the first zymogen granules appear. The experiments of George E. Palade and Philip Siekevitz and their collaborators at Rockefeller University had suggested that newly synthesized enzymes first appear within the endoplasmic reticulum and then proceed to the larger cavities of the Golgi apparatus, where they are packaged in zymogen granules. In none of our experiments have we ever observed high levels of specific enzyme formation without the development of these intracellular organelles; the accelerated synthesis of specific proteins appears always to be coupled with the formation of new structures involved in their synthesis, storage and secretion.

Turning to the endocrine cells, we first established that the insulin content of the adult mouse or rat pancreas is remarkably high: more than a billion molecules per *B* cell. (Very low but significant levels were found in other adult tissues, presumably reflecting the normal dissemination of the hormone from the pancreas.) Our findings in the early embryo were more unexpected: we were unable to detect any insulin before the ninth day, although our assay could have detected one molecule of insulin in 10 cells. Clearly there is little, if any, transfer of

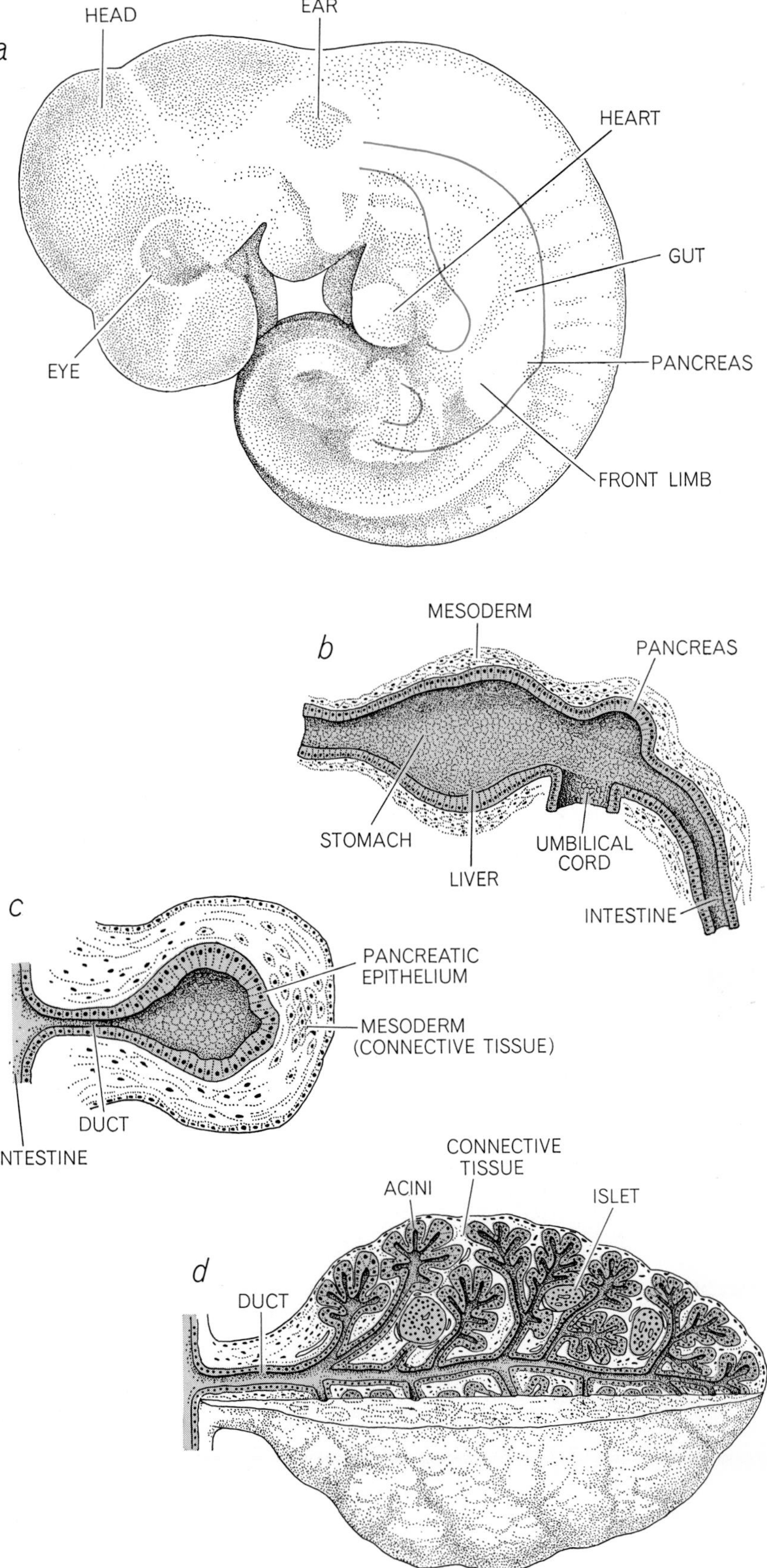


**DEVELOPMENT of the mouse pancreas begins when the embryo is in its ninth day of gestation (*a*). The organ forms from the central part of the gut by a process of evagination (*b*): a group of endodermal cells (*color*) bulge into adjacent mesoderm (*black*). By the 11th day the pancreas forms a pouchlike diverticulum (*c*). By the 15th day some pancreas cells have stopped dividing and begun to synthesize large quantities of specific proteins (*d*).**

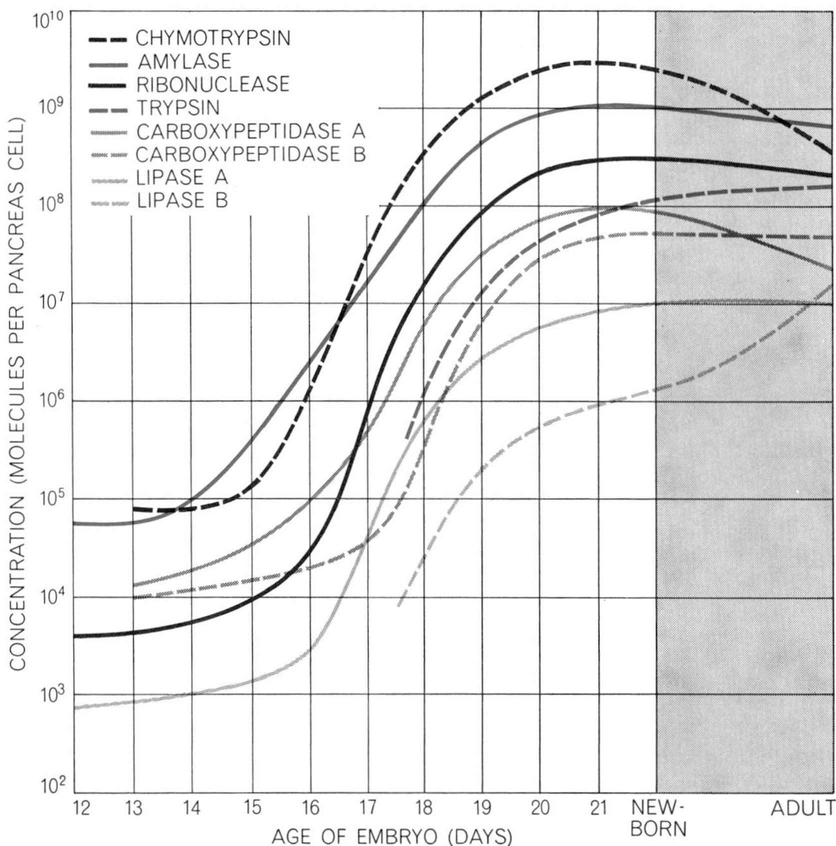


**PATTERN OF ACCUMULATION of the digestive enzymes is an index of exocrine-cell development. Extracts of pancreatic rudiments of various ages were prepared and assayed for enzyme content on the basis of specific catalytic activity. The midpoints between low and high levels differ for most of the proteins. Clearly they are not regulated as a single group.**

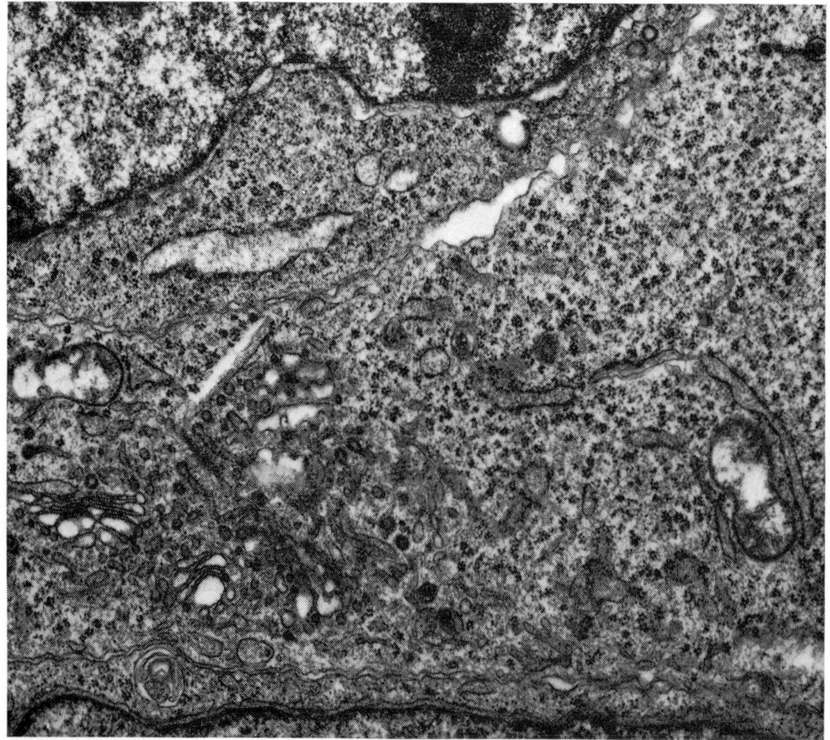

**CYTOPLASM of an exocrine cell from a 10-day embryo reflects the cell's differentiation. Some ribosomes are organized into endoplasmic reticulum but there are no zymogen granules yet, perhaps because there is not enough specific protein to require such structures.**

insulin from the mother to the embryo and the early embryo itself must synthesize insignificant quantities. Early embryonic development apparently proceeds in the absence of this hormone. There must be remarkably stringent regulation of specific protein synthesis in embryonic cells. Until a given time in development, proteins such as insulin simply are not made.

Late on the ninth day insulin is detected in the gut region, where it is presumably present in primitive pancreas cells [*see top illustration on opposite page*]. The concentration then increases until, during the period of State I, there may be as many as a million insulin molecules per *B* cell. Later, at the time the exocrine cells make the transition from State I to State II, the insulin content of the endocrine cells increases another several hundredfold. By applying an antigen-antibody assay for glucagon we have shown that glucagon too is present at the earliest stages of pancreatic development. Electron micrographs show that there are secretory granules in the endocrine cells at this early stage of differentiation. In the exocrine cells, on the other hand, zymogen granules only appear later, during the large rise in specific activity from State I to State II. It seems that the details of the differentiation processes in these two cell types are different.

These correlated studies of pancreatic development suggest a model for the differentiation of pancreatic exocrine and endocrine cells [*see illustration on page 206*]. The model recognizes several distinct stages of development. The protodifferentiated state corresponds to the State I we described earlier. The differentiative state (State II) is found in the late embryo, and the modulated levels found in the adult characterize State III. This implies that there are three regulatory transitions, each introducing a new stage of differentiation.

It is difficult, if not impossible, to identify the factors that control these transitions in the intact embryo, because one cannot effectively manipulate the cellular environment or the physical relations among cells. The cultivation of cells or tissues in laboratory vessels seemed to offer a way to obtain further information about the differentiative process, particularly the nature of the regulatory events. We found that a pancreas isolated from a 10-day or 11-day mouse embryo would develop normally in culture, and that both the pattern of increase in enzyme levels and the appearance of zymogen granules closely mimic the developmental sequence in intact embryos. This developmental competence in tissue culture opened up attractive experimental vistas. We could investigate relations among different cell types and also monitor developmental events within cells much more effectively.

First we tested the ability of progressively younger pancreatic tissues and more primitive gut tissues to develop into normal pancreas. We found that even the youngest pancreatic rudiment differentiates normally, provided that both mesodermal and epithelial elements are present. Then we found that gut tissue from the eight-day mouse embryo, excised 18 hours before the pancreas would become visible, develops into normal pancreatic tissue (plus some other tissues) in culture; gut material taken from earlier embryos forms liver-like and stomach-like tissues in culture but no pancreatic cells. It seems, therefore, that some significant developmental event must occur on the eighth embryonic day that confers pancreatic potential on a group of precursor cells. This event presumably requires an input from some part of the embryonic system that is absent in the laboratory culture. After this event, however, the tissue is somehow "determined," so that when it is removed from the embryo, it will develop just as it would in the intact embryo.

The simplest way to test for an interaction between two tissues in an organ is to separate the tissues and see whether each is capable of differentiating normally alone or whether recombination is necessary for normal development. We found we could carry out such an experiment with the pancreatic rudiment in the protodifferentiated state (the 11-day mouse embryo or the 13-day rat embryo). At this point the epithelial cells exist as a bulbar structure encased in mesoderm. We separated the two tissues by first incubating the intact rudiment for a short time in a mixture of protein-digesting enzymes and then stripping off the mesoderm by drawing the rudiment up into a micropipette. We then tested for an interaction between the two components with a technique, perfected by Clifford Grobstein of the University of California at San Diego, in which a thin, porous membrane serves as a platform to support the individual tissues or as a barrier to separate them [*see bottom illustration on page 207*]. We found that the epithelium would not develop normally alone but that it did differentiate if it was cultured directly across the filter from mesoderm. To our surprise,

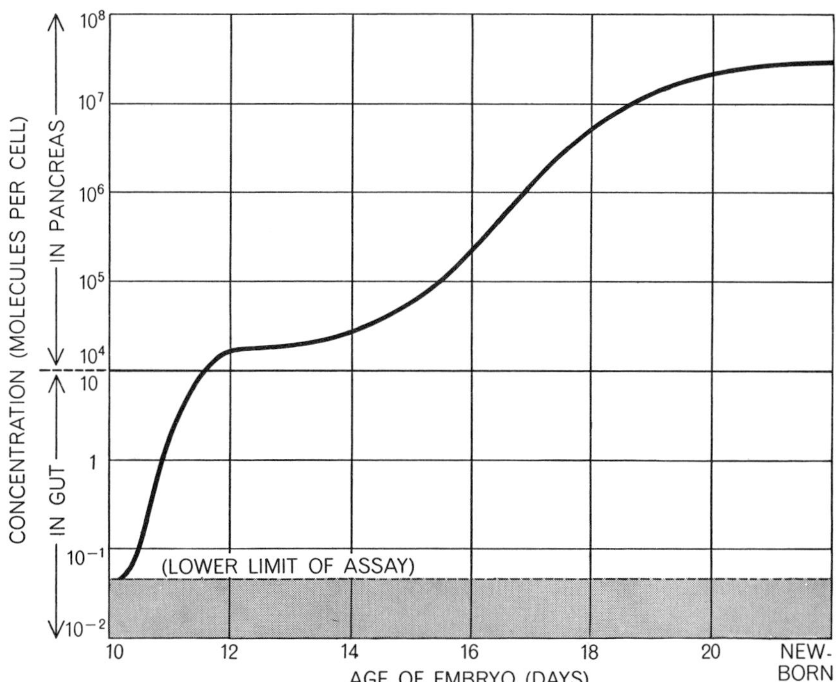


**INSULIN CONCENTRATION** in tissue extracts was measured by an antigen-antibody assay and is given here in molecules per gut cell and (when the embryo is large enough for dissection) molecules per pancreas cell. The sensitivity of the assay makes it possible to detect insulin at an earlier developmental stage than one can detect any of the exocrine proteins.

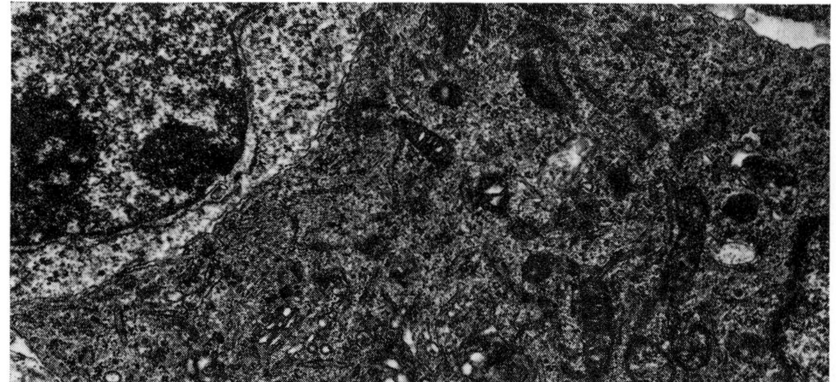

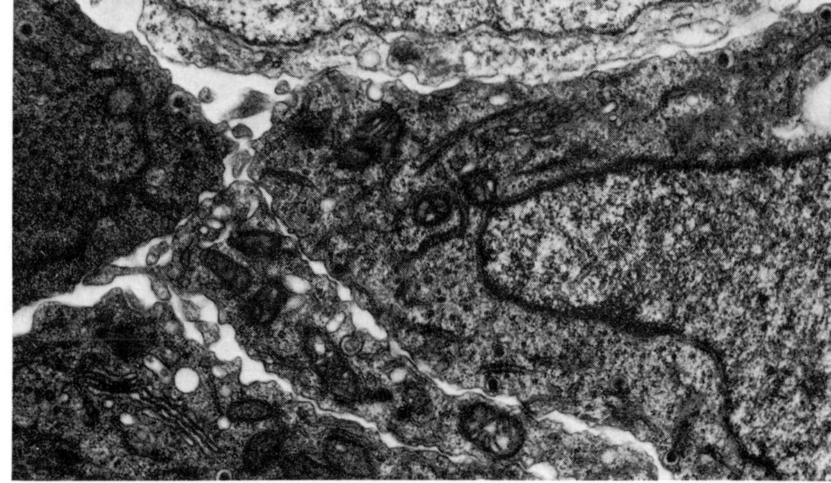

**SECRETORY GRANULES** appear in endocrine cells as soon as hormones begin to be synthesized. There are no granules in the cytoplasm of future endocrine cells when the pancreas begins to take shape (*top*), but in pancreas cells from an early 10-day embryo, when insulin is first detected (*see illustration at top of page*), a number of granules can be seen (*bottom*).

the mesoderm did not have to be from the pancreas; we learned that mesodermal tissues from salivary gland, kidney, stomach, lung or spleen were just as effective in promoting epithelial development.

Apparently the mesoderm contributed some factor that is required for epithelial differentiation and that is small enough to pass through the filter. To get enough material from which to attempt to isolate the factor, we first prepared a crude homogenate of an entire embryo, which turned out to contain enough of the mesodermal factor to promote pancreatic differentiation. We were able to remove most of the active material from the homogenate by low-speed centrifugation, indicating that the factor was part of (or was bound to) some rather large particulate structure. A number of experiments suggest that it is a protein, or at least that a protein is involved in the activity. The active material has now been solubilized and partially purified.

Is the mesodermal factor required at all times during development of the pancreas or just at a certain time? We have noted an epithelial requirement for the presence of mesoderm as early as the ninth day, when a restricted region of the gut responds to mesodermal tissue by forming pancreatic acini. Nearby epithelium, treated similarly, forms other gut derivatives such as stomach or intestine but does not form pancreatic tissue. Several hours before the pancreas can be seen, then, a discrete group of gut cells has the capacity to form the organ, provided that mesodermal tissue is present. Presumably this ability reflects the determination event that occurred some eight hours earlier. Similar experiments with older pancreatic epithelium indicate that the mesodermal requirement ceases just before the accelerated synthesis of the specific proteins; the mesoderm can be removed from the epithelium late in the protodifferentiated state without any effect on subsequent development. In summary, the mesodermal factor must be present from a time before the pancreas can be seen until about five days later, when cell division ceases and enzymes begin to accumulate at high levels.

How does the mesodermal factor act? It could promote growth, cause the formation of acini or actually cause differentiation in the epithelial cells. Recent results suggest that a major action of the mesodermal factor is to stimulate DNA synthesis and thus the proliferation of epithelial cells. (As we shall see, such DNA synthesis may be required for differentiation to occur.) The eventual availability of pure preparations of the factor may give us clear answers to these questions.

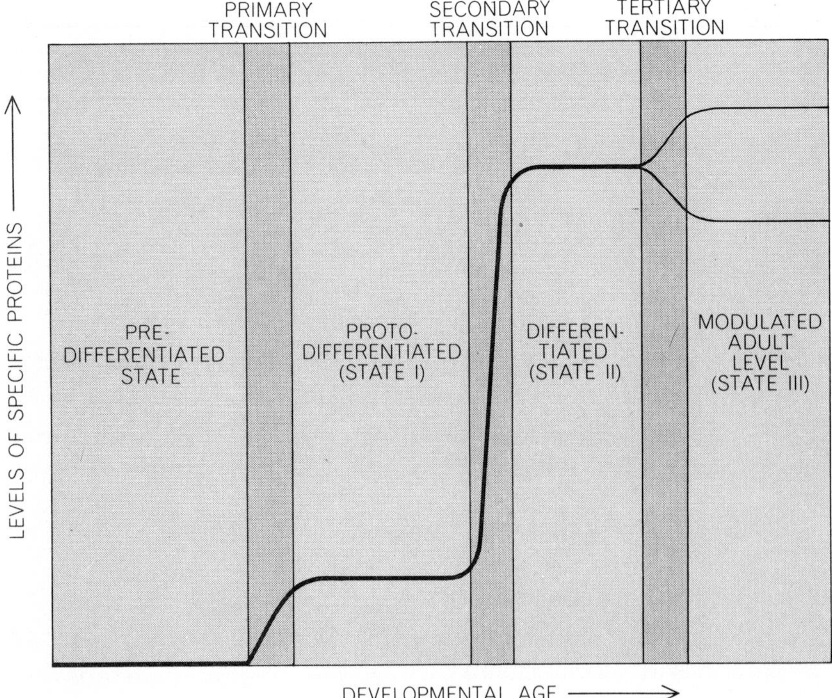


**POSSIBLE REGULATORY PHASES in cell differentiation are indicated by this schematic curve. The three stages of differentiation correspond to the numbered states associated with the exocrine enzyme levels. Each of them is introduced by a regulatory transition phase.**

Our efforts to describe and analyze some of the changes that take place at the cellular and molecular level during development have been focused on the secondary transition leading to the differentiated state. First we sought to trace changes in the pattern of protein synthesis during the transition. (What we had measured in intact embryos was enzyme content, not the rate at which enzymes were being synthesized.) We incubated 14-day and 19-day pancreatic rudiments in a medium containing radioactively labeled amino acids—the building blocks of proteins—and then sorted out the resulting radioactively labeled proteins by electrophoresis in polyacrylamide gel. In this technique specific proteins are identified by the rate at which they move through the gel in the presence of an electric field, and the amount of radioactivity at each point shows how much of each protein has been synthesized.

There was a remarkable difference between the two patterns of protein synthesis [*see illustration on page 208*]. Similar experiments with rudiments at various stages of the secondary transition confirmed that there is about a fiftyfold increase in the rate of synthesis of the enzymes during the transition. This increase in rate accounts quantitatively for the increase in enzyme content between State I and State II noted in the earlier assays. There must therefore be little, if any, secretion of these proteins by exocrine cells during this period of development; the proteins that are synthesized simply accumulate within the cells.

Protein synthesis depends on the activity of a number of species of RNA. We decided to see if the increase in specific enzyme synthesis in pancreas depends on the synthesis of new RNA. For these studies we employed actinomycin *D*, a compound that is bound to the DNA template and thus blocks RNA synthesis. Very low levels of actinomycin applied to dividing cells in the protodifferentiated state inhibited only part (about 70 percent) of the RNA synthesis in those cells but completely prevented the accelerated synthesis of the differentiative proteins. This was a specific effect, since the cells remained healthy and continued to synthesize cellular (as opposed to secretory) proteins at nearly the normal rate. If the tissue was treated with actinomycin at a later time, when the central group of cells in the culture had stopped dividing, those cells differentiated more or less normally and accumulated zymogen granules; cells located on the periphery of the same

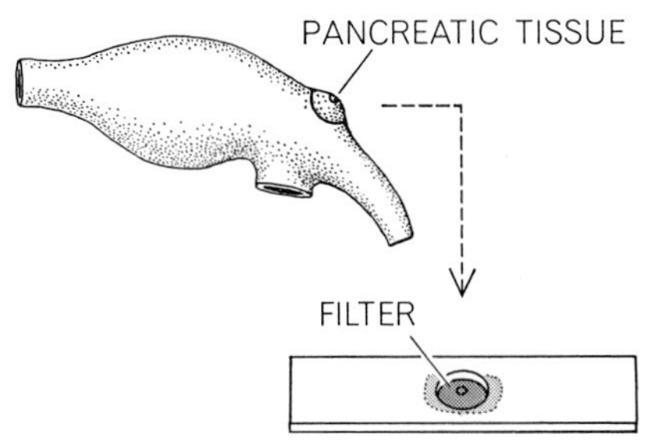


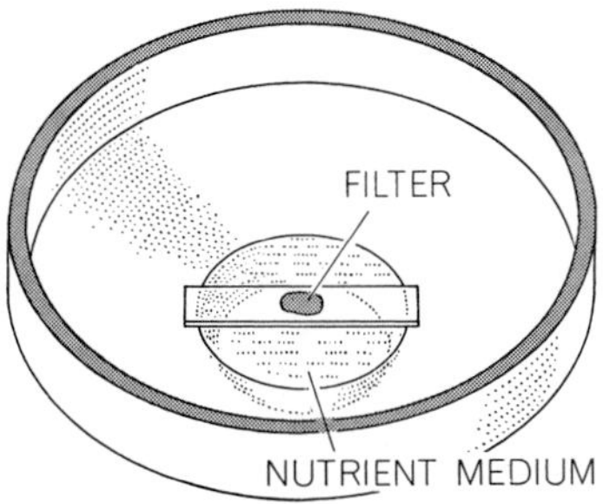


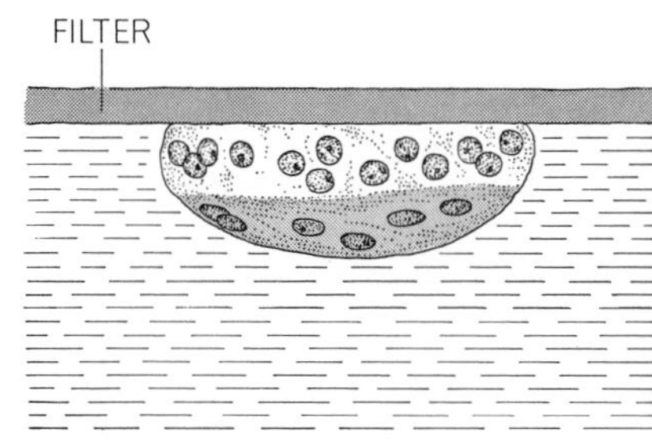


**PANCREATIC TISSUE** is grown in the laboratory. Nine-day gut tissue is excised (*left*) and placed in a culture dish (*middle*). The tissue, mesoderm (*color*) as well as endoderm (*black*), is suspended under a plastic support and incubated in nutrient (*right*).

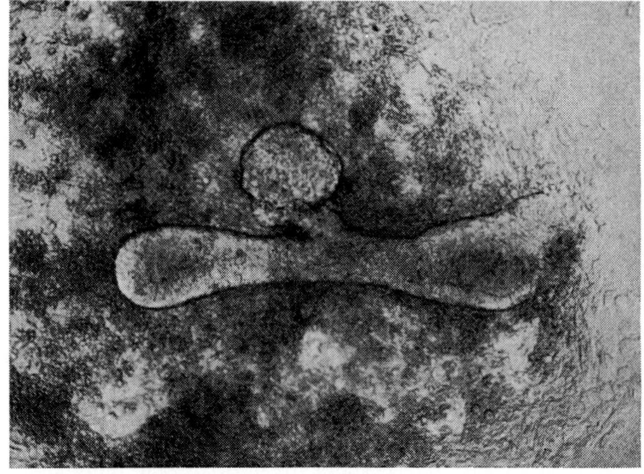

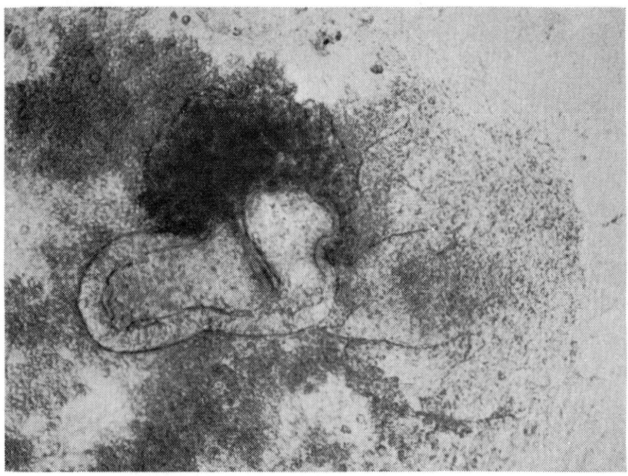

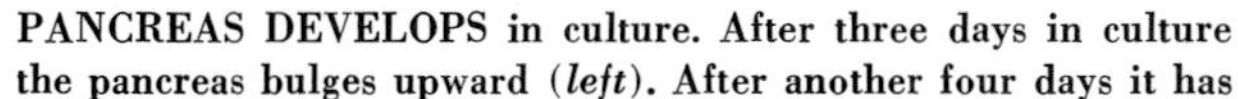

**PANCREAS DEVELOPS** in culture. After three days in culture the pancreas bulges upward (*left*). After another four days it has grown, acini have formed and the acinar tissue has been darkened by zymogen granules (*right*). The enlargement is 50 diameters.

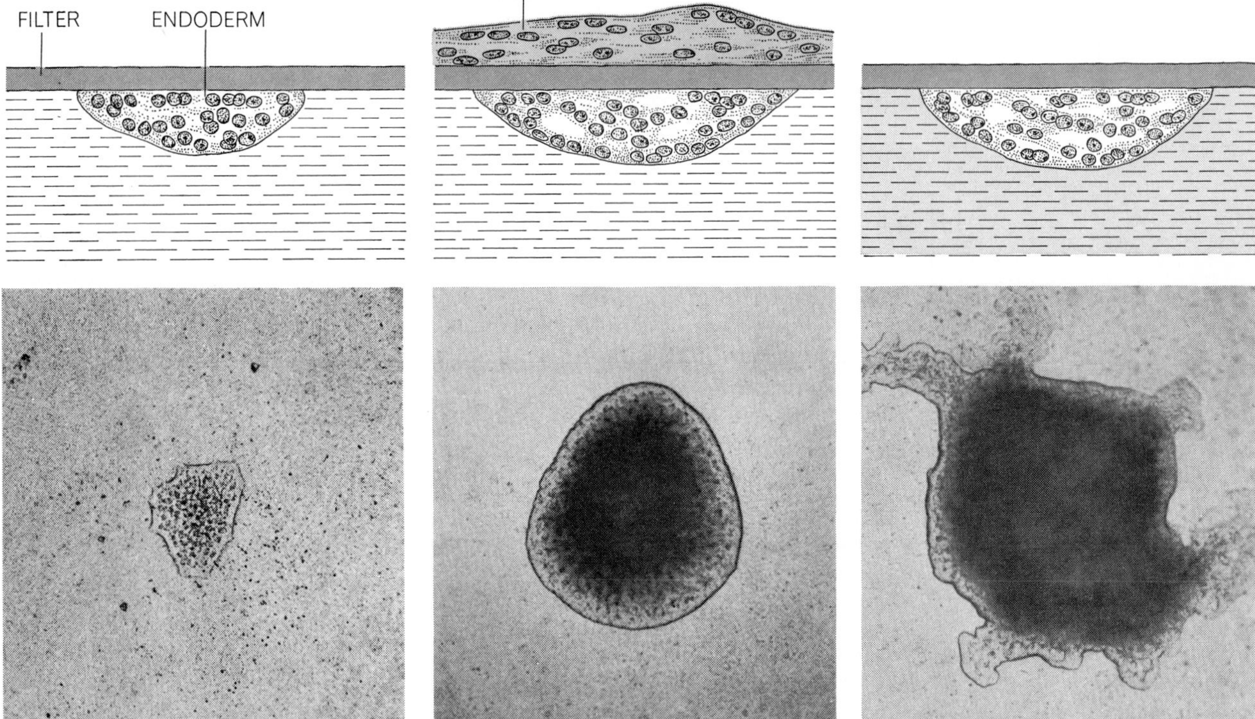


**TRANSFILTER CULTURE** tests for mesoderm-endoderm interaction. When pancreatic epithelium (endoderm) is cultured alone (*top left*), the cells fail to differentiate in five days (*bottom left*). When epithelium and mesoderm (*color*) are separated by a thin filter and cultured together, there is substantial growth in five days, acini are visible and zymogen granules darken the cells (*center*). Epithelium that is cultured without mesoderm, but with an embryo extract included in the nutrient medium, does equally well (*right*).

rudiments were still dividing and they (like all cells treated in the protodifferentiated state) never made the secondary transition. If cultures were treated still later, when most of the cells had ceased to divide, there was no significant effect of actinomycin on the synthesis or accumulation of the specific proteins. In collaboration with Fred H. Wilt of the University of California at Berkeley we showed that actinomycin has the same inhibitory effect on RNA synthesis in early pancreatic epithelial cells as it does in older cells. The results were therefore not due to a difference in the susceptibility of the two populations of cells to actinomycin but to the fact that the synthesis of new RNA is required for the accelerated synthesis of the pancreatic proteins.

One further experiment with actinomycin is of special significance in relation to our earlier observations about different enzymes being regulated together in groups. When actinomycin *D* is added to cultures early in the secondary transition, the synthesis of the various specific proteins is inhibited to different degrees. This sequential inhibition suggests that the "messenger" RNA's that encode the various proteins are synthesized at slightly different times, supporting our earlier contention that the exocrine proteins are not all regulated as a single set.

Inhibition of DNA synthesis also blocks exocrine cell differentiation. A culture treated with fluorodeoxyuridine (a strong inhibitor of DNA synthesis and hence of cell division) during the protodifferentiated state does not differentiate, although it appears otherwise healthy; cells treated after they have ceased to divide, however, develop normally. Similar results are obtained with bromodeoxyuridine, a structural analogue of thymidine, one of the four nucleosides of DNA. Bromodeoxyuridine is actually incorporated into the DNA during the synthesis period; by forming a faulty DNA that does not allow the synthesis of functional RNA, bromodeoxyuridine blocks the differentiative process. In contrast to the action of actinomycin *D* (which has a sequential inhibitory effect on the synthesis of certain proteins), bromodeoxyuridine exhibits an "all or none" effect, depending on just when it is administered: differentiation is either blocked completely or is unaffected. These experiments suggest that some event that occurs only in dividing cells or is facilitated by DNA synthesis must precede the subsequent changes in RNA synthesis that bring about the new pattern of protein synthesis.

Our present studies have provided some insight into the basic mechanisms underlying the formation of organs. They demonstrate the importance of cell interaction in the developmental process. Specifically, after the determination event, a mesodermal-epithelial interaction is required to maintain the protodifferentiated state and initiate the secondary transition. It may be that interactions between the exocrine and endocrine cells are also involved in the developmental process.

Perhaps the most striking thing about the molecular events within maturing pancreatic cells is the fact that many complex regulatory events are apparently coordinated at a few distinct times. This is most obvious in the exocrine cells. There some early initiating event in a population of cells starts the synthesis of specific proteins and the morphogenesis of the pancreas. The ensuing protodifferentiated state is primarily a phase of growth and acinus formation. Then there is a transition to the differentiated state, when cells stop dividing and undergo intracellular differentiation, which is marked by the accelerated synthesis and the accumulation of specific proteins in zymogen granules.

The two major developmental phases must involve hundreds of genes. We believe sets of genes are activated and other sets are inactivated simultaneously, rather than each of the hundreds of individual genes' being activated in sequence. One possible mechanism to accomplish such a concerted transition is an alteration of chromosomal structure so that some sets of genes are chemically unmasked, and thus activated, while others are chemically sequestered and thus repressed. These experiments have not yielded a precise description of the molecular mechanisms involved, but they have simplified our view of the problem and suggested a hypothesis that will be the basis for continuing explorations.

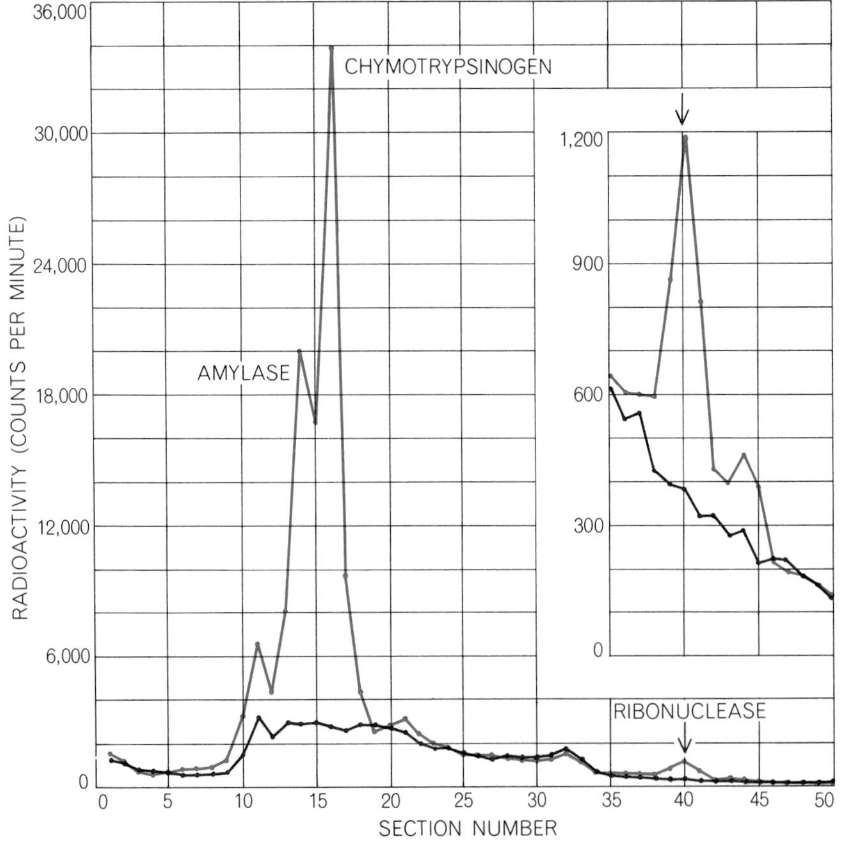


**PROTEIN SYNTHESIS in 14-day and 19-day rat pancreatic tissue is compared. The tissues were incubated with a radioactive amino acid, which was incorporated in the proteins being synthesized. When the proteins were subjected to electrophoresis in a gel, they migrated to different points in the gel. The gel was cut into sections and the amount of radioactivity in each section was measured. In the 19-day tissue three new peaks corresponding to specific exocrine proteins are identified. An expanded scale is used (*inset*) to visualize the ribonuclease peak. If 14-day rudiments are cultured for five days, they exhibit a synthetic pattern like that of the 19-day pancreas, unless an inhibiting agent such as actinomycin *D* or bromodeoxyuridine (*see text*) is in the medium during the early part of the culture period.**

# IV
# *Genetics and Evolution*

# IV

# *Genetics and Evolution*

## INTRODUCTION

Evolution, established conceptually by Darwin and his disciples without their having any knowledge of genetic systems as they exist, cannot be really understood in the absence of knowledge of the genetic systems that make evolution inevitable. In his remarkable analysis of what turned out to be chromosomal heredity, Mendel could not have been greatly helped by extensive exposure to the Darwinism of his century. But Darwin's contribution must be viewed with reference to his lack of the informational base which might have helped him most, and which accounts for weaknesses in his formulation. Considered in this context, Darwin must only rise in our esteem.

The major point is that evolution demands heritable variability, and genetic systems supply it, primarily by means of mutation and recombination. The capacities of DNA to replicate and to be altered into an enormous number of functionally meaningful and also replicable forms endow it with those properties that meet the fundamental requirements of a genetic-evolutionary system. Given the DNA-type genetic system, evolution is inevitable. Genetic systems themselves, as well as the organisms they define, are subject to evolution.

And so today, in a biological context of any breadth, it is difficult to consider either genetics or evolution without considering the other. Evolutionary studies may be made by other than genetic means, but outside of a framework of genetic thinking and interpretation they are unlikely to have meaning. On the other hand, significant genetic phenomena, and indeed all sorts of biological phenomena, are better understood when their evolutionary implications emerge.

Dobzhansky's article "The Genetic Basis of Evolution" serves as an admirable introduction to this section. Notable for its breadth of scope and for the historical context it provides, the article is especially valuable because it shows how particular experiments and specific kinds of information elaborate our understanding of speciation—a process that under natural conditions moves so slowly that it is seldom directly observed.

A natural sequel is Kettlewell's "Darwin's Missing Evidence." By inference, protective coloration in organisms is one of the more obvious evolutionary adaptations. A classical example is provided by certain species of moths, known before the industrial revolution in England to have been mostly light in color. Concomitant with changes in their background environment—the blackening of tree trunks by industrial soot—members of these species of moths have become mostly dark in color. Kettlewell has demonstrated by direct experiment that birds capture resting moths *selectively* on the basis of color. Knowing that genetic variability for color existed in the moth populations, as indeed it still does, Kettlewell was able to provide a convincing example of adaptive selection in action. The simple elegance of the primary observations should not obscure other significant aspects of the studies—for example, the implications of the fact that with time the allele for the melanic form has become increasingly dominant over the allele for the light-colored form.

Without genetic variability in the moth populations, selection for the darker forms could not have occurred. In the ultimate sense, gene mutation may be viewed as providing the raw material for evolution, with recombination and selection working to shape the raw material into evolving products. In a real way, then, the factors accounting for gene mutation are ingredients of the evolutionary process. In "Ionizing Radiation and Evolution," an article rich in content, Crow considers mutation rates in relation to evolution and, in particular, the effects of mutagenic ionizing radiations on plant and animal evolution. Crow's conclusions that ionizing radiation is probably not an important factor in evolution, and that for organisms like man is more probably harmful than potentially beneficial, gives pertinence to the article by Muller, which focuses on "Radiation and Human Mutation." Muller was one of the leading exponents of the thesis that radiation can be a potential hazard to man's genetic well-being, and he presents his arguments in this article. We know now that certain organic chemicals are even more effectively mutagenic than ionizing radiation, and Muller's general point of view on mutagenic hazards is equally relevant to chemicals that induce mutation. The closing passages of the article, typical of Muller and his attitude toward genetics, constitute a passionate and idealistic appeal for the use of genetic knowledge toward the betterment of man.

The possibility of a certain degree of human control over mutation frequencies is suggested by recent findings that alterations in DNA can be reversed. In "The Repair of DNA," Hanawalt and Haynes describe the cellular mechanisms for rectifying damaged genetic material. The work they describe on the molecular biology of a microorganism immediately raises questions that are of significance for evolution in general and perhaps of great importance to human biology and medicine.

The final articles in this section also deal with molecules in relation to evolution. Articles in earlier sections have shown that polypeptides are the molecular basis for enzymatic, structural, and other functions that define the attributes of organisms; that polypeptides are colinear with the DNA that codes for them, with the amino acid sequences reflecting sequences of coding units; and that mutation may result in substitutions of amino acids in polypeptides. A well-known example of heterozygote advantage in human populations involves the mutant hemoglobin designated S, which differs from normal hemoglobin by the substitution of an amino acid and appears to confer resistance to malaria (see "Sickle Cells and Evolution," by A. C. Allison; SCIENTIFIC AMERICAN, August 1956). In "The Evolution of Hemoglobin," Zuckerkandl deals with various amino acid sequences found in hemoglobin, both within and among species, and views them as living records of mutational change and of the evolutionary process. Comparison of common features of the sequences and of the differences among them permits inferences to be drawn about ancestral relationships and the time scale of evolutionary processes. In "Computer Analysis of Protein Evolution," Dayhoff continues the kind of approach used by Zuckerkandl. On firm genetic grounds, and based upon data that reflect accurately the genetic material itself, studies of this kind clearly provide another dimension to our view of evolution. Eventually we must come to understand not only gene evolution, as explored in these studies, but the evolution of regulatory systems as well.

# 22

# *The Genetic Basis of Evolution*

THEODOSIUS DOBZHANSKY

*January 1950*

THE living beings on our planet come in an incredibly rich diversity of forms. Biologists have identified about a million species of animals and some 267,000 species of plants, and the number of species actually in existence may be more than twice as large as the number known. In addition the earth has been inhabited in the past by huge numbers of other species that are now extinct, though some are preserved as fossils. The organisms of the earth range in size from viruses so minute that they are barely visible in electron microscopes to giants like elephants and sequoia trees. In appearance, body structure and ways of life they exhibit an endlessly fascinating variety.

What is the meaning of this bewildering diversity? Superficially considered, it may seem to reflect nothing more than the whims of some playful deity, but one soon finds that it is not fortuitous. The more one studies living beings the more one is impressed by the wonderfully effective adjustment of their multifarious body structures and functions to their varying ways of life. From the simplest to the most complex, all organisms are constructed to function efficiently in the environments in which they live. The body of a green plant can build itself from food consisting merely of water, certain gases in the air and some mineral salts taken from the soil. A fish is a highly efficient machine for exploiting the organic food resources of water, and a bird is built to get the most from its air environment. The human body is a complex, finely coordinated machine of marvelously precise engineering, and through the inventive abilities of his brain man is able to control his environment. Every species, even the most humble, occupies a certain place in the economy of nature, a certain adaptive niche which it exploits to stay alive.

The diversity and adaptedness of living beings were so difficult to explain that during most of his history man took the easy way out of assuming that every species was created by God, who contrived the body structures and functions of each kind of organism to fit it to a predestined place in nature. This idea has now been generally replaced by the less easy but intellectually more satisfying explanation that the living things we see around us were not always what they are now, but evolved gradually from very different-looking ancestors; that these ancestors were in general less complex, less perfect and less diversified than the organisms now living; that the evolutionary process is still under way, and that its causes can therefore be studied by observation and experiment in the field and in the laboratory.

The origins and development of this theory, and the facts that finally convinced most people of its truth beyond reasonable doubt, are too long a story to be presented here. After Charles Darwin published his convincing exposition and proof of the theory of evolution in 1859, two main currents developed in evolutionary thought. Like any historical process, organic evolution may be studied in two ways. One may attempt to infer the general features and the direction of the process from comparative studies of the sequence of events in the past; this is the method of paleontologists, comparative anatomists and others. Or one may attempt to reconstruct the causes of evolution, primarily through a study of the causes and mechanisms that operate in the world at present; this approach, which uses experimental rather than observational methods, is that of the geneticist and the ecologist. This article will consider what has been learned about the causes of organic evolution through the second approach.

Darwin attempted to describe the causes of evolution in his theory of natural selection. The work of later biologists has borne out most of his basic contentions. Nevertheless, the modern theory of evolution, developed by a century of new discoveries in biology, differs greatly from Darwin's. His theory has not been overthrown; it has evolved. The authorship of the modern theory can be credited to no single person. Next to Darwin, Gregor Mendel of Austria, who first stated the laws of heredity, made the greatest contribution. Within the past two decades the study of evolutionary genetics has developed very rapidly on the basis of the work of Thomas Hunt Morgan and Hermann J. Muller of the U. S. In these developments the principal contributors have been C. D.

Darlington, R. A. Fisher, J. B. S. Haldane, J. S. Huxley and R. Mather in England; B. Rensch and N. W. Timofeeff-Ressovsky in Germany; S. S. Chetverikov, N. P. Dubinin and I. I. Schmalhausen in the U.S.S.R.; E. Mayr, J. T. Patterson, C. G. Simpson, G. L. Stebbins and Sewall Wright in the U. S., and some others.

### Evolution in the Laboratory

Evolution is generally so slow a process that during the few centuries of recorded observations man has been able to detect very few evolutionary changes among animals and plants in their natural habitats. Darwin had to deduce the theory of evolution mostly from indirect evidence, for he had no means of observing the process in action. Today, however, we can study and even produce evolutionary changes at will in the laboratory. The experimental subjects of these studies are bacteria and other low forms of life which come to birth, mature and yield a new generation within a matter of minutes or hours, instead of months or years as in most higher beings. Like a greatly speeded-up motion picture, these observations compress into a few days evolutionary events that would take thousands of years in the higher animals.

One of the most useful bacteria for this study is an organism that grows, usually harmlessly, in the intestines of practically every human being: *Escherichia coli,* or colon bacteria. These organisms can easily be cultured on a nutritive broth or nutritive agar. At about 98 degrees Fahrenheit, bacterial cells placed in a fresh culture medium divide about every 20 minutes. Their numbers increase rapidly until the nutrients in the culture medium are exhausted; a single cell may yield billions of progeny in a day. If a few cells are placed on a plate covered with nutritive agar, each cell by the end of the day produces a whitish speck representing a colony of its offspring.

Now most colon bacteria are easily killed by the antibiotic drug streptomycin. It takes only a tiny amount of streptomycin, 25 milligrams in a liter of a nutrient medium, to stop the growth of the bacteria. Recently, however, the geneticist Milislav Demerec and his collaborators at the Carnegie Institution in Cold Spring Harbor, N. Y., have shown that if several billion colon bacteria are placed on the streptomycin-containing medium, a few cells will survive and form colonies on the plate. The offspring of these hardy survivors are able to multiply freely on a medium containing streptomycin. A mutation has evidently taken place in the bacteria; they have now become resistant to the streptomycin that was poisonous to their sensitive ancestors.

How do the bacteria acquire their

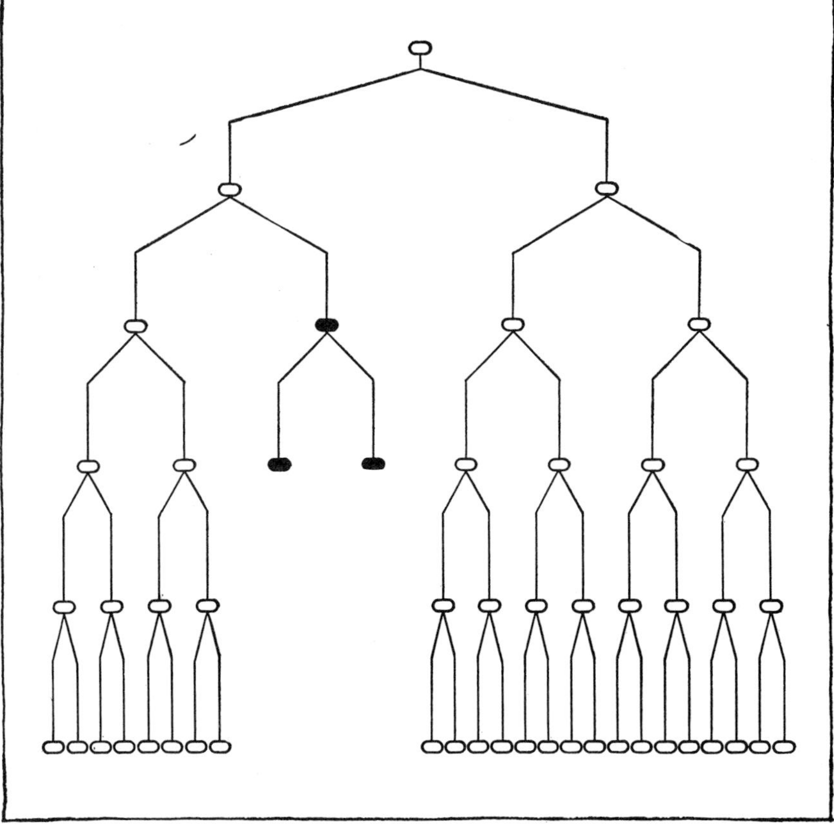

**IN NORMAL ENVIRONMENT the common strain of the bacterium *Escherichia coli* (*white bacteria*) multiplies. A mutant strain resistant to streptomycin (*black bacteria*) remains rare because the mutation is not useful.**

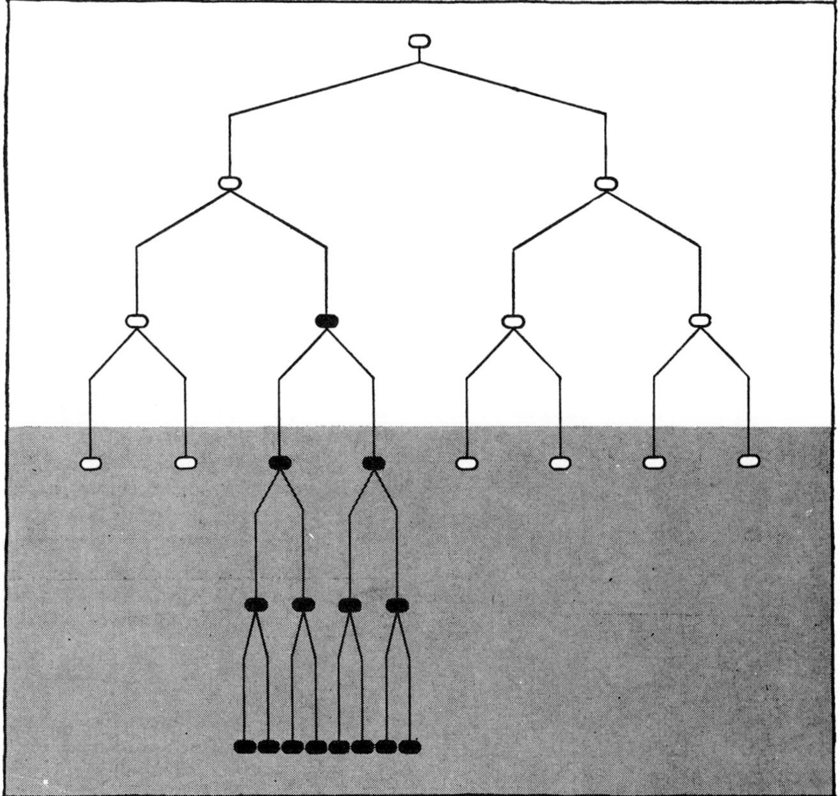

**IN CHANGED ENVIRONMENT produced by the addition of streptomycin (*gray area*) the streptomycin-resistant strain is better adapted than the common strain. The mutant strain then multiplies and the common one dies.**

**CONTROLLED ENVIRONMENT** for the study of fruit-fly populations is a glass-covered box. In bottom of the box are cups of food that are filled in rotation to keep food a constant factor in environment.

resistance? Is the mutation caused by their exposure to streptomycin? Demerec has shown by experimental tests that this is not so; in any large culture a few resistant mutants appear even when the culture has not been exposed to streptomycin. Some cells in the culture undergo mutations from sensitivity to resistance regardless of the presence or absence of streptomycin in the medium. Demerec found that the frequency of mutation was about one per billion; *i.e.*, one cell in a billion becomes resistant in every generation. Streptomycin does not induce the mutations; its role in the production of resistant strains is merely that of a selecting agent. When streptomycin is added to the culture, all the normal sensitive cells are killed, and only the few resistant mutants that happened to be present before the streptomycin was added survive and reproduce. Evolutionary changes are controlled by the environment, but the control is indirect, through the agency of natural or artificial selection.

What governs the selection? If resistant bacteria arise in the absence of streptomycin, why do sensitive forms predominate in all normal cultures; why has not the whole species of colon bacteria become resistant? The answer is that the mutants resistant to streptomycin are at a disadvantage on media free from this drug. Indeed, Demerec has discovered the remarkable fact that about 60 per cent of the bacterial strains derived from streptomycin-resistant mutants become dependent on streptomycin; they are unable to grow on media free from it!

On the other hand one can reverse the process and obtain strains of bacteria that can live without streptomycin from cultures predominantly dependent on the drug. If some billions of dependent bacteria are plated on nutrient media free of the drug, all dependent cells cease to multiply and only the few mutants independent of the drug reproduce and form colonies. Demerec estimates the frequency of this "reverse" mutation at about 37 per billion cells in each generation.

Evolutionary changes of the type described in colon bacteria have been found in recent years in many other bacterial species. The increasing use of antibiotic drugs in medical practice has made such changes a matter of considerable concern in public health. As penicillin, for example, is used on a large scale against bacterial infections, the strains of bacteria that are resistant to penicillin survive and multiply, and the probability that they will infect new victims is increased. The mass application of antibiotic drugs may lead in the long run to increased incidence of cases refractory to treatment. Indications exist that this has already happened in some instances: in certain cities penicillin-resistant gonorrhea has become more frequent than it was.

The same type of evolutionary change has also been noted in some larger organisms. A good example is the case of DDT and the common housefly, *Musca domestica*. DDT was a remarkably effective poison for houseflies when first introduced less than 10 years ago. But already reports have come from places as widely separated as New Hampshire, New York, Florida, Texas, Italy and Sweden that DDT sprays in certain localities have lost their effectiveness. What has happened, of course, is that strains of houseflies relatively resistant to DDT have become established in these localities. Man has unwittingly become an agent of a selection process which has led to evolutionary changes in housefly populations. Similar changes are known to have occurred in other insects; *e.g.*, in some orchards of California where hydrocyanic gas has long been used as a fumigant to control scale insects that prey on citrus fruits, strains of these insects that are resistant to hydrocyanic gas have developed.

Obviously evolutionary selection can take place only if nature provides a supply of mutants to choose from. Thus no bacteria will survive to start a new strain resistant to streptomycin in a culture in which no resistant mutant cells were present in the first place, and housefly races resistant to DDT have not appeared everywhere that DDT is used. Adaptive changes are not mechanically forced upon the organism by the environment. Many species of past geological epochs died out because they did not have a supply of mutants which fitted changing environments. The process of mutation furnishes the raw materials from which evolutionary changes are built.

## Mutations

Mutations arise from time to time in all organisms, from viruses to man. Perhaps the best organism for the study of mutations is the now-famous fruit fly, Drosophila. It can be bred easily and rapidly in laboratories, and it has a large number of bodily traits and functions that are easy to observe. Mutations affect the color of its eyes and body, the size and shape of the body and of its parts, its internal anatomical structures, its fecundity, its rate of growth, its behavior, and so on. Some mutations produce differences so minute that they can be detected only by careful measurements; others are easily seen even by beginners; still others produce changes so drastic that death occurs before the development is completed. The latter are called lethal mutations.

The frequency of any specific mutation is usually low. We have seen that in colon bacteria a mutation to resistance to streptomycin occurs in only about one cell per billion in every generation, and the reverse mutation to independence of streptomycin is about 37 times more frequent. In Drosophila and in the corn plant mutations have been found to range in frequency from one in 100,000 to one in a million per generation. In man, according to estimates by Haldane in England and James Neel in the U. S., mutations that produce certain hereditary diseases, such as hemophilia and Cooley's anemia, arise in one in 2,500 to one in 100,000 sex cells in each generation. From this it may appear that man is more mutable than flies and bacteria, but it should be remembered that a generation in man takes some 25 years, in flies two weeks, and in bacteria 25 minutes. The frequency of mutations per unit of time is actually greater in bacteria than in man.

A single organism may of course produce several mutations, affecting different features of the body. How frequent are all mutations combined? For technical reasons, this is difficult to determine; for example, most mutants produce small changes that are not detected unless especially looked for. In Drosophila it is estimated that new mutants affecting one part of the body or another are present in between one and 10 per cent of the sex cells in every generation.

In all organisms the majority of mutations are more or less harmful. This may seem a very serious objection against the theory which regards them as the mainspring of evolution. If mutations produce incapacitating changes, how can adaptive evolution be compounded of them? The answer is: a mutation that is harmful in the environment in which the species or race lives may become useful, even essential, if the environment changes. Actually it would be strange if we found mutations that improve the adaptation of the organism in the environment in which it normally lives. Every kind of mutation that we observe has occurred numerous times under natural conditions, and the useful ones have become incorporated into what we call the "normal" constitution of the species. But when the environment changes, some of the previously rejected mutations become advantageous and produce an evolutionary change in the species. The writer and B. A. Spassky have carried out certain experiments in which we intentionally disturbed the harmony between an artificial environment and the fruit flies living in it. At first the change in environment killed most of the flies, but during 50 consecutive generations most strains showed a gradual improvement of viability, evidently owing to the environment's selection of the better-adapted variants.

This is not to say that every mutation will be found useful in some environment somewhere. It would be difficult to imagine environments in which such human mutants as hemophilia or the absence of hands and feet might be useful. Most mutants that arise in any species are, in effect, degenerative changes; but some, perhaps a small minority, may be beneficial in some environments. If the environment were forever constant, a species might conceivably reach a summit of adaptedness and ultimately suppress the mutation process. But the environment is never constant; it varies not only from place to place but from time to time. If no mutations occur in a species, it can no longer become adapted to changes and is headed for eventual extinction. Mutation is the price that organisms pay for survival. They do not possess a miraculous ability to produce only useful mutations where and when needed. Mutations arise at random, regardless of whether they will be useful at the moment, or ever; nevertheless, they make the species rich in adaptive possibilities.

## The Genes

To understand the nature of the mutation process we must inquire into the nature of heredity. A man begins his individual existence when an egg cell is fertilized by a spermatozoon. From an egg cell weighing only about a 20-millionth of an ounce, he grows to an average weight at maturity of some 150 pounds—a 48-billionfold increase. The material for this stupendous increase in mass evidently comes from the food consumed, first by the mother and then by the individual himself. But the food becomes a constituent part of the body only after it is digested and assimilated, *i.e.*, transformed into a likeness of the assimilating body. This body, in turn, is a likeness of the bodies of the individual's ancestors. Heredity is, then, a process whereby the organism reproduces itself in its progeny from food materials taken in from the environment. In short, heredity is self-reproduction.

The units of self-reproduction are called genes. The genes are borne chiefly in chromosomes of the cell nucleus, but certain types of genes called plasmagenes are present in the cytoplasm, the part of the cell outside the nucleus. The chemical details of the process of self-reproduction are unknown. Apparently a gene enters into some set of chemical reactions with materials in its surroundings; the outcome of these reactions is the appearance of two genes in the place of one. In other words, a gene synthesizes a copy of itself from nongenic materials. The genes are considered to be stable because the copy is a true likeness of the original in the overwhelming majority of cases; but occasionally the copying process is faulty, and the new gene that emerges differs from its model. This is a mutation. We can increase the frequency of mutations in experimental animals by treating the genes with X-rays, ultraviolet rays, high temperature or certain chemical substances.

Can a gene be changed by the environment? Assuredly it can. But the important point is the kind of change produced. The change that is easiest to make is to treat the gene with poisons or heat in such a way that it no longer reproduces itself. But a gene that cannot produce a copy of itself from other materials is no longer a gene; it is dead. A mutation is a change of a very special kind: the altered gene can reproduce itself, and the copy produced is like the changed structure, not like the original. Changes of this kind are relatively rare. Their rarity is not due to any imperviousness of the genes to influences of the environment, for genic materials are probably the most active chemical constituents of the body; it is due to the fact that genes are by nature self-reproducing, and only the rare changes that preserve the genes' ability to reproduce can effect a lasting alteration of the organism.

Changes in heredity should not be confused, as they often are, with changes in the manifestations of heredity. Such expressions as "gene for eye color" or "inheritance of musical ability" are figures of speech. The sex cells that transmit heredity have no eyes and no musical ability. What genes determine are patterns of development which result in the emergence of eyes of a certain color and of individuals with some musical abilities. When genes reproduce themselves from different food materials and in different environments, they engender the development of different "characters" or "traits" in the body. The outcome of the development is influenced both by heredity and by environment.

In the popular imagination, heredity is transmitted from parents to offspring through "blood." The heredity of a child is supposed to be a kind of alloy or solution, resulting from the mixture of the paternal and maternal "bloods." This blood theory became scientifically untenable as long ago as Mendel's discovery of the laws of heredity in 1865. Heredity is transmitted not through miscible bloods but through genes. When different variants of a gene are brought together in a single organism, a hybrid, they do not fuse or contaminate one another; the genes keep their integrity and separate when the hybrid forms sex cells.

## Genetics and Mathematics

Although the number of genes in a single organism is not known with precision, it is certainly in the thousands, at least in the higher organisms. For Drosophila, 5,000 to 12,000 seems a reasonable estimate, and for man the figure is, if anything, higher. Since most or all genes suffer mutational changes from time to time, populations of virtually every species must contain mutant variants of many genes. For example, in the human species there are variations in the skin, hair and eye colors, in the shape and distribution of hair, in the form of the head, nose and lips, in stature, in body proportions, in the chemical composition of the blood, in psychological traits, and so on. Each of these traits is influenced by several or by many genes. To be conservative, let us assume that the human species has only 1,000 genes and that each gene has only two variants. Even on this conservative basis, Mendelian segregation and recombination would be capable of producing $2^{1000}$ different gene combinations in human beings.

The number $2^{1000}$ is easy to write but is utterly beyond comprehension. Compared with it, the total number of electrons and protons estimated by physicists to exist in the universe is negligibly small! It means that except in the case of identical twins no two persons now living, dead, or to live in the future are at all likely to carry the same complement of genes. Dogs, mice and flies are as individual and unrepeatable as men are. The mechanism of sexual reproduction, of which the recombination of genes is a part, creates ever new genetic constitutions on a prodigious scale.

One might object that the number of possible combinations does not greatly matter; after all, they will still be combinations of the same thousand gene variants, and the way they are combined is not significant. Actually it is: the same gene may have different effects in combinations with different genes. Thus Timofeeff-Ressovsky showed that two mutants in Drosophila, each of which reduced the viability of the fly when it was present alone, were harmless when combined in the same individual by hybridization. Natural selection tests the fitness in certain environments not of single genes but of constellations of genes present in living individuals.

Sexual reproduction generates, therefore, an immense diversity of genetic constitutions, some of which, perhaps a small minority, may prove well attuned to the demands of certain environments. The biological function of sexual reproduction consists in providing a highly efficient trial-and-error mechanism for the operation of natural selection. It is a reasonable conjecture that sex became established as the prevalent method of reproduction because it gave organisms the greatest potentialities for adaptive and progressive evolution.

Let us try to imagine a world providing a completely uniform environment. Suppose that the surface of our planet were absolutely flat, covered everywhere with the same soil; that instead of summer and winter seasons we had eternally constant temperature and humidity; that instead of the existing diversity of foods there was only one kind of energy-yielding substance to serve as nourishment. The Russian biologist Gause has pointed out that only a single kind of organism could inhabit such a tedious world. If two or more kinds appeared in it, the most efficient form would gradually crowd out and finally eliminate the less efficient ones, remaining the sole inhabitant. In the world of reality, however, the environment changes at every step. Oceans, plains, hills, mountain ranges, regions where summer heat alternates with winter cold, lands that are permanently warm, dry deserts, humid jungles—these diverse environments have engendered a multitude of responses by protoplasm and a vast proliferation of distinct species of life through the evolutionary process.

### Some Adaptations

Many animal and plant species are polymorphic, *i.e.*, represented in nature by two or more clearly distinguishable kinds of individuals. For example, some individuals of the ladybird beetle *Adalia bipunctata* are red with black spots while others are black with red spots. The color difference is hereditary, the black color behaving as a Mendelian dominant and red as a recessive. The red and black forms live side by side and interbreed freely. Timofeeff-Ressovsky observed that near Berlin, Germany, the black form predominates from spring to autumn, and the red form is more numerous during the winter. What is the origin of these changes? It is quite improbable that the genes for color are transformed by the seasonal variations in temperature; that would mean epidemics of directed mutations on a scale never observed. A much more plausible view is that the changes are produced by natural selection. The black form is, for some reason, more successful than the red in survival and reproduction during summer, but the red is superior to the black under winter conditions. Since the beetles produce several generations during a single season, the species undergoes cyclic changes in its genetic composition in response to the seasonal alterations in the environment. This hypothesis was confirmed by the discovery that black individuals are more frequent among the beetles that die during the rigors of winter than among those that survive.

The writer has observed seasonal changes in some localities in California in the fly *Drosophila pseudoobscura.* Flies of this species in nature are rather uniform in coloration and other external traits, but they are very variable in the structure of their chromosomes, which can be seen in microscopic preparations. In the locality called Piñon Flats, on Mount San Jacinto in southern Califor-

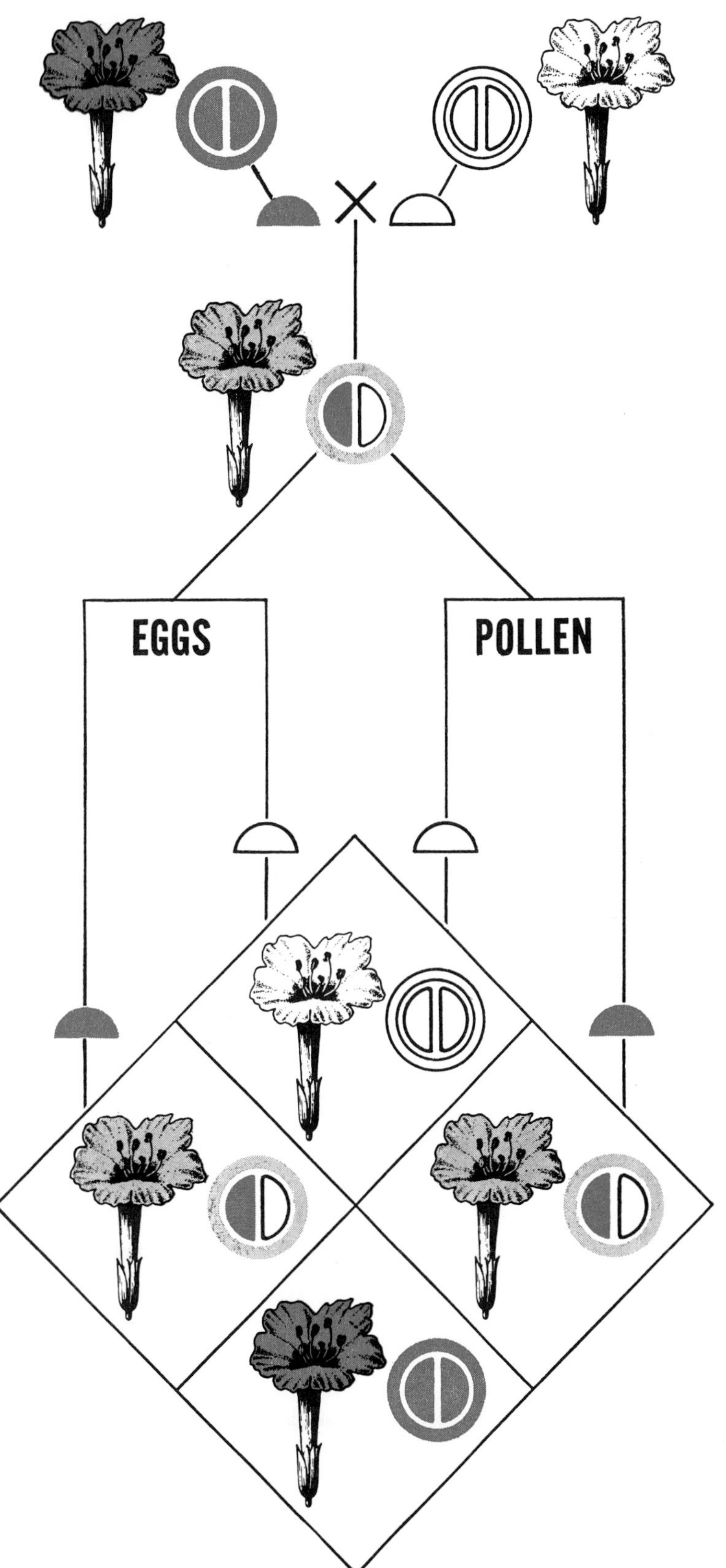


**MENDELIAN SEGREGATION is illustrated by the four o'clock (*Mirabilis jalapa*). The genes of red and white flowers combine in a pink hybrid. Genes are segregated in the cross-fertilized descendants of pink flowers.**

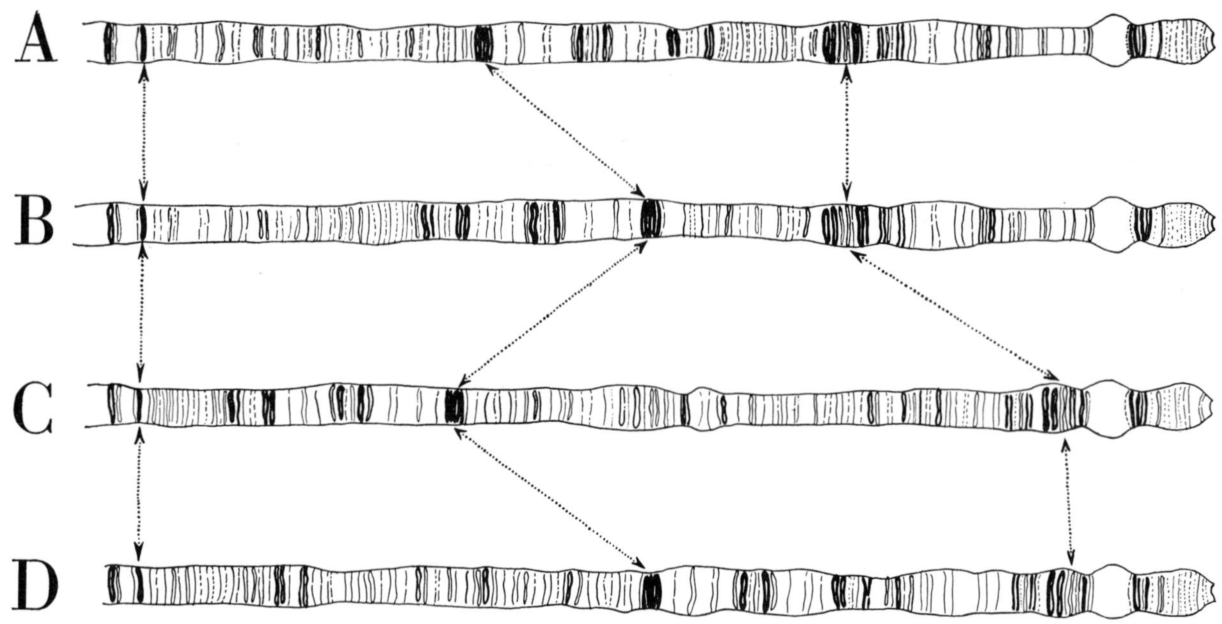


**FOUR VARIETIES** of the species *Drosophila pseudoobscura* are revealed by differences in the structure of their chromosomes. Under the microscope similar markings may be observed at different locations (*arrows*).

nia, the fruit-fly population has four common types of chromosome structure, which we may, for simplicity, designate as types A, B, C and D. From 1939 to 1946, samples of flies were taken from this population in various months of the year, and the chromosomes of these flies were examined. The relative frequencies of the chromosomal types, expressed in percentages of the total, varied with the seasons as follows:

| Month | A | B | C | D |
|---|---|---|---|---|
| March | 52 | 18 | 23 | 7 |
| April | 40 | 28 | 28 | 4 |
| May | 34 | 29 | 31 | 6 |
| June | 28 | 28 | 39 | 5 |
| July | 42 | 22 | 31 | 5 |
| Aug. | 42 | 28 | 26 | 4 |
| Sept. | 48 | 23 | 26 | 3 |
| Oct.-Dec. | 50 | 26 | 20 | 4 |

Thus type A was common in winter but declined in the spring, while type C waxed in the spring and waned in summer. Evidently flies carrying chromosomes of type C are somehow better adapted than type A to the spring climate; hence from March to June, type A decreases and type C increases in frequency. Contrariwise, in the summer type A is superior to type C. Types B and D show little consistent seasonal variation.

Similar changes can be observed under controlled laboratory conditions. Populations of Drosophila flies were kept in a very simple apparatus consisting of a wood and glass box, with openings in the bottom for replenishing the nutrient medium on which the flies lived—a kind of pudding made of Cream of Wheat, molasses and yeast. A mixture of flies of which 33 per cent were type A and 67 per cent type C was introduced into the apparatus and left to multiply freely, up to the limit imposed by the quantity of food given. If one of the types was better adapted to the environment than the other, it was to be expected that the better-adapted type would increase and the other decrease in relative numbers. This is exactly what happened. During the first six months the type A flies rose from 33 to 77 per cent of the population, and type C fell from 67 to 23 per cent. But then came an unexpected leveling off: during the next seven months there was no further change in the relative proportions of the flies, the frequencies of types A and C oscillating around 75 and 25 per cent respectively.

If type A was better than type C under the conditions of the experiment, why were not the flies with C chromosomes crowded out completely by the carriers of A? Sewall Wright of the University of Chicago solved the puzzle by mathematical analysis. The flies of these types interbreed freely, in natural as well as in experimental populations. The populations therefore consist of three kinds of individuals: 1) those that obtained chromosome A from father as well as from mother, and thus carry two A chromosomes (AA); 2) those with two C chromosomes (CC); 3) those that received chromosomes of different types from their parents (AC). The mixed type, AC, possesses the highest adaptive value; it has what is called "hybrid vigor." As for the pure types, under the conditions that obtain in nature AA is superior to CC in the summer. Natural selection then increases the frequency of A chromosomes in the population and diminishes the C chromosomes. In the spring, when CC is better than AA, the reverse is true. But note now that in a population of mixed types neither the A nor the C chromosomes can ever be entirely eliminated from the population, even if the flies are kept in a constant environment where type AA is definitely superior to type CC. This, of course, is highly favorable to the flies as a species, for the loss of one of the chromosome types, though it might be temporarily advantageous, would be prejudicial in the long run, when conditions favoring the lost type would return. Thus a polymorphic population is better able than a uniform one to adjust itself to environmental changes and to exploit a variety of habitats.

## Races

Populations of the same species which inhabit different environments become genetically different. This is what a geneticist means when he speaks of races. Races are populations within a species that differ in the frequencies of some genes. According to the old concept of race, which is based on the notion that

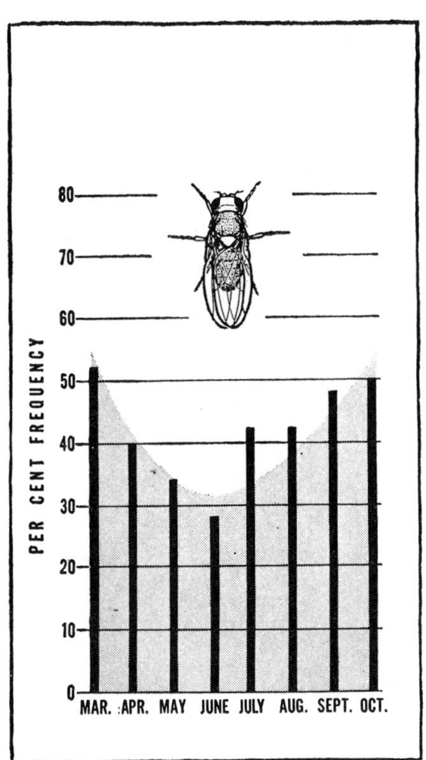


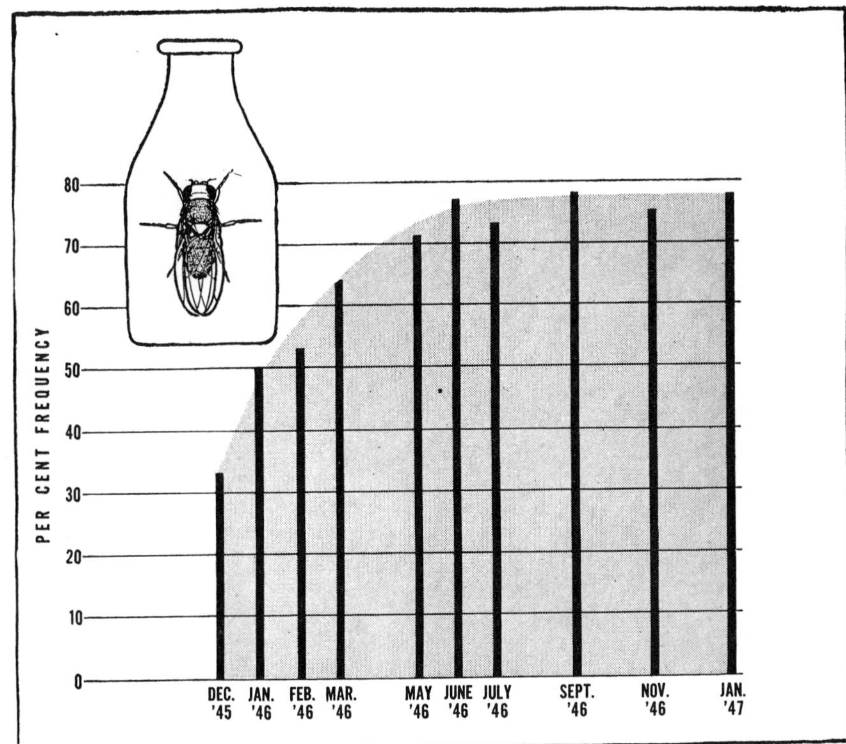


**NUMBER OF FLIES** of one chromosomal type varies in nature (*left*) and in the laboratory. In seasonal environment of nature the type increases and decreases regularly; in constant environment of laboratory it levels off.

heredity is transmitted through "blood" and which still prevails among those ignorant of modern biology, the hereditary endowment of an isolated population would become more and more uniform with each generation, provided there was no interbreeding with other tribes or populations. The tribe would eventually become a "pure" race, all members of which would be genetically uniform. Scientists misled by this notion used to think that at some time in the past the human species consisted of an unspecified number of "pure" races, and that intermarriage between them gave rise to the present "mixed" populations.

In reality, "pure" races never existed, nor can they possibly exist in any species, such as man, that reproduces by sexual combination. We have seen that all human beings except identical twins differ in heredity. In widely differing climatic environments the genetic differences may be substantial. Thus populations native in central Africa have much higher frequencies of genes that produce dark skin than do European populations. The frequency of the gene for blue eye color progressively diminishes southward from Scandinavia through central Europe to the Mediterranean and Africa. Nonetheless some blue-eyed individuals occur in the Mediterranean region and even in Africa, and some brown-eyed ones in Norway and Sweden.

It is important to keep in mind that races are populations, not individuals. Race differences are relative and not absolute, since only in very remote races do all members of one population possess a set of genes that is lacking in all members of another population. It is often difficult to tell how many races exist in a species. For example, some anthropologists recognize only two human races while others list more than 100. The difficulty is to know where to draw the line. If, for example, the Norwegians are a "Nordic race" and the southern Italians a "Mediterranean race," to what race do the inhabitants of Denmark, northern Germany, southern Germany, Switzerland and northern Italy belong? The frequencies of most differentiating traits change rather gradually from Norway to southern Italy. Calling the intermediate populations separate races may be technically correct, but this confuses the race classification even more, because nowhere can sharp lines of demarcation between these "races" be drawn. It is quite arbitrary whether we recognize 2, 4, 10, or more than 100 races—or finally refuse to make any rigid racial labels at all.

The differences between human races are, after all, rather small, since the geographic separation between them is nowhere very marked. When a species is distributed over diversified territories, the process of adaptation to the different environments leads to the gradual accumulation of more numerous and biologically more and more important differences between races. The races gradually diverge. There is, of course, nothing fatal about this divergence, and under some circumstances the divergence may stop or even be turned into convergence. This is particularly true of the human species. The human races were somewhat more sharply separated in the past than they are today. Although the species inhabits almost every variety of environment on earth, the development of communications and the increase of mobility, especially in modern times, has led to much intermarriage and to some genetic convergence of the human races.

The diverging races become more and more distinct with time, and the process of divergence may finally culminate in transformation of races into species. Although the splitting of species is a gradual process, and it is often impossible to tell exactly when races cease to be races and become species, there exist some important differences between race and species which make the process of species formation one of the most important biological processes. Indeed, Darwin called his chief work *The Origin of Species*.

Races of sexually reproducing organisms are fully capable of intercrossing; they maintain their distinction as races only by geographical isolation. As a rule in most organisms no more than a single race of any one species inhabits the same territory. If representatives of two or more races come to live in the same territory, they interbreed, exchange genes, and eventually become fused into a single population. The human species, however, is an exception. Marriages are influenced by linguistic, religious, social, economic and other cultural factors.

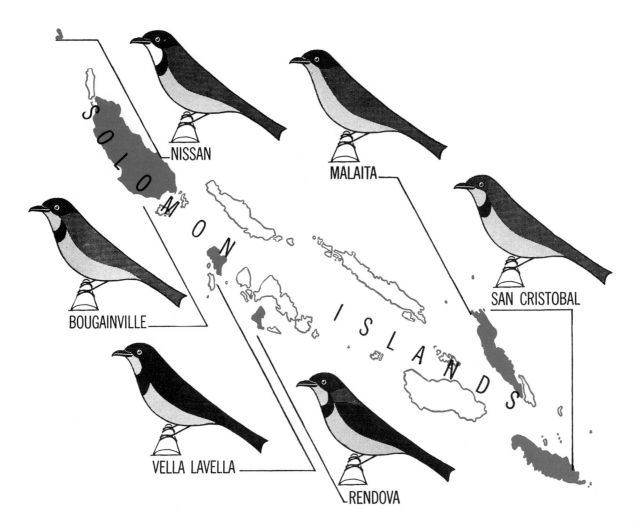

**CONCEPT OF RACE** is illustrated by the varieties of the golden whistler (*Pachycephala pectoralis*) of the Solomon Islands. The races are kept distinct principally by geographical isolation. They differ in their black and white and colored markings. Dark gray areas are symbol for green markings; light gray for yellow.

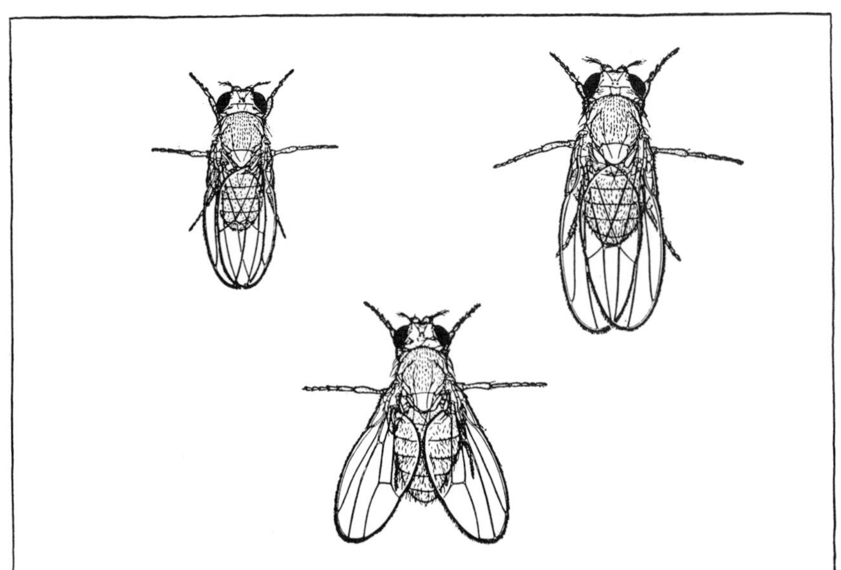

**SPECIES OF DROSOPHILA** and some other organisms tend to remain separate because their hybrid offspring are often weak and sterile. At left is *D. pseudoobscura*; at right *D. miranda*. Their hybrid descendant is at bottom.

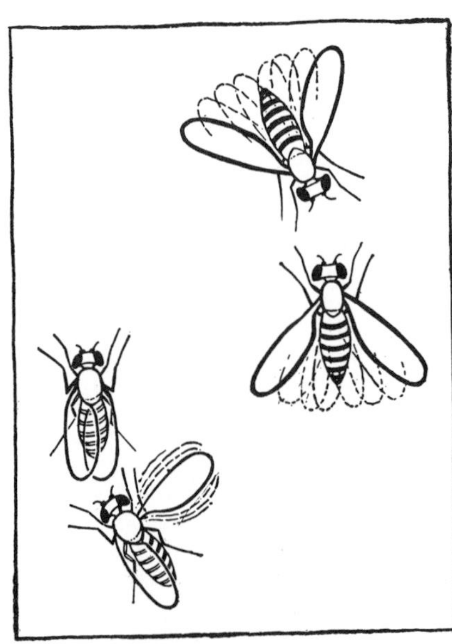

**RITUALS OF MATING** in *D. nebulosa* (*top*) and *D. willistoni* are example of factor that separates species.

Hence cultural isolation may keep populations apart for a time and slow down the exchange of genes even though the populations live in the same country. Nevertheless, the biological relationship proves stronger than cultural isolation, and interbreeding is everywhere in the process of breaking down such barriers. Unrestricted interbreeding would not mean, as often supposed, that all people would become alike. Mankind would continue to include at least as great a diversity of hereditary endowments as it contains today. However, the same types could be found anywhere in the world, and races as more or less discrete populations would cease to exist.

## The Isolationism of Species

Species, on the contrary, can live in the same territory without losing their identity. F. Lutz of the American Museum of Natural History found 1,402 species of insects in the 75-by-200-foot yard of his home in a New Jersey suburb. This does not mean that representatives of distinct species never cross. Closely related species occasionally do interbreed in nature, especially among plants, but these cases are so rare that the discovery of one usually merits a note in a scientific journal.

The reason distinct species do not interbreed is that they are more or less completely kept apart by isolating mechanisms connected with reproduction, which exist in great variety. For example, the botanist Carl C. Epling of the University of California found that two species of sage which are common in southern California are generally separated by ecological factors, one preferring a dry site, the other a more humid one. When the two sages do grow side by side, they occasionally produce hybrids. The hybrids are quite vigorous, but their seed set amounts to less than two per cent of normal; *i.e.*, they are partially sterile. Hybrid sterility is a very common and effective isolating mechanism. A classic example is the mule, hybrid of the horse and donkey. Male mules are always sterile, females usually so. There are, however, some species, notably certain ducks, that produce quite fertile hybrids, not in nature but in captivity.

Two species of Drosophila, *pseudoobscura* and *persimilis*, are so close together biologically that they cannot be distinguished by inspection of their external characteristics. They differ, however, in the structure of their chromosomes and in many physiological traits. If a mixed group of females of the two species is exposed to a group of males of one species, copulations occur much more frequently between members of the same species than between those of different species, though some of the latter do take place. Among plants, the flowers of related species may differ so much in structure that they cannot be pollinated by the same insects, or they may have such differences in smell, color and shape that they attract different insects. Finally, even when cross-copulation or cross-pollination can occur, the union may fail to result in fertilization or may produce offspring that cannot live. Often several isolating mechanisms, no one of which is effective separately, combine to prevent interbreeding. In the case of the two fruit-fly species, at least three such mechanisms are at work: 1) the above-mentioned disposition to mate only with their own kind, even when they are together; 2) different preferences in climate, one preferring warmer and drier places than the other; 3) the fact that when they do interbreed the hybrid males that result are completely sterile and the hybrid females, though fertile, produce offspring that are poorly viable. There is good evidence that no gene exchange occurs between these species in nature.

The fact that distinct species can coexist in the same territory, while races generally cannot, is highly significant. It permits the formation of communities of diversified living beings which exploit the variety of habitats present in a territory more fully than any single species, no matter how polymorphic, could. It is responsible for the richness and colorfulness of life that is so impressive to biologists and non-biologists alike.

## Evolution *v.* Predestination

Our discussion of the essentials of the modern theory of evolution may be concluded with a consideration of the objections raised against it. The most serious objection is that since mutations occur by "chance" and are undirected, and since selection is a "blind" force, it is difficult to see how mutation and selection can add up to the formation of such complex and beautifully balanced organs as, for example, the human eye. This, say critics of the theory, is like believing that a monkey pounding a typewriter might accidentally type out Dante's *Divine Comedy*. Some biologists have preferred to suppose that evolution is directed by an "inner urge toward perfection," or by a "combining power which amounts to intentionality," or by "telefinalism" or the like. The fatal weakness of these alternative "explanations" of evolution is that they do not explain anything. To say that evolution is directed by an urge, a combining power, or a telefinalism is like saying that a railroad engine is moved by a "locomotive power."

The objection that the modern theory of evolution makes undue demands on chance is based on a failure to appreciate the historical character of the evolutionary process. It would indeed strain credulity to suppose that a lucky sudden combination of chance mutations produced the eye in all its perfection. But the eye did not appear suddenly in the offspring of an eyeless creature; it is the result of an evolutionary development that took many millions of years. Along the way the evolving rudiments of the eye passed through innumerable stages, all of which were useful to their possessors, and therefore all adjusted to the demands of the environment by natural selection. Amphioxus, the primitive fish-like darling of comparative anatomists, has no eyes, but it has certain pigment cells in its brain by means of which it perceives light. Such pigment cells may have been the starting point of the development of eyes in our ancestors.

We have seen that the "combining power" of the sexual process is staggering, that on the most conservative estimate the number of possible gene combinations in the human species alone is far greater than that of the electrons and protons in the universe. When life developed sex, it acquired a trial-and-error mechanism of prodigious efficiency. This mechanism is not called upon to produce a completely new creature in one spectacular burst of creation; it is sufficient that it produces slight changes that improve the organism's chances of survival or reproduction in some habitat. In terms of the monkey-and-typewriter analogy, the theory does not require that the monkey sit down and compose the *Divine Comedy* from beginning to end by a lucky series of hits. All we need is that the monkey occasionally form a single word, or a single line; over the course of eons of time the environment shapes this growing text into the eventual masterpiece. Mutations occur by "chance" only in the sense that they appear regardless of their usefulness at the time and place of their origin. It should be kept in mind that the structure of a gene, like that of the whole organism, is the outcome of a long evolutionary development; the ways in which the genes can mutate are, consequently, by no means indeterminate.

Theories that ascribe evolution to "urges" and "telefinalisms" imply that there is some kind of predestination about the whole business, that evolution has produced nothing more than was potentially present at the beginning of life. The modern evolutionists believe that, on the contrary, evolution is a creative response of the living matter to the challenges of the environment. The role of the environment is to provide opportunities for biological inventions. Evolution is due neither to chance nor to design; it is due to a natural creative process.

# 23

# *Darwin's Missing Evidence*

H. B. D. KETTLEWELL

*March 1959*

Charles Darwin's *Origin of Species*, the centenary of which we celebrate in 1959, was the fruit of 26 years of laborious accumulation of facts from nature. Others before Darwin had believed in evolution, but he alone produced a cataclysm of data in support of it. Yet there were two fundamental gaps in his chain of evidence. First, Darwin had no knowledge of the mechanism of heredity. Second, he had no visible example of evolution at work in nature.

It is a curious fact that both of these gaps could have been filled during Darwin's lifetime. Although Gregor Mendel's laws of inheritance were not discovered by the community of biologists until 1900, they had first been published in 1866. And before Darwin died in 1882, the most striking evolutionary change ever witnessed by man was taking place around him in his own country.

The change was simply this. Less than a century ago moths of certain species were characterized by their light coloration, which matched such backgrounds as light tree trunks and lichen-covered rocks, on which the moths passed the daylight hours sitting motionless. Today in many areas the same species are predominantly dark! We now call this reversal "industrial melanism."

It happens that Darwin's lifetime coincided with the first great man-made change of environment on earth. Ever since the Industrial Revolution commenced in the latter half of the 18th century, large areas of the earth's surface have been contaminated by an insidious and largely unrecognized fallout of smoke particles. In and around industrial areas the fallout is measured in tons per square mile per month; in places like Sheffield in England it may reach 50 tons or more. It is only recently that we have begun to realize how widely the lighter smoke particles are dispersed, and to what extent they affect the flora and fauna of the countryside.

In the case of the flora the smoke particles not only pollute foliage but also kill vegetative lichens on the trunks and boughs of trees. Rain washes the pollutants down the boughs and trunks until they are bare and black. In heavily polluted districts rocks and the very ground itself are darkened.

Now in England there are some 760 species of larger moths. Of these more than 70 have exchanged their light color and pattern for dark or even all-black coloration. Similar changes have occurred in the moths of industrial areas of other countries: France, Germany, Poland, Czechoslovakia, Canada and the U. S. So far, however, such changes have not been observed anywhere in the tropics. It is important to note here that industrial melanism has occurred only among those moths that fly at night and spend the day resting against a background such as a tree trunk.

These, then, are the facts. A profound change of color has occurred among hundreds of species of moths in industrial areas in different parts of the world. How has the change come about? What underlying laws of nature have produced it? Has it any connection with one of the normal mechanisms by which one species evolves into another?

In 1926 the British biologist Heslop Harrison reported that the industrial melanism of moths was caused by a special substance which he alleged was present in polluted air. He called this substance a "melanogen," and suggested that it was manganous sulfate or lead nitrate. Harrison claimed that when he fed foliage impregnated with these salts to the larvae of certain species of light-colored moths, a proportion of their offspring were black. He also stated that this "induced melanism" was inherited according to the laws of Mendel.

Darwin, always searching for missing evidence, might well have accepted Harrison's Lamarckian interpretation, but in 1926 biologists were skeptical. Although the rate of mutation of a hereditary characteristic can be increased in the laboratory by many methods, Harrison's figures inferred a mutation rate of 8 per cent. One of the most frequent mutations in nature is that which causes the disease hemophilia in man; its rate is in the region of .0005 per cent, that is, the mutation occurs about once in 50,000 births. It is, in fact, unlikely that an increased mutation rate has played any part in industrial melanism.

At the University of Oxford during the past seven years we have been attempting to analyze the phenomenon of industrial melanism. We have used many different approaches. We are in the process of making a survey of the present frequency of light and dark forms of each species of moth in Britain that exhibits industrial melanism. We are critically examining each of the two forms to see if between them there are any differences in behavior. We have fed large numbers of larvae of both forms on foliage impregnated with substances in polluted air. We have observed under various conditions the mating preferences and relative mortality of the two forms. Finally we have accumulated much information about the melanism of moths in parts of the world that are far removed from industrial centers, and we have sought to link industrial melanism with the melanics of the past.

Our main guinea pig, both in the field and in the laboratory, has been the peppered moth *Biston betularia* and its melanic form *carbonaria*. This species occurs throughout Europe, and is probably identical with the North American *Amphidasis cognataria*. It has a one-year life cycle; the moth appears from May to August. The moth flies at night and passes the day resting on the trunks or on the underside of the boughs of rough-barked deciduous trees such as the oak. Its larvae feed on the foliage of such trees from June to late October; its pupae pass the winter in the soil.

The dark form of the peppered moth was first recorded in 1848 at Manchester in England. Both the light and dark forms appear in each of the photographs at right and on the next page. The background of each photograph is noteworthy. In the photograph on the next page the background is a lichen-encrusted oak trunk of the sort that today is found only in unpolluted rural districts. Against this background the light form is almost invisible and the dark form is conspicuous. In the photograph at right the background is a bare and blackened oak trunk in the heavily polluted area of Birmingham. Here it is the dark form which is almost invisible, and the light form which is conspicuous. Of 621 wild moths caught in these Birmingham woods in 1953, 90 per cent were the dark form and only 10 per cent the light. Today this same ratio applies in nearly all British industrial areas and far outside them.

We decided to test the rate of survival of the two forms in the contrasting types of woodland. We did this by releasing known numbers of moths of both forms. Each moth was marked on its underside with a spot of quick-drying cellulose paint; a different color was used for each day. Thus when we subsequently trapped large numbers of moths we could identify those we had released and established the length of time they had been exposed to predators in nature.

In an unpolluted forest we released 984 moths: 488 dark and 496 light. We recaptured 34 dark and 62 light, indicating that in these woods the light form had a clear advantage over the dark. We then repeated the experiment in the polluted Birmingham woods, releasing 630 moths: 493 dark and 137 light. The result of the first experiment was completely reversed; we recaptured proportionately twice as many of the dark form as of the light.

For the first time, moreover, we had witnessed birds in the act of taking moths from the trunks. Although Britain has more ornithologists and bird watch-

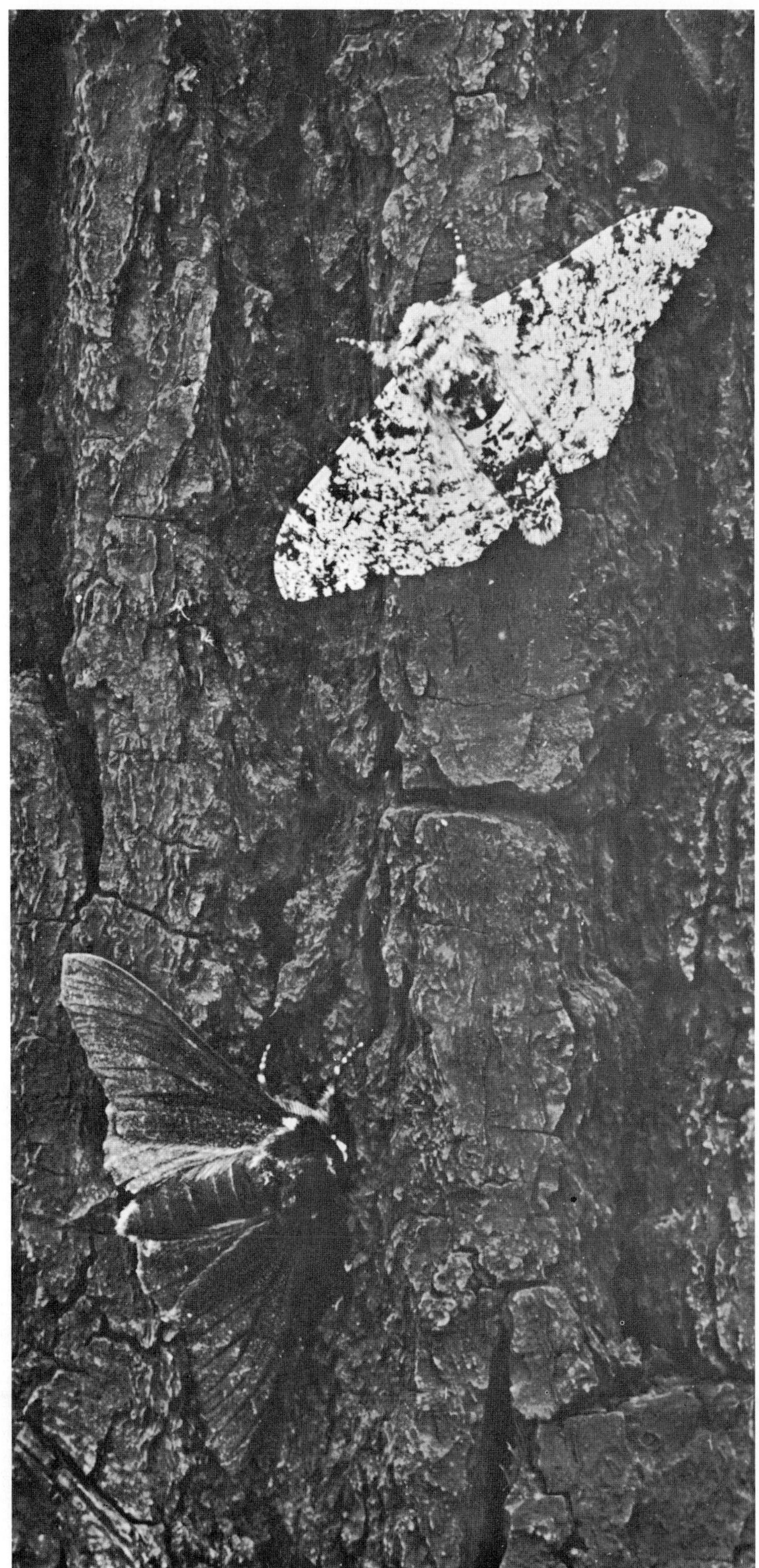

**DARK AND LIGHT FORMS** of the peppered moth were photographed on the trunk of an oak blackened by the polluted air of the English industrial city of Birmingham. The light form (*Biston betularia*) is clearly visible; the dark form (*carbonaria*) is well camouflaged.

ers than any other country, there had been absolutely no record of birds actually capturing resting moths. Indeed, many ornithologists doubted that this happened on any large scale.

The reason for the oversight soon became obvious. The bird usually seizes the insect and carries it away so rapidly that the observer sees nothing unless he is keeping a constant watch on the insect. This is just what we were doing in the course of some of our experiments. When I first published our findings, the editor of a certain journal was sufficiently rash as to question whether birds took resting moths at all. There was only one thing to do, and in 1955 Niko Tinbergen of the University of Oxford filmed a repeat of my experiments. The film not only shows that birds capture and eat resting moths, but also that they do so selectively.

These experiments lead to the following conclusions. First, when the environment of a moth such as *Biston betularia* changes so that the moth cannot hide by day, the moth is ruthlessly eliminated by predators unless it mutates to a form that is better suited to its new environment. Second, we now have visible proof that, once a mutation has occurred, natural selection alone can be responsible for its rapid spread. Third, the very fact that one form of moth has replaced another in a comparatively short span of years indicates that this evolutionary mechanism is remarkably flexible.

The present status of the peppered moth is shown in the map on page 300. This map was built up from more than 20,000 observations made by 170 voluntary observers living in various parts of Britain. The map makes the following points. First, there is a strong correlation between industrial centers and a high percentage of the dark form of the moth. Second, populations consisting entirely of the light form are found today only in western England and northern Scotland. Third, though the counties of eastern England are far removed from industrial centers, a surprisingly high percentage of the dark form is found in them. This, in my opinion, is due to the long-standing fall-out of smoke particles carried from central England by the prevailing southwesterly winds.

Now in order for the dark form of a moth to spread, a mutation from the light form must first occur. It appears that the frequency with which this happens—that is, the mutation rate—varies according to the species. The rate at which the light form of the peppered

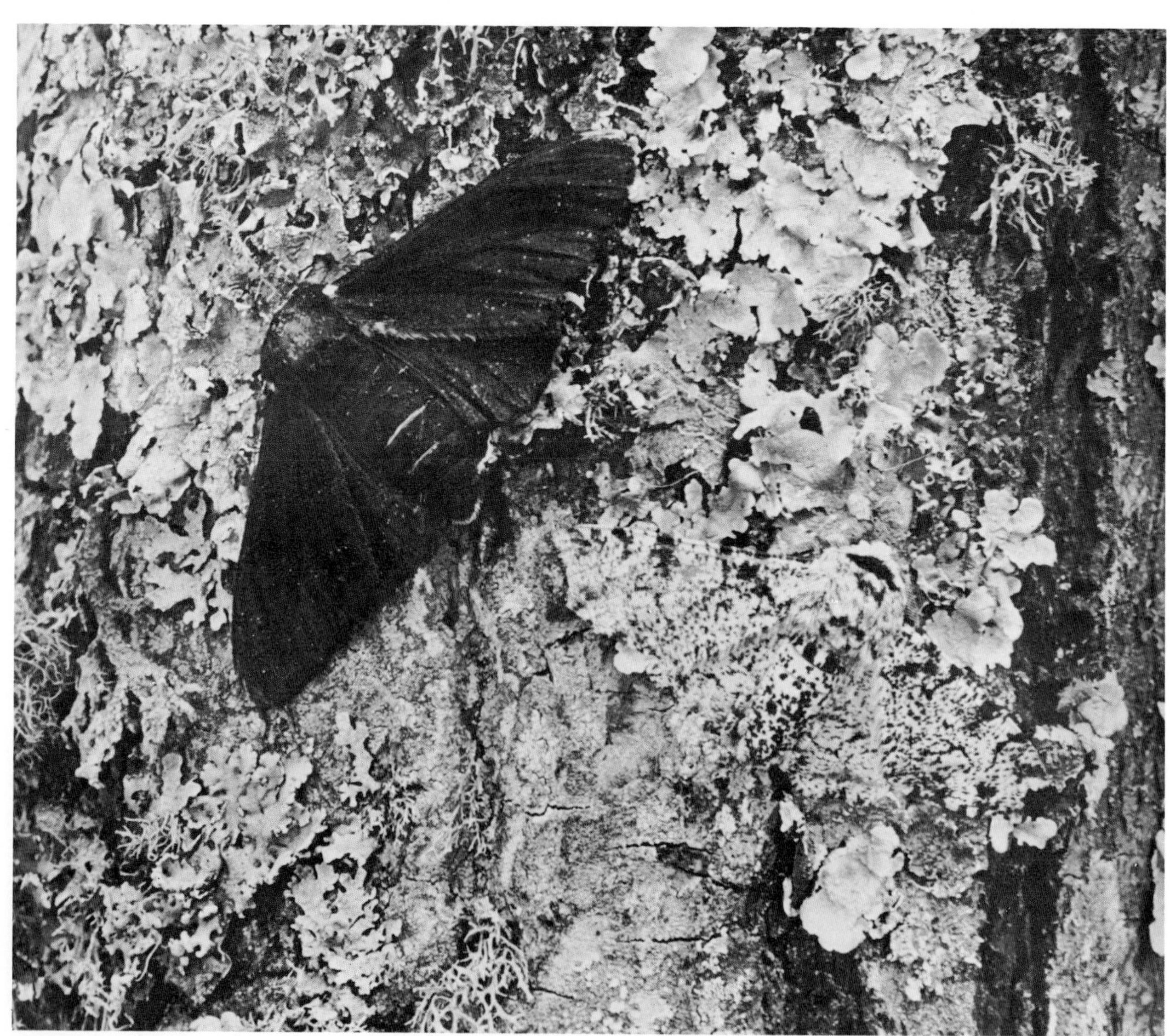

**SAME TWO FORMS of the peppered moth were photographed against the lichen-encrusted trunk of an oak in an unpolluted area. Here it is the dark form which may be clearly seen. The light form, almost invisible, is just below and to the right of the dark form.**

moth mutates to the dark form seems to be fairly high; the rate at which the mutation occurs in other species may be very low. For example, the light form of the moth *Procus literosa* disappeared from the Sheffield area many years ago, but it has now reappeared in its dark form. It would seem that a belated mutation has permitted the species to regain lost territory. Another significant example is provided by the moth *Tethea ocularis.* Prior to 1947 the dark form of this species was unknown in England. In that year, however, many specimens of the dark form were for the first time collected in various parts of Britain; in some districts today the dark form now comprises more than 50 per cent of the species. There is little doubt that this melanic arrived in Britain not by mutation but by migration. It had been known for a considerable time in the industrial areas of northern Europe, where presumably the original mutation occurred.

The mutation that is responsible for industrial melanism in moths is in the majority of cases controlled by a single gene. A moth, like any other organism that reproduces sexually, has two genes for each of its hereditary characteristics: one gene from each parent. The mutant gene of a melanic moth is inherited as a Mendelian dominant; that is, the effect of the mutant gene is expressed and the effect of the other gene in the pair is not. Thus a moth that inherits the mutant gene from only one of its parents is melanic.

The mutant gene, however, does more than simply control the coloration of the moth. The same gene (or others closely linked with it in the hereditary material) also gives rise to physiological and even behavioral traits. For example, it appears that in some species of moths the caterpillars of the dark form are hardier than the caterpillars of the light form. Genetic differences are also reflected in mating preference. On cold nights more males of the light form of the peppered moth appear to be attracted to light females than to dark. On warm nights, on the other hand, significantly more light males are attracted to dark females.

There is evidence that, in a population of peppered moths that inhabits an industrial area, caterpillars of the light form attain full growth earlier than caterpillars of the dark form. This may be due to the fact that the precipitation of pollutants on leaves greatly increases late in the autumn. Caterpillars of the dark form may be hardier in the presence

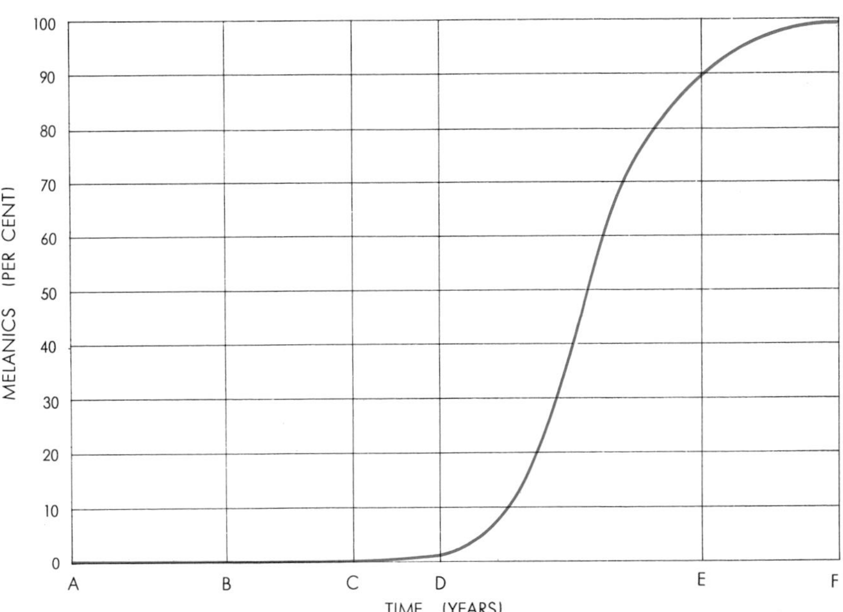


**SPREAD OF MUTATION from the light form to the dark (melanic) is expressed by this curve, discussed in detail in the text. The mutation occurs in the period AB, spreads slowly during BD and spreads rapidly during DE. During EF the light form is either gradually eliminated, as indicated by the curve, or remains at a level of about 5 per cent of the population.**

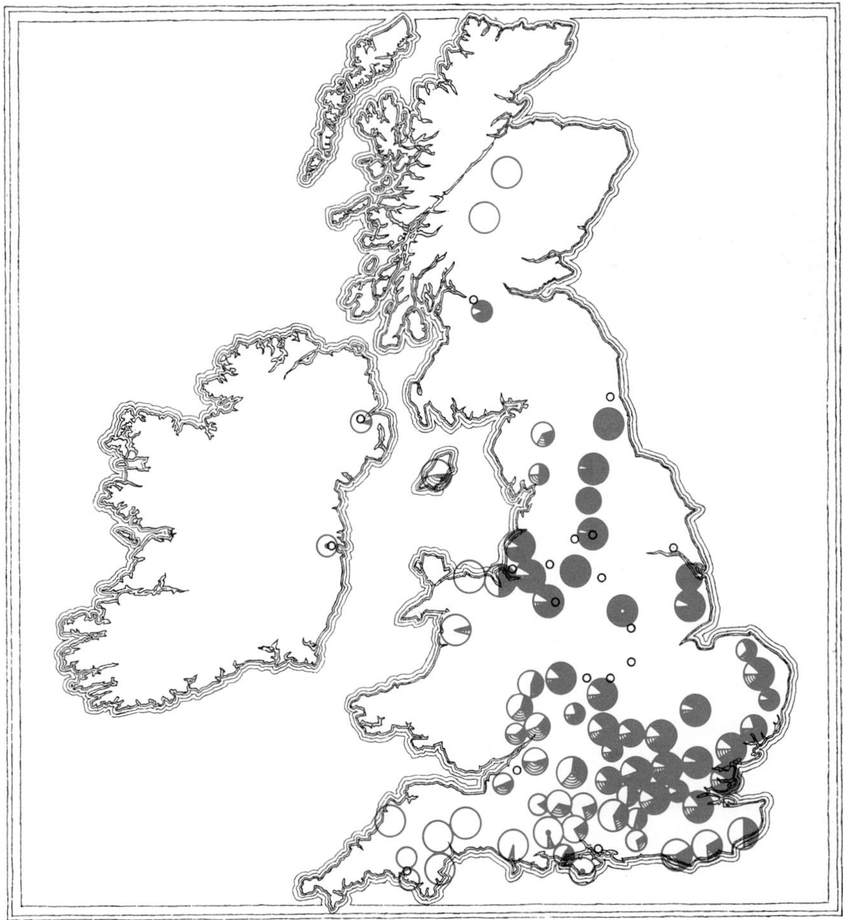

**PROPORTION OF FORMS of the peppered moth at various locations in the British Isles is indicated on this map. The open area within a colored circle represents the proportion of the light form *Biston betularia* recorded; the solid colored area, the proportion of the dark form *carbonaria;* the hatched colored area, the proportion of another dark form, *insularia.* Small black circles on the map indicate the location of major industrial centers.**

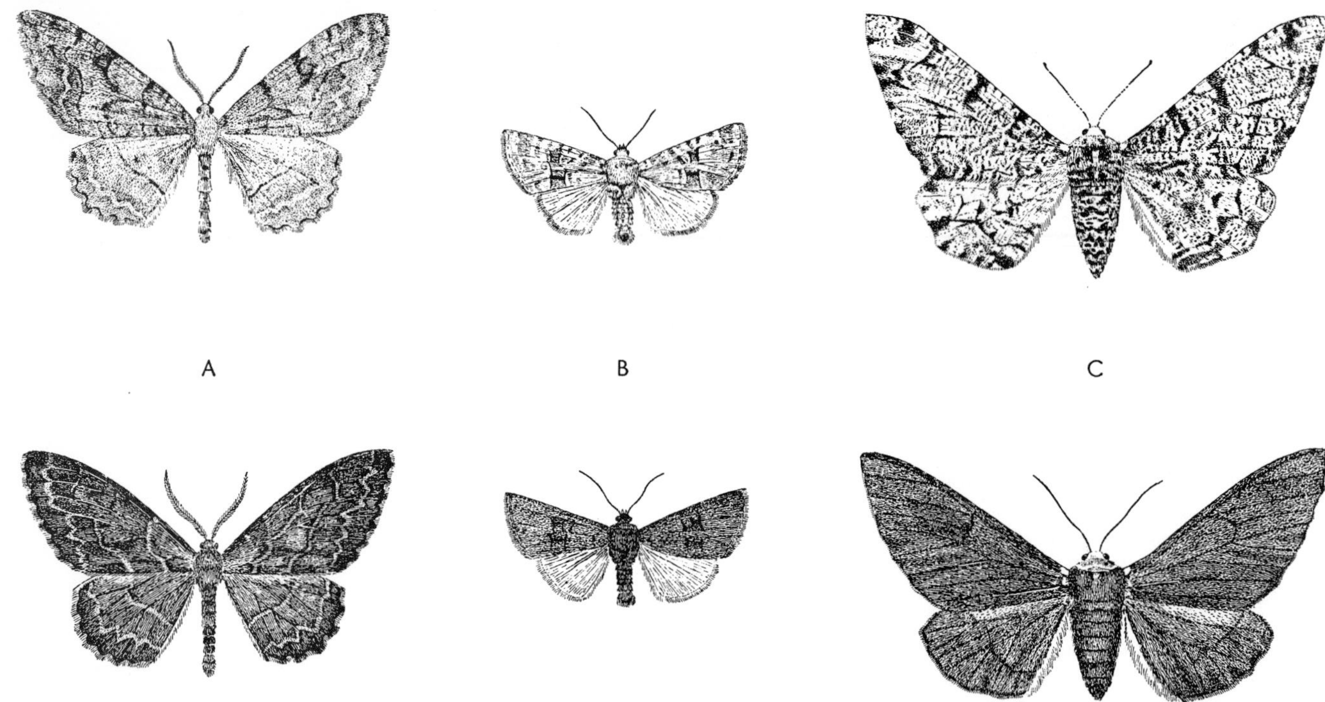


**FIVE SPECIES OF MOTH** that have both light and dark forms are depicted in their actual size. In each case the light form is at the top and the dark form is at the bottom. The species are: *Cleora repandata* (A), *Procus literosa* (B), *Biston betularia* (C), *Ectropis*

of such pollution than caterpillars of the light form. In that case natural selection would favor light-form caterpillars which mature early over light-form caterpillars which mature late. For the hardier caterpillars of the dark form, on the other hand, the advantages of later feeding and longer larval life might outweigh the disadvantages of feeding on increasingly polluted leaves. Then natural selection would favor those caterpillars which mature late.

Another difference between the behavior of *B. betularia* and that of its dark form *carbonaria* is suggested by our experiments on the question of whether each form can choose the "correct" background on which to rest during the day. We offered light and dark backgrounds of equal area to moths of both forms, and discovered that a significantly large proportion of each form rested on the correct background. Before these results can be accepted as proven, the experiments must be repeated on a larger scale. If they are proven, the behavior of both forms could be explained by the single mechanism of "contrast appreciation." This mechanism assumes that one segment of the eye of a moth senses the color of the background and that another segment senses the moth's own color; thus the two colors could be compared. Presumably if they were the same, the moth would remain on its background; but if they were different, "contrast conflict" would result and the moth would move off again. That moths tend to be restless when the colors conflict is certainly borne out by recent field observations.

It is evident, then, that industrial melanism is much more than a simple change from light to dark. Such a change must profoundly upset the balance of hereditary traits in a species, and the species must be a long time in restoring that balance. Taking into account all the favorable and unfavorable factors at work in this process, let us examine the spread of a mutation similar to the dark form of the peppered moth. To do so we must consult the diagram at the top of the preceding page.

According to the mutation rate and the size of the population, the new mutation may not appear in a population for a period varying from one to 50 years. This is represented by AB on the diagram. Let us now assume the following: that the original successful mutation took place in 1900, that subsequent new mutations failed to survive, that the total local population was one million, and that the mutant had a 30-per-cent advantage over the light form. (By a 30-per-cent advantage for the dark form we mean that, if in one generation there were 100 light moths and 100 dark, in the next generation there would be 85 light moths and 115 dark.)

On the basis of these assumptions there would be one melanic moth in 1,000 only in 1929 (BC). Not until 1938 would there be one in 100 (BD). Once the melanics attain this level, their rate of increase greatly accelerates.

In the period between 1900 and 1938 (BD) natural selection is complicated by other forces. Though the color of the dark form gives it an advantage over the light, the new trait is introduced into a system of other traits balanced for the light form; thus the dark form is at first at a considerable physiological disadvantage. In fact, when moths of the dark form were crossed with moths of the light form 50 years ago, the resulting broods were significantly deficient in the dark form. When the same cross is made today, the broods contain more of the dark form than one would expect. The system of hereditary traits has become adjusted to the new trait.

There is evidence that other changes take place during the period BD. Specimens of the peppered moth from old collections indicate that the earliest melanics were not so dark as the modern dark form: they retained some of the white spots of the light form. Today a large proportion of the moths around a city such as Manchester are jet black. Evidently when the early melanics inherited one gene for melanism, the gene was not entirely dominant with respect to the gene for light coloration. As the

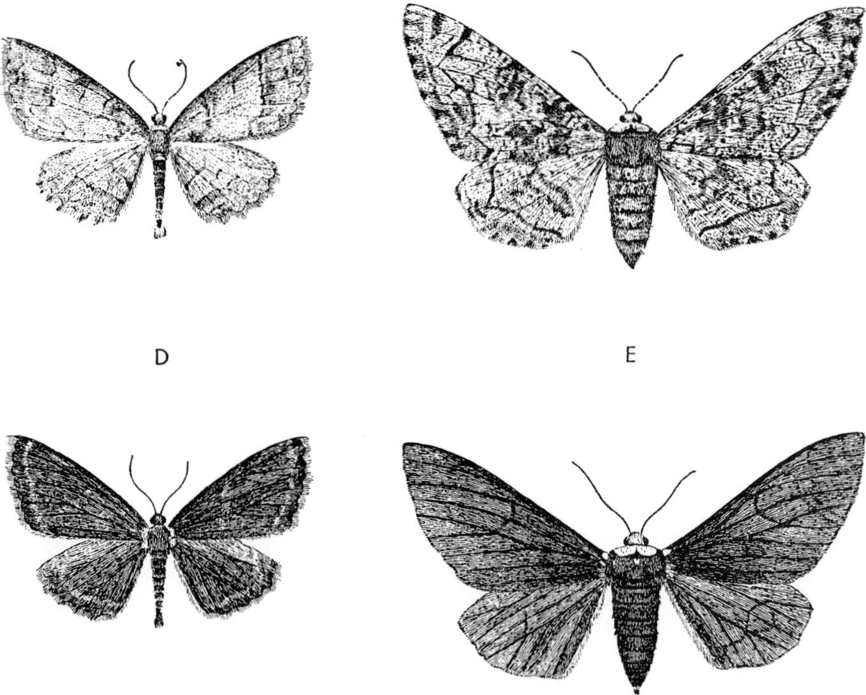


*consonaria* **(D) and** ***Amphidasis cognataria*** **(E). All of the species are European except the last, which occurs in North America and may be identical with** ***Biston betularia.***

gene complex adjusted to the mutation, however, the new gene became almost entirely dominant.

When the dark form comprises about 10 per cent of the population, it may jump to 90 per cent in as little as 15 or 20 years. This is represented by period DE on the graph. Thereafter the proportion of the dark form increases at a greatly reduced rate.

Eventually one of two things must happen: either the light form will slowly be eliminated altogether, or a balance will be struck so that the light form continues to appear as a small but definite proportion of the population. This is due to the fact that the moths which inherit one gene for dark coloration and one for light (heterozygotes) have an advantage over the moths which inherit two genes for dark coloration (homozygotes). And when two heterozygotes mate, a quarter of their offspring will have two genes for light coloration, *i.e.*, they will be light. Only after a very long period of time, therefore, could the light forms (and with them the gene for light coloration) be entirely eliminated. This period of removal, represented by EF on the diagram, might be more than 1,000 years. Indications so far suggest, however, that complete removal is unlikely, and that a balance of the two forms would probably occur. In this balance the light form would represent about 5 per cent of the population.

The mechanisms I have described are without doubt the explanation of industrial melanism: normal mutation followed by natural selection resulting in an insect of different color, physiology and behavior. Industrial melanism involves no new laws of nature; it is governed by the same mechanisms which have brought about the evolution of new species in the past.

There remains, however, one major unsolved problem. Why is it that, in almost all industrial melanics, the gene for melanism is dominant? Many geneticists would agree that dominance is achieved by natural selection, that it is somehow related to a successful mutation in the distant past. With these thoughts in mind I recently turned my attention away from industrial centers and collected moths in one of the few remaining pieces of ancient Caledonian pine forest in Britain: the Black Wood of Rannoch. Located in central Scotland far from industrial centers, the Black Wood is probably very similar to the forests that covered Britain some 4,000 years ago. The huge pines of this forest are only partly covered with lichens. Here I found no fewer than seven species of moths with melanic forms.

I decided to concentrate on the species *Cleora repandata,* the dark form of which is similar to the dark form of the same species that has swept through central England. This dark form, like the industrial melanics, is inherited as a Mendelian dominant. Of just under 500 specimens of *C. repandata* observed, 10 per cent were dark.

*C. repandata* spends the day on pine trunks, where the light form is almost invisible. The dark form is somewhat more easily seen. By noting at dawn the spot where an insect had come to rest, and then revisiting the tree later in the day, we were able to show that on some days more than 50 per cent of the insects had moved. Subsequently we found that because of disturbances such as ants or hot sunshine they had had to fly to another tree trunk, usually about 50 yards away. I saw large numbers of these moths on the wing, and three other observers and I agreed that the dark form was practically invisible at a distance of more than 20 yards, and that the light form could be followed with ease at a distance of up to 100 yards. In fact, we saw birds catch three moths of the light form in flight. It is my belief that when it is on the wing in these woods the dark form has an advantage over the light, and that when it is at rest the reverse is true.

This may be one of many ways in which melanism was useful in the past. It may also explain the balance between the light and dark forms of *Cleora repandata* in the Black Wood of Rannoch. In this case a melanic may have been preserved for one evolutionary reason but then have spread widely for another.

The melanism of moths occurs in many parts of the world that are not industrialized, and in environments that are quite different. It is found in the mountain rain forest of New Zealand's South Island, which is wet and dark. It has been observed in arctic and subarctic regions where in summer moths must fly in daylight. It is known in very high mountains, where dark coloration may permit the absorption of heat and make possible increased activity. In each case recurrent mutation has provided the source of the change, and natural selection, as postulated by Darwin, has decided its destiny.

Melanism is not a recent phenomenon but a very old one. It enables us to appreciate the vast reserves of genetic variability which are contained within each species, and which can be summoned when the occasion arises. Had Darwin observed industrial melanism he would have seen evolution occurring not in thousands of years but in thousands of days—well within his lifetime. He would have witnessed the consummation and confirmation of his life's work.

24

# *Ionizing Radiation and Evolution*

JAMES F. CROW
*September 1959*

The mutant gene—a fundamental unit in the mechanism of heredity that has been altered by some cause, thereby changing some characteristic of its bearers—is the raw material of evolution. Ionizing radiation produces mutation. Ergo, ionizing radiation is an important cause of evolution.

At first this seems like a compelling argument. But is it really? To what extent is the natural rate at which mutations occur dependent upon radiation? Would evolution have been much the same without radiation? Is it possible that increased exposure to radiation, rising from human activities, has a significant effect upon the future course of human evolution?

Radiation has been of tremendous value for genetic research, and therefore for research in evolution. To clarify the issue at the outset, however, it is likely that ionizing radiation has played only a minor role in the recent evolutionary history of most organisms. As for the earliest stages of evolution, starting with the origin of life, it is problematical whether ionizing radiation played a significant role even then. Laboratory experiments show that organic compounds, including the amino acid units of the protein molecule, may be formed from simple nitrogen and carbon compounds upon exposure to ionizing radiation or an electric discharge. But ultraviolet radiation serves as well in these demonstrations, and was probably present in large amounts at the earliest ages of life. In the subsequent history of life the development of photosynthesis in plants and of vision in animals indicates the much greater importance of nonionizing radiation in the terrestrial environment.

Nonetheless, as will be seen, man presents an important special case in any discussion of the mutation-inducing effects of ionizing radiation. The mutation rate affects not only the evolution of the human species but also the life of the individual. Almost every mutation is harmful, and it is the individual who pays the price. Any human activity that tends to increase the mutation rate must therefore raise serious health and moral problems for man.

H. J. Muller's great discovery that radiation induces mutations in the fruit fly *Drosophila*, and its independent confirmation in plants by L. J. Stadler, was made more than 30 years ago. One of the first things to be noticed by the Drosophila workers was that most of the radiation-induced changes in the characteristics of their flies were familiar, the same changes having occurred repeatedly as a result of natural mutation. Not only were the external appearances of the mutant flies recognizable (white eyes, yellow body, forked bristles, missing wing-veins); it could also be shown that the mutant genes were located at the same sites on the chromosomes as their naturally occurring predecessors. Thus it seems that radiation-induced mutants are not unique; they are the same types that occur anyhow at a lower rate and that we call spontaneous because we do not know their causes.

The prevailing hypothesis is that the hereditary information is encoded in various permutations in the arrangement of subunits of the molecule of deoxyribonucleic acid (DNA) in the chromosomes. A mutation occurs when this molecule fails, for some reason, to replicate itself exactly. The mutated gene is thereafter reproduced until its bearer dies without reproducing, or until, by chance, the altered gene mutates again.

A variety of treatments other than radiation will induce mutation. Most agents, including radiation, seem to affect the genes indiscriminately, increasing the mutation rate for all genes in the same proportion. But some chemical mutagens are fairly selective in the genes they affect. It is even conceivable that investigators may find a chemical that will regularly mutate a particular gene and no other. This will probably be very difficult to achieve: A mutagen capable of affecting a specific gene would probably have to have the same order of informational complexity as the gene itself in order to recognize the gene and influence it.

Moreover, different mutant genes known to be at different locations on the chromosome frequently produce effects that mimic each other. Geneticists who work with fruit flies are familiar with several different genes that result in indistinguishable eye colors. This is not surprising when we consider the complexity of the relationship between the genes and the characteristics that they produce. There are many different paths by which any particular end-point may be reached; it is to be expected that many different gene changes can lead to the same result, though the chemical pathways by which this is accomplished may be greatly different.

To consider another kind of example, many insects have come to survive insecticides. They do so by such diverse means as behavior patterns that enable them to avoid the insecticide, mechanisms that interfere with the entrance of the insecticide into the body, enzymes that detoxify the insecticide, and by somehow becoming able to tolerate more of the insecticide. All these modes of survival develop by selection of mutant genes that are already present in low frequency in the population. This is one of the characteristic features of evolution: its opportunism. It makes use of the raw materials—that is, the mutant genes—that happen to be available.

As a first conclusion it appears that radiation-induced mutations do not play any unique role in evolution. The same gene mutations would probably occur anyhow. Even if they did not, the same end result could be achieved with other mutants.

One might still suspect that spontaneous mutations are caused by natural radiations. This was quickly ruled out as a possibility in Drosophila. For one thing, the spontaneous-mutation rate is strongly dependent upon temperature, which would be surprising if mutation were a simple and direct consequence of radiation. But much more decisive evidence comes from the fact that the amount of natural radiation is entirely inadequate to account for the rate of spontaneous mutation. The natural-mutation rate in Drosophila, if it were caused solely by radiation, would require 50 or more r. Yet the amount of radiation received by a fly in the 12-day interval between egg and mature adult is about .004 r. Natural radiation would have to be increased more than 10,000 times to account for the natural Drosophila mutation rate! Thus, in Drosophila at least, radiation accounts for only a trivial part of the spontaneous rate. The same is true for mice. Although the fraction of spontaneous mutations that owe their origin to radiation is higher than that in Drosophila, it is still less than .1 per cent.

It is clear that natural-mutation rates, if they are measured in absolute time-units, cannot be the same in all organisms. One example will demonstrate this decisively. The spontaneous-mutation rate in Drosophila is such that about one embryo in 50 carries a lethal mutation (a mutation harmful enough to cause death) that occurred during the preceding generation. Most such mutations are recessive, that is, they have a lethal effect only when they are in a double dose, but this does not prevent the death; it only postpones it. The reproductive life-cycle in man is approximately 1,000 times as long as that of a fly: some 30 years as compared with 12 days. If the absolute mutation-rate in man were the same as that in Drosophila, each human embryo would bear an average of 20

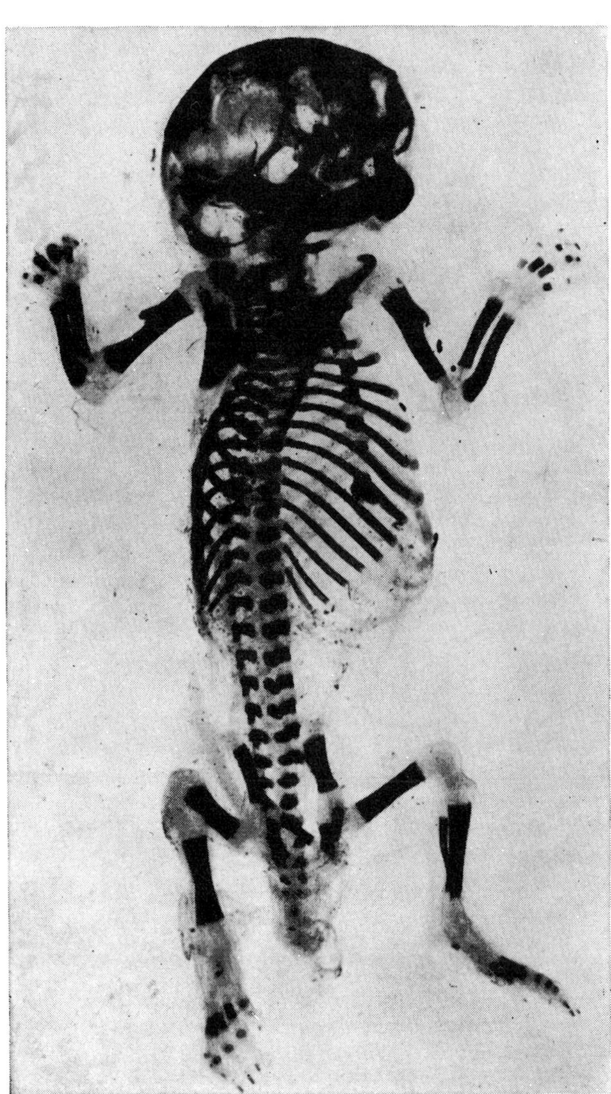

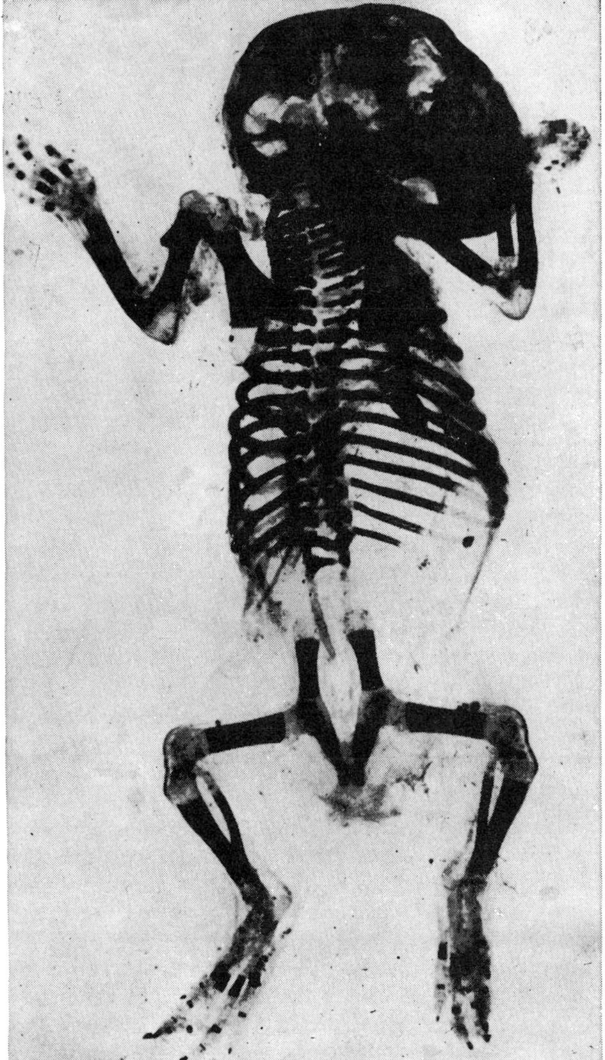

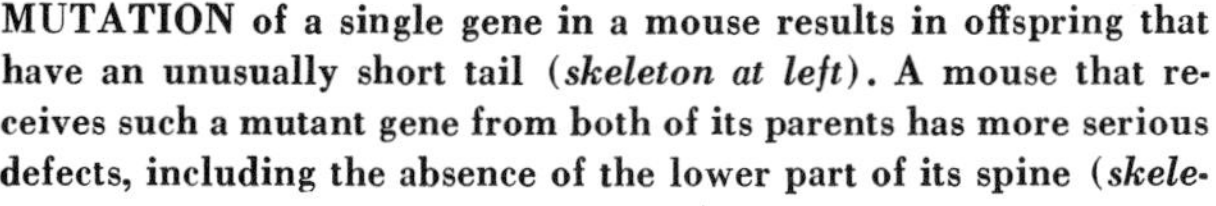

**MUTATION of a single gene in a mouse results in offspring that have an unusually short tail (*skeleton at left*). A mouse that receives such a mutant gene from both of its parents has more serious defects, including the absence of the lower part of its spine (*skeleton at right*); it dies soon after birth. These specimens were prepared in the laboratory of L. C. Dunn of Columbia University. The soft tissues were treated with strong alkali and glycerin to make them transparent; the bones were stained with a red dye.**

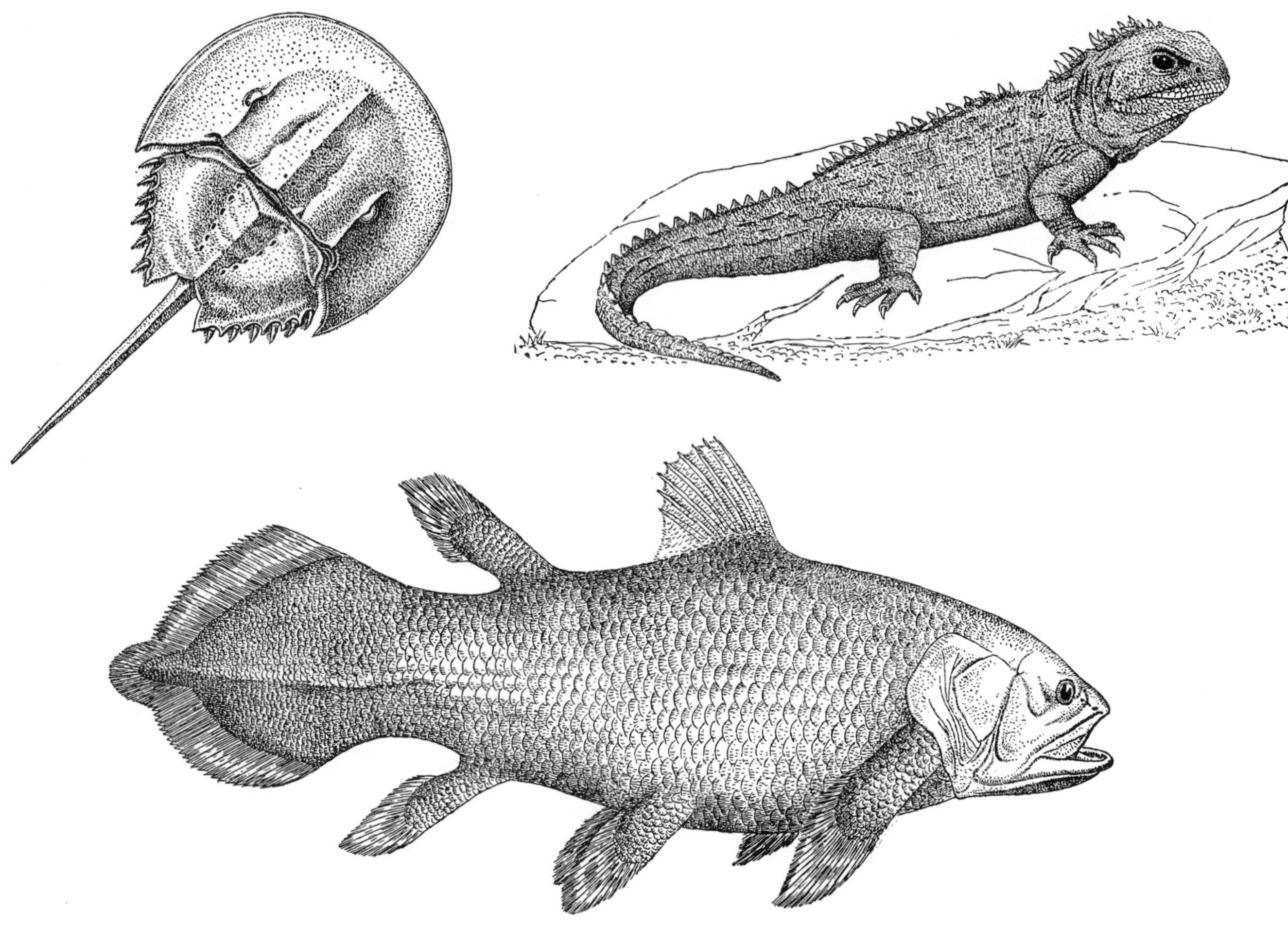

**RATE OF EVOLUTION** in certain organisms is extremely slow. Depicted in these drawings are four animals that have not changed appreciably over a period of millions or even hundreds of millions of years. At top left is the horseshoe crab; at top center, the tuatara;

new lethal mutations; man would quickly become extinct.

Actually, if the rate is measured in mutations per generation, the spontaneous-mutation rate of those few human genes whose mutation rates can be effectively measured is roughly the same as that of Drosophila genes. The rule seems to be that the absolute mutation-rate divided by the generation period is much more nearly constant from one species to another than the rate itself. The implication of this is exciting. It can only mean that the mutation rate itself is capable of modification; that it, too, is changed by natural selection. By being brought into adjustment with the life cycle, the underlying mechanism of evolution itself is undergoing evolution!

It appears, however, that the rates of radiation-induced mutations are not as readily modified. If these rates were being adjusted along with the rates of mutations from other causes, we would expect that mice would show a lesser response to radiation than do fruit flies because their life cycle is longer. The facts are otherwise; mouse genes average some 15 times as many mutations per unit of radiation as Drosophila genes.

This strongly suggests that the process whereby mutations are produced by radiation is less capable of evolutionary adjustment than other mutation-producing processes. The implication is that organisms with a long life-cycle—Sequoia trees, elephants, men—have a larger fraction of mutations due to natural radiations. It may be that the majority of mutations in a very long-lived organism are due to radiations. Possibly this sets an upper limit on the length of the life cycle.

Unfortunately the rate of radiation-induced mutations is not known for any organism with a long life-cycle. If we assume that the susceptibility of human genes to mutation by radiation is the same as that of mouse genes, we would conclude that less than 10 per cent of human mutations are due to natural radiations. But the exact fraction is quite uncertain, because of inaccurate knowledge of the rate of spontaneous human mutations. So it cannot be ruled out that even a majority of human mutations owe their origin to radiation.

Thus the over-all mutation rate is not determined by radiation to any significant extent, except possibly in some very long-lived organisms. In general, radiation does not seem to play an important role in evolution, either by supplying qualitatively unique types of mutations or by supplying quantitatively significant numbers of mutations.

There is another question: To what extent is the rate of evolution dependent on the mutation rate? Will an organism with a high mutation-rate have a correspondingly higher rate of evolution? In general, the answer is probably no.

The measurement of evolutionary rates is fraught with difficulty and doubt. There is always some uncertainty about the time-scale, despite steady improvements in paleontology and new techniques of dating by means of the decay

at bottom center, the recently discovered coelacanth; at the far right, the opossum.

of radioactive isotopes. In addition there is the difficulty of determining what a comparable rate of advance in different species is. Is the difference between a leopard and a tiger more or less than that between a field mouse and a house mouse? Or has man changed more in developing his brain than the elephant has by growing a trunk? How is it possible for one to devise a suitable scale by which such different things can be compared?

Despite these difficulties, the differences in rates of evolution in various lines of descent are so enormous as to be clear by any standard. One criterion is size. Some animals have changed greatly. The horse has grown from a fox-sized ancestor to its present size while many other animals have not changed appreciably. (The British biologist J. B. S. Haldane has suggested that the word "darwin" be used as a unit for rate of size change. One darwin is taken to be an increase by a factor *e* per million years. By this criterion the rate of evolution of tooth size in the ancestors of the horse was about 40 millidarwins.)

More difficult to quantify, but more significant, are changes in structure. Some animals have changed very little in enormous lengths of time. There are several examples of "living fossils," some of which are depicted on the preceding two pages. During the time the coelacanth remained practically unchanged whole new classes appeared: the birds and mammals, with such innovations as feathers, hair, hoofs, beaks, mammary glands, internal temperature-regulation and the ability to think conceptually. What accounts for such differences in rate?

The general picture of how evolution works is now clear. The basic raw material is the mutant gene. Among these mutants most will be deleterious, but a minority will be beneficial. These few will be retained by what Muller has called the sieve of natural selection. As the British statistician R. A. Fisher has said, natural selection is a "mechanism for generating an exceedingly high level of improbability." It is Maxwell's famous demon superimposed on the random process of mutation.

Despite the clarity and simplicity of the general idea, the details are difficult and obscure. Selection operates at many levels—between cells, between individual organisms, between families, between species. What is advantageous in the short run may be ruinous in the long run. What may be good in one environment may be bad in another. What is good this year may be bad next year. What is good for an individual may be bad for the species.

A certain amount of variability is necessary for evolution to occur. But the comparison of fossils reveals no consistent correlation between the measured variability at any one time and the rate of evolutionary change. George Gaylord Simpson, now at Harvard University, has measured contemporary representatives of low-rate groups (crocodiles, tapirs, armadillos, opossums) and high-rate ones (lizards, horses, kangaroos) and has found no tendency for the latter to be more variable. Additional evidence comes from domesticated animals and plants. G. Ledyard Stebbins, Jr., of the University of California has noted that domesticated representatives of slowly evolving plant groups produce new horticultural varieties just as readily as those from more rapidly evolving groups. Finally, the rates of "evolutionary" change in domestic animals and plants under artificial selection by man are tremendous compared with even the more rapidly evolving natural forms—perhaps thousands of times as fast. This can only mean that the genetic variability available for selection to work on is available in much greater abundance than the variability actually utilized in nature. The reasons why evolution is so slow in some groups must lie elsewhere than in insufficient genetic variability.

Instances of rapid evolution are probably the result of the opening-up of a new, unfilled ecological niche or environmental opportunity. This may be because of a change in the organism itself, as when the birds developed the power of flight with all the new possibilities this offered. It may also be due to opening-up of a new area, as when a few fortunate colonists land on a new continent.

One of the best-known examples of rapid change is to be found in the finches of the Galápagos Islands that Charles Darwin noted during his voyage on the *Beagle*. Darwin's finches are all descendants of what must have been a small number of chance migrants. They are a particularly striking group, especially in their adaptations to different feeding habits. Some have sparrow-like beaks for seed feeding; others have slender beaks and eat insects; some resemble the woodpeckers and feed on insect larvae in wood; still others feed with parrot-like beaks on fruits. The woodpecker type is of special interest. Lacking the woodpecker's long tongue it has evolved the habit of using a cactus spine or a twig as a substitute—it is the only example of a tool-using bird. In the rest of the world the finches are a relatively homogeneous group. Their tremendous diversity on these islands must be due to their isolation and the availability of new, unexploited ecological opportunities. A similar example of multiple adaptations is found in the honey creepers of the Hawaiian Islands [*see illustration on page 233*].

When paleontologists look into the ancestry of present-day animals, they find that only a minute fraction of the species present 100 million years ago is represented by descendants now. It is estimated that 98 per cent of the living vertebrate families trace their ancestry to eight of the species present in the Mesozoic Era, and that only two dozen of the tens of thousands of vertebrate species that were then present have left any descendants at all. The overwhelmingly probable future of any species is extinction. The history of

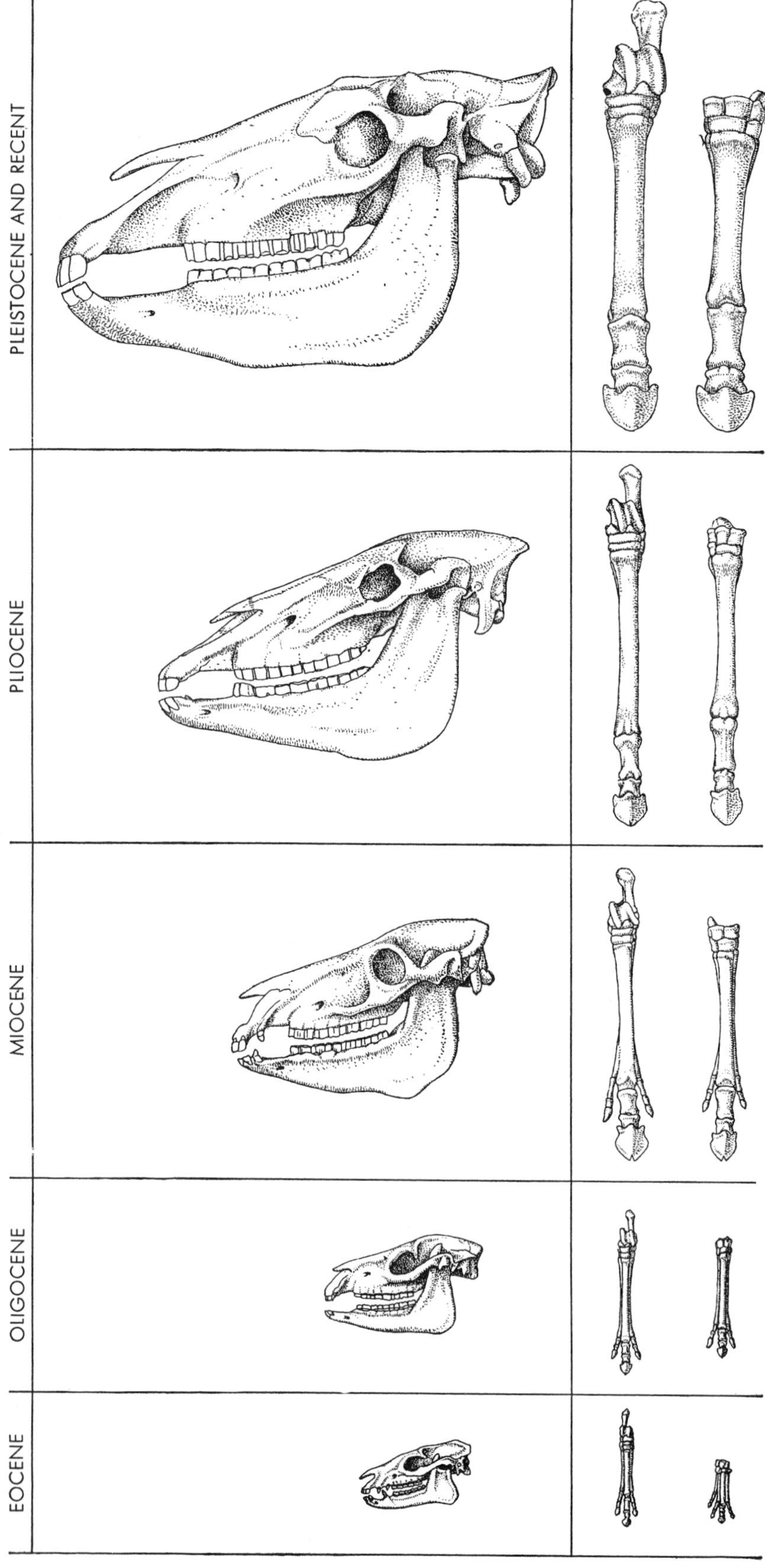


**RAPID RATE OF EVOLUTION is illustrated by the horse. Comparisons of the skulls, hindfeet and forefeet of animals of successive periods indicate the changes which led from a fox-sized ancestor in the Eocene period, 60 million years ago, to the modern horse (*top*).**

evolution is a succession of extinctions along with a tremendous expansion of a few fortunate types.

The causes of extinctions must be many. A change of environment—due to a flood, a volcanic eruption or a succession of less dramatic instances—may alter the ecological niche into which the species formerly fitted. Other animals and plants are probably the most important environmental variables: one species is part of the environment of another. There may be a more efficient predator, or a species that competes for shelter or food supply, or a new disease vector or parasite. As Theodosius Dobzhansky of Columbia University has said, "Extinction occurs either because the ecological niche disappears, or because it is wrested away by competitors."

Extinction may also arise from natural selection itself. Natural selection is a short-sighted, opportunistic process. All that matters is Darwinian fitness, that is, survival ability and reproductive capacity. A population is always in danger of becoming extinct through "criminal" genes—genes that perpetuate themselves at the expense of the rest of the population. An interesting example is the so-called SD gene, found in a wild Drosophila population near the University of Wisconsin by Yuichiro Hiraizumi. Ordinarily the segregation ratio—the ratio of two alternative genes in the progeny of any individual that carries these genes —is 1:1. In this strain the segregation is grossly distorted to something like 10:1 in favor of the SD gene; hence its name, SD, for segregation distorter. The SD gene somehow causes the chromosome opposite its own chromosome to break just prior to the formation of sperm cells. As a result the cells containing the broken chromosome usually fail to develop into functional sperm. The SD-bearing chromosome thus tends to increase itself rapidly in the population by effectively killing off its normal counterpart. Such a gene is obviously harmful to the population. But it cannot be eliminated except by the extinction of the population, or by the occurrence of a gene that is immune to the SD effect.

A similar gene exists in many mouse populations. Causing tail abnormalities, it is transmitted to much more than the usual fraction of the offspring, though in this case the detailed mechanism is not known. L. C. Dunn of Columbia University has found this gene in many wild-mouse populations. Most of these genes are highly deleterious to the mouse (a mouse needs a tail!), and some are lethal when they are borne in a double dose, so that the gene never completely takes over despite its segregational advantage. The result is an equilibrium frequency determined by the magnitude of the two opposing selective forces: the harmfulness of the gene and its segregational advantage. The population as a whole suffers, a striking illustration of the fact that natural selection does not necessarily improve the fitness of the species.

A gene causing extremely selfish, antisocial behavior—for example genes for cannibalism or social parasitism—could have similar effects in the human population. Any species must be in constant danger from the short-sightedness of the process of natural selection. Those that are still here are presumably the descendants of those that were able to avoid such pitfalls.

Aside from mutation itself, the most important evolutionary invention is sexual reproduction, which makes possible Mendelian heredity. The fact that such an elaborate mechanism exists throughout the whole living world—in viruses and bacteria, and in every major group of plants and animals—attests to its significance.

In a nonsexual population, if two potentially beneficial mutations arise in separate individuals, these individuals and their descendants can only compete with each other until one or the other type is eliminated, or until a second mutation occurs in one or the other

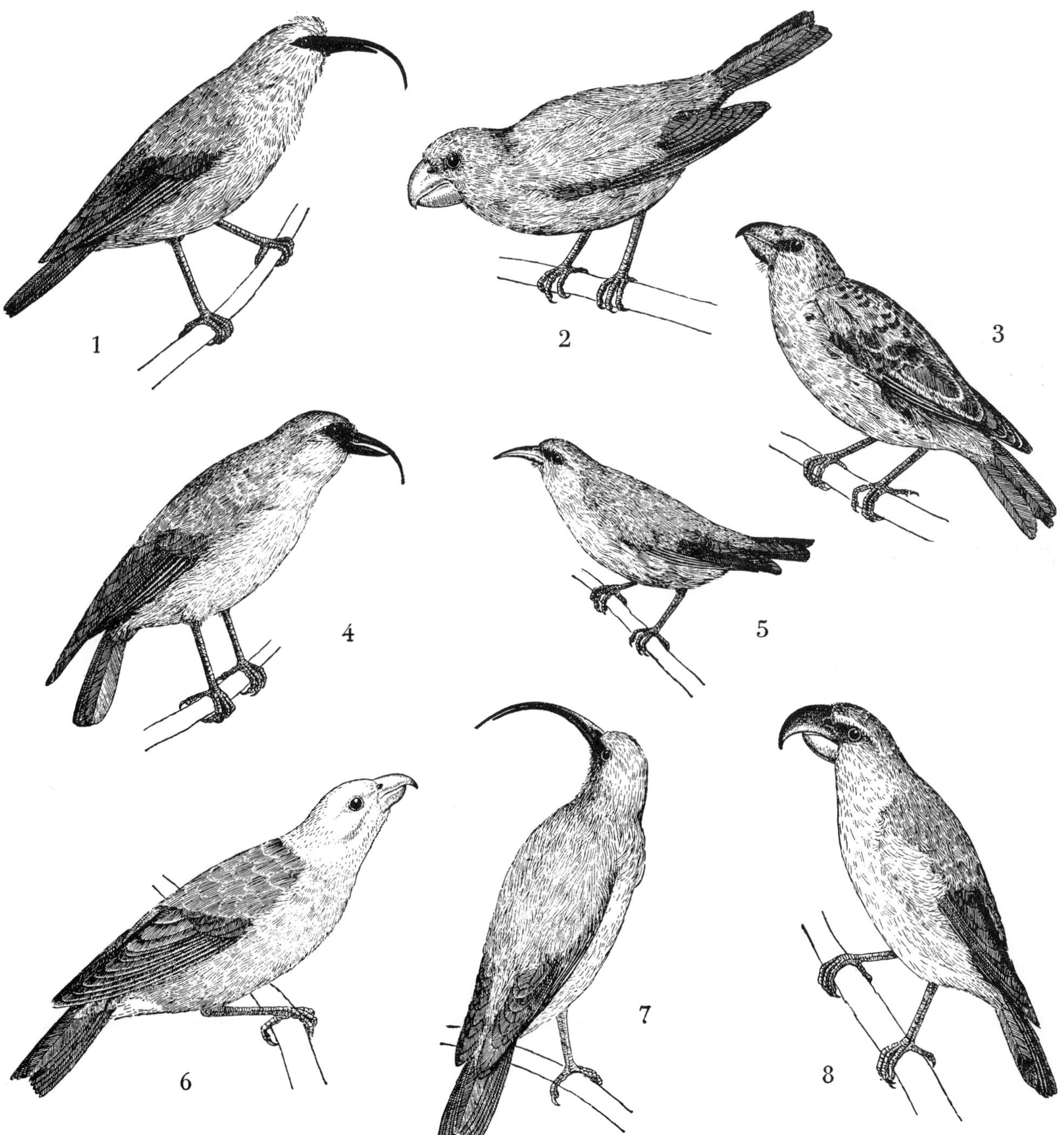


**HONEY CREEPERS of Hawaii are a striking example of rapid evolution. Dozens of varieties, adapted to different diets, arose from a common ancestor. The long bills of *Hemignathus lucidus affinis* (1), *Hemignathus wilsoni* (4) and *Hemignathus o. obscurus* (7) are used to seek insects as woodpeckers do or to suck nectar. The beaks of *Psittirostra kona* (2), *Psittirostra cantans* (3), and *Psittirostra psittacea* (6) are adapted to a diet of berries and seeds. *Loxops v. virens* (5) sucks nectar and probes for insects with its sharp bill. *Pseudonestor xanthrophrys* (8) wrenches at hard wood to get at burrowing insects. Some of these species are presently extinct.**

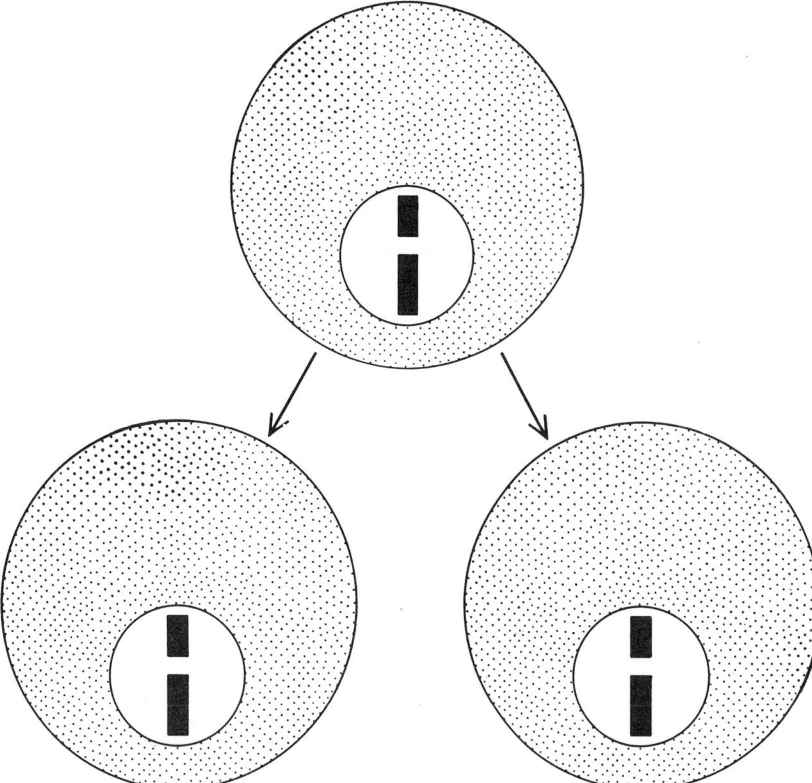

**ASEXUAL REPRODUCTION permits little variability in a population; the evolution of asexual organisms depends entirely on mutation. At top is a schematic diagram of an asexual cell. Within its nucleus (*open circle*) are chromosomes (*rectangles*). When the cell divides (*bottom*), its two daughter cells have replicas of the original chromosomes and are exactly like the parent. To simplify the diagram only two of the chromosomes of a cell are shown.**

group. A Mendelian species, with its biparental reproduction and consequent gene-scrambling, permits both mutants to be combined in the same individual. An asexual species must depend upon newly occurring mutations to provide it with variability. Sexual reproduction permits the combination and recombination of a whole series of mutants from the common pool of the species. Before they enter the pool the mutants have been to some extent pretested and the most harmful ones have already been eliminated.

Suppose that a population has 50 pairs of genes: Aa, Bb, Cc, etc. Suppose that each large-letter gene adds one unit of size. The difference between the smallest possible individual in this population (aa bb cc . . .) and the largest possible (AA BB CC . . .) is 100 units. Yet if all the genes are equally frequent, the size of 99.7 per cent of the population will be between 35 and 65 units. Only one individual in $2^{100}$ (roughly 1 followed by 30 zeros) would have 100 size units. Yet if the population is sexual, this type can be produced by selection utilizing only the genes already in the population.

This example illustrates the evolutionary power of a system that permits Mendelian gene-assortment. In an asexual system the organism would have to wait for new mutations. Most of the size-increasing mutations would probably have deleterious side effects. If the asexual organism had a mutation rate sufficient to give it the potential variability of a sexual population, it would probably become extinct from harmful mutations.

The very existence of sexual reproduction throughout the animal and plant kingdoms argues strongly for the necessity of an optimum genetic variability. This is plain, even though there is no consistent correlation between evolution rates as observed in the fossil record and the variability of the animals observed. It must be that a certain level of genetic variability is a necessary, but by no means sufficient, condition for progressive evolution.

Evolution is an exceedingly complex process, and it is obviously impossible, even in particular cases, to assign relative magnitudes to the various causal factors. Perhaps species have become extinct through mutation rates that are too high or too low, though there is no direct observational evidence for this. Probably many asexual forms have become extinct because they were unable to adapt to changes. Other forms have become overspecialized, developing structures that fit them for only a particular habitat; when the habitat disappears, they are lost. The population size and structure are also important. To be successful in evolution, by the criterion of having left descendants many generations later, a species has the best chance if it has an optimum genetic system. But among a number of potentially successful candidates, only a few will succeed, and probably the main reason is simply the good luck of having been at the right place at the right time.

Of all the natural selection that occurs, only a small fraction leads to any progressive or directional change. Most selection is devoted to maintaining the status quo, to eliminating recurrent harmful mutations, or to adjusting to transitory changes in the environment. Thus much of the theory of natural selection must be a theory of statics rather than dynamics.

The processes that are necessary for evolution demand a certain price from the population in the form of reduced fitness. This might be said to be the price that a species pays for the privilege of evolving. The process of sexual reproduction, with its Mendelian gene-shuffling in each generation, produces a number of ill-adapted gene combinations, and therefore a reduction in average fitness. This can be avoided by a nonsexual system, but at the expense of genetic variability for evolutionary change in a changing environment.

The plant breeder knows that it is easy to maintain high-yielding varieties of the potato or sugar cane because they can be propagated asexually. But then his potential improvement is limited to the varieties that he has on hand. Only by combining various germ plasms sexually can he obtain varieties better than the existing best. But the potato and cane breeder can have his cake and eat it. He gets his new variants by sexual crosses; when he gets a superior combination of genes, he carries this strain on by an asexual process so that the combination is not broken up by Mendelian assortment.

Many species appear to have sacrificed long-term survival for immediate fitness. Some are highly successful. A familiar example is the dandelion; any lawn-keeper can testify to its survival value. In dandelion reproduction, although seeds are produced by what looks superficially like the usual method, the chromosome-assorting features of the sexual process are bypassed and the seed contains exactly the same combination of genes as the plant that produced it. In the long run the dandelion will probably become extinct, but in the present environment it is highly successful.

The process of mutation also produces ill-adapted types. The result is a lowering of the average fitness of the population, the price that asexual, as well as sexual, species pay for the privilege of evolution. Intuition tells us that the effect of mutation on fitness should be proportional to the mutation rate; Haldane has shown that the reduction in fitness is, in fact, exactly equal to the mutation rate.

The environment is never constant, so any species must find itself continually having to adjust to transitory or permanent changes in the environment. The most rapid environmental changes are usually brought about by the continuing evolution of other species. A simple example was given by Darwin himself. He noted that a certain number of rabbits in every generation are killed by wolves, and that in general these will be the rabbits that run the slowest. Thus by gradual selection the running speed of rabbits would increase. At the same time the slower wolves would starve, so that this species too has a selective premium on speed. As a result both improve, but the position of one with respect to the other does not change. It is like the treadmill situation in Lewis Carroll's famous story: "It takes all the running *you* can do, to keep in the same place."

A change in the environment will cause some genes that were previously favored to become harmful, and some that were harmful to become beneficial. At first it might seem that this does not make any net difference to the species. However, when the change occurs, the previously favored genes will be common as a result of natural selection in the past. The ones that were previously deleterious will be rare. The population will not return to its original fitness until the gene numbers are adjusted by natural selection, and this has its costs.

Just how much does it cost to exchange genes in this way? Let us ask the question for a single gene-pair. If we start with a rare dominant gene present in .01 per cent of the population, it requires the equivalent of about 10 selective "genetic deaths" (*i.e.*, failure to survive or reproduce) per surviving individual in the population to substitute this gene for its predecessor. This means that if the population is to share an average of one gene substitution per generation, it must have a sufficient reproductive capacity to survive even though nine out of 10 individuals can die without offspring in each generation. The cost is considerably greater if the gene is recessive. The surprising part of this result is that the number does not depend on the selective value of the gene. As long as the difference between two alternative genes is small, the cost of replacing one by the other depends only on the initial frequency and dominance of the gene and not at all on its fitness.

If the gene is less rare, the cost is lowered. Thus the species can lower the price of keeping up with the environment by having a higher frequency of deleterious genes that are potentially favorable. One way to accomplish this is to have a higher mutation rate, but this, too, has its price. Once again there is the conflict between the short-term objective of high fitness and the longer-term objective of ability to change with

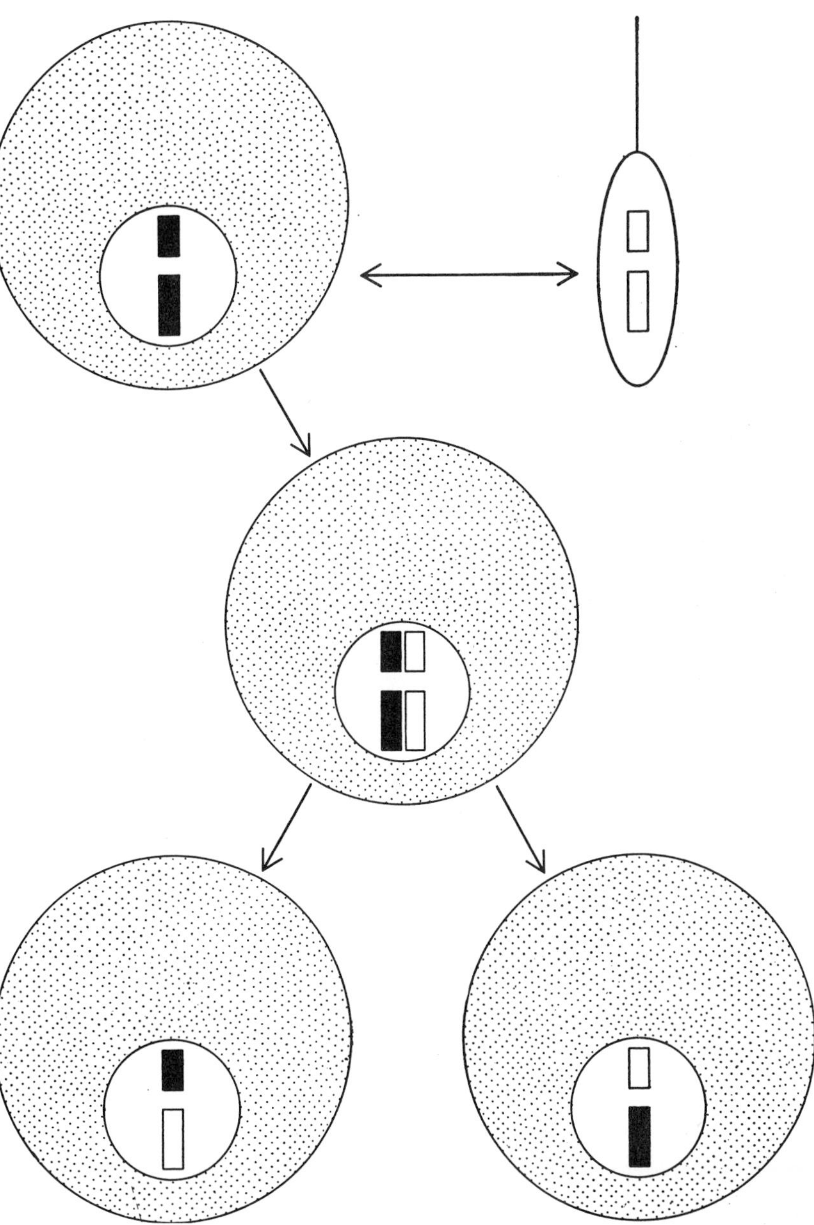

**SEXUAL REPRODUCTION, also depicted in highly schematic form, depends on the union of two cells that may be from different lines of descent. At top left is an egg cell; at top right, a sperm cell. Each contains one set of chromosomes (*rectangles*). They join to form a new individual with two sets of chromosomes in its body cells (*center*). When this organism produces its own sperm or egg cells (*bottom*), each receives a full set of chromosomes. Some, however, may be from the father and some from the mother. Evolution in sexual organisms thus does not depend entirely on new mutations, because the genes already present in the population may be combined in different ways leading to improvements in the species.**

the environment.

There are known to be genes that affect the mutation rate. Some of them are quite specific and affect only the mutation rate of a particular gene; others seem to enhance or depress the over-all rate. A gene whose effect is to lower the mutation rate certainly has selective advantage: it will cause an increase in fitness. This means that in most populations there should be a steady decrease in the mutation rate, possibly so far as to reduce the evolutionary adaptability of the organism. This is offset by selection for ability to cope with fluctuating environments.

How does man fare in this respect? From the standpoint of his biological evolution, is his mutation rate too high, is it too low or is it just right? It is not possible to say.

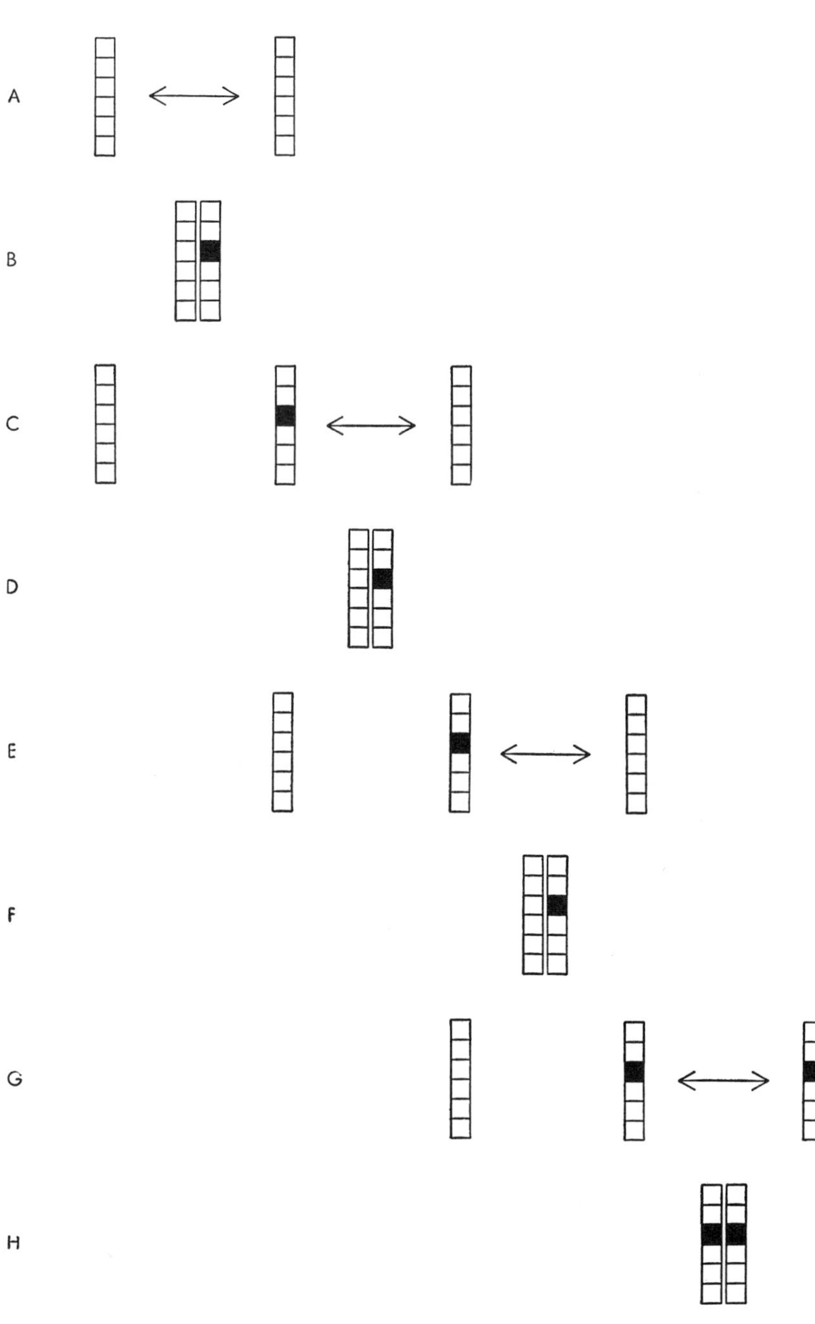


**RECESSIVE MUTATION cannot be expressed in a sexual organism unless the corresponding gene of the paired chromosome has the same defect. Thus a harmful mutation can remain latent for generations. In this diagram the bars representing chromosomes are divided into squares, each symbolizing the gene for a different character. In row A the chromosomes of normal parents are paired (*arrows*). In row B a gene of their offspring mutates (*black*). In row C the mutant gene has been transmitted to the next generation but is still latent because the other parent has supplied a normal gene. This continues (*rows E and F*) until two individuals harboring the mutant gene mate (*row G*). The full effect of the mutation, harmful or otherwise, is expressed in offspring receiving a pair of mutant genes (*row H*).**

In general one would expect the evolutionary processes to work so that the mutation rate would usually be below the optimum from the standpoint of long-term evolutionary progress, because selection to reduce the mutation load has an immediate beneficial effect. Yet selection for a mutation-rate-adjusting gene is secondary to selection for whatever direct effects this gene has on the organism, such as an effect on size or metabolic rate. So we do not have much idea about how rapidly such selection would work.

Our early simian ancestors matured much more rapidly than we do now. This means that the mutation rate would have to be lowered in order to be brought into adjustment with the lengthening of our life cycle. To whatever extent the adjustment is behind the times, the mutation rate is too high. Furthermore, if it is true (as it appears to be) that radiation-induced mutation rates are less susceptible to evolutionary adjustment than those due to other causes, then an animal with a life cycle so long that it receives considerable radiation per generation may have too many from this cause.

I think it is impossible, however, to say whether man has a mutation rate that is too high or too low from the viewpoint of evolutionary advantage. It is worth noting that man, like any sexually reproducing organism, already has a tremendous store of genetic variability available for recombination. If all mutation were to stop, the possibilities for evolution would not be appreciably altered for a tremendous length of time—perhaps thousands of generations. Consider the immense range of human variability now found, for example the difference between Mozart, Newton, da Vinci and some of our best athletes, in contrast to a moron or the genetically impaired. Haldane once said: "A selector of sufficient knowledge and power might perhaps obtain from the genes at present available in the human species a race combining an average intellect equal to that of Shakespeare with the stature of Carnera." He goes on to say: "He could not produce a race of angels. For the moral character or for the wings he would have to await or produce suitable mutations." Surely the most hopefully naive eugenist would settle for considerably less!

I would argue that from any practical standpoint the question of whether man's mutation-rate is too high or too low for long-term evolution is irrelevant. From any other standpoint the present mutation rate is certainly too high. I suspect that man's expectations for future progress depend much more on what he does to his environment than on how he changes his genes. His practice has been to change the environment to suit his genes rather than vice versa. If he should someday decide on a program of conscious selection for genetic improvement, the store of genes already in the population will probably be adequate. If by some remote chance it is not, there will be plenty of ways to increase the number of mutations at that time.

There can be little doubt that man would be better off if he had a lower mutation-rate. I would argue, in our present ignorance, that the ideal rate for the foreseeable future would be zero. The effects produced by mutations are of all sorts, and are mostly harmful. Some cause embryonic death, some severe disease, some physical abnormalities, and probably many more cause minor impairments of body function that bring an increased susceptibility to the various vicissitudes of life. Some have an immediate effect; others lie hidden to cause their harm many generations later. All in all, mutations must be responsible for a substantial fraction of human premature death, illness and misery in general.

At the present time there is not much that can be done to lower the spontaneous-mutation rate. But at least we can do everything possible to keep the rate from getting any higher as the result of human activities. This is especially true for radiation-induced mutations; if anything they are probably more deleterious and less likely to be potentially useful than those from other causes. It is also important to remember that there are very possibly things in the environment other than radiation that increase the mutation rate. Among all the new compounds to which man is exposed as a result of our complex chemical technology there may well be a number of mutagenic substances. It is important that these be discovered and treated with caution.

The general conclusion, then, is that ionizing radiation is probably not an important factor in animal and plant evolution. If it is important anywhere it is probably in those species, such as man, that have a long life span, and at least for man it is a harmful rather than a potentially beneficial factor.

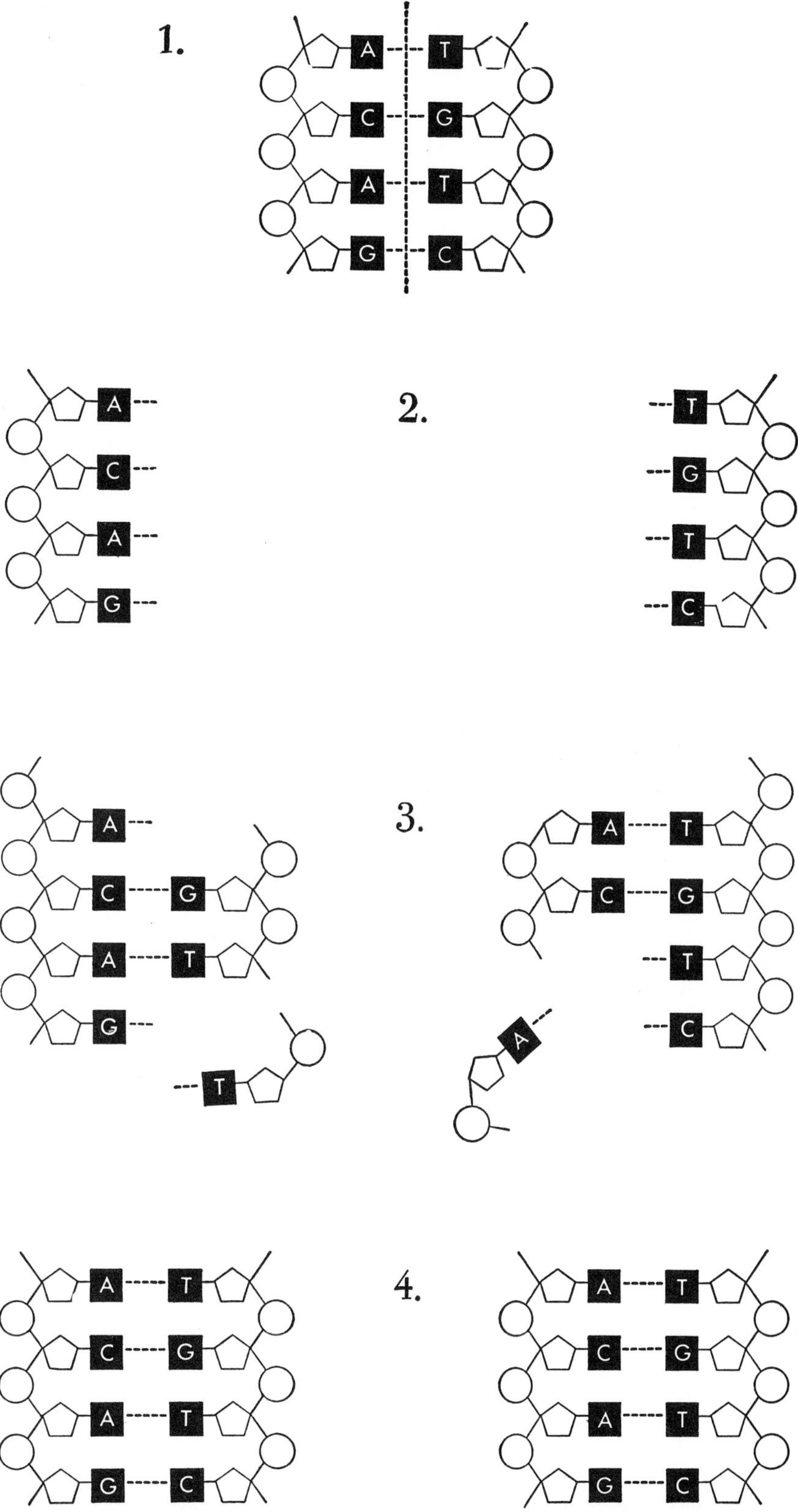


**THE GENETIC MATERIAL,** deoxyribonucleic acid (DNA), consists of a chain of sugar units (*pentagons*) and phosphate groups (*circles*) with side chains of bases: adenine (A), cytosine (C), thymine (T) and guanine (G). One possible mechanism for replication postulates that DNA normally contains two complementary strands, linked A—T and C—G (1). The chains then separate (2), and separate precursor units assemble along each chain (3). When they are completed, the double strands are identical to the original (4).

# 25

# *Radiation and Human Mutation*

H. J. MULLER
*November 1955*

The revolutionary impact on men's minds brought about by the development of ways of manipulating nuclear energy, both for destructive and for constructive purposes, is causing a public awakening in many directions: physical, biological and social. Among the biological subjects attracting wide interest is the effect of radiation upon the hereditary constitution of mankind. This article will consider the part which may be played by radiation in altering man's biological nature, and also the no less important effects that may be produced on our descendants by certain other pertinent influences under modern civilization.

At the cost of being too elementary for readers who are already well informed on biological matters it must first be explained that each cell of the body contains a great collection—10,000 or more—of diverse hereditary units, called genes, which are strung together in a single-file arrangement to form the tiny threads, visible under the microscope, called chromosomes. It is by the interactions of the chemical products of these genes that the composition and structure of every living thing is determined. Before any cell divides, each of its genes reproduces itself exactly or, as we say, duplicates itself. Thus each chromosome thread becomes two, both structurally identical. Then when the cell divides, each of the two resulting cells has chromosomes exactly alike. In this way the descendant cells formed by successive divisions, and, finally, the individuals of subsequent generations derived from such cells, tend to inherit genes like those originally present.

However, the genes are subject to rare chemical accidents, called gene mutations. Mutation usually strikes but one gene at a time. A gene changed by mutation thereafter produces daughter genes having the mutant composition. Thus descendants arise that have some abnormal characteristic. Since each gene is capable of mutating in numerous more or less different ways, the mutant characteristics are of many thousands of diverse kinds, chemically at least.

Very rarely a mutant gene happens to have an advantageous effect. This allows the descendants who inherit it to multiply more than other individuals in the population, until finally individuals with that mutant gene become so numerous as to establish the new type as the normal type, replacing the old. This process, continued step after step, constitutes evolution.

But in more than 99 per cent of cases the mutation of a gene produces some kind of harmful effect, some disturbance of function. This disturbance is sometimes enough to kill with certainty any individual who has inherited a mutant gene of the same kind from both his parents. Such a mutant gene is called a lethal. More often the effect is not fully lethal but only somewhat detrimental, giving rise to some risk of premature death or failure to reproduce.

Now in the great majority of cases an individual who receives a mutant gene from one of his parents receives from the other parent a corresponding gene that is "normal." He is said to be heterozygous, in contrast to the homozygous individual who receives like genes from both parents. In a heterozygous individual the normal gene is usually dominant, the mutant gene recessive. That is, the normal gene usually has much more influence than the mutant gene in determining the characteristics of the individual. However, exact studies show that the mutant gene is seldom completely recessive. It does usually have some slight detrimental effect on the heterozygous individual, subjecting him to some risk of premature death or failure to reproduce or, as we may term it, a risk of genetic extinction. This risk is commonly of the order of a few per cent, down to a fraction of 1 per cent.

If a mutant gene causes an average risk of extinction of, for instance, 5 per cent, that means there is one chance in 20 that an individual possessing it will die without passing on the same gene to offspring. Thus such a mutant gene will, on the average, pass down through about 20 generations before the line of descent containing it is extinguished. It is therefore said that the "persistence" of that particular gene is 20 generations. There is some reason to estimate that the average persistence of mutant genes in general may be something like 40 generations, although there are vast differences between genes in this respect.

## The Human Store of Mutations

Observations on the frequency of certain mutant characteristics in man, supported by recent more exact observations on mice by W. L. Russell at Oak Ridge, indicate that, on the average, the chance of any given human gene or chromosome region undergoing a mutation of a given type is one in 50,000 to

100,000 per generation. Moreover, studies on the fruit fly *Drosophila* show that for every mutation of a given type there are at least 10,000 times as many other mutations occurring. Now since it is very likely that man is at least as complicated genetically as Drosophila, we must multiply our figure of 1/100,000, representing our more conservative estimate of the frequency of a given type of mutation, by at least 10,000 to obtain a minimum estimate of the total number of mutations arising in each generation among human germ cells. Thus we find that at least every tenth egg or sperm has a newly arisen mutant gene. Taking the less conservative estimate of 1/50,000 for the frequency of a given type of mutation, our figure would become two in 10.

Every person, however, arises from both an egg and a sperm and therefore contains twice as many newly arisen mutant genes as the mature germ cells do, so the figure becomes two to four in 10. When we say that the per capita frequency of newly arisen mutations is .2 to .4, we mean that there are, among every 10 of us, some two to four mutant genes which arose among the germ cells of our parents. This is the frequency of so-called spontaneous mutation, which occurs even without exposure to radiation or other special treatment.

Far more frequent than the mutant genes that have newly arisen are those that have been handed down from earlier generations and have not yet been eliminated from the population by causing death or failure to reproduce. The average per capita frequency of all the mutant genes present, new and old, is calculated by multiplying the frequency of newly arisen mutations by the persistence figure.

The greatly simplified diagram on page 8, in which we suppose the frequency of new mutations in each generation to be .2 and the persistence to be only four generations, shows why this relation holds. We start with 10 individuals. Let us suppose that in this first generation eight persons contain no mutant genes while each of the other two has one newly arisen mutant gene. In the second generation these two mutant genes are passed along and two new ones are added to the group, making the total frequency 4/10. By the fourth generation the frequency is 8/10. After that the frequency remains constant because each mutant gene lasts only four generations and is assumed to be replaced by a normal gene.

Of course in any actual case neither the multiplication nor the distribution of mutant genes among individuals is as regular as in this simplified illustration, but the general principle holds. However, as previously mentioned, the persistence of mutant genes is of the order of 40 generations, instead of only four. Thus the equilibrium frequency becomes not 8/10 but 8. In other words, each person would on the average contain, by this reckoning, an accumulation of about eight detrimental mutant genes.

It happens that this very rough, "conservative" estimate, made by the present writer six years ago, agrees well with the estimate arrived at a few months ago by Herman Slatis, in a study carried out in Montreal by a more direct method. His method was based on the frequency with which homozygous abnormalities appeared among the children of marriages between cousins.

The eight mutations estimated above,

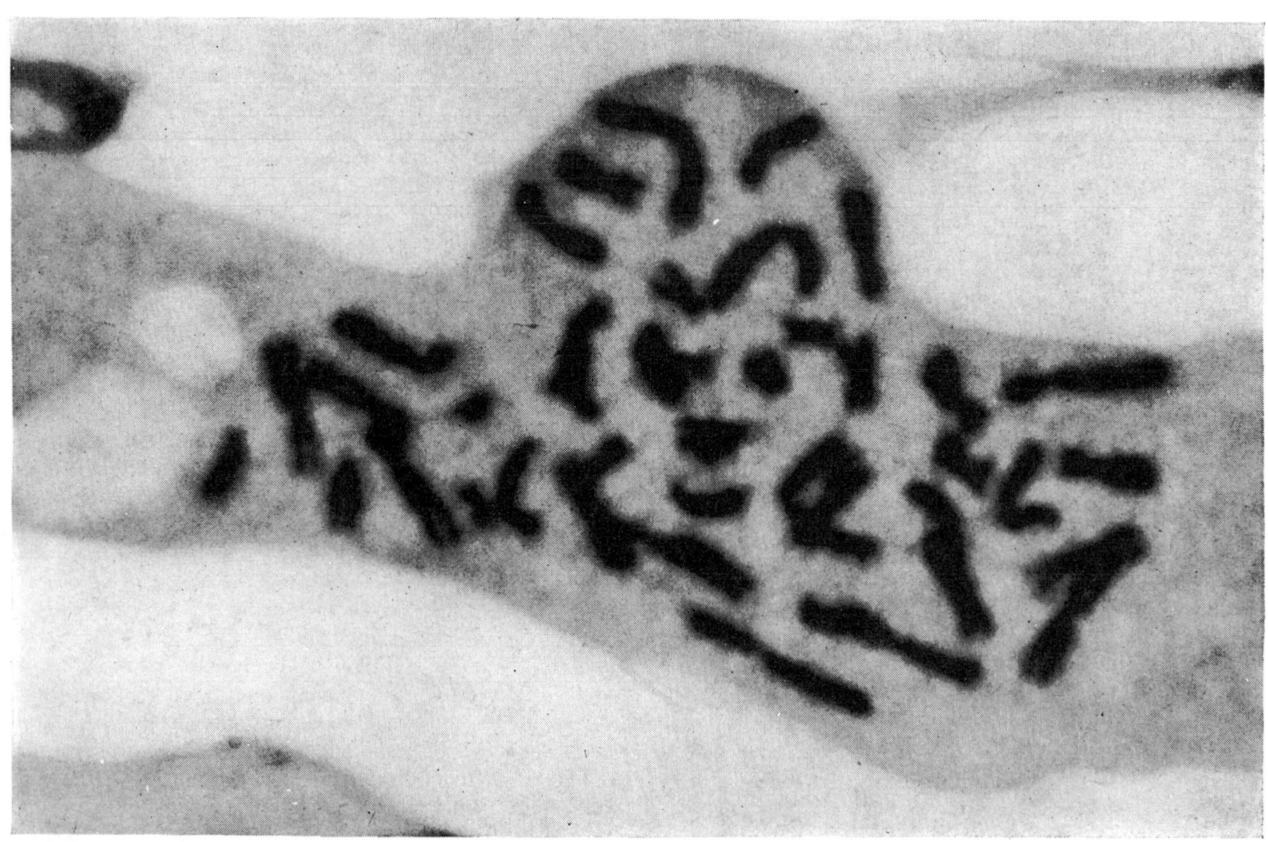

**HUMAN CHROMOSOMES, which are much more difficult to photograph than those of fruit flies, are clearly revealed as dark bodies in this photomicrograph by T. C. Hsu of the M. D. Anderson Hospital and Tumor Institute in Houston, Tex. The chromosomes, which are enlarged approximately 3,000 diameters, are in a human spleen cell. The cell was grown in a laboratory culture after spleen tissue had been removed from a four-month-old fetus. Human body cells normally contain 48 chromosomes; human germ cells, 24.**

it should be understood, do not include most of the multitude of more or less superficial differences, sometimes conspicuous but very minor in the conduct of life, whereby, in the main, we recognize one another. The latter mutations probably arise relatively seldom yet become inordinately numerous because of their very high persistence. Thus the value that we arrive at for the frequency of mutant genes depends very much upon just where the line is drawn in excluding this mutational "froth." As yet little attention has been given to this point. The number eight, at any rate, includes only mutant genes which when homozygous give rather definite abnormalities. In the great majority of cases these genes are only heterozygous and usually are but slightly expressed. Yet they do become enough expressed to cause, in each individual, his distinctive pattern of functional weaknesses, depending upon which of these mutant genes he contains and what his environment has been. The influence of environment on gene expression is often important.

Even the genes that give only a trace of detrimental effect, or are detrimental only when homozygous, play an important role, because of their high persistence and consequent high frequency. When conditions change, certain combinations of these genes may occasionally happen to be more advantageous than the type previously prevailing, and so tend to become established.

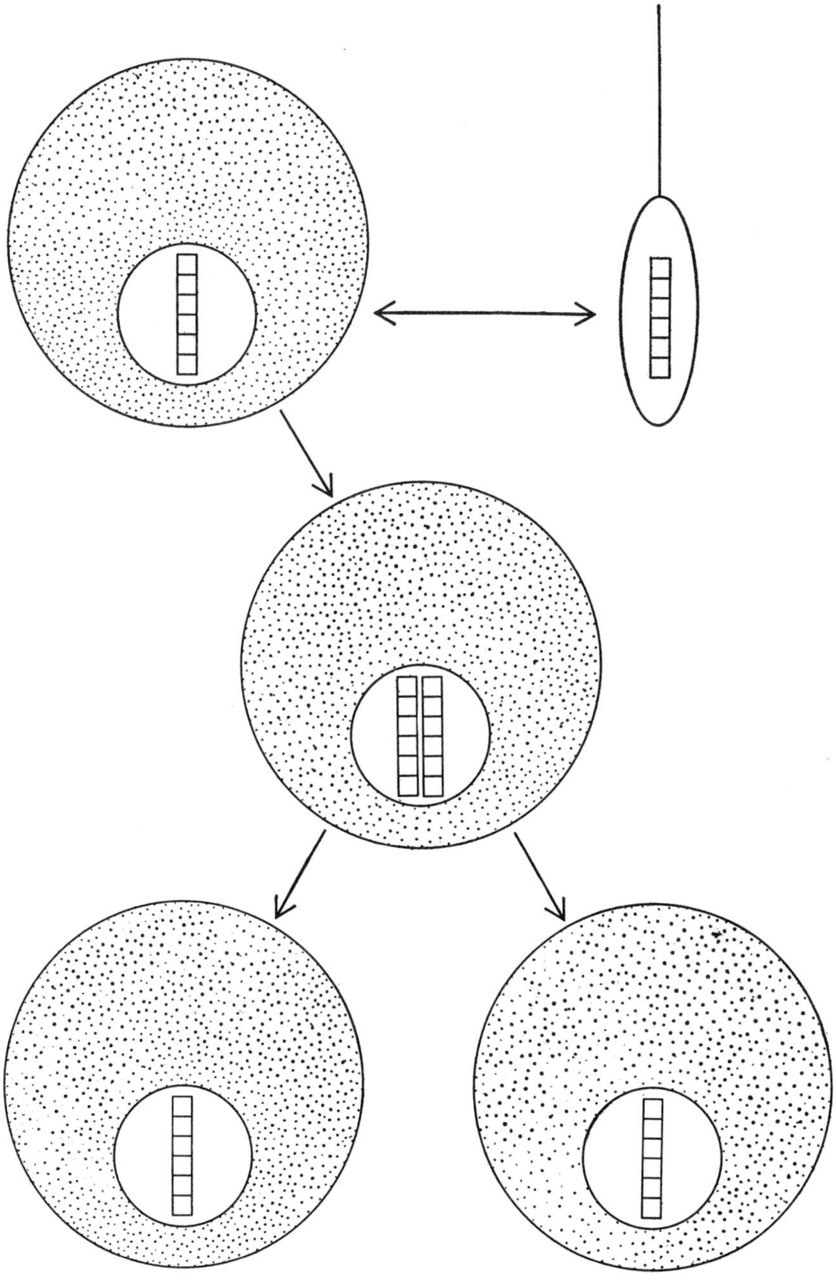

FUNDAMENTAL PROCESS of heredity is depicted in highly schematic form. At the upper left is an egg cell; at the upper right, a sperm cell. Each contains a single chromosome bearing only six genes (*square segments of chromosome*). The chromosomes are paired in the fertilized egg (*center*), resulting in an organism with a complete set of genes from each parent. When the organism produces its own germ cells (*bottom*), the chromosomes are separated, leaving one set of genes to combine with those of a mate in the new generation.

## The Effects of Mutant Genes

In general each detrimental mutant gene gives rise to a succession of more or less slight impairments in the generations that carry it. Even if only slightly detrimental, it must finally result in extinction. Moreover, even though an individual suffers less from a slightly detrimental gene than from a markedly detrimental or lethal one, nevertheless the slightly detrimental gene, being passed down to a number of individuals which is inversely proportional to the amount of harm done to each individual, occasions a total amount of damage comparable to that produced by the very detrimental gene. Although each of us may be handicapped very little by any one of our detrimental genes, the sum of all of them causes a noticeable amount of disability, which is usually felt more as we grow older.

The frequency of mutant genes levels off at an equilibrium only when conditions for both mutation frequency and gene elimination have remained stable (or have at any rate fluctuated about a given average) for many generations. During such a period about as many mutations must be eliminated as are arising per generation. If, however, the mutation rate or the average persistence or both changed significantly because of increased radiation or a change in environmental conditions which made mutant genes more or less harmful than previously, then the frequency would move toward a new level. But it would be a long time before the new equilibrium was reached. If the average persistence of mutant genes was 40 generations, the new equilibrium would still be very incompletely attained after 1,000 years.

## The Effects of Radiation

We may next consider how a given dose of ionizing radiation would affect the population. Such radiation, when absorbed by the germ cells of animals, usually induces mutations which are similar to the spontaneous ones. They are induced at a frequency which is proportional to the total amount of the radiation received, regardless of the duration or time-distribution of the exposure. Russell's data on mice—the nearest experimental object to man that has been used in such studies—show that it would take about 40 roentgen units of radiation to produce mutations in them at a frequency equal to their spontaneous frequency. If the frequency of spontaneous mutations is two new mutations per generation among each 10 individuals, a dose of 40 roentgens, by adding two induced mutations to the two spontaneous ones, would result in a total mutation frequency of four new mutations per generation among 10 individuals. Now assuming the total mutant gene content is eight per individual to begin with, the radiation dose would raise this figure from 8 to 8.2, an increase of only 2.5 per cent. This effect on the population would ordinarily be too small to produce noticeable changes in important characteristics. One must also bear in mind that an actual mean change in a population may be masked by the great genetic differences among individuals and by differences in environment between two groups that are to be compared. These considerations explain why even Hiroshima survivors who had been relatively close to the blast, and who may have absorbed several hundred roentgens, showed no statistically significant increase in genetic defects among their children. However, offspring of U. S. radiologists who (judging by the incidence of leukemia) probably were exposed during their work over a long period to about as much radiation as these Hiroshima survivors, do show a statistically significant increase in congenital abnormalities, as compared with the offspring of other medical specialists. This was recently established in a study by Stanley Macht and Philip Lawrence.

The toll taken by mutant genes upon the descendants of exposed individuals is spread out over more than a thousand years—40 generations. It is too small to be demonstrable in any one generation of descendants. In the first generation of offspring of a population exposed to 40 roentgens, where the induced mutation rate is .2 per individual and the average risk of extinction for any given mutant gene is 1/40, the frequency of extinctions occasioned by these mutant genes would be .2/40 or 1/200. This would mean, for example, that in a total population of 100 million some 500,000 persons would die prematurely or fail to reproduce as a result of having mutant genes that had been induced in their parents by the exposure. Moreover, a much larger number would be damaged to a lesser extent. The total of induced extinctions in all generations subsequent to the exposure would be .2 times 100 million, or 20 million, and the disabilities short of extinction would be numbered in the hundreds of millions. And yet the amount of genetic deterioration in the population due to the exposure would be small in a relative sense, for the induced mutations would have added only 2.5 per cent to the load of mutant genes already accumulated by spontaneous mutation.

The situation would be very different if the doubling of the mutation frequency by irradiation in each generation were continued for many generations, say for 1,500 years. For after 1,500 years the mutant gene content would have been raised from eight to nearly 16 per individual. Along with this doubled frequency of detrimental genes there would of course be a corresponding increase in the amount of disability and in the frequency of genetically occasioned extinction of individuals.

It is possible that all this would be ruinous to a modern human population, even though in most kinds of animals it could probably be tolerated. For, in the first place, human beings multiply at a low rate which does not allow nearly as rapid replacement of mutant genes by normal ones as can occur in the great majority of species. Secondly, under modern conditions the rate of human multiplication is reduced much below its potential. Thirdly, the pressure of natural selection toward eliminating detrimental genes is greatly diminished, under present conditions at least, through the artificial saving of lives. Under these circumstances a long-continued doubling of the mutation frequency might eventually mean, if the situation persisted, total extinction of the population. However, we do not now have nearly enough knowledge of the strength of the various factors here involved to pass a quantitative judgment as to how high the critical mutation frequency would have to be, and how low the levels of multiplication and selection, to bring about this denouement. We can only see that danger lies in this direction, and call for further study of the whole matter.

## Bomb Effects

In the light of the facts reviewed, we are prepared to come to some conclusions concerning the problem of the genetic effects of nuclear explosions. Let us start with the test explosions. J. Rotblat of London has estimated that the tests of the past year approximately doubled the background radiation for the year, in regions of the earth remote from the explosions. In the U. S. they raised the background radiation from about .1 to about .2 of a roentgen for the year. The natural background radiation of about .1 roentgen per year causes, we estimate, about 5 per cent of the spontaneous mutations in man. Hence a doubling of it would cause a rise of the same amount in the occurrence of new mutant genes. Although this influence, if continued over a generation, would induce an enormous number of mutations—of the order of 20 million in the world population of some two billion—nevertheless the effect, in relation to the already accumulated store of detrimental mutations, would be comparatively small. It would raise the per capita content of mutant genes at most by only a few tenths of 1 per cent.

Much more serious genetic consequences would follow from atomic warfare itself, in the regions subject to the fall-outs of the first few days. As for regions remote from the explosions (say the Southern Hemisphere), Rotblat and Ralph Lapp have reckoned that a hydrogen-uranium bomb like those tested in the Pacific would deliver an effective dose of about .04 roentgen throughout the whole period of radioactive disintegration. Thus 1,500 such bombs would deliver about 60 roentgens—an amount which might somewhat more than double the mutation frequency for one generation. Since there would be relatively little residual radioactivity in these remote regions after the passage of a generation, and since it is scarcely conceivable that such bombing would be repeated in many successive generations, it seems probable that most of the world's inhabitants below about the Tropic of Cancer would escape serious genetic damage. However, they would be likely in the course of centuries to become contaminated by extensive interbreeding with the survivors of the heavy irradiations in the North. For although an attempt might be made to establish a genetic quarantine, this would, for psychological reasons, be unlikely to be maintained with sufficient strictness for the hundreds of years required for the success of such a program.

In the regions subject to the more immediate fall-outs, pattern bombing could have resulted in practically all populous areas receiving several thousand roentgens of gamma radiation. Even persons well protected in shelters during the first week might subsequently be subjected to a protracted exposure

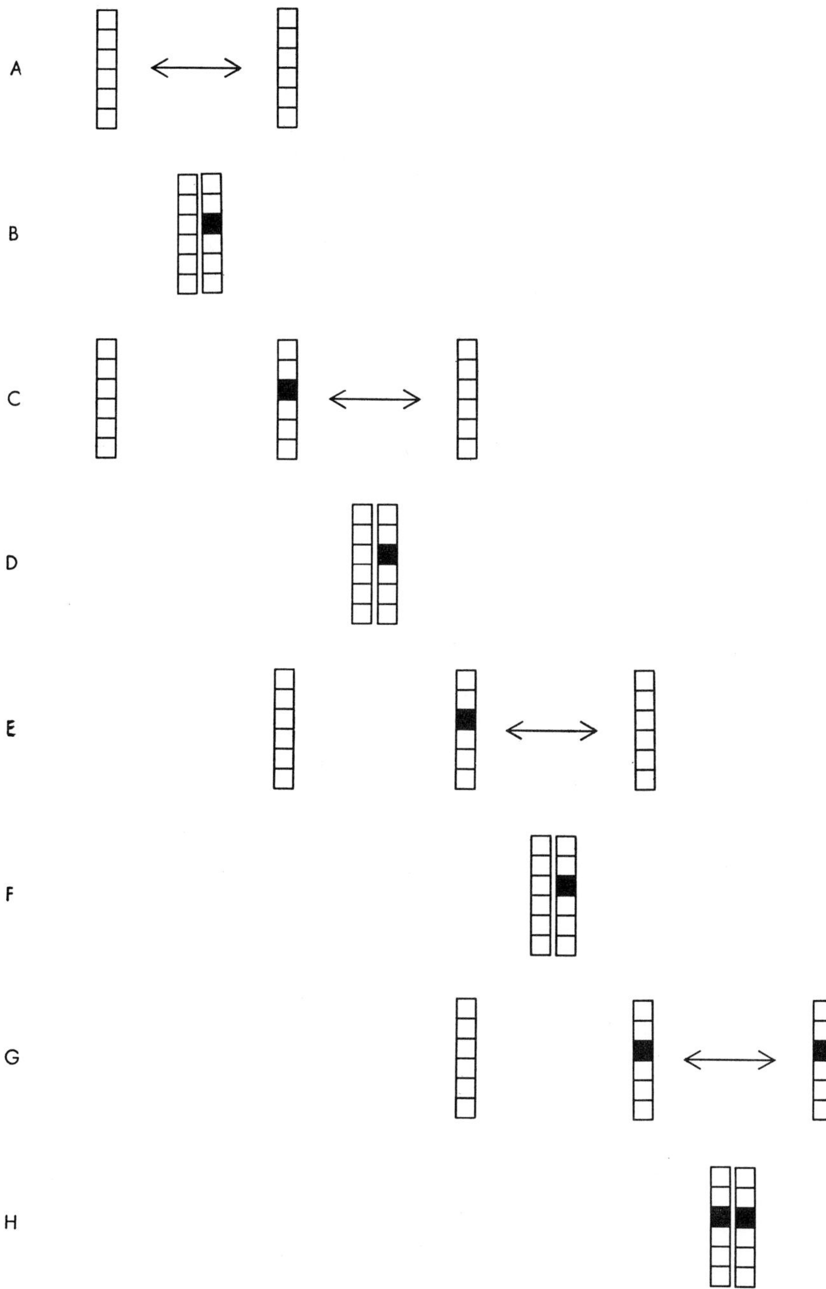


**HARMFUL RECESSIVE MUTATION may persist for generations before it is fully expressed. This diagram is based on the schematic chromosomes depicted on page 241. In row A the chromosomes of two parents are paired (*arrow*). In row B a gene of their offspring mutates (*black*). In row C the mutant gene has been transmitted to the next generation. If the mutant gene is recessive (*i.e.*, if the corresponding gene of the paired chromosome has a dominant effect), it is masked. Here a new set of genes is introduced (*second arrow*) from another line of descent. In rows D, E and F the mutant gene is passed along. In row G a mutant gene of the same character is introduced from still another line of descent. In offspring of this union (*row H*), the harmful effect of the paired recessive genes is expressed.**

adding up to some 2,500 roentgens. Moreover, this estimate fails to take into account the soft radiation from inhaled and ingested materials which under some circumstances, as yet insufficiently dealt with in open publications, may become concentrated in the air, water or food and find fairly permanent lodgment in the body. Now although some 400 roentgens is the semilethal dose (that killing half its recipients) if received within a short time, a considerably higher dose can be tolerated if spread out over a long period. Thus it is quite possible that a large proportion of those who survive and reproduce will have received a dose of some 1,000 to 1,500 roentgens or even more. This would cause a 12-fold to 40-fold rise in the mutation frequency of that generation.

Such an increase, assuming that the population was already loaded with an accumulation of mutant genes amounting to 40 times the annual spontaneous mutation rate, would at one step cause a 30 to 100 per cent increase in the mutant gene content. In fact, the detrimental effect would be considerably greater than that indicated by these figures, because the newly added mutant genes, unlike those being "stored" at an equilibrium level, would not yet have been subjected to any selective elimination in favor of the less detrimental ones. It can be estimated that this circumstance might cause the total detrimental influence to be twice as strong for each new mutant gene, on the average, as for each old one. Therefore the increase in detrimental effect would be between 60 and 200 per cent.

Owing to these circumstances, an effect would be produced similar to that of a doubled accumulation of genes, such as we saw would ensue from a doubled mutation frequency after about a thousand years of repetition. Thus offspring of the fall-out survivors might have genetic ills twice or even three times as onerous as ours.

The worst of the matter is that the effects of this enormous sudden increase in the genetic load would by no means be confined to just one or two generations. Here is where the inertia of mutant-gene content, which in the case of a moderately increased mutation frequency works to spread out and thus to soften its impact, now shows the reverse side of its nature: its extreme prolongation of the effect. That is, the gene content is difficult to raise, but once raised, it is equally resistant to being reduced.

Supposing the average content of markedly detrimental genes per person to be only doubled, from 8 to 16, more than 50 per cent of the population would come to contain a number of these mutant genes (16 or more) that was as great or greater than that now present in the most afflicted 1 per cent, if the distribution followed the Poisson principle. When we consider the extent to which we are already troubled with ills of partly or wholly genetic origin, especially as we grow older, the prospect of so great an increase in them in the future is far from reassuring.

It is fortunate, in the long run, that sterility and death ensue when the accumulated dose has risen beyond about 1,000 to 3,000 roentgens. For the frequency of mutations received by the descendants of an exposed population is thereby prevented from rising much beyond the amount which we have here considered. This being the case, it is probable that the offspring of the survivors, even though considerably weakened genetically, would nevertheless—some of them—be able to struggle through and reestablish a population which could continue to survive.

Yet, supposing the population were able to re-establish its stability of numbers within, say, a couple of centuries, what would be the toll among the later generations in terms of premature death and failure to reproduce? If 40 roentgens produce .2 new mutant genes per person, then 1,000 roentgens must on the average add five mutant genes to each person's composition. All of these five genes must ultimately lead to genetic extinction. But if, to be conservative, we suppose that two to three genes, on the average, work together in causing extinction, we reach the conclusion that, in a population whose numbers remain stable after the first generation following the exposure, there will ultimately be about two cases of premature death or failure to reproduce for each first-generation offspring of an exposed individual.

If, however, the descendants multiply and re-establish the original population size in a century or two, then the number of extinctions will be multiplied also. Over the long run the number of "genetic deaths" will be approximately twice as large, altogether, as the population total in any one generation. The future extinctions would in this situation be several times as numerous as the deaths that had occurred in the directly exposed generation.

Even though it is probable that mankind would revive ultimately after exposure to radiation, large or small, let us not make the all-too-common mistake of gauging whether or not such an exposure is genetically "permissible" merely by the criterion of whether or not humanity would be completely ruined by it. The instigation of nuclear war, or indeed of any other form of war, can hardly find a valid defense in the proposition, even though true, that it will probably not wipe out the whole of mankind.

## Radiation from Other Sources

It is by the standard of whether individuals are harmed, rather than whether the human race will be wiped out, that we should judge the propriety of everyday practices that may affect the human genetic constitution. We have to consider, for one thing, the amount of radiation which the population should be allowed to receive as a result of the peacetime uses of atomic energy.

How much effort, inconvenience and money are we willing to expend in the avoidance of one genetic extinction, one frustrated life and other partially frustrated lives, not to be beheld by us? Shall we accept the present official view that the "permissible" dose for industrially exposed personnel may be as high as .3 roentgen per week, that is, 300 roentgens in 20 years—a dose which would cause such a worker to transmit somewhere between .5 and 1.5 mutations per offspring conceived after that time?

Exactly the same questions apply in medical practice. A U. S. Public Health survey conducted three years ago showed that at that time Americans were receiving a skin dose of radiation averaging about two roentgens per year per person from diagnostic examinations alone. Of course only a small part of this could have reached the germ cells, but if the relative frequencies of the different types and amounts of exposure were similar to those enumerated in studies recently carried out in British hospitals, we may calculate that the total germ-cell dose was about a thirtieth of the total skin dose, namely, about .06 roentgen per person per year. This is about 12 times as much as the dose that had previously been estimated to reach the reproductive organs of the general population (not the hospital population) in England. However, the U. S. is notoriously riding "the wave of the future" in regard to the employment of X-rays; it is still expanding their use rapidly, while other countries are following as fast as they can.

Now this dose of .06 roentgen per year, the only estimate for the U. S. that we have, is of the same order of magnitude (perhaps twice as large) as the annual dose received in the U. S. over the past four years from all the nuclear test explosions. It seems rather disproportionate that so much furor should be raised about the genetic effects of the latter and so little about the former.

The writer's personal conviction is that, at the present stage of international relations or at least at the stage of the past several years, the tests have been fully justified as warnings and defensive measures against totalitarianism, despite the future sacrifices that they inexorably bring in their train, although it is to be hoped that this stage is now about to become obsolete. On the other hand, it seems impossible to find justification for the large doses to which the germ cells of patients are exposed in medical practice. It would involve comparatively little care or expense to shield the gonads or take other precautions to reduce the dose being received by the reproductive organs and other parts not being examined. And the deliberate irradiation of the ovaries to induce ovulation, and of the testes to provide an admittedly temporary means of avoiding pregnancies, should be regarded as malpractice.

We must remember that nuclear weapons tests and possibly nuclear warfare may be dangers of our own turbulent times only, whereas physicians will always be with us. It is easier and better to establish salutary policies with regard to any given practice early than late in its development. If we continue neglectful of the genetic damage from medical irradiations, the dose received by the germ cells will tend to creep higher and higher. It will also be joined by a rising dose from industrial uses of radioactivity. For the industrial and administrative powers-that-be will tend to take their cue in such matters from the physicians, not from the biologists, even as they do today. It should be our generation's concern to take note of this situation and to make further efforts to start off the expected age of radiation, if there is to be one, in a rational way as regards protection from this insidious agent, so

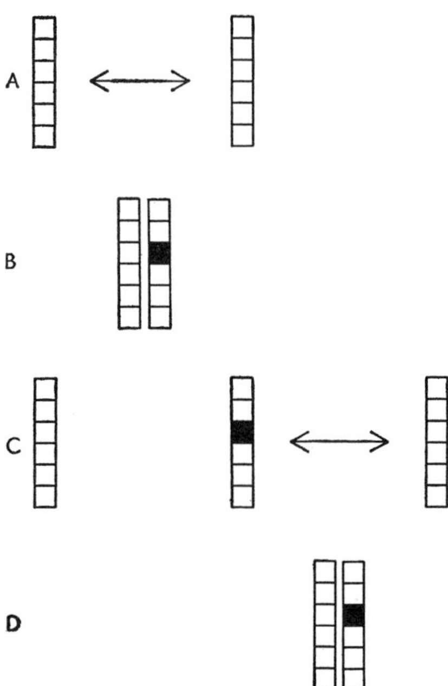


**HARMFUL DOMINANT MUTATION, as opposed to a recessive one, is quickly eliminated. Here the mutation (*black*) occurs at the same stage as in the diagram on page 243. It occurs in a germ cell, so its effect is not expressed in that generation. When the chromosome bearing the mutant gene is paired with another, however, the mutation is expressed in the offspring of the union.**

as to avoid that permanent, significant raising of the mutation frequency which in the course of ages could do even more genetic damage than a nuclear war.

## Chemical Agents

Radiation is by no means the only agent that is capable of drastically increasing the frequency of mutation. Diverse organic substances, such as the mustard gas group, some peroxides, epoxides, triazene, carbamates, ethyl sulfate, formaldehyde and so forth, can raise the mutation frequency as much as radiation.

The important practical question is: to what extent may man be unwittingly raising his mutation frequency by the ingestion or inhalation of such substances, or of substances which, after entering the body, may induce or result in the formation of mutagens that penetrate to the genes of the germ cells? As yet far too little is known of the extent to which our genes, under modern conditions of exposure to unusual chemicals, are being subjected to such mutagenic influences.

A surprising recent finding by Aaron Novick and Leo Szilard at the University of Chicago is that in coli bacteria the feeding of ordinary purines normal to the organism more than doubled the spontaneous mutation frequency, while methylated purines, and more especially caffeine (as had been found by other workers), had a much stronger mutagenic effect. Thus far, however, caffeine has not proved mutagenic in fruit flies, although it is possible that it is destroyed in their gut. In Novick and Szilard's work compounds of purines with ribose (*e.g.* adenosine) counteracted the mutagenic effect of the purines. Furthermore, adenosine and guanosine even acted as "antimutagens" when there had been no addition of purines to the nutrient medium, as though a considerable part, about a third, of the spontaneous mutations were being caused by the purines naturally present in the cells. This work, then, indicates both the imminence of the mutagenic risks to which we may be subject and also the fact that means of controlling these risks and, to some extent, even of controlling the processes of spontaneous mutation themselves, are already coming into view.

Other large differences in the frequency of so-called spontaneous mutations were found in my studies in 1946 on the mutation frequencies characterizing different stages in the germ-cell cycle of the fruit fly. Moreover, J. B. S. Haldane, dealing with data of others,

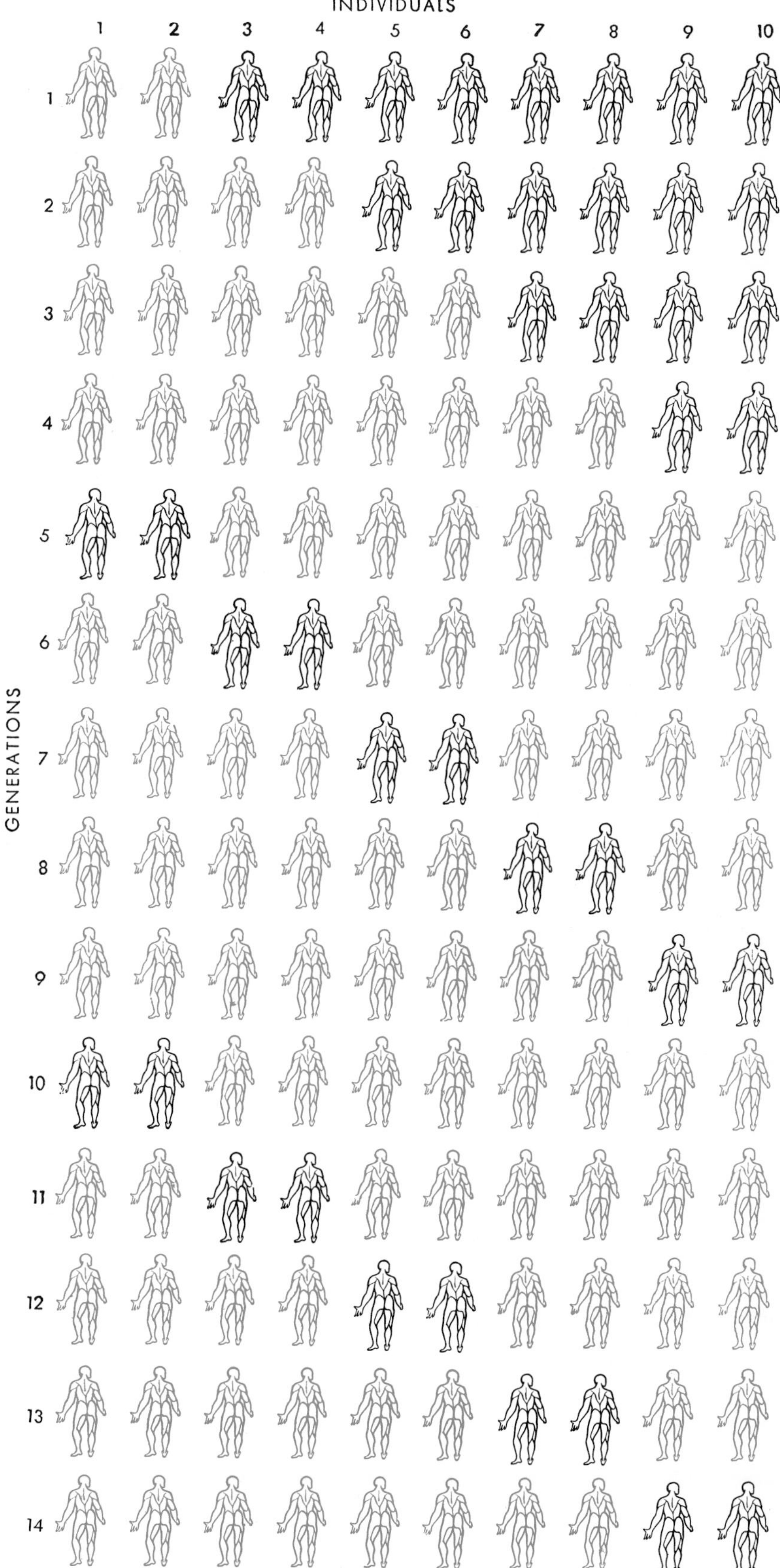


**EQUILIBRIUM IS ATTAINED** by recessive mutant genes under the conditions assumed here. The first assumption is that in each generation (*horizontal rows*) two new mutant genes (*colored figures*) arise among 10 individuals. The second assumption is that each line of descent bearing the mutant gene dies out four generations after the mutation has occurred. In the diagram this extinct line is then replaced by a new one. Thus after three generations the number of individuals bearing a mutant gene is stabilized at eight out of 10.

adduced some evidence that the germ cells of older men have a much higher frequency of newly arisen mutant genes than those of young men. If this result for man, so different from what we have just noted for fruit flies, should be confirmed, it might prove to be more damaging, genetically, for a human population to have the habit of reproduction at a relatively advanced age than for its members to be exposed regularly to some 50 roentgens of ionizing radiation in each generation.

It is evident from these varied examples that the problem of maintaining the integrity of the genetic constitution is a much wider one than that of avoiding the irradiation of the germ cells, inasmuch as diverse other influences may play a mutagenic role equal to or greater in importance than that of radiation.

The view has been expressed that, since some chemical mutagenesis and even radiation mutagenesis occur naturally, the effects of such normal processes should cause us no great concern. Aside from the fact that not everything that is natural is desirable, we must always be conscious of the hazards added by civilization. Certain civilized practices, such as the use of X-rays and radioactivity (and possibly reproduction at an advanced age or the drinking of coffee and tea), are causing genetic damage to be done at a significantly more rapid rate than in olden times.

## Relaxed Selection

It is evident that the rate of elimination of mutant genes is just as important as the mutation frequency in the determination of the human genetic constitution. What we really mean here, of course, is "selective" elimination. The importance of this distinction is seen in the fact that in the ancestors of both men and mice much the same mutations must have occurred, but that the different conditions of their existence—the ever more mousy living of the mouse progenitors and the manlier living of the pre-men—caused a different group of genes to become selected from out of their common store.

A very distinctive feature of our modern industrial civilization is the tremendous saving of human lives which would have been sacrificed under primitive conditions. This is accomplished in part by medicine and sanitation but also by the abundant and diverse artificial aids to living supplied by industry and widely disseminated through the operation of modern social practices. The proportion of those who die prematurely is now so small that it must be considerably below the proportion who would have to be eliminated in order to extinguish mutant genes as fast as new ones arise. In other words, many of the saved lives must represent persons who under more primitive conditions would have died as a result of genetic disabilities. Moreover, the genetically less capable survivors apparently do not have a much lower rate of multiplication than the more capable; in fact, there are certain oppositely working tendencies.

It is probably a considerable underestimate to say that half of the detrimental genes which under primitive conditions would have met genetic extinction, today survive and are passed on. On the basis of this conservative estimate we can calculate that in some 10 generations, or 250 to 300 years, the accumulated genetic effect would be much like that from exposure of a population to a sudden heavy dose of 200 to 400 roentgens, such as was received by the most heavily exposed survivors of Hiroshima. If the techniques of saving life in our civilization continue to advance, the accumulation of mutant genes will rise to ever higher levels. After 1,000 years the population in all likelihood would be as heavily loaded with mutant genes as though it were descended from the survivors of hydrogen-uranium bomb fall-outs, and the passage of 2,000 years would continue the story until the system fell of its own weight or changed.

The process just depicted is a slow, invisible, secular one, like the damage resulting from many generations of exposure to overdoses of diagnostic X-rays. Therefore it is much less likely to gain credence or even attention than the sensational process of being overdosed by fall-outs from bombs. This situation, then, even more than the danger of fall-outs, calls for basic education of the public and publicists, if they are to reshape their deep-rooted attitudes and practices as required.

It is necessary for mankind to realize that a species rises no higher, genetically, and stays no higher, than the pressure of selection forces it to, and that it responds to any relaxation of that pressure by sinking correspondingly. It will in fact take as much rope in sinking as we pay out to it. The policy of saving all possible genetic defectives for reproduction must, if continued, defeat its own purposes. The reason for this is evident as soon as we consider that when, by artificial devices, a moderately detrimental gene is made less detrimental, its frequency will gradually creep upward toward a new equilibrium level, at which it is finally being eliminated anyway at the same rate as that at which it had been eliminated originally, namely, at the rate at which it arises by mutation. This rate of elimination, being once more just as high as before medicine began, will at the same time reflect the fact that as much suffering and frustration (except insofar as we may deaden them with opiates) will then be existing, in consequence of that detrimental gene, as existed under primitive conditions. Thus, with all our medicine and other techniques, we will be as badly off as when we started out.

Not all genetic disabilities, however, would simply be made less detrimental. Some of them would be rendered not detrimental at all under the circumstances of a highly artificial civilization, in the sense that they were enabled to persist indefinitely and thus to become established as the new norm of our descendants. The number of these disabilities would increase up to such a level that no more of them could be supported and compensated for by the technical means available and by the resources of the social system. The burden of the individual cases, up to that level, would have become largely shifted from the given individuals themselves to the whole community, through its social services (a form of insurance), yet the total cost would be divided among all individuals and that cost would keep on rising as far as it was allowed to rise.

Ultimately, in that Utopia of Inferiority in the direction of which we are at the moment headed, people would be spending all their leisure time in having their ailments nursed, and as much of their working time as possible in providing the means whereby the ailments of people in general were cared for. Thus we should have reached the acme of the benefits of modern medicine, modern industrialization and modern socialization. But, because of the secular time scale of evolutionary change and the inertia which retards changes in gene frequency, this condition would come upon the world with such insensible slowness that, except for a few long-haired cranks who took genetics seriously, and perhaps some archaeologists, no one would be conscious of the transformation. If it were called to their attention, they would be likely to rationalize it off as progress. It is hard to think of such a system not at length collapsing, as people lost the capabilities and the incentives needed to keep it going. Such a collapse could not be into barbarism, however, since the population would have become unable to survive primitive conditions; thus a collapse at this point would mean annihilation.

## Countermeasures

There is an alternative policy, and I am hopeful it will be adopted. The alternative does not by any means abandon modern social techniques or call for a return to the fabulous golden age of noble savages or even of rugged individualism. It makes use of all the science, skills and genuine arts we have, to ameliorate, improve and ennoble human life, and, so far as is consistent with its quality and well-being, to extend its quantity and range. Medicine, especially that of a far-seeing and a promoting kind, seeking actively to foster health, vigor and ability, becomes, on this policy, more developed than ever. Persons who nevertheless had defects would certainly have them treated and compensated for, so as to help them to lead useful, satisfying lives. But—and here is the crux of the matter—those who were relatively heavily loaded with genetic defects would consider it their obligation, even if these defects had been largely counteracted, to refrain from transmitting their genes, except when they also possessed genes of such unusual value that the gain for the descendants was likely to outweigh the loss. Only by the adoption of such an attitude towards genetics and reproduction, an attitude seldom encountered as yet, will it be possible for posterity indefinitely to sustain and extend the benefits of medicine, of technology, of science and of civilization in general.

With advance in realistic education should come a better realization of man's place in the great sweep of evolution, and of the risks and the opportunities, genetic as well as nongenetic, which are increasingly opening up for him.

It is evident from these considerations that the same change in viewpoint that leads to the policy of voluntary elimination of detrimental genes would carry with it the recognition that there is no reason to stop short at the arrested norm of today. For all goods, genetic or otherwise, are relative, and, so far as the genetic side of things is concerned, our own highest fulfillment is attained by enabling the next generation to receive the best possible genetic equipment. What the implementation of this viewpoint involves, by way of techniques on the one hand, and of wisdom in regard to values on the other hand, is too large a matter for treatment here. Nevertheless, certain points regarding the genetic objectives to be more immediately sought do deserve our present notice.

For one thing, the trite assertion that one cannot recognize anything better than oneself, or in imagination rise above oneself, is merely a foolish vanity on the part of the self-complacent. Among the important objectives to be sought for mankind are all-around health and vigor, joy of life and longevity. Yet they are far from the supreme aims. For these aims we must search through the most rational and humane thought of those who have gone before us, and integrate with it thinking based on our present vantage point of knowledge and experience. In the light of such a survey it becomes clear that man's present paramount requirements are, on the one hand, a deeper and more integrated understanding and, on the other hand, a more heartfelt, keener sympathy, that is, a deeper fellow-feeling, leading to a stronger impulse to cooperation—more, in a word, of love.

It is wishful thinking on the part of some psychologists to assert that these qualities result purely from conditioning or education. For although conditioning certainly plays a vital role, nevertheless *Homo sapiens* is both an intelligent and a cooperating animal. It is these two complex genetic characteristics, working in combination and serviced by the deftness of his hands, which above all others have brought man to his present estate. Moreover, there still exist great, diverse and numerous genetic differences in the biological bases of these traits within any human population. Although our means of recognition of these genetic differences are today very faulty and tend to confound differences of genetic origin with those derived from the environment, these means can be improved. Thus we can be enabled to recognize our betters. Yet even today our techniques are doubtless more accurate than the trials and errors whereby, after all, nature did manage to evolve us up to this point where we have become effective in counteracting nature. Certainly then it would be possible, if people once became aware of the genetic road that is open, to bring into existence a population most of whose members were as highly developed in regard to the genetic bases of both intelligence and social behavior as are those scattered individuals of today who stand highest in either separate respect.

If the dread of the misuse of nuclear energy awakens mankind not only to the genetic dangers confronting him but also to the genetic opportunities, then this will have been the greatest peacetime benefit that radioactivity could bestow upon us.

# 26
# *The Repair of DNA*

PHILIP C. HANAWALT and
ROBERT H. HAYNES
*February 1967*

One of the most impressive achievements of modern industry is its ability to mass-produce units that are virtually identical. This ability is based not solely on the inherent precision of the production facilities. It also involves intensive application of quality-control procedures for the correction of manufacturing errors, since even the best assembly lines can introduce faulty parts at an unacceptable rate. In addition industry provides replacement parts for the repair of a product that is subsequently damaged by exposure to the hazards of its natural environment. Recent studies have demonstrated that living organisms employ analogous processes for repairing defective parts in their genetic material: deoxyribonucleic acid (DNA). This giant molecule must be replicated with extraordinary fidelity if the organism is to survive and make successful copies of itself. Thus the existence of quality-control mechanisms in living cells may account in large part for the fact that "like produces like" over many generations.

Until recently it had been thought that if the DNA in a living cell were damaged or altered, for example by ionizing radiation, the cell might give rise either to mutant "daughter" cells or to no daughter cells at all. Now it appears that many cells are equipped to deal with some of the most serious hazards the environment can present. In this article we shall describe the experimental results that have given rise to this important new concept.

The instructions for the production of new cells are encoded in the sequences of molecular subunits called bases that are strung together along a backbone of phosphate and sugar groups to form the chainlike molecules of DNA. A sequence of a few thousand bases constitutes a single gene, and each DNA molecule comprises several thousand genes. Before a cell can divide and give rise to two daughter cells, the DNA molecule (or molecules) in the parent cell must be duplicated so that each daughter cell can be supplied with a complete set of genes. On the basis of experiments made with the "chemostat"—a device for maintaining a constant number of bacteria in a steady state of growth—Aaron Novick and the late Leo Szilard estimated that bacterial genes may be duplicated as many as 100 million times before there is a 50 percent chance that even one gene will be altered. This is a remarkable record for any process, and it seems unlikely that it could be achieved without the help of an error-correcting mechanism.

The ability of cells to repair defects in their DNA may well have been a significant factor in biological evolution. On the one hand, repair would be advantageous in enabling a species to maintain its genetic stability in an environment that caused mutations at a high rate. On the other hand, without mutations there would be no evolution, mutations being the changes that allow variation among the individuals of a population. The individuals whose characteristics are best adapted to their environment will leave more offspring than those that are less well adapted. Presumably even the efficiency of genetic repair mechanisms may be subject to selection by evolution. If the repair mechanism were too efficient, it might reduce the natural mutation frequency to such a low level that a population could become trapped in an evolutionary dead end.

Although the error-correcting mechanism cannot yet be described in detail, one can see in the molecular architecture of DNA certain features that should facilitate both recognition of damage and repair of damage. The genetic material of all cells consists of two complementary strands of DNA linked side by side by hydrogen bonds to form a double helix [*see upper illustration on page 250*]. Normally DNA contains four chemically distinct bases: two purines (adenine and guanine) and two pyrimidines (thymine and cytosine). The two strands of DNA are complementary because adenine in one strand is always hydrogen-bonded to thymine in the other, and guanine is similarly paired with cytosine [*see lower illustration on page 250*]. Thus the sequence of bases that constitute the code letters of the cell's genetic message is supplied in redundant form. Redundancy is a familiar stratagem to designers of error-detecting and error-correcting codes. If a portion of one strand of the DNA helix were damaged, the information in that portion could be retrieved from the complementary strand. That is, the cell could use the undamaged strand of DNA as a template for the reconstruction of a damaged segment in the complementary strand. Recent experimental evidence indicates that this is precisely what happens in many species of bacteria, particularly those that are known to be highly resistant to radiation.

The ability to recover from injury is a characteristic feature of living organisms. There is a fundamental difficulty, however, in detecting repair processes in bacteria. For example, when a population of bacteria is exposed to a dose of ultraviolet radiation or X rays, there is no way to determine in advance what proportion of the population will die. How can one tell whether the observed mortality accurately reflects all the damage sustained by the irradiated cells or whether some of the damaged cells have repaired themselves? Fortunately it is possible to turn the repair mechanism on or off at will.

A striking example can be found in the process called photoreactivation [*see bottom illustration on page 252*]. Although hints of its existence can be traced back to 1904, photoreactivation was not adequately appreciated until Albert Kelner rediscovered the effect in 1948 at the Carnegie Institution of Washington's Department of Genetics in Cold Spring Harbor, N.Y. Kelner was puzzled to find that the number of soil organisms (actinomycetes) that survived large doses of ultraviolet radiation could be increased by a factor of several hundred thousand if the irradiated bacteria were subsequently exposed to an intense source of visible light. He concluded that ultraviolet radiation had its principal effect on the nucleic acid of the cell, but he had no inkling what the effect was. In an article published before the genetic significance of DNA was generally appreciated, Kelner wrote: "Perhaps the real stumbling block [to understanding photoreactivation] is that we do not yet understand at all well the biological role of that omnipresent and important substance—nucleic acid" [see "Revival by Light," by Albert Kelner; SCIENTIFIC AMERICAN, May, 1951].

It is now known that the germicidal action of ultraviolet radiation arises chiefly from the formation of two unwanted chemical bonds between pyrimidine bases that are adjacent to each other on one strand of the DNA molecule. Two molecules bonded in this way are called dimers; of the three possible types of pyrimidine dimer in DNA, the thymine dimer is the one that forms most readily [*see upper illustration on page 251*]. It is therefore not surprising that a given dose of ultraviolet radiation will create more dimers in DNA molecules that contain a high proportion of thymine bases than in DNA molecules with fewer such bases. Consequently bacteria whose DNA is rich in thymine tend to be more sensitive to ultraviolet radiation than those whose DNA is not.

Richard B. Setlow, his wife Jane K. Setlow and their co-workers at the Oak Ridge National Laboratory have shown that pyrimidine dimers block normal replication of DNA and that bacteria with even a few such defects are unable to divide and form colonies [see "Ultraviolet Radiation and Nucleic Acid," by R. A. Deering; SCIENTIFIC AMERICAN Offprint No. 143]. In the normal replication of DNA each parental DNA strand serves as a template for the synthesis of a complementary daughter strand. This mode of replication is termed semiconservative because the parental strands separate in the course of DNA synthesis; each daughter cell receives a "hybrid" DNA molecule that consists of one parental strand and one newly synthesized complementary strand. The effect of a pyrimidine dimer on DNA replication may be analogous to the effect on a zipper of fusing two adjacent teeth.

Claud S. Rupert and his associates at

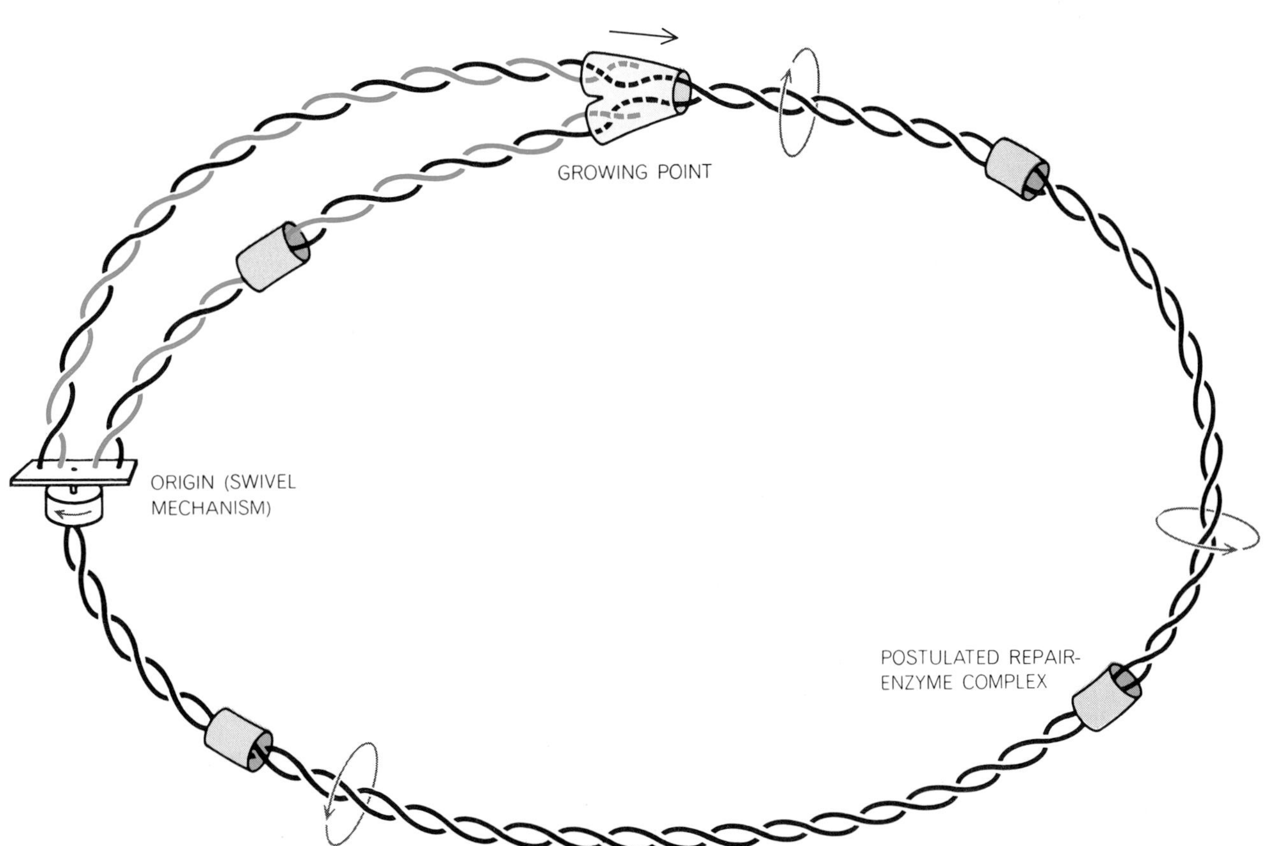


**REPLICATION OF BACTERIAL CHROMOSOME, a ring-shaped molecule of deoxyribonucleic acid (DNA), has now been shown to take two forms: normal replication and repair replication. In the former process the two strands that constitute the double helix of DNA are unwound and a daughter strand (*color*) is synthesized against each of them. In this way the genetic "message" is transmitted from generation to generation. The pairing of complementary subunits that underlies this process is illustrated on the next page. In repair replication, defects that arise in individual strands of DNA are removed and replaced by good segments. It is hypothesized that "repair complexes," composed of enzymes, are responsible for the quality control of the DNA structure. Although this diagram shows the swivel mechanism for unwinding the parent strands to be at the origin, it may in fact be located at the growing point.**

Johns Hopkins University have shown that photoreactivation involves the action of an enzyme that is selectively bound to DNA that has been irradiated with ultraviolet. When this enzyme is activated by visible light (which simply serves as a source of energy), it cleaves the pyrimidine dimers, thereby restoring the two bases to their original form. Photoreactivation is thus a repair process that can be turned on or off merely by flicking a light switch.

Let us now consider another kind of repair mechanism in which light plays no role and that is therefore termed dark reactivation. This type of repair process can be turned off genetically, by finding mutant strains of bacteria that lack the repair capabilities of the original radiation-resistant strain. The "$B/r$" strain of the bacterium *Escherichia coli,* first isolated in 1946 by Evelyn Witkin of Columbia University, is an example of a microorganism that is particularly resistant to radiation. The first radiation-sensitive mutants of this strain, known as $B_{s-1}$, were discovered in 1958 by Ruth Hill, also of Columbia.

Not long after the discovery of the $B_{s-1}$ strain a number of people suggested that its sensitivity to radiation might be due to the malfunction of a particular enzyme system that enabled resistant bacteria such as $B/r$ to repair DNA that had been damaged by radiation. This was a reasonable suggestion in view of the steadily accumulating evidence that DNA is the principal target for many kinds of radiobiological damage. Experiments conducted by Howard I. Adler at Oak Ridge and by Paul Howard-Flanders at Yale University lent further support to this hypothesis. It had been known for some years that bacteria can exchange genes by direct transfer through a primitive form of sexual mating [see "Viruses and Genes," by François Jacob and Elie L. Wollman, beginning on page 42 in this book]. Howard-Flanders and his co-workers found that bacteria of a certain radiation-resistant strain of *E. coli* (strain *K*-12) have at least three genes that can be transferred by bacterial mating to radiation-sensitive cells, thereby making them radiation-resistant. Since genes direct the synthesis of all enzymes in the living cell, these experiments supported the hypothesis that $B_{s-1}$ and other radiation-sensitive bacteria lack one or more enzymes needed for the repair of radiation-damaged DNA.

The question now arises: Do the enzymes involved in dark reactivation operate in the same way as the enzyme that

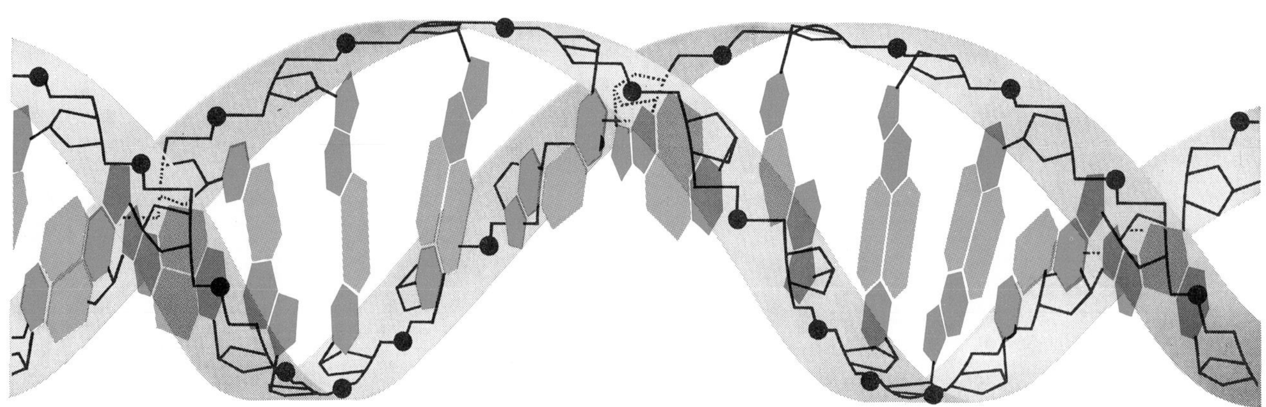

**DNA MOLECULE is a double helix that carries the genetic message in redundant form. The backbone of each helix consists of repeating units of deoxyribose sugar (*pentagons*) and phosphate (*black dots*). The backbones are linked by hydrogen bonds between pairs of four kinds of base: adenine, guanine, thymine and cytosine. The bases are the "letters" in which the genetic message is written. Because adenine invariably pairs with thymine and guanine with cytosine, the two strands carry equivalent information.**

**DNA BASES are held together in pairs by hydrogen bonds. The cytosine-guanine pair (*left*) involve three hydrogen bonds, the thymine-adenine pair (*right*) two bonds. If the $CH_3$ group in thymine is replaced by an atom of bromine (*Br*), the resulting molecule is called 5-bromouracil. Thymine and 5-bromouracil are so similar that bacteria will incorporate either in synthesizing DNA. Because the bromine compound is so much heavier than thymine its presence can be detected by its effect on the weight of the DNA.**

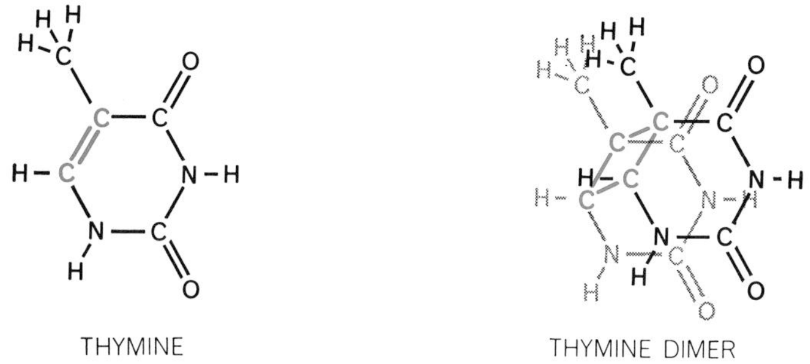

is known to split pyrimidine dimers in the photoreactivation process? Another possibility is that the resistant cells might somehow bypass the dimers during replication of DNA and leave them permanently present, although harmless, in their descendant molecules.

The actual mechanism is even more elegant than either of these possibilities; it exploits the redundancy inherent in the genetic message. The radiation-resistant strains of bacteria possess several enzymes that operate sequentially in removing the dimers and replacing the defective bases with the proper complements of the bases in the adjacent "good" strand. We shall recount the two key observations that substantiate this postulated repair scheme.

THYMINE

THYMINE DIMER

**EFFECT OF ULTRAVIOLET RADIATION on DNA is to fuse adjacent pyrimidine units: thymine or cytosine. The commonest linkage involves two units of thymine, which are coupled by the opening of double bonds. The resulting structure is known as a dimer.**

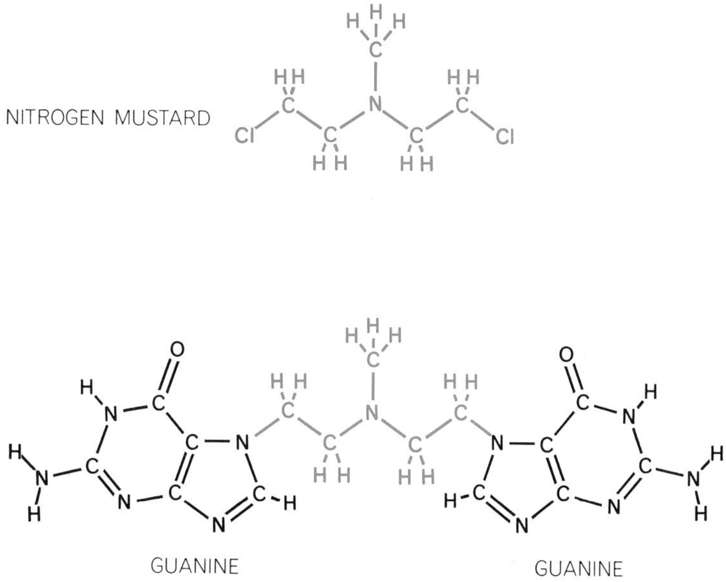

NITROGEN MUSTARD

GUANINE

GUANINE

**EFFECT OF NITROGEN MUSTARD, a compound related to mustard gas, is to cross-link units of guanine within the DNA molecule. Unless repaired, the structural defects caused by nitrogen mustard and ultraviolet radiation can prevent the normal replication of DNA.**

The excision of dimers was first demonstrated by Richard Setlow and William L. Carrier at Oak Ridge and was soon confirmed by Richard P. Boyce and Howard-Flanders at Yale. In their studies cultures of ultraviolet-resistant and ultraviolet-sensitive bacteria were grown separately in the presence of radioactive thymine, which was thereupon incorporated into the newly synthesized DNA. The cells were then exposed to ultraviolet radiation. After about 30 minutes they were broken open so that the fate of the labeled thymine could be traced. In the ultraviolet-sensitive strains all the thymine that had been incorporated into DNA was associated with the intact DNA molecules. Therefore any thymine dimers formed by ultraviolet radiation remained within the DNA. In the ultraviolet-resistant strains, however, dimers originally formed in the DNA were found to be associated with small molecular fragments consisting of no more than three bases each. (Thymine dimers can easily be distinguished from the individual bases or combinations of bases by paper chromatography, the technique by which substances are separated by their characteristic rate of travel along a piece of paper that has been wetted with a solvent.) These experiments provided strong evidence that dark repair of ultraviolet-damaged DNA does not involve the splitting of dimers in place, as it does in photoreactivation, but does depend on their actual removal from the DNA molecule.

Direct evidence for the repair step was not long in coming. At Stanford University one of us (Hanawalt), together with a graduate student, David Pettijohn, had been studying the replication of DNA after ultraviolet irradiation of a radiation-resistant strain of *E. coli.* In these experiments we used as a tracer a chemical analogue of thymine called 5-bromouracil. This compound is so similar to the natural base thymine that a bacterium cannot easily tell the two apart [*see right half of lower illustration on opposite page*]. When 5-bromouracil is substituted for thymine in the growth medium of certain strains of bacteria that are unable to synthesize thymine, it is incorporated into the newly replicated DNA. The fate of 5-bromouracil can be traced because the bromine atom in it is more than five times heavier than the methyl ($CH_3$) group in normal thymine that it replaces. Therefore DNA fragments containing 5-bromouracil are denser than normal fragments containing thymine. The density difference can be detected by density-gradient centrifugation, a technique introduced in 1957 by Matthew S. Meselson, Franklin W. Stahl and Jerome Vinograd at the California Institute of Technology.

In density-gradient centrifugation DNA fragments are suspended in a solution of the heavy salt cesium chloride and are spun in a high-speed centrifuge for several days. When equilibrium is reached, the density of the solution varies from 1.5 grams per milliliter at the top of the tube to two grams per milliliter at the bottom. If normal DNA from *E. coli* is also present, it will eventually concentrate in a band corresponding to a density of 1.71 grams per milliliter. A DNA containing 5-bromouracil instead of thymine has a density of 1.8 and so will form a band closer to the bottom of the tube.

The entire genetic message of *E. coli* is contained in a single two-strand molecule of DNA whose length is nearly 1,000 times as long as the cell itself. This long molecule must be coiled up like a

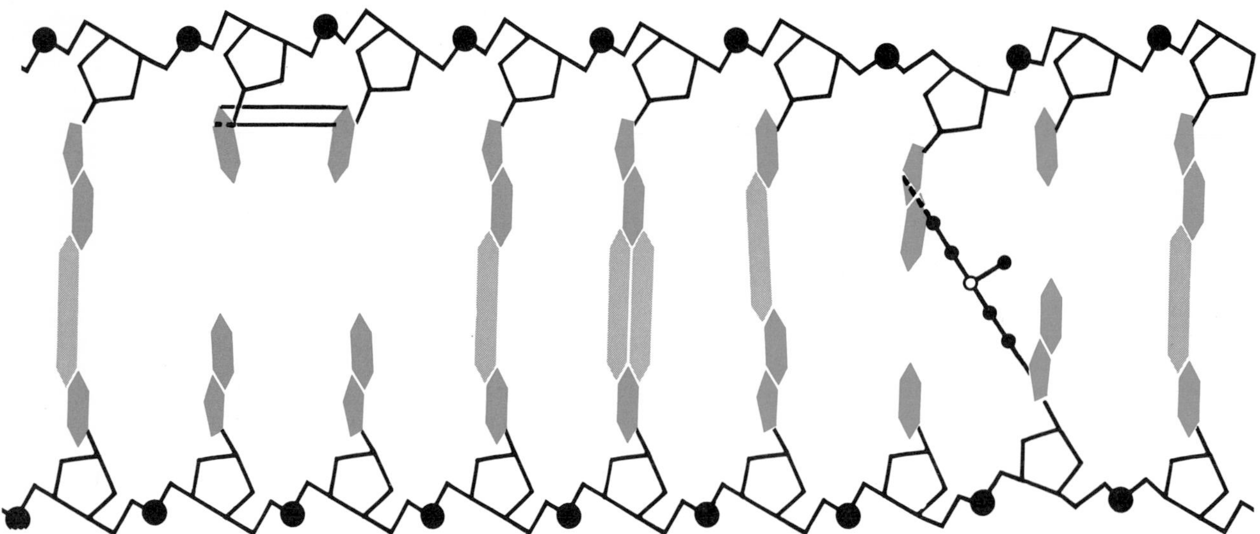

DEFECTS IN DNA probably distort the symmetry of its helical structure. To make the distortions more apparent this diagram shows the DNA in flattened form. A thymine dimer appears in the left half of the structure; a guanine-guanine cross-link is shown in the right half. The authors believe the repair complex recognizes the distortions rather than the actual defects in the bases.

skein of yarn to be accommodated within the cell [see "The Bacterial Chromosome," by John Cairns, which begins on page 21 in this book]. Such a molecule is extremely sensitive to fluid shearing forces; in the course of being extracted from the cell it is usually broken into several hundred pieces.

If 5-bromouracil is added to a culture of growing bacteria for a few minutes (a small fraction of one generation), the DNA fragments isolated from the cells fall into several categories of differing density, each of which forms a distinct band in a cesium chloride density gradient. The lightest band will consist of unlabeled fragments: regions of the DNA molecule that were not replicated during the period when 5-bromouracil was present. A distinctly heavier band will contain fragments from regions that have undergone replication during the labeling period. This is the band containing hybrid DNA: molecules made up of one old strand containing thymine and one new strand containing 5-bromouracil in place of thymine. If synthesis proceeds until the chromosome has completed one cycle of replication and has started on the next cycle, some DNA fragments will have 5-bromouracil in both strands and therefore will form a band still heavier than the band containing the hybrid fragments. Finally, one fragment from each chromosome will include the "growing point" where the new strands are being synthesized on the pattern of the old ones, and thus will consist of a mixture of replicated and unreplicated DNA. This fragment, which is presumably shaped like a Y, will show up in the density gradient at a position

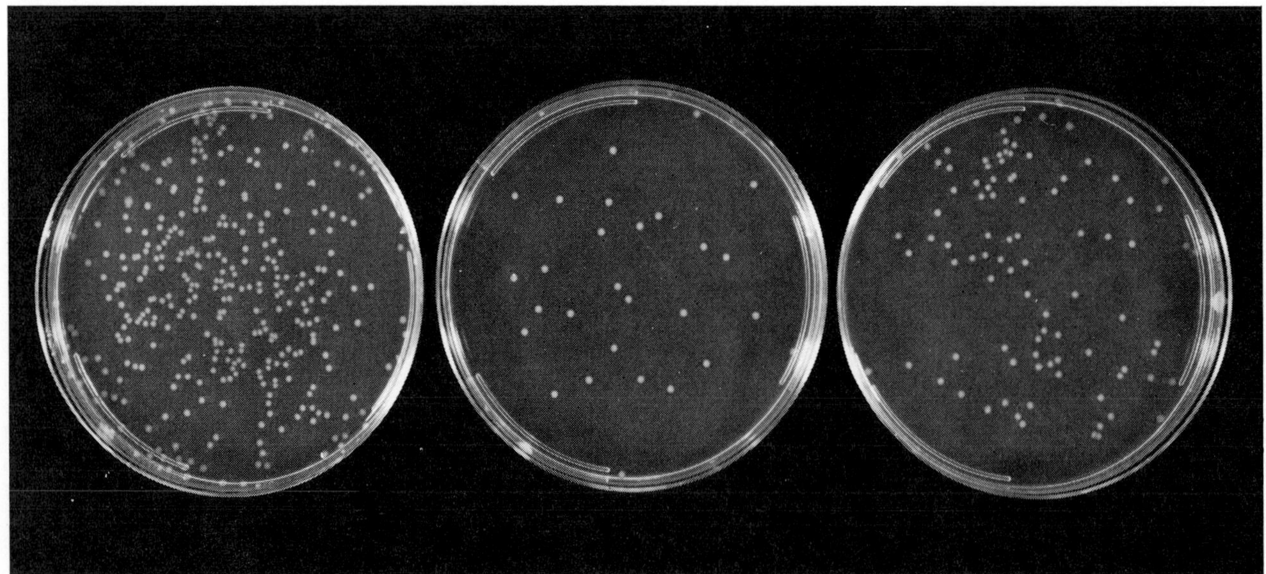

PHOTOREACTIVATION, a type of DNA repair process, is demonstrated in this photograph of three bacterial culture dishes. The dish at the left is a control: it contains 368 colonies of *B/r* strain of *Escherichia coli.* The middle dish contains bacteria exposed to ultraviolet radiation; only 35 cells have survived to form colonies. The bacteria in the dish at right were exposed to visible light following ultraviolet irradiation; it contains 93 colonies. Thus exposure to visible light increased the survival rate nearly threefold.

intermediate between the unlabeled DNA and the hybrid fragments containing 5-bromouracil.

When we used this technique to study DNA replication in bacteria exposed to ultraviolet radiation, we observed a pattern quite different from the one expected for normal replication. The DNA fragments containing the 5-bromouracil appeared in the gradient at the same position as normal fragments containing thymine! There could be no doubt of this because in these experiments we used 5-bromouracil labeled with tritium (the radioactive isotope of hydrogen) and thymine labeled with carbon 14 [*see illustration on next page*].

This pattern, which at first seems puzzling, is just the one to be expected if many thymine dimers—created at random throughout the DNA by ultraviolet radiation—had been excised and if 5-bromouracil had been substituted for thymine in the repaired regions. As a result many DNA fragments would contain 5-bromouracil, but no one fragment would contain enough 5-bromouracil to affect its density appreciably.

How can we be sure that the density distribution of 5-bromouracil observed in the foregoing experiment arises from "repair replication" rather than from normal replication? A variety of tests confirmed the repair interpretation. By using enzymes to break down the DNA molecule and separating the bases by paper chromatography we verified that the radioactive label was still in 5-bromouracil and had not been transferred to some other base. Various physical studies showed that the 5-bromouracil had been incorporated into extremely short segments that were distributed randomly throughout both DNA strands. This mode of DNA replication was not observed in the $B_{s-1}$ strain of *E. coli*, the radiation-sensitive mutant that cannot excise pyrimidine dimers and therefore could not be expected to perform repair replication. Moreover, repair replication was not observed in the radiation-resistant bacteria in which visible light had triggered the splitting of pyrimidine dimers by photoreactivation; this indicates that repair replication is not necessary if the dimers are otherwise repaired *in situ*.

Finally it was shown that DNA repaired by dimer excision and strand reconstruction could ultimately replicate in the normal semiconservative fashion. This is rather compelling evidence for the idea that biologically functional DNA results from repair replication and that the process is not some aberrant form of synthesis with no biological importance.

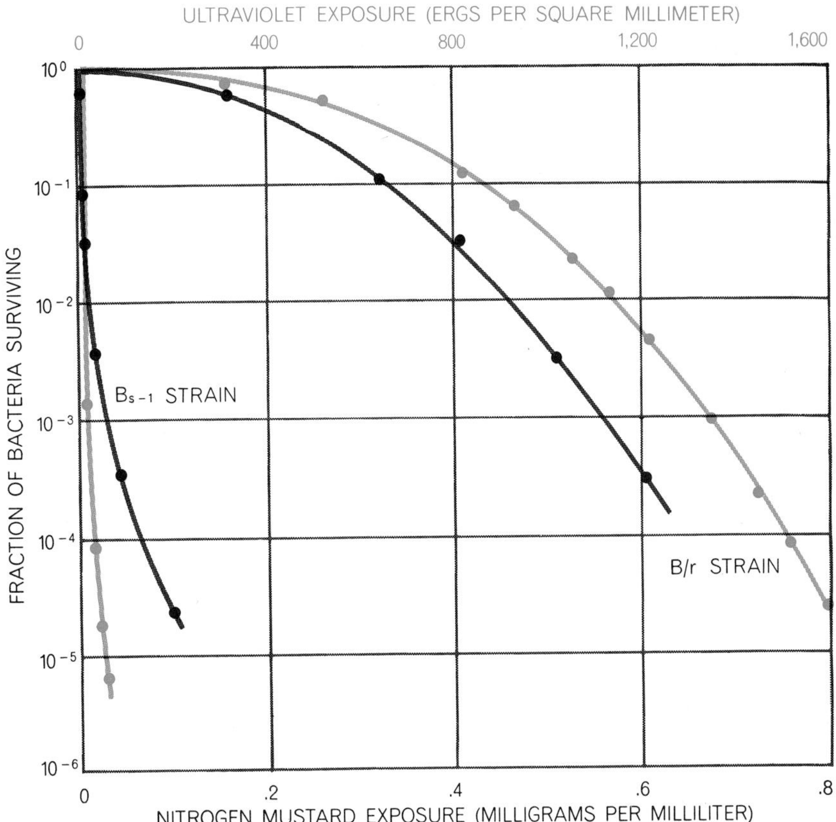


**RESISTANCE TO LETHAL AGENTS is demonstrated by certain strains of *E. coli* but not by others. The *B/r* strain, for example, shows a high tolerance to doses of ultraviolet radiation (*colored curves*) and nitrogen mustard (*black curves*) that kill a large percentage of the sensitive $B_{s-1}$ strain. This result suggests that the DNA repair mechanism of the *B/r* strain is effective in removing guanine-guanine cross-links produced by nitrogen mustard as well as in removing thymine dimers formed by exposure to ultraviolet radiation.**

How can one visualize the detailed sequence of events that must be involved in this type of repair? Two models have been suggested, and the present experimental data seem to be equally compatible with each. The two models are distinguished colloquially by the terms "cut and patch" and "patch and cut" [*see illustrations on page 255*]. The former refers to the model originally proposed by Richard Setlow, Howard-Flanders and others. The latter refers to a model that took form during a discussion at a recent conference on DNA repair mechanisms held in Chicago.

The cut-and-patch scheme postulates an enzyme that excises a short, single-strand segment of the damaged DNA. The resulting gap is enlarged by further enzyme attack and then the missing bases are replaced by repair replication in the genetically correct sequence according to the rules that govern the pairing of bases.

In the patch-and-cut scheme the process is assumed to be initiated by a single incision that cuts the strand of DNA near the defective bases. Repair replication begins immediately at this point and is accompanied by a "peeling back" of the defective strand as the new bases are inserted. This patch-and-cut scheme is attractive because it could conceivably be carried out by a single enzyme complex or particle that moves in one direction along the DNA molecule, repairing defects as it goes. Furthermore, it does not involve the introduction of long, vulnerable single-strand regions into the DNA molecule while the repair is taking place. Both models are undoubtedly oversimplifications of the actual molecular events inside the living cell, but they have great intuitive appeal and are helpful in planning further studies of the DNA repair process.

Repair replication would be of interest only to radiation specialists if it were not for the evidence that DNA structural defects other than pyrimidine dimers can be repaired and that similar repair

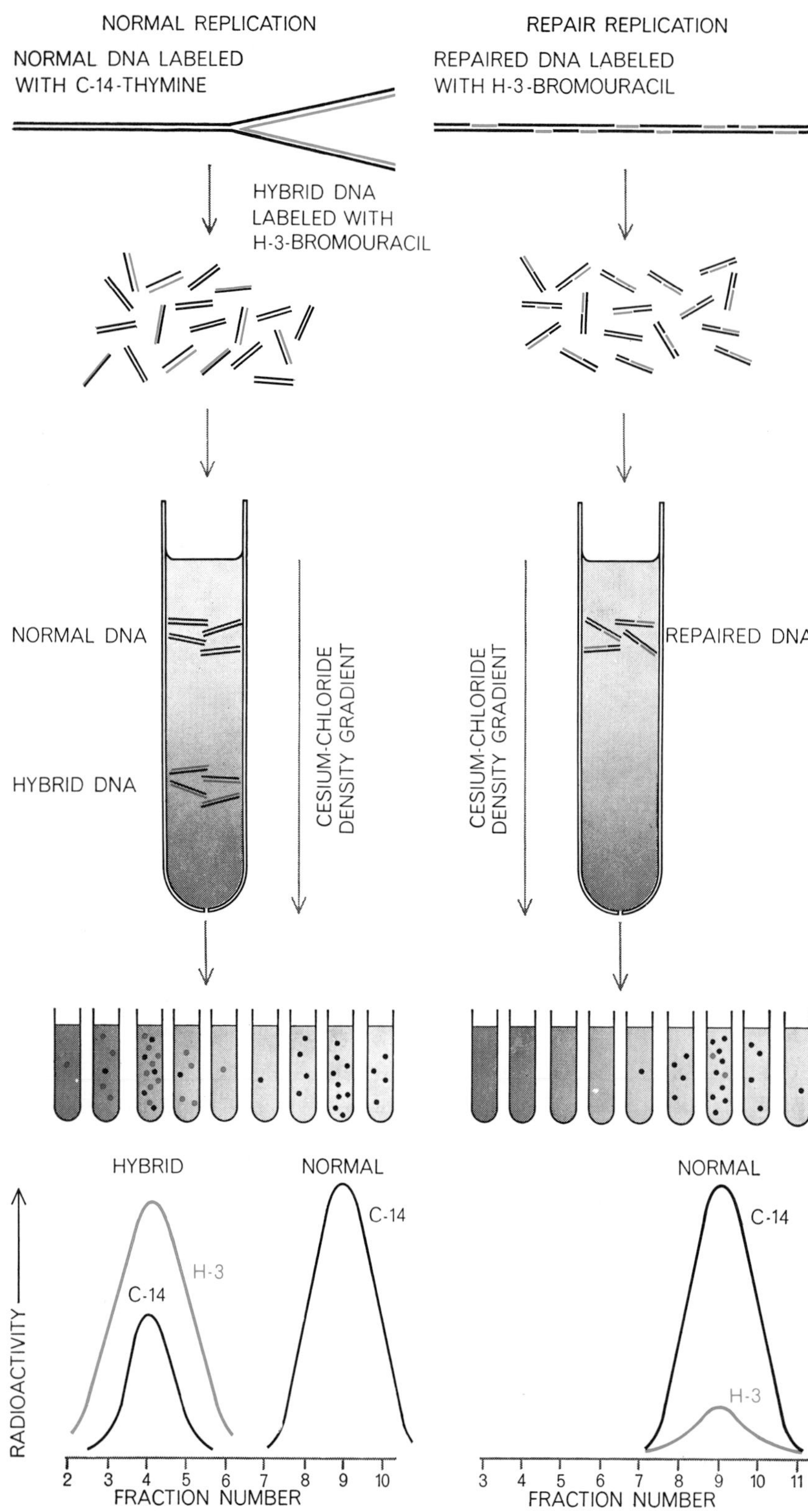


**TWO KINDS OF REPLICATION** can be demonstrated by growing bacteria first in a culture containing thymine labeled with radioactive carbon 14 and then in a culture containing 5-bromouracil labeled with radioactive hydrogen 3. In normal replication (*left*), also known as semiconservative replication, daughter strands of "hybrid" DNA incorporate the 5-bromouracil (*color*). Because 5-bromouracil is much heavier than the thymine for which it substitutes, fragments of hybrid DNA form a separate heavier layer when they have been centrifuged and have reached equilibrium in a density gradient of cesium chloride. When the radioactivity in the various fractions is analyzed (*bottom left*), carbon 14 appears in two peaks but hydrogen 3 occurs in only one peak. If the experiment is repeated with DNA fragments that have undergone repair replication (*right*), they all appear to be of normal density. This implies that relatively little 5-bromouracil has been incorporated and also that the repaired segments are randomly scattered throughout the DNA molecule.

phenomena occur in organisms other than *E. coli.* We shall review some of the evidence indicating that repair replication is of general biological significance.

Just as strains of *E. coli* vary considerably in their sensitivity to ultraviolet radiation, so they vary considerably in their sensitivity to other mutagenic agents. One such agent is nitrogen mustard, so named because it is chemically related to the mustard gas used in World War I. It was the first chemical agent known to be capable of producing mutations and chromosome breaks in fruit flies and other organisms. Its biological action arises primarily from its ability to react with neighboring guanine bases in DNA, thereby producing guanine-guanine cross-links [*see lower illustration on page 251*].

If one compares the survival curves of different strains of bacteria treated with nitrogen mustard with survival curves for bacteria subjected to ultraviolet radiation, one finds that the curves are almost identical [*see illustration on preceding page*]. This similarity led us to suggest that it is not the altered bases themselves that are "recognized" by the repair enzymes but rather the associated distortions, or kinks, that the alteration of the bases produces in the backbone of the DNA molecule. On this hypothesis one would predict that a wide variety of chemically different structural defects in DNA might be repaired by a common mechanism.

A substantial amount of biochemical evidence has now accumulated in support of this idea. We have established, for example, that repair replication of DNA takes place in *E. coli* that have been treated with nitrogen mustard. Others have found evidence that defects produced by agents as diverse as X rays, the chemical mutagen nitrosoguanidine and the antibiotic mitomycin C can all be repaired in radiation-resistant strains of *E. coli.* Walter Doerfler and David Hogness of Stanford have even found evidence that simple mispairing of bases between two strands of DNA can be corrected.

Finally, it now seems that certain steps in repair replication may also be involved in such phenomena as genetic recombination and the reading of the DNA code in preparation for protein synthesis. Evidence for these exciting possibilities has begun to appear in the work of Howard-Flanders, Meselson (who is now at Harvard University), Alvin J. Clark, of the University of California at Berkeley, Crellin Pauling of the University of California at Riverside and other investigators.

Repair replication has also been observed in a number of bacterial species other than *E. coli*. For example, Douglas Smith, a graduate student at Stanford, has demonstrated the repair of DNA in the pleuropneumonia-like organisms, which are probably the smallest living cells. These organisms, which are even smaller than some viruses, are thought to possess only the minimum number of structures needed for self-replication and independent existence [see "The Smallest Living Cells," by Harold J. Morowitz and Mark E. Tourtellotte; SCIENTIFIC AMERICAN Offprint 1005]. This suggests that repair replication may be a fundamental requirement for the evolution of free-living organisms.

In view of the impressive versatility of the repair replication process it is natural to ask if there is any type of DNA damage that cannot be mended by the cell. The evidence so far is limited and indirect, but William Rodger Inch, working at the Lawrence Radiation Laboratory of the University of California, has found that the *B/r* strain of *E. coli* is unable to repair all the damage caused when it is exposed to certain energetic beams of atomic nuclei produced by the heavy-ion linear accelerator (HILAC). Considering the extensive damage that must be done to cells by a beam of such intensely ionizing radiation, the result is not too surprising.

The discovery that cells have the facility to repair defects in DNA is a recent one. It is already apparent, however, that the process of repair replication could have broad significance for biology and medicine. Many questions remain to be investigated: What is the structure of the various repair enzymes? Are they organized into particulate units within the cell? What range of DNA defects can be recognized and repaired? Does DNA repair, as we now understand it, take place in the cells of mammals, or do even more complicated processes underlie the recovery phenomena that are observed after these higher types of cells are exposed to radiation? Might it be possible to increase the radiosensitivity of tumors by inhibiting the DNA repair mechanisms that may operate in cancer cells? If so, the idea could be of great practical value in the treatment of cancer.

These and many related questions are now being investigated in many laboratories around the world. Once again it has been demonstrated that the study of what may appear to be rather obscure properties of the simplest forms of life can yield rich dividends of much intellectual and practical value.

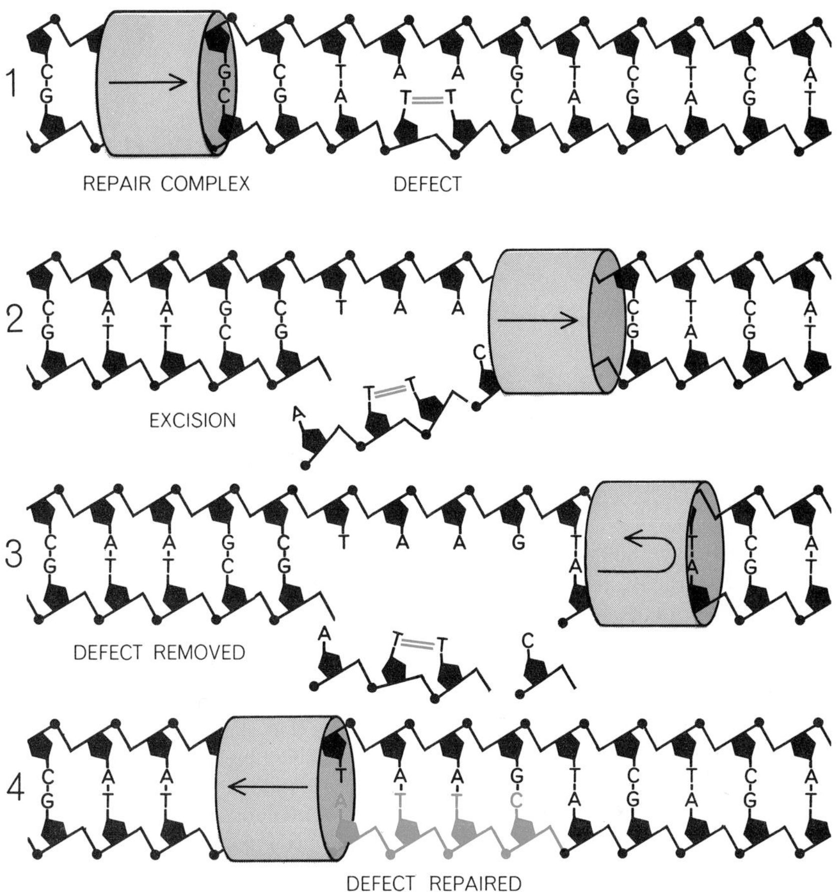


**"CUT AND PATCH" repair mechanism was the first one proposed to explain how bacterial cells might remove thymine dimers (*1*) and similar defects from a DNA molecule. The hypothetical repair complex severs the defective strand (*2*) and removes the defective region (*3*). Retracing its path, it inserts new bases according to the rules of base pairing (*4*).**

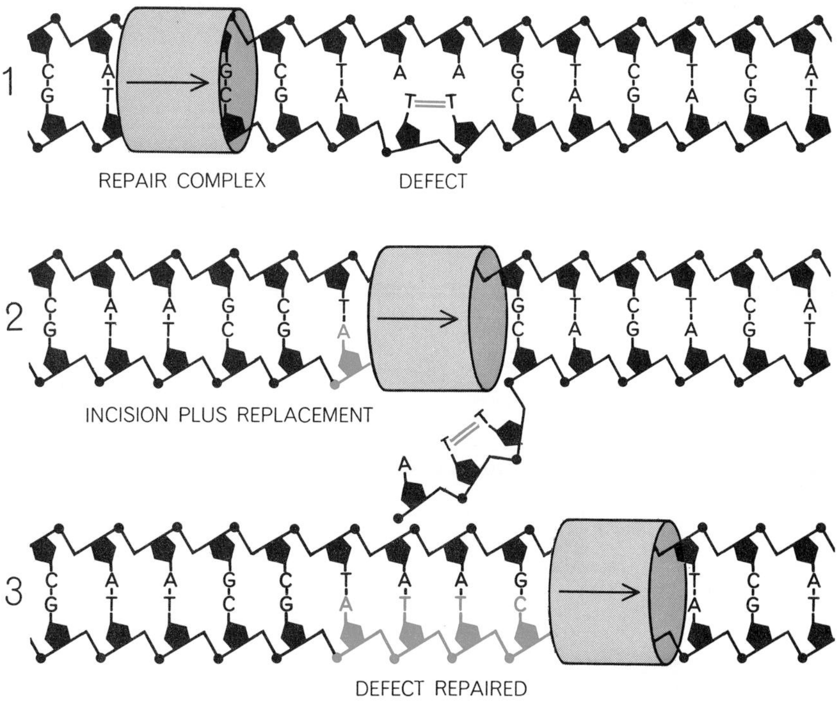


**"PATCH AND CUT" mechanism has been proposed as an alternative to the cut-and-patch scheme. On the new model the repair complex inserts new bases as it removes defective ones.**

# 27
# *The Evolution of Hemoglobin*

EMILE ZUCKERKANDL
*May 1965*

Every living thing carries within itself a richly detailed record of its antecedents from the beginning of life on earth. This record is preserved in coded form in the giant molecules of deoxyribonucleic acid (DNA) that constitute the organism's genome, or total stock of genetic information. The genetic record is also expressed more tangibly in the protein molecules that endow the organism with its form and function.

These two kinds of molecule—DNA and protein—are living documents of evolutionary history. Although chemically very different, they have in common a fundamental characteristic: they are both made up of a one-dimensional succession of slightly differing subunits, like differently colored beads on a string. Each colored bead occupies a place specifically assigned to it unless the heritable changes called mutations either change the color of a bead or displace, eliminate or add a bead (or several beads) at a time. In addition the protein molecules are folded in a specific way that enables them to carry out their specific functions.

To examine these molecular documents of evolutionary history a new discipline has emerged: chemical paleogenetics. It sets itself the ambitious goal of reconstructing, insofar as possible, how evolution proceeds at the molecular level. The new discipline is still in its infancy because almost nothing is yet known about the linear sequence of subunits that embody the code for a single gene in a molecule of DNA. Viruses, the smallest structures containing the blueprints for their own replication, possess from a few to several hundred genes. Each gene, in turn, consists of a string of several hundred code "letters." It has not been possible to isolate a single gene from any organism for chemical analysis.

It has been possible, however, to study and determine the chemical structure of a number of individual polypeptide chains that embody the coded information contained in individual genes. The term "polypeptide" refers to the principal chain of a protein molecule; it describes a sequence of amino acid molecules that are held together by peptide bonds. Such bonds are formed when two amino acid molecules link up with the release of a molecule of water; when they are linked in this way, the amino acids are called residues.

Because three code letters in DNA

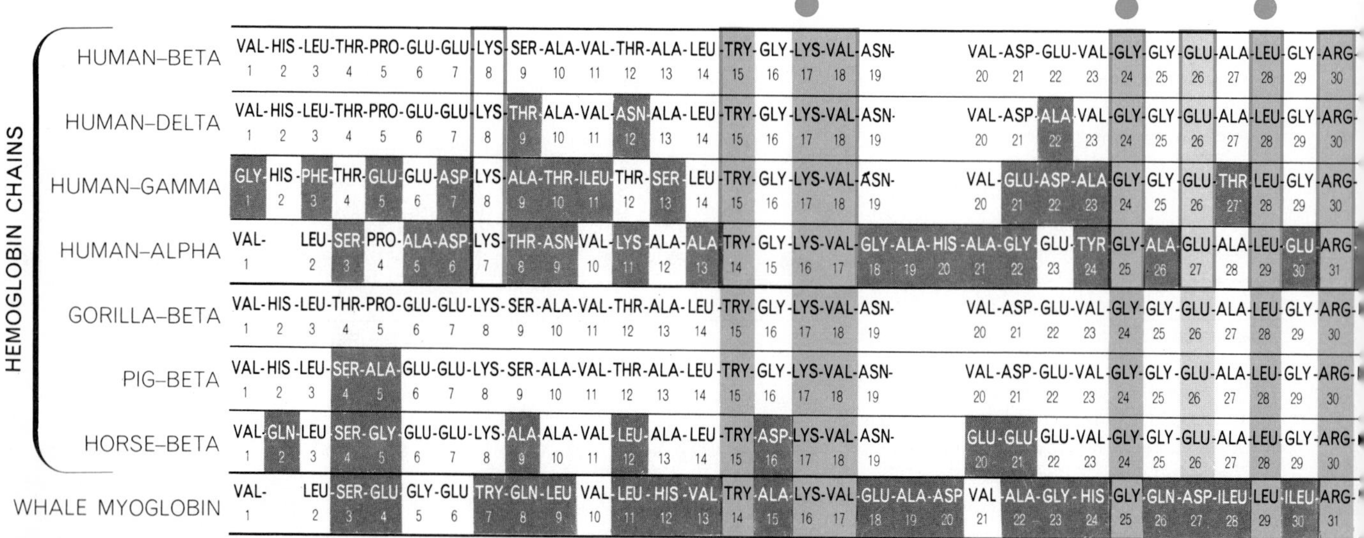


**FAMILY RESEMBLANCES are exhibited by the polypeptide chains found in several kinds of hemoglobin and by the polypeptide chain of sperm whale myoglobin. Hemoglobin is the oxygen-carrying molecule of the blood; myoglobin stores oxygen in muscle. Polypeptides are molecular chains whose links are amino acid units, usually called residues. The hemoglobin chains comprise either 141 or 146 residues; the myoglobin chain, 153. (The illustration is continued on pages 258 and 259.) Each molecule of**

are required to make a "word" specifying one amino acid molecule, there is a certain compression of information between the gene and the polypeptide chain it encodes. A "structural" gene containing 600 code letters is required to specify a polypeptide containing 200 amino acid residues. The reason for the three-to-one ratio is that there are 20 kinds of amino acid and only four kinds of DNA code letters, embodied in subunits called bases, to identify them; a minimum of three code letters, or bases, is needed to specify each amino acid. (In fact, three code units can specify 64 different items, and there is evidence that more than one DNA triplet exists for some of the amino acids.)

Enough is now known about the amino acid sequence of certain polypeptides to enable the chemical paleontologist to test the validity of three basic postulates. The first asserts that polypeptide chains in present-day organisms have arisen by evolutionary divergence from similar polypeptide chains that existed in the past. The present and past chains would be similar in that many of their amino acid residues match; such chains are said to be homologous. The second postulate is that a gene existing at some past epoch can occasionally be duplicated so that it appears at two or more sites in the genome of descendent organisms. Thus a contemporary organism can have two or more homologous genes represented by two or more homologous polypeptide chains, which have mutated independently and are therefore no longer identical in structure. The third postulate holds that the mutational events most commonly retained by natural selection are those that lead to the replacement of a single amino acid residue in a polypeptide chain.

In addition to these three postulates I would like to suggest a fourth that is much more controversial: Contemporary organisms that look much like ancient ancestral organisms probably contain a majority of polypeptide chains that resemble quite closely those of the ancient organisms. In other words, certain animals said to be "living fossils," such as the cockroach, the horseshoe crab, the shark and, among mammals, the lemur, probably manufacture a great many polypeptide molecules that differ only slightly from those manufactured by their ancestors millions of years ago. This postulate is controversial because it is often said that evolution has been just as long for organisms that appear to have changed little as for those that have changed much; consequently it is held that the biochemistry of living fossils is probably very different from that of their remote ancestors. My own view is that it is unlikely that selective forces would favor the stability of morphological characteristics without at the same time favoring the stability of biochemical characteristics, which are more fundamental.

As an example of the application of chemical paleogenetics I shall describe how evolutionary changes are reflected in the molecular structure of hemoglobin, the oxygen-carrying protein of the blood. Hemoglobin is the most complex protein whose detailed molecular composition and structure are known in man, in his near relatives among the primates and in his more distant relatives such as horses and cattle. The composition and structure of hemoglobin molecules in more primitive organisms such as fishes are rapidly being worked out.

Hemoglobin is a particularly good

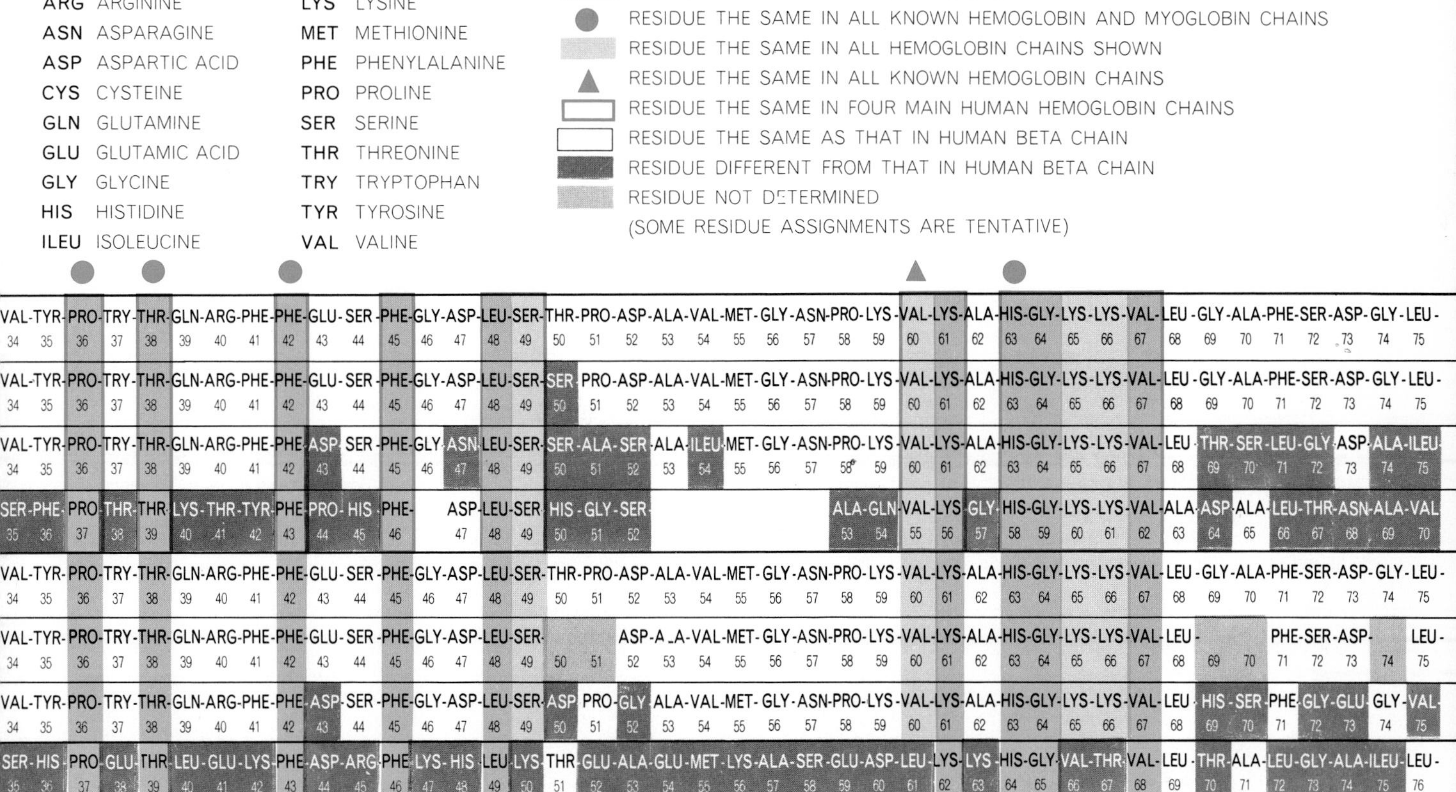


**hemoglobin contains two subunits of a polypeptide chain called alpha ($\alpha$) and two of a chain called beta ($\beta$). In human adults about 2 percent of the hemoglobin molecules contain delta ($\delta$) chains in place of beta chains. Two other chains, gamma ($\gamma$) and epsilon ($\varepsilon$, *not shown*), are manufactured during fetal life and can also serve in place of the $\beta$-chain. The illustration enables one to compare the four principal chains ($\alpha$, $\beta$, $\gamma$, $\delta$) found in human hemoglobin with the $\beta$-chains found** **(*caption continued on next page*)**

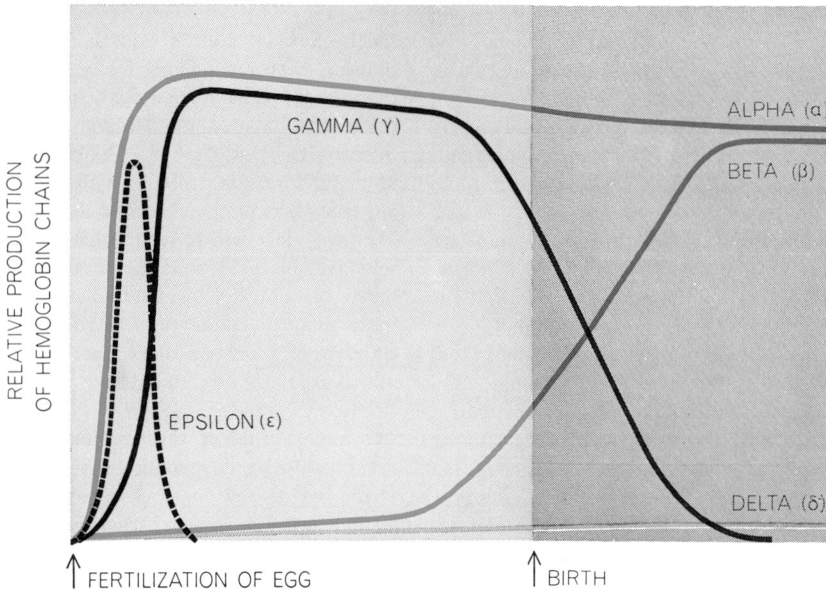


**OUTPUT OF HUMAN HEMOGLOBIN CHAINS shifts abruptly during fetal development. Throughout life two of the four subunits in normal hemoglobin are $\alpha$-chains. These chains pair first with epsilon ($\varepsilon$) chains, then with $\gamma$-chains. Just before birth $\beta$-chains begin to replace $\gamma$-chains. Simultaneously $\delta$-chains appear and also pair with some $\alpha$-chains.**

subject for chemical paleogenetics because it is produced in several slightly variant forms even within an individual organism, and the study of these variants suggests how their components may have descended from a common ancestral form. A molecule of hemoglobin is composed of four large subunits, each a polypeptide chain. Each chain enfolds an iron-containing "heme" group that can pick up an atom of oxygen as hemoglobin passes through the lungs and release it in tissues where oxygen is needed.

The principal kind of hemoglobin found in the human adult is composed of two alpha chains and two beta chains, and it is believed that they too have a common ancestry. The alpha chain comprises 141 amino acid residues; the beta chain, 146. Although the two chains are quite similar in their three-dimensional conformation, they differ considerably in composition. When the two chains are placed side by side, there are 77 sites where the residues in the two chains are different and only 64 sites where the residues are the same. Both the similarities and the differences are of interest to the chemical paleontologist.

The reader may wonder at this point how one can assume that the alpha and beta chains of human hemoglobin have a common ancestry if they are now more different than they are alike. The answer is that it seems most improbable that two different and unrelated polypeptide chains could evolve in such a way as to have the same function, the same conformation and a substantial number of identical amino acid residues at corresponding molecular sites. Consequently the chemical paleontologist interprets their marked difference in amino acid sequence as evidence that a long time has elapsed since they diverged from a common ancestor.

The argument for a common ancestry is strengthened by the fact that in the hemoglobins of man the beta chain is sometimes replaced by chains with still other amino acid sequences known as gamma, delta and epsilon chains. The epsilon chain is manufactured only for a brief period early in fetal life. The gamma chain replaces the beta chain during most of embryonic development and disappears almost entirely shortly after birth. Throughout adult life a small fraction of the hemoglobin in circulation contains delta chains rather than beta chains [*see top illustration on this page*]. The beta, gamma and delta chains are all 146 units long and closely resemble one another in amino acid sequence. There are only 39 differences in amino acid residues between the beta

| | | | ▲ | | | | | | | | | | ▲ | | | ● | | | | |
|---|---|---|---|---|---|---|---|---|---|---|---|---|---|---|---|---|---|---|---|---|
| ALA | HIS | LEU | ASP | ASN | LEU | LYS | GLY | THR | PHE | ALA | THR | LEU | SER | GLU | LEU | HIS | CYS | ASP | LYS | LEU |
| 76 | 77 | 78 | 79 | 80 | 81 | 82 | 83 | 84 | 85 | 86 | 87 | 88 | 89 | 90 | 91 | 92 | 93 | 94 | 95 | 96 |
| ALA | HIS | LEU | ASP | ASN | LEU | LYS | GLY | THR | PHE | SER | GLN | LEU | SER | GLU | LEU | HIS | CYS | ASP | LYS | LEU |
| 76 | 77 | 78 | 79 | 80 | 81 | 82 | 83 | 84 | 85 | 86 | 87 | 88 | 89 | 90 | 91 | 92 | 93 | 94 | 95 | 96 |
| LYS | HIS | LEU | ASP | ASP | LEU | LYS | GLY | THR | PHE | ALA | GLN | LEU | SER | GLU | LEU | HIS | CYS | ASP | LYS | LEU |
| 76 | 77 | 78 | 79 | 80 | 81 | 82 | 83 | 84 | 85 | 86 | 87 | 88 | 89 | 90 | 91 | 92 | 93 | 94 | 95 | 96 |
| ALA | HIS | VAL | ASP | ASP | MET | PRO | ASN | ALA | LEU | SER | ALA | LEU | SER | ASP | LEU | HIS | ALA | HIS | LYS | LEU |
| 71 | 72 | 73 | 74 | 75 | 76 | 77 | 78 | 79 | 80 | 81 | 82 | 83 | 84 | 85 | 86 | 87 | 88 | 89 | 90 | 91 |
| ALA | HIS | LEU | ASP | ASN | LEU | LYS | GLY | THR | PHE | ALA | THR | LEU | SER | GLU | LEU | HIS | CYS | ASP | LYS | LEU |
| 76 | 77 | 78 | 79 | 80 | 81 | 82 | 83 | 84 | 85 | 86 | 87 | 88 | 89 | 90 | 91 | 92 | 93 | 94 | 95 | 96 |
| LYS | HIS | LEU | ASP | ASN | LEU | LYS | GLY | THR | PHE | ALA | LYS | LEU | SER | GLU | LEU | HIS | CYS | ASP | GLU | LEU |
| 76 | 77 | 78 | 79 | 80 | 81 | 82 | 83 | 84 | 85 | 86 | 87 | 88 | 89 | 90 | 91 | 92 | 93 | 94 | 95 | 96 |
| HIS | HIS | LEU | ASP | ASN | LEU | LYS | GLY | THR | PHE | ALA | ALA | LEU | SER | GLU | LEU | HIS | CYS | ASP | LYS | LEU |
| 76 | 77 | 78 | 79 | 80 | 81 | 82 | 83 | 84 | 85 | 86 | 87 | 88 | 89 | 90 | 91 | 92 | 93 | 94 | 95 | 96 |
| LYS | LYS | LYS | GLY | HIS | HIS | GLU | ALA | GLU | LEU | LYS | PRO | LEU | ALA | GLN | SER | HIS | ALA | THR | LYS | HIS |
| 77 | 78 | 79 | 80 | 81 | 82 | 83 | 84 | 85 | 86 | 87 | 88 | 89 | 90 | 91 | 92 | 93 | 94 | 95 | 96 | 97 |

| | ▲ | ▲ | ▲ | | | ▲ | | ▲ | ▲ | | | | ▲ | | | | | ▲ | |
|---|---|---|---|---|---|---|---|---|---|---|---|---|---|---|---|---|---|---|---|
| HIS | VAL | ASP | PRO | GLU | ASN | PHE | ARG | LEU | LEU | GLY | ASN | VAL | LEU | VAL | CYS | VAL | LEU | ALA | HIS |
| 97 | 98 | 99 | 100 | 101 | 102 | 103 | 104 | 105 | 106 | 107 | 108 | 109 | 110 | 111 | 112 | 113 | 114 | 115 | 116 |
| HIS | VAL | ASP | PRO | GLU | ASN | PHE | ARG | LEU | LEU | GLY | ASN | VAL | LEU | VAL | CYS | VAL | LEU | ALA | ARG- |
| 97 | 98 | 99 | 100 | 101 | 102 | 103 | 104 | 105 | 106 | 107 | 108 | 109 | 110 | 111 | 112 | 113 | 114 | 115 | 116 |
| HIS | VAL | ASP | PRO | GLU | ASN | PHE | LYS | LEU | LEU | GLY | ASN | VAL | LEU | VAL | THR | VAL | LEU | ALA | ILEU |
| 97 | 98 | 99 | 100 | 101 | 102 | 103 | 104 | 105 | 106 | 107 | 108 | 109 | 110 | 111 | 112 | 113 | 114 | 115 | 116 |
| ARG | VAL | ASP | PRO | VAL | ASN | PHE | LYS | LEU | LEU | SER | HIS | CYS | LEU | LEU | VAL | THR | LEU | ALA | ALA |
| 92 | 93 | 94 | 95 | 96 | 97 | 98 | 99 | 100 | 101 | 102 | 103 | 104 | 105 | 106 | 107 | 108 | 109 | 110 | 111 |
| HIS | VAL | ASP | PRO | GLU | ASN | PHE | LYS | LEU | LEU | GLY | ASN | VAL | LEU | VAL | CYS | VAL | LEU | ALA | HIS |
| 97 | 98 | 99 | 100 | 101 | 102 | 103 | 104 | 105 | 106 | 107 | 108 | 109 | 110 | 111 | 112 | 113 | 114 | 115 | 116 |
| HIS | VAL | ASP | PRO | GLU | ASN | PHE | ARG | | | GLY | ASN | VAL | | VAL | | VAL | LEU | ALA | ARG- |
| 97 | 98 | 99 | 100 | 101 | 102 | 103 | 104 | 105 | 106 | 107 | 108 | 109 | 110 | 111 | 112 | 113 | 114 | 115 | 116 |
| HIS | VAL | ASP | PRO | GLU | ASN | PHE | ARG | LEU | LEU | GLY | ASN | VAL | LEU | ALA | LEU | VAL | VAL | ALA | ARG |
| 97 | 98 | 99 | 100 | 101 | 102 | 103 | 104 | 105 | 106 | 107 | 108 | 109 | 110 | 111 | 112 | 113 | 114 | 115 | 116 |
| LYS | ILEU | PRO | ILEU | LYS | TYR | LEU | GLU | PHE | ILEU | SER | GLU | ALA | ILEU | ILEU | HIS | VAL | LEU | HIS | SER- |
| 98 | 99 | 100 | 101 | 102 | 103 | 104 | 105 | 106 | 107 | 108 | 109 | 110 | 111 | 112 | 113 | 114 | 115 | 116 | 117 |

**in the hemoglobin molecules of gorillas, pigs and horses. The $\delta$-, $\gamma$- and $\alpha$-chains are ranked below the human $\beta$-chain in order of increasing number of differences. The gorilla $\beta$-chain differs from the human $\beta$-chain at only one site. The pig $\beta$-chain appears to differ at about 17 sites (based on the known differences), and the horse $\beta$-chain at 26 sites. The number of differences indicates roughly how far these animals are separated from man on the phylogenetic tree. Relatively few sites have been completely re-**

and the gamma chains and only 10 between the beta and the delta chains. The sequence of the human epsilon chain has not yet been established.

One other oxygen-carrying protein molecule figures in this discussion of hemoglobin evolution: the protein known as myoglobin, which does not circulate in the blood but acts as an oxygen repository in muscle. Myoglobin is a single polypeptide chain of 153 amino acid residues that has nearly the same three-dimensional configuration as the various hemoglobin chains. In fact, the unraveling of the three-dimensional structure of sperm whale myoglobin in 1958 by John C. Kendrew and his colleagues at the University of Cambridge marked the first complete determination of the structure of any protein molecule. Two years later Kendrew's colleague M. F. Perutz announced the three-dimensional conformation of the alpha and beta chains of horse hemoglobin; their topological similarity to myoglobin was immediately apparent [see "The Three-dimensional Structure of a Protein Molecule," by John C. Kendrew, SCIENTIFIC AMERICAN Offprint 121, and "The Hemoglobin Molecule," by M. F. Perutz, SCIENTIFIC AMERICAN, Offprint 196].

In amino acid sequence whale myoglobin and the alpha and beta chains of human hemoglobin are far apart. The sequence for human myoglobin is only now being determined, and it is apparent that it will be much closer to the sequence of whale myoglobin than to that of any of the human hemoglobin chains. Whale myoglobin and the alpha chain of human hemoglobin have the same residues at 37 sites; whale myoglobin and the human beta chain are alike at 35 sites. Again the chemical paleontologist regards it as probable that myoglobin and the various hemoglobin chains have descended from a remote common ancestor and are therefore homologous.

Although I have been speaking loosely of the evolution and descent of polypeptide chains, the reader should keep in mind that the molecular mutations underlying the evolutionary process take place not in polypeptide molecules but in the structural genes of DNA that carry the blueprint for each polypeptide chain. The effect of a single mutation of the most common kind is to change a single base in a structural gene, with the result that one triplet code word is changed into a different code word. Unless the new code word happens to specify the same amino acid as the old code word (which is sometimes the case) the altered gene will specify a polypeptide chain in which one of the amino acid residues is replaced by a different one. The effect of such a substitution is usually harmful to the organism, but from time to time a one-unit alteration in a polypeptide chain will increase the organism's chances of survival in a particular environment and the organism will transmit its altered gene to its progeny. This is the basic mechanism of natural selection.

As I have mentioned, there are also types of mutation that produce deletions or additions in a polypeptide chain. And there are the still more complex genetic events in which it is believed a structural gene is duplicated. One of the duplicates may later be shifted to a different location so that copies appear at two or more places in the genome. Such gene duplication, followed by independent mutation, would seem to account for the various homologues of hemoglobin found in all vertebrates.

Duplicate genes may have several values for an organism. For example, they may provide the organism with twice as much of a given polypeptide chain as it had before the duplication. They may also have subtler and more important values. It may be that the gamma chain found in fetal hemoglobin is particularly adapted to the needs of prenatal existence whereas the beta chain that replaces the gamma chain soon after birth is more suitable for life outside the womb. The precise value to the organism of having these two kinds of hemoglobin chain available at different stages of development remains to be discovered. It is somewhat puzzling that adult humans who have a certain genetically controlled abnormality go through life with gamma chains rather than beta chains in a significant fraction of their hemoglobin and show no ill effects.

Even without detailed knowledge of the role of duplicate genes it is clear that they are valuable both for the evolution of species and for the development of the individual organism. For purposes of evolution they provide two (or more) copies of genetic material that

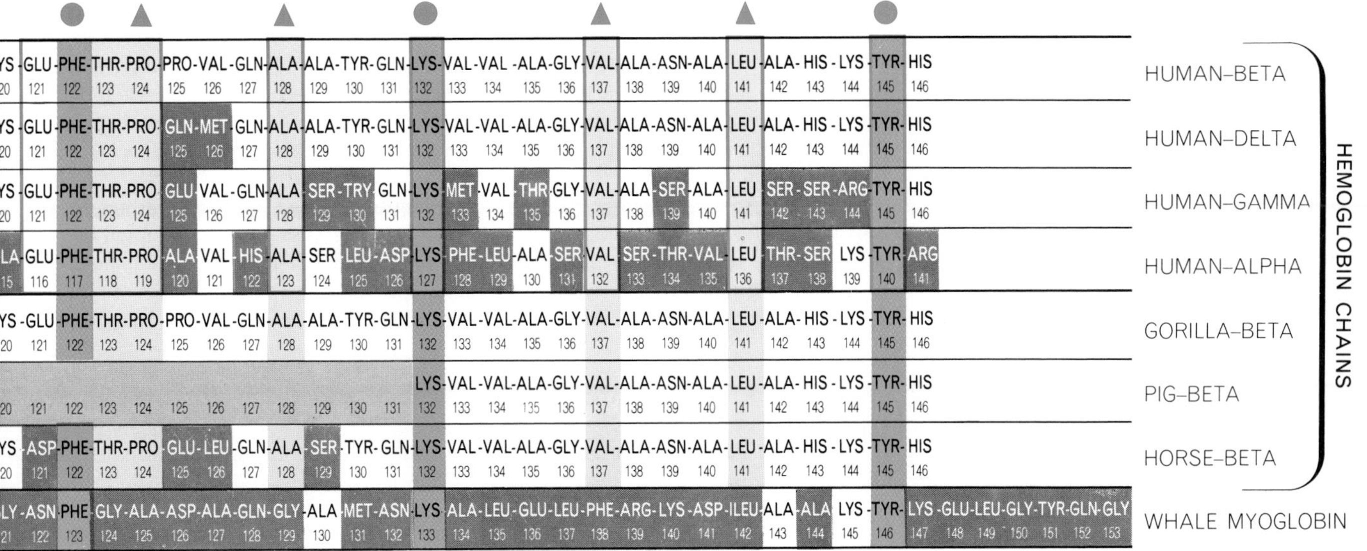


**sistant to evolutionary change. Only 11 of the sites (*colored circles*) have the same residues in all known hemoglobin and myoglobin chains, and only 15 more sites (*colored triangles*) have the same residues in all known chains of hemoglobin. Among the four principal chains of normal human hemoglobin the same residues are found at 49 sites. The $\beta$-, $\delta$- and $\gamma$-chains, which are closely related, have 103 sites in common. The three-dimensional conformation of all these chains is illustrated at the top of the next page.**

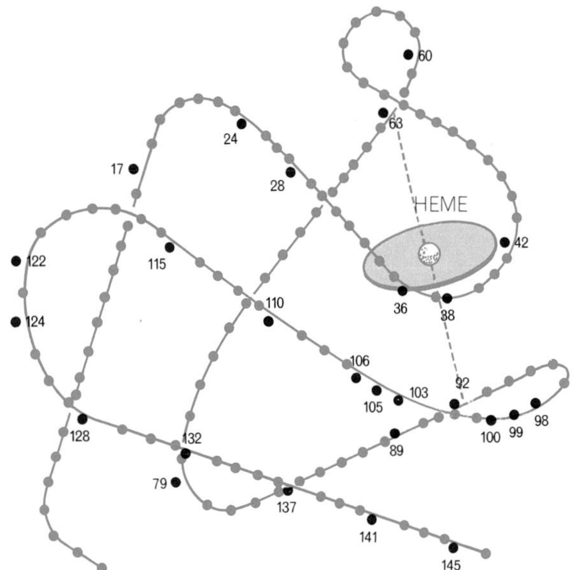


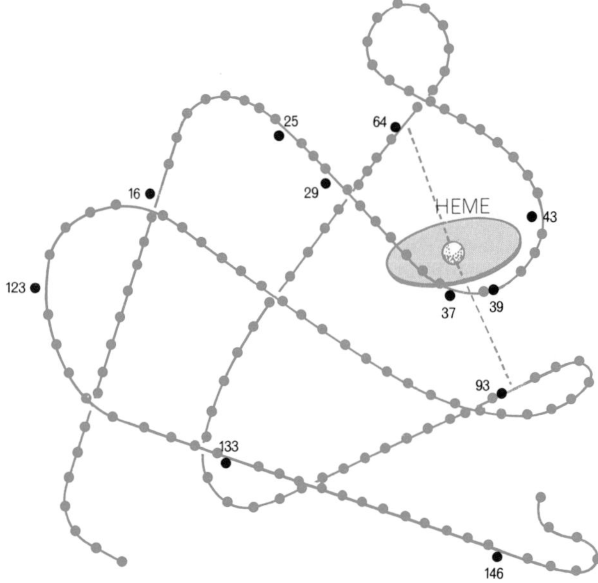


**INVARIANT SITES** are identified on knotlike shapes representing the three-dimensional structure of the polypeptide chains of hemoglobin and myoglobin. The 26 numbered sites at left are occupied by the same residues in all known hemoglobin chains. Eleven of these same sites (*assigned slightly different numbers at right*) are occupied by the same residues in all known chains of hemoglobin and myoglobin. Presumably the invariant sites are important in establishing the structure and function of these polypeptides.

are free to evolve separately. Thus a duplicate gene may be transformed so completely that it gives rise to a new type of polypeptide chain with a function entirely different from that of its ancestor. In the life history of an individual organism the existence of duplicate genes at different sites in the genome enables the organism to obtain a supply of an essential polypeptide without activation of the whole genome. In this way gene duplication makes possible a more complex pattern of gene activation and inactivation during an organism's development.

It is not always easy to decide when two polypeptide chains are homologous and when they are not. As long as one is dealing with rather similar chains that serve the same function—as in the case of the various hemoglobin chains—there is a strong *prima facie* case for homology. As the amino acid sequences of more and more polypeptides are deciphered, however, one can expect ambiguities to arise.

One potential source of ambiguity arises in the identification of "corresponding" molecular sites. Such sites are often made to correspond by shifting parts of one chain with respect to the homologous chain [*see illustration below*]. The shifts are justified on the grounds that deletions or additions of one to several residues in a row have occurred during the evolution of certain polypeptide chains. A shift is considered successful when it maximizes the number of identities between the segments of two chains. The argument, therefore, is somewhat circular in that the shifts are justified by the presumed deletions (or additions) and the deletions (or additions) by the shifts. The argument that breaks the circle is that by invoking a small number of shifts, homologous polypeptide chains can be brought to display remarkable coincidences, whereas nonhomologous chains cannot be. There remains, however, the problem of placing the concept of homology on an objective basis. An effort is being made to do this with the help of a computer analysis of real and hypothetical polypeptide chains.

Now that the reader has this background I can provide a more detailed statement of the aims and methods of chemical paleogenetics. Fundamentally it attempts to discover the probable amino acid sequence of ancestral polypeptide chains and also the probable base sequence in the genes that controlled them. It is concerned with the fate of the descendent line of each gene. It inquires whether gene duplication has occurred and, if so, when it occurred; it asks what became of the duplicate genes, how they may have been shifted to various parts of the genome and how they have mutated. Finally it is concerned with the factors that regulate the rate and timing of the synthesis of the various polypeptide chains.

Present evidence suggests (although exceptions are known) that the number of differences between homologous polypeptide chains of a certain type found in different animals is roughly

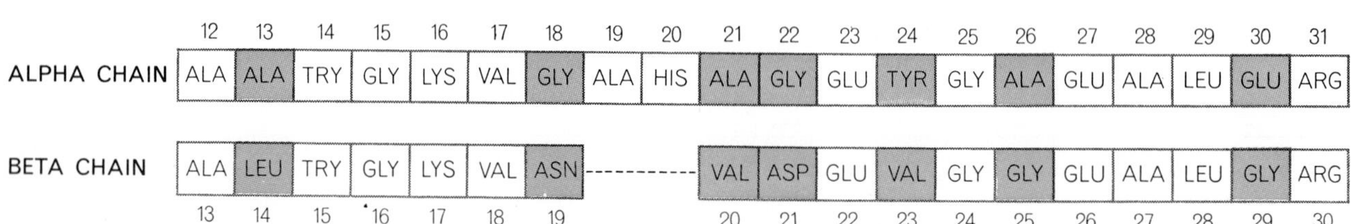


**CORRESPONDING REGIONS** of the $\alpha$- and $\beta$-chains of human hemoglobin show how a short deletion must be inferred in the $\beta$-chain to produce a good match at corresponding sites. An earlier one-unit deletion in the $\alpha$-chain explains why $\alpha$-site 12 corresponds to $\beta$-site 13. By postulating the two-unit deletion shown here the two chains can be made to have the same residues at 11 sites.

proportional to the relatedness of these animals as established by standard methods of phylogenetic classification. Indeed, the readily observable differences among living things must be to a significant extent the expression of differences in their enzymes—the proteins that catalyze the chemical reactions of life—and therefore of differences in the amino acid sequences of the polypeptide chains that form the enzymes. It is probable that observable differences also reflect differences in the regulation of rate and timing of the synthesis of polypeptide chains rather than differences in the amino acid sequence of these chains.

On the other hand, a difference in sequence may express itself primarily as a difference in rate and timing. It is quite probable that regulatory enzymes play an important role, and less obvious regulatory mechanisms may also exist. It has been suggested, for example, that differences in rate and timing may be attributable to certain sequences of bases in DNA that never find expression in a polypeptide chain. It seems in the last analysis, however, that the differences between organisms, if the environment is kept constant, boil down to differences in molecular sequences. These differences may reside in base sequences in genes, which are then expressed in amino acid sequences in polypeptide chains; they may reside in other base sequences that are not so expressed; finally they may reside in the sequential order in which genes are distributed within the genome.

Although chemical paleogenetics will ultimately have taxonomic value in providing a fundamental way of measuring the distance between living things on the evolutionary scale, this is not its prime objective. A major value of analyzing evolutionary changes at the molecular level will be to provide a deeper understanding of natural selection in relation to different types of mutation.

Let me proceed, then, to apply the methods of chemical paleogenetics to the myoglobin-hemoglobin family of polypeptide chains. The top illustration at the right shows the number of differences in amino acid sequence between four animal-hemoglobin alpha and beta chains and the corresponding human chains. For purposes of rough computation let us assume that the alpha and beta chains evolve at the same rate and pool the number of differences they exhibit. The reason for doing this is to

| ANIMAL | NUMBER OF DIFFERENCES: ALPHA CHAIN | NUMBER OF DIFFERENCES: BETA CHAIN | MEAN NUMBER OF DIFFERENCES, ALL CHAINS | ESTIMATED TIME SINCE COMMON ANCESTOR |
|---|---|---|---|---|
| HORSE | 17 | 26 | ~22 | 80 MILLION YEARS |
| PIG | ~18 | ~17 | | |
| CATTLE | ~27 | | | |
| RABBIT | ~27 | | | |

**COMPARISON OF HEMOGLOBIN CHAINS offers a way to estimate the number of years required to produce an evolutionarily effective change at one site. The values given here for the number of differences represent a comparison with the $\alpha$- and $\beta$-chains of human hemoglobin. The mean of 22 differences between any pair of human and animal chains implies an average of 11 mutations per chain, or about one change per seven million years.**

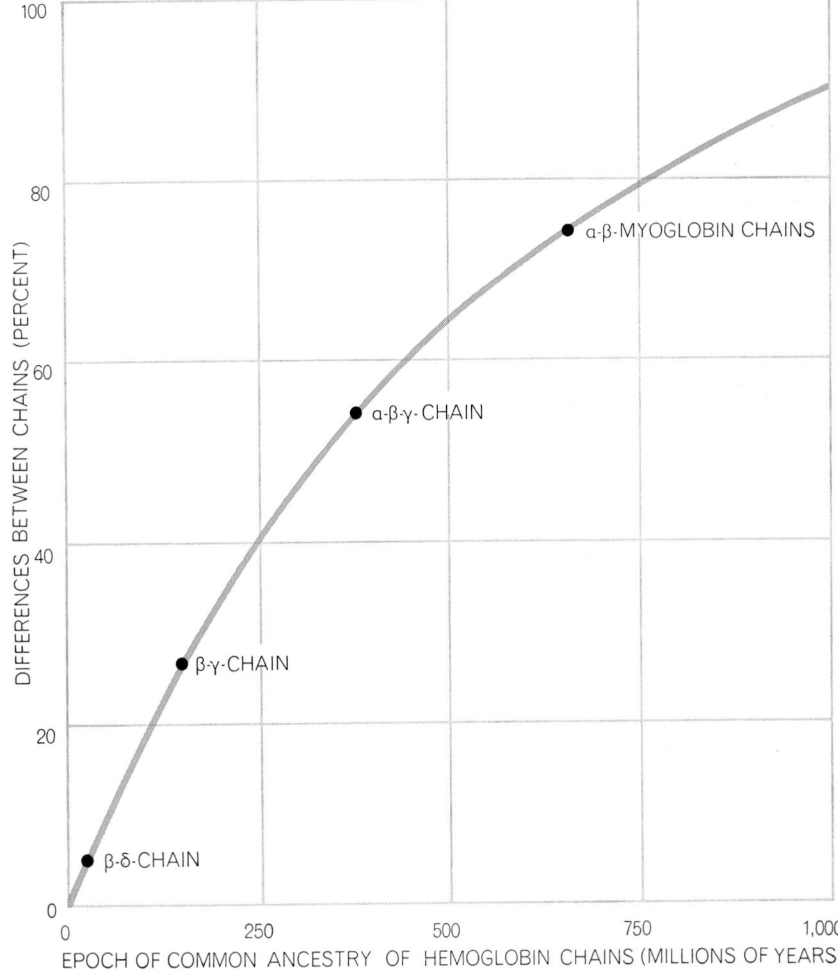


**AGE OF ANCESTRAL HEMOGLOBIN-MYOGLOBIN CHAINS is plotted on a curve computed by Linus Pauling. Except for myoglobin the chains represented are those of humans. Where only a few differences are observed it is assumed that about seven million years are needed to establish an effective mutation. But where chains show large differences today it can be assumed that more than one mutation occurred at a given site in the course of evolution. For example, the $\alpha$-chain and $\beta$-chain each differ from the myoglobin chain at about 110 sites. Thus the ancestral $\alpha$-$\beta$-myoglobin chain appears on the curve where the vertical axis reads 75 percent (110/146 is about 3/4). This point corresponds to a period about 650 million years ago rather than the 385 million years that would be obtained if 55 mutations per gene line (110 ÷ 2) were simply multiplied by seven million.**

establish a mean value for the number of apparent amino acid substitutions that have occurred in the alpha and beta chains of the four animal species (horse, pig, cattle and rabbit) since the time when the four species and man had a common ancestor. The mean difference is 22 apparent changes in the two chains, or an average of 11 changes per chain. If the common ancestor of man and the four other animals lived about 80 million years ago, as is thought to be the case, the average time required to establish a successful amino acid substitution in any species is about seven million years. Until more chains have been analyzed, however, 10 million years per substitution is a good order-of-magnitude figure.

Such a figure can now be used for a different purpose: to estimate very roughly the time elapsed since the four principal types of chain in human hemoglobin had a common ancestor. In making such a calculation one must employ statistical principles to allow for the following fact. The greater the number of differences in sequence between two homologous chains, the greater the chances that at some molecular sites more than one amino acid substitution will have been retained temporarily by evolution since the time of the common ancestor. An appropriate calculation was recently performed by Linus Pauling, with the result shown in the bottom illustration on the preceding page. The curve in the illustration allows one to read off the probable time of existence of the common molecular ancestor of various polypeptide chains as a function of the percentage of differences in amino acid sequence between the chains.

The two chains that are most nearly alike—the beta and delta chains—differ at only 10 sites and presumably were the most recent to arise by duplication of a common genetic ancestor. To exhibit 10 differences each gene line would have to undergo only five changes, which implies an elapsed time of roughly 35 million years on Pauling's chart. The beta and gamma chains are different at 37 sites and thus seem to have arisen by gene duplication about 150 million years ago. The beta and alpha chains are different at 76 sites and therefore their common ancestor goes back some 380 million years. If the calculation is valid as a rough approximation, the common genetic ancestor of the hemoglobin chains now circulating in the human bloodstream dates back to the Devonian period and to the appearance of the first amphibians.

The curve also indicates very roughly how long it has been since the chains of hemoglobin and myoglobin may have arisen as the result of duplication of a common ancestral gene. The differences in amino acid sequence between hemoglobin chains and myoglobin are so numerous that their common molecular ancestor may date back about 650 million years to the end of the Precambrian era, long before the appearance of the vertebrates. This suggests, in turn, that it may be possible to find in living invertebrates a distant relative of the vertebrate hemoglobins and myoglobins.

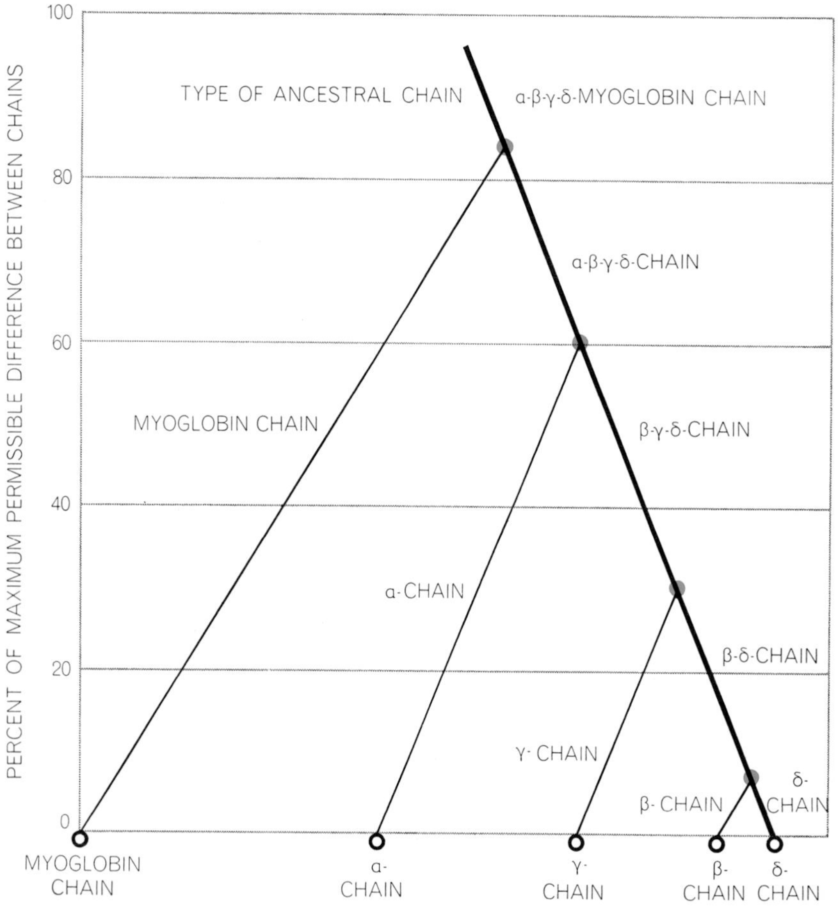

Let me turn now from discussing the overall differences between homologous polypeptide chains to the question of how one might construct a molecular "phylogenetic tree." Such a tree would show an evolutionary line of descent for an entire family of polypeptide molecules. One can also construct individual trees for individual molecular sites. Later this site-by-site information can be synthesized to obtain probable residue sequences for complete ancestral chains.

If the amino acid residue is the same in two homologous chains at a given molecular site, there is a certain probability that the same residue was also present in the common ancestor of the two chains. There is also a chance, of course, that the ancestral residue was different and that the identity observed in the two homologous existing chains was produced by molecular convergence or simply by coincidence. Traditional paleontology reveals many examples of convergences at the level of large-scale morphology. In chemical paleogenetics molecular convergence or coincidence is particularly troublesome because the path from difference to similarity runs directly counter to that needed to trace a molecular phylogenetic tree. About all one can say at this stage in the development of the new discipline is that convergence or coincidence do not seem to occur often enough to vitiate the effort of constructing such trees.

The illustration at the left shows



**HEMOGLOBIN-MYOGLOBIN RELATIONSHIP is traced back through evolution, based on the number of differences in the various chains. The four colored dots indicate where ancestral genes were presumably duplicated, giving rise each time to a new gene line.**

schematically, in the form of an inverted tree, the probable evolutionary relationships for the known chains of human hemoglobin and myoglobin. The tree also represents the relationships of any given molecular site in these chains. The epsilon chain has been omitted because too little is known about it. The vertical axis is not an absolute time scale but shows how chain differences rate on a scale in which the maximum permissible difference is 100 percent. Some of the branching points in the tree are assumed to coincide with a gene duplication. Following such duplication the resulting independent genes (and their polypeptide chains) evolve separately. The most ancient duplication presumably separated the myoglobin gene from the gene that ultimately gave rise, by repeated duplications, to the alpha, beta, gamma and delta genes of hemoglobin. Additional gene duplications will surely have to be postulated along various lines.

The next molecular phylogenetic tree [*see top illustration at right*] attempts to reconstruct the evolutionary changes at one particular site (site No. 4 in the human alpha chain) that led to the amino acid residues now observed at that site in various animal species, including man. As the genetic code is being worked out, it is becoming possible to distinguish amino acid substitutions that may have occurred in one step from those requiring two or more steps. It is a principle of chemical paleogenetics that in postulating a possible ancestral amino acid residue one should prefer the residue that can be reached by invoking the fewest number of mutations in the genetic message. In the tree just referred to the residue of the amino acid alanine has been selected as the residue at site No. 4 in the ancestral polypeptide chain from which the 17 present-day hemoglobin chains are descended. This selection may seem odd; among the 17 chains eight have proline in the No. 4 position and only four have alanine. (The remaining five chains have glycine, glutamic acid or serine in the No. 4 position.) The explanation is that if proline is assumed to be in the No. 4 position in the most remote ancestral chain, one has to postulate nine or 10 evolutionarily effective amino acid substitutions in the various descendent chains to reach the residues actually observed in the 17 present-day chains, but if alanine is selected as the ancestral residue, only eight effective substitutions are needed.

The choice of alanine becomes more impressive when it is shown that no

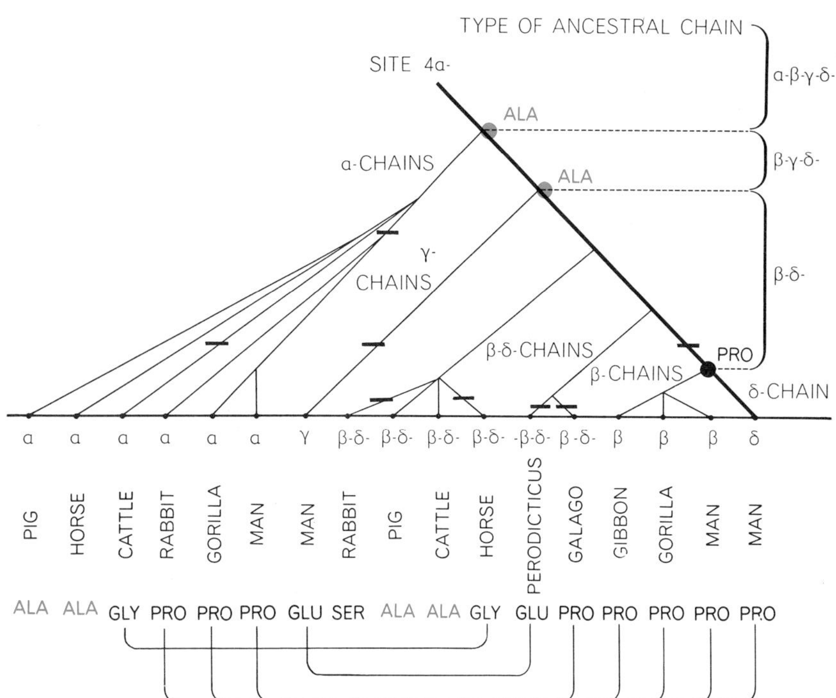


**ANCESTRAL RESIDUE can be traced by trying to establish the simplest lines of descent for residues now found at a particular site in polypeptide chains of hemoglobin. The residues shown across the bottom occupy site No. 4 in the human $\alpha$-chain. (*Perodicticus* and *Galago* are small monkeys commonly known as the potto and the bush baby.) Alanine (*ala*) is selected as the probable residue in the earliest ancestral chain because it provides a line of descent requiring fewer mutations than any other that might be selected: eight. They are represented by short horizontal bars. The lines at the bottom identify convergences or coincidences: identical residues that presumably resulted from independent mutations.**

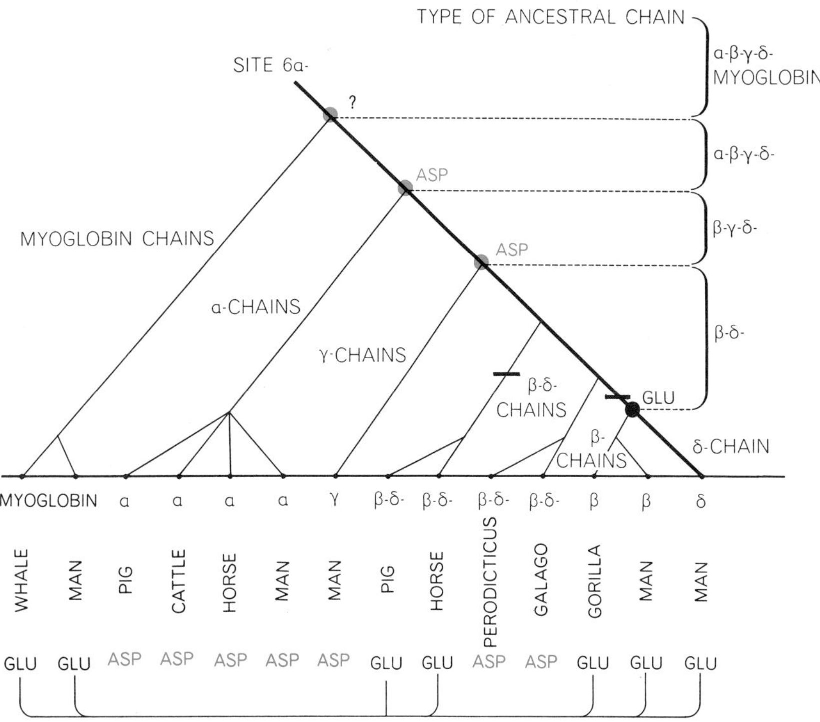


**AMBIGUOUS ANCESTRAL RELATIONSHIP is encountered when the present-day residues at a particular site are those of two amino acids that frequently replace each other, such as glutamic acid (*glu*) and aspartic acid (*asp*). The sites compared are No. 6 in the $\alpha$-chain of human hemoglobin. Note that myoglobin has been included in this evolutionary tree.**

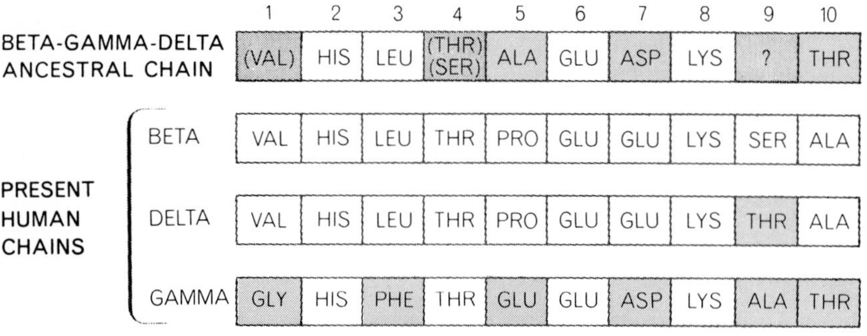

1 2 3 4 5 6 7 8 9 10
BETA-GAMMA-DELTA ANCESTRAL CHAIN
(VAL) HIS LEU (THR) (SER) ALA GLU ASP LYS ? THR
PRESENT HUMAN CHAINS
BETA
VAL HIS LEU THR PRO GLU GLU LYS SER ALA
DELTA
VAL HIS LEU THR PRO GLU GLU LYS THR ALA
GAMMA
GLY HIS PHE THR GLU GLU ASP LYS ALA THR


**RECONSTRUCTION OF ANCESTRAL CHAIN represents a synthesis of evolutionary trees for individual sites as illustrated on the preceding page. This chart shows the first 10 sites in the ancestral $\beta$-$\gamma$-$\delta$-chain and the corresponding region in its three present-day descendants. Residues in the $\delta$-, $\gamma$- and ancestral chains that differ from those in the contemporary $\beta$-chain are shown in color. Gray indicates uncertain or unknown residues.**

more than one amino acid substitution is needed in any single line of descent to explain the residues currently observed. If proline is made the ancestral choice, double substitutions—the ones least likely to occur—must be postulated in three of the lines of descent. The choice of any other amino acid for the ancestral position would necessitate many more substitutions. Alanine is therefore adopted as the most probable ancestral residue—a conclusion that is not likely to need revision unless the genetic code is revised with regard to proline. In this particular example molecular coincidence is represented in some of the chains that now contain proline, glutamic acid and glycine.

The problem of identifying a probable ancestral residue is often difficult. At a site where there are frequent interchanges between residues that seem to be more or less functionally equivalent, any conclusion about ancestry becomes doubtful. This is demonstrated at site No. 6, as numbered in the human alpha chain, where there is a frequent interchange between aspartic acid and glutamic acid [*see bottom illustration on preceding page*].

The information from a series of molecular phylogenetic trees can finally be synthesized to produce a complete sequence of residues representing the composition of an ancestral polypeptide chain. The illustration above shows such a postulated sequence for the first 10 sites of the polypeptide chain that is presumed to be the ancestor of the beta, gamma and delta chains now present in human hemoglobin.

In order to establish by chemical paleogenetics the evolutionary relationship between two different organisms it should not be necessary to know the sequential composition of thousands or even hundreds of homologous polypeptide chains. To require such knowledge would be discouraging indeed. It can reasonably be predicted, however, that a comparison of relatively few chains—perhaps a few dozen—should yield a large fraction of the maximum amount of information that polypeptide chains can provide. The reason is that even relatively few chains should yield a good statistical sample of the evolutionary behavior of many chains.

Chemical paleogenetics offers many new possibilities. For example, after one has reconstructed a number of ancestral polypeptides for some ancient organism one should be able to make various deductions about some of its physiological functions. One might be able to decide, for instance, whether it could live successfully in an atmosphere composed as we know it today or whether it was designed for a different atmosphere.

From similar polypeptide reconstructions it may be possible to make informed guesses about organisms, such as soft-bodied animals, that have left no fossil record. In this way the state of living matter in past evolutionary times can be pieced together, at least in part, without the help of fossil remains. But one of the main attractions of chemical paleogenetics is the possibility of deriving strictly from molecular sequences a phylogenetic tree that is entirely independent of phylogenetic evidence gathered by traditional methods. If this can be accomplished, one can compare the two kinds of phylogenetic tree—the molecular and the traditional—and see if they tell the same story of evolution. If they do, chemical paleogenetics will have provided a powerful and independent confirmation of the already well-documented theory of evolution.

# 28

# *Computer Analysis of Protein Evolution*

MARGARET OAKLEY DAYHOFF

*July 1969*

The protein molecules that determine the form and function of every living thing are intricately folded chains of amino acid units. The primary structure of each protein—the sequence in which its amino acid units are linked together—is governed by the sequence of subunits in the nucleic acid of the genetic material. The proteins of an organism are therefore the immediate manifestation of its genetic endowment. From a biochemical point of view a fungus and a man are different primarily because each of them has a different complement of proteins.

Yet human beings and fungi and organisms of intermediate biological complexity have some proteins in common. These homologous proteins are quite similar in structure, reflecting the ultimate common ancestry of all living things and the remarkable extent to which proteins have been conserved throughout geologic time. Because of this conservation the millions of proteins existing today are in effect living fossils: they contain information about their own origin and history and about the ancestry and evolution of the organisms in which they are found. The comparative study of proteins therefore provides an approach to critical issues in biology: the exact relation and order of origin of the major groups of organisms, the evolution of the genetic and metabolic complexity of present-day organisms and the nature of biochemical processes. A new discipline, chemical paleogenetics, concerns itself with such studies [see "The Evolution of Hemoglobin," by Emile Zuckerkandl, which begins on page 256 of this book]. In order to exploit the possibilities of this new field we have developed a computer technique for analyzing the relations among protein sequences.

The body of data available in protein sequences is something fundamentally new in biology and biochemistry, unprecedented in quantity, in concentrated information content and in conceptual simplicity. The data give direct information about the chemical linkage of atoms, and that linkage determines how protein chains coil, fold and cross-link—and thus establishes the three-dimensional structure of proteins. Because of our interest in the theoretical aspects of protein structure our group at the National Biomedical Research Foundation has long maintained a collection of known sequences. For the past four years we have published an annual *Atlas of Protein Sequence and Structure,* the latest volume of which contains nearly 500 sequences or partial sequences established by several hundred workers in various laboratories. In addition to the sequences, we include in the *Atlas* theoretical inferences and the results of computer-aided analyses that illuminate such inferences. This article is based in part on that material, to which contributions have been made by Chan Mo Park, Minnie R. Sochard, Lois T. Hunt and Patricia J. McLaughlin, and by Richard V. Eck, now of the University of Georgia.

Basic metabolic processes are similar in all living cells. Many identical structures, mechanisms, compounds and chemical pathways are found in widely diverse organisms; even the genetic code is almost the same in all species. It is by this code that the sequence of nucleotides, or nucleic acid subunits, that constitutes a gene is translated into the amino acid sequence of the protein derived from it. It is therefore not surprising that a large number of proteins have been found to have identifiable counterparts in most living things. These homologues appear to perform the same functions in the organisms in which they are found, and they can often be substituted for one another in laboratory experiments. Being complex substances, they are only rarely identical, but in the past 15 years homologous proteins have been shown to have nearly the same amino acid sequences and quite similar three-dimensional structures.

One such protein whose amino acid sequence has been established for a number of species is the protein of cytochrome *c,* a complex substance that in animals and higher plants is found in the cellular organelles called mitochondria, where it plays a role in biological oxidation. Twenty different sequences of cytochrome *c* have been identified and analyzed by a number of investigators, including Emil L. Smith of the University of California at Los Angeles, Emanuel Margoliash of the Abbott Laboratories and Shung Kai Chan and I. Tulloss of the University of Kentucky. Recently Karl M. Dus, Knut Sletten and Martin D. Kamen of the University of California at San Diego found a clearly related protein in a bacterium, *Rhodospirillum rubrum,* which lacks mitochondria.

The correspondence in amino acid sequence among these proteins is clear when the sequences are arrayed below one another [*see top illustration on next two pages*]. There are differences in length, reflecting additions or deletions of nucleotides in the corresponding genes. These changes are at the ends of sequences except for the internal deletions or additions revealed by the bacterial protein. Once the sequences have been adjusted to allow for these changes there is no question about the correct alignment. In man and the gray kangaroo, for example, the amino acids are the same in 94 out of 104 positions; in the less similar human and baker's yeast sequences, 64 positions conform, or some

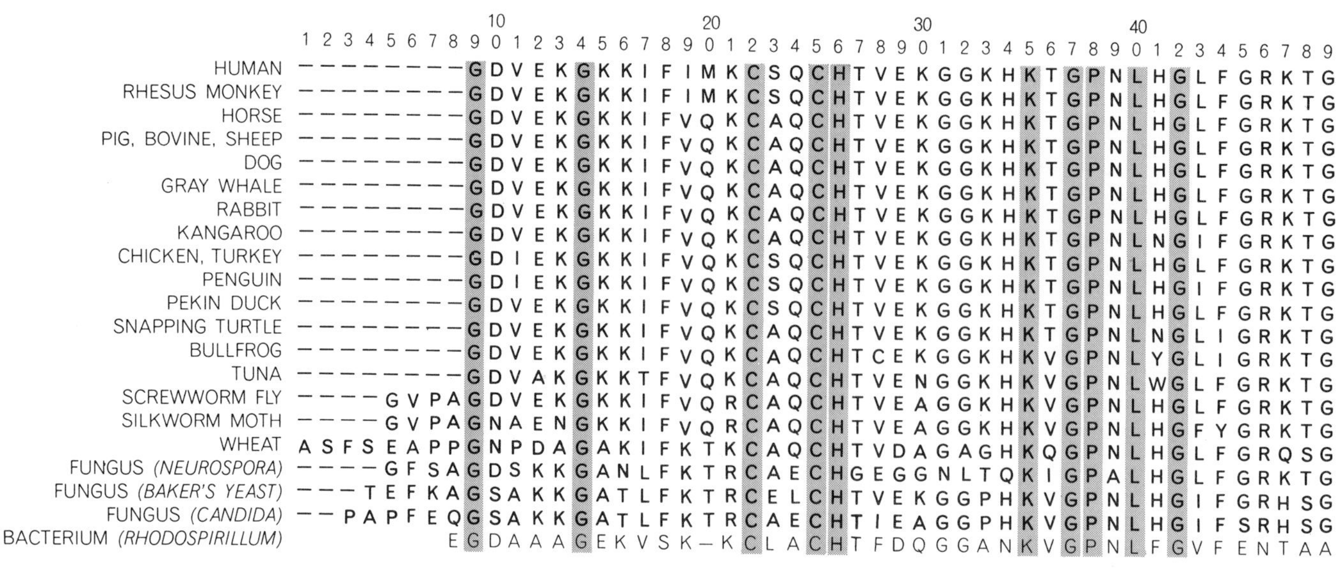


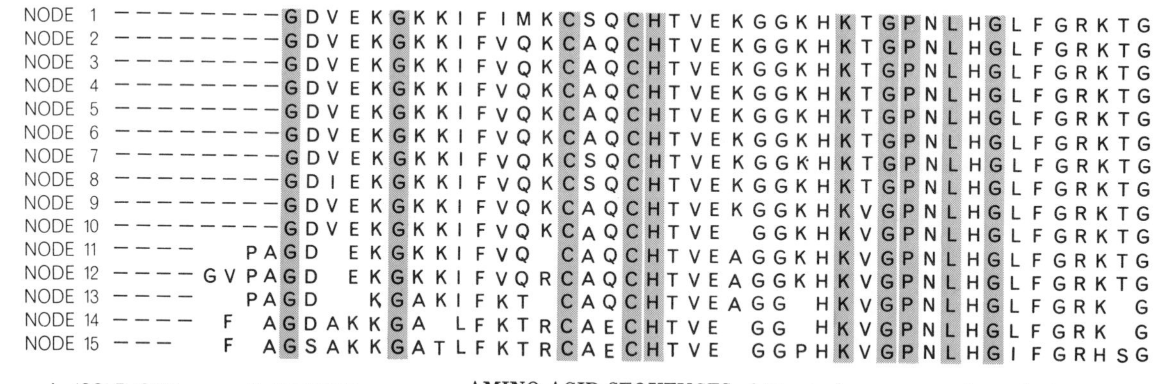


A ALANINE
C CYSTEINE
D ASPARTIC ACID
E GLUTAMIC ACID
F PHENYLALANINE
G GLYCINE
H HISTIDINE
I ISOLEUCINE
K LYSINE
L LEUCINE
M METHIONINE
N ASPARAGINE
P PROLINE
Q GLUTAMINE
R ARGININE
S SERINE
T THREONINE
V VALINE
W TRYPTOPHAN
Y TYROSINE

**AMINO ACID SEQUENCES of 20 cytochrome *c* proteins and of a related bacterial protein are arrayed below one another. For the purposes of the computer each amino acid is represented by a single letter (*see key at left*) instead of the usual three-letter symbol. The proteins differ in length, and dashes have been inserted in order to preserve the correct alignment; these differences come at**

three-fifths of the total length. All 21 sequences, including the bacterial one, have the same amino acid in 20 positions. When the amino acids at a given position are not the same, they usually have similar shapes or chemical properties.

Such similarity of sequence is impressive testimony to the evolution of all these organisms from common ancestors, confirming earlier morphological, embryological and fossil evidence. The alternative to common ancestry—that the similar cytochrome *c* proteins originated independently in different organisms—is not plausible. Consider the probability of duplicating the sequence of amino acids in just one chain 100 units long. Since any of 20 amino acids can occur in every position, the number of different possible chains is $20^{100}$. With so many possibilities it is improbable that two unrelated organisms would happen independently to have manufactured—

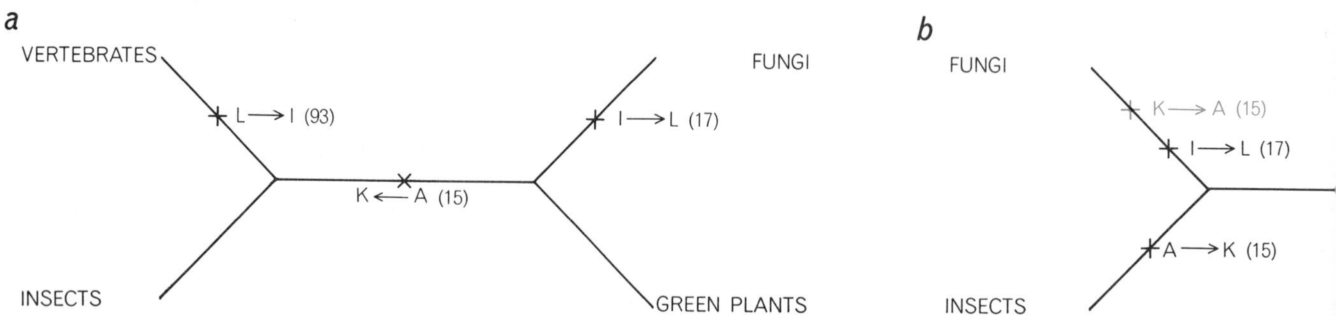


**LINEAGES of major groups are constructed from the evidence at three positions in the sequences in order to illustrate the principles involved in constructing a phylogenetic tree. At Position 15 vertebrates and insects have a *K* (lysine); fungi and wheat have an *A* (alanine). This suggests that a single lineage connected the animals and plants and that a single mutation from *A* to *K* took**

```
QAPGYSYTAANKNKGITWGEDTLMEYLENPKKYIPGTKMIFVGIKKKEERADLIAYLKKATNE
QAPGFSYTDANKNKGITWGEDTLMEYLENPKKYIPGTKMIFAGIKKKGERADLIAYLKKATNE
QAPGFSYTDANKNKGITWGEETLMEYLENPKKYIPGTKMIFAGIKKKGERADLIAYLKKATNE
QAPGFSYTDANKNKGITWGEETLMEYLENPKKYIPGTKMIFAGIKKKGEREDLIAYLKKATNE
QA GFSYTDANKNKGITWGEDTLMEYLENPKKYIPGTKMIFAGIKKKGERADLIAYLK ATSK
QAEGFSYTDANKNKGITWGEDTLMEYLENPKKYIPGTKMIFAGIKKK ERADLIAYLKDATSK
QAEGFSYTDANKNKGITWGEDTLMEYLENPKKYIPGTKMIFAGIKKKSERADLIAYLKDATSK
QAEGFSYTDANKNKGITWGEDTLMEYLENPKKYIPGTKMIFAGIKKKSERADLIAYLKDATSK
QA GFSYTDANKNKGITWGEDTLMEYLENPKKYIPGTKMIFAGIKKKGER DLIAYLKSATSK
QA GYSYTDANKNKGITWGEDTLMEYLENPKKYIPGTKMIFAGIKKKGER DLIAYLKSATS–
QAAGYSYT ANKNKGITWGEDTLFEYLENPKKYIPGTKM FAGLKK  ERADLIAYLK AT –
QAAGFSYTNANKAKGITWGDDTLFEYLENPKKYIPGTKM FAGLKK NERADLIAYLK ATK–
QAAGYSYT ANKNKG  W ENTLFEYLENPKKYIPGTKM F GLKK  DRADLIAYLK AT –
QA GYSYTDANK KG  WDENTLFEYLENPKKYIPGTKMAFGGLKK KDRNDLITY KEAT –
QAQGYSYTDANK KGV WDENTMSEYLENPKKYIPGTKMAFGGLKKAKDRNDLITY KEA  –
```

**the ends of sequences except in the case of the bacterium, where there are internal differences in length. The amino acid positions are numbered according to the wheat sequence, which has 112 amino acids. At 20 positions (*color*) the same amino acid is found in all the sequences, and the degree of identity is far greater among related species. These observed sequences constitute the raw data that are fed into the computer. The computer determines the ancestral sequences that can best account for the relations among observed sequences. These ancestral sequences establish the nodes: locations at which the branches of the phylogenetic tree diverge. Node 1, for example, is the ancestor of the primates, Node 2 is the mammalian ancestor and Node 10 is the vertebrate ancestor.**

and to have preserved through natural selection—such similar structures. On the other hand, gradual evolution from a common ancestor through millions of generations provides a convincing explanation for both the similarities and the differences among present-day cytochrome sequences.

The evolutionary process is made possible by mutations: errors in the copying and passing along of genetic material from generation to generation. The most frequently accepted mutation within a gene is the exchange of a single nucleotide for another, which may yield a protein that has one amino acid changed. A second kind of error is the duplication of a portion of a gene. This can yield an elongated gene or, often, two almost complete copies of the original genetic material that proceed to mutate independently. Finally, nucleotides can be deleted or inserted, resulting in a protein of altered length.

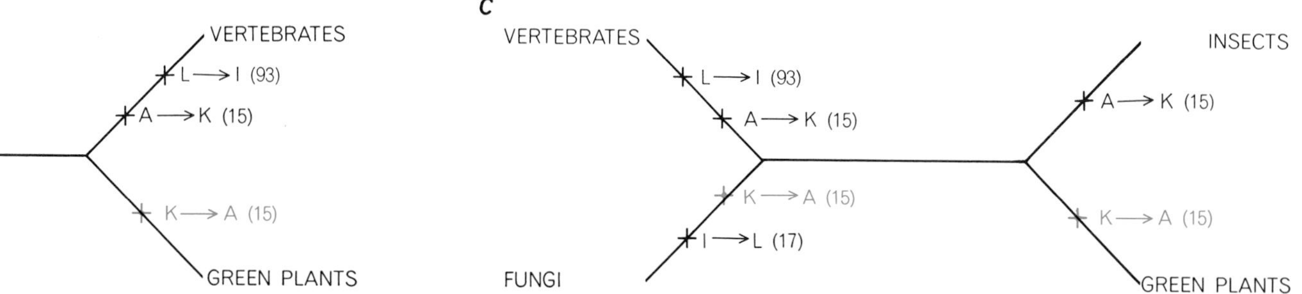


**place between them. This reasoning, together with similar reasoning from evidence at Position 17 and Position 93 (*see text*), suggests a certain topological relation of lineages (*a*). It includes three mutations. There are two other possible configurations (*b, c*) but they each require four mutations, two of which have alternative forms (*color*). The first topology is therefore the most probable one.**

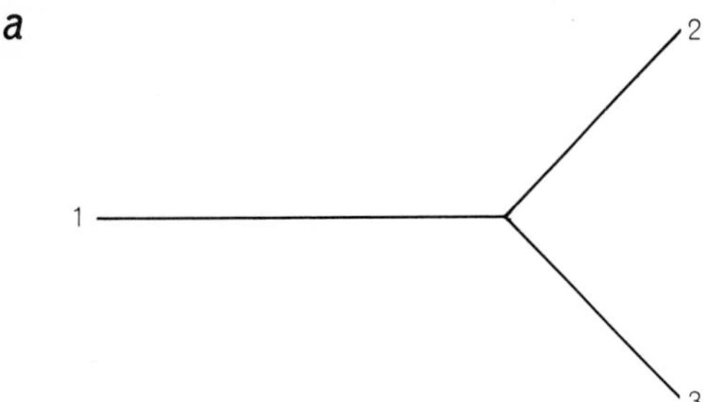


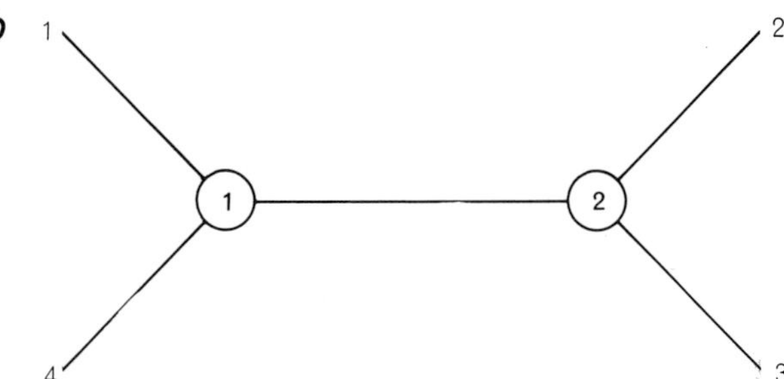


**COMPUTER PROGRAM 2 builds an approximate topology by beginning with three observed sequences, which can only be related by a simple three-branch topology (*a*). It then adds a fourth sequence to each of the three original branches in turn, establish-**

Over billions of years many such errors have occurred in individual organisms. A few have been selected as beneficial and perpetuated in the species; most have been deleterious and have been eliminated. One pressure against their selection is the biochemical conservatism that results from the interdependence of the various cell components. A protein must automatically fold into a precise three-dimensional shape when it is synthesized, and the shape is predetermined by the sequence. Each protein becomes adapted to performing a particular function in which it must interact with other components, whether through its chemical action, through the complexes it forms, through the rate at which its reactions proceed or through its structural properties. Moreover, all these capabilities must be little disturbed by changes or extremes in the environment.

Under such circumstances as these, most changes in protein sequence—even if they are advantageous for a particular function—are likely to disturb so many other interactions as to be almost always deleterious on balance. So severe are these constraints that an identical sequence of each protein is found in most individuals of a species, and a given sequence may be predominant in a species for several million years. Occasionally a minor variant may be tolerated, persist and eventually become preferable because of a change in other cell components or in the external environment. In other cases a rarely occurring error may immediately prove to be beneficial. Sometimes the environmental circumstances are so strained or the beneficial error is so profound that two separate populations or even two species develop. Subsequent changes arise independently in the two separate groups.

The degree of difference among present-day species and the order of their derivation from common ancestors are commonly represented by a phylogenetic tree. It is possible to derive such a tree from protein-sequence data. The basic method is to infer from observed sequences the ancestral sequences of the proteins from which they diverged, and thus to establish a series of nodes that define the connections of twigs to branches and of branches to limbs. Then all the observed and reconstructed sequences and the topology, or order of branching, that connects them are considered at once, and the configuration that is most consistent with the known characteristics of the mutational process is chosen. Within the limitations of the small quantity of data available so far, a tree constructed in this way has the same topology as trees that have been derived from conventional morphological or other biological considerations. When the structure of a large number of proteins has been worked out, there will be enough evidence to establish the order of divergence of the major living groups of organisms and even a relative time scale for these divergences. The detailed nature and order of acceptance of mutations that occurred in the distant past may then become clear.

Each point on a phylogenetic tree derived from protein sequences represents a definite time, a particular species and a predominant protein structure for the individuals of the species. For any such tree there is a "point of earliest time"; radiating from this point, time increases along all branches, with protein sequences from present-day organisms at the ends of the branches. The location of the point of earliest time—the connection to the trunk of the tree—cannot be inferred directly from the sequences; it must be estimated from other evidence.

To illustrate some of the general considerations in building a phylogenetic tree let us consider just three amino acid positions in the cytochrome *c* sequences (excluding the bacterial one). It is clear, first of all, that biologically similar organisms tend to have the same amino acid in a given position. In Position 15 the plants all have the amino acid alanine (*A*), whereas the animals have lysine (*K*). In Position 17 the fungi (*Neurospora*, yeast and *Candida*) have leucine (*L*), whereas the wheat and most animal sequences have isoleucine (*I*); only a fish (the tuna) has threonine (*T*). In Position 93 the insects and plants have leucine, whereas the vertebrates have isoleucine.

Changes arise so seldom that an observed change almost always reflects a mutation in a single ancestral organism.

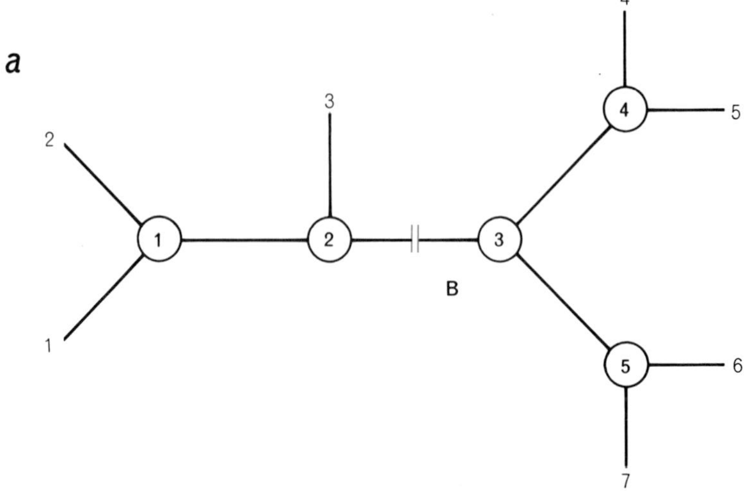


**PROGRAM 3 improves on the topology established by Program 2 by trying alternative configurations. It does this by cutting each branch of the tree and grafting the resulting pieces**

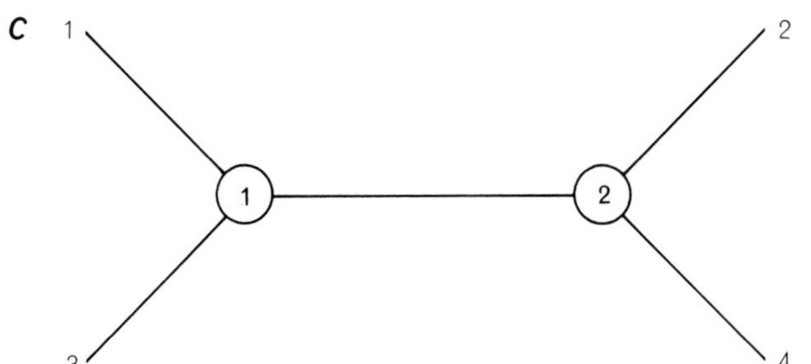


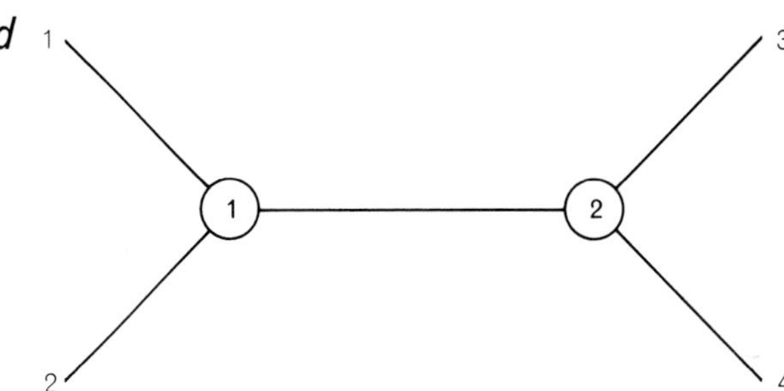


**ing three possible topologies for a four-branch tree (*b, c, d*). Program 1 establishes the sequences for the nodes (*numbered circles*) and evaluates each topology. The best one is accepted. In this way all the sequences are added until a complete tree has been formed.**

The evidence at Position 15 favors the hypothesis that there was a single mutation in a single lineage connecting the animal group with the plant group. The mutation at Position 17 indicates a single lineage between fungi and the other species; the one at Position 93 indicates a single lineage between vertebrates on the one hand and insects and plants on the other. Taken together, these pieces of evidence yield a topology that accommodates all the information from the three sites and requires that only three changes occurred in three ancestral organisms. There are two other possible topologies, but they require that at least four changes must have taken place [*see bottom illustration, pages 266 and 267*]. Since changes in sequence are so rare, we assume that the first configuration is the one most likely to be correct.

It is necessary, of course, to consider all the evidence, not just that found at three amino acid positions. Evidence

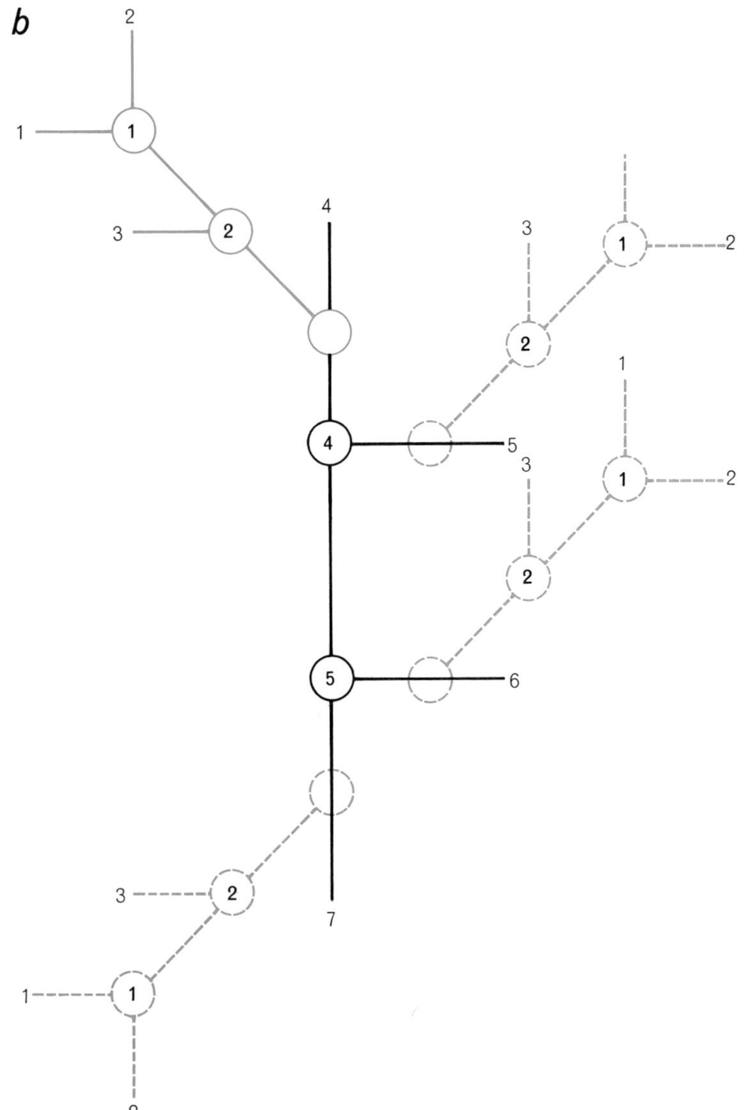


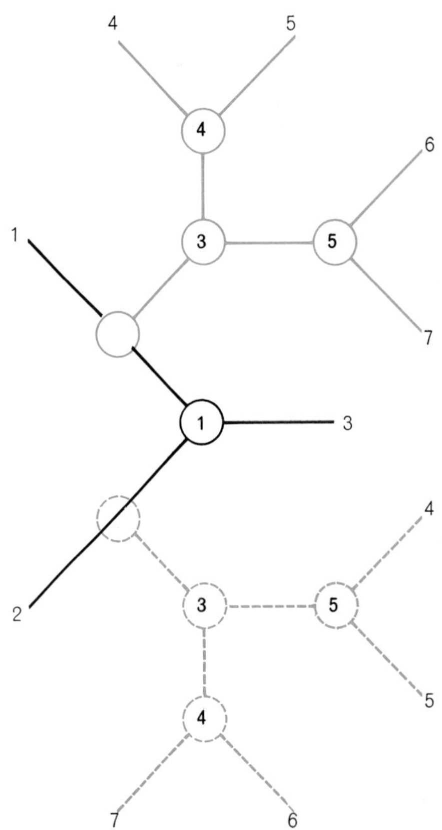

**in different ways. In this example a small tree (*a*) is divided into parts *A* and *B*. Four new topologies are created (*b*) when *A* is grafted to *B* at four points (*color*). Two new structures result (*c*) when *B* is grafted to *A*. The procedure is repeated for each branch.**

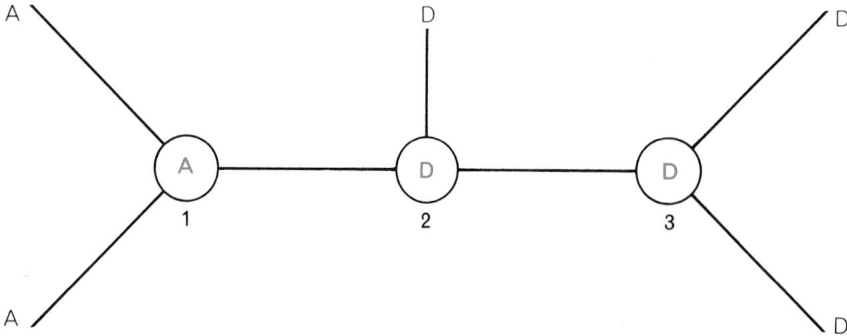


**PROGRAM 1, which evaluates topologies, infers the ancestral sequences at each node. It does this one amino acid at a time. In this tree the amino acids at a certain position in five observed sequences are shown (*black letters*). From this information the amino acids at that position in ancestral sequences at Node 1, Node 2 and Node 3 are inferred (*color*).**

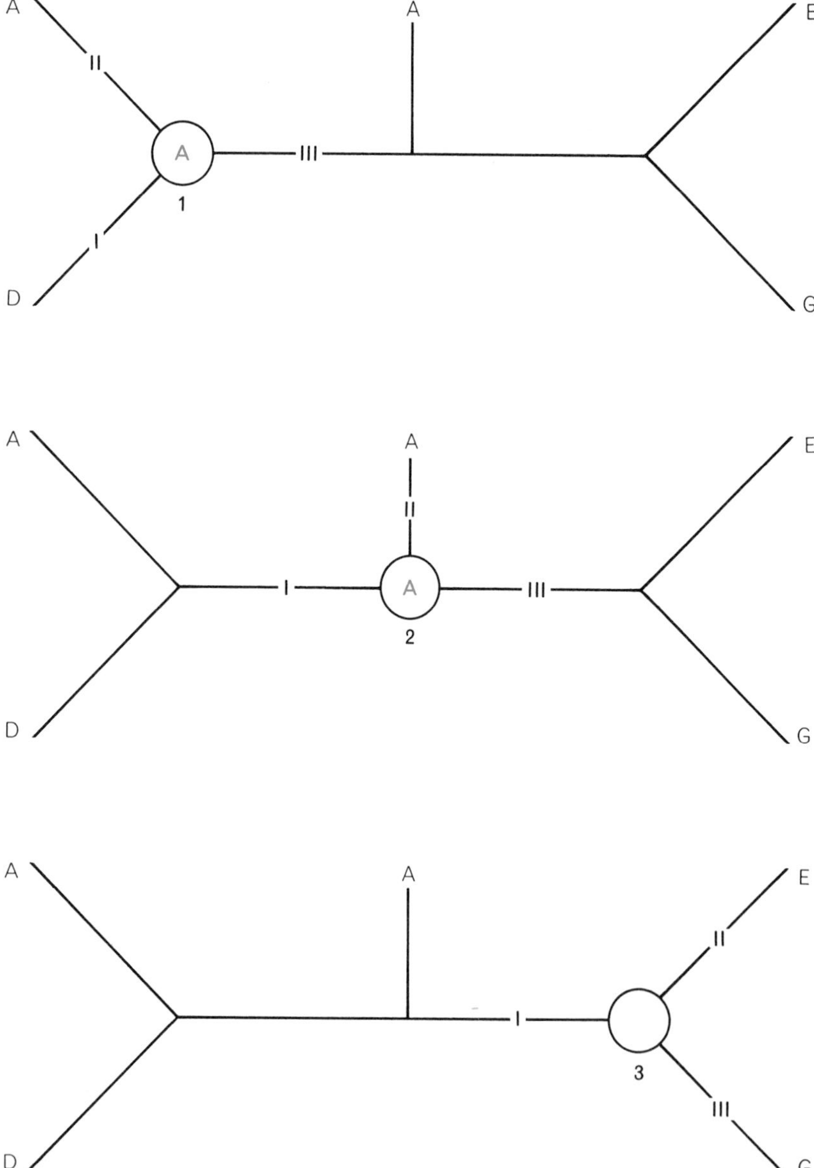


**PROCEDURE followed by the computer is to make a list, for each node, of the amino acids on each branch (*Roman numerals*). The amino acid that is on more lists than any other one is assigned to the node. Here the lists would read, for Node 1, *D, A* and *AEG* (*top*); for Node 2, *DA, A* and *EG* (*middle*); for Node 3, *DAA, E* and *G* (*bottom*). Amino acid *A* appears on two lists for Node 1 and so it is assigned to the node. For the same reason it is assigned to Node 2. No amino acid is clearly the best for Node 3 and so it is left blank.**

from a number of other positions confirms the choice of the first topology described above, but occasionally there is conflicting evidence. For example, at Position 74 there is evidence that wheat and *Candida* are in one group that is connected by a single lineage to all other species. Since this is contrary to the weight of all the other evidence, we must assume that in this position there were two distinct mutations, in two different groups, that by coincidence yielded the same amino acid.

In constructing a phylogenetic tree the quantity of data to be considered is so large and objectivity is so essential that processing the information is clearly an appropriate task for a computer. Our approach is to make an approximation of the topology and then try a large number of small changes in order to find the best possible tree. We have developed three computer programs to do this. Program 1 evaluates a topology. It does this, as I shall explain in more detail below, by first determining the ancestral sequences at all the nodes in a given topology and then counting the total number of amino acid changes that must have occurred in order to derive all the present-day sequences from the ancestral ones. The lower the number, the better the topology is assumed to be. The other two programs use Program 1 to build an approximate tree and then to improve it.

Program 2 starts with three observed present-day sequences, which can only have a simple three-branch topology. It then adds a fourth sequence to each of the three branches in turn [*see top illustration on preceding two pages*] and applies Program 1 to evaluate each resulting topology. The best one is chosen. Then a fifth sequence is added, and then, one at a time, the rest of the sequences. Since each placement is decided without regard to the sequences to be located later, at least one wrong decision is very likely to be made, producing a tree that is almost but not quite the best one. Program 3 is therefore applied to shift each of the branches to other parts of the tree, thus testing all likely alternative configurations. This can be done systematically by cutting each branch or group of branches and grafting it to every other branch or limb of the tree [*see bottom illustration on preceding two pages*]. Again the resulting topologies are evaluated by Program 1, and the best one is finally chosen.

Program 2 and Program 3 are straightforward in logic, although they were intricate programming problems. Our major decisions were made in designing Program 1, which evaluates the topologies proposed by the other two programs.

Program 1 begins by making inferences about the ancestral sequences to be assigned to the nodes. It does this by considering, one at a time, the amino acid positions along the chain. Where only one amino acid is found in all the observed sequences, almost certainly it was present in all the ancestors at all times. In less clear-cut situations a number of reasonable conjectures can be made regarding the ancestral sequences. Consider the case of the amino acids at a certain position in five sequences connected by a definite topology [*see top illustration on opposite page*]. What was the ancestral amino acid at that position in the three nodal sequences? At Node 1 it is most likely to have been *A*, and at Node 2 and Node 3 it is most likely to have been *D*. There was, then, one mutation between Node 2 and Node 1. Any other assignment of amino acids would require two or more mutations.

Let us now see how the computer handles such a problem in practice. The computer must treat all possible topologies, not just one particular case. For this purpose any tree can be thought of as being made up entirely of nodes connecting three branches [*see bottom illustration on opposite page*]. More complex branching simply involves two or more such nodes with zero distance between them. Each of the three branches connects the node either with an observed sequence or with another node and, through it, ultimately with two or more other observed sequences. The computer makes a list of the amino acids that lie on each branch. For example, the lists for Node 1 would show *D* on Branch I, *A* on Branch II and *A*, *E* and *G* on Branch III. Then the amino acid that is found on more lists than any other is assigned to each node. If no single first choice exists, the position is left blank. By this procedure *A* would be selected for Node 1 and Node 2; Node 3 would be left blank.

In a number of situations this simple program gives an equivocal assignment when it need not [*see top illustration on this page*]. The procedure I have described would assign blanks to Node 1 and Node 3 although it is intuitively clear that the choice of *A* for all three nodes is best, necessitating two independent mutations from *A* to *D*. Any other choice would require at least three mutations, a less likely history.

We therefore added further steps to enable the computer to fill in unnecessary blanks. The first assignment of nodal amino acids is examined. If at least two of the three positions adjacent to a blank node contain the same amino acid, that amino acid can be inferred also for the blank node. This second assignment may supply the information required to fill in other blanks, and so the procedure is repeated until no more changes occur. Finally any node that does not have the same amino acid as two of its neighbors is changed to a blank. The entire process yields a definite assignment of ancestral amino acids wherever one choice is clearly preferable and leaves blanks where there is reasonable doubt. By applying these procedures to each position the program eventually spells out all the ancestral sequences.

The nodal sequences for cytochrome *c* are displayed along with the observed sequences [*see top illustrations on pages 266 and 267*]. The very small number of blanks indicates how few of the positions remain doubtful. These computed ancestral sequences, incidentally, may take

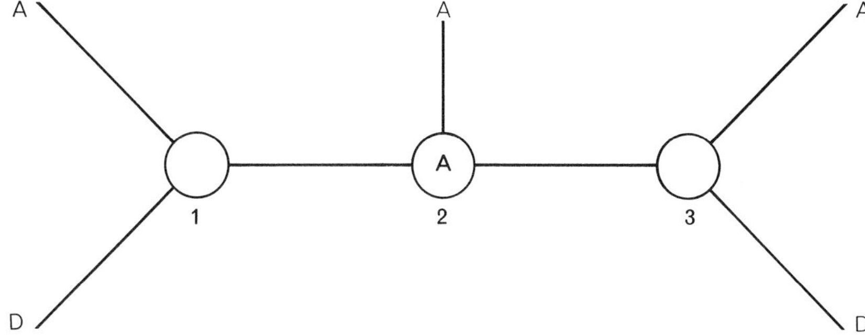


**FURTHER STEPS are required of Program 1 to avoid leaving unnecessary blanks. In this example the basic procedure would leave Node 1 and Node 3 blank (because at both nodes *A* and *D* would each appear on two lists. The program therefore examines the first nodal assignments and, if at least two of the three positions adjacent to a blank node (including another node) have the same amino acid, assigns that amino acid to the blank node. Thus *A* is assigned to Node 1 and Node 3. Ultimately each node must agree with two of its neighbors.**

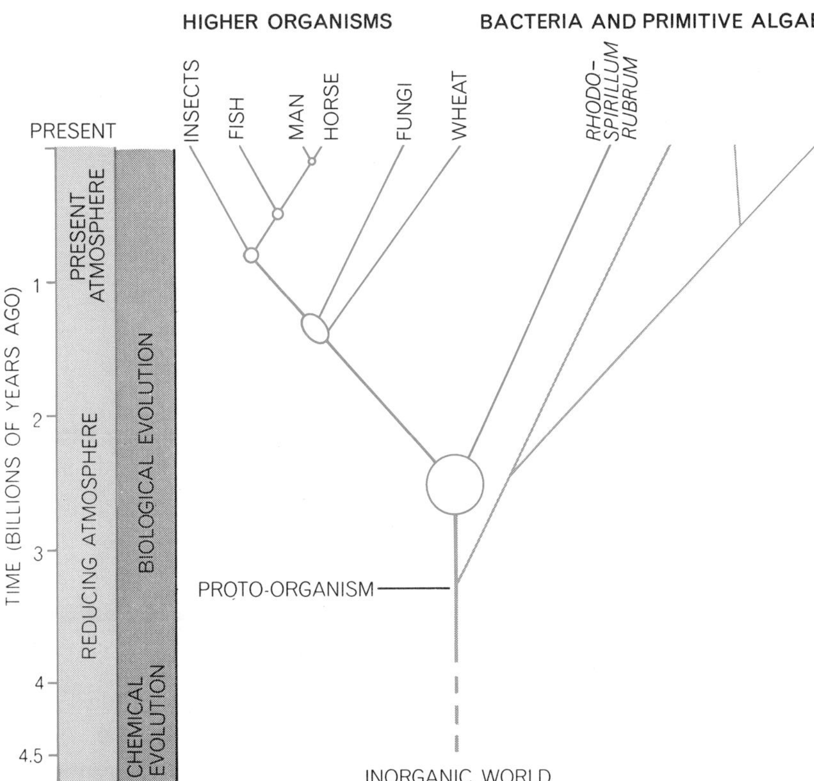


**CYTOCHROME *C* TREE (*dark color*) is redrawn on an absolute time scale and in the context of earth history. The bacterial branch has been added and the "point of earliest time" is taken to be equidistant from the bacterial and the other present-day sequences (*see text*). The size of each node reflects the degree of uncertainty in determining its position.**

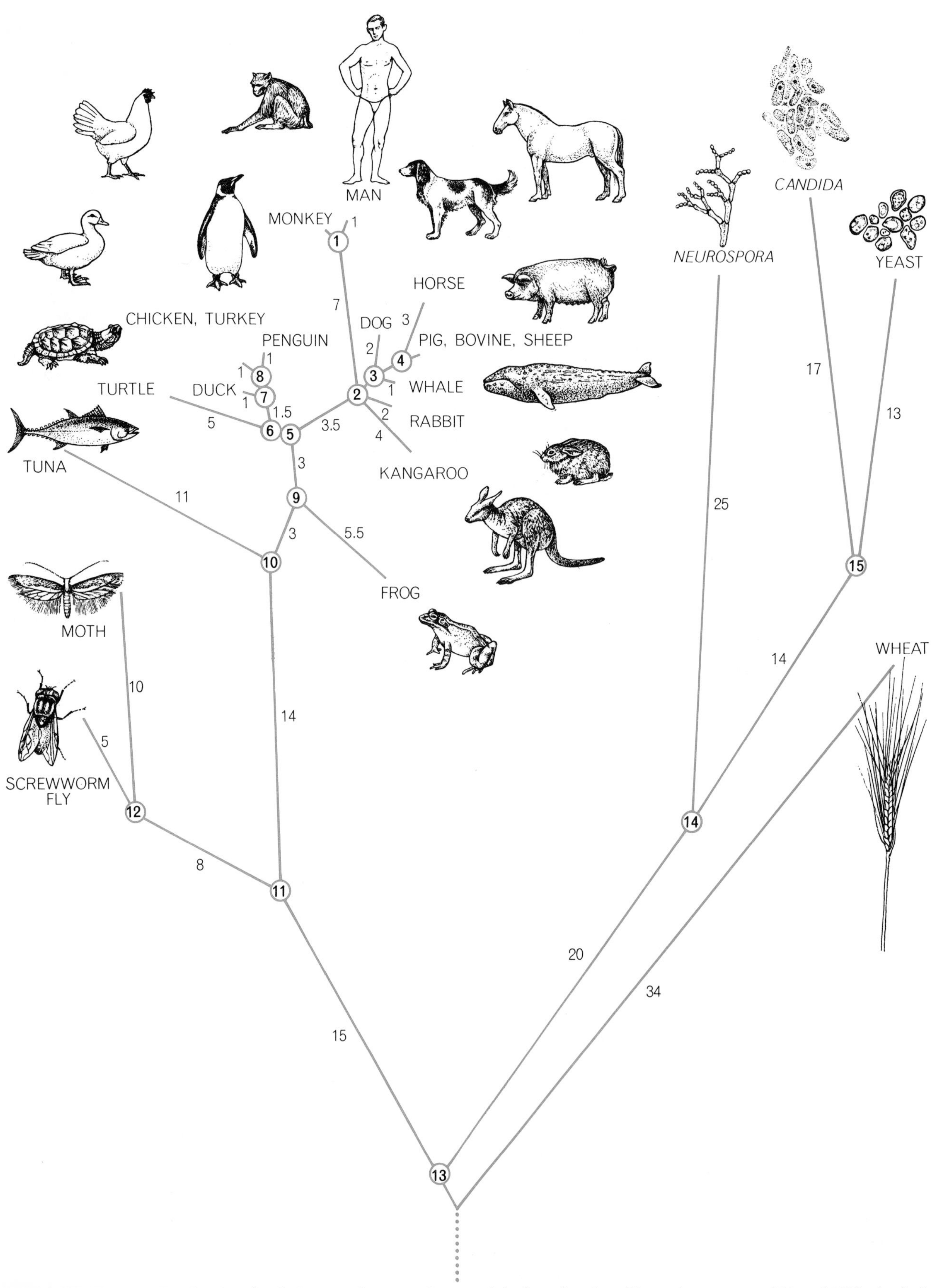


**PHYLOGENETIC TREE** showing the derivation of present-day organisms was constructed on the basis of a computer analysis of homologous proteins of cytochrome *c*, a complex substance that is found in similar versions in different species. The sequence of the amino acids that constitute the homologous protein chains is slightly different in each of the organisms shown at the ends of the branches (*see illustration on pages 266 and 267*). Analysis of the differences reveals the ancestral relations that dictate the topology of the tree. The computer programs determine the sequences of the unknown ancestral proteins shown at the nodes of the tree (*numbered circles*) and compute the number of mutations that must have taken place along the way (*numbers on branches*).

on real meaning in view of the increasing possibility of synthesizing proteins in the laboratory. As investigators succeed in duplicating the sequences we may learn a great deal about the chemical capabilities of ancient organisms.

Once the ancestral sequences have been established, the amino acid changes along each branch of the tree are totaled. (Even when a position is left blank, it is possible to determine the number of changes that must have taken place there.) The sum, representing the total number of changes on the tree, is the final score for that tree. In this way each of the alternative topologies proposed by Program 2 and Program 3 is evaluated in turn.

To make the best topology into the best phylogenetic tree one needs a measure of branch length. We use the number of mutations between nodes. The figures for the observed amino acid changes, however, understate the actual number of mutations because mutations can be superimposed: in a long enough time interval, for example, an *A* might change to a *D* and the *D* to an *S* and the *S* to an *A*, eradicating the evidence of change. We correct for superimposed mutations by applying factors based on the known probability that any amino acid will change to any other given amino acid. That provides our unit of branch length: accepted point mutations per 100 amino acid positions (PAM's). Now it is possible to draw the tree [*see illustration on page 272*]. The major groups fall clearly into the topology shown, but some of the details are still uncertain. It is hard to establish the exact sequence of events in the short interval during which the lines to the kangaroo, the rabbit, the ungulates and the primates diverged. For some divergences, such as the one to the dog and the gray whale, the topology depends on a single amino acid position, and there is perhaps one chance in five or 10 that the branching point is incorrect by one unit. In time other protein sequences from these animals should clear up the uncertainties.

It remains only to establish a time scale for the tree and, by establishing a point of earliest time, to relate the history of cytochrome *c* to geologic time. Our impression is that a protein such as cytochrome *c*, once its function is well established, is subject to about the same risk of mutation in a given time interval no matter what species it is in. It may well turn out that this is not true—that the risk varies in major groups and that occasionally a species may undergo a large change. For the time being, however, we assume that the mutation rate is constant over long intervals, and we define a time scale in terms of the number of mutations.

The bacterial sequence provides information with which to establish the point of earliest time. Because the *Rhodospirillum* sequence is so different from the other sequences it is not shown on the cytochrome *c* tree. There is evidence for its placement, however. At Position 13 and Position 29 *Rhodospirillum* and wheat are different from all the other sequences but are like each other. This indicates that the bacterium should be attached to the wheat branch. Then the fungi and the animals must have diverged from each other after the line to higher plants diverged from the bacteria. To allow for its many differences, the bacterial branch must be very long—about 95 PAM's. That being the case, the point of earliest time must be well back on the bacterial branch.

Now it is possible to redraw the cytochrome tree in simplified form with a time scale in years. The translation from PAM's to years is derived from geological evidence, the best of which dates the divergence of the lines to the bony fishes and the mammals at about 400 million years ago. The cytochrome tree puts that divergence at 11.5 PAM's, on the aver-

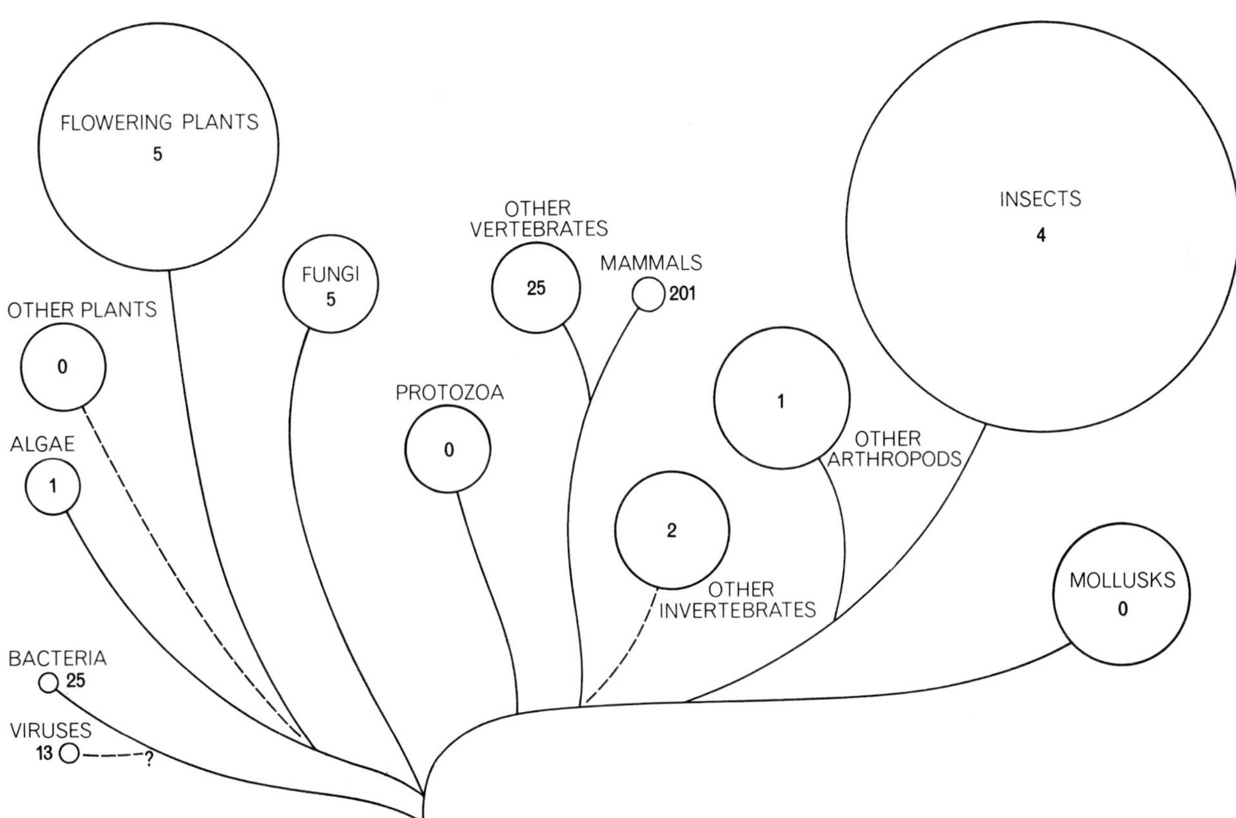


**SEQUENCE DATA are accumulating rapidly. The numbers indicate how many sequences of 30 amino acid positions or more have been determined in each biological group; the area of the circles is proportional to the number of species described in each group (except in bacteria and viruses, where species are not clearly defined). Data from a wide range of groups are needed for paleogenetics.**

age. Therefore 11.5 PAM's corresponds to 400 million years. Now the major nodes can be plotted on the basis of the average number of mutations on the branches above each node. We assume, moreover, that the point of earliest time—the connection to the trunk of the tree—is equidistant from the bacterial and the other present-day sequences. Thus from one family of related proteins we can estimate the temporal relations for an extensive tree of life [*see bottom illustration on page 271*].

The tree is shown in the context of current theory from geological and other biochemical considerations. Elso S. Barghoorn of Harvard University has reported fossil evidence that at least two kinds of organism existed more than three billion years ago, one resembling a bacterium and the other a blue-green alga. Their common ancestor, the "proto-organism," must already have had a complex cell chemistry; it contained many related proteins presumably descended by evolutionary processes from fewer ancestral proteins. Before the time of the proto-organism there was an era of chemical evolution through which life emerged from an inorganic world [see "Chemical Fossils," by Geoffrey Eglinton and Melvin Calvin; SCIENTIFIC AMERICAN Offprint 308].

The rate of change over geologic time varies greatly from one protein to another. Cytochrome *c* protein is the most slowly changing one that has been studied so far in a wide variety of organisms; it appears to have changed at the rate of about 30 PAM's every billion years. The comparable figure for insulin is 40; for the enzyme glyceraldehyde-3-phosphate dehydrogenase, 20; for histones (proteins that are bound to DNA) only .6 PAM. On the other hand, hemoglobin has undergone about 120 PAM's per billion years, ribonuclease 300 and the fibrinopeptides, which are involved in the chemistry of blood clotting, 900. It seems likely that some of the slowly changing proteins will provide the best information on long-term evolution because they have undergone fewer superimposed mutations. Proteins that change more rapidly will provide higher resolution for sorting out closely related species.

Each protein sequence that is established, each evolutionary mechanism that is illuminated, each major innovation in phylogenetic history that is revealed will improve our understanding of the history of life. Surely insight into the biochemistry of man will be obtained from better understanding of his origins.

# V
# *Genetics and Man*

# V

# *Genetics and Man*

## INTRODUCTION

The articles that follow admittedly form a conglomerate rather than a structured intellectual or historical sequence. They are put together in recognition of the fact that certain aspects of genetics have special interest to us primarily because we are humans. We are naturally intrigued by any study of the genetics of man, even when the general scientific contribution is relatively minor. And we have a powerful interest in information that relates directly to man's health or other well-being, to the components of his culture, or to his self-understanding. The foregoing remarks are not intended to imply that genetic studies that appeal to the special interests of man fall outside the generally unified genetic-evolutionary picture that has emerged at every level of biological organization. They do help to explain, however, why it is that the answers to certain genetic problems are pursued in man even though general solutions would come more easily with organisms more amenable to the required kinds of experimentation.

Human genetics had auspicious beginnings in comparison with the science as a whole. Early work by Sir Archibald Garrod (see "The Genes of Men and Molds," by G. W. Beadle, page 78 of this book) established fundamental principles of relationship between genes and metabolic pathways long before these principles were studied in depth in microorganisms. Today, in the context of detailed knowledge obtained from microbial genetic studies, Garrod's principles are widely used in the flourishing field of medical genetics. Immediately after Garrod, however, there was something of a hiatus in human genetic studies of general significance. The emphasis of the science as a whole was on the identification of genes, the establishment of their patterns of transmission, and the analysis of the mechanisms by which they are transmitted. Because of the length of time between generations and the infeasibility of controlling either matings or environments, human genetic studies were confined to family pedigree analysis, inevitably inefficient, often inaccurate, and relatively unproductive of information of generalized significance. And when the experimental induction of mutations became one of the more powerful tools in all sorts of genetic inquiry, including studies of the nature of the gene and of gene action, human genetics remained outside the mainstream. Nevertheless, persistent and compelling interest in the genetics of man has led both to the effective use of those of his attributes that favor certain kinds of genetic study and to the development of techniques that will surmount, or even bypass, the difficulties inherent in the use of man as an object of investigation. Similarly, the principles and materials of genetics have been applied to the improvement of domestic plants and animals—not the easiest organisms to work with—in such ingenious ways as to inspire intellectual admiration.

The first article in this Section, "Porphyria and King George III," by Macalpine and Hunter, represents, to an accentuated degree, both the fascination and the difficulties of certain kinds of human genetic study. Porphyria belongs to the class of heritable metabolic deficiency diseases first perceived with insight by Garrod. The literary detective work done by the authors of this article was feasible only because King George's blood relatives were important enough to be closely observed and written about. Because the available medical records are incomplete, the pattern of transmission for porphyria cannot be established with certainty. This is a difficulty that plagues many pedigree studies, even those that go back only a generation or two. But porphyria is known from other studies to be heritable, and besides, the article is chiefly of human interest. Like the pedigree for another disease that afflicted the royal houses of Europe (see "The Royal Hemophilia," by V. A. McKusick; SCIENTIFIC AMERICAN, August 1965), this one for porphyria can lead to fascinating speculations on interactions between DNA and major events in human history. Moreover, the misunderstanding about George III in his time and the subsequent historical misinterpretations lead naturally into reflections on the advantages that genetic knowledge may offer when man attempts to interpret man.

Macalpine and Hunter point out that present-day medical knowledge might have ameliorated the condition of George III. Actually, the challenge of several metabolic deficiency diseases is currently being met; for example, the adverse effects of the gene for galactosemia can be prevented by dietary means, and phenylpyruvic amentia may be similarly controllable. In "The Prevention of Rhesus Babies," C. A. Clarke, pursuing a scientific trail that begins with butterflies and ends with man, presents a sophisticated example of the use of environmental means for preventing a biological defect that arises out of complex but frequently recurring genetic interactions.

The difficulties encountered in pedigree analysis do not arise—at least to the same degree—in the studies of the population genetics of man. In "The Genetics of the Dunkers" Bentley Glass reports on a study in which groups rather than individuals are considered. His article, addressed to the question of why different groups of humans are characterized by different frequencies of genes, shows the rather drastic changes in gene frequency that occur, merely as a function of chance, in small isolated populations.

Application of genetic knowledge for man's direct benefit is seen most strikingly in plant and animal breeding, where perhaps pre-eminent among many major achievements has been the development of hybrid corn (see "Hybrid Corn," by P. C. Mangelsdorf; SCIENTIFIC AMERICAN, August 1951). In "Hybrid Wheat," Curtis and Johnston tell the beginnings of a story that may come to parallel that of hybrid corn. Note the multiple considerations that go into a plant-breeding program, the complexity of the materials, and the ingenuity of the techniques, which in this instance include putting both chromosomal and cytoplasmic genetic elements into particular sets of relationships. The major successes in plant and animal improvement, achieved through intelligent manipulation of genetic materials, have led to highly controversial suggestions that similar means should be taken to improve mankind.

Genetic problems of several kinds are studied most advantageously in organisms that reproduce vegetatively as well as through sexual or other forms of reproduction that permit genetic recombination. Man and other higher animals that are biologically similar to man, and therefore of special interest, do not reproduce vegetatively; thus they lack some of the experimental advantages offered by such organisms as Neurospora or bacteria. But the development of methods for culturing cells of higher organisms in vitro has broadened the scope of genetic study considerably. Little more than a decade ago, significant cytological or cytogenetic studies of man scarcely existed. Not long before that, even the chromosome number for man was in doubt. The major difficulty was that actively dividing cells are needed in considerable numbers in order to provide an accurate picture of a chromsome complement. With cell-culture techniques available, a trivially small number of cells from an individual provide a potentially endless supply of growing cells, available for colchicine treatment, autoradiography, and all the devices of cytological research. A good sense of what all this has meant to human genetics is found in Ursula Mittwoch's article "Sex Differences in Cells," which includes some of the early definitive work in human cytology that focused on the X and Y chromosomes. Note that even relatively simple cytological work almost immediately uncovered information of significance both to medical genetics and to general problems of human development.

Much more sophisticated uses of tissue culture appear in Dulbecco's

"Induction of Cancer by Viruses." Although the cells used in the experiments are hamster cells or mouse cells, the implication for human cells are immediate. An intriguing aspect of the work is the nearly direct application of techniques and concepts developed in studies of the molecular genetics of microorganisms. Dulbecco could not have obtained the information that enabled him to expand his insight if he had confined himself to studies of viruses in relation to whole mice or hamsters.

Not long ago, descriptions of man-mouse hybridization—the kind of genetics represented in "Hybrid Somatic Cells," by Ephrussi and Weiss—would have been considered science fiction. The work clearly presages a revolution in the means of studying the genetics of higher organisms, including man. Again, animal cells in culture are treated like microorganisms. They can be subjected to the kinds of selective devices that are the basis for coping with rare events; they can be mutagenized; they can be hybridized, and recombinants can be sought; and potentially they offer unique means of exploring morphogenesis through the hybridization of differentiated and undifferentiated cells.

Who can tell what possibilities in the genetics of somatic cells may be revealed as facts ten years from now—possibilities that today might seem beyond the limits of science fiction? Whatever these facts may be, scientists and nonscientists alike will have to assimilate them.

# 29
# *Porphyria and King George III*

IDA MACALPINE and
RICHARD HUNTER
*July 1969*

King George III, who is held in low regard on both sides of the Atlantic as the stubborn monarch whom the American colonies fought for their independence, was not a well man. His putative "madness" affected the course of Britain's history and, among other things, led to the establishment of psychiatry (then called the "mad-business") as a serious branch of medicine. Oddly enough, it has now become clear, a century and a half after his death, that George III was by no means psychotic. The much maligned king suffered spells of a painful and delirious metabolic disease that has only recently been recognized.

While working on a history of psychiatry, we learned with considerable interest how greatly its origins and development had been influenced by George III's alarming attacks, and we decided to find out what we could about the illness itself. Fortunately we were able to round up the notes and records of physicians who examined the king at the time; their manuscripts were preserved in Windsor Castle, the British Museum, the Lambeth Palace Library and in the hands of descendants. The recently published correspondence of George III also was helpful. The physicians' descriptions of the king's illness (not previously examined in recent times), together with the other available evidence, enabled us to arrive at a firm diagnosis of his disease in the light of present medical knowledge.

Let us begin with a review of his medical history as it was reported at the time. The first severe attack came in the fall of 1788, when the king was 50 years old. He had had a seizure of acute abdominal pain in June; his physician, Sir George Baker, diagnosed the cause as "biliary concretions in the gall duct" and sent him to Cheltenham Spa to drink the waters. The episode subsided, but in October the pains returned, accompanied by constipation, darkening of the urine, weakness of the limbs, hoarseness and a fast pulse. In the ensuing weeks the king was afflicted with insomnia, headache, visual disturbances and increasing restlessness. By the third week he became delirious, and over the weekend of November 8–9 he had convulsions, followed by prolonged stupor. His doctors feared that a fever had "settled on the brain" and that his life was in imminent danger. For a week he apparently hovered between life and death. Then his physical condition began to improve, but his mind was "deranged." There were periods of great excitement, interspersed with moments of lucidity and calm. "Wrong ideas" took hold of the king, and his physicians found him increasingly difficult to manage.

During all this time, although he was attended by coveys of physicians, the patient was not really examined in the present sense of the word. The doctors looked at the king's tongue, felt his pulse, inquired about his excretory functions, listened to his complaints and attempted to pronounce a diagnosis by "an estimate of symptoms and appearances." There were, indeed, no tools for examination to speak of in that era—no stethoscope, not even a reliable clinical thermometer to measure fever. The physicians often could not agree on the pulse rate, presumably because their timepieces differed. Doctors did not listen to the chest; even if they had, they would not have known what to make of what they heard. They were also handicapped by the fact that they did not dare to question the king about his symptoms unless he addressed them first. (After one session of fruitless silence the physicians plaintively reported: "His Majesty appears to be very quiet this morning, but not having been addressed we know nothing more of His Majesty's condition of mind or body than what is obvious in his external appearance.")

In contrast to the obscurity and vagueness of the physical symptoms, the king's mental symptoms spoke loudly and clearly. His physicians needed no modern aids to observe that his behavior was excited and irrational and his mind confused. Moreover, his mental state caused much concern about his fitness to rule and the dangers to the nation and the empire. The mental symptoms therefore overshadowed the physical complaints. Thus it came about that the king's physical sufferings were minimized (and later disregarded), whereas his mental derangement was magnified as if it were the whole illness. Physicians who specialized in "intellectual maladies" were called in, took up residence in the palace and took charge of the sickroom.

One of these practitioners was the Reverend Dr. Francis Willis, called "Doctor Duplicate" because he was a doctor of medicine as well as of theology. Dr. Willis, who managed a madhouse, arrived at Kew Palace with the aids and tools of his establishment, including attendants and a straitjacket. He applied to the king the usual treatment of the day for insane persons: coercion and restraint. The king was put in the straitjacket for infringements of discipline such as throwing off his necktie and wig when he had attacks of sweating, or refusing to eat when he had difficulty swallowing, or walking about the room when he became too restless to lie down. The king's unpredictable and obstreperous behavior was taken to be the ebullition of furious mania, and his fierce (and understandable) dislike of his doc-

**GEORGE WILLIAM FREDERICK**, the third Hanoverian king of England, is seen in an official portrait by Allan Ramsay, painted when the monarch was 30 years old. Born in 1738, George III ruled from 1760 to 1811, when the fourth in a series ot misdiagnosed bouts of "madness," apparently a hereditary enzyme imbalance known today as porphyria, required the appointment of a regent.

tors and keepers was attributed to delusions.

His illness precipitated a historic party struggle in Parliament known as the "Regency Crisis." The Whigs, led by Charles James Fox, Edmund Burke and Richard Brinsley Sheridan (who was a member of Parliament as well as a celebrated playwright), tried to oust the king's prime minister, William Pitt, and the other members of his cabinet. For four months Parliament gave its entire attention to the king's illness and the constitutional issues it raised. Members of Parliament interrogated the physicians exhaustively on the question of whether the king was suffering merely a prolonged delirium, from which he could be expected to recover with unimpaired mind, or was actually afflicted with "a lunacy" that would permanently cloud his judgment.

Then, just as Parliament was about to pass a bill setting up a regency, George's mind suddenly began to clear. At the end of February, 1789, his doctors announced "the entire cessation" of his illness. Although Willis claimed the credit for the cure, in retrospect it is clear that the king's recovery must have been spontaneous. He was soon well enough to leave his confinement in Kew Palace and return to Windsor Castle, his favorite residence. His recovery was celebrated with demonstrations of national rejoicing the like of which had never before been witnessed.

In 1801 and again in 1804 George III had recurrences of the same illness. Each time he was at first dangerously ill with identical physical symptoms and then deranged mentally for only a brief period. Eventually, in 1810, he fell into an illness that incapacitated him to the point where he was replaced by the Prince of Wales under the regency act of 1811. For at least a year there were hopes that he would again recover, and his son, as regent, refrained from dismissing George's ministers to avoid embarrassing him in case he became able to resume the throne. The king did experience periods of recovery, but each time he relapsed. He was then well past 70, blind and much reduced physically and mentally by the repeated onslaughts of his illness. Senility had set in. During his last years George was on the whole tranquil, played the harpsichord and had intervals of good humor and cheerfulness; however, he was often "sullen and lost in mind," tears and laughter would come in quick succession and from time to time he was stricken with the old, painful paroxysms. A month before his death, in the last of these attacks, he spent 58 turbulent hours without sleep or rest and "gave other remarkable proof of the extraordinary energies of his constitution." He died quietly on January 29, 1820, at the age of 81.

After his death political bias and professional opinion developed an image of George III as a "mad king" who was more or less deranged throughout his life. A spell of sickness he had experienced in 1765, when he was 26, was taken to have been an early sign of his madness. There is not a shred of evidence that any mental disturbance accompanied that early illness, but it was popularly believed the king must have been insane to permit the 1765 enactment of the infamous Stamp Act that sowed the seeds of the American War of Independence. Furthermore, psychiatrists who later diagnosed George III's illnesses as primarily mental also adopted the lunacy interpretation of the king's 1765 illness to support their theory; it would not make sense to suppose that the king, if mentally unstable, would have come through the first 28 stormy years of his reign without any sign of psychological distress.

The great prominence given to George III's supposed insanity aroused wide public and professional interest in mental illness and generated the first systematic attempts to deal with it as a medical problem. William Black, a contemporary teacher of medicine who was intrigued by the physicians' fumbling efforts to forecast the prospects for the king's recovery, looked into the question statistically and thus became the founder of psychiatric statistics. Studying the records of people who had been pronounced insane, he came to the conclusion, which may be called "Black's law," that a third of such patients could be expected to recover to full mental health, a third recovered somewhat but did not regain all their former mental ca-

PORPHIN

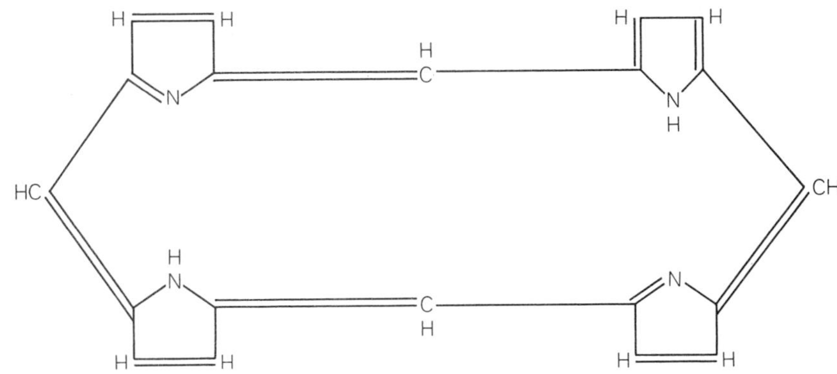


HEMIN

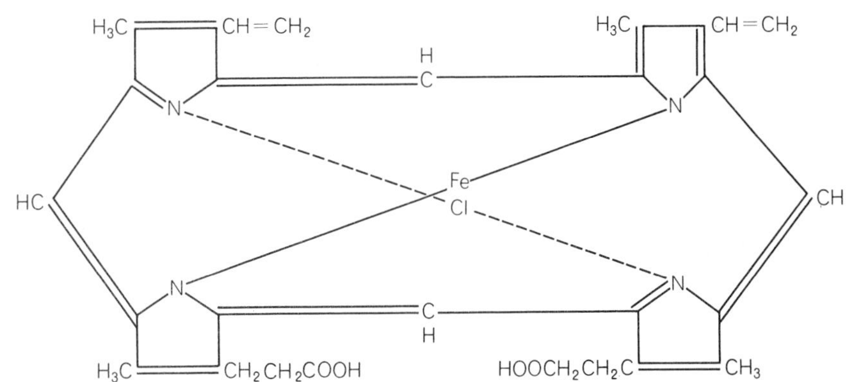


**PORPHYRIN BUILDING BLOCK, the porphin molecule, is comprised of four pyrrole rings joined together by four methene bridges. The pigments that may be constructed from porphin molecules include three that are essential to animal and plant life: hemoglobin, chlorophyll and cytochrome. For comparison a hemin molecule is illustrated; it differs from porphin mainly in having an iron and a chlorine atom attached to the pyrrole rings.**

pacities and a third did not improve at all or sank into deeper illness.

Richard Powell, another physician with a statistical bent, found that in the years immediately following the king's 1788 attack there was a big increase in the number of insane persons admitted to private asylums. He presented his findings graphically in a histogram, introducing this device into medical reporting for the first time. Dr. Powell attributed the apparent rise in mental illness to the mounting complexities of civilization, and his social interpretation is still widely put forward as an explanation of increases in the incidence of mental disorders.

Two of George III's sons, the dukes of Kent and Sussex, set up the first fund for research in psychiatry and initiated the first controlled trial of a treatment for insanity. The trial was conducted by two laymen who had developed a secret remedy they hoped would be used on the king. A London physician named Edward Sutleffe also offered a remedy; he called it a "herbaceous tranquillizer," thereby introducing the term that describes the dominant treatment of mental illness with drugs today.

Parliament, prodded by demands for better care of the mentally ill, particularly among the poor ("pauper lunatics"), set up a committee "to enquire into Madhouses." Under the chairmanship of George III's personal friend George Rose, the committee took evidence for two years and published reports that paved the way to the system of caring for mental patients in "asylums," which lasted well into our century and whose memorials are still with us. This advance had some unfortunate consequences. It isolated patients from society, often in remote establishments, and it created an artificial separation between mental and physical disease, each with its own specialists. Thus psychiatry was unhappily set apart from the mainstream of medicine, and physicians and psychiatrists became two separate breeds.

In view of the historic importance of George III's illness, it is remarkable that so little inquiry has since been made, either by psychiatrists or by physicians, into what was really the matter with the king. Astonishingly, only two medical studies have ever been attempted. Both were made by individual U.S.

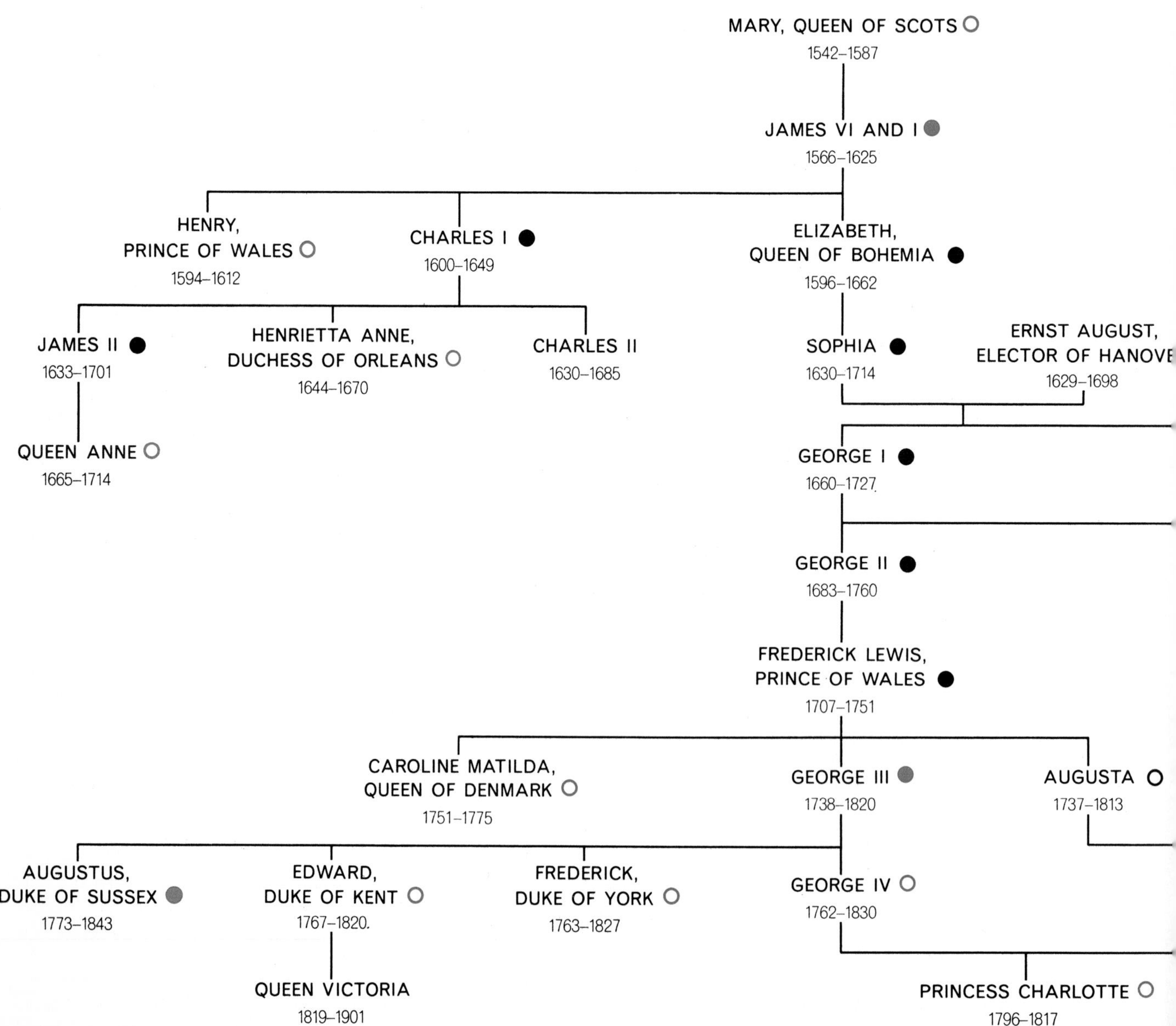


**THREE ROYAL HOUSES suffered from porphyria. First to be afflicted was Mary, Queen of Scots, a Stuart. From her descendants the disorder spread to both the Hanoverian and the Prussian royal lines. Colored circles mark those who showed some signs of the**

psychiatrists, almost a century apart, and both completely missed the medical complexities of the case.

In 1855 Isaac Ray, the distinguished president of the Association of Medical Superintendents of American Institutions for the Insane (since renamed the American Psychiatric Association), reviewed the then available information on George's sickness. He was surprised by the lack of background for the king's attacks of mental derangement. Dr. Ray wrote: "Few men would have seemed less likely to be visited by insanity. His general health had always been good; his powers were impaired by none of those indulgences almost inseparable from the kingly station; he was remarkably abstemious at the table, and took much exercise in the open air. Insanity had never appeared in his family, and he was quite free from those eccentricities and peculiarities which indicate an ill-balanced mind." Nevertheless, on the basis of the reports to which he had access Dr. Ray diagnosed George III's malady as "mania" (which is as unspecific for mental illness as "fever" is for a physical complaint). Ray's attempt at diagnosis was severely handicapped by the paucity of facts he had on the case and by the comparatively primitive state of medical knowledge in the 19th century.

In 1941 the eminent Baltimore psychiatrist Manfred S. Guttmacher reexamined George III's case from the angle of modern psychoanalysis. It is characteristic of the psychoanalytic point of view that, given a case of mental aberration, it attaches little weight to physical symptoms and causes. Guttmacher dismissed the king's physical complaints, attributing them in part to efforts by the court to cover up the king's madness and in part to neurotic imaginings by the king himself. Describing the illness in modern terms as manic-depressive psychosis, Guttmacher added: "Self-blame, indecision and frustration destroyed the sanity of George III.... A vulnerable individual, this unstable man ...could not tolerate his own timorous uncertainty [and] broke under the strain. [Had the king] been a country squire, he would in all probability not have been psychotic." (Actually the king was known to his subjects as Farmer George because of his interest in agriculture.)

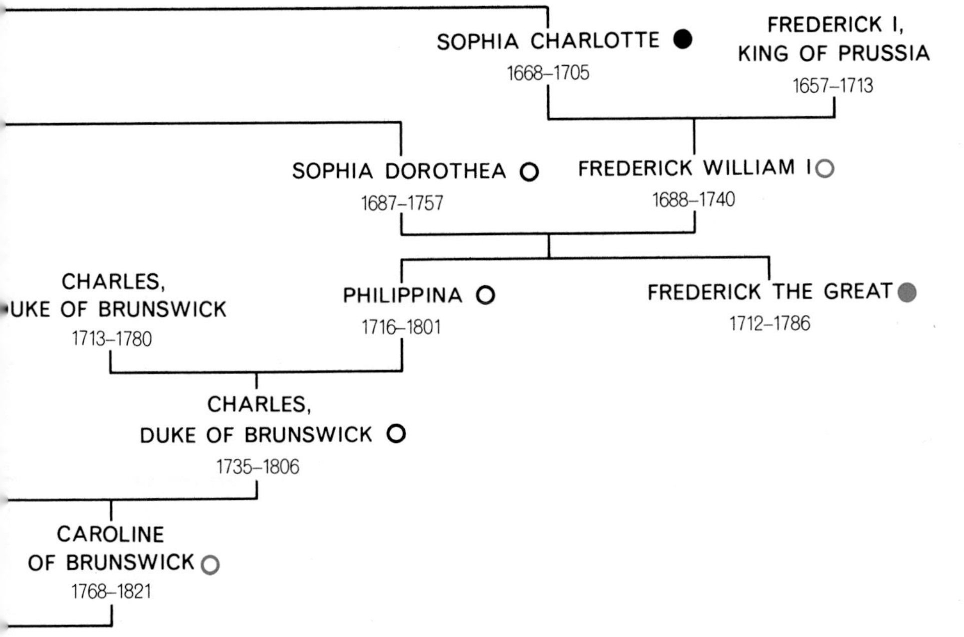


**disorder, colored dots those whose urine was also abnormal. Black dots mark transmitters who did not suffer from the disorder; black circles, those who may have been transmitters.**

When we came to our detailed study of George III's career and illnesses, we found no grounds for support of this psychoanalytic interpretation of his case. George's contemporaries and early biographers described him as one of Britain's most devoted and best-informed rulers; he was a musician, a book collector (whose collection forms an important part of the British Museum), a patron of the arts and sciences, fond of country life and his family. If he had been emotionally and mentally unbalanced, how could he have lived through the disastrous period of his reign—the loss of the American colonies and the 18-year struggle leading up to it—without a suspicion of breakdown? In view of the political troubles that beset him, not to mention his large and unruly family, one should be surprised that he was ever sane at all, if the psychoanalytic diagnosis of his personality were correct.

The fact is that, before physical illness and senility finally incapacitated him, George had only three attacks of mental derangement, and all together these periods did not add up to more than six months. In each instance the nervous disorder was ushered in by serious physical symptoms that perplexed his physicians and brought him to the brink of death. "It is not merely the delirium of fever, nor is it any common form of insanity," said one of his doctors, William Heberden, Jr. "The whole frame has been more or less disordered, both body and mind... [due to] a peculiarity of constitution of which I can give no distinct account." Sir Henry Halford, another eminent physician of the time, remarked: "The King's case appears to have no exact precedent in the records of insanity."

There were clues to the root of George's illness, if the physicians had

**"DOCTOR DUPLICATE" WILLIS**, the chief physician during George III's first porphyria attack, maintained a private asylum, shown here in a cartoon by Thomas Rowlandson. Willis' nickname derived from his being both a physician and a doctor of divinity. His method of treatment emphasized restraint and discipline for any disorderly patient; two here are in "winding sheets." Use of his method, including straitjacketing when his delirious royal patient failed to follow his orders, earned all doctors the king's enmity.

**SAVAGE CARICATURE** of the royal family was published in the year preceding the king's first seizure. The work, by James Gillray and titled "Monstrous Craws at a New Coalition Feast," was inspired by popular belief that "foreign" Hanoverians squandered the nation's funds for their own benefit. The queen (*left*), the Prince of Wales (*center*) and the king (*right*) are shown seated outside the Treasury, gorging themselves on gold. The king was in fact frugal, but the Prince of Wales (later his regent) was a notorious spender.

only known how to interpret them. The doctors reported, for instance, that his attacks appeared to be caused by "the force of a humour" that first showed itself in the legs, then drove "into the bowels" and finally was projected "upon the brain." Quaint as this description now sounds, it was a significant account of the course of the king's attacks, involving a progression of symptoms from the limbs to the abdomen to the brain. Of all the king's symptoms, the most revealing one, which has led us now to the discovery of the true nature of his illness, is the color of his urine. At least half a dozen times the doctors who examined him noted that the king's urine was "dark," red or discolored.

Considered in connection with the king's other symptoms and the character of his attacks, it is quite clear today that this discoloration of his urine must have been due to the presence of porphyrins. Porphyrin is a pigment, contained in the hemoglobin of the blood, that normally is metabolized in the body cells. Hence its presence in the urine is a signal of faulty metabolism: namely, inability of the cells to convert porphyrin, presumably because of the absence of a necessary enzyme. The clinical seriousness of such defects was first called to the attention of the medical world in 1908 by the London physician Sir Archibald Garrod, who discovered that "inborn errors of metabolism" could cause profound disorders [see "The Chemistry of Hereditary Disease," by A. G. Bearn; SCIENTIFIC AMERICAN, December, 1956]. It has since been found that the inability to metabolize porphyrin produces a disease, called porphyria, that attacks the nervous system [see "Pursuit of a Disease," by Geoffrey Dean; SCIENTIFIC AMERICAN, March, 1957]. The attack usually begins in the autonomic system, then advances to the peripheral nerves, the cranial nerves and finally the brain itself. At the height of the attack the patient is paralyzed, delirious and in agonizing pain.

George III's symptoms, the sequence of their development and the climaxes of his illness read like a textbook case of porphyria. His attacks started with colic, constipation and nausea; there followed a painful weakness of the limbs, so that he could not walk or stand, a speedup of his pulse, attacks of sweating, hoarseness, visual disturbances, difficulty in swallowing, intractable insomnia, mounting excitement, nonstop rambling, dizziness, headache, tremors, stupor and convulsions. The physicians described his climactic mental state thus: "Delirious

**LIKENESS OF WILLIS appeared on one side and a patriotic injunction on the other of a medal that "Doctor Duplicate" distributed to the public when the king, in February, 1789, spontaneously recovered four months after the onset of his chief porphyria attack.**

**PHYSICIANS' JOURNAL, a chronological record of the king's illness, notes events of December 23 and 24, 1788, as follows: "The waistcoat was taken off at nine–& blisters drefs'd—discharg'd well—very sore—Pulse 96—perspir'd through the night profusely—but little sleep—& very quiet & in good humour for the most part—Tongue white. Copy of the letter to the Prince of Wales—not sign'd by Dr. Willis—The straight waistcoat was taken off from his Majesty at noon yesterday, but was put on again soon after two oClock & was not taken off till nine this morning. His Majesty has not had more than an hours sleep in the night, is good humour'd but as incoherent as ever. Mr. Keate is of opinion that the blisters on his legs are in a healing state—Bulletin—His Majesty pafs'd the night quietly but with little sleep—& is quiet this morning—Sometime betwixt 10 & 11—fell fast a sleep upon a Sophy—nearly an hour—awak'd & lay very compos'd. Before He fell asleep He had a very pertinent conversation with my Father [Dr. Willis] concerning Mr. Smelt & religion—his sense [struck out] & worthinefs but too much refinement—&c. [conclusion of page]."**

**WINE-COLORED URINE**, shown in the middle test tube, was produced by a patient during an acute attack of porphyria. For comparison, normal urine is at left, port wine at right. James I, who also had porphyria, commented that his urine resembled port.

all day . . . impressed by false images . . . continually addressed people dead or alive as if they were present . . . engrossed in visionary scenery . . . his conversation like the details of a dream in its extravagant confusion."

These mental symptoms are the hallmarks of a state in which the brain is disordered by toxin. Other aspects of the king's attacks also were characteristic of porphyria: they were usually precipitated by mild infections; his condition fluctuated rapidly, and each attack was followed by a protracted convalescence. Porphyria is usually accompanied by high blood pressure; there were of course no measurements of the king's blood pressure, but the repeated crises that made his doctors fear "a paralytic stroke" may well have been hypertensive. As for the illness of 1765, that was probably a mild attack of porphyria that did not go on to involve the brain.

Since porphyria is a hereditary disease, we looked into the medical histories of George III's blood relatives. The available records showed that signs of porphyria in his family went as far back as his 16th-century ancestor Mary, Queen of Scots. Her son, King James, suffered from colics (which he told his physician he had inherited from his mother) and described his urine as the color of his favorite Alicante wine. George III's sister, Queen Caroline Matilda of Norway and Denmark (who is the subject of many novels and of a Verdi opera), died at 23 of a mysterious illness that was featured by rapidly progressive paralysis. Some of George's children were afflicted with his disorder. The son who succeeded him on the throne, George IV, had a disease that his physicians called "unformed gout" but that must certainly have been porphyria. George IV's daughter, Princess Charlotte, showed characteristic symptoms of the disease and died in childbirth, apparently during an acute attack. George III's son Augustus, the Duke of Sussex, had severe attacks of illness accompanied by discoloration of his urine. Another son, the Duke of Kent (who was the father of Queen Victoria), suffered severely from colics and died of an attack a week before the death of George III. Porphyria, introduced into the House of Brandenburg-Prussia by George I's sister, also claimed Frederick the Great as a victim. The disease has persisted in descendants of George III up to the present day. We examined some of them and found the characteristic signs, including the discoloration of the urine. Our laboratory tests showed that the family had a form of porphyria that makes the skin sensitive to the sun and to injury.

**PORCELAIN PLAQUE, made by Josiah Wedgwood in commemoration of the king's recovery, shows George III crowned with laurels. It bears the inscription "Health restored."**

We see, then, from the perspective of 20th-century medical knowledge, that George III's image, like his pain-racked body, has been the victim of a cruel misunderstanding. His episodes of derangement were merely the mutterings of a delirious mind temporarily disordered by an intoxicated brain. The royal malady was not "insanity" or "mania" or "manic-depressive psychosis"—whatever meaning these nebulous terms may retain in the modern era of diagnostic and investigative medicine. Partly because of the backwardness of medical knowledge at the time and partly because of the king's position, the bodily disorder he suffered was pushed into the shadows. With a good diet, avoidance of medication with drugs and generally rational treatment his attacks of delirium might have been curtailed.

# 30

# *The Prevention of "Rhesus" Babies*

C. A. CLARKE

*November 1968*

As a child I was fascinated by butterflies, particularly a yellow swallowtail butterfly that flies in a marshy area of the east coast of England known as the Norfolk Broads. After World War II, I wanted to breed these insects but found it easier said than done. Swallowtails usually will not mate in a cage, as they need an elaborate courtship flight to stimulate pairing. My interest in the insect did not wane, and by persevering I learned a simple trick to make captive swallowtails mate. Holding the female in the left hand and the male in the right, one brings the pair close together, pries open the male's claspers with the nail of the left-hand middle finger and thereby induces the male to lock onto the female, after which mating follows naturally.

The happy acquisition of this technique in 1952 led me on to experiments in crossbreeding butterflies that turned up some surprising results and fruitful genetic findings. Thus, by the pleasant route of pursuing idle curiosity, my colleagues and I were led unexpectedly to a solution for the well-known medical problem having to do with the inheritance of the Rh factor in human blood! A clue suggested by the butterfly work has enabled us to develop a successful method of preventing the anemia hazard for babies born of an Rh-positive father and an Rh-negative mother. The method can best be explained by describing both the butterfly work and the blood-group investigation from the beginning.

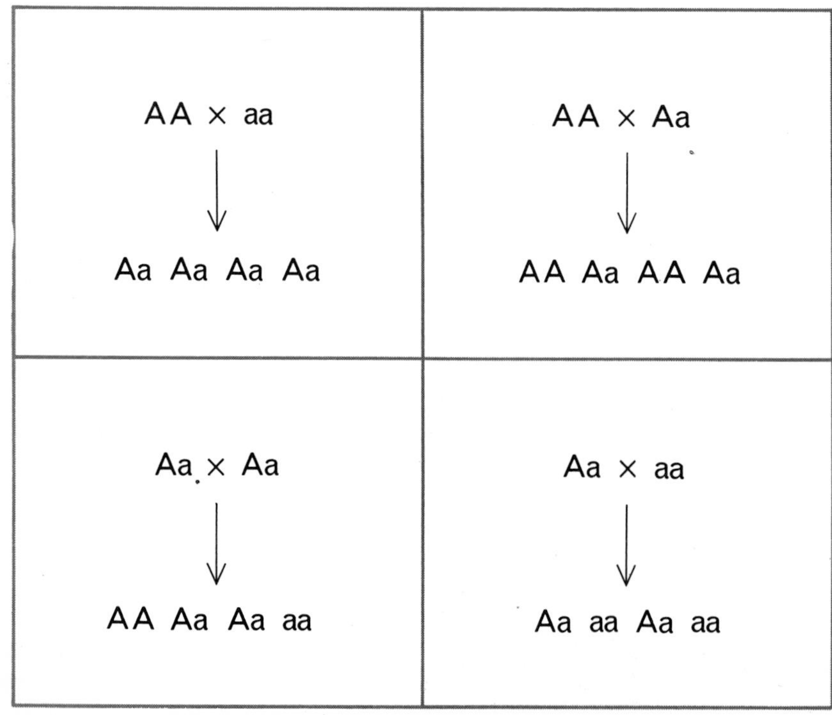


DOMINANT AND RECESSIVE GENES figure in the Rh problem. Each box shows in the top line a parental arrangement of dominant genes (*capital letters*) and recessive ones (*small letters*) and in bottom line the combination of genes to be expected in offspring.

In 1952 I happened to acquire a female butterfly of a black swallowtail species common in America (*Papilio asterias*), and in an idle moment one Sunday afternoon I hand-mated her to a male of the yellow British species (*Papilio machaon*). Since the two species are related, the mating was successful, and their first-generation offspring turned out to be like the American parent (showing that black and American were dominant to yellow and British!). When the hybrid was back-crossed to the recessive yellow parent species, however, the new (second) generation segregated for the ground color again: half of the offspring were black and half were yellow. Clearly, then, the ground color of the wings must be controlled by a single gene. A butterfly that inherited the dominant gene for black from either hybrid parent would have black wings, whereas an offspring that received yellow genes from both parents (each of which possessed the yellow as the recessive gene) would be yellow.

It was this experiment that aroused my interest in genetics. Soon afterward I met P. M. Sheppard, who is now a colleague of mine at the University of Liverpool but then was at the University of Oxford. We decided to use the mating technique to investigate the genetic aspects of mimicry in certain butterfly species. In wing coloring and form the mim-

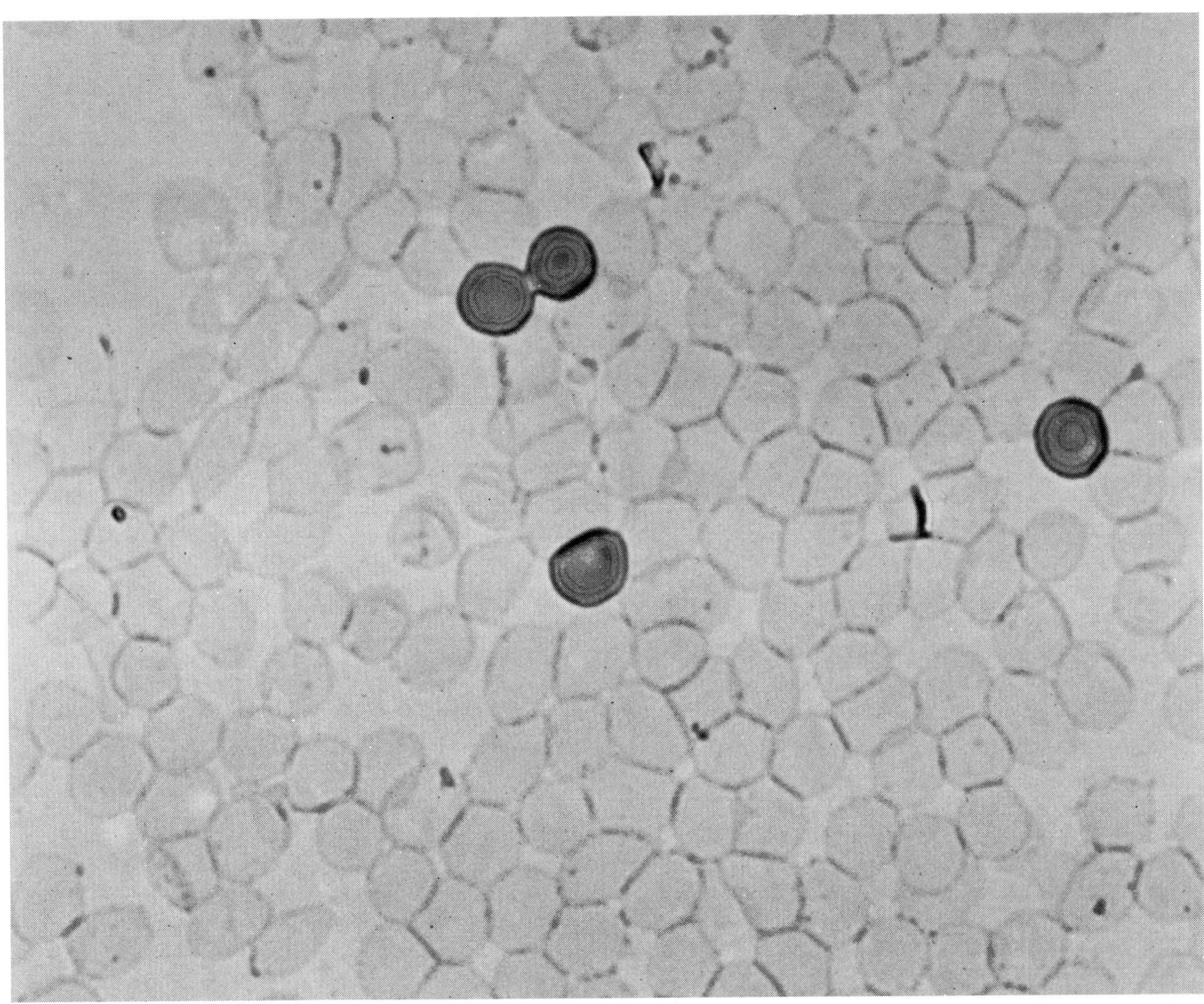

BABY'S CELLS in the mother's circulation give rise to the Rh problem. In this photomicrograph, made by Flossie Cohen of the Child Research Center of Michigan, the red blood cells of the baby are darker than the mother's red blood cells because of a technique that has washed the hemoglobin out of the adult cells while leaving hemoglobin in the fetal cells that are more resistant to the technique. The steps whereby fetal cells in the maternal circulation can create an Rh problem are shown at top of next two pages.

icking butterflies copy butterflies of different species or genera and even of different families. The "model" butterflies are nauseous to predators (particularly birds), and thus the mimics avoid attack although they are themselves edible. A single species of mimic may have several different forms, each imitating a different model; such a species is said to be polymorphic. In formal terms polymorphism refers to distinctly different types (such as blood types in human beings) that persist in an interbreeding population living in the same habitat, and that occur in frequencies such that the rarest form could not be maintained by recurrent mutation. In both mimic butterflies and their models the relative proportion of edible and inedible species in any one area is kept in balance by natural selection. For example, if the edible species mimicking a particular wing pattern comes to outnumber the inedible species, that pattern becomes attractive to predators. The pressure of natural selection is then unfavorable to the mimic species instead of being favorable to it.

How does the mimicry arise in the first place? Does it come about by the mutation of a single gene, causing a butterfly to acquire a mimetic pattern at one jump? This seemed to us a tall order, considering the complexity of colors and configurations in a butterfly's wings. For an answer to our question we investigated an African butterfly (*Papilio dardanus*) and one from southeast Asia (*Papilio memnon*). In both cases we found that the "gene" controlling the wing pattern is really a group of closely linked genes behaving as a single unit—what is known as a supergene. We found evidence for this in crossovers (exchanging of genes) within the genetic group. For example, in a southeast Asian butterfly we bred we were able to see a crossover in the chromosomal unit responsible for the change in wing pattern [*see illustration on page 293*].

There is another significant point about these two species of butterfly that is relevant to the blood-group work. In these species mimicry occurs only in the female; in both species the males do not show mimetic patterns although they carry the genes that are responsible for the patterns in the female. Evidently there is an interaction between sex and the mimetic supergene, so that the supergene is somehow switched off in the male.

By the time we had got this far in the genetic research on butterflies we could not help noticing certain striking parallels between the inheritance of their wing patterns and the inheritance of blood types in man. In man the blood-

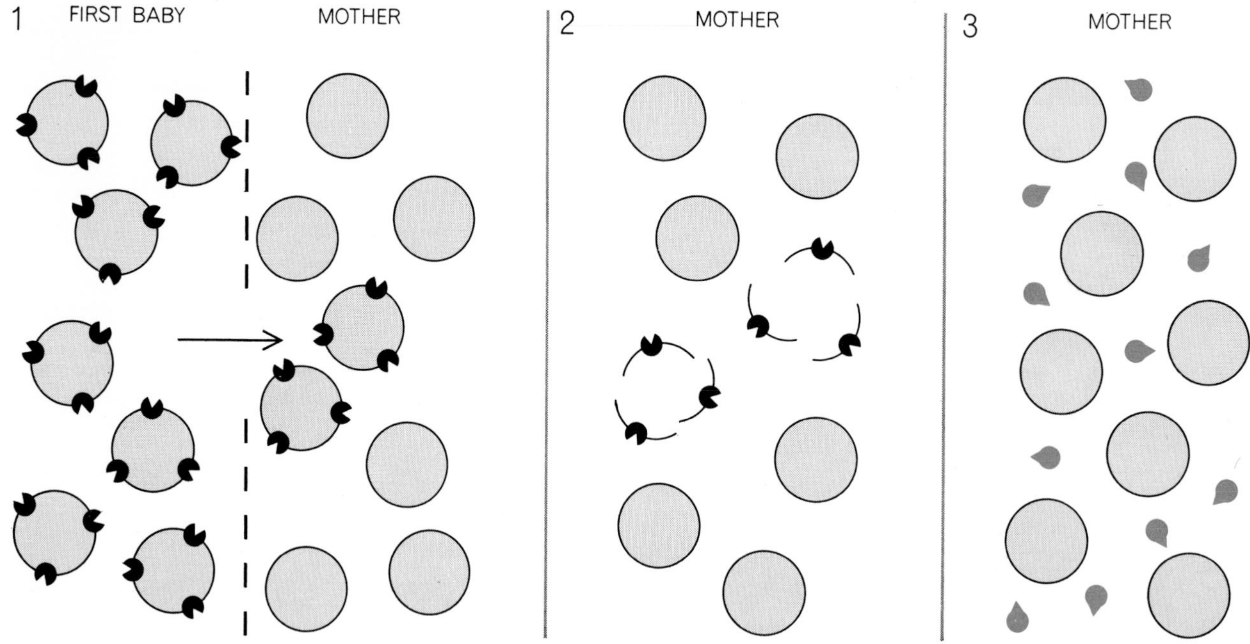


**STEPS IN DEVELOPMENT** of the Rh problem begin (*1*) when an Rh-negative mother has an Rh-positive baby and some of the baby's red blood cells get into her circulation. Although the baby's cells soon disappear naturally (*2*), the mother may manufacture antibody to the Rh antigen (*3*). The first baby is not affected, because it has been born by the time the antibody

group differences between individuals —Rh-positive, Rh-negative, O, A, B, AB and so on—are genuine manifestations of polymorphism. The Rh genetic units are supergenes, composed of three closely linked genes, with alternative recessive forms in many different combinations. Moreover, as in the male butterfly, in man there is an interaction whereby another blood-group system can interfere with the formation of antibodies against the antigens controlled by the Rh supergene. To appreciate the significance of these facts we must take a fresh look at the particulars of the Rh problem.

About 85 percent of the population in Britain and in the U.S. are Rh-positive, which means their red blood corpuscles contain the "rhesus" factor or substance, so named because it was originally detected in rhesus monkeys. The Rh factor is an antigen; if Rh-positive blood gets into the bloodstream of an Rh-negative individual, the person may produce an antibody (called anti-Rh or anti-D) that destroys the Rh-positive red blood cells. Therein lies the hazard for babies of an Rh-negative mother. The hazard, when it arises, usually comes about in the following way. If an Rh-positive father and an Rh-negative mother produce an Rh-positive baby (inheriting the Rh factor from the father), and if some of the baby's Rh-positive red blood cells get into the mother's circulation at the time of delivery, the mother may subsequently manufacture the Rh antibody. The antibody does not harm the mother, and a first baby is not affected because the antibody is produced after the baby's birth. The antibody in the mother's blood remains as a threat, however, to any subsequent Rh-positive baby, because it will enter the circulation of the fetus in the womb and destroy red blood cells, thereby causing possibly fatal anemia. The baby may be stillborn or be born with hemolytic disease.

The risk of this happening is not very high; although 15 percent of women are Rh-negative, among the 850,000 births each year in Britain, for instance, the number of "rhesus" babies is probably not more than about 5,000. Several factors operate to limit the risk. First, leakage of the baby's blood through the placenta into the mother's circulation in sufficient quantity to stimulate the production of Rh antibodies does not occur often. Second, some women do not produce antibodies even though there is leakage. Third, when the Rh-positive father is heterozygous (having received an Rh-negative gene from one of his parents), there is only a 50 percent chance that the baby will be Rh-positive. Fourth, and this is what particularly interested us, in about 20 percent of all the potential cases the formation of antibodies is prevented by the protective mechanism arising from interaction with other blood-group genes.

As an example, the mechanism operates in cases where the blood of the Rh-negative mother is of the type known as Group O. Blood of the O type always contains naturally occurring antibodies against A-type or B-type blood. The antibodies, called anti-A and anti-B, attack the red cells in blood of Group A or Group B. Thus if a Group O Rh-negative mother bears an Rh-positive baby whose blood is of the A type, her anti-A will rapidly get rid of any red cells that leak into her circulation from the baby at delivery, thereby removing the stimulus for the production of anti-Rh antibodies [*see illustration at right*]. This situation, technically called ABO incompatibility between the mother and the fetus, is almost always effective in preventing immunization of an Rh-negative mother against an Rh-positive baby.

Here, then, was an intriguing analogy to our findings about butterflies. The mode of inheritance of the blood groups and the interaction of the Rh and ABO systems were remarkably similar to what we had observed in the insects, particularly the interaction of sex and the mimetic supergene that in male butterflies prevents wing mimicry. Could we somehow devise a protective system for unprotected Rh-negative mothers, that is, for cases where there is no ABO incompatibility between the mother and the fetus?

For months I puzzled over the problem with my colleagues Sheppard and Richard B. McConnell. One night my

SECOND BABY MOTHER

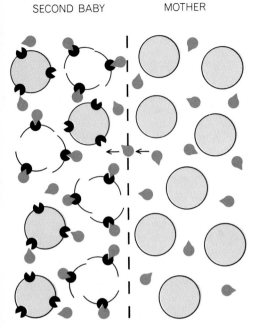

**appears. If the mother has a subsequent Rh-positive baby, however, the antibody may attack the baby's red blood cells (*4*), thereby giving rise to a possibly fatal anemia.**

wife, who had taken a keen interest in our work, woke me from a sound sleep and said: "Give them anti-Rh." Now, nothing is more irritating to a physician than to be awakened in the middle of the night and told how to manage his medical affairs. In a huff I replied, "It is anti-Rh we are trying to *prevent* them from making," and turned over and went to sleep again. In the clear light of morning, infuriatingly, the idea began to make sense. Giving the mother antibody to get rid of incompatible Rh-positive cells before her own antibody machinery went into production was obviously similar to the way nature accomplished the same objective, for instance in Group O mothers with a Group A baby or Group B one. I discussed the proposal with my colleagues, and we decided to test it.

The first experiments consisted in injecting Rh-positive red blood cells, labeled with radioactive chromium atoms, into Rh-negative male volunteers and then giving anti-Rh to half of the subjects, the other half serving as controls. We put Ronald Finn in charge of the experiments, and Dermot Lehane, director of the Liverpool Blood Transfusion Service, provided and injected the radioactively labeled infusions. The immediate results were exciting: the injected anti-Rh did indeed knock down a high proportion of the Rh-positive cells. Alas, the initial effect did not stand up. After six months we found that instead of suppressing the formation of antibodies the treatment had actually enhanced it. Confident nonetheless that our reasoning was basically sound, we persisted and discovered that we had given the wrong type of anti-Rh: the "complete" form of the antibody (which acts in saline solution). This material, we found, destroys the Rh-positive cells but still leaves the residue antigenic. We therefore did a second series of experiments with "incomplete" anti-Rh, which coats the antigen so that it does not make contact with the antibody-forming cells. These injections were much more successful: they prevented the production of Rh antibody in most of the subjects.

We now proceeded to find out if the treatment would work in Rh-negative mothers who received the antibody injection after delivery of their first baby. First of all, Finn and Joseph C. Woodrow of our group determined that the likelihood of production of Rh antibody by mothers generally depended on the number of Rh-positive cells that had leaked into the maternal circulation from the fetus: the more such cells in the mother's blood just after delivery, the greater the risk she would produce antibody. On the strength of this information we felt justified in testing the preventive effect of antibody injection in Rh-negative mothers who, after delivery of their first baby, showed a fair number of fetal red cells in their blood (five or more per 50 fields in a low-power microscope). Our obstetrical liaison was with Shoma H. Towers of the department at Liverpool headed by T. N. A. Jeffcoate. In these clinical trials we used a new form of the antibody preparation that was similar to one that had been

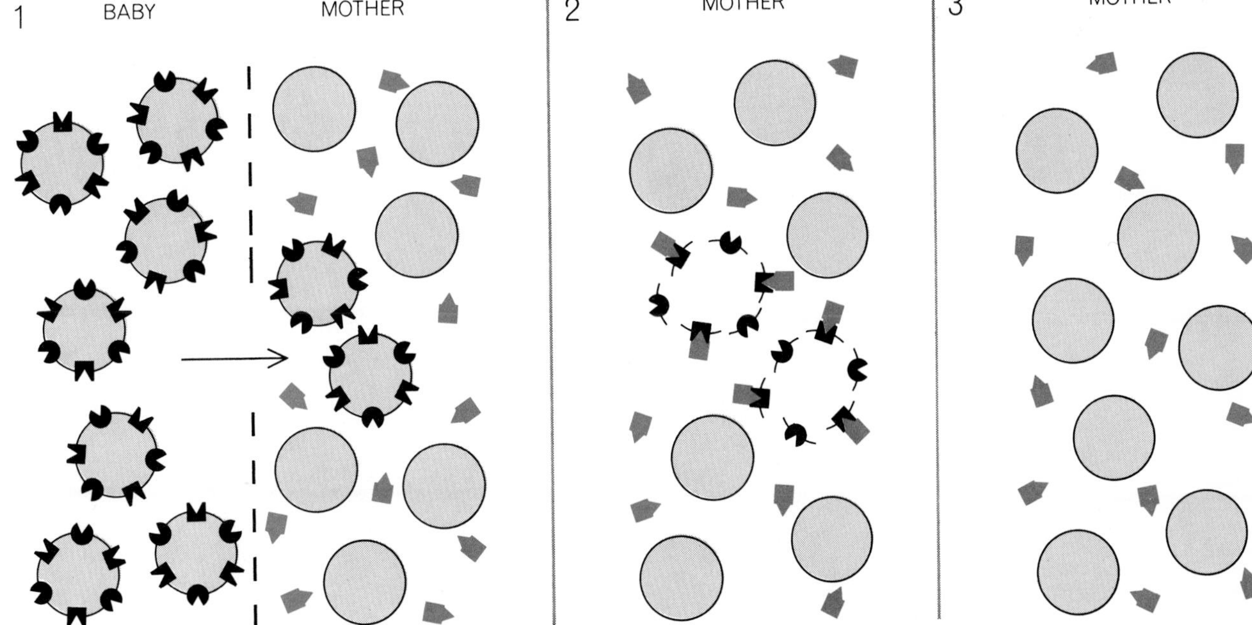


**NATURAL IMMUNITY to the Rh problem can result from what is called ABO incompatibility. Blood of the O type always has antibodies against blood of the A and B. Hence if an Rh-negative woman with O-type blood has an Rh-positive baby with A-type blood (*1*), her anti-A factor will attack any fetal red blood cells that enter her circulation (*2*). The cells are thus made nonantigenic, and the mother's body does not make antibodies against them. As a result she is not immunized against Rh-positive blood, so her subsequent Rh-positive babies will not be harmed.**

developed by a team at the Sloane Hospital for Women in New York: John G. Gorman, Vincent J. Freda and William Pollack, who had arrived independently at the anti-Rh concept. The preparation consists of anti-Rh gamma globulin instead of anti-Rh serum; its great advantage is that it avoids the risk of jaundice, which is always a potential danger in blood transfusions. Employing anti-Rh gamma globulin prepared for us from the serum of our immunized male volunteers by William d'A. Maycock and his colleagues at the Lister Institute in London, we gave the antibody to 131 selected first-baby mothers within 48 hours after delivery of the baby. Six months later W. T. A. Donohoe and his staff, who have carried out all our serological tests, assayed the mothers' blood for the presence of Rh antibody. Only one of the 131 mothers produced this antibody, whereas in a comparable control group of 136 first-baby mothers who had not received the anti-Rh injection 21 percent proved to be anti-Rh producers at the six-month examination.

This result, suggesting that the treatment gave almost complete protection, was far better than we had anticipated on the basis of the anti-immunization results in our male subjects. Furthermore, clinical trials of essentially the same method in the U.S., Canada, West Germany, Australia and elsewhere produced similarly successful outcomes for first-baby mothers. Critics objected that the six-month test was not necessarily conclusive as to protection: the treated mothers might start to make anti-Rh antibody under the stimulus of a second pregnancy. This objection proved to be groundless: subsequent tests at various centers on treated women who had a second Rh-positive baby showed that these mothers very rarely produced anti-Rh antibody. Immunologically speaking, the treated mothers entered their second pregnancy as if it were their first. The possibility remains, of course, that a "bleed" from the baby across the placenta at the second delivery or any subsequent one may stimulate the mother to begin producing antibody. In such cases the mother will require a new injection of the protective treatment whenever her blood after the birth carries fetal Rh-positive cells.

We are currently conducting studies to settle on a standard, minimum effective dose of antibody for the treatment. Tentatively it appears that about 200 micrograms of anti-Rh (about a fifth of what we gave in the original trials) is effective in most cases.

We have, of course, given a great deal of study to the possible risks involved in the anti-Rh injection. Occasionally we have noted a local swelling afterward at the site of the injection (made intramuscularly), but this disappears within a day or two. Will the injected gamma globulin produce harmful effects later? It disappears from the blood after a few months, but some women (about 5 percent in our trials) do produce antibodies to the gamma globulin that perhaps might cause a reaction to an anti-Rh injection given after a later pregnancy. This, however, does not appear to be a substantial hazard, as is evidenced by the fact that after a person has received an ordinary blood transfusion (which can generate gamma globulin antibodies) it is not considered necessary as a general practice to test the person for sensitivity before giving him a second transfusion. All in all, it appears that the risks of giving anti-Rh gamma globulin to an Rh-negative mother after her first pregnancy are negligible.

On the other hand, there is some risk for the donors of the anti-Rh: the male Rh-negative volunteers who must be injected with whole Rh-positive blood to

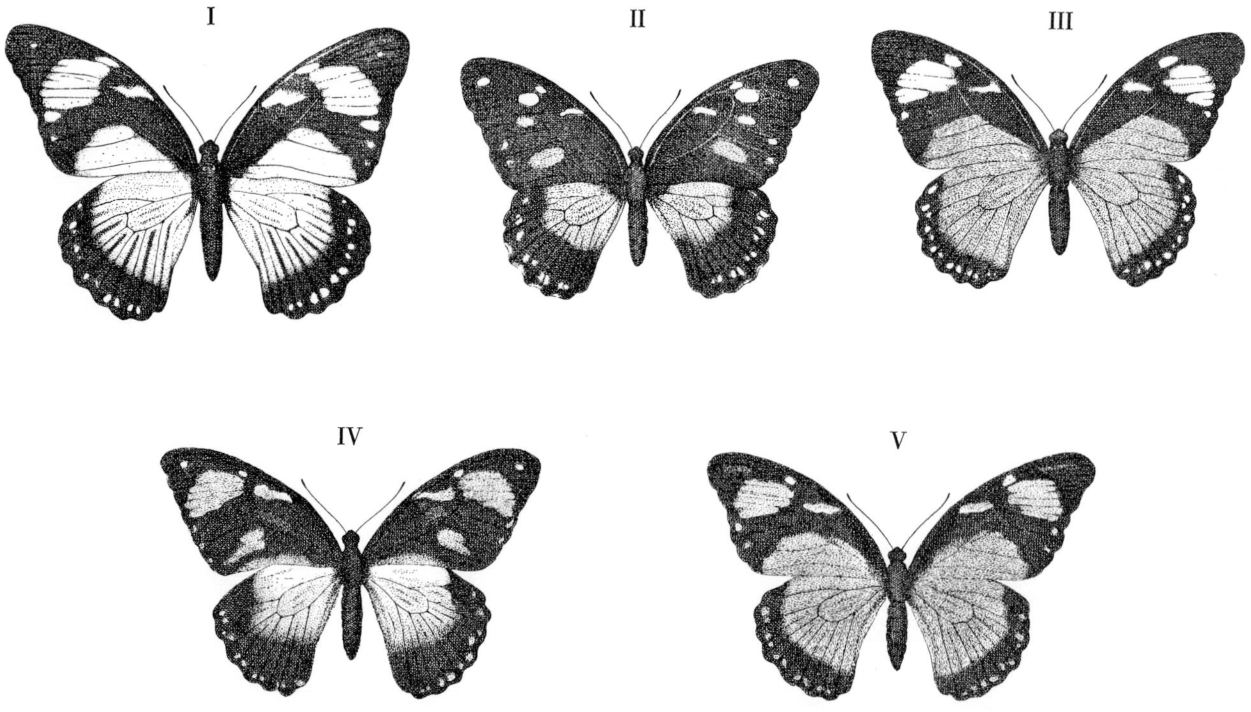


**WORK WITH BUTTERFLIES** provided a lead to the solution of the Rh problem. The work involved an investigation of mimicry, in which a form of butterfly that is palatable to birds comes to resemble a form the birds find distasteful. Mimicry occurs only in females. A single difference in genes can turn one mimetic pattern into another. Here five female forms of the South African butterfly *Papilio dardanus* are arranged in order from the bottom recessive (*I*) to the two top dominants (*III and IV*), which form a recognizable heterozygote (*V*). The human blood types equivalent to these mimetic patterns are respectively O, B, $A_2$, $A_1$ and AB.

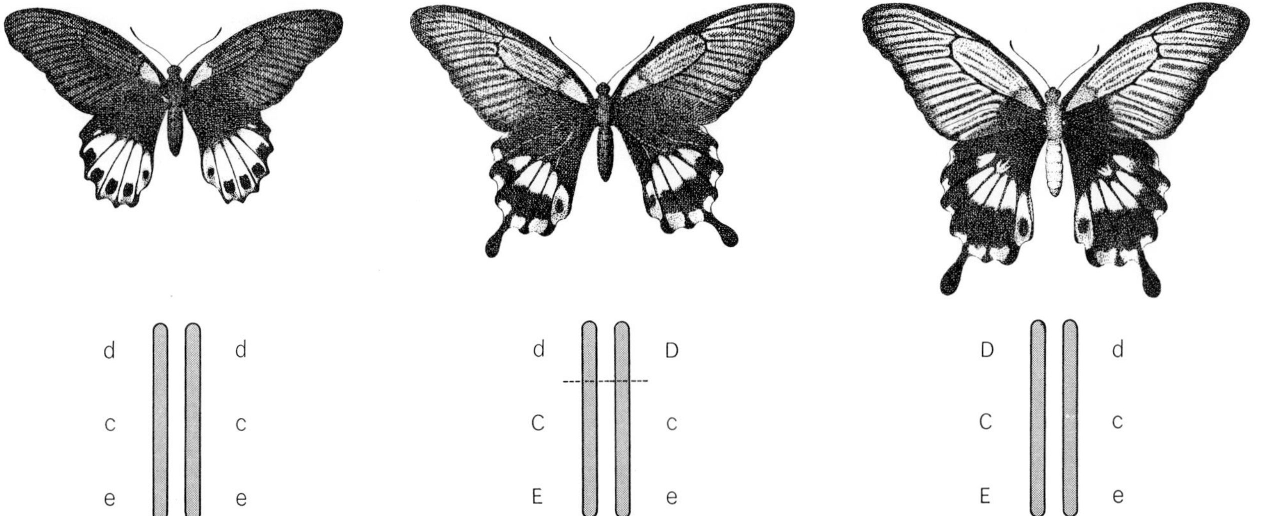


**BREEDING EXPERIMENTS** involving a nonmimetic (*left*) and a mimetic form (*right*) of the butterfly *Papilio memnon* of southeast Asia produced a crossover insect (*center*). At bottom are the respective chromosomal arrangements; *D* is yellow body color, *C* is small white wing window, *E* is long tails, *d* is black body color, *c* is large white wing window and *e* is absence of tails. The patterns are controlled by a group of closely linked genes called a supergene. There is interaction of sex and the mimetic supergene, reflected by the fact that mimicry does not occur in male butterflies. The Rh factor in human genetics is also controlled by a supergene.

manufacture the antibody. This risk is the possibility of the virus of jaundice being present in the injected blood. It can be minimized by making sure that the blood is obtained only from donors whose contributions have never induced jaundice in recipients. We have used blood from a few carefully selected donors and so far have not had a case of transmitted jaundice. There is, of course, the possibility of using serum from Rh-negative women who have been naturally immunized by an Rh-positive pregnancy; the Canadian workers use this source exclusively. The number of volunteers can also be greatly reduced with the new technique called plasmapheresis, which makes it possible to bleed the donors of anti-Rh every few weeks; a liter of blood is taken, the plasma containing the anti-Rh is skimmed off, and the red blood cells are reinjected into the donor.

Some authorities feel that, because of the jaundice risk to the donors and possible long-term harm to the women who receive anti-Rh, it is unwise to administer the antibody on a wholesale basis to all Rh-negative mothers who conceivably might have "rhesus" babies. Considering the small proportion of cases in which this actually happens, they suggest that anti-Rh should be given only to mothers who show a high probability of producing Rh antibody. The trouble is that it is impossible to identify these "high risk" cases with precision. The number of red cells from the baby found in the mother's circulation after the birth is not always a reliable measure of the risk. Many women produce antibodies after receiving only a very tiny bleed from the baby, and the indications are that the cell-count basis for selection of women to be treated would catch only about a third of the risky cases. It seems to us that if proper precautions are taken, the hazards involved in the treatment are so small, for the donors as well as the mothers, that they are far outweighed by the benefits. All vaccination programs entail giving the injection to great numbers of people who do not necessarily need it but who take it nonetheless for safety's sake. In this case the inoculation would banish anxiety for Rh-negative women, who would no longer need to worry about the possibility of endangering their babies.

The discussion of the risks has, however, called attention to the fact that the need for anti-Rh treatment could be obviated in some cases by doing more than is now done to keep the baby from bleeding into the mother's circulation in the first place. It appears that the situations most likely to produce this untoward happening (called transplacental hemorrhage) are the following: peeling the placenta off the womb by hand when the afterbirth is delayed, attempts to turn around a poorly positioned baby in the womb before it is delivered, an excessive number of abdominal examinations of the pregnant mother, toxemia of pregnancy, Cesarean delivery and abortion. Therapeutic abortions are particularly likely to give rise to transplacental bleeding, and when this occurs in an Rh-negative mother with an Rh-positive fetus, all the Rh-positive children she may deliver subsequently are apt to be exposed to anti-Rh attack. It therefore seems prudent to give anti-Rh to all Rh-negative women who have had an abortion even if the husband is Rh-negative, as one cannot always be absolutely certain that the husband is the father.

Among the thousands of mothers who have received the anti-Rh treatment since the trials began in 1964, there have been a few failures—cases in which the mother produced antibody in spite of the treatment. Two interesting possible explanations suggest themselves. The general rule has been to select for the treatment only women who apparently are not immunized, that is, who do not show evidence of producing Rh antibodies. One can suppose, however, that the antibodies may be present in the blood but not detectable by the usual methods, and that they manifest their presence only by going into action against Rh-positive cells when the woman becomes pregnant with an Rh-positive child. This may happen in the case of a mother who had a previous Rh-positive pregnancy but showed no sign of immunization afterward. In short, she may have been "primed" by the earlier pregnancy. We have actually encountered a case of priming in an Rh-negative man who volunteered for one of our experiments. After testing his blood and finding no evidence of the presence of

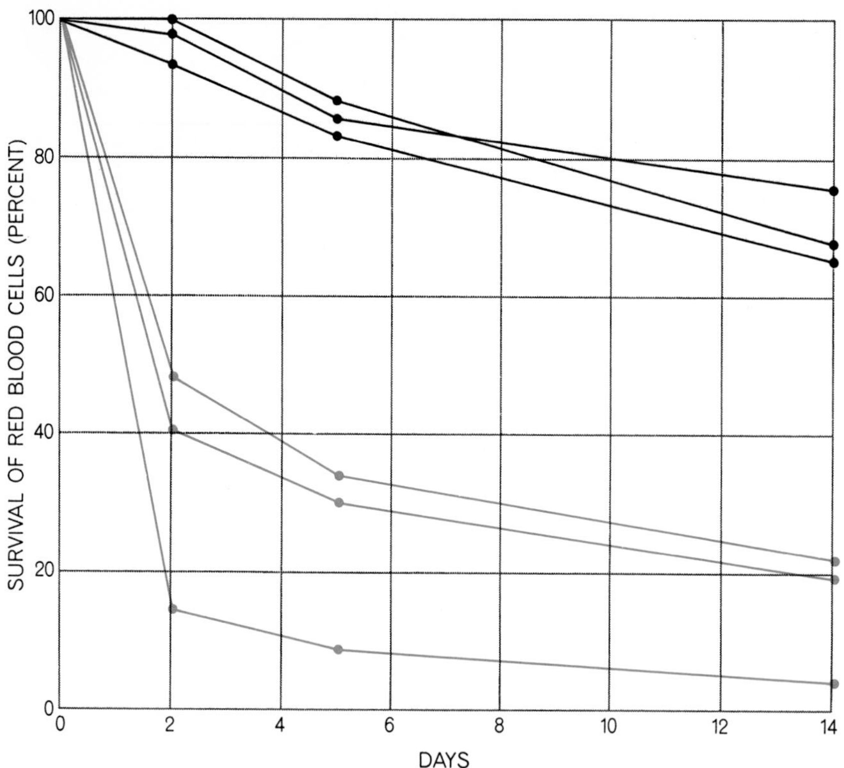


**FIRST TEST of plan to attack the Rh problem by administering Rh antibody produced these results. Male volunteers were injected with Rh-positive cells and then half of them (*color*) were given anti-Rh. Technique produced the same result as natural ABO incompatibility.**

Rh antibodies, we injected five milliliters of Rh-positive fetal blood. Within 48 hours all the fetal cells disappeared from his circulation, and we found traces of Rh antibodies in his blood. Since they could not have been produced so soon after the injection of the Rh-positive blood, we had to assume that he had been primed earlier. We then learned that the subject, who was 70 years old, had received a blood transfusion in World War I, which evidently was the source of his immunization against Rh-positive red cells.

This clear case of unsuspected priming suggests the exciting possibility that the anti-Rh treatment may work even for some mothers who are immunized, that is, whose antibody-producing mechanism has already been set in motion. It may be that a number of the mothers who have been given anti-Rh in the trials had actually been primed by previous exposure to Rh-positive cells although they showed no sign of being immunized when they were accepted for the treatment. If that is the case, the nearly perfect success in eliminating immunization in the thousands of women who have received anti-Rh is all the more remarkable.

Undetected priming is one of the two interesting explanations that have been offered for the few failures. The other is that occasionally the Rh-positive fetus's red cells may get into the Rh-negative mother's circulation early in her pregnancy, in which case the mother may develop the antibody-producing capacity before the baby is delivered and treatment with anti-Rh at that time will be too late. We believe such cases must be very rare, because first babies are seldom exposed to antibody from the mother. Bruce Chown and his colleagues at the University of Manitoba have, however, found evidence of the presence of Rh antibodies in some mothers during the first pregnancy or immediately after delivery. They have therefore begun giving anti-Rh to the mother during pregnancy and find that in the doses they use it does no harm to the baby. In cases where a bleed is known to have occurred early in pregnancy this method may be extremely valuable.

So, starting with experiments in the breeding of butterflies, the research has grown into a project that has enlisted the enthusiastic interest of a large team at Liverpool, stimulated workers in other laboratories around the world and produced a helpful advance in medicine and that may be found to have still wider applications.

# 31

# *The Genetics of the Dunkers*

H. BENTLEY GLASS

*August 1953*

THE DUNKERS, a small, sober religious sect that originally settled in eastern Pennsylvania, possess a characteristic of great interest to human geneticists. They are a group genetically isolated by strict marriage customs in the midst of a much larger community. Their beliefs have kept them distinct for more than two centuries. When Milton S. Sacks, director of the Baltimore Rh Typing Laboratory, and I were searching for a group in which to study how present blood-type and other hereditary differences arose in human populations, we turned to the Dunkers. Not only were they ideal for the study but also they opened up a larger area of inquiry. In them it is possible to perceive how racial differences which today distinguish millions of men first emerged and became set, so to speak, in small populations.

For 20 years anthropologists and geneticists have been increasingly aware of the importance in evolutionary processes of population size. In prehistoric times, when man was a hunter and gatherer, the world was sparsely inhabited. Hunting tribes are never large—1,000 people or so at most—and each tribe must range a rather large area to secure food and clothing. To keep this area inviolate and the tribe intact, primitive social customs include a simple avoidance of outsiders as well as active hostility, headhunting and cannibalism. Such customs tend to keep each tribe to itself and to make intermating between tribes quite rare. This was the period in human history when racial differences must have arisen and become set.

Such traits were probably established either by natural selection or by genetic drift, a term which may be explained as follows. In large populations chances in opposite directions tend to balance—*e.g.*, boy babies would generally equal girls in number—and the frequency of any genetic trait is expected to repeat itself generation after generation. But in small groups chance fluctuates more around the expected norm. No single family, for instance, is certain to have an equal number of boys and girls, and a predictable number of families will have all one or the other. If, in a small population, the initial proportion of brown eyes and blue were equal, the following generation might by chance have 45 per cent brown eyes and 55 per cent blue. This variation would then be expected to repeat itself by genetic law. However, the next generation might again by chance return to the 50-50 ratio or might shift to 40 per cent brown and 60 blue. If the latter were to happen, the expectation for the next generation would again be shifted, and this cumulative change by pure chance is what is meant by genetic drift. The phenomenon might continue until the whole population was either blue or brown eyed. Which ever way it went, once it had gone all the way it could not easily drift back. At that point it would be fixed in character until some mutation of the gene for eye color occurred. The chance of this in a small population is low, because the probability that a single gene will mutate is generally only one in 100,000 or even one in a million.

The actual operations of genetic drift are hard to pin down at this late date. Some present-day primitive peoples have been observed to have sharp divergences of inherited traits. The Eskimos of Thule in northern Greenland, where we are building a big air base, have a much higher frequency of the gene for blood group O (83.7 per cent) and a lower frequency of the gene for blood group A (9 per cent) than other Eskimos. They live far to the north and rarely mingle with the more populous Eskimo communities of the south. The Thule tribe numbers only 57. While its differences are probably due to genetic drift, we cannot rule out the possibility that some pressure of natural selection has operated more rigorously on blood-group genes in the far north than in the milder south.

A parallel case occurs in the aborigines of South Australia, whose traits have been studied for more than a decade by Joseph B. Birdsell of the University of California at Los Angeles. Among the South Australian desert tribes is the Pitjandjara, whose frequency of the gene for blood group A exceeds 45 per cent. To the west is a tribe of the same stock but having a much lower frequency of the group A gene (27.7 per cent); to the east is an apparently unrelated tribe with a frequency of this gene still lower (20 to 25 per cent). All three tribes are much more nearly alike in frequency of genes for the blood group MN. This is a situation to be expected from genetic drift, for it is not likely that in small populations chance will act on unrelated genes in the same way. But here again it is impossible to prove that genetic drift is the real agent. Each tribe has its own slightly different territory, customs and way of life: who is to say that it is not natural selection in a particular environment which has favored the marked increase of blood-group A in the Pitjandjara?

To get at the elusive drift in the frequencies of alternative kinds of genes, a small community of known origin is required, existing as nearly inviolate as possible in a larger civilization. After some discussion, it became apparent to Sacks and me that these requirements might happily be met right here among various German religious sects which immigrated to the U. S. in the 18th and 19th centuries. Not only are many of these sects still held together by strict rules regarding marriage, but also we have precise knowledge of their racial origins. This gives a starting point for comparison, since the present genetic composition of their homeland is also known. Moreover, they are now isolated in a much larger population, which provides an even more important basis for comparison. If natural selection or intermarriage were to influence their gene frequencies, these frequencies would shift from those of the homeland toward those characteristic of the large surrounding population. But if any sharp divergencies showed up, they would be attributable not to natural selection but to genetic drift.

After some exploration, a Dunker community in Pennsylvania's Franklin

**GENETIC DRIFT is depicted by disembodied eyes. The group labeled 1 represents a small population in which 50 per cent of the people have brown eyes (*dark in drawing*) and 50 per cent have blue eyes (*light*). Group 2 shows a second generation in which 45 per cent of the people have brown eyes and 55 per cent have blue eyes. Group 3 indicates a third generation in which, if the proportion of brown eyes to blue did not return to that of the first generation, the percentages might be 40 for brown eyes and 60 for blue.**

County was settled upon. No small part in influencing the choice was played by Charles Hess, a young medical student at the University of Maryland and a Dunker. He became interested in the project and gave invaluable help in collecting information and blood samples and in winning the cooperation of the Dunkers, without which the project could not have been carried out.

THE DUNKER sect, more formally known as the German Baptist Brethren, was founded in the province of Hesse in 1708, with a second center arising at Krefeld in the Rhineland. Between 1719 and 1729 some 20 families from the latter place and 30 from the former completely transplanted the sect to the New World, settling around Germantown, Pa. Over the next century the Dunkers doubled in number, and nearly all of them were descended from the original 50 families. To marry outside the church was a grave offense followed by either voluntary withdrawal or expulsion from the community. By 1882 the sect had grown to 58,000 and spread to the Pacific Coast. Under this steady drain to the frontier the Pennsylvania groups remained fairly stationary in numbers. The Franklin County or Antietam Church community, the subject of this study, seems never to have grown larger than a few hundred persons up to 1882.

In that year a schism split the church and further contracted the size of this and other communities. For some time there had been trouble between those who wished to retain all the old tenets of the sect and those who wanted some relaxation of the more restrictive rules governing baptism, foot-washing, love feasts, head coverings for women, sober dress for men and the like. An open rupture finally separated the strict Old Order, as well as a Progressive group, from the main body, which went on to form the present-day Church of the Brethren. The Franklin County community studied by us remained in the Old Order, which from that day to this has numbered only about 3,000 people scattered in 55 communities over the country.

In all visible respects the Franklin County Dunkers live as their neighbors live. Most of them are farmers, though some have moved to the county's two medium-sized towns. They own cars and farm machinery; most have modest but comfortable frame houses; their food is typically American; the children attend public schools; medical care is good. Distinctions of dress are not conspicuous to the degree seen, for example, in the better-known Amish sect. Except for strict adherence to their religion, the Dunkers are typical rural and small-town Americans. In marriage pattern the community is not wholly self-contained. Over the last three generations about 13 per cent of their marriages were with members of other Old Order communities and about 24 per cent with converts, a factor taken into account in our study. Thus in each generation somewhat more than 12 per cent of the parents in the community came in from the general population. The equalizing force of this "gene flow" from outside would of course tend to make the Dunkers more like everyone else in hereditary makeup. The effects of genetic drift, if perceptible, would have to be large enough to overcome the equalizing tendency. Altogether the community now numbers 298 persons, or 350, if children who have left the church are included. For several generations the number of parents has been about 90. This, then, is exactly the type of small "isolate" in which the phenomenon of genetic drift might be expected to occur.

THE CHARACTERISTICS chosen for study in the Dunker group were limited to those in which inheritance is clearly established and in which alternative types are clear-cut, stable and, so far as is known, non-adaptive. The frequency records of these characteristics were available for both the West German and North American white populations, or at least the latter. Complete comparisons could be made for frequencies of the four ABO blood groups (O, A, B and AB) and the three MN types (M, MN and N). The Rh blood groups were almost as good. Although no Rh frequencies are available from West Germany, other European peoples have been studied, the English very extensively, and it is evident that all West Europeans are quite similar in this respect. Four other traits were examined in which the Dunkers could at least be compared with the surrounding American population. These were: (1) the presence or absence of hair on the middle segment of one or more fingers, known as "mid-digital hair"; (2) hitchhiker's thumb, technically called "distal hyperextensibility," the ability to bend the tip of the thumb back to form more than a 50-degree angle with the basal segment; (3) the nature of the ear lobes, which may be attached to the side of the head or hang free; and (4) right- or left-handedness.

The findings were clear-cut. They show that in a majority of all these factors the Dunkers are neither like the West Germans nor like the Americans surrounding them, nor like anything in between. Instead, the frequencies of particular traits have deviated far to one extreme or the other. Whereas in the U. S. the frequency of blood group A is 40 per cent and in West Germany 45 per cent, among the Dunkers it has risen to nearly 60 per cent, instead of being intermediate as would be expected in the absence of genetic drift. On the other hand, frequencies of groups B and AB, which together amount to about 15 per cent in both major populations, have declined among the Dunkers to scarcely more than 5 per cent. These differences are statistically significant and unlike any ever found in a racial group of West European origin. One would have to go to the American Indians, Polynesians or Eskimos to find the like. The ancestry of all 12 persons with blood groups B and AB was checked to find out whether their B genes had been inherited from within the community. Only one had inherited this blood factor from a Dunker ancestor in the Franklin community; all the others had either been converts or married in from other Dunker groups. Evidently this gene was nearly extinct in the group before its recent reintroduction.

The three MN types showed even more unexpected trends. These have almost identical frequencies in West Germany and the U. S.: 30 per cent for type M, 50 for MN and 20 for N. In the Dunkers the MN percentage had diminished slightly, but frequencies for the other two types had deviated radically. M had jumped to 44.5 per cent and N had dropped to 13.5. One would have to go to the Near East or look in Finland, Russia or the Caucasus to find any whites with hereditary MN distributions like these. Only in the Rh blood groups do the Dunkers not differ greatly from their parent stock or adopted land. As against an average of close to 15 per cent for the Rh-negative type in both English and U. S. populations, the Dunkers show 11 per cent.

From the other traits in which comparisons were made almost equally striking conclusions can be drawn. Without going into details, the Dunkers had fewer persons with mid-digital hair or hitchhiker's thumb or an attached ear lobe than other U. S. communities. Only in right- and left-handedness, as in the Rh blood types, do the Dunkers agree well with the major populations used for comparison.

THERE SEEMS to be no explanation for these novel combinations of hereditary features except the supposition that genetic drift has been at work. To clinch the matter, the Dunkers were divided into three age groups—3 to 27 years, 28 to 55 years and 56 years and older—roughly corresponding to three successive generations. When the ABO blood types were singled out, it was at once apparent that their frequencies were the same in all three generations. It follows that the unusual ABO distribution is of fairly long standing and antedates the birth of anyone still living. When the same analysis of MN blood types was made, however, a very different story emerged. In the oldest generation the M and N genes were exactly the same in frequency as in the surrounding population. In the second generation the frequency of M had risen to 66 per cent and N had dropped to 34. In the third generation this trend continued, M going to 74, N sinking to 26. While other genes remained unaltered in frequency, these genes were apparently caught in the act of drifting.

Let us consider the phenomenon a little more deeply. There can be no doubt from these instances that genetic drift does occur in small, reproductively isolated human groups in which the parents in any one generation number less than 100. Drift is probably somewhat effective, though slower, in populations two or three times that size. Such were the tribes of man before the dawn of agriculture. How inevitable it was, then, that numerous hereditary differences, perhaps of a quite noticeable but really unimportant kind, became established in different tribes. It is my opinion, no doubt open to dispute, that most inherited racial differences are of this kind and were not materially aided by natural selection.

Some traits, of course, must originally have been fixed by selection. Dark skin is probably a biological advantage in the

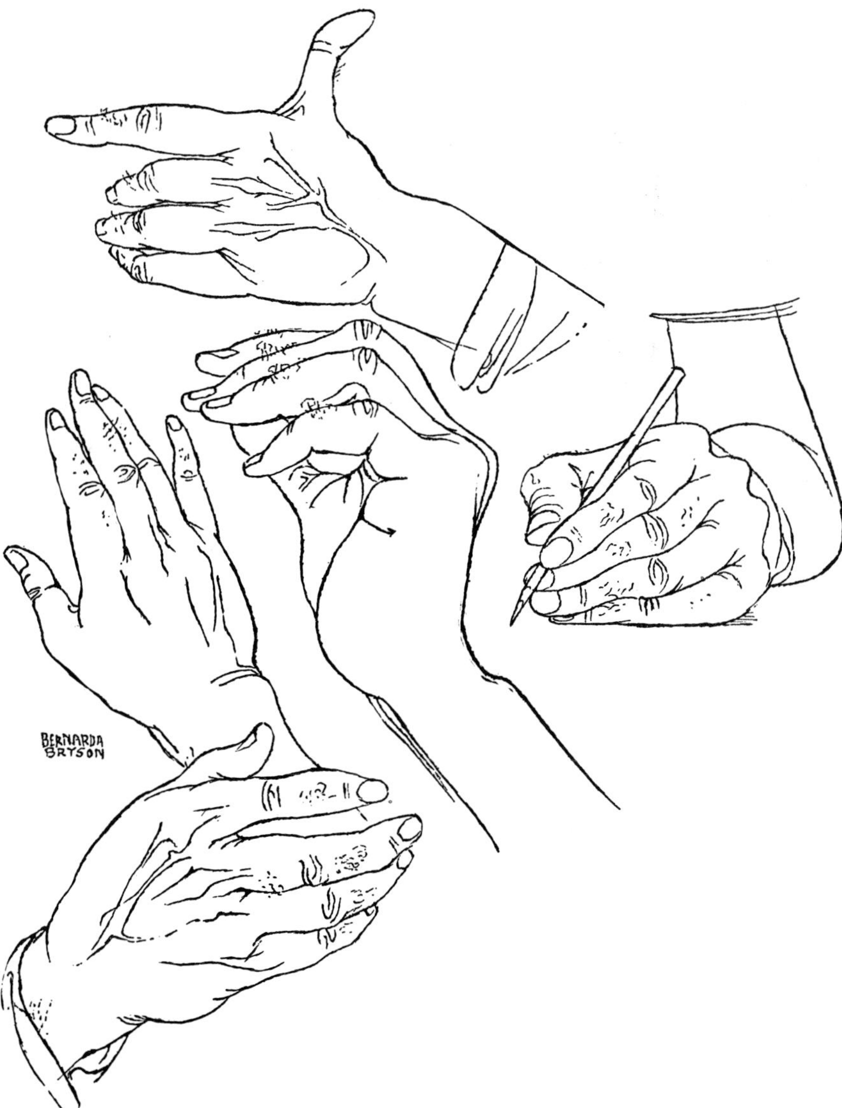


**THREE OTHER CHARACTERISTICS were studied. One was "hitchhiker's thumb" (*hand at top*), the ability to bend the end of the thumb backward at an angle of more than 50 degrees. The second was "mid-digital hair" (*hands at lower left*), or hair on the middle segments of the fingers. The third was right-handedness *v.* left-handedness (*hand at right*).**

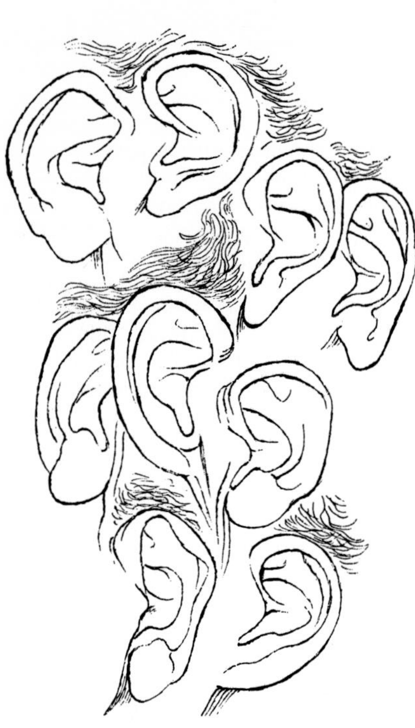

**EAR LOBES either hang free or are attached to the head. Among Dunkers there are fewer attached lobes than among the U.S. population as a whole.**

tropics, while pale skin may be an advantage in weaker northern light. The same may hold for kinky hair as against straight hair, for dark eyes as against light eyes, and the like. Many possibilities of this kind have been suggested in a recent book, *Races: A Study of the Problems of Race Formation in Man*, by Carleton S. Coon, Stanley M. Garn and Joseph B. Birdsell. I remain skeptical when I think of the prevailing hairlessness of man in many regions where more body hair would have helped to keep him warm. I am particularly skeptical of any selective advantage in blondness, "the most distinctive physical trait or group of traits shown by Europeans." It seems more likely that these traits confer no advantage to speak of, in Europe or elsewhere. I would add that if blonds had been eliminated by selection in other parts of the world, and if a blond type happened by genetic drift to become established in Europe, then it could have persisted and spread in large populations and given rise to the present racial distribution of the blond caucasoid or "Nordic" man.

THE STUDY of the Dunker community confirms the suspicion of many anthropologists that genetic drift is responsible for not a few racial characteristics. Further studies along these lines, together with studies of mortality and fertility in contrasted hereditary types, may in time tell us which racial traits were established because of selective advantage and which owe their presence solely to chance and genetic drift.

# 32
# *Hybrid Wheat*

BYRD C. CURTIS and
DAVID R. JOHNSTON
*May 1969*

So far in this century the production of two major grain crops, corn and sorghum, has been revolutionized by the technique of mass hybridization: the crossing of two dissimilar inbred lines or varieties to obtain offspring with more desirable qualities than those possessed by either of the parent lines. It now appears that another important grain, wheat, is on the verge of a similar revolution. The problems associated with the production of hybrid wheat on a commercially feasible scale have been particularly difficult, but enough progress has been made in the past few years to predict with some assurance that the eventual widespread introduction of hybrid wheat varieties will have a far greater economic and nutritional impact than the introduction of any other hybrid crop grown in the world today.

The basis of all such attempts at genetic manipulation is the phenomenon of hybrid vigor, the tendency for the offspring of crossed varieties to have greater vitality than the offspring of inbred varieties. Hybrid vigor can be manifested in a number of ways: increased yield, greater resistance to disease or insects or harsh climate, a shorter growing season and better milling or baking qualities. In the case of wheat the primary benefit being sought is increased yield.

It turns out that the extra vigor of hybrids is expressed at a maximum in the first generation after the cross; later generations show a drastic reduction in vigor. Hence the object of any hybridization program is to perfect a technique for producing enough hybrid seed to grow first-generation plants on a large scale.

This was easy to accomplish with corn, because in a broad sense all corn is hybrid. Corn is a cross-pollinated species in which the male sex parts (in the tassel) and the female sex parts (in the ear) are located in quite separate parts of the same plant. Removing the tassel by hand makes the plant female and therefore incapable of self-fertilization; thus all the seed produced on the ear will be hybrid since it must be fertilized by pollen from other plants. Enough hybrid seed was produced by this method to plant the entire corn acreage of the U.S. within the first 20 years after the first hybrid varieties were introduced in the 1930's.

In the early 1950's a new technique for producing hybrid corn seed was developed. It involves a sophisticated genetic procedure for inducing male sterility in a generation of corn plants; these plants are then crossed with a variety that is capable of restoring full fertility to the first-generation offspring of the cross. This approach eliminated the need for the laborious hand-detasseling operation and is currently employed in the production of a large percentage of the world's crop of hybrid corn. A similar technique has proved successful for producing hybrid sorghum; in only a few years it has resulted in the hybridization of the entire sorghum crop of the U.S. It is basically a variation of this general approach that has been applied to the hybridization of wheat.

Unlike corn, wheat is almost 100 percent self-pollinated. Both the male and the female sex parts—the male stamens containing the pollen and the female pistil containing the egg—are located in the same floret, or flower [*see illustration on next page*]. Normally the anthers (the elongated pollen-bearing portions of the stamens) supply pollen to the stigma (the feathery portion of the pistil) before the floret opens enough to allow the entrance of pollen from other plants. To obtain a single hybrid seed it is necessary to remove the three anthers in a floret with small forceps and later apply pollen from another plant to the stigma by hand. To ensure success all these operations must be timed precisely and executed with great care. Obviously commercial quantities of hybrid seed cannot be produced in this manner.

Here it should be noted that many varieties of wheat already under cultivation are often referred to incorrectly as hybrid wheat. This misnomer arises from the fact that it is possible to derive improved varieties from a handmade hybrid several generations removed. The best descendants of the original hybrid are selected and inbred for five or six successive self-pollinated generations until a "pure line" or "true breeding" plant is obtained. Such a plant, combining the good traits of the original parents, will

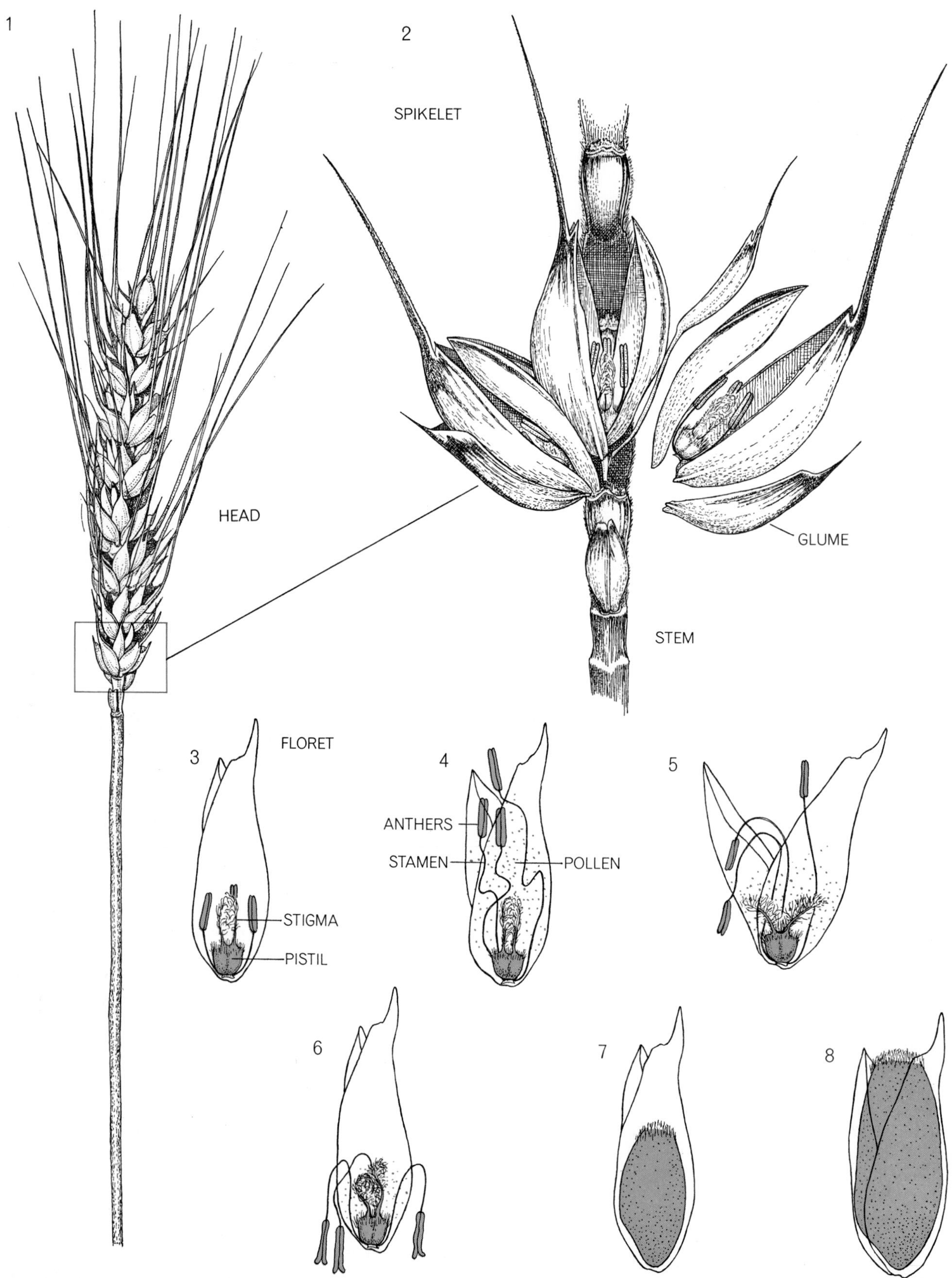


**SELF-FERTILIZATION** of an individual floret, or flower, of wheat is represented in this series of drawings. Drawing *1* shows a typical mature wheat spike; drawing *2* shows how the wheat head is composed of alternating rows of spikelets, each of which contains several florets (in this case three are shown). Drawings *3* through *6* show how the anthers (the elongated pollen-bearing portions of the male stamens) normally supply pollen on the stigma (the feathery portion of the female pistil) before the floret opens enough to allow the entrance of pollen from other plants. Drawings *7* and *8* show how the fertilized egg develops into the seed, or grain, of wheat. Threshing separates the full-grown grains from the rest of the plant, which is referred to collectively as chaff.

produce similar plants in all subsequent generations, provided that it is self-pollinated and that mutation does not take place. Most modern wheat varieties were obtained by some such procedure, but they are clearly not true hybrids.

The development of a technique for producing true hybrid wheat seed on a large scale began in the early 1950's, at about the same time that the male-sterile mechanism for producing hybrid corn was being perfected. The Japanese investigator H. Kihara reported in 1951 that he had succeeded in inducing "cytoplasmic male sterility" in wheat; this type of male sterility sometimes results when the nucleus of one cell interacts with the cytoplasm of an unrelated cell. Kihara had transferred the nucleus of a common bread wheat (*Triticum aestivum* subspecies *vulgare*) into the cytoplasm of a wild relative of wheat called goat grass (*Aegilops caudata*) and had found that the progeny were female-fertile but male-sterile. In 1953 H. Fukasawa of Japan obtained similar results with a different wild species (*Aegilops ovata*) as the female parent in a cross with durum wheat (*Triticum durum*), a species used primarily for making macaroni. Later Kihara developed still another male-sterile variety by crossing emmer wheat (*Triticum dicoccum*) with a species known only as *Triticum timopheevi.*

None of these developments proved successful in providing a commercially useful male-sterile mechanism for hybrid wheat, owing to adverse side effects produced by the particular lines that were crossed. The findings did, however, stimulate later and more successful experiments. In 1961 J. A. Wilson and W. M. Ross, working at the Kansas Agricultural Experiment Station, obtained stable cytoplasmic-male-sterile bread wheats by crossing *T. timopheevi* as the female or seed parent with a variety of *T. aestivum* called Bison wheat. Repeated backcrossing of the Bison variety with the male-sterile progeny plants resulted in stable, cytoplasmic-male-sterile Bison lines. These lines were widely distributed internationally and are the source of many of the male-sterile wheat varieties currently available.

For the cytoplasmic-male-sterile system to be useful in producing hybrid wheat a corresponding fertility-restoring mechanism must be found. A male-sterile Bison line, when pollinated by normal fertile Bison wheat or other normal, nonrestoring varieties, will produce offspring that are also male-sterile. This

**EMASCULATION of a wheat floret is accomplished by removing the three anthers in the floret with small forceps. Pollen from another plant can later be applied to the stigma by hand in order to obtain a single seed of hybrid wheat. Such a procedure is obviously not commercially feasible as a method of producing large quantities of hybrid wheat seed.**

**MALE-STERILE SEED on the wheat head at center was obtained by hand-fertilizing the florets of an emasculated plant called *Triticum timopheevi* with pollen from a common bread-wheat variety of *Triticum aestivum*. Normal spikes of the female and male parents are shown at left and right respectively. The "cytoplasmic male sterility" of the crossed variety results from a little-understood interaction of the chromosomal genes in the nuclei of the male parent's cells with an unknown heredity factor in the cytoplasm of the female parent's cells. Repeated backcrossing of fully fertile plants of the common bread-wheat variety with the male-sterile progeny plants results in a stable cytoplasmic-male-sterile line.**

outcome is of course an important and integral step in maintaining and increasing seed of the male-sterile variety, but seed used by the farmer to plant his hybrid crop must have, in the resulting plants, the capacity for male fertility. This is achieved by pollinating the male-sterile plants with a variety that restores fertility. Such varieties have in them dominant restorer genes capable of overcoming the cytoplasm-nucleus reaction that causes male sterility. Finding highly effective dominant gene systems for fertility-restoration has proved difficult indeed, but a few such systems that restore fertility well under certain environments have been reported.

In 1960 Wilson and Ross announced the discovery of partial fertility-restoring factors in bread wheat for one of the cytoplasms on which Fukasawa had worked. Then in February, 1962, Wilson suggested that restorer genes must exist in *T. timopheevi,* since it carried the sterile cytoplasm; otherwise *T. timopheevi* would be male-sterile and unable to reproduce itself. Several months later John W. Schmidt, V. A. Johnson and S. D. Mann of the Nebraska Agricultural Experiment Station demonstrated that a bread wheat derived from *T. timopheevi* was effective in restoring reasonable fertility to male-sterile Bison wheat. Shortly thereafter Wilson, working independently, reported similar results. Subsequent studies by these same investigators and others have shown that the original restorer sources were not completely effective in restoring fertility to the Bison plants and other sterile wheats in all environments.

In the years since the discovery of these partial fertility-restoring factors it has become obvious that restorer genes and modifiers of restorer genes are distributed among several of the world's existing wheat varieties. For example, Ronald W. Livers of the Kansas Agricultural Experiment Station has reported a number of common varieties that carry genes for partial restoration. An important restorer gene was also found in a variety called Primepi wheat by E. Oehler and M. Ingold of France. Much research is currently under way to collect enough of the restorer genes into single agronomically desirable varieties in order to make hybrids completely fertile when crossed with agronomically sound male-sterile varieties. One private seed company has announced success in this effort and has distributed several varieties of hybrid wheat seed for test planting by farmers.

The cytoplasm-nucleus reaction that leads to male sterility is not a well-understood process. In fertile plants there is apparently a good balance between the chromosomal genes and an unknown heredity mechanism carried in the cytoplasm. In cytoplasmic-male-sterile plants this delicate balance is undoubtedly upset in some way, giving rise to deformed anthers and empty or sterile pollen grains [*see illustration on page 304*]. In contrast to normal anthers the anthers of male-sterile plants are more slender and tend to curl at the base into the shape of an arrowhead. Such anthers produce little pollen, and what little is produced cannot effect fertilization. Apparently the induced imbalance has no effect on the pistil. Normal pollen applied to the stigma will function properly, and a seed will be produced. In fact, there is scant evidence that male-sterile plants are morphologically different from normal plants except for defective anthers and pollen.

Cytoplasmic-male-sterile lines can be developed, starting with a stable line such as Bison wheat or Wichita wheat, by means of the backcross method [*see top illustration on page 305*]. The final progeny will be identical with the recurrent parent in most characteristics except that it will be male-sterile instead of fully fertile. In order to prevent contamination by foreign pollen and to ensure that only pollen from the recurrent parent effects fertilization, some stratagem must be found for isolating the individual plant, such as placing a small plastic bag over its head.

A few wheat varieties have proved difficult to sterilize. These plants, which usually have a restorer gene that prevents sterilization, include several important commercial varieties. Just why these varieties have such genes is not known. Perhaps they arose by mutation or are carry-over genes from natural crosses to *T. timopheevi* in past generations.

Male-sterile varieties are maintained and increased by growing the male-sterile, or *A*-line, plants in "drill strips" in the field. These are situated between drill strips of the normal fertile, or *B*-line, plants. The male-sterile plants are pollinated by windblown pollen from the fertile plants. Seed from the male-sterile plants is harvested (by combine) separately from seed from the fertile plants; after a sufficient increase is attained the male-sterile seed is ready to be planted as the female next to a restorer line as the male for hybrid-seed production.

Developing additional restorer, or *R*-line, plants from a given source is much more difficult and laborious than developing additional male-sterile lines. For example, in order to produce a restorer line of the Scout variety one must select and backcross the dominant restorer plants in later generations to the parent Scout line. After several repetitions of the cycle the characteristics of the Scout plants are combined with the dominant restorer genes, resulting in a Scout restorer variety [*see bottom illustration on page 305*].

The same general procedure can be used to produce a new restorer variety with the desirable characteristics of the Scout line combined with the desirable characteristics of the original restorer source. Here, as before, the second-generation restorer plants would be selected for producing the third and succeeding generations. In each generation, however, the best agronomic plants would be selected for continuation. Thus after the fifth generation it should be possible to select plants approaching the pureline status.

An important requirement for any breeding scheme designed to produce restorer lines is that the selected plants of each generation be crossed to a cytoplasmic-male-sterile variety to determine if the selected plants actually carry the restorer genes. The first-generation plants from these test crosses will be fully restored to male fertility if the selected plants carry the necessary fertility-restoring genes.

After a number of male-sterile and restorer lines are produced, various combinations of hybrids are made and tested under field conditions to determine which hybrids display the greatest hybrid vigor for a given production area. Once this is established the hybrid seed can be produced on a large scale for sale to farmers. Such seed will produce first-generation hybrid plants that one hopes will possess the desired amount of hybrid vigor. Obviously hybrid vigor must be manifested or the farmer gains nothing from planting the hybrid.

The amount of hybrid vigor to be gained from hybrid wheat in large-scale field plantings has not yet been established, owing to the unavailability of sufficient seed stocks for such plantings. Many reports on the gain in hybrid vigor

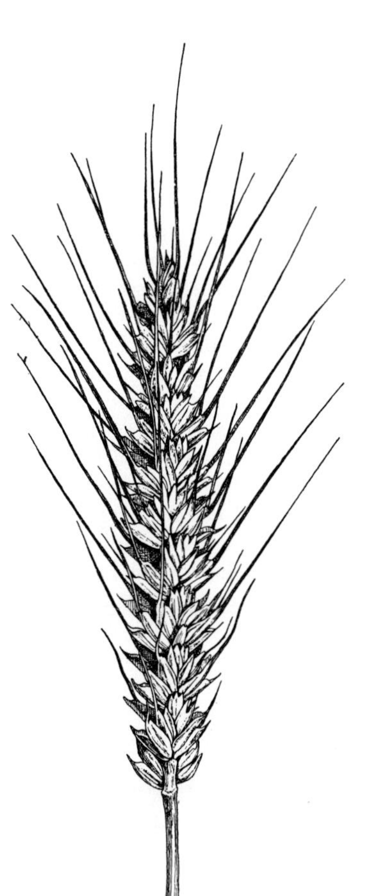

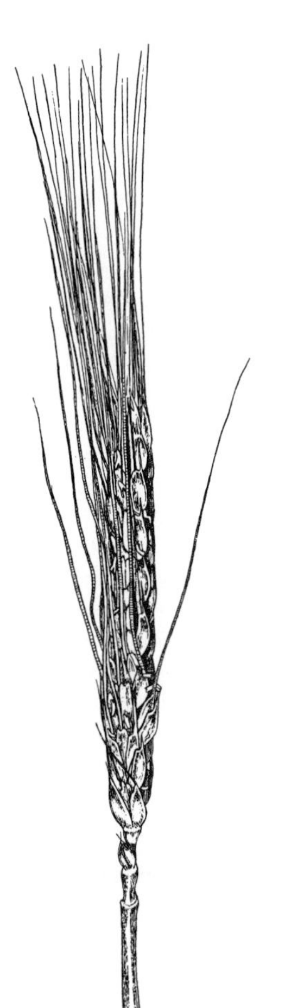

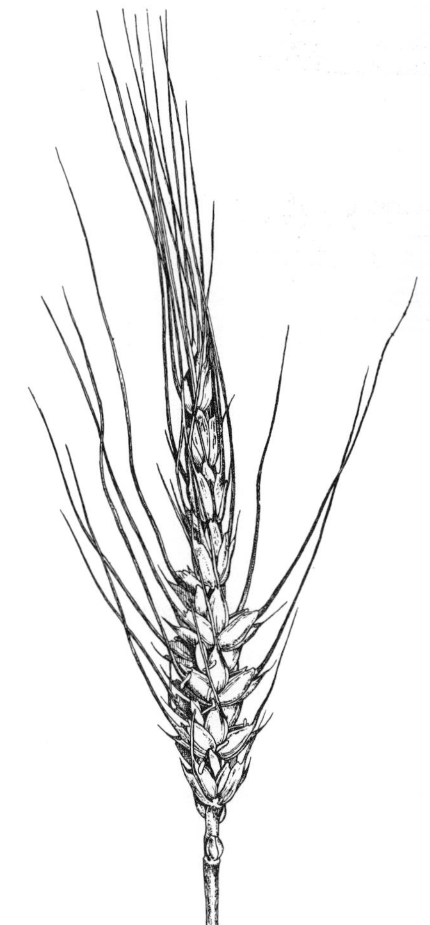

**THREE MATURE SPIKES** of Bison wheat are compared in this illustration. At left is a cytoplasmic-male-sterile spike open and ready for cross-pollination. At center is a fully fertile spike undergoing self-pollination. At right is a partially fertile spike derived from a cytoplasmic-male-sterile line; its fertility has been partially restored by crossing its male-sterile parent as the female with a variety of bread wheat derived from *T. timopheevi* as the male. The discovery of the fertility-restoring factors in bread wheat in the early 1960's stimulated a major expansion in the research effort devoted to the development of new hybrid wheat varieties.

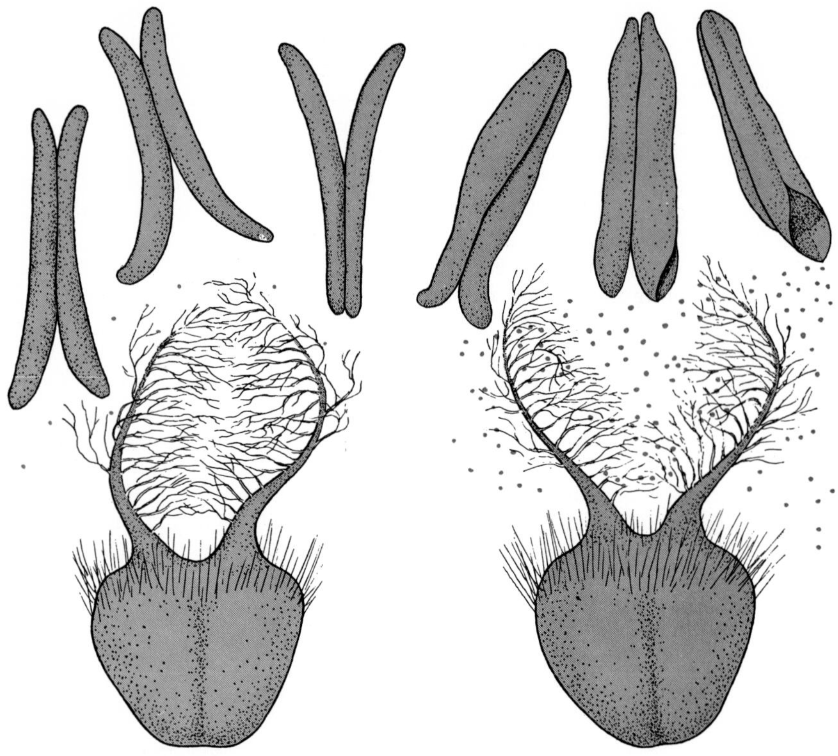

**SEX PARTS from a male-sterile plant of Bison wheat (*left*) are compared with the sex parts from a normal plant of Wichita wheat (*right*). The anthers of the male-sterile plant (*top left*) are more slender than the normal anthers (*top right*) and tend to curl at the base into the shape of an arrowhead. The little pollen produced by the anthers of the male-sterile wheat plant is sterile and cannot effect fertilization. The pistil of the male-sterile plant (*bottom left*) is morphologically identical with that of the normal plant (*bottom right*).**

from small experimental plantings in various parts of the world are in the literature on wheat. One can surmise from these reports that a 20 to 30 percent increase in yield should not be difficult to obtain from large-scale field plantings. Data from such plantings should be available in two to four years; it will then be possible to ascertain the degree of hybrid vigor to be realized from wheat.

So far we have discussed only the mechanics of producing hybrid wheat plants experimentally. No mention has been made of the difficulties usually encountered in increasing the male-sterile lines or in growing hybrid seed under field conditions.

The major difficulty in producing economically feasible hybrids, other than developing adequate restorer lines, has been the failure to obtain consistently good cross-pollination in the field. Cross-pollination is affected by many factors, including the synchronization of the flowering times of the male-sterile plants and the pollinator plants. The vagaries of the weather also have a strong bearing on cross-pollination. The problem of providing receptive female florets at the time of maximum pollen dispersal depends on the relative maturity of the two parent varieties. Concurrent plantings of male-sterile and pollinator varieties, with the pollinator plants reaching the "heading" stage one to three days later than the male-sterile plants, will usually provide adequate synchronization.

Under conditions of cool weather and an adequate supply of moisture, the female sex parts will remain receptive for as long as eight to 10 days. Maximum receptivity is attained three to five days after the wheat heads are fully formed. If the pollinator variety happens to mature earlier than the male-sterile one, maximum pollen dispersal may precede maximum floret receptivity, and the result will be a reduced yield of seed. Solutions to such problems may be found in adjusting the planting times or in varying the seeding rates. Such activities will, however, add considerably to production costs and thereby increase the cost of hybrid seed. Extra seed costs to the farmer reduce the value he gains from hybrid vigor.

Wheat pollen is short-lived and is adversely affected by hot and dry conditions. Richard E. Watkins of Colorado State University has found under laboratory conditions that the life-span of pollen from typical wheat varieties is less than five minutes after anthesis (the opening of the anther) at a temperature of 95 degrees Fahrenheit and a relative humidity of 20 percent. At 65 degrees F. and 80 percent relative humidity some pollen remained alive for as long as an hour, with more than 60 percent still viable 20 minutes after anthesis. In addition to affecting the viability of pollen, high temperatures and low humidity reduce the pollen load in the air, since the wilting of the plant parts impedes the release of pollen.

In general the best cross-pollination has been achieved at cool (but not too cool) temperatures and medium humidity. Too much rainfall or fog also impedes pollen dispersal. Under these conditions pollen is either washed down or made so soggy that it fails to become airborne. Such weather conditions may also cause "lodging," or bending of the entire wheat plant toward the ground, another situation that could prevent cross-pollination. Gentle to strong breezes are necessary to move pollen from one plant to another, but too much wind may result in the loss of pollen. Wheat pollen is heavier than air, but it is easily borne aloft by air movement. Some workers have envisioned the use of wind machines to enhance cross-pollination, but no reports of the successful application of this technique are at hand.

Cross-pollination of more than 90 percent of the male-sterile plants has been achieved under field conditions, but much lower percentages are usually the case. It is believed that at least 70 percent cross-pollination is needed to keep the cost of hybrid wheat seed at an acceptable level. Efforts are under way to enhance cross-pollination by incorporating (through breeding) larger and more prominent anther types into pollinators and wider-opening florets with larger stigmas into male-sterile lines.

What will the advent of hybrid wheat mean to farmers, millers, bakers and the ultimate consumers of wheat products? To the farmer hybrid wheat should provide greater returns per acre of land, resulting not only from hybrid vigor but also from the more intensive and efficient management practices that seem to accompany the introduction of a hybrid crop. When hybrid corn was introduced, improved management accounted for as much or more of the increase in yield as hybrid vigor did. The production prob-

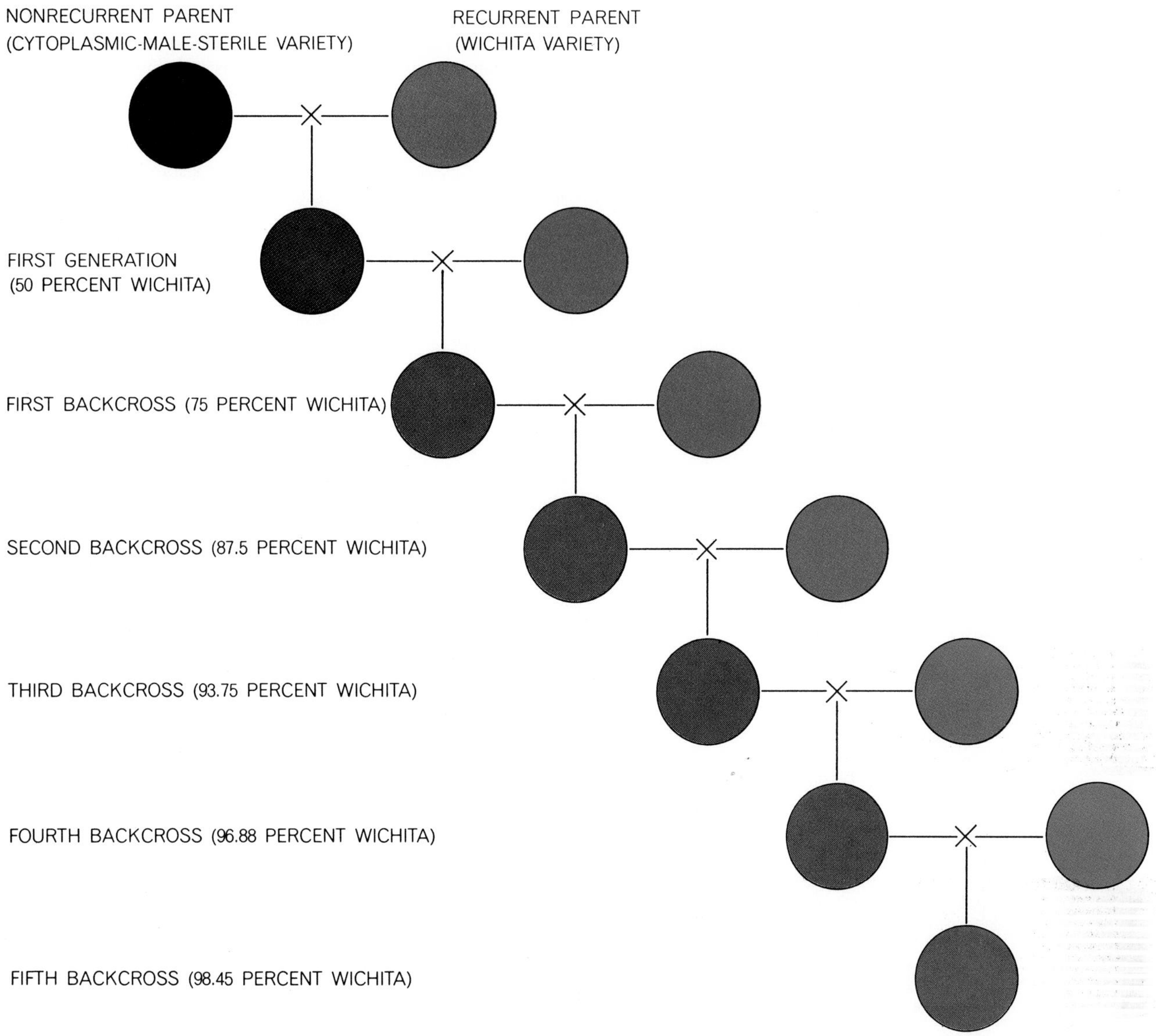

NONRECURRENT PARENT
(CYTOPLASMIC-MALE-STERILE VARIETY)

RECURRENT PARENT
(WICHITA VARIETY)

FIRST GENERATION
(50 PERCENT WICHITA)

FIRST BACKCROSS (75 PERCENT WICHITA)

SECOND BACKCROSS (87.5 PERCENT WICHITA)

THIRD BACKCROSS (93.75 PERCENT WICHITA)

FOURTH BACKCROSS (96.88 PERCENT WICHITA)

FIFTH BACKCROSS (98.45 PERCENT WICHITA)

**BACKCROSS METHOD is used to develop and increase a cytoplasmic-male-sterile line of Wichita wheat. The final progeny will be identical with the recurrent Wichita parent in most characteristics except that it will be male-sterile instead of fully fertile. Theoretically the genetic content of the nonrecurrent male-sterile parent will be reduced by half for each generation of backcrossing.**

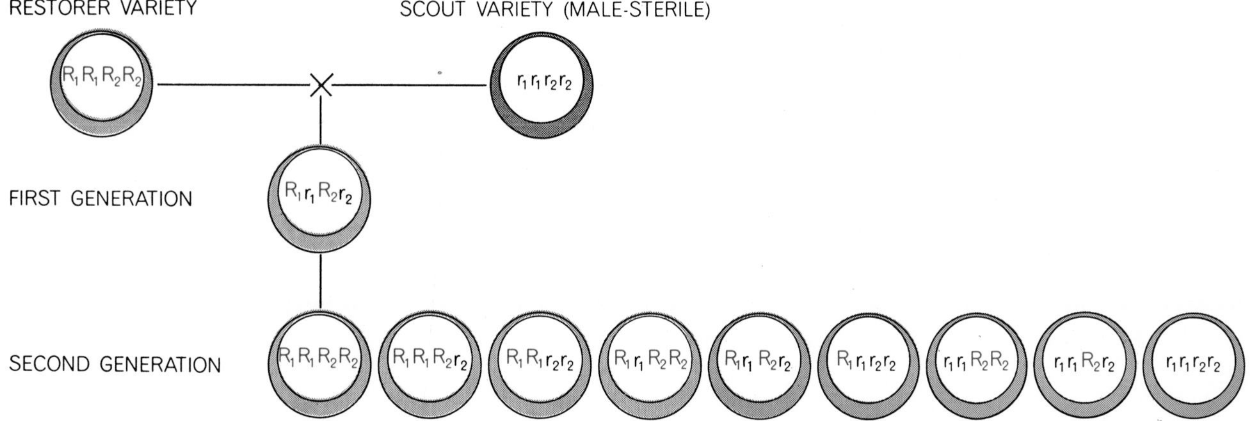


**RESTORER LINE, consisting of wheat plants of the Scout variety, is developed by selecting and backcrossing the dominant restorer plants of later generations to the parent Scout line. After several repetitions of the cycle the characteristics of the Scout plants are combined with the dominant restorer genes, resulting in a Scout restorer variety. In this schematic representation of one such cycle the dominant restorer genes, located in the nuclei of plants derived from the original restorer plants, are denoted by $R_1$ and $R_2$; the recessive, or nonrestorer, genes are denoted by $r_1$ and $r_2$. The cytoplasm from the male-sterile variety is colored; the normal cytoplasm from the Scout variety is gray. The completely dominant restorer plant in the second generation is designated $R_1R_1R_2R_2$.**

lems associated with hybrid wheat should be no greater than those encountered with high-yield pure-line wheat varieties. Hybrids must of course be acceptable in all the agronomic characteristics to which the wheat farmer is accustomed.

One specific trait that hybrid wheats must possess is good resistance to lodging, particularly in the high-yield areas. Increased grain yields of 25 to 50 percent impose a proportionately increased weight load on the wheat straw and may result in severe lodging. Heavily lodged plants result in high harvest losses in grain, manpower and harvesting-machine time. One way of increasing resistance to lodging is to breed hybrids with shorter and stiffer stems. Most breeding programs aimed at developing hybrid wheat are taking this approach to attain resistance to lodging. Sources for these hybrids include the highly productive semidwarf Mexican wheat varieties and semidwarf wheat varieties from the northwestern U.S.

Hybrids with shorter stems may not be necessary in some of the important wheat areas that normally do not support the development of tall straw. Much of the high plains of the U.S. is such an area. Irrigation is on the increase in some of the high plains areas, however, and there are indications that better lodging resistance will be needed.

Resistance to disease is an important attribute for stabilizing wheat production in many parts of the world. Hybrid wheat should offer more flexibility than pure-line varieties in the control of diseases, particularly parasitic diseases such as the rusts. When a pure-line variety that is resistant to the prevalent rust fungi is released, new rust species usually arise to attack the new variety. The same pattern will probably hold true for hybrids, except that other hybrids, resistant to the new rusts, can be easily substituted for the old hybrid. The reason is that farmers must obtain new seed for each crop from the producer of the original hybrid. Farmers growing pure-line varieties would be more likely to plant subsequent crops from the seed of preceding crops and thus perpetuate the rust-susceptible varieties. The producers of hybrid wheat have a great responsibility to keep abreast of the rust situation and to develop resistant hybrids to combat the disease.

A large number of wheat varieties are required to fit the varied ecological conditions of the areas where wheat is grown. This suggests that many different hybrids will also be required to fit many ecological niches. There is some hope, however, that one of the benefits of hybrid vigor will be increased adaptability. Some experiments have indicated that the root system of a wheat plant may be improved to allow the plant to perform better under drought conditions. Hybrids so designed will help to stabilize production in areas of the world where rainfall is highly variable from season to season.

To millers, bakers and consumers the quality of wheat produced by hybrids

**IN THE FIELD hybrid wheat seed is produced by growing alternating "drill strips" of the male-sterile line and the fertility-restoring line. Cross-pollination of the male-sterile plants is effected by windborne pollen from the restorer plants. The hybrid seed from the**

is of great importance. Approximately three-fourths of the wheat produced in the world is destined for human consumption. This being so, a close relation has developed between breeders and cereal chemists seeking to maintain or improve quality as new varieties are developed.

The quality of a particular variety of wheat is defined in terms of its ultimate use. Some flours require a high protein content; others do not. Some call for the ability to absorb more water than others. The large number of wheat products results in an equally large number of flour specifications. Most of the wheat produced in North America is classed as bread wheat. These wheats are generally high in protein content and have good water-absorption and gas-retention properties. The best of them are blended with poorer wheats to improve the flour quality in the making of bread.

Wheat varieties differ in their ability to confer favorable characteristics on their progeny. Studies have shown that the protein content of first-generation hybrids can be higher than that of the superior parent. This is somewhat at variance with regular varieties, in which high yield has been associated with low protein content. Conversely, some hybrids have been lower in protein (and in yield) than the inferior parent. Other crosses have produced hybrids whose protein content is intermediate between the protein content of the parents. Similar results have been obtained for water-absorption and gas-retention properties. There is still very little information on the quality characteristics of hybrid wheat, but the information available indicates that with the proper selection of parents hybrids of any desired quality can be obtained.

Research on the hybridization of wheat has been expanded on all fronts in the past decade. Much of the expansion was triggered by the discovery of the male-sterility and fertility-restoring system. Until the early 1960's wheat research was centered in the land-grant institutions and supported primarily by tax funds, both state and Federal. Credit for the discovery of the mechanism to produce hybrid wheat belongs to those institutions. The prospect of an extremely large and continuing market for hybrid wheat seed has prompted several private seed companies, most of them experienced in breeding other crops, to initiate research programs aimed at the development of hybrid wheat varieties. The combined efforts of public and private breeders have already produced dramatic results, but there is a clear need for continued research in many areas. The value of hybrid wheat can only be maintained and improved by further development of improved parental lines. These are the backbone of a successful hybridization program. The development of high-yielding, strong-strawed, disease-resistant parental lines producing grain of good milling and baking quality will ensure the success of hybrid wheat for the future.

**male-sterile strips is harvested (by combine) separately from the inbred seed from the restorer strips. The photograph on these two pages shows a typical drill-strip wheat field that is located at the Colorado State University Agronomy Farm in Fort Collins, Colo.**

# 33

# *Sex Differences in Cells*

URSULA MITTWOCH
*July 1963*

"Male and female created He them." Until the turn of the century this statement had to stand as a full description of what was known about the mechanism that brings about the divergence of the sexes. With the development of the discipline of cell genetics it has become clear that the chromosomes, the cellular structures that encode the hereditary plan, differ in male and female animals. Sex is determined at conception by the chromosome complement with which the fertilized egg is endowed by the union of the parental sex cells. In recent years other sex differences in cells have been discovered; it is now possible to tell whether a piece of skin or a drop of blood has come from a man or a woman by examining the nuclei of cells of these tissues under a microscope. More important, the recognition of these sex indicators in the nuclei of body cells has opened the way to study of the still largely unknown processes by which the parental chromosomes commit the fertilized egg to development as a male or a female of its species.

The first step in the understanding of the determination of sex came with the discovery in insects that the male sex cells—the spermatozoa—do not all carry the same complement of chromosomes. Half of them were found to contain a chromosome that stands out from the rest because of its characteristic form; it was named the X chromosome. In many species the spermatozoa that lack an X chromosome were found to contain another, usually smaller chromosome; this was labeled the Y chromosome. The chromosome complements of insect eggs, on the other hand, turned out to be uniformly the same; each contains an X chromosome and never a Y.

The body cells of animals, as distinguished from their sex cells, have pairs of homologous, or similar, chromosomes, one member of each pair being contributed by the parental sex cells at conception. When investigators first examined the body cells of insects in the early years of this century, they found that the cells of females contain two X chromosomes. In the body cells of male insects they found instead one X and one Y. They concluded that the union of an egg with an X-bearing spermatozoon would give rise to a female offspring with two X chromosomes; correspondingly, a sperm bearing a Y chromosome would give rise to an individual with an X and a Y chromosome, or a male [*see top figure on page 310*]. So it became established that sex is determined at the moment of conception, that it is determined by the sex-chromosome content of the germ cells and that in most species the differentiating chromosome is carried by the spermatozoon.

The early period of cell genetics can be said to have closed with the appearance in 1925 of Edmund B. Wilson's book *The Cell in Development and Heredity,* which set out the fundamental principles of sex determination by chromosomes. A second period of fruitful investigation of cellular genetics in its bearing on the determination of sex, particularly in human beings, began about 1950 and has gained momentum steadily. Two lines of research—chromosome studies and the work that led to the discovery of the nuclear sex indicators—have joined to bring about an unexpect-

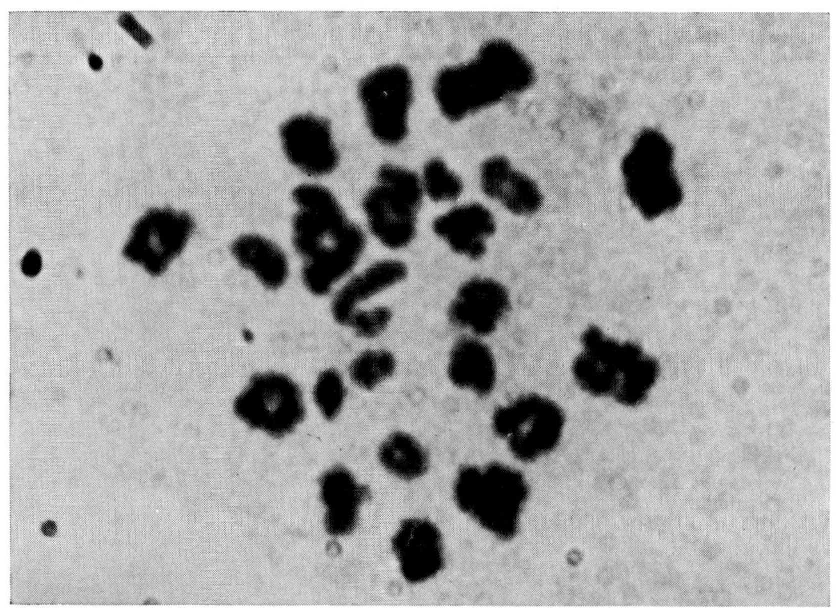

**SEX CHROMOSOMES are clearly identifiable in a male gonadal cell dividing to form spermatozoa. In this photomicrograph from the author's laboratory the homologous chromosomes are paired. All pairs but one are composed of equal chromosomes and are roughly symmetrical. The one hook-shaped pair is composed of X and Y sex chromosomes.**

edly rapid advance in the understanding of sexual development in the embryo.

In spite of man's great interest in matters regarding his own species, geneticists for many years knew far less about human chromosomes than about those of many animals and plants. The number of chromosomes in a human cell is large and they tend, under ordinary circumstances, to cluster together in the nucleus; it is practically impossible to count them accurately, let alone study their individual shapes. In the past decade workers in many countries have developed important new techniques for overcoming these difficulties [see "Chromosomes and Disease," by A. G. Bearn and James L. German III; SCIENTIFIC AMERICAN, Offprint 150]. The turning point came in 1956, when J. H. Tjio and Albert Levan of the University of Lund in Sweden announced that they had counted 46 chromosomes in cells grown from aborted human embryos—not 48, as had been supposed. The correctness of the new number was soon confirmed by Charles E. Ford and John L. Hamerton of the Radiobiological Research Unit at Harwell in England. Once the number of chromosomes in the normal human cell was determined, the stage was set for more detailed investigations.

Chromosomes can be seen only in cells that are dividing. They can be seen in gonadal cells, which divide by the special process called meiosis to give rise to sex cells, as well as in cells that divide by the more common process of mitosis, whereby single-celled organisms and the body cells of higher organisms reproduce themselves. The gonadal cells offer certain advantages for study. In mitosis all the chromosomes in the cell replicate before cell division and each of the daughter cells is furnished with a full complement of chromosomes. In meiosis, on the other hand, the homologous chromosomes separate, one member of each pair going to each of the sex cells; the resulting sperm or egg cell is thus endowed with a half-complement of chromosomes, representing a mixture of the

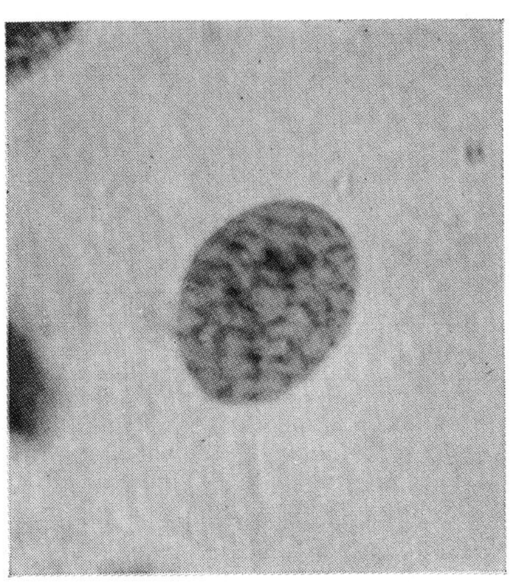

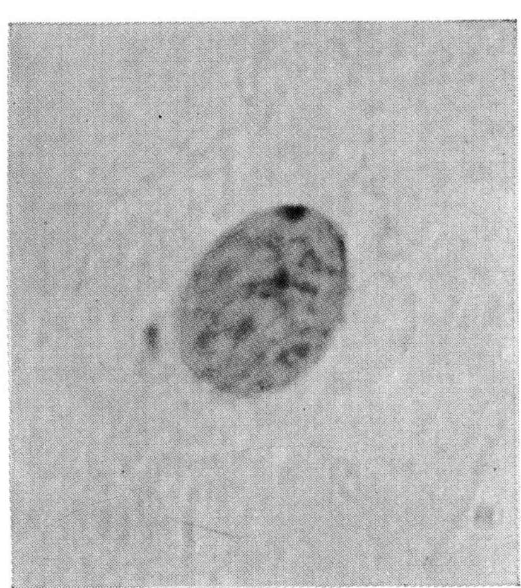

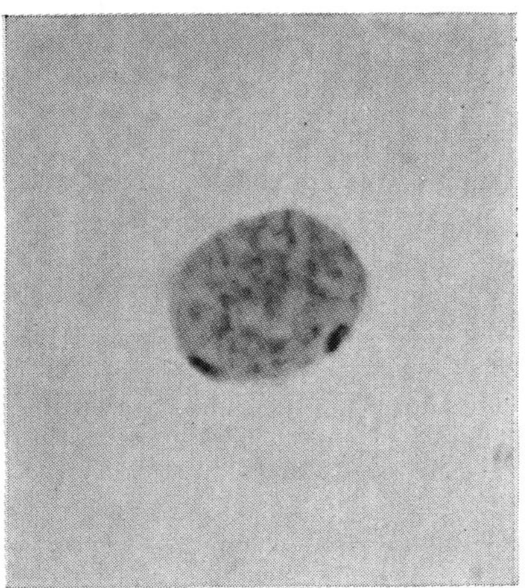

**SEX CHROMATIN, or Barr body, is one cellular indicator of sex. It is a darkly staining body that is seen in the nuclei of certain female cells but not in male cells. The photomicrograph at left shows the nucleus of a cell with no sex chromatin. The next photomicrograph shows a similar cell from a normal woman, with one Barr body at the periphery of the nucleus. The cell at right, taken from a woman with three X chromosomes instead of the normal two, has two Barr bodies. The magnification is 2,100 diameters.**

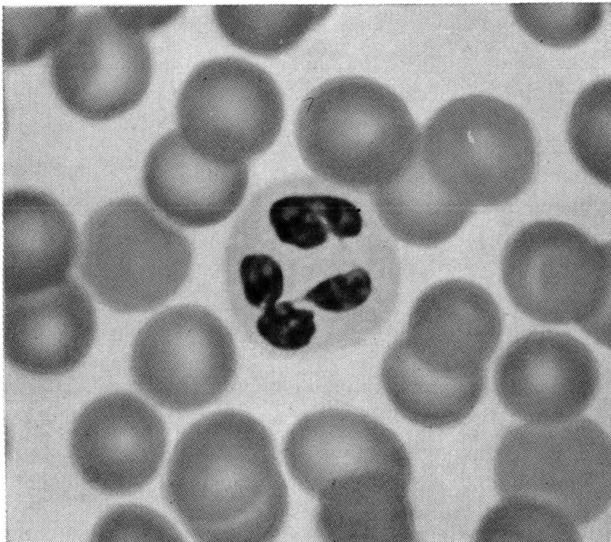

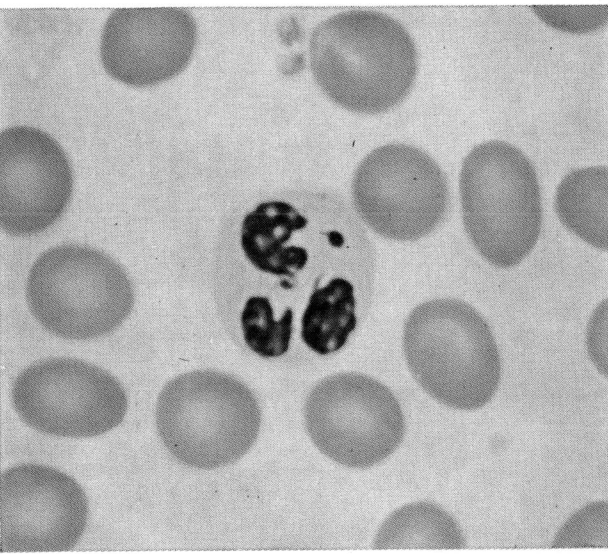

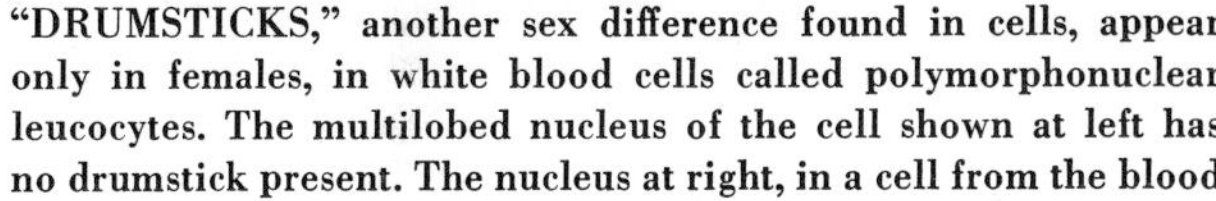

**"DRUMSTICKS," another sex difference found in cells, appear only in females, in white blood cells called polymorphonuclear leucocytes. The multilobed nucleus of the cell shown at left has no drumstick present. The nucleus at right, in a cell from the blood of a normal woman, has one: the small round body, which is attached to the nucleus by a faint filament. The magnification is 2,200 diameters. The photomicrographs on this page were made by Anthony Bligh of the University College Hospital Medical School.**

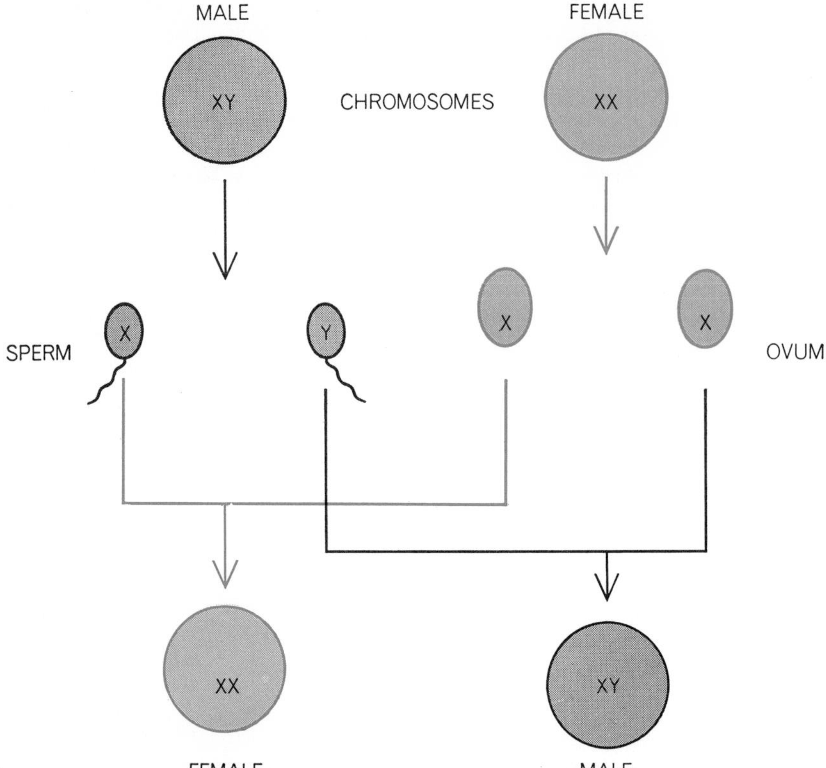


**SEX IS DETERMINED** by the sex chromosomes of the fertilized egg. Male cells have an X and a Y chromosome; female cells, two X chromosomes. Half of the sperm therefore carry X chromosomes and half Y; all ova, or eggs, have X chromosomes. Fertilization by an X-carrying sperm produces a female; fertilization by a sperm with a Y produces a male.

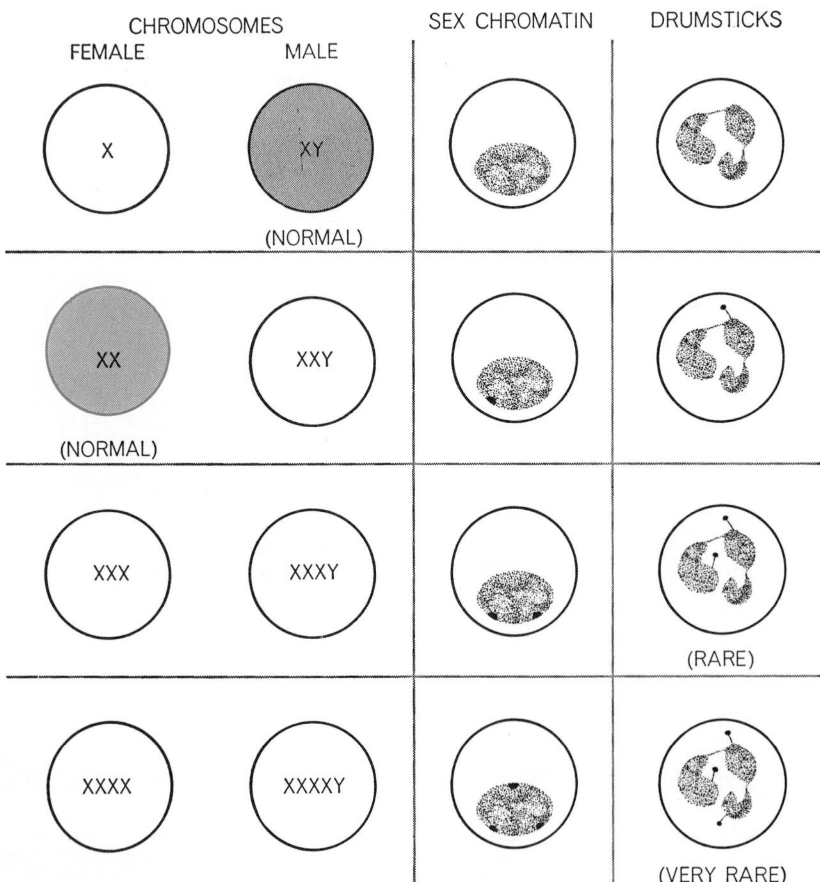


**SEX DIFFERENCES** in cells are related as shown here. The maximum number of sex-chromatin bodies and of drumsticks is one less than the number of X chromosomes in the cell.

chromosomes with which the gonadal cell was endowed by the parents of the organism. Just before they part company the homologous chromosomes of the gonadal cell are intimately associated with each other; in the cells of the human gonad the 46 chromosomes appear as 23 pairs.

A photomicrograph of such a cell from the male gonad [*see illustration on page 308*] shows that all but one of the pairs are symmetrical since they are composed of two similar chromosomes—the autosomes, or nonsex chromosomes. The asymmetric pair is made up of the X and Y chromosomes. The X is considerably larger than the Y; the two are associated end to end. The trouble is that male gonadal cells are rarely available for study; the stage of chromosome division in the female when the ova, or egg cells, are formed has never been seen.

In human beings chromosomes are most commonly studied in body cells taken from the bone marrow, skin or blood. Cells in bone marrow are normally in a state of active division, since it is their function to replace blood cells as they become worn out. Dividing cells can be seen, therefore, in bone marrow samples immediately after the material has been taken from the body. Cells from the lower layer of skin also divide, although slowly, to replace those in the upper layer that are sloughed off; these cells must be cultured for at least two weeks before chromosome studies can be undertaken.

The white cells of the blood are, of course, more conveniently sampled, from both the subject's and the investigator's point of view. Although they do not normally divide once they have been liberated from the bone marrow, it was shown by Edwin E. Osgood and John Brooke of the University of Oregon Medical School that a reagent extracted from beans, called phytohemagglutinin, can be used to induce them to divide in a culture medium. Addition of the plant alkaloid colchicine to the medium inhibits cell division at the stage known as metaphase, when the chromosomes are most clearly seen. Then the cells are bathed in a solution of sodium citrate at a concentration lower than that of the salts inside the cells. Osmotic pressure causes water to enter the cell, swell the nucleus and disperse the chromosomes. Finally the cells are killed, placed on a slide and usually stained.

Human chromosomes obtained from blood cells by this procedure are discrete structures of different sizes, symmetrical along a longitudinal axis and roughly cross-shaped or V-shaped. Their symmetry results from the fact that they have already replicated; each has formed a copy of itself, to which it is joined at a constriction called the centromere. (In the normal course of cell division the centromere would have divided and the two halves of the doubled chromosome would have moved off in opposite directions to two daughter cells.) Both the over-all size and the position of the centromere are characteristic for each chromosome. In this type of cell, unlike a germ cell, the chromosomes are not naturally paired with their homologues. But a trained person can cut the individual chromosomes from a photograph and arrange them in pairs according to their size and the position of their centromeres.

If the cell comes from a man, one of the pairs will be composed of unequal partners; these are the X and Y chromosomes. The X chromosome is of medium size and is cross-shaped, whereas the Y is very small and is V-shaped. The other chromosomes—the autosomes—are numbered from one to 22 in order of decreasing size. If the chromosomes of a similar cell from a woman are arranged in the same manner, all the pairs will be composed of equal partners because a woman has two X chromosomes [*see illustration below*].

By comparing cells from males and females it was possible to learn to identify the X chromosome fairly reliably as one of medium size with its centromere near the middle. But there are a number of other chromosomes (those labeled six to 12) that look rather similar, so at present no one can be sure that the one la-

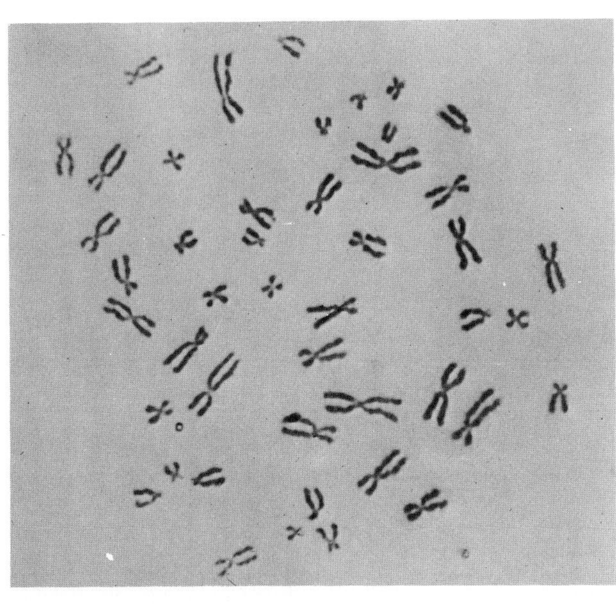

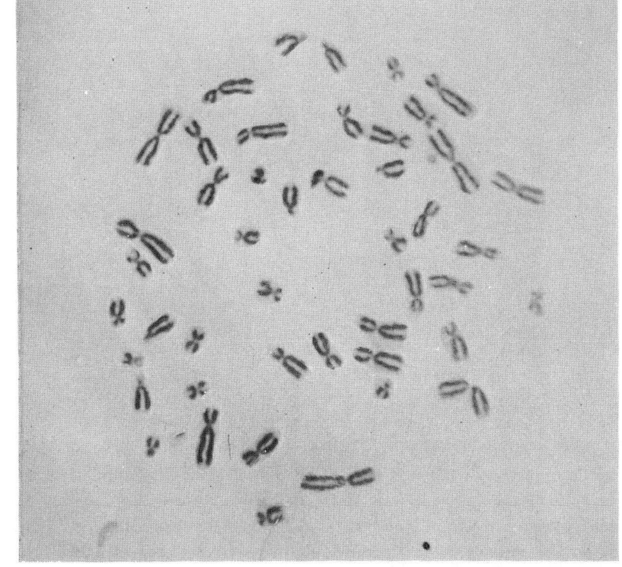

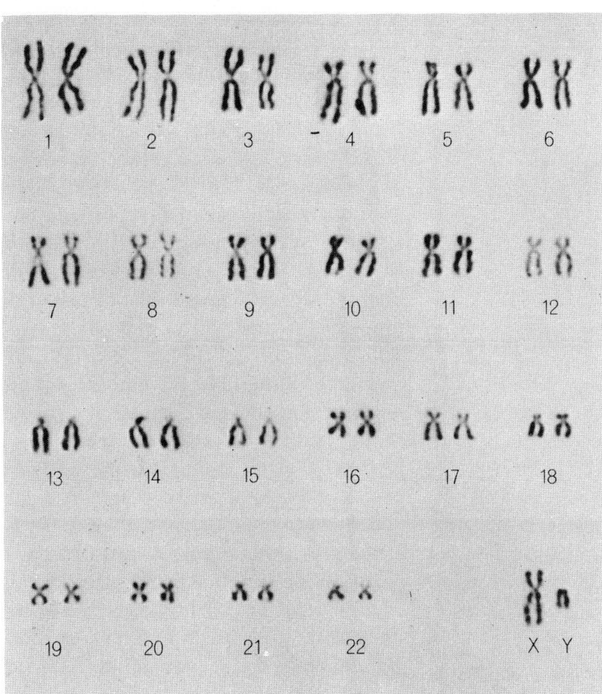


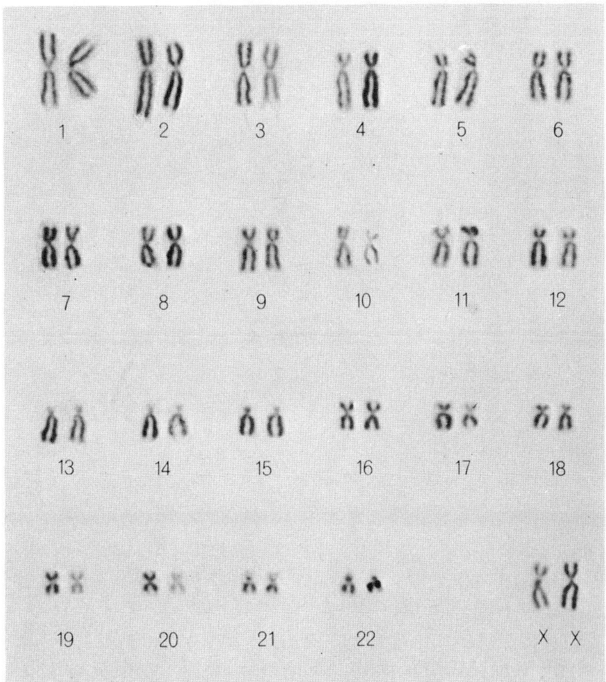


**HUMAN CHROMOSOMES from blood cells, prepared as described in the text, are enlarged 1,400 diameters in the two upper photomicrographs. The cell at left is from a man, the one at right from a woman. In the lower part of the illustration the chromosomes have been arranged in pairs. There are 22 pairs of autosomes in each cell and one pair of sex chromosomes: XY in the male and XX in the female. The cells were prepared by Ruth Marshall in the author's laboratory; the photomicrographs were made by Bligh.**

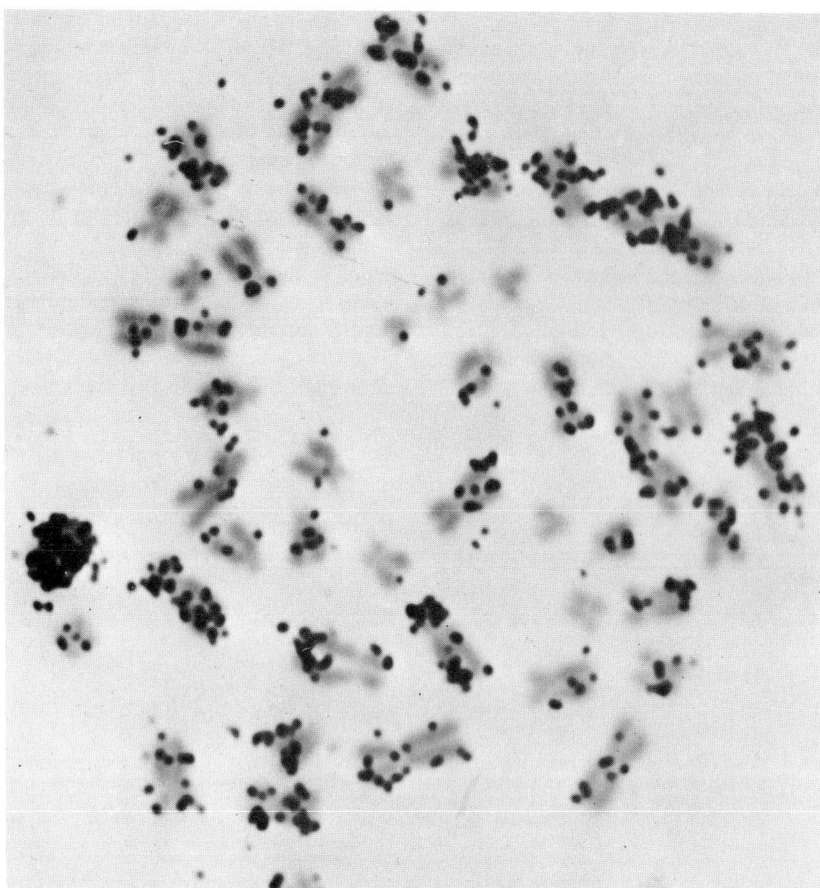

**AUTORADIOGRAPHY reveals a late-replicating X chromosome in this photomicrograph made by James L. German III of the Rockefeller Institute. Radioactive thymidine was added to a culture of female blood cells when this cell had almost completed DNA synthesis, and was taken up only by late-replicating chromosomal regions. Now, at metaphase, the late X is identified by the concentration of dark spots (*left*) on a radiosensitive emulsion.**

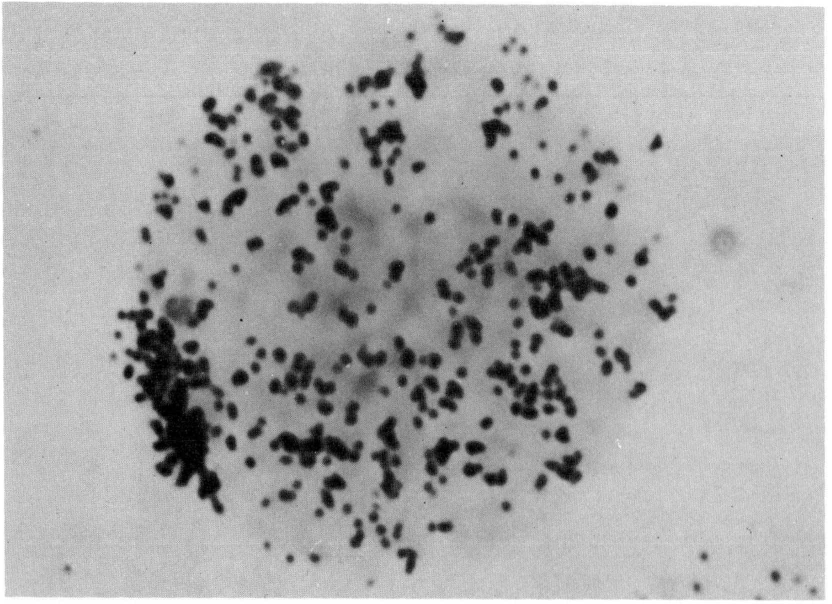

**LATE-REPLICATING MATERIAL is seen at another stage of the cell-division cycle in this photomicrograph, also made by German. This cell, a neighbor on the microscope slide of the one at the top of the page, was also far advanced in DNA synthesis when exposed to radioactive thymidine; unlike the other cell, it has not yet reached metaphase. The chromosomal material that took up the largest quantity of radioactive thymidine, and which would at metaphase presumably be seen as a late-replicating X like the one at the top of the page, is visible (*left*) at the edge of the nucleus, the area in which Barr bodies are usually found.**

beled X in any specific instance is in fact the X chromosome. The Y chromosome is easier to identify; although it looks quite like chromosomes No. 21 and No. 22, it can usually be distinguished from them because it is a bit larger and its two arms lie closer together.

The second line of research on the cellular basis of sex stems from a fortuitous discovery in 1949 by Murray L. Barr and E. G. Bertram of the Medical School of the University of Western Ontario. While studying the nerve cells of a cat they noted a body in the nucleus of each cell that stained deeply with certain dyes. When they looked for this body in the nerve cells of a number of cats, they could see it in the cells of some animals but not in those of others. On investigation they found that it was the sex of the cats that made the difference: the nuclei of female cells contained a special body that was missing in male cells. The same sex difference turned up in other cat cells and then in other animals and in humans. The differentiating structure is called sex chromatin, or, after its discoverer, a Barr body. It is found in a variety of tissues, but the simplest way to demonstrate it is in cells scraped off the buccal mucosa: the inside surface of the cheek. The cells are placed on a slide, fixed in alcohol and stained. If the cells come from a woman, many will contain a darkly stained Barr body at the periphery of the nucleus [*see upper illustration on page 309*].

A few years after Barr and Bertram announced their findings William M. Davidson and David Robertson Smith of King's College Hospital in London decided to conduct a systematic search for sex differences in blood cells. They found one in the cells called polymorphonuclear leucocytes, white cells in which the nuclei, as the name implies, assume many different shapes. In the nuclei of some of these cells from females Davidson and Smith observed a unique appendage: a round body attached to one of the lobes of the nucleus by a thin stalk. They called it a "drumstick" and showed that although it is present in a small percentage of the polymorphonuclear leucocytes of a woman, it is entirely absent in normal males and therefore provides an additional indicator of sex. To demonstrate drumsticks one needs only to spread a drop of blood on a slide, fix and stain it and examine it under the microscope [*see lower illustration on page 309*]. In view of the great number of such blood films studied over the years in medical practice and research, it seems strange that until 1954 no one

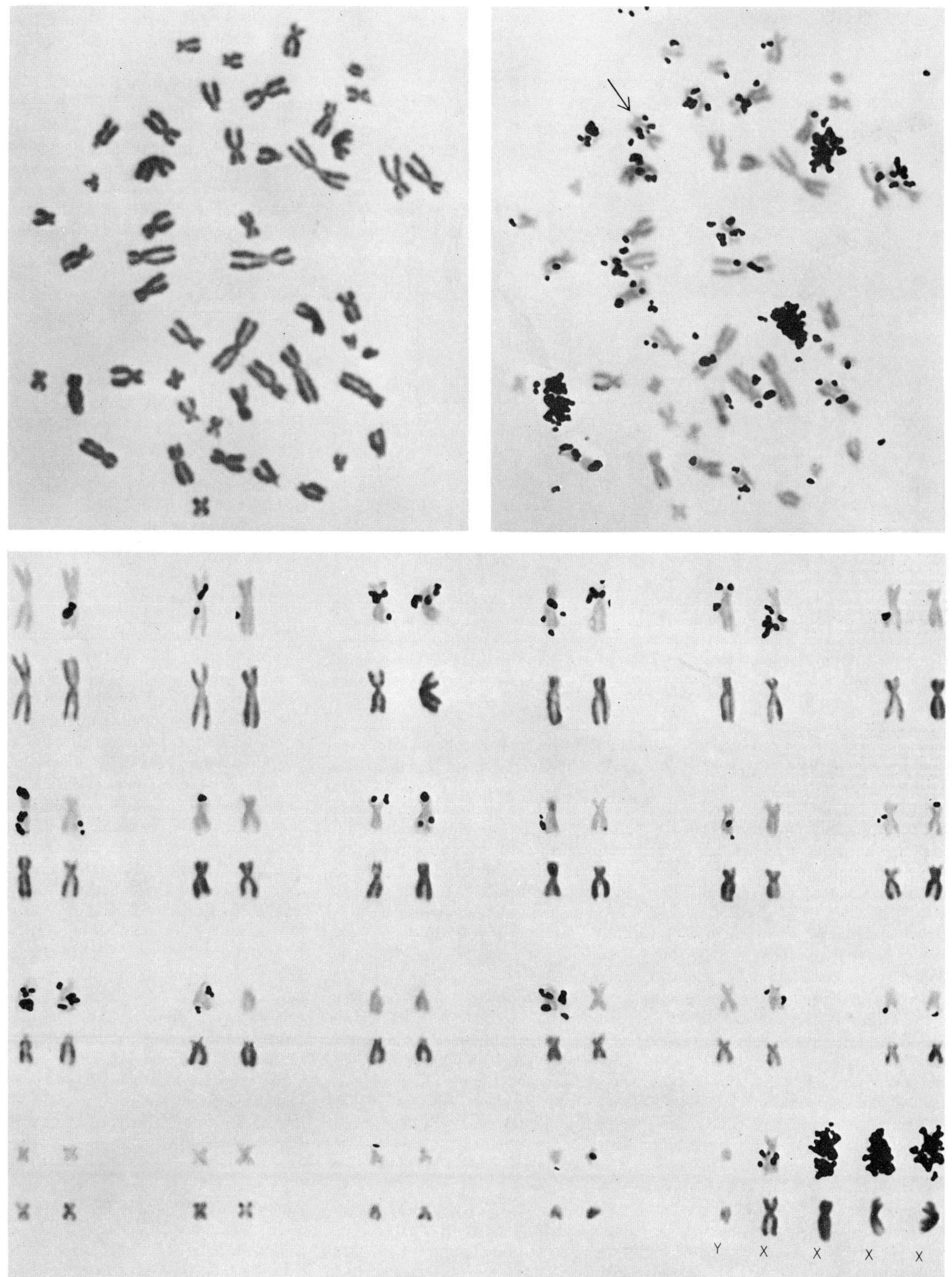


**ABNORMAL CELL** with four X chromosomes and a Y shows three late-replicating X chromosomes. The chromosomes were photographed before (*top left*) and after (*top right*) the application of a radiosensitive emulsion. Pairing the chromosomes of the unlabeled photomicrograph according to the usual standards (*bottom rows in lower part of illustration*) identified the X chromosomes. When the labeled chromosomes were similarly paired (*top rows*), the heavily radioactive late replicators were shown to be X's. The arrow (*top right*) indicates the "first" X chromosome. These photomicrographs are from a study by B. B. Mukherjee, O. J. Miller and Saul Bader of the Columbia University College of Physicians and Surgeons and W. Roy Breg of the Southbury Training School.

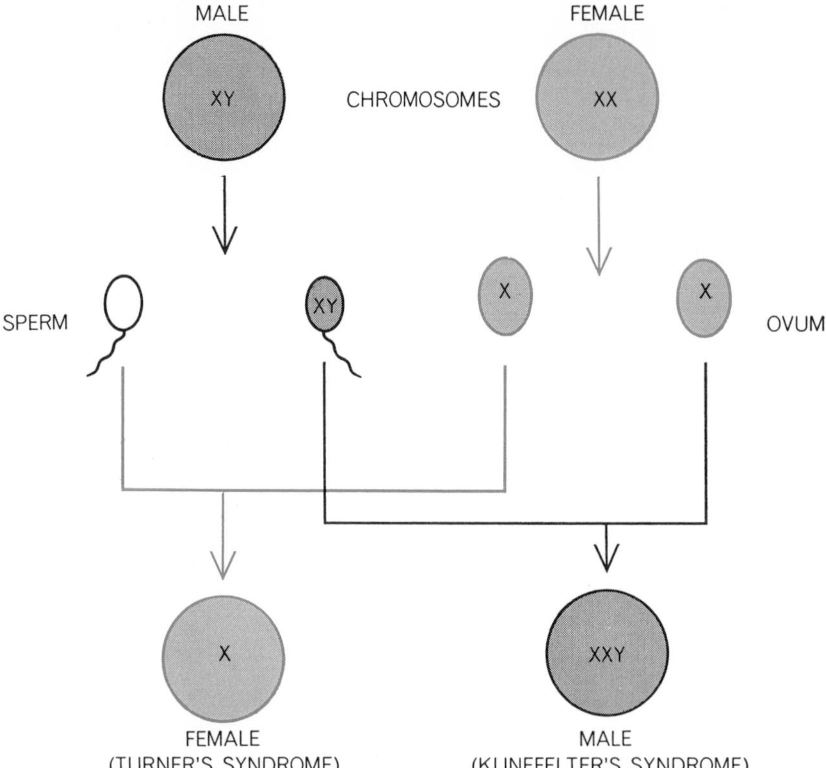


**NONDISJUNCTION, the failure of homologous chromosomes to separate during cell division, can cause sexual abnormalities. This diagram shows the possible effects of nondisjunction in the father during sperm formation. The fertilized egg will have two X chromosomes and a Y or only one X. Both chromosome constitutions lead to known intersex syndromes.**

had noticed that female cells can be distinguished from those of males so easily. But in science nothing seems more obvious than what has just been discovered; the problem is to know what to look for.

Both Barr bodies and drumsticks occur in cells that are not dividing, whereas chromosomes are visible only in dividing cells. There can be no doubt, however, that sex chromatin and drumsticks are somehow related to the X chromosome. Some of the first indications that this is so emerged from studies of certain abnormalities in sexual development.

Although the great majority of people are clearly either male or female, with the normal characteristics of their sex, a few individuals suffer from various types of errors in sexual differentiation: their apparent sex is not the same as the "actual" sex of their cells. One such condition in men is Klinefelter's syndrome, named after Harry F. Klinefelter, an American physician who first described it in 1942. It is characterized by underdevelopment of the male sex glands, enlargement of the breasts and sometimes mental retardation. When the new techniques of "nuclear sexing" became available, it developed that in most of these men there are Barr bodies in cells from the buccal mucosa and drumsticks in the polymorphonuclear leucocytes. In other words, the nuclear sex of these individuals is female. A little later Patricia A. Jacobs and J. A. Strong of the Western General Hospital in Edinburgh counted 47 chromosomes in the cells of Klinefelter's-syndrome men instead of 46; the additional chromosome appeared to be an X. The existence of an XXY chromosome constitution in this condition has since been confirmed by many workers.

An abnormality in women had been described in 1938 by Henry H. Turner of the University of Oklahoma School of Medicine. Individuals with Turner's syndrome are usually abnormally short, have tiny ovaries and lack many sexual characteristics that normally develop at puberty. In 1954 the British investigators Paul E. Polani, W. F. Hunter and Bernard Lennox and the U.S. workers Lawson Wilkins, Melvin M. Grumbach and Judson J. Van Wyk reported that most of these women are "chromatin-negative": there are no Barr bodies in their buccal mucosa cells and no drumsticks in their blood cells. A few years later the chromosome count of these women was found to be only 45: one X chromosome and, of course, no Y. More recently Jacobs and her colleagues discovered that certain women have three X chromosomes—a total count of 47. The triple-X condition does not seem to be associated with a distinct clinical abnormality, although many of these women are mentally retarded. Their cells show not one but two Barr bodies. So do the cells of the few men who have three X chromosomes and a Y, along with the characteristics of Klinefelter's syndrome. More rarely still, investigators have found individuals with four X chromosomes—males (XXXXY) as well as females (XXXX). Whether from male or female, cells with four X chromosomes have a maximum of three Barr bodies.

All this made it evident that the formation of sex chromatin bears a direct relation to the number of X chromosomes present in a cell. Here was a clue to the behavior of the chromosomes from one cell generation to the next and to their function in shaping the destiny of the cell. During cell division, when the chromosomes are visible, they split to form two daughter nuclei. But the molecular agent of heredity in the chromosomes—deoxyribonucleic acid (DNA)—performs its genetic work, including its own replication, during interphase: between cell divisions. It is then, when the chromosomes as such are invisible, that Barr bodies and drumsticks are seen.

The technique of autoradiography has provided some clues to what goes on during interphase and has thereby thrown some light on the relation of Barr bodies to the X chromosome. Cells are supplied with a radioactive component of DNA, usually thymidine labeled with radioactive hydrogen (tritium). The radioactive thymidine, incorporated into the replicated chromosomes as DNA is synthesized, reveals itself by producing dark spots on a photographic emulsion placed in contact with the cell. When J. Herbert Taylor of Columbia University applied this technique to hamster cells, he learned that the various chromosomes of a cell synthesize DNA at different times in the course of interphase [see "The Duplication of Chromosomes," by J. Herbert Taylor, beginning page 14 of this book]. In the past few years a number of investigators in the U.S. and England have autoradiographed human cells. Their results show that if a cell is exposed to tritiated thymidine late in interphase, most of the chromosomes will already have completed their DNA synthesis. But in female cells one of the chromosomes, which in shape and size resembles an X, replicates late and therefore incorporates a great deal of the radioactive thymidine, and stands out from the others because it is so heavily labeled [*see top illustration on page 312*]. No such late-replicating X chromosome is found in normal male cells. In individuals with more than two X chromosomes, the number that replicate late is regularly one less than the total number of X chromosomes [*see illustration on page 313*]. The incidence of late-replicating X chromosomes is strikingly parallel to that of Barr bodies. This suggests that any X chromosome in excess of one behaves differently from the rest of the chromosomes and is responsible for the presence of sex chromatin.

There are a number of other observations, some of them quite tentative, that support this conclusion. For one thing, when the chromosomes first become visible as delicate threads in the stage of cell division called prophase, one thread may be more highly condensed than the rest in female body cells. Susumu Ohno of the City of Hope Medical Center in Duarte, Calif., suggested in 1959 that this thread is the chromosome responsible for the formation of the Barr body. Moreover, some investigators have reported that the late X chromosomes tend to be situated toward the periphery of the nucleus—as are Barr bodies.

If Barr bodies are clearly related to the X chromosome, the same must be true of drumsticks, since there is such good agreement between the presence of Barr bodies and of drumsticks in normal people and in people with abnormalities of the X chromosome. The formation of drumsticks, however, is dependent on the rather peculiar maturation process of the polymorphonuclear leucocytes, during which their nuclei divide into a number of lobes. In some chromosomal anomalies lobe formation is not quite normal, and drumsticks may for this reason be reduced in number or even entirely absent. The interpretation of drumsticks is therefore sometimes less straightforward than that of Barr bodies; although the maximum number of drumsticks is one less than the number of X chromosomes, this maximum may rarely be reached. But there is no longer any doubt that both structures are the expressions of any X chromosomes present in excess of one.

The numerical relation between X chromosomes and Barr bodies is not only of theoretical interest; it has already proved to be of great value in studies of various intersex conditions. Since the techniques of nuclear sexing are much simpler and less time-consuming than an analysis of chromosomes, one can investigate the presence or absence of Barr bodies in a large number of people. The results allow one to forecast the number of X's in any patient before undertaking a chromosome study. The determination of the nuclear sex is also an indispensable adjunct to correct chromosome analysis because, as I have indicated, the X chromosome cannot yet be distinguished with certainty.

It nows appears that, regardless of the number of X chromosomes present, an individual who bears a Y chromosome in his cells is a male and one without this chromosome is a female. Although normal women are XX and normal men are XY, the addition of one, two or even three X chromosomes to an XY constitution still results in development that is essentially male. The male-determining effect of the little Y chromosome in humans has been recognized only within the past few years, and the discovery came as a surprise. This was because a

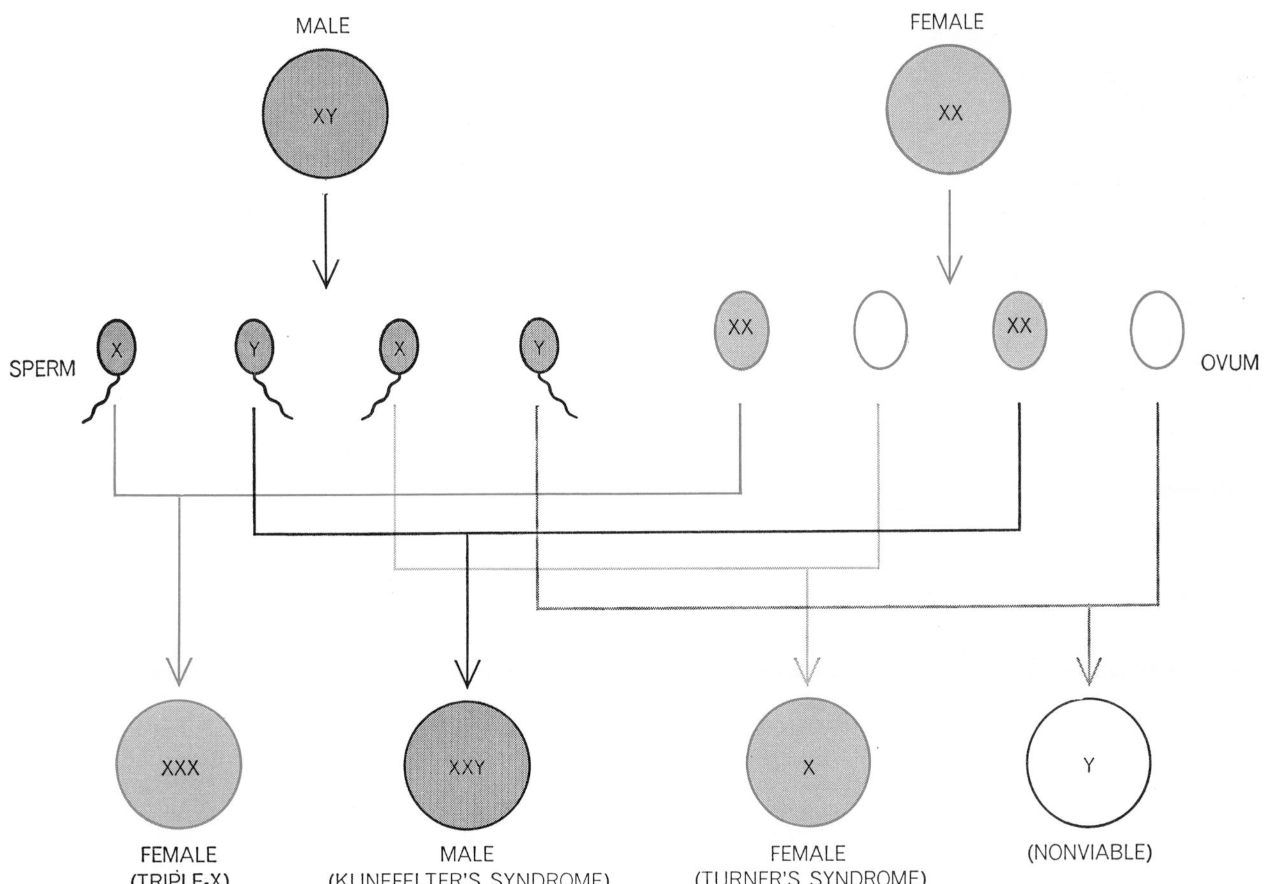


**NONDISJUNCTION IN MOTHER produces ova with either two X chromosomes or none. Depending on the sperm by which they are fertilized, such ova can give rise to any one of four possible chromosome constitutions. Three are abnormal; one is not viable.**

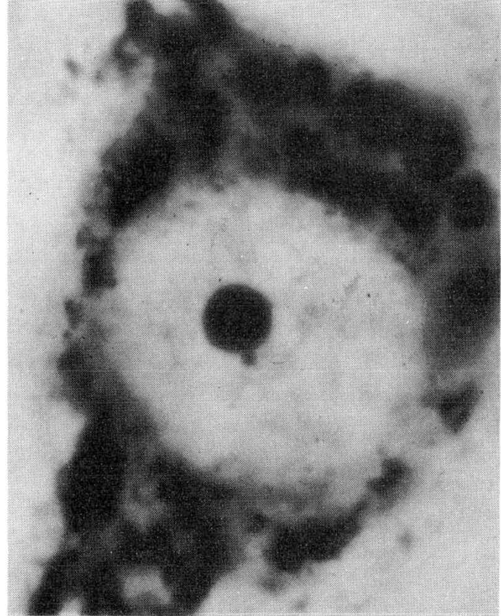

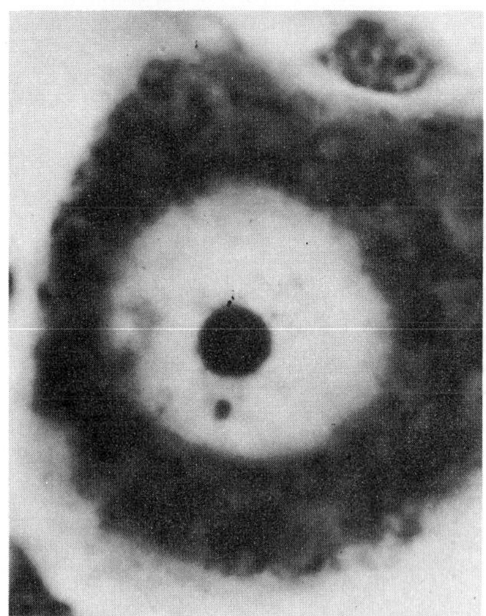

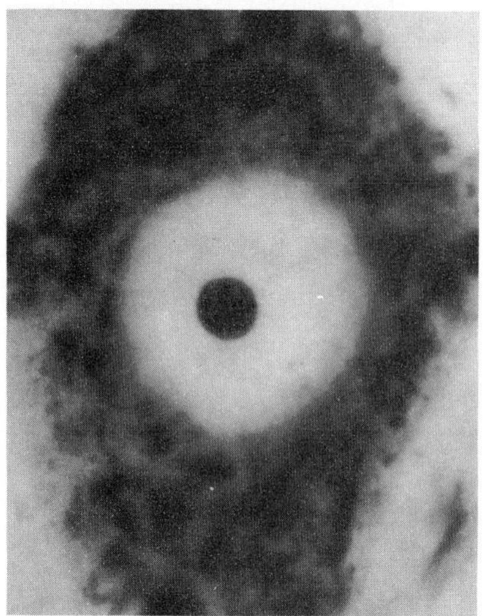

**BARR BODY was observed for the first time in the nerve cells of cats, seen in these photomicrographs made by Murray L. Barr and E. G. Bertram of the Medical School of the University of Western Ontario. Their attention was drawn to a small, dense body near the nucleolus of female cells (*top*) that migrated away from the nucleolus (*middle*) after electrical stimulation. No such body appeared in male cells (*bottom*).**

great deal of work had been done on the determination of sex in the fruit fly *Drosophila melanogaster*. In flies as in humans males are XY and females are XX but the Y chromosome plays only a small part in development. XXY flies are ordinary females, and flies with one X and no Y are males, although they produce nonmotile spermatozoa and are therefore sterile. In *Drosophila* the X chromosome has a tendency to produce female characteristics; it is the autosomes that tend to produce male characteristics. It is now certain that the details of sex determination in humans are different: the Y chromosome has a strong tendency to switch development in the direction of maleness. The exact function of the X chromosome in humans is less certain. For example, although a single X chromosome will in the absence of a Y give rise to a female, such a female is not normal.

The sex of an embryo is determined at conception, but the structural differentiation of the sexes does not become apparent until the seventh week of embryonic life. It seems likely that in the beginning the sex glands have the potentialities of both sexes. At some critical stage the presence or absence of a normal Y chromosome is probably all-important. If a Y chromosome is present, the sex glands develop into testes; if there is no Y chromosome, ovaries form. Once testes are present in an embryo they produce hormones under the influence of which further male characteristics develop. Alfred Jost of the University of Paris has been able to demonstrate that male development occurs in rabbit embryos only if testes are present. But ovaries are not necessary for female development; this can occur even in the absence of sex glands.

It would seem that the function of the sex chromosomes in man is to switch the embryo into one or the other channel of sexual development. Thereafter the hormones take over the work of further differentiation. If the switch mechanism is faulty, the production of hormones will become abnormal and give rise to an aberrant sexual condition.

The discovery that errors in sexual development may be associated with abnormalities of the chromosomes has provided new leads to understanding of the process of sex determination. Many of these errors result from accidents affecting the production of the ovum or the spermatozoon that goes into the making of the organism. Normally the two partners of a pair of sex chromosomes separate before a germ cell is formed. In the great majority of cases this "disjunction" comes about without a hitch. Sometimes, however, the chromosomes fail to disjoin. Klinefelter's syndrome and Turner's syndrome can arise from nondisjunction in the germ cells of either the father or the mother. Nondisjunction in the mother can also produce a triple-X female. It might even produce a fertilized ovum with only a Y chromosome and no X, but such individuals have never been found. It is reasonable to assume that cells need at least one X chromosome in order to function and that embryos with none at all fail to develop. An ovum or spermatozoon lacking a sex chromosome might also be formed if during cell division an X or a Y moved too slowly and failed to become included in one of the daughter cells. Such a loss of a chromosome may be an additional cause of Turner's syndrome.

Errors can occur in the development of the embryo as well as in the formation of germ cells. In young embryos the cells are engaged in active division, with each daughter cell ordinarily receiving one longitudinal half of each replicated chromosome. Very rarely, however, a cell may receive both halves or neither—and such a cell may then produce a whole line of cells with one chromosome too many or one too few. The organism to which such an embryo gives rise may have more than one type of chromosome constitution. Such "mosaic" individuals have, in fact, been encountered among patients with sexual abnormalities. A few patients with Klinefelter's syndrome have some XX cells as well as XXY; others have XY and XXY. Similarly, some women with Turner's syndrome have both XX and X cells, whereas others are mosaics of XXX and X cells. The number of possible combinations is large and new ones are continually being found.

The study of the cellular basis of sex in man is in a state of rapid progress. The knowledge recently gained has already yielded its first fruits in medicine by ascribing physical causes to a number of hitherto unexplained conditions; further studies may well uncover even more subtle forms of abnormality in human beings. The next step, the prevention or cure of these afflictions, is undoubtedly still some way ahead, but it will surely follow. There is a special fascination in these new insights into the relation between the structures of cells as seen under the microscope and the characteristics of the two sexes—unless, of course, one agrees with Hamlet that "man delights not me; no, nor woman neither."

# 34

# *The Induction of Cancer by Viruses*

RENATO DULBECCO

*April 1967*

Cancer, one of the major problems of modern medicine, is also a fascinating biological problem. In biological terms it is the manifestation of changes in one of the more general properties of the cells of higher organisms: their ability to adjust their growth rate to the architectural requirements of the organism. To learn more about cancer is therefore to learn more about this basic control mechanism. Over the past decade dramatic advances in our knowledge of cancer have resulted from the use of viruses to elicit the disease in simple model systems. A certain understanding of the molecular aspects of cancer has been attained, and the foundation has been laid for rapid progress in the foreseeable future.

A cancer arises from a single cell that undergoes permanent hereditary changes and consequently multiplies, giving rise to billions of similarly altered cells. The development of the cancer may require other conditions, such as failure of the immunological defenses of the organism. The fundamental event, however, is the alteration of that one initial cell.

There are two main changes in a cancer cell. One change can be defined as being of a regulatory nature. The multiplication of the cells of an animal is carefully regulated; multiplication takes place only when it is required, for example by the healing of a wound. The cancer cell, on the other hand, escapes the regulatory mechanisms of the body and is continuously in a multiplication cycle.

The other change of the cancer cell concerns its relations with neighboring cells in the body. Normal cells are confined to certain tissues, according to rules on which the body's overall architecture depends. The cancer cell is not confined to its original tissue but invades other tissues, where it proliferates.

The basic biological problem of cancer is to identify the molecular changes that occur in the initial cancer cell and determine what causes the changes. The particular site in the cell affected by the changes can be approximately inferred from the nature of the changes themselves. For example, a change in the regulation of cell growth and multiplication must arise from a change in the regulation of a basic process in the cell, such as the synthesis of the genetic material deoxyribonucleic acid (DNA). The alterations in relations with neighboring cells are likely to flow from changes in the outside surface of the cell, which normally recognizes and responds to its immediate environment.

Experimental work directed toward the solution of this central problem makes use of cancers induced artificially rather than cancers that occur spontaneously. Spontaneous cancers are not suitable for experiments because by definition their occurrence cannot be controlled; moreover, when a spontaneous cancer becomes observable, its cells have often undergone numerous changes in addition to the initial one. In recent years model systems for studying cancers have been developed by taking advantage of the fact that animal cells can easily be grown in vitro—in test tubes or boxes of glass or plastic filled with a suitable liquid medium. This is the technique of tissue culture.

Since the use of tissue culture has many obvious experimental advantages, methods for the induction of cancer in vitro have been developed. The most successful and most widely employed systems use viruses as the cancer-inducing agent. In these systems the initial cellular changes take place under controlled conditions and can be followed closely by using an array of technical tools: genetic, biochemical, physical and immunological.

It may seem strange that viruses, which are chemically complex structures, would be preferable for experimental work to simple cancer-inducing chemicals, of which many are available. The fact is that the action of cancer-inducing chemicals is difficult to elucidate; they have complex chemical effects on a large number of cell constituents. Furthermore, even if one were to make the simple and reasonable assumption that chemicals cause cancer by inducing mutations in the genetic material of the cells, the problem would remain enormously difficult. It would still be almost impossible to know which genes are affected, owing to the large number of genes in which the cancer-causing mutation could occur. It is estimated that there are millions of genes in an animal cell, and the function of most of them is unknown. With viruses the situation can be much simpler. As I shall show, cancer is induced by the genes of the virus, which, like the genes of animal cells, are embodied in the structure of DNA. Since the number of viral genes is small (probably fewer than 10 in the system discussed in this article), it should be possible to identify those responsible for cancer induction and to discover how they function in the infected cells. The problem can thus be reduced from one of cellular genetics to one of viral genetics. The reduction is of several orders of magnitude.

A number of different viruses have the ability to change normal cells into cancer cells in vitro. In our work at the Salk Institute for Biological Studies we employ two small, DNA-containing viruses called the polyoma virus and simian virus 40 (SV 40), both of which induce cancer when they are inoculated into newborn rodents, particularly hamsters, rats and, in the case of the polyoma virus, mice. Together these viruses are referred to as the small papovaviruses.

In tissue-culture studies two types of host cell are employed with each virus. In one cell type—the "productive" host cell—the virus causes what is known as a productive infection: the virus multiplies unchecked within the cell and finally kills it. In another type of cell—the "transformable" host cell—the virus causes little or no productive infection but induces changes similar to those in cancer cells. This effect of infection is called transformation rather than cancer induction because operationally it is recognized from the altered morphology of the cells in vitro rather than from the production of cancer in an animal.

In the experimental work it is convenient to employ as host cells, particularly for transformation studies, permanent lines of cellular descent, known as clonal lines, that are derived from a single cell and are therefore uniform in composition. By using these clonal lines the changes caused by the virus can be studied without interference from other forms of cellular variation; one simply compares the transformed cells with their normal counterparts. Two lines that are widely employed are the "BHK" line, which was obtained from a hamster by Ian A. Macpherson and Michael G. P. Stoker of the Institute of Virology of the University of Glasgow, and the "3T3" line, which was obtained from a mouse by George J. Todaro and Howard Green of the New York University School of Medicine. BHK cells are particularly suitable for transformation by the polyoma virus; 3T3 cells are readily transformed by SV 40 and less easily by the polyoma virus.

In a typical transformation experiment a suspension of cells in a suitable liquid medium is mixed with the virus [*see illustration at left*]. The cells are incubated at 37 degrees centigrade for an hour; they are stirred constantly to prevent them from settling and clumping together. A sample of the cells is then distributed in a number of sterile dishes of glass or plastic that contain a suitable nutrient medium. The dishes are incubated at 37 degrees C. for a period ranging from one to three weeks. The cells placed in the dishes settle and adhere to the bottom. There they divide, each cell giving rise to a colony [*see illustration at left on opposite page*]. If the number of cells is sufficiently small, the colonies remain distinct from one another and are recognizable to the unaided eye after about 10 days' incubation; they can be studied sooner with a low-power microscope. When the colonies are fully developed, they are usually fixed and stained. Colonies of normal and transformed cells can be recognized on the basis of morphological characters I shall describe. By picking a colony of transformed cells and reseeding its cells in a fresh culture, clonal lines of transformed cells can be easily prepared. The transformation of BHK cells can also be studied by a selective method that involves suspending the cells in melted agar, which then sets. Transformed cells give rise to spherical colonies, visible to the unaided eye, whereas normal cells grow little or not at all [*see illustration at right on opposite page*].

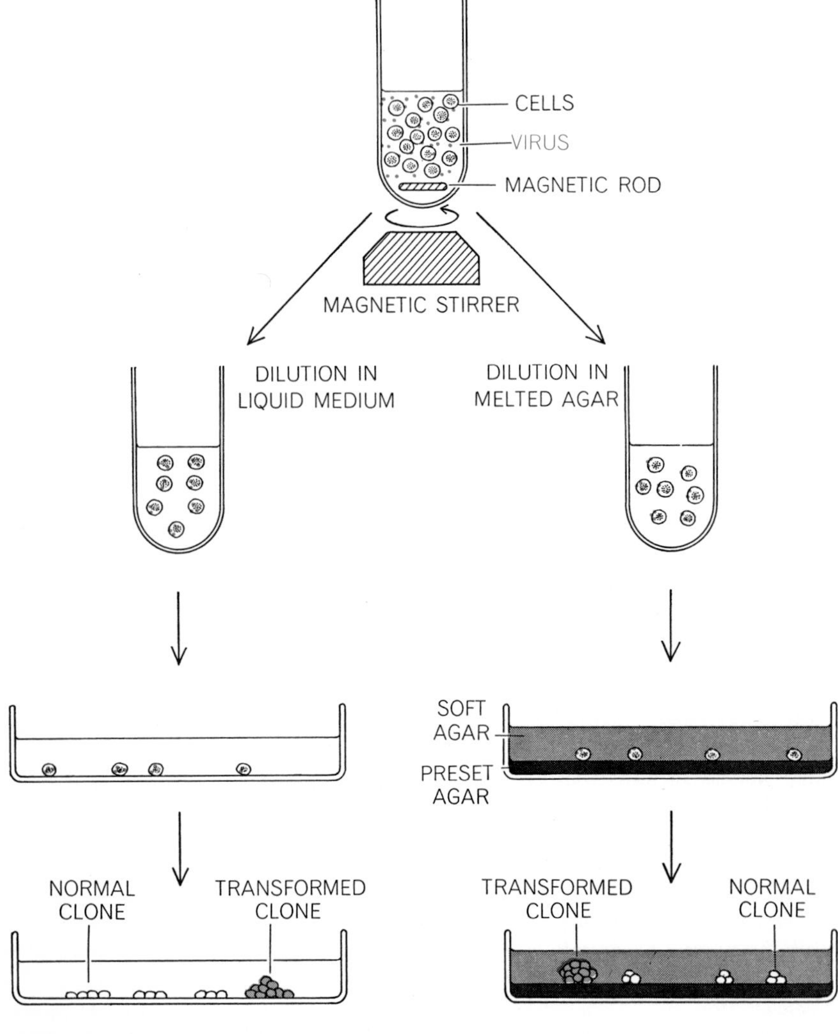


**TRANSFORMATION EXPERIMENT produces cell colonies that differ in appearance, depending on the nature of the culture medium. BHK (hamster) cells are first incubated with polyoma virus for about an hour at 37 degrees centigrade, being stirred constantly. During this time viral particles enter the cells. The infected cells are then diluted either in a liquid medium or in melted agar and transferred to culture dishes. In the agar system the melted agar is poured on a layer of preset agar. The dishes are incubated at 37 degrees C. Cell colonies in the liquid medium develop in contact with the bottom of the container, whereas those in agar form spherical colonies above the preset agar layer. The results of using the two kinds of media are shown in the photographs on the opposite page.**

Colonies of transformed cells, and cultures derived from such colonies, differ morphologically from their normal counterparts in two obvious ways; these differences show that changes have occurred in the regulatory properties of the cells and also in the way they relate to their neighbors. The transformed cultures are thicker because they continue to grow rapidly, whereas normal cultures slow down or stop; in addition the transformed cells are not regularly oriented with respect to each other because they fail to respond to cell-to-cell contact. The altered response to contacts can be best appreciated in time-lapse motion pictures of living cultures.

In sparse BHK cultures the cells move

around actively; if a cell meets another cell in its path, it usually stops moving and slowly arranges itself in contact with and parallel to the other cell. In this way a characteristic pattern of parallel lines and whorls is generated, since the cells do not climb over each other. In a culture of a derivative of the BHK line transformed by polyoma virus the same active movement of the cells is observed. When a cell meets another in its path, however, it continues to move, climbing over the other. In this way the arrangement of the cells becomes chaotic, without any discernible pattern [*see illustration on page 321*].

These alterations of the transformed cells indicate their intimate relatedness to cancer cells. The relatedness is shown in a more dramatic way by the ability of the transformed cells to grow into a cancer when injected, in sufficient number, into a live host that does not present an insurmountable immunological barrier to their survival. For example, BHK cells, which were originally obtained from a hamster, can be transplanted into hamsters; similarly, cells of inbred strains of mice can easily be transplanted into mice of the same strain. The injection of roughly a million transformed cells into a hamster or mouse will be followed by the development of a walnut-sized tumor at the site of inoculation in about three weeks. Untransformed cells, on the other hand, fail to produce tumors.

A crucial finding is that the transformation of healthy cells is attributable to the genes present in the viral DNA that penetrates the cells at infection. The viral genes are the units of information that determine the consequences of infection. Each viral particle contains a long, threadlike molecule of DNA wrapped in a protein coat. Each of these molecules is made up of two strands twisted around each other. Attached to the molecular backbone of each strand is the sequence of nitrogenous bases that contains the genetic information of the virus in coded form. There are four kinds of base, and the DNA molecule of a papovavirus has some 5,000 bases on each strand. Each species of virus has a unique sequence of bases in its DNA; all members of a species have the same base sequence, except for isolated differences caused by mutations.

The double-strand molecule of DNA is so constructed that a given base in one strand always pairs with a particular base in the other strand; these two associated bases are called complementary. Thus the two DNA strands are also complementary in base sequence. Complementary bases form bonds with each other; the bonds hold the two strands firmly together. The two strands fall apart if a solution of DNA is heated to a fairly high temperature, a process called denaturation. If the heated solution is then slowly cooled, a process called annealing, the complementary strands unite again and form double-strand molecules identical with the original ones.

When suitable cells are exposed to a virus, a large number of viral particles are taken up into the cells in many small vesicles, or sacs, which then accumulate around the nucleus of the cell. Most of the viral particles remain inert, but the protein coat of some is removed and their naked DNA enters the inner compartments of the cell, ultimately reaching the nucleus. Evidence that cell transformation is caused by the viral DNA, and by the genes it carries, is supplied by two experimental results.

The first result is that cell transformation can be produced by purified viral DNA, obtained by removing the protein coat from viral particles; this was first shown by G. P. Di Mayorca and his col-

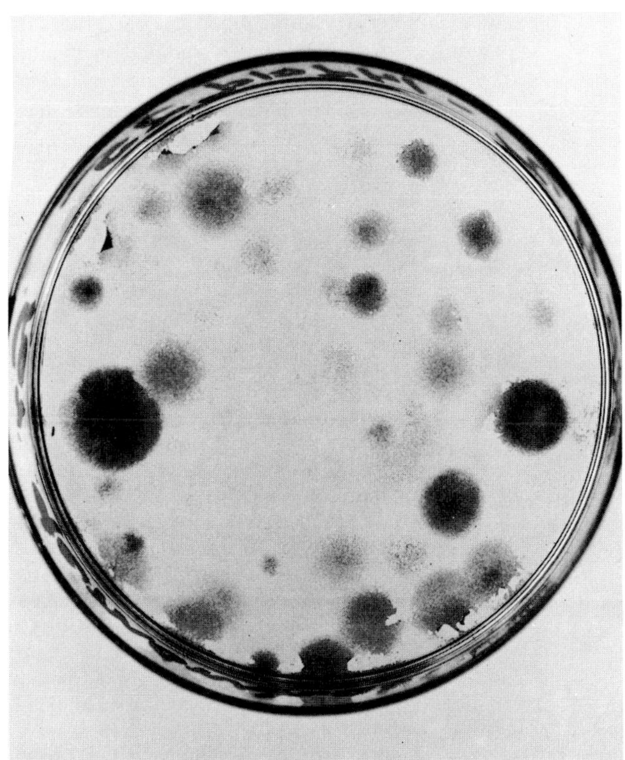

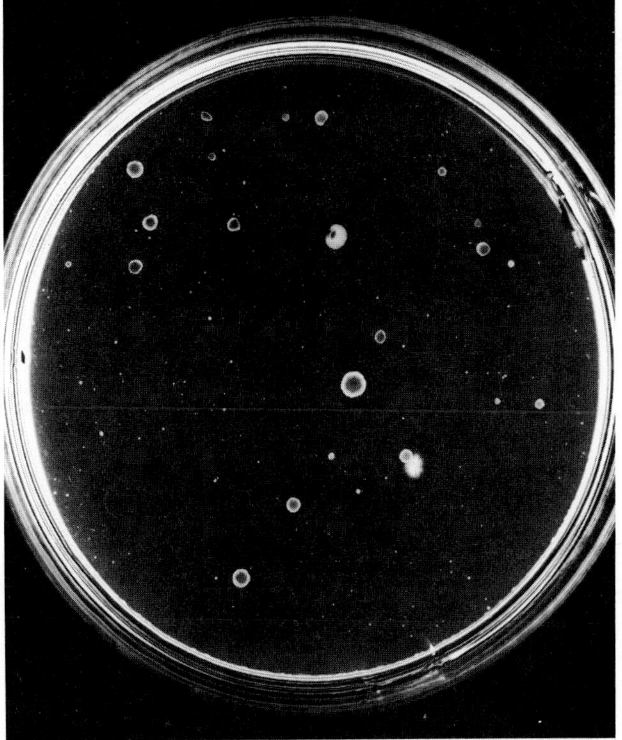

**TWO KINDS OF CELL COLONY are depicted in these photographs made in the author's laboratory. Colonies formed by 3T3 (mouse) cells exposed to SV 40 particles grow on the bottom of a plastic dish under a liquid nutrient medium (*left*). The two large, dark colonies on opposite sides of the culture dish consist of transformed cells. The other colonies are made up of normal cells. Colonies formed by BHK (hamster) cells exposed to polyoma virus are suspended in agar (*right*). Transformed cells create large spherical colonies, which appear as white disks with gray centers. Colonies of normal cells are small or invisible.**

leagues at the Sloan-Kettering Institute. The extraction of the DNA is usually accomplished by shaking the virus in concentrated phenol. In contrast, the empty viral coats do not cause transformation. These DNA-less particles are available for experimentation because they are synthesized in productively infected cells together with the regular DNA-containing particles. The empty coats have a lower density than the complete viral particles; hence the two can be separated if they are spun at high speed in a heavy salt solution, the technique known as density-gradient centrifugation.

A more sophisticated experiment performed at the University of Glasgow also rules out the possibility that the transforming activity resides in contaminant molecules present in the extracted DNA. The basis for this experiment is the shape of the DNA molecules of papovaviruses. The ends of each molecule are joined together to form a ring. When the double-strand filaments that constitute these ring molecules are in solution, they form densely packed supercoils. If one of the strands should suffer a single break, the supercoil disappears and the molecule becomes a stretched ring. Supercoiled molecules, because of their compactness, settle faster than stretched-ring molecules when they are centrifuged. Thus the two molecular types can be separated in two distinct bands.

By this technique polyoma virus DNA containing both molecular types can be separated into fractions, each of which contains just one type. Examination of the biological properties of these fractions shows that the transforming efficiency is strictly limited to the two bands of the viral DNA. Similarly, only the material in the two bands will give rise to productive infection. This result, among others, rules out the possibility that transformation is due to fragments of cellular DNA, which are known to be present in some particles of polyoma virus and therefore contaminate the preparations of viral DNA. The contaminant molecules have a very different distribution in the gradient.

The second result demonstrates directly that the function of a viral gene is required for transformation, by showing that a mutation in the viral genetic material can abolish the ability of the virus to transform. This important finding was made by Mike Fried of the California Institute of Technology, who studied a temperature-sensitive mutant line of polyoma virus called Ts-a. The virus of this line behaves like normal virus in cells at 31 degrees C., causing either transformation or productive infection, depending on the cells it infects. At 39 degrees C., however, the effect of the mutation is manifest, and the virus is unable to cause either transformation or infection; it is simply inactive [*see top illustration on page 322*].

We can now inquire whether the viral gene functions needed to effect transformation are transient or continuous. In other words, do the genes act only once and produce a permanent transformation of the cell line or must they act continuously to keep the cell and its descendants transformed?

A result pertinent to this question is that the transformed cells contain functional viral genes many cell generations after transformation has occurred, although they never contain, or spontaneously produce, infectious virus. The presence of viral genes has been demonstrated particularly well by T. L. Benjamin at Cal Tech, who has shown that the transformed cells contain virus-specific

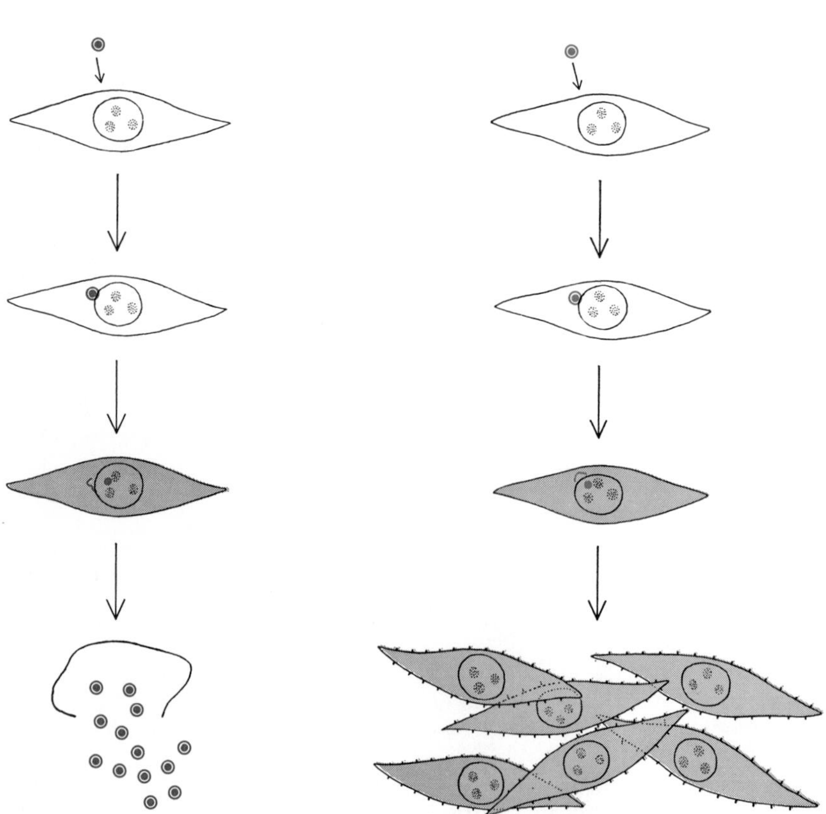


**VIRAL INVASION OF CELLS can have two different results. One result is "productive infection" (*left*), in which viral particles (*color*) mobilize the machinery of the cell for making new viral particles, complete with protein coats. The cell eventually dies, releasing the particles. The other result is transformation (*right*), in which the virus alters the cell so that it reproduces without restraint and does not respond to the presence of neighboring cells. Viral particles cannot be found in the transformed cells. The tint of color in these cells indicates the presence of new functions induced by the genes of the virus. The change in the cell membrane (*lower right*) denotes the presence of a virus-specific antigen.**

**NORMAL AND TRANSFORMED cells are shown on the opposite page in three stages of density, or growth, increasing to the right in each row. Normal cells (*A*, *C*) tend to adhere to one another and form either a pattern of bundles (*A*) or a mosaic-like arrangement referred to as pavement (*C*). Cells that have been transformed by viruses (*B*, *D*, *E*) generally overlap one another and form irregular patterns. The cellular bodies are dark gray and contain a lighter round or oval nucleus in which two or more dark nuclei are embedded. Two of the cultures (*A*, *B*) are a strain of hamster cells identified as "BHK." The other three cultures are "3T3" cells derived from a mouse. Two of the cultures (*B*, *E*) have been transformed by polyoma virus; one (*D*) has been transformed by simian virus 40, also known as SV 40. The cells were photographed in the living state by the author, using a phase-contrast microscope, at the Salk Institute for Biological Studies.**

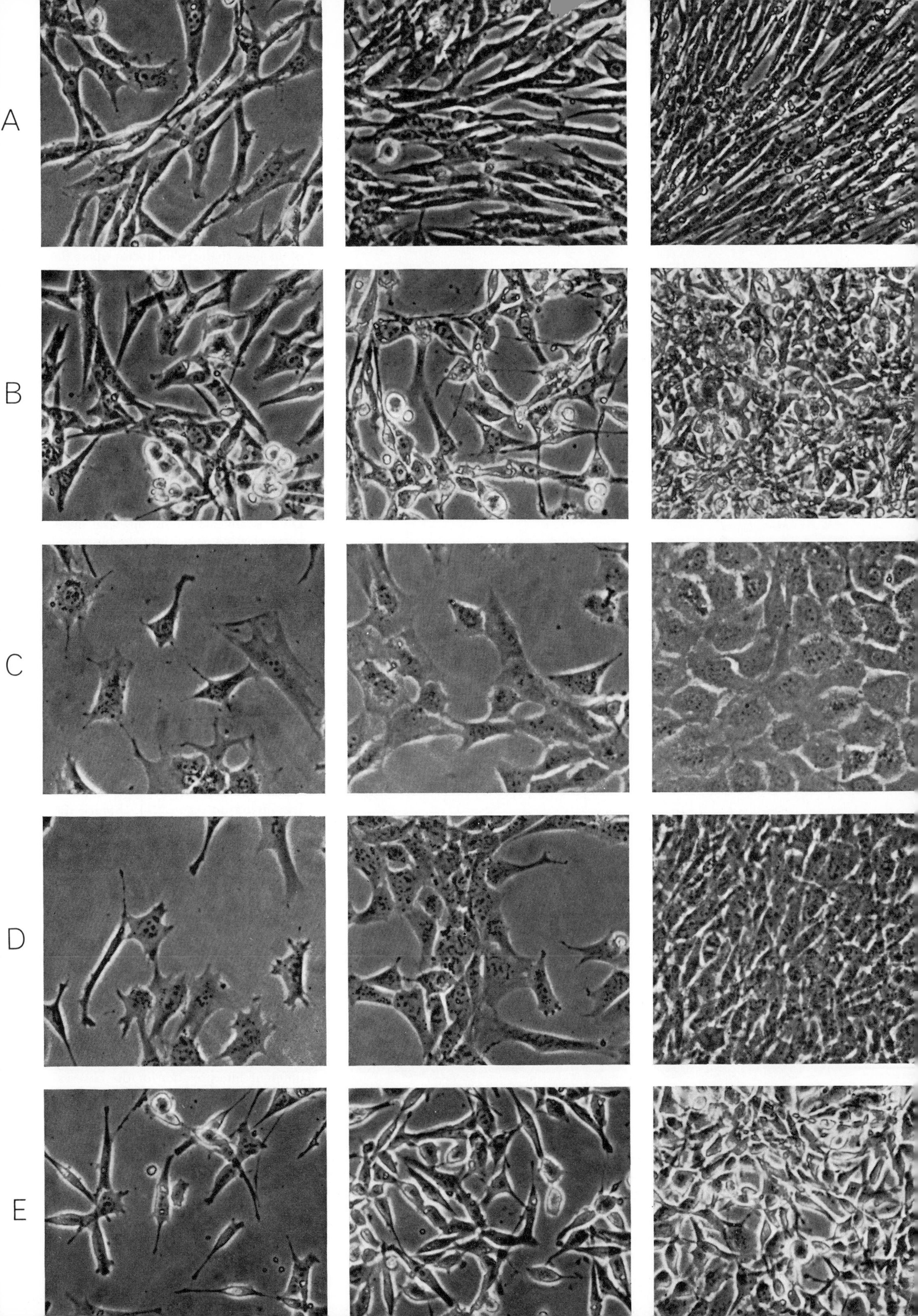
A
B
C
D
E

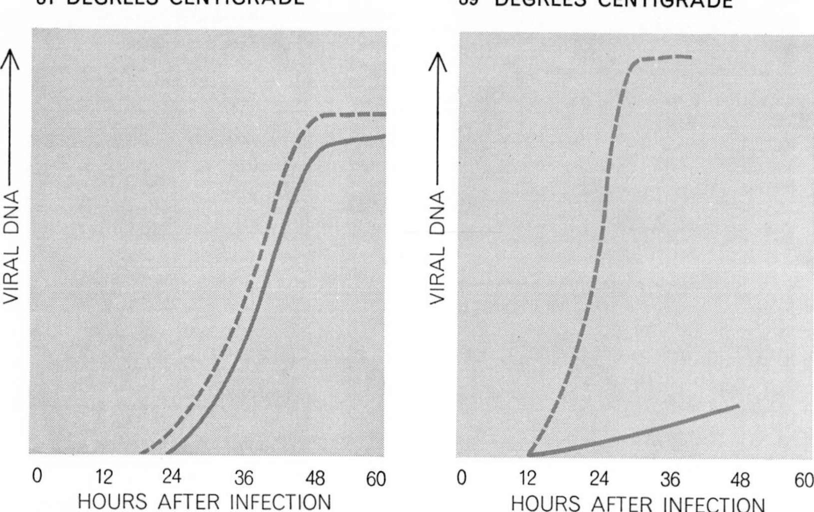


**TEMPERATURE-DEPENDENT STRAINS OF POLYOMA VIRUS act normally at a temperature of 31 degrees C. (*left*) but exhibit mutated behavior at 39 degrees C. (*right*). The solid curves show the amount of viral deoxyribonucleic acid (DNA) synthesized in productive host cells containing the mutant virus. Broken curves show the viral DNA output in cells containing "wild type" (ordinary) polyoma virus. The mutant virus is called Ts-a.**

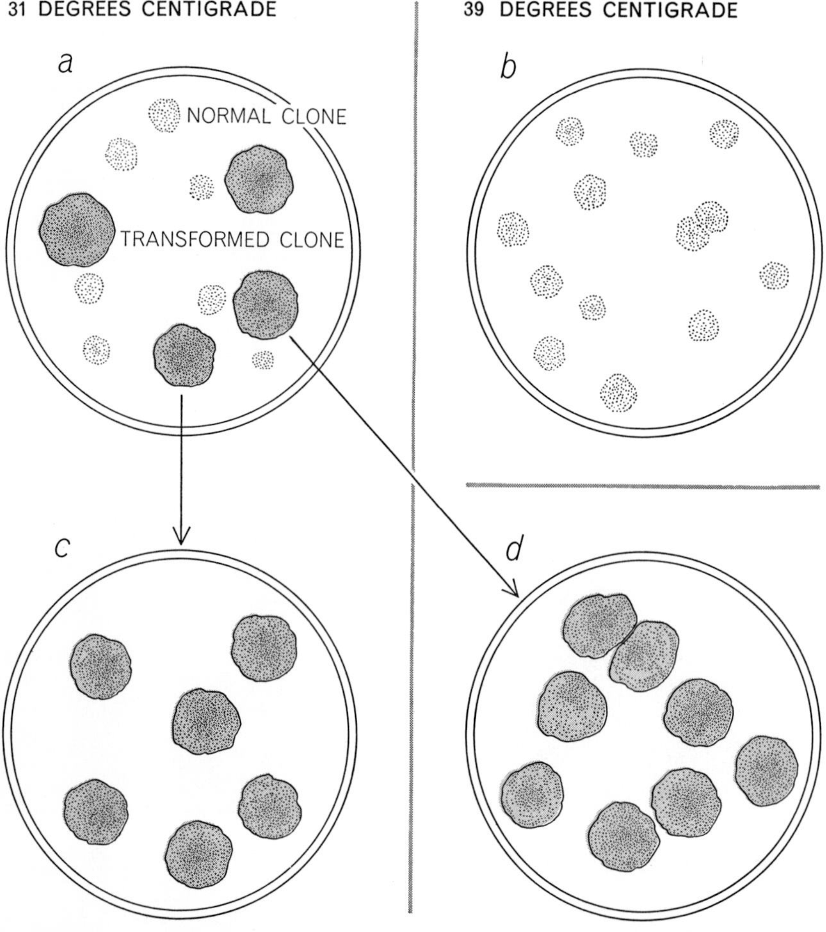


**TRANSIENT ROLE OF TS-A GENE, which gives rise to temperature-dependent mutants of polyoma virus, can be demonstrated by raising the temperature of experimental cultures after the cells have been transformed by the virus. The mutant virus is able to transform cells at low temperature (*a*) but not at high temperature (*b*). Transformed cell colonies, or clones, remain transformed, as expected, at low temperature (*c*), but they also remain transformed when the temperature is raised (*d*). This experiment provides evidence that the Ts-a gene is needed for the initial transformation of the cell but is not needed thereafter.**

ribonucleic acid (RNA). To make the significance of this finding clear it should be mentioned that the instructions contained in the base sequence of the DNA in cells or viruses are executed by first making a strand of RNA with a base sequence complementary to that of one of the DNA strands. This RNA, called messenger RNA, carries the information of the gene to the cellular sites where the proteins specified by the genetic information are synthesized. Each gene gives rise to its own specific messenger RNA. If one could show that viral messenger RNA were present in transformed cells, one would have evidence not only that the cells contain viral genes but also that these genes are active. The viral RNA molecules can be recognized among those extracted from transformed cells, which are mostly cellular RNA, by adding to the mixture of RNA molecules heat-denatured, single-strand viral DNA. When the mixture of RNA and DNA is annealed, only the viral molecules of RNA enter into double-strand molecules with the viral DNA. The reaction is extremely sensitive and specific.

It is likely, therefore, that the viral genes persisting in the transformed cells are instrumental in maintaining the transformed state of the cells. This idea is supported by the observation that the form of the transformed cells is controlled by the transforming virus. This is seen clearly in cells of the line 3T3, which can be transformed by either SV 40 or polyoma virus. The transformed cells, although descended from the same clonal cell line, are strikingly different [*see figure on page 321*]. Similar differences are also observed in other cell types transformed by the two viruses. Since the cells were identical before infection, the differences that accompany transformation by two different viruses can be most simply explained as the result of the continuing function of the different viral genes in the same type of cell. In fact, it is difficult to think of a satisfactory alternative hypothesis.

It must be clear, however, that there is no conclusive evidence for this continuing role of the viral genes. It is therefore impossible to exclude an entirely different interpretation of the observation. One can argue, for example, that the persistence of the viral genes is irrelevant for transformation, and that the genes remain in the cells as an accidental result of the previous exposure of the cells, or of their ancestors, to the virus. Indeed, under many other circumstances viruses are often found in association with cells without noticeably affecting them. A conclusive clarification of the role of the persisting viral genes is being sought by using temperature-dependent viral mutants analogous to the Ts-a mutant I have mentioned. A virus bearing a temperature-dependent mutation in a gene whose function is required for maintaining the cells in the transformed state would cause transformation at low temperature. The cells, however, would revert to normality if the ambient temperature were raised. A small-scale search for mutants with these properties has already been carried out in our laboratory but without success; a large-scale search is being planned in several laboratories.

It should be remarked that no protein of the outer coat of the viral particles is ever found in the transformed cells. Thus the gene responsible for the coat protein is always nonfunctional. This could be either because the transformed cells have an incomplete set of viral genes and the coat gene is absent or because some genes remain "silent." The silence of these genes in turn could be attributed to failure either of transcription of the DNA of the gene into messenger RNA or of translation of the messenger RNA into protein. If failure of transcription were the mechanism, transformed cells would be similar to lysogenic bacteria. Such bacteria have a complete set of genes of a bacteriophage (a virus that infects bacteria), but most of the viral genes are not transcribed into RNA. No other significant similarities exist, however, between ordinary lysogenic bacteria and transformed cells; therefore it is more likely that the coat gene is either absent or, if it is present, produces messenger RNA that is not translated into coat protein. Whatever the mechanism, the lack of expression of the coat-protein gene, and probably of other genes as well, is essential for the survival of the transformed cells, since it prevents productive infection that would otherwise kill the cells.

So far we have considered the genes of the virus in abstract terms. Let us now consider them in concrete ones by asking how many genes each viral DNA molecule possesses and what their functions are. The function of a viral gene is the specification, through its particular messenger RNA, of a polypeptide chain, which by folding generates a protein subunit; the subunits associate to form

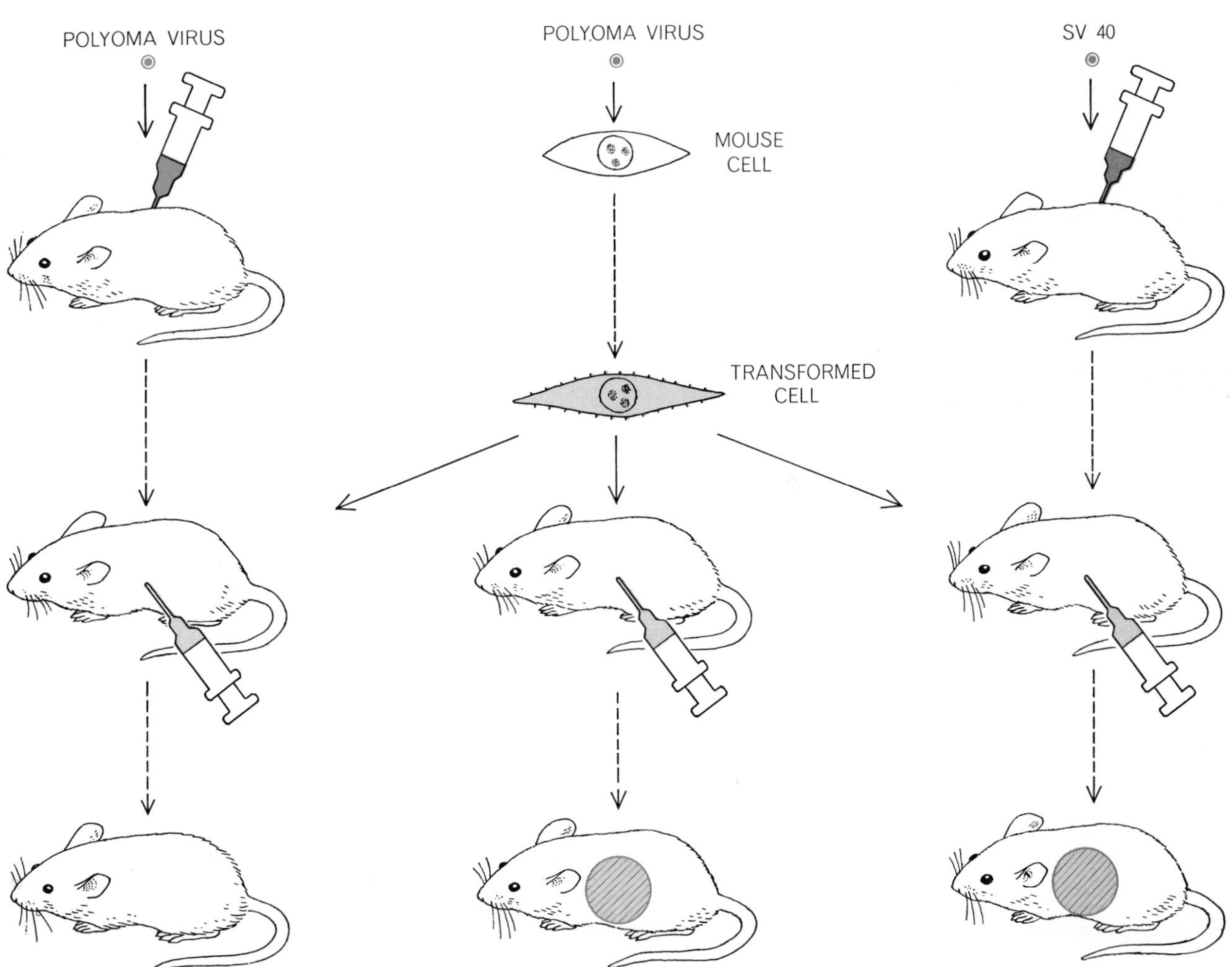


**IMMUNIZATION EXPERIMENT** shows that an animal will not develop a tumor after receiving a massive injection of transformed cells if it has previously received a mild inoculation of the virus used to transform the cells. Thus the animal at left, which has been immunized by an injection of polyoma virus, does not develop a tumor when injected with cells transformed by polyoma virus. The animal in the middle, not so immunized, develops a tumor following the injection of polyoma-transformed cells. The animal at right, which has received an injection of a different virus, SV 40, is not immunized against cells transformed by polyoma virus, hence it too develops a tumor. It would not develop a tumor, however, if injected with cells transformed by SV 40. Cells transformed by either polyoma virus or SV 40 contain a new antigen in their surface that makes them foreign to the animal strain from which they derive and therefore subject to its immunological defenses. These defenses can be mobilized by direct injection of the virus.

a functional viral protein. The final product can be an enzyme or a regulator molecule that can control the function of other genes (viral or cellular), or it can be a structural protein such as the coat protein of the viral particles.

As I have said, each strand of the DNA of the small papovaviruses contains about 5,000 bases. Three bases are required to specify one amino acid, or one building block, in a polypeptide chain; therefore 5,000 bases can specify some 1,700 amino acids. It can be calculated from the total molecular weight of the coat protein of the viral particles and from the number of subunits it has that between a third and a fourth of the genetic information of the virus is tied up in specifying the coat protein. This genetic information is irrelevant for transformation, because no coat protein is made in the transformed cells. What remains, therefore, is enough genetic information to specify about 1,200 amino acids, which can constitute from four to eight small protein molecules, depending on their size. This is the maximum number of viral genetic functions that can be involved in the transformation of a host cell.

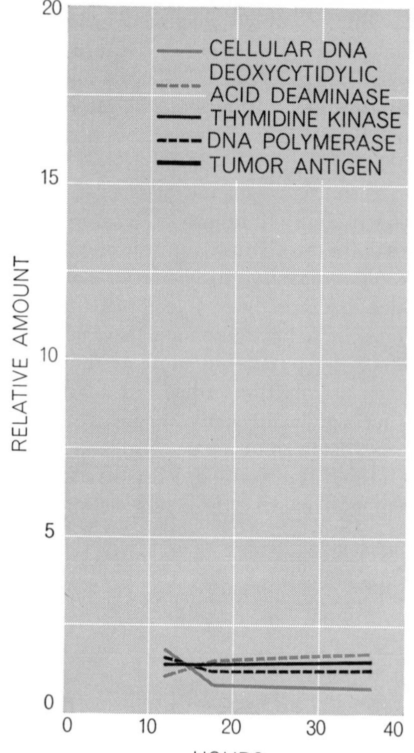


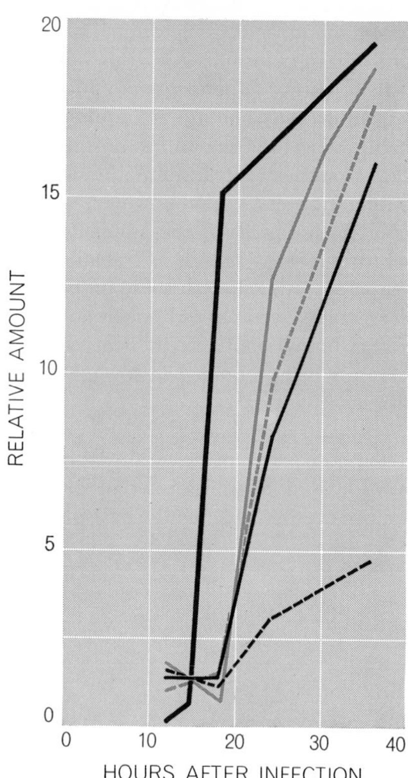


**ACTIVATION OF DNA SYNTHESIS, along with activation of enzymes needed for its production, is a major consequence of viral infection of animal cells. Resting, uninfected cells make little DNA or enzymes associated with its synthesis (*left*). The values plotted are for kidney cells of an African monkey. When the cells are infected with SV 40, the output of DNA and associated enzymes rises steeply (*right*). Before these cellular syntheses are activated a new virus-specific protein, the "T antigen," whose role is unknown, appears.**

In order to discover these viral genetic functions the properties of normal cells have been carefully compared with the properties of cells that have been either transformed or productively infected. Characteristics present in the infected cultures can be considered to result, directly or indirectly, from the action of viral genes. We shall call these new characteristics "new functions." In this way six new functions have been discovered in the infected cells, in addition to the specification of the viral coat protein [*see illustration on page 326*]. Some of the new functions can be recognized biochemically; others can be shown by immunological tests to act as new cellular antigens.

The genetic studies with the papovaviruses have not gone far enough to reveal whether each of the new functions indeed represents the function of a separate viral gene, or whether all the gene functions have been identified. On the basis of the possible number of genes and the number of new functions it is likely that most, if not all, of the gene functions have been detected. At present the new functions are being attributed to the genes, and a large-scale effort is being made to produce temperature-sensitive mutants that will affect each of the genes separately. By studying the effect of such mutations on transformation it will be possible to establish the role of each gene in an unambiguous way.

For the moment we must limit ourselves to examining the various new functions and making educated guesses about their possible role in transformation. If transformation is continuously maintained by the function of viral genes, two new functions are particularly suspect as agents of transformation. One function involves a virus-specific antigen present on the surface of transformed cells; the other is the activation of the synthesis of cellular DNA and of cellular enzymes required for the manufacture of DNA by productively infected cells.

The induction of a virus-specific antigen on the cell surface was detected independently by Hans Olof Sjögren of the Royal Caroline Institute in Stockholm and by Karl Habel of the National Institutes of Health. They have shown that if an animal is inoculated with a mild dose of SV 40 or polyoma virus, it will develop an immune response that will enable it to reject cells transformed by the virus. Whereas the cells grow to form a tumor in the untreated animals, they are immunologically rejected by and form no tumor in the immunized animals [*see the illustration on page 323*]. Rejection occurs only if the animals were immunized by the same virus used for transforming the cells. For instance, immunity against cells transformed by polyoma virus is induced by polyoma virus but not by SV 40, and vice versa. This shows that the antigen is virus-specific. The antigenic change is an indication of structural changes in the cellular surface, which may be responsible for the altered relations of transformed cells and their neighbors.

The activation of cellular syntheses, discovered independently in several laboratories, can be demonstrated in crowded cultures. If the cells in the culture are uninfected, they tend to remain in a resting stage. In these cells the synthesis of DNA, and of enzymes whose operation is required for DNA synthesis (such as deoxycytidylic acid deaminase, DNA polymerase and thymidine kinase), proceeds at a much lower rate than it does in growing cells. After infection by a small papovavirus a burst of new synthesis of both DNA and enzymes occurs; a viral function thus activates a group of cellular genes that were previously inactive [*see illustration on preceding page*]. If the infection of the cells is productive, the activation of cellular syntheses occurs before the cells are killed. The activating viral function must act centrally, presumably at the level of transcription or translation of cellular genes that receive regulatory signals from the periphery of the cell; the signals themselves should be unchanged, since the cell's environment, in which the signals originate, is not changed. If the viral gene responsible for the activating function persists and operates in the transformed cells, it will make the cells insensitive to regulation of growth. Direct evidence for the operation of this mechanism in the transformed cells, however, has not yet been obtained.

A third viral function may be connected with such activation. This is the synthesis of a protein detected as a virus-specific antigen and called the T antigen (for tumor antigen). This antigen, discovered by Robert J. Huebner and his colleagues at the National Institute of Allergy and Infectious Diseases, differs in immunological specificity from either the protein of the viral coat or the transplantation antigen [*see lower illustration on this page*]. The T antigen is present in the nucleus of both productively infected and transformed cells. In productive infection the T antigen appears before the induction of the cellular syntheses begins, and before the viral DNA replicates. Therefore the T antigen may represent a protein with a control function; for instance, it may be the agent that activates the cellular syntheses. For this assumption also direct evidence is lacking.

A fourth viral function relevant to transformation is the function of the gene bearing the Ts-a mutation, which we can call the Ts-a gene. The reader will recall that a virus line carrying this mutation transforms cells at low temperature but not at high temperature. Cells transformed at low temperature, however, remain transformed when they are subjected to the higher temperature, in spite of the inactivation of the gene [*see bottom illustration on page 322*]. Thus the function of the Ts-a gene is only transiently required for transformation. In order to evaluate the significance of this result we must also recall that in productive infection the function of the Ts-a gene is required for the synthesis of the viral DNA. Therefore the transient requirement of this function in transformation may simply mean that the viral DNA must replicate before transformation takes place. If so, the Ts-a gene is not directly involved in transformation.

Another interpretation is possible. The function of the Ts-a gene is likely to be the specification of an enzyme involved in the replication of the viral DNA, for example a DNA polymerase, or a nuclease able to break the viral DNA at specific points, or even an enzyme with both properties. The action of a specific nuclease seems to be required for the replication of the viral DNA because the viral DNA molecules are in the form of closed rings. A nuclease, in breaking one of the strands, could provide a swivel around which the remainder of the molecule could rotate freely, allowing the two strands to unwind. The enzyme, although required for the replication of the viral DNA, may also affect the cellular DNA, for instance by causing breaks and consequently mutations. Such breaks have been observed in the DNA of cells that have been either productively infected or transformed by papovaviruses. If the Ts-a gene indeed acts on the DNA of the host cell, it could play a direct role in the transformation of the cell. Its actions would appear to

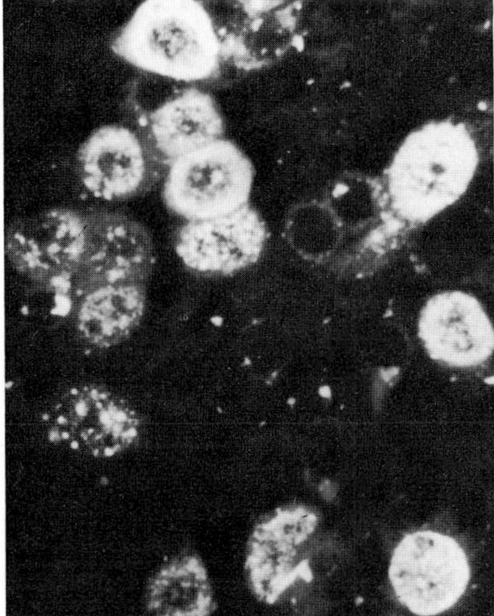
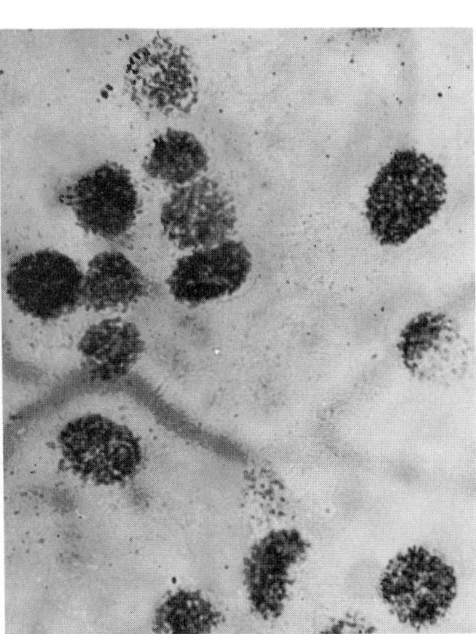

**SITE OF VIRAL-COAT PROTEIN SYNTHESIS in infected cells is found to coincide with the site of DNA synthesis. A culture of mouse kidney cells was exposed to radioactive thymidine (needed in DNA synthesis) some 20 hours after infection with polyoma virus. Six hours later the culture was fixed and stained with antibodies coupled to a fluorescent dye that are specific for the coat protein of the virus. When the culture was photographed in ultraviolet light (*left*), the brilliant fluorescence of the bound antibodies showed that some of the cell nuclei were rich in coat protein. Then the culture was coated with a photographic film to disclose where beta rays emitted by the radioactive thymidine would expose grains of silver. The result (*right*) shows that the nuclei rich in coat protein were the same ones that had accumulated thymidine and were thus the site of DNA synthesis.**

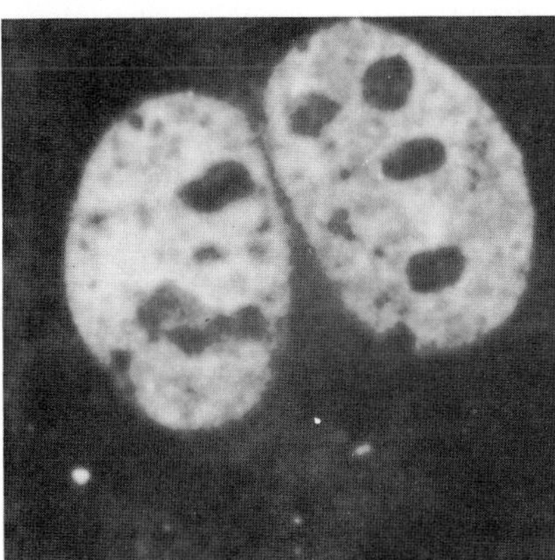
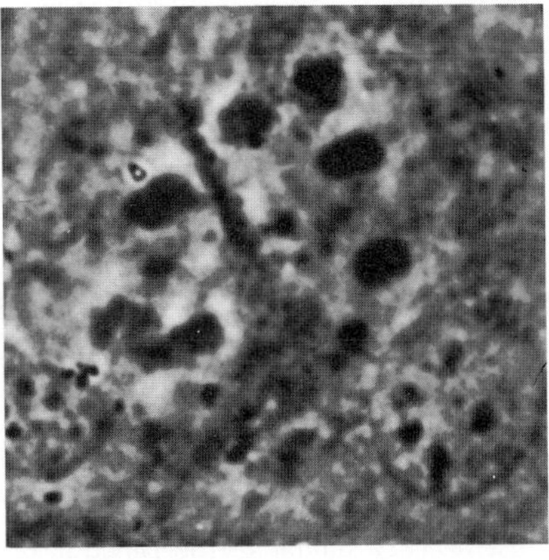

**LOCATION OF T ANTIGEN in mouse cells transformed by SV 40 can be established by staining the cells with fluorescent antibodies that are specific for the T antigen (*left*). The same cells were also photographed in the phase-contrast microscope to show details of cell structure (*right*). It can be seen that the fluorescent antibodies nearly fill two large nuclei.**

be transient, however, since mutations in the cellular DNA would not be undone if the Ts-a gene were subsequently inactivated by raising the temperature of the system. A more definite interpretation of the Ts-a results must await the completion of the biochemical and genetic studies now in progress in several laboratories.

The last two of the six new viral functions observed in infected cells are not sufficiently well known to permit evaluation of their possible roles in cell transformation. One of these two functions is the induction of a thymidine kinase enzyme that is different from the enzyme of the same type normally made by the host cell. Thymidine kinase participates at an early stage in a synthetic pathway leading to the production of a building block required in DNA synthesis. There are reasons to believe, however, that the thymidine kinase induced by the virus may have a general regulatory effect in activating the DNA-synthesizing machinery of the cell after infection. One reason is that the viral thymidine kinase has not been found in transformed cells. Since this enzyme is induced by many viruses containing DNA, whether or not they cause transformation, its induction by the papovaviruses may be connected exclusively with productive infection.

The last new function is one observed so far only with SV 40. After cells have been productively infected with this virus they are changed in some way so that they become productive hosts for a completely different kind of virus, an adenovirus, even though they are normally not a suitable host for such viruses. Little is known about the biochemical steps involved.

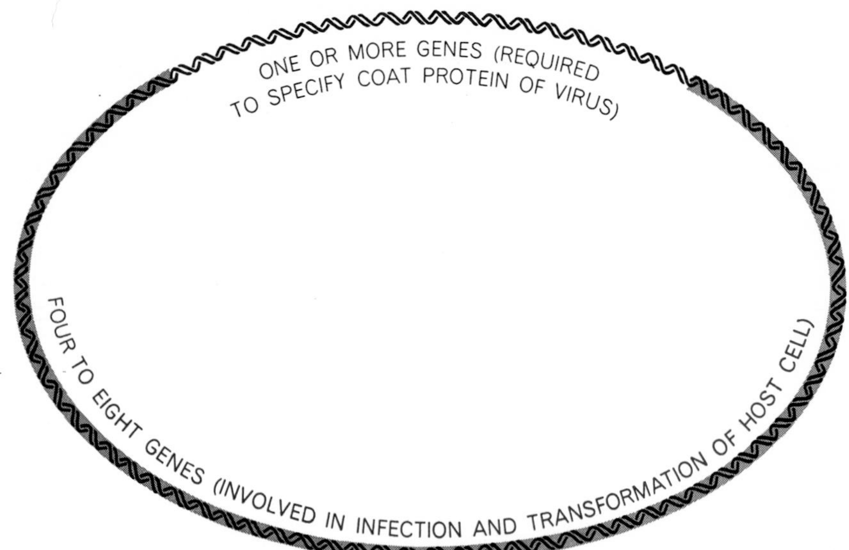


SEVEN FUNCTIONS IDENTIFIED WITH VIRUS ACTIVITY

1. Specification of antigen found on surface of transformed cells.
2. Specification of factor that activates synthesis of cellular DNA.
3. Specification of antigen (T antigen) found in nuclei of infected and transformed cells.
4. Specification of enzyme involved in initial replication of viral DNA. (Attributed to the Ts-a gene.)
5. (Facilitation of cell infection by other viruses.)
6. (Induction of thymidine kinase enzyme.)
7. (Specification of coat protein of virus.)

**SEVEN VIRAL FUNCTIONS have been identified in the infection and transformation of cells. The DNA present in the polyoma virus and SV 40 takes the form of a single ring-shaped molecule consisting of two helically intertwined strands (*top*). Each strand contains some 5,000 molecular subunits called bases that embody the genetic information of the virus in coded form. These bases, in groups of three, specify the amino acids that link together to form protein molecules. Thus 5,000 bases can specify some 1,700 amino acids, or enough to construct some six to 12 proteins. By definition it takes one gene to specify one protein. It is estimated that a third to a fourth of the bases in the viral DNA are needed to specify the protein in the coat of the virus. The remaining bases, enough for four to eight genes, specify the proteins involved in infection and transformation. Little is yet known about the fifth function in this list of seven. Functions 6 and 7 are not involved in cell transformation.**

The central mechanism of cell transformation and cancer induction would appear to be contained within the half-dozen viral functions I have discussed, perhaps together with a few others as yet unknown. Thus the problem is narrowly restricted. It is likely that the dubious points still remaining will be resolved in the near future, since the dramatic advances of the past several years have set the stage for rapid further progress.

This article should not be concluded without an attempt's being made to answer a question that will undoubtedly have arisen in the minds of many readers: Why are viruses able to induce cancer at all? For the two viruses discussed in this article, at least, it seems likely that the viral functions that are probably responsible for cell transformation have been selected by evolutionary processes to further the multiplication of the virus. Because the virus is small and cannot contain much genetic information it must exploit the synthetic mechanisms of the cell, including a large number of cellular enzymes, to achieve its own replication. Furthermore, in the animal hosts in which these viruses normally multiply, most cells that can undergo productive infection are in a resting stage and have their DNA-synthesizing machinery turned off. Thus the evolution of a viral function capable of switching on this machinery is obviously quite advantageous to the virus. This function must be very similar to the function of the cellular gene that regulates cellular DNA synthesis (and overall growth) in the absence of viral infection. The functions of the viral gene and of the cellular gene, however, must differ in one point, again for selective reasons: the cellular function must be subject to control by external signals, whereas the viral function must not be. The virus-induced alteration of the cellular surface seems also to be connected, in a way not yet understood, with viral multiplication, since in many viral infections viral proteins appear on the surface of cells.

The cancer-producing action of the papovaviruses can therefore be considered a by-product of viral functions developed for the requirements of viral multiplication. These viral functions lead to cancer development because they are similar to cellular functions that control cell multiplication, but they somehow escape the regulatory mechanisms that normally operate within the cell.

# 35

# *Hybrid Somatic Cells*

BORIS EPHRUSSI and
MARY C. WEISS
*April 1969*

One of the most powerful tools available to the biologist is genetic analysis, through which the structure of the hereditary material and its relation to the functions of cells and organisms are revealed in great detail and with high resolution by the manner in which various characteristics are passed from one generation to the next. At first this analysis was made through sexual breeding, which is easily accomplished in such organisms as the fruit fly *Drosophila* or the mold *Neurospora.* Much of modern genetics, however, rests on information gained by taking advantage of processes that represent alternatives to sexual breeding. These alternatives were first exploited for the genetic analysis of bacteria and viruses. Now such an alternative to sexual breeding has been developed for the genetic analysis of higher animals, including man. The new procedure stems from the discovery that somatic cells (body cells, as opposed to eggs and sperm) can be crossed to form hybrid cells that live and multiply.

The technique was discovered in 1960 by George Barski, Serge Sorieul and Francine Cornefert of the Institut Gustave Roussy in Paris. They had mixed together cultures of two different mouse-cancer cells that could be distinguished by differences in cell morphology and in the shapes of some of their chromosomes, the threadlike structures in which the genes are arranged. After a few months they saw that a few cells of a new type had appeared, containing in a single nucleus the chromosomes of both parents. They were hybrid cells; they had arisen by the fusion of pairs of cells of the two different types. Barski and his colleagues went on to produce pure cultures of these hybrid cells and grow them successfully. Soon other somatic hybrids were produced by crossing various pairs of cell lines.

It became clear that the hybrids were not mere curiosities. They have two properties that make them suitable for genetic analysis. First, both sets of chromosomes are functional, and the hybrids therefore exhibit the hereditary characteristics of both parents. Second, as the hybrids multiply they lose some of their chromosomes, and this process produces cells with many different constellations of "parental" genes.

The genetic analysis of hybrid cells has been brought to bear on a number of biological problems. One problem is the formal genetics of higher animals, including man: the location of genes on specific chromosomes. Breeding analysis, which is effective for this purpose in lower animals and plants, is too slow in mammals (even in mice the generation time is about three months), and it is impossible in man. The analysis of somatic cells, on the other hand, proceeds without any breeding of individuals, and the generation time for somatic cells (from one cell division to another) is generally between 12 and 24 hours. Another problem under study is that of cellular differentiation, the process whereby cells that presumably have the same genetic equipment become differently specialized in form and function. To investigate the mechanisms that bring this about one must work with cells that have undergone differentiation; somatic hybridization makes it possible for the first time to apply genetic analysis to differentiated cells. We shall give an account of the first results in both areas of investigation after describing the methods of somatic hybridization and the relevant properties of hybrid cells.

The basic techniques of cell hybridization are those of standard cell culture, in which cells taken from a bit of tissue are allowed to settle out of a suspension and proliferate to form a layer on the bottom of a laboratory dish [*see illustration on page 4*]. To give the cells room to grow the layer is periodically broken up, and a few of the cells are transferred to a new dish; these "serial passages" can be repeated many times, and some cells go on multiplying indefinitely. Such a culture is heterogeneous, having been derived from a fragment containing several cell types. In order to obtain uniform populations of cells one must inoculate a culture dish with a very small number of highly dispersed cells, so that each produces a discrete "clone," or a colony that represents the progeny of a single cell.

Experiments in hybridization begin with a mixed culture of two parent cell lines, each characterized by the presence of marker chromosomes not found in the other. After a few days or weeks of growth one can identify hybrid cells by examining chromosome preparations. The drug colchicine is added to the medium to arrest cell division in metaphase, a stage that is favorable for observation of the chromosomes. The cells are fixed, pipetted onto a slide and stained. Under the microscope such preparations show many normal metaphase chromosome sets of both parental types, each recognizable by the number and shape of the chromosomes. If any hybrid cells have been formed, there will also be hybrid metaphases, recognizable by the large number of chromosomes and by the presence of marker chromosomes from both parents [*see illustration on page 329*].

In Barski's first hybridization experiment, performed in this manner, the hybrids turned out to have a selective advantage over the parental cells and grew

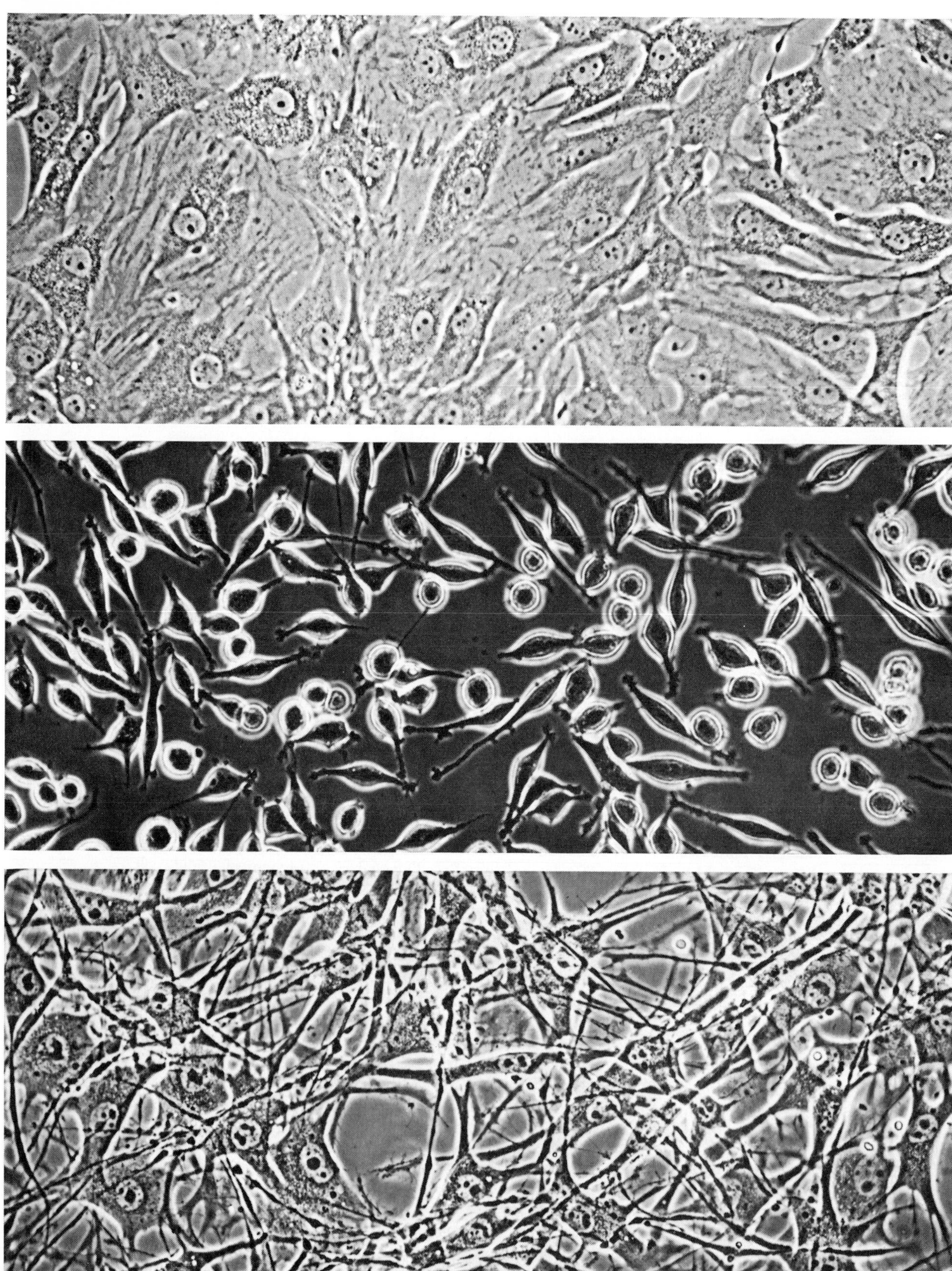

HYBRID CELLS and their parent cells are seen in a set of phase-contrast photomicrographs made by the authors. Rat cells (*top*) and mouse cells (*middle*) are grown together in a cell culture. Some of them fuse and form hybrids, which are isolated and grown in a pure culture (*bottom*). The rat cells tend to flatten out on the surface of the glass culture dish and so they appear thin. The mouse cells attach more loosely to the glass and therefore appear more refractile. The hybrid cells combine the morphological characteristics of both parent lines: they are more refractile than the mouse cells but they attach more completely than the rat cells.

more rapidly, and so they soon constituted a large enough fraction of the total cell population to be isolated by cloning. One of us (Ephrussi), together with co-workers first at the Laboratoire de Génétique Physiologique at Gif in France and then at Western Reserve University, isolated a number of other somatic hybrids. In each case success depended on the hybrid's having a selective advantage in the mixed culture. Since hybrids were rare and it took a long time for enough of them to accumulate so that they could be isolated, and since some crosses were almost surely failing only because the hybrids had no selective advantage, it was clear that a method conferring a decisive selective advantage on hybrid cells would be extremely valuable.

In 1964 John W. Littlefield of the Harvard Medical School devised a system for the selective isolation of hybrid cells [*see illustrations on page 332 and at top of page 333*]. He cultured, in a medium containing the drug aminopterin, two kinds of mutant cells, each of which lacked a different enzyme necessary for growth in the presence of aminopterin. When hybrid cells were formed, they were able to grow in the aminopterin medium by virtue of mutual complementation: each parent supplied the gene for the enzyme the other parent lacked. Littlefield's selective medium therefore kills cells of both parent lines but allows the survival and unhampered growth of hybrid cells. In the course of his experiments Littlefield established that one hybrid cell was produced per 200,000 parental cells (half of each kind).

With Richard L. Davidson, one of us (Ephrussi) then modified Littlefield's method to devise a "half-selective" system that is more generally applicable, in which only one parent cell line lacks one of the enzymes necessary for growth in Littlefield's medium. The other parent can be a normal line, carrying no selective marker, provided that it grows slowly or is inoculated in small numbers. Assume, for example, that a dish is inoculated with a million enzyme-deficient cells and only 100 normal cells. In the selective medium the majority parent degenerates, leaving discrete colonies of the minority parent and of hybrid cells. The hybrids can usually be recognized by their shape, and their hybrid nature

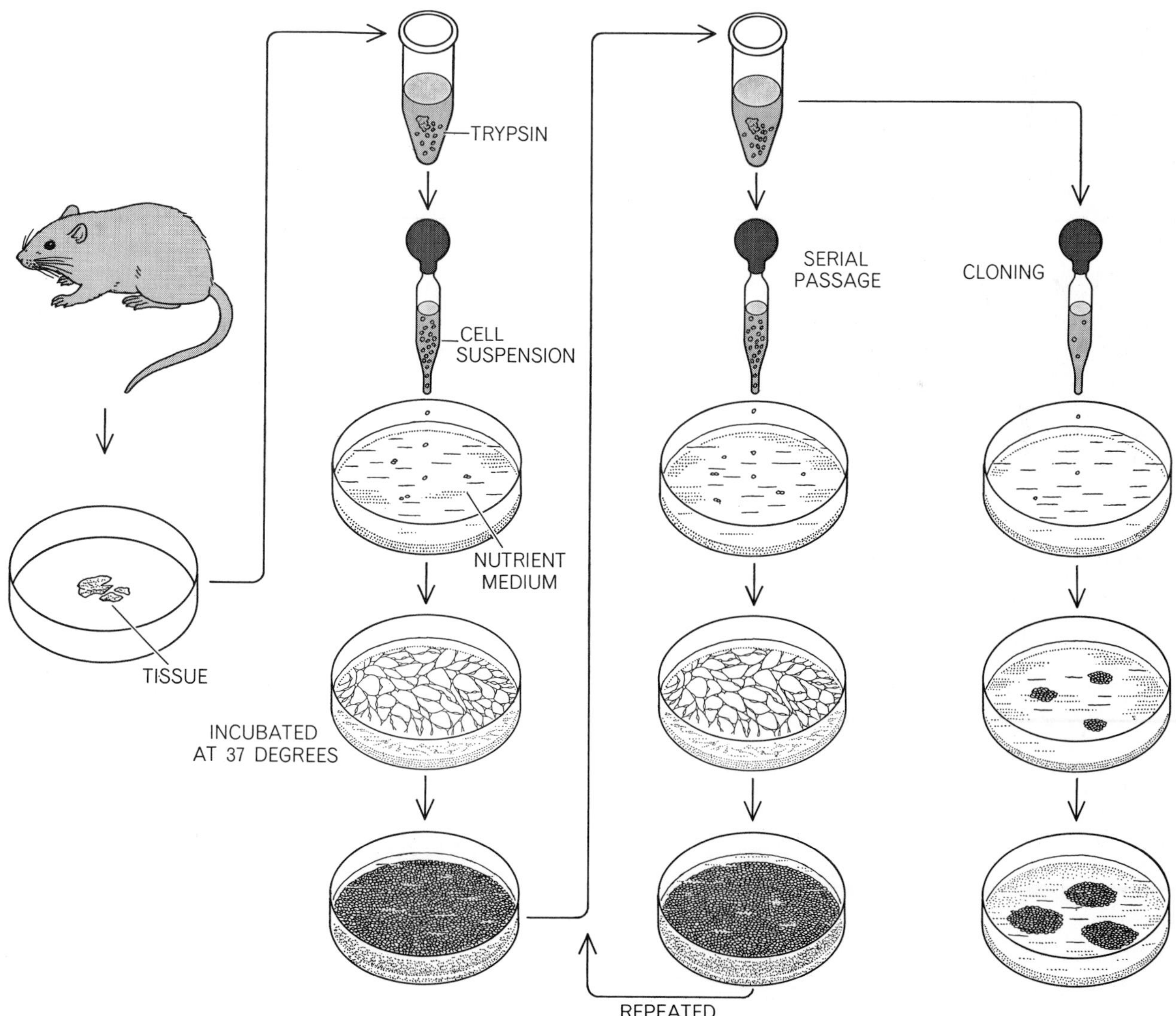


**CELL CULTURE begins with minced pieces of tissue from an adult animal or embryo. Incubating the pieces in a digestive enzyme such as trypsin breaks them up into a suspension of single cells. Inoculated into a liquid nutrient medium and incubated, the cells attach to the bottom of the culture dish and divide, producing a continuous sheet. The sheet is again digested with trypsin. If a large fraction of the cells are thereupon inoculated in new medium, the process is repeated. Such serial passages yield heterogeneous cultures because the tissue pieces contained several kinds of cells. If instead a few dispersed cells are inoculated (*right*), they divide to form discrete "clones," or colonies of the progeny of one cell. One clone is then selected with which to start a pure culture.**

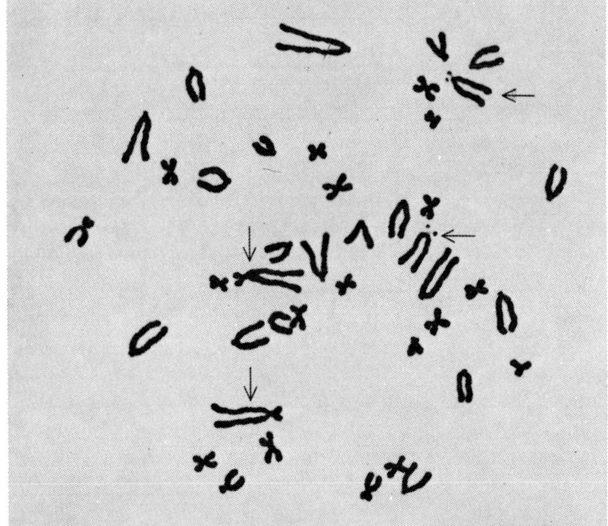

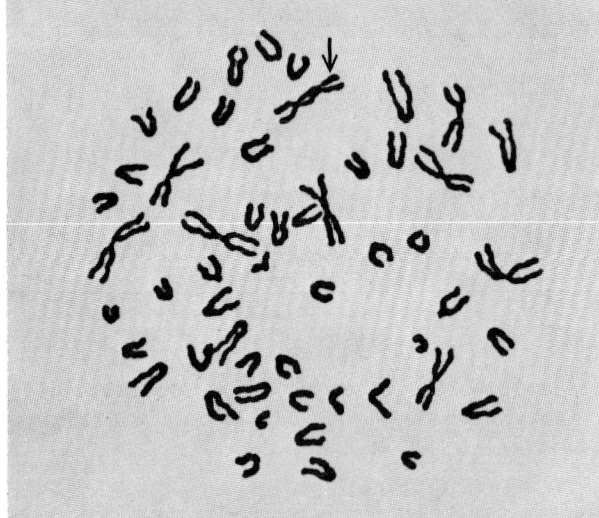

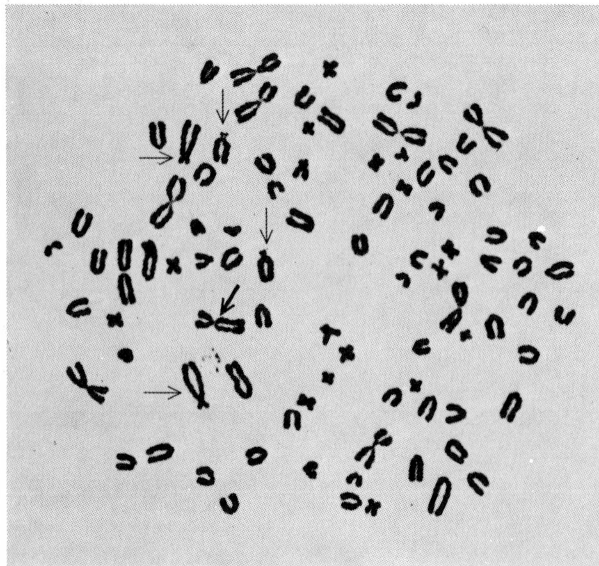

CHROMOSOMES of the rat, mouse and rat-mouse hybrid cells illustrated on page 3 are displayed here in two forms: metaphase figures (*left*) and karyograms (*right*), in which the chromosomes are arranged in groups on the basis of shape and size. The 42 rat chromosomes (*top*) include large ones like unbalanced *X*'s (*thin arrows*) and small *X*-shaped ones. The mouse parental cell (*middle*) has an abnormal number of chromosomes, 54 instead of the usual 40, and a distinctive *X*-shaped one (*heavy arrow*) as well as the usual *V*-shaped ones. The hybrid from which this preparation was made (*bottom*) has 89 chromosomes, of both parental types.

is confirmed by examination of the chromosomes.

These selective systems were first applied to intraspecific crosses (between different cell lines of the same species, usually the mouse). Then we managed to cross cells of different species. The first of these interspecific crosses involved rat and mouse cells; later hamster-mouse hybrids and finally mouse-human hybrids were obtained. Most of the current experimentation is being done with interspecific hybrid cells. The reason is that they fulfill the two requirements for genetic analysis much better than the intraspecific hybrids do.

Since a gene can be recognized only when it mutates to an alternative form that is recognizable in the progeny, the existence of distinguishing genetic markers in the two parents that can be traced in the progeny is a *sine qua non* of genetic analysis. In microbial genetics the most valuable markers have been enzymes that exist in a normal form and also in a form altered by mutation. Such enzyme markers could serve in mammalian-cell studies too, but the trouble is that the necessary mutations are rare and are more difficult to induce in mammalian cultured cells than they are, for example, in bacteria. This paucity of markers led us in 1965 to undertake crosses between different species. We knew that as a consequence of evolution many animal species have come to possess variants of the same enzyme that differ in their structure. These differences affect the physical properties of the enzymes, so that the two variants can be distinguished by such methods as electrophoresis and chromatography. We could expect, therefore, that crossing cells of different species would yield hybrid cells endowed with many built-in enzyme markers.

The only question remaining was whether or not the genes of both parents would in such cases be functional in the hybrids. Fortunately that proved to be the case. For example, in rat-mouse hybrid cells both of the parental forms of the enzyme lactate dehydrogenase (LDH) are synthesized [*see bottom illustration on page 333*]. Malate dehydrogenase and beta-glucuronidase are two other enzymes that have been found in both parental forms in interspecific hybrid cells.

The wealth of genetic markers in interspecific hybrids fulfilled the first prerequisite for genetic analysis. We soon found that the second requirement was fulfilled too: successive generations of

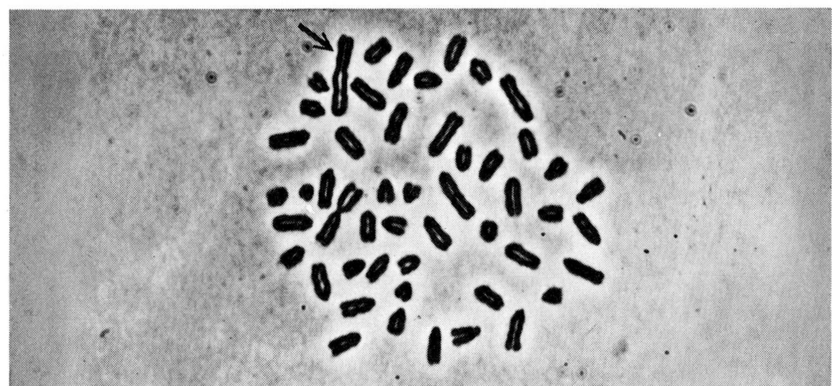

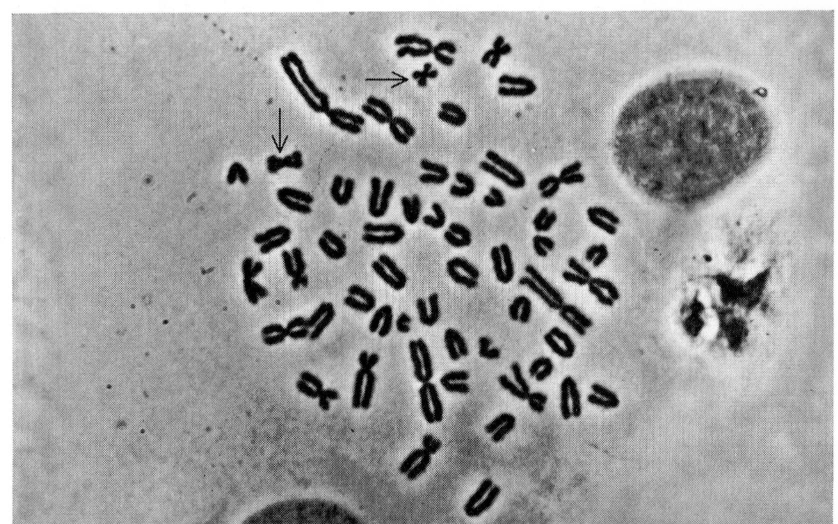

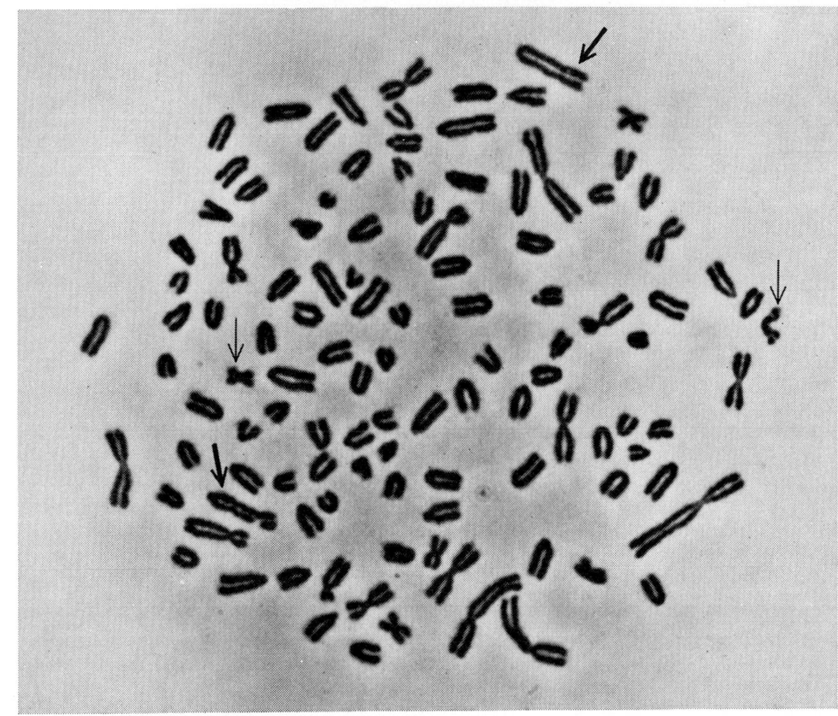

**METAPHASE** chromosome preparations of the two mouse tumor cells first used in somatic hybridization are shown (*top and middle*) along with a metaphase of the hybrid cell (*bottom*). The chromosomal DNA has been replicated and the two copies of each chromosome are joined to form *V*-shaped or X-shaped double chromosomes. The cells of one line (*top*) have two extra-long *V*-shaped chromosomes, one of which is seen in this metaphase (*heavy arrow*). The other line has many X-shaped chromosomes (*middle*), two of which are distinctive (*thin arrows*). The hybrid cell (*bottom*) contains the chromosomes of both parent lines.

the interspecific hybrid cells contain decreasing numbers of chromosomes as a result of the loss of chromosomes during cell division. The rate and the extent of the loss vary in different hybrids, but they are generally greater than they are in the intraspecific hybrids.

As an example of an intraspecific cross, consider the hybrids derived from two lines of mouse cells. Detailed analysis of the chromosomal changes is not possible because all normal mouse chromosomes are about the same shape and most cultured mouse cells contain only a few marker chromosomes. In general the karyotype, or chromosomal constitution, of hybrid mouse cells is rather stable. There is some loss—as one might expect, since the hybrids contain an excess of most genes and can therefore survive the loss of some chromosomes. The loss usually does not exceed 10 to 20 percent of the number of chromosomes in the original fused nucleus, however, so that even after hundreds of generations the hybrid cells retain most of the chromosomes of both parents.

In contrast to the intraspecific hybrids, those produced by the fusion of cells of different species have a great many marker chromosomes, since there are large differences in the shapes and sizes of the two species' chromosomes [*see illustration on page 330*]. Analysis of the numbers and kinds of chromosomes in successive generations of rat-mouse hybrids showed that although the decrease in the total number of chromosomes is not much greater than it is in mouse-mouse hybrids, the loss is slightly preferential: significantly more of the rat chromosomes than of the mouse chromosomes disappear. In hamster-mouse hybrids there is greater loss of the mouse chromosomes than of the hamster chromosomes.

In 1967 one of us (Weiss) and Howard Green, working at the New York University School of Medicine, succeeded in crossing mouse cells with human cells and found that the hybrids presented an extreme case of preferential chromosome loss. The mouse-human hybrid cells appeared from the beginning to be different from other interspecific hybrids. Instead of having some of the characteristics of each parent, they looked much more like the mouse cells than the human cells [*see illustration on page 334*]. The reason became clear when their karyotype was examined: the cells contained all the expected mouse chromosomes but only from two to 15 of the 46 human chromosomes. Apparently these hybrids had lost most of their human chromosomes soon after being formed. They continued to lose them as they were cultivated. After 100 generations some of the clones had lost all their human chromosomes; the others retained no more than 10 of them. Because it is possible to obtain cells that contain all the mouse chromosomes and either no human ones or very few, the mouse-human hybrids lend themselves, as we shall see, to studies of human genetics.

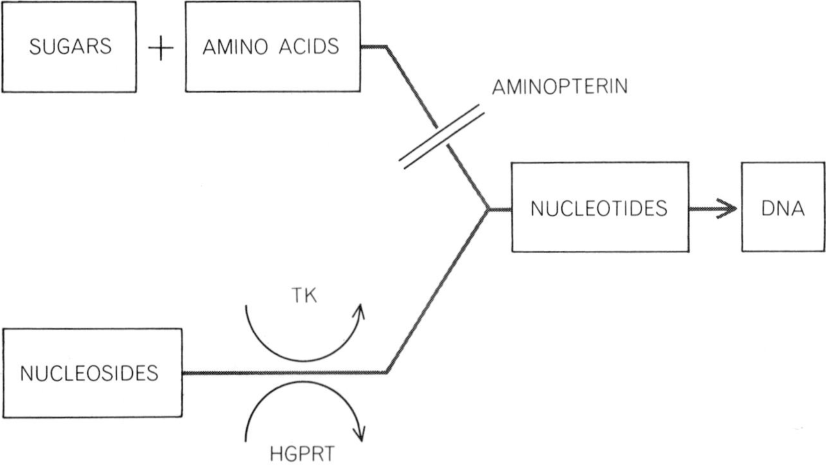


**DNA SYNTHESIS from sugars and amino acids is blocked by aminopterin (*top*). An alternative pathway depends on preformed nucleosides (DNA precursors) and the enzymes thymidine kinase (TK) and hypoxanthine guanine phosphoribosyl transferase (HGPRT). Cells with these enzymes can grow in a medium containing aminopterin and nucleosides.**

**MUTATIONS produce some cells that lack TK but produce HGPRT (*black dots*) and some that lack HGPRT but produce TK (*colored dots*). If such cells are crossed, the hybrid cells contain the genes from both parent lines and therefore produce both parental enzymes.**

One would like to be able to cross any two cell types at will. Although Littlefield's system and the half-selective system are in principle applicable to a large number of crosses, they depend on the introduction of specific mutations into the cells to be crossed. In mammalian cells this is a difficult and time-consuming process, and for some kinds of cells it may be impossible. During the time it takes to select cells with the mutation resulting in the required enzyme deficiency, for example, other changes may occur that alter other cell properties one would like to retain. There are two ways in which one might bypass this difficulty: either by devising a selective system based on naturally occurring markers or by somehow increasing the frequency of cell fusion to such an extent that the hybrids no longer need to be selected. The second of these approaches has proved to be rewarding.

Some years ago Y. Okada of Osaka University reported that the Sendai strain of parainfluenza virus causes animal cells in suspension to clump together and that many of the clumped cells undergo multiple fusions. With this observation in mind Henry Harris and J. F. Watkins of the University of Oxford were able in 1965 to bring about the fusion of different types of cells with Sendai virus that had been rendered noninfectious by ultraviolet irradiation. The virus treatment induced the formation of giant cells with from two to 10 or more nuclei—in some cases heterokaryons, or cells with nuclei from different parents. Some of these nuclei fused, yielding hybrid karyotypes, but Harris and Watkins saw no hybrids capable of more than a few divisions. Yet the occurrence of nuclear fusion, hybridization and some cell division, and the absence of viral infection (since the virus was inactivated), implied that the method

a

MUTANT LACKING IN ENZYME TK

MUTANT LACKING IN ENZYME HGPRT

NUTRIENT MEDIUM WITH AMINOPTERIN AND NUCLEOSIDES

KILLED CELLS

HYBRID

b

MUTANT LACKING IN ENZYME TK

NORMAL CELL

NUTRIENT MEDIUM WITH AMINOPTERIN AND NUCLEOSIDES

KILLED CELLS

HYBRID

NORMAL CELL

c

VIRUS-TREATED CELL A

VIRUS-TREATED CELL B

NUTRIENT MEDIUM

CELL A

HYBRID

CELL B

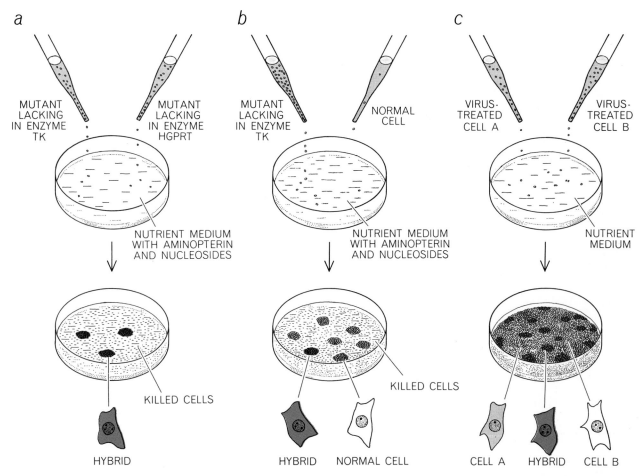

**SELECTIVE SYSTEMS** are required in order to isolate the rare hybrid cells from the proliferating parent lines. One selective system (*a*) depends on the enzyme activity outlined in the illustration on the opposite page. Cells lacking either enzyme are killed; hybrids, which have both enzymes, live and form colonies (*color*). In the half-selective system (*b*) only one parent cell lacks an enzyme, but since only a few of the other (normal) cells are inoculated, the rare but discrete hybrid colonies can be isolated. In virus-induced hybridization (*c*) any two cells can be crossed. The virus causes them to clump and fuse, promoting hybrid formation.

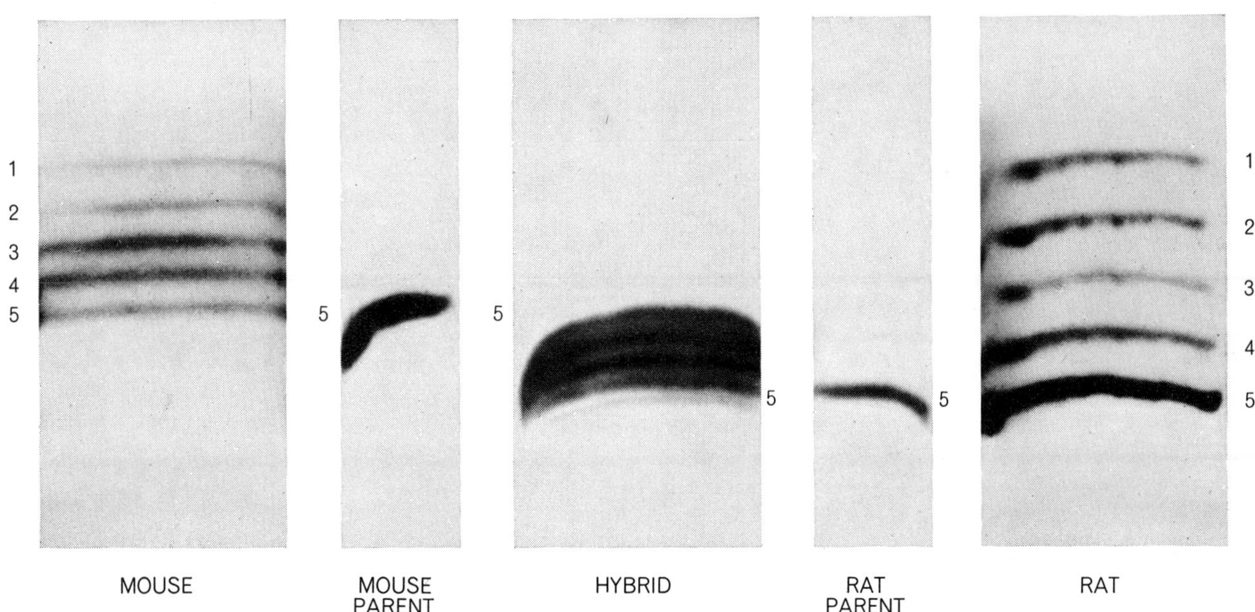


**ENZYME MARKERS** are available in somatic cells because different species contain slightly different forms of the same enzyme. These forms can be separated on the basis of their mobility in an electric field. Electrophoresis of the LDH found in mouse and rat diaphragm yields five different LDH bands in each (*left and right*). The two No. 5 bands are present in the parental cells used in hybridization. The hybrid has both parental bands and also has three intermediate bands representing hybrid molecules (*center*).

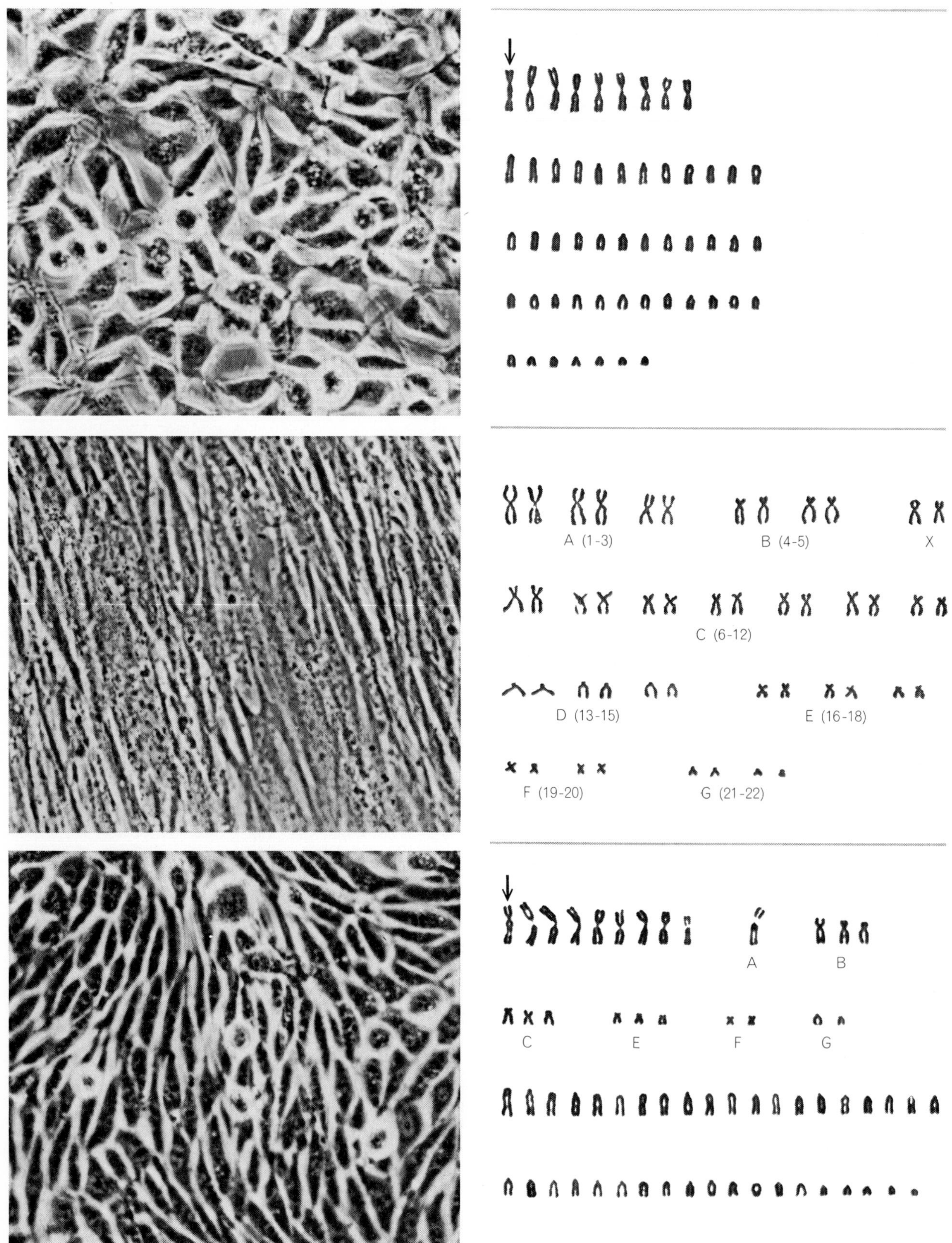


**MOUSE-HUMAN** hybrids are illustrated by the cell cultures (*left*) and the karyograms (*right*) of the mouse parent line (*top*), the human parent (*middle*) and the hybrid (*bottom*). The human cells, derived from embryonic lung tissue, contain the normal number of chromosomes (46, or 23 pairs), arranged here in the usual seven groups (plus the two female sex chromosomes). Except for a tendency to align in parallel, the hybrid cells look more like the mouse cells than the human ones. This is in keeping with the fact that the hybrid karyogram contains only 14 of the 46 human chromosomes, which are readily distinguished from mouse chromosomes.

could be developed to provide large numbers of hybrids capable of prolonged multiplication.

This was achieved by George Yerganian and M. B. Nell of the Children's Cancer Research Foundation in Boston and then by Hayden Coon and one of us (Weiss) at the Department of Embryology of the Carnegie Institution of Washington. In the latter case the cell lines involved were such that the frequency of virus-induced hybrids could be compared with that of spontaneous hybrids isolated by the selective techniques, and so it was possible to determine the effectiveness of the virus in promoting hybridization. Viable hybrids appeared from 100 to 1,000 times more frequently in cultures treated with inactivated virus than they did in cultures left to spontaneous hybridization. The virus-induced hybrids proved to have the same properties as the spontaneous ones and can therefore be used for the same kinds of experiments. It is now possible to make crosses between cells to which no selective system is applicable, and this should mean that almost any two cells can be crossed and the resulting hybrid can be isolated.

One of the most interesting findings of somatic-hybridization studies is the very fact that somatic cells of different origins are compatible. The incompatibility between the sperm of one species and the egg of another is well established; in extreme cases an egg fertilized by a sperm of another species immediately expels the nucleus of the sperm. It is therefore surprising to see that the nuclei of cells of two different species fuse and in most cases at once begin to function harmoniously. This implies that the intracellular signals that dictate the sequence of biochemical events in one parent's division cycle must be "understood" by the components of the other cell—in spite of the millions of years during which mammalian species have diverged from their common ancestors by accumulating gene mutations.

A related finding is that the hybrid cells synthesize hybrid enzymes that function satisfactorily. As we mentioned before, rat-mouse hybrids synthesize both parental forms of the enzyme LDH. We have found, moreover, that in the hybrids some of the active enzyme molecules, which are composed of four subunits, are themselves hybrid in nature, formed by the random association of rat subunits and mouse subunits. Several other examples of hybrid enzyme molecules have been reported in interspecific hybrids; since many enzymes are composed of subunits, molecules of this kind are probably common in hybrid cells. The existence of functional hybrid enzymes was not in itself a surprise. Clement L. Markert of Johns Hopkins University had earlier produced such LDH molecules by chemical methods. What is surprising is to find that homologous genes that have diverged as widely as is suggested by the different structures of the enzymes they specify can still produce proteins similar enough to associate into molecules whose enzyme activity can fully satisfy the requirements of a living cell.

The two examples we shall give of current experiments utilizing somatic hybridization have to do with the formal genetics of man and with the study of gene expression and its control in cellular differentiation.

Study of the formal genetics of any organism begins with the determination of linkage groups and the assignment of genes to specific chromosomes. Then genes are localized in specific segments of the chromosomes, establishing genetic maps for the species. The required data are ordinarily obtained by sexual breeding over many generations, each involving segregation and recombination: the processes, which occur during formation of the germ cells, by which the various parental genes are distributed in different assortments to different daughter cells, resulting in progeny with a range of different characteristics. In effect, the loss of chromosomes by successive generations of somatic hybrid cells takes the place of the segregation and recombination that occur in germ cells, making it possible to begin to determine human linkage groups.

Mouse-human hybrids eventually lose all their human chromosomes. By studying them at a stage when they contain only a few, one can correlate the loss of a specific human gene product with the loss of a specific chromosome. So far this has been done for one enzyme. The mouse cells Weiss and Green used were deficient in thymidine kinase, one of the enzymes required for growth in Littlefield's medium. The survival and growth of hybrid cells maintained in the selective medium therefore depend on the presence of the human gene for thymidine kinase. In several clones of hybrid cells only from one to three human chromosomes remained after 100 to 150 generations [*see illustration on page 336*]. In each of them one specific human chromosome was still present: a small one of the group designated *E*. Presumably this chromosome carries the thymidine kinase gene. This was confirmed when the clones were removed from the selective medium and exposed to bromodeoxyuridine, which kills cells containing thymidine kinase. None of the cells that survived this treatment contained the chromosome in question.

The mouse-human hybrids can surely be used to locate other genes in specific chromosomes. The gene need not be for an enzyme that is missing from the mouse parent as in the case of thymidine kinase. Because of the physical differences between mouse and human forms of many enzymes, one can find out which human enzymes are retained in various hybrid populations and so correlate the presence of the enzyme with that of a specific chromosome. Similar experiments should reveal the location in human chromosomes of genes that determine the presence of antigens. Somatic hybridization has already shown that genes for human antigens must be widely distributed among the human chromosomes since the human antigenic activity of the hybrid cells is proportional to the number of human chromosomes they contain. Adding a purified antiserum that acts against a specific antigen should make it possible to trace that antigen's gene to a single chromosome. In short, it appears likely that within a few years a number of human genes will have been assigned to specific chromosomes. Whether more refined genetic analysis and mapping will be possible with somatic hybrids depends on whether genetic events comparable to those that produce recombination within the chromosomes of germ cells occur also in mammalian somatic cells.

Crosses between differentiated and undifferentiated cells, or between differently differentiated cells, can provide information on the nature of the regulatory processes involved in cell specialization. The activities of somatic cells can be divided into two general categories. There are "essential" functions that are indispensable for the cells' own maintenance and growth and there are specialized "luxury" functions, such as the formation of muscle fiber, the secretion of hormones and the production of pigment, that are necessary for the survival of the organism but not for the survival of isolated cells. The essential functions are expressed in all growing somatic cells. Each luxury function, on the other hand, is expressed in a different line of specialized cells. In 1961

François Jacob and Jacques Monod of the Pasteur Institute in Paris discovered mechanisms that regulate the activity of genes in bacteria. It is generally believed that similar mechanisms are responsible for specialization of somatic cells—that sets of genes governing the various luxury functions are activated or repressed as required in the cells of the different tissues. Since bacterial and mammalian cells have very different properties and requirements, however, it would not be surprising if somatic cells have evolved some unique mechanisms for the regulation of gene activity.

Working with Davidson and Kohtaro Yamamoto at Western Reserve, one of us (Ephrussi) crossed two cells, one of which synthesizes a certain luxury product. This differentiated parent is a line of hamster tumor cells that produce melanin, a dark pigment. When cells of this melanoma line are crossed with cells of a number of different mouse lines that have never produced melanin, the hybrids have a rather stable karyotype and contain most of the chromosomes of both parents. Yet among the many hybrid colonies obtained, none synthesizes melanin or contains the enzyme dopa oxidase, which is required for the synthesis of the pigment and which is present in large quantities in the melanoma cells. Obviously, when the melanoma cells fuse with the unpigmented ones, the synthesis of the enzyme is halted by some regulatory substance produced by the normal cells. Whether this block occurs at the level of the gene specifying dopa oxidase (which was active in the melanoma cell) or at some later step in the process leading to dopa oxidase production is something that remains to be determined.

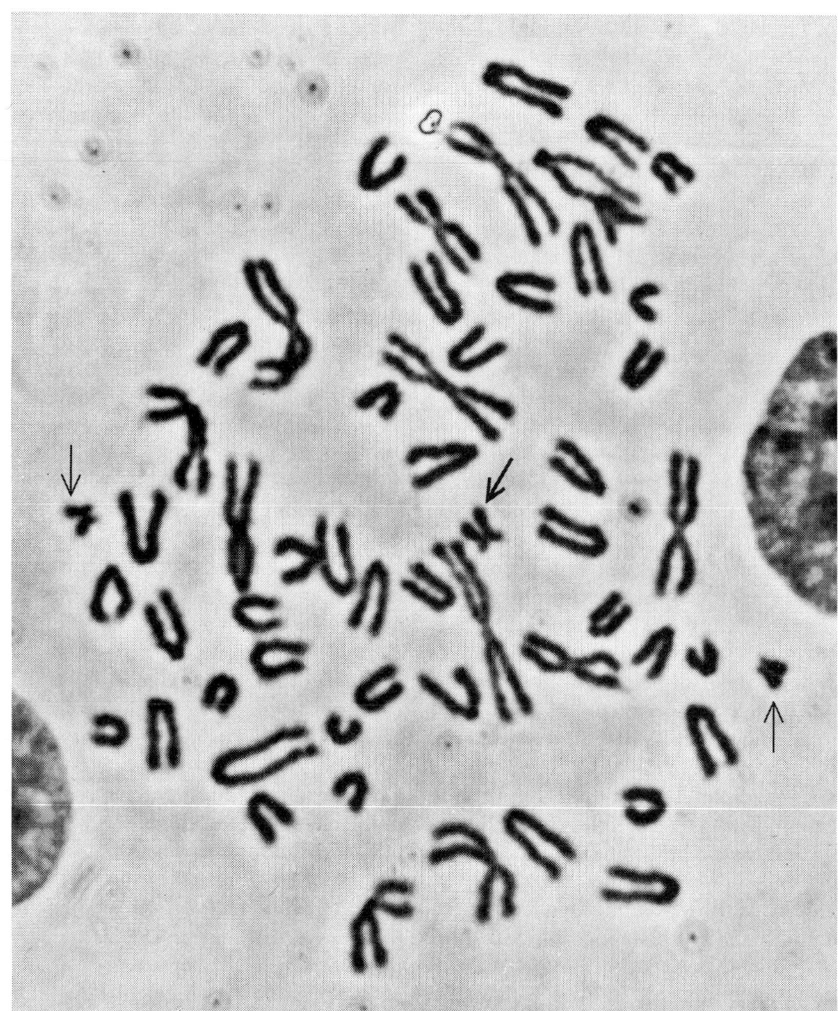

**ENZYME AND CHROMOSOME can be correlated in the clone from which this metaphase was taken. All but three human chromosomes have been lost: two of Group *G* (*thin arrows*) and one of Group *E* (*heavy arrow*) remain. The Group *E* chromosome is always retained in clones grown, as this one was, in a medium in which cells survive only if they produce the enzyme TK. Therefore that chromosome must carry the gene for TK.**

Some light may be shed on the mechanism of regulation of gene activity in mammalian cells by examination of highly segregated hybrids between melanoma cells and cells that do not produce the pigment. If the genes responsible for producing melanin are active in the melanoma parent because of some stable change at the chromosome level, and if they are only temporarily repressed in the hybrid cells, then the loss of the repressing chromosomal material should bring a resumption of melanin synthesis. If synthesis does not resume, one will have to conclude that the continuous production of melanin in the melanoma parent was not due to a stable cellular change in the first place.

The melanoma cell is a differentiated cell that synthesizes both a specialized product and all the enzymes necessary for continuous and rapid growth. Some other highly differentiated cells no longer grow at all; the essential syntheses that occur in growing cells are arrested. Does this reflect a total and definitive inactivation of the nucleus associated with extreme differentiation? Harris and his co-workers have shown that it does not. They produced heterokaryons between nucleated red blood cells taken directly from the blood of a hen and actively growing, undifferentiated human cells. The human cells are characterized by rapid synthesis of deoxyribonucleic acid (DNA) and ribonucleic acid (RNA); the hen blood cells synthesize neither and they have small, highly condensed nuclei characteristic of quiescent cells. In the heterokaryons the red-cell nuclei undergo dramatic changes: they swell up, their chromosomal material becomes less condensed and they resume the synthesis of both DNA and RNA. Clearly the red-cell nucleus is reactivated; it must not have been irreversibly inactivated to begin with.

It will be most interesting to determine precisely which of the numerous possible functions are resumed in these reactivated nuclei. Can the nuclei of highly differentiated cells such as the red blood cells be induced to perform not only some essential functions but also specialized functions characteristic of other differentiated cell types? It should be possible to answer this question by determining the nature of the products synthesized by red-cell nuclei that are reactivated by fusion with other differentiated cells. A positive answer would open a new line of biochemical investigation of the factors that control cellular differentiation in embryonic development.

# *Biographical Notes and Bibliographies*

## *Part* I *The Elements of Inheritance*

### 1. The Gene

#### *The Author*

NORMAN H. HOROWITZ is professor of biology at the California Institute of Technology, where he took his Ph.D. in 1939. He has long been associated with George W. Beadle, first at Stanford and then at Cal Tech, in research on the biochemical genetics of *Neurospora*. In 1955 he worked at the *Institut de Biologie* of the University of Paris as a Fulbright and Guggenheim fellow.

#### *Bibliography*


LIFE OF MENDEL. Hugo Iltis. W. W. Norton & Company, Inc., 1932.

ON THE EVOLUTION OF BIOCHEMICAL SYNTHESES. N. H. Horowitz in *Proceedings of the National Academy of Sciences,* Vol. 31, pages 153–157; June, 1945.

WHAT IS LIFE? Erwin Schrödinger. Cambridge University Press, 1948.


### 2. The Duplication of Chromosomes

#### *The Author*

J. HERBERT TAYLOR is professor of biology at Florida State University. He has been using isotopes in biological research ever since high-resolution autoradiography was first developed in 1950. At that time he was teaching at the University of Tennessee and working as a consultant to the Oak Ridge National Laboratory. Taylor comes from Corsicana, Texas, was graduated from Southeastern State College in Oklahoma and acquired a Ph.D. in biology from the University of Virginia.

#### *Bibliography*


THE BIOCHEMISTRY OF THE NUCLEIC ACIDS. J. N. Davidson. John Wiley & Sons, Inc., 1953.

THE NUCLEIC ACIDS. Edited by Erwin Chargaff and J. N. Davidson. Academic Press, Inc., 1955.

NUCLEIC ACIDS. F. H. C. Crick in *Scientific American,* Vol. 197, No. 3, pages 188–200; September, 1957.

THE ORGANIZATION AND DUPLICATION OF CHROMOSOMES AS REVEALED BY AUTORADIOGRAPHIC STUDIES USING TRITIUM-LABELED THYMIDINE. J. Herbert Taylor, Philip S. Woods and Walter L. Hughes in *Proceedings of the National Academy of Sciences,* Vol. 43, No. 1, pages 122–128; January, 1957.

PHYSICAL TECHNIQUES IN BIOCHEMICAL RESEARCH. VOL. III: CELLS AND TISSUES. Edited by Gerald Oster and Arthur W. Pollister. Academic Press, Inc., 1956.

A SYMPOSIUM ON THE CHEMICAL BASIS OF HEREDITY. William D. McElroy and Bentley Glass. Johns Hopkins Press, 1957.


### 3. The Bacterial Chromosome

#### *The Author*

JOHN CAIRNS has recently served as director of the Cold Spring Harbor Laboratory of Quantitative Biology. He was born in England and obtained a medical degree at the University of Oxford. For several years he did research in Australia on the multiplication of influenza virus and vaccinia virus. Later he worked on the visualization of DNA molecules by autoradiography, a project he describes in part in the present article.

#### *Bibliography*


THE BACTERIAL CHROMOSOME AND ITS MANNER OF REPLICATION AS SEEN BY AUTORADIOGRAPHY. John Cairns in *Journal of Molecular Biology,* Vol. 6, No. 3, pages 208–213; March, 1963.

COLD SPRING HARBOR SYMPOSIA ON QUANTITATIVE BIOLOGY, VOLUME XXVIII: SYNTHESES AND STRUCTURE OF MACROMOLECULES. Cold Spring Harbor Laboratory of Quantitative Biology, 1963.

MOLECULAR BIOLOGY OF THE GENE. James D. Watson. W. A. Benjamin, Inc., 1965. See pages 255–296.

## 4. Transformed Bacteria

### *The Authors*

ROLLIN D. HOTCHKISS AND ESTHER WEISS together performed the first transformations producing drug-resistant bacteria at the Rockefeller Institute for Medical Research in New York, now Rockefeller University, at which Hotchkiss is a professor. Hotchkiss graduated from Yale University's Sheffield Scientific School and received a Ph.D. from Yale in organic chemistry. After a year as an instructor at Yale he went to the Rockefeller Institute in 1935. He has worked there since, except for a year of study in Copenhagen and a wartime tour of scientific duty with the Navy. At the Institute he was associated with René J. Dubos in determining the composition of the antibiotics tyrocidine and gramicidin. This led him to study the metabolism of bacteria as the object of attack in chemotherapy, which in turn led to his work in genetic transformation. Miss Weiss holds a B.A. in biology from Smith College. She worked with Hotchkiss at the Institute from 1950 to 1952, and later was with the Olin Mathieson laboratories in New Haven and Children's Hospital in Boston. She was formerly a member of the Board of Editors of SCIENTIFIC AMERICAN.

### *Bibliography*

THE GENETIC CHEMISTRY OF THE PNEUMOCOCCAL TRANSFORMATIONS. Rollin D. Hotchkiss in *The Harvey Lectures (1953–54)*, Series 49, pages 124–144. Academic Press Inc., 1955.

## 5. "Transduction" in Bacteria

### *The Author*

NORTON D. ZINDER was born in New York City and there pursued an accelerated course of studies which carried him through the Bronx High School of Science and Columbia University by the time he was 18. Proceeding to the University of Wisconsin, he acquired a Ph.D. in medical microbiology in 1952. By that time Zinder—then 23 years old—had not only shared in the discovery of "transduction" but had married and become a father. In 1952 Zinder returned to New York to join the staff of Rockefeller University, where he is a professor.

### *Bibliography*

BACTERIAL TRANSDUCTION. Norton D. Zinder in *Journal of Cellular and Comparative Physiology*. Vol. 45, Supplement 2, pages 23–49; May, 1955.

GENETIC EXCHANGE IN SALMONELLA. Norton D. Zinder and Joshua Lederberg in *Journal of Bacteriology*, Vol. 64, No. 5, pages 679–699; November, 1952.

GENETIC STUDIES WITH BACTERIA. M. Demerec, Zlata Hartman, Philip E. Hartman, Takashi Yura, Joseph S. Gots, Haruo Ozeki and S. W. Glover. Publication 612, Carnegie Institution of Washington, 1956.

TRANSDUCTION IN ESCHERICHIA COLI K-12. M. L. Morse, Esther M. Lederberg and Joshua Lederberg in *Genetics*, Vol. 41, No. 1, pages 142–156; January, 1956.

TRANSDUCTION OF FLAGELLAR CHARACTERS IN SALMONELLA. B. A. D. Stocker, N. D. Zinder and J. Lederberg in *The Journal of General Microbiology*, Vol. 9, No. 3, pages 410–433; December, 1953.

TRANSDUCTION OF LINKED GENETIC CHARACTERS OF THE HOST BY BACTERIOPHAGE P1. E. S. Lennox in *Virology*, Vol. 1, Nos. 1–5, pages 190–206; 1955.

## 6. Viruses and Genes

### *The Authors*

FRANÇOIS JACOB and ELIE L. WOLLMAN both work at the Pasteur Institute in Paris. Jacob's medical studies at the University of Paris were interrupted by World War II. In 1940 he escaped to England to join the Free French forces there and he later fought in both Africa and France. Jacob completed his M.D. degree after the war and went to the Pasteur Institute in 1950. In 1954 he received a D.Sc. degree from the Sorbonne. He was a corecipient of the Nobel Prize in 1965. Wollman, whose parents were microbiologists at the Pasteur Institute, studied medicine and biology at the University of Paris until war intervened. During the German occupation of France, in which both of his parents were killed, Wollman finished his medical degree at the University of Lyon and served as a physician with the resistance forces and later with the French army. He became a staff member of the Pasteur Institute in 1945.

### *Bibliography*

THE CONCEPT OF VIRUS. A. Lwoff in *The Journal of General Microbiology*, Vol. 17, No. 2, pages 239–253; October, 1957.

GENETIC CONTROL OF VIRAL FUNCTIONS. François Jacob in *The Harvey Lectures*, Series LIV, 1958–1959, pages 1–39. Academic Press Inc., 1960.

MICROBIAL GENETICS. Tenth Symposium of the Society for General Microbiology. Cambridge University Press, 1960.

PHYSIOLOGICAL ASPECTS OF BACTERIOPHAGE GENETICS. S. Brenner in *Advances in Virus Research*, Vol. 6, pages 137–158; 1959.

A SYMPOSIUM ON THE CHEMICAL BASIS OF HEREDITY. Edited by William D. McElroy. Johns Hopkins Press, 1957.

VIRUSES AS INFECTIVE GENETIC MATERIAL. S. E. Luria in *Immunity and Virus Infection*, edited by Victor A. Najjar, pages 188–195. John Wiley & Sons, Inc., 1959.

THE VIRUSES: BIOCHEMICAL, BIOLOGICAL, AND BIOPHYSICAL PROPERTIES. Edited by F. M. Burnet and W. M. Stanley. Academic Press Inc., 1956.

## 7. Infectious Drug Resistance

### *The Author*

TSUTOMU WATANABE is associate professor of microbiology at the Keio University School of Medicine in Tokyo, where he received his M.D. degree in 1948. He spent a year as an exchange student in radiobiology and bacteriology at the University of Utah in 1951 and 1952 and also worked as a research associate in the department of zoology at Columbia University in 1957 and 1958. He writes that his special field of study is "microbial genetics, particularly the genetics of bacterial drug resistance. Through the studies on infectious drug resistance I have become interested in various episomes and plasmids (parasitic or symbiotic agents) of bacteria and also in the evolution of microorganisms."

### *Bibliography*

EVOLUTIONARY RELATIONSHIPS OF R FACTORS WITH OTHER EPISOMES AND PLASMIDS. T. Watanabe in *Federation Proceedings,* Vol. 26, No. 1, pages 23–28; January–February, 1967.

INFECTIVE HEREDITARY OF MULTIPLE DRUG RESISTANCE IN BACTERIA. Tsutomu Watanabe in *Bacteriological Reviews,* Vol. 27, No. 1, pages 87–115; March, 1963.

## 8. Genes Outside the Chromosomes

### *The Author*

RUTH SAGER is professor of biology at Hunter College, New York. She was born in Chicago and was graduated from the University of Chicago in 1938, obtaining an M.S. from Rutgers University in 1944 and a Ph.D. from Columbia University in 1948. Thereafter she worked for several years at the Rockefeller Institute before returning to Columbia in 1955. She is coauthor with Francis J. Ryan of *Cell Heredity.* In describing how she became interested in nonchromosomal inheritance, she writes: "I have always been intrigued by the physicist's approach to scientific inquiry, particularly in the fact that the way to find out something really new is to question a basic tenet of existing theory. In this case, the theory that chromosomal genes constitute the total genetic apparatus of cells was already under fire. The serious possibility of a second genetic system, however, was rarely discussed. I found this situation challenging."

### *Bibliography*

CELL HEREDITY. Ruth Sager and Francis J. Ryan. John Wiley & Sons, Inc., 1961.

NUCLEO-CYTOPLASMIC RELATIONS IN MICROORGANISMS. Boris Ephrussi. Oxford University Press, 1953.

ON NON-CHROMOSOMAL HEREDITY IN MICRO-ORGANISMS. Ruth Sager in *Function and Structure of Micro-Organisms,* edited by M. R. Pollock and M. H. Richmond. Cambridge University Press, in press.

THE PARTICULATE NATURE OF NON-CHROMOSOMAL GENES IN CHLAMYDOMONAS. Ruth Sager and Zenta Ramanis in *Proceedings of the National Academy of Sciences,* Vol. 50, No. 2, pages 260–268; August, 1963.

# *Part* II *The Nature of the Gene*

## 9. The Genes of Men and Molds

### *The Author*

GEORGE BEADLE served for many years as head of the California Institute of Technology Division of Biology. He is a geneticist, whose most notable work has been done with the bread mold *Neurospora.* In 1958, while on a visiting professorship at the University of Oxford, Beadle was summoned to Stockholm to receive a Nobel prize jointly with his collaborator Edward L. Tatum for the work on Neurospora described in this article. In 1961 Beadle left Cal Tech to become president of the University of Chicago. He is presently Director of the AMA Institute for Biomedical Research.

### *Bibliography*

GENES AND THE CHEMISTRY OF ORGANISM. G. W. Beadle in *American Scientist,* Vol. 34, pages 31–53; 1946.

THE PRINCIPLES OF HEREDITY, Third Edition. Laurence H. Snyder. D. C. Heath, 1946.

THE PHYSIOLOGY OF THE GENE. S. Wright in *Physiological Reviews,* Vol. 21, pages 487–527; 1941.

## 10. The Fine Structure of the Gene

### *The Author*

SEYMOUR BENZER is professor of biology at the California Institute of Technology. After taking his B.A. at Brooklyn College in 1942, Benzer went to Purdue, where he acquired an M.S. the following year and a Ph.D. in physics in 1947. He then spent a year at the Oak Ridge National Laboratory, two years at the California Institute of Technology and a year at the Pasteur Institute in Paris.

### *Bibliography*

THE ELEMENTARY UNITS OF HEREDITY. Seymour Benzer in *The Chemical Basis of Heredity,* edited by William D. McElroy and Bentley Glass, pages 70–93. The Johns Hopkins Press, 1957.

GENETIC RECOMBINATION BETWEEN HOST-RANGE AND PLAQUE-TYPE MUTANTS OF BACTERIOPHAGE IN SINGLE BACTERIAL CELLS. A. D. Hershey and Raquel Rotman in *Genetics,* Vol. 34, No. 1, pages 44–71; January, 1949.

INDUCTION OF SPECIFIC MUTATIONS WITH 5-BROMOURACIL. Seymour Benzer and Ernst Freese in *Proceedings of the National Academy of Sciences,* Vol. 44, No. 2, pages 112–119; February, 1958.

THE STRUCTURE OF THE HEREDITARY MATERIAL. F. H. C. Crick in *Scientific American,* Vol. 191, No. 4, pages 54–61; October, 1954.

ON THE TOPOGRAPHY OF THE GENETIC FINE STRUCTURE. Seymour Benzer in *Proceedings of the National Academy of Sciences,* Vol. 47, No. 3, pages 403–415; March, 1961.

## 11. The Genetic Code

### *The Author*

F. H. C. CRICK is, with James D. Watson and Maurice H. F. Wilkins, winner of the 1962 Nobel Prize in physiology and medicine for the discovery of the molecular structure of the genetic material deoxyribonucleic acid (DNA). Originally a physicist, Crick made significant contributions to the development of radar; after World War II he turned to basic research on the structure of viruses, collagen and nucleic acids. Working at Cambridge in the early 1950's, Crick and Watson conceived and built the now-classic double spiral model of the DNA molecule, confirming the results of earlier X-ray diffraction studies made by Wilkins at Kings College, London. Crick is carrying on his researches into the nature of the genetic code under the auspices of the Medical Research Council Laboratory of Molecular Biology at the University Postgraduate Medical School in Cambridge.

### *Bibliography*

THE FINE STRUCTURE OF THE GENE. Seymour Benzer in *Scientific American,* Vol. 206, No. 1, pages 70–84; January, 1962.

GENERAL NATURE OF THE GENETIC CODE FOR PROTEINS. F. H. C. Crick, Leslie Barnett, S. Brenner and R. J. Watts-Tobin in *Nature,* Vol. 192, No. 4809, pages 1227–1232; December 30, 1961.

MESSENGER RNA. Jerard Hurwitz and J. J. Furth in *Scientific American,* Vol. 206, No. 2, pages 41–49; February, 1962.

THE NUCLEIC ACIDS: Vol. III. Edited by Erwin Chargaff and J. N. Davidson. Academic Press Inc., 1960.

## 12. The Genetic Code: III

### *The Author*

F. H. C. CRICK is a molecular biologist who works for the British Medical Research Council's Laboratory of Molecular Biology at the University Postgraduate Medical School in Cambridge. For additional information about him, see biographical note 11.

### *Bibliography*

THE GENETIC CODE, VOL. XXXI: 1966 COLD SPRING HARBOR SYMPOSIA ON QUANTITATIVE BIOLOGY. Cold Spring Harbor Laboratory of Quantitative Biology, in press.

MOLECULAR BIOLOGY OF THE GENE. James D. Watson. W. A. Benjamin, Inc., 1965.

RNA CODEWORDS AND PROTEIN SYNTHESIS, VII: ON THE GENERAL NATURE OF THE RNA CODE. M. Nirenberg, P. Leder, M. Bernfield, R. Brimacombe, J. Trupin, F. Rottman and C. O'Neal in *Proceedings of the National Academy of Sciences,* Vol. 53, No. 5, pages 1161–1168; May, 1965.

STUDIES ON POLYNUCLEOTIDES, LVI: FURTHER SYNTHESIS, IN VITRO, OF COPOLYPEPTIDES CONTAINING TWO AMINO ACIDS IN ALTERNATING SEQUENCE DEPENDENT UPON DNA-LIKE POLYMERS CONTAINING TWO NUCLEOTIDES IN ALTERNATING SEQUENCE. D. S. Jones, S. Nishimura and H. G. Khorana in *Journal of Molecular Biology,* Vol. 16, No. 2, pages 454–472; April, 1966.

## 13. Gene Structure and Protein Structure

### *The Author*

CHARLES YANOFSKY is professor of biology at Stanford University. After his graduation from the City College of the City of New York in 1948 he did graduate work in the department of microbiology at Yale University, receiving a Ph. D. there is 1951. He remained at Yale until 1954, when he joined the faculty of the Western Reserve University School of Medicine. Four years later he went to the department of biological sciences at Stanford. Yanofsky has received several awards for his work in molecular biology. He is a fellow of the American Academy of Arts and Sciences and a member of the National Academy of Sciences.

### *Bibliography*

CO-LINEARITY OF $\beta$-GALACTOSIDASE WITH ITS GENE BY IMMUNOLOGICAL DETECTION OF INCOMPLETE POLYPEPTIDE CHAINS. Audree V. Fowler and Irving Zabin in *Science,* Vol. 154, No. 3752, pages 1027–1029; November 25, 1966.

CO-LINEARITY OF THE GENE WITH THE POLYPEPTIDE CHAIN. A. S. Sarabhai, A. O. W. Stretton, S. Brenner and A. Bolle in *Nature,* Vol. 201, No. 4914, pages 13–17; January 4, 1964.

THE COMPLETE AMINO ACID SEQUENCE OF THE TRYPTOPHAN SYNTHETASE A PROTEIN ($\alpha$ SUBUNIT) AND ITS CO-LINEAR RELATIONSHIP WITH THE GENETIC MAP OF THE A GENE. Charles Yanofsky, Gabriel R. Drapeau, John R. Guest and Bruce C. Carlton in *Proceedings of the National Academy of Sciences,* Vol. 57, No. 2, pages 296–298; February, 1967.

MUTATIONALLY INDUCED AMINO ACID SUBSTITUTIONS IN A TRYPTIC PEPTIDE OF THE TRYPTOPHAN SYNTHETASE A PROTEIN. John R. Guest and Charles Yanofsky in *Journal of Biological Chemistry,* Vol. 240, No. 2, pages 679–689; February, 1965.

ON THE COLINEARITY OF GENE STRUCTURE AND PROTEIN STRUCTURE. C. Yanofsky, B. C. Carlton, J. R. Guest, D. R. Helinski and U. Henning in *Proceedings of the National Academy of Sciences,* Vol. 51, No. 2, pages 266–272; February, 1964.

## 14. The Genetics of a Bacterial Virus

### *The Authors*

R. S. EDGAR and R. H. EPSTEIN are respectively professor of biology at the California Institute of Technology and professor at the Institute of Molecular Biology of the University of Geneva. They were graduate students together at the University of Rochester, where Epstein, who was born in Rochester, had done his undergraduate work; Edgar, a native of Canada, went to Rochester after undergraduate work at McGill University. Both obtained a Ph.D. at Rochester and then, in the late 1950's, went to the California Institute of Technology to study under Max Delbrück. Of the work reported in their article Edgar writes: "Epstein started the 'embers' here and then went to Geneva to continue that work while I developed the temperature-sensitive system. The two systems were developed independently, with communication going on through an intermediary, an associate of Epstein's, since he is a notoriously bad letter writer (he doesn't)."

### *Bibliography*

BACTERIOPHAGE REPRODUCTION. Sewell P. Champe in *Annual Review of Microbiology, Vol. 17, 1963.*

MOLECULAR BIOLOGY OF BACTERIAL VIRUSES. Gunther S. Stent. W. H. Freeman and Company, 1963.

PHYSIOLOGICAL STUDIES OF CONDITIONAL LETHAL MUTANTS OF BACTERIOPHAGE T4D. R. H. Epstein, A. Bolle, C. M. Steinberg, E. Kellenberger, E. Boy de la Tour, R. Chevalley, R. S. Edgar, M. Susman, G. H. Denhardt and A. Lielausis in *Cold Spring Harbor Symposia on Quantitative Biology, Vol. XXVIII.* 1963.

# *Part* III *From Gene to Organism*

## 15. The Control of Biochemical Reactions

### *The Author*

JEAN-PIERRE CHANGEUX, when he wrote this article in 1965, was *maître-assistant* in biochemistry at the University of Paris. It was in 1959, at the Pasteur Institute, that he began an investigation of the mechanism by which the activity of enzymes is regulated. That work led him into other investigations of cellular regulatory processes in an effort to elucidate the mechanisms by which a metabolite, or regulatory signal, controls a chemical reaction at the molecular level. After an interval at the University of California, he returned to the Pasteur Institute, where he is now *maître de recherche.*

### *Bibliography*

ALLOSTERIC PROTEINS AND CELLULAR CONTROL SYSTEMS. Jacques Monod, Jean-Pierre Changeux and François Jacob in *Journal of Molecular Biology,* Vol. 6, No. 4, pages 306–329; April, 1963.

GENETIC REGULATORY MECHANISMS IN THE SYNTHESIS OF PROTEINS. François Jacob and Jacques Monod in *Journal of Molecular Biology,* Vol. 3, No. 3, pages 318–356; June, 1961.

ON THE REGULATION OF DNA REPLICATION IN BACTERIA. François Jacob, Sydney Brenner and François Cuzin in *Cold Spring Harbor Symposia on Quantitative Biology, Vol. XXVIII.* 1963.

A PLAUSIBLE MODEL OF ALLOSTERIC TRANSITION. Jacques Monod, Jeffries Wyman and Jean-Pierre Changeux in *Journal of Molecular Biology,* Vol. 12, No. 1, pages 88–118; May, 1965.

## 16. Chromosome Puffs

### *The Authors*

WOLFGANG BEERMANN and ULRICH CLEVER work at the Max Planck Institute for Biology in Tübingen. They are also members of the faculty of the University of Tübingen. Beermann is a director of one of the Max Planck Institutes in Tübingen and professor of zoology at the university. A native of Hanover, he received his doctorate from the University of Göttingen in 1952. He did research at the Max Planck Institute for Marine Biology in Wilhelmshaven from 1952 to 1954, when he was appointed assistant professor at the Zoological Institute of the University of Marburg. He took up his present post in 1958. Clever is a research associate at the Max Planck Institute in Tübingen and lecturer in zoology and genetics at the university. He received his doctorate from Göttingen in 1957 and did research for a year at the Federal Research Institute for Viticulture before going to Tübingen in 1958.

### *Bibliography*

CHROMOSOMES AND CYTODIFFERENTIATION. Joseph G. Gall in *Cytodifferentiation and Macromolecular Synthesis,* edited by Michael Locke. Academic Press, 1963.

NUCLEIC ACIDS AND CELL MORPHOLOGY IN DIPTERAN SALIVARY GLANDS. Hewson Swift in *The Molecular Control of Cellular Activity,* edited by John M. Allen. McGraw-Hill Book Company, 1962.

RIESENCHROMOSOMEN. Wolfgang Beermann in *Protoplasmatologia,* Vol. VI/D. Springer-Verlag, 1962.

UNTERSUCHUNGEN AN RIESENCHROMOSOMEN ÜBER DIE WIRKUNGSWEISE DER GENE. Ulrich Clever in *Materia Medica Nordmark,* Vol. 15, No. 10, pages 438–452; July, 1962.

## 17. Hormones and Genes

### *The Author*

ERIC H. DAVIDSON is assistant professor at Rockefeller University, working in cell biology. As a high school student in Nyack, N.Y., he worked summers at the Marine Biological Laboratory in Woods Hole, Mass., and was one of the national winners of the Westinghouse Science Talent Search. Davidson was graduated from the University of Pennsylvania in 1958, having majored in zoology. For the next five years he was a graduate fellow at the Rockefeller Institute, obtaining a doctor's degree there in 1963. He collaborated with Alfred E. Mirsky in the study of gene action in the initiation and control of embryological development.

### *Bibliography*

EFFECT OF ACTINOMYCIN AND INSULIN ON THE METABOLISM OF ISOLATED RAT DIAPHRAGM. Ira G. Wool and Arthur N. Moyer in *Biochimica et Biophysica Acta,* Vol. 91, No. 2, pages 248–256; October 16, 1964.

ON THE MECHANISM OF ACTION OF ALDOSTERONE ON SODIUM TRANSPORT: THE ROLE OF RNA SYNTHESIS. George A. Porter, Rita Bogoroch and Isidore S. Edelman in *Proceedings of the National Academy of Sciences,* Vol. 52, No. 6, pages 1326–1333; December, 1964.

PREVENTION OF HORMONE ACTION BY LOCAL APPLICATION OF ACTINOMYCIN D. G. P. Talwar and Sheldon J. Segal in *Proceedings of the National Academy of Sciences,* Vol. 50, No. 1, pages 226–230; July 15, 1963.

SELECTIVE ALTERATIONS OF MAMMALIAN MESSENGER-RNA SYNTHESIS: EVIDENCE FOR DIFFERENTIAL ACTION OF HORMONES ON GENE TRANSCRIPTION. Chev Kidson and K. S. Kirby in *Nature,* Vol. 203, No. 4945, pages 599–603; August 8, 1964.

TRANSFER RIBONUCLEIC ACIDS. E. N. Carlsen, G. J. Trelle and O. A. Schjeide in *Nature,* Vol. 202, No. 4936, pages 984–986; June 6, 1964.

## 18. Building a Bacterial Virus

### *The Authors*

WILLIAM B. WOOD and R. S. EDGAR are in the division of biology of the California Institute of Technology; Wood is associate professor and Edgar is professor. Wood, who did his undergraduate work at Harvard College, received a doctorate in biochemistry from Stanford University in 1963 and spent a year and a half as a postdoctoral fellow in Switzerland before joining the Cal Tech faculty. Edgar, a graduate of McGill University, obtained his Ph.D. from the University of Rochester. Wood writes that they began discussing the experiments described in their article in 1963 and started work in 1965. "I suspect," he adds, "that either of us alone might never have initiated these experiments."

### *Bibliography*

CONDITIONAL MUTATIONS IN BACTERIOPHAGE T4. R. S. Edgar and R. H. Epstein in *Genetics Today,* edited by S. J. Geerts. Pergamon Press, 1963.

GENE ACTION IN THE CONTROL OF BACTERIOPHAGE T4 MORPHOGENESIS. W. B. Wood in *Genetics and Developmental Biology,* edited by H. J. Teas, University of Kentucky Press, 1969.

SOME STEPS IN THE MORPHOGENESIS OF BACTERIOPHAGE T4. R. S. Edgar and I. Lielausis in *Journal of Molecular Biology,* Vol. 32, No. 2, pages 263–276; March, 1968.

## 19. Transplanted Nuclei and Cell Differentiation

### *The Author*

J. B. GURDON is a lecturer at the University of Oxford. He was educated at Eton, where he studied Greek and Latin, and at Christ Church College, Oxford, where he took up biology. Before his graduation in 1960 he was mainly interested in entomology, and his first published paper dealt with the discovery of a species of sawfly new to the British Isles. During graduate work at Oxford he became interested in embryology. Gurdon writes that aside from his work his "main interests include ski-mountaineering and horticulture."

### *Bibliography*

NUCLEOCYTOPLASMIC INTERACTIONS IN EGGS AND EMBRYOS. Robert Briggs and Thomas J. King in *The Cell: Biochemistry, Physiology, Morphology, Vol I,* edited by Jean Brachet and Alfred E. Mirsky. Academic Press, 1959.

NUCLEAR TRANSPLANTATION IN AMPHIBIA AND THE IMPORTANCE OF STABLE NUCLEAR CHANGES IN PROMOTING CELLULAR DIFFERENTIATION. J. B. Gurdon in *The Quarterly Review of Biology,* Vol. 38, No. 1, pages 54–78; March, 1963.

INTERACTING SYSTEMS IN DEVELOPMENT. James D. Ebert. Holt, Rinehart and Winston, 1965.

NUCLEAR TRANSPLANTATION IN AMPHIBIA. Thomas J. King in *Methods in Cell Physiology, Vol. II,* edited by David M. Prescott. Academic Press, 1966.

THE CYTOPLASMIC CONTROL OF NUCLEAR ACTIVITY IN ANIMAL DEVELOPMENT. J. B. Gurdon and H. R. Woodland in *Biological Reviews of the Cambridge Philosophical Society,* Vol. 43, No. 2, pages 233–267; May, 1968.

## 20. Transdetermination in Cells

### *The Author*

ERNST HADORN is professor of zoology and comparative anatomy at the University of Zurich, with which he has been affiliated since 1939. From 1962 to 1964 he acted as rector of the university. Hadorn received his Ph.D.

from the University of Bern in 1931 and lectured there for several years. He then spent a year as a Rockefeller fellow at Harvard University and the University of Rochester before taking up his work at Zurich. His research deals with problems of experimental embryology and physiological and biochemical genetics. Among his writings are a book, *Developmental Genetics and Lethal Factors,* and the article "Fractionating the Fruit Fly" in the April 1962 issue of SCIENTIFIC AMERICAN.

*Bibliography*


DEVELOPMENTAL GENETICS AND LETHAL FACTORS. Ernst Hadorn. Methuen & Co., Ltd., 1961.

PROBLEMS OF DETERMINATION AND TRANSDETERMINATION. Ernst Hadorn in *Brookhaven Symposia in Biology, No. 18: Genetic Control of Differentiation,* pages 148–161; December, 1965.

DYNAMICS OF DETERMINATION. Ernst Hadorn in *Major Problems in Developmental Biology: The Twenty-fifth Symposium,* edited by Michael Locke. Academic Press, 1966.

KONSTANZ, WECHSEL UND TYPUS DER DETERMINATION UND DIFFERENZIERUNG IN ZELLEN AUS MÄNNLICHEN GENITALANLAGEN VON DROSOPHILA MELANOGASTER NACH DAUERKULTUR IN VIVO. Ernst Hadorn in *Developmental Biology,* Vol. 13, No. 3, pages 424–509; June, 1966.


## 21. Phases in Cell Differentiation

*The Authors*

NORMAN K. WESSELLS and WILLIAM J. RUTTER are respectively associate professor of biological sciences at Stanford University and professor of biochemistry and genetics at the University of Washington. Wessells did both his undergraduate and his graduate work at Yale University and began teaching at Stanford in 1962. Rutter was graduated from Harvard College in 1949, went to the University of Utah for his master's degree and received his Ph.D. from the University of Illinois, where he taught from 1955 to 1965. In 1969 he became chairman of the department of biochemistry and biophysics at the San Francisco Medical Center of the University of California. The authors wish to thank a number of associates in the work they describe: Thomas G. Sanders, William R. Clark, William S. Bradshaw, John D. Kemp, William D. Ball, Leslie C. Brock, Julia H. Cohen and Jean Evans.

*Bibliography*


EARLY PANCREAS ORGANOGENESIS: MORPHOGENESIS, TISSUE INTERACTIONS, AND MASS EFFECTS. Norman K. Wessells and Julia H. Cohen in *Developmental Biology,* Vol. 15, No. 3, pages 237–270; March, 1967.

MULTIPHASIC REGULATION IN CYTODIFFERENTIATION. William J. Rutter, William R. Clark, John D. Kemp, William S. Bradshaw, Thomas G. Sanders and William D. Ball in *Epithelial-Mesenchymal Interactions,* edited by Raul Fleischmajer and Rupert E. Billingham. The Williams & Wilkins Company, 1968.

ULTRASTRUCTURAL STUDIES OF EARLY MORPHOGENESIS AND CYTODIFFERENTIATION IN THE EMBRYONIC MAMMALIAN PANCREAS. Norman K. Wessells and Jean Evans in *Developmental Biology,* Vol. 17, No. 4, pages 413–446; April, 1968.

REGULATION OF SPECIFIC PROTEIN SYNTHESIS IN CYTODIFFERENTIATION. W. J. Rutter, J. D. Kemp, W. S. Bradshaw, W. R. Clark, R. A. Ronzio and T. G. Sanders in *Journal of Cellular Physiology,* Vol. 72, No. 2, Part II, pages 1–18; October, 1968.


# *Part* IV *Genetics and Evolution*

## 22. The Genetic Basis of Evolution

*The Author*

THEODOSIUS DOBZHANSKY, professor at the Rockefeller University, has spent much of his career unraveling the genetics of fruit flies. Born and educated in Russia, he "got excited" about biology at the age of ten, and dates his decision to specialize in genetics and evolution from his reading of *Origin of Species* when he was fifteen. After teaching at the University of Leningrad for a number of years, he came to the United States as a Rockefeller Foundation Research Fellow to study with Thomas Hunt Morgan. At the end of his fellowship Morgan invited him to stay permanently. Biology in America has been enriched in consequence, not only by Dobzhansky's own significant researches but by the many fruitful careers his teaching has inspired. Before accepting his present position, he was for many years professor of zoology at Columbia University.

*Bibliography*


GENETICS AND THE ORIGIN OF SPECIES. Th. Dobzhansky. Columbia University Press, 1937.


## 23. Darwin's Missing Evidence

*The Author*

H. B. D. KETTLEWELL visited Brazil last year to retrace the routes traveled by Charles Darwin in 1832. He has recently participated in the production of two Darwin centenary films, "Evolution in Progress" and "Darwin and the Insect Adaptations of Brazil." Educated at the

Charterhouse School and the University of Cambridge (Gonville and Caius College), Kettlewell qualified as a physician at St. Bartholomew's Hospital, London, in 1933, then practiced medicine for 15 years. He has, however, been a lepidopterist since childhood. Long convinced that the genetics of evolution could be seen at work among moths, Kettlewell quit medical practice in 1948 in order to prove his point. As Nuffield Research Fellow in genetics at the University of Oxford he has pursued his quarry in the Belgian Congo, Uganda, the southern Sudan, Kenya, Portuguese East Africa, South Africa, Namaqualand, Norway, Corsica and Canada as well as Great Britain.

### *Bibliography*

THE CONTRIBUTION OF INDUSTRIAL MELANISM IN THE LEPIDOPTERA TO OUR KNOWLEDGE OF EVOLUTION. H. B. D. Kettlewell in *The Advancement of Science,* Vol. 13, No. 52, pages 245–252; March, 1957.

FURTHER SELECTION EXPERIMENTS ON INDUSTRIAL MELANISM IN THE LEPIDOPTERA. H. B. D. Kettlewell in *Heredity,* Vol. 10, Part 3, pages 287–301; December, 1956.

A RÉSUMÉ OF INVESTIGATIONS ON THE EVOLUTION OF MELANISM IN THE LEPIDOPTERA. H. B. D. Kettlewell in *Proceedings of the Royal Society of London,* Series B, Vol. 145, No. 920, pages 297–303; July 24, 1956.

SELECTION EXPERIMENTS ON INDUSTRIAL MELANISM IN THE LEPIDOPTERA. H. B. D. Kettlewell in *Heredity,* Vol. 9, Part 3, pages 323–342; December, 1955.

A SURVEY OF THE FREQUENCIES OF BISTON BETULARIA (L.) (LEP.) AND ITS MELANIC FORMS IN GREAT BRITAIN. H. B. D. Kettlewell in *Heredity,* Vol. 12, Part 1, pages 51–72; February, 1958.

## 24. Ionizing Radiation and Evolution

### *The Author*

JAMES F. CROW is professor of genetics and chairman of the Department of Medical Genetics at the University of Wisconsin. He grew up in Kansas and attended Friends University at Wichita. "I started out to major in music, but discovered I wasn't talented enough, and shifted successively to physics, then to chemistry and finally to biology. I took a course in genetics . . . and decided that this was the most interesting of the biology courses and something that I would like to pursue." He took his Ph.D. in genetics at the University of Texas in 1941, and taught at Dartmouth College until he joined the Wisconsin faculty in 1948.

### *Bibliography*

THE CAUSES OF EVOLUTION. J. B. S. Haldane. Harper & Brothers, 1932.

THE DARWINIAN AND MODERN CONCEPTIONS OF NATURAL SELECTION. H. J. Muller in *Proceedings of the American Philosophical Society,* Vol. 93, No. 6, pages 459–479; December 29, 1949.

EVOLUTION, GENETICS AND MAN. Theodosius Dobzhansky. John Wiley & Sons, Inc., 1955.

GENETICS, PALEONTOLOGY AND EVOLUTION. Edited by Glenn L. Jepsen, Ernst Mayr and George Gaylord Simpson. Princeton University Press, 1949.

RADIATION AND THE ORIGIN OF THE GENE. Carl Sagan in *Evolution,* Vol. 11, No. 1, pages 40–55; March, 1957.

## 25. Radiation and Human Mutation

### *The Author*

H. J. MULLER, professor of genetics at Indiana University, is the man who found that mutations could be accelerated by X-rays. This discovery, which he made in 1927, won him the Nobel prize in physiology and medicine for 1946. Muller was born in New York City in 1890 and was educated at Columbia University, where he took a Ph.D. in zoology in 1916. From 1915 to 1936 he taught zoology, first at the Rice Institute, then at the University of Texas. In 1933 to 1937 he worked at the Institute of Genetics in Moscow as senior geneticist, but he later became a fierce foe of the Soviet system. After spending some years at the University of Edinburgh and at Amherst College, he joined Indiana University in 1945, remaining there until his death in 1967.

### *Bibliography*

NATIONAL SURVEY OF CONGENITAL MALFORMATIONS RESULTING FROM EXPOSURE TO ROENTGEN RADIATION. Stanley H. Macht and Philip S. Lawrence in *American Journal of Roentgenology and Radiation Therapy.* Vol. 73, No. 3, pages 442–446; March, 1955.

OUR LOAD OF MUTATIONS. H. J. Muller in *American Journal of Human Genetics,* Vol. 2, No. 2, pages 111–176; June, 1950.

X-RAY INDUCED MUTATIONS IN MICE. W. L. Russell in *Cold Spring Harbor Symposia on Quantitative Biology,* Vol. 16, pages 327–336; 1951.

## 26. The Repair of DNA

### *The Authors*

PHILIP C. HANAWALT and ROBERT H. HAYNES are respectively associate professor of biological sciences at Stanford University and associate professor of biophysics and medical physics at the University of California at Berkeley. Hanawalt majored in physics at Oberlin College, from which he was graduated in 1954, and did graduate work in physics and biophysics at Yale University, from which he received a Ph.D. in 1959. Haynes obtained a bachelor's degree in physics and a Ph.D. in biophysics from the University of Western Ontario. Haynes writes that he was "raised a true-blue Canadian Tory" but during a year as an exchange fellow in England "was subverted by the success of the British National Health Service and began drinking in workingmen's pubs." Looking back, he says, "it is clear that it was this experience that ensured I would later be on the side of the angels and the Free Speech Movement in Berkeley." Both Hanawalt and Haynes are teaching introductory courses in biology. Haynes writes: "Although I have always spent most of my time in research, I am convinced that recent events at Berkeley will accelerate the swing back to teaching in academia; and somewhat to my surprise I found teaching Biology 1 to be a rather exhilarating experience. In spite of my political stance at Berkeley, my interest in fine food and wine, poetry and ballet appears to be disconcertingly nonproletarian. However, it was in a scruffy Oxford pub that Hanawalt and I began our continuing collaboration."

### *Bibliography*

THE DISAPPEARANCE OF THYMINE DIMERS FROM DNA: AN ERROR-CORRECTING MECHANISM. R. B. Setlow and W. L. Carrier in *Proceedings of the National Academy of Sciences,* Vol. 51, No. 2, pages 226–231; February, 1964.

EVIDENCE FOR REPAIR REPLICATION OF ULTRAVIOLET-DAMAGED DNA IN BACTERIA. David Pettijohn and Philip C. Hanawalt in *Journal of Molecular Biology,* Vol. 9, No. 2, pages 395–410; August, 1964.

A GENETIC LOCUS IN E. COLI K 12 THAT CONTROLS THE REACTIVATION OF UV-PHOTOPRODUCTS ASSOCIATED WITH THYMINE IN DNA. P. Howard-Flanders, Richard P. Boyce, Eva Simson and Lee Theriot in *Proceedings of the National Academy of Sciences,* Vol. 48, No. 12, pages 2109–2115; December 15, 1962.

STRUCTURAL DEFECTS IN DNA AND THEIR REPAIR IN MICROORGANISMS. Radiation Research, Supplement 6, edited by Robert H. Haynes, Sheldon Wolff and James E. Till. Academic Press, in press.

## 27. The Evolution of Hemoglobin

### *The Author*

EMILE ZUCKERKANDL is an investigator with the French National Center for Scientific Research, working at the Physico-Chemical Colloidal Laboratory in Montpellier. A native of Vienna, he became a French citizen in 1938. After he was graduated from the Sorbonne, he obtained a master's degree at the University of Illinois and then returned to the Sorbonne for his doctorate. For several years he served at a marine biological station in Brittany, investigating proteins. From 1959 to 1964 he worked with Linus Pauling at the California Institute of Technology, investigating hemoglobin. He is now at work in "the new field of chemical paleogenics," attempting "to elucidate questions related to the evolutionary succession of major and minor components of a polypeptide chain and to the correlation, in hemoglobins, between structure and function."

### *Bibliography*

EVOLUTIONARY DIVERGENCE AND CONVERGENCE IN PROTEINS. Emile Zuckerkandl and Linus Pauling in *Evolving Genes and Proteins,* edited by Henry J. Vogel. Academic Press, in press.

GENE EVOLUTION AND THE HAEMOGLOBINS. Vernon M. Ingram in *Nature,* Vol. 189, No. 4766, pages 704–708; March 4, 1961.

THE HEMOGLOBINS. G. Braunitzer, K. Hilse, V. Rudloff and N. Hilschmann in *Advances in Protein Chemistry: Vol. XIX,* edited by C. B. Anfinsen, Jr., John T. Edsall, M. L. Anson and Frederic M. Richards. Academic Press, 1964.

MOLECULAR DISEASE, EVOLUTION, AND GENIC HETEROGENEITY. Emile Zuckerkandl and Linus Pauling in *Horizons in Biochemistry,* edited by Michael Kasha and Bernard Pullman. Academic Press, 1962.

## 28. Computer Analysis of Protein Evolution

### *The Author*

MARGARET OAKLEY DAYHOFF ("Computer Analysis of Protein Evolution") is head of the chemical biology department and associate director of research at the National Biomedical Research Foundation. She writes: "I received my B.A. from New York University in 1945 and my Ph.D. in chemistry from Columbia University in 1948. I then joined the staff of the Rockefeller Institute, where I remained for three years as a physical chemist. For the next few years my main interests centered around my husband, who is a physicist, and my two daughters. The development of the accurate programmed high-speed computer has paralleled and greatly influenced my scientific career. The recent accumulation of fundamental structural data in biochemistry has sparked my hope that the computer's memory, its capacity for handling details and its many features analogous to living systems would make possible a new level of human understanding of biological structure, evolution and function. As our contribution to the realization of this hope, my colleagues and I have set out to collect, organize and analyze protein and nucleic acid sequence data, and to publish this collection for the convenience of workers in biomedical research."

### *Bibliography*

PRINCIPLES OF ANIMAL TAXONOMY. George Gaylord Simpson. Columbia University Press, 1961.

COMPUTER AIDS TO PROTEIN SEQUENCE DETERMINATION. M. O. Dayhoff in *Journal of Theoretical Biology,* Vol. 8, No. 1, pages 97–112; January, 1965.

ATLAS OF PROTEIN SEQUENCE AND STRUCTURE, 1969. Margaret O. Dayhoff. National Biomedical Research Foundation, 1969.

# Part V Genetics and Man

## 29. Porphyria and King George III

### *The Authors*

IDA MACALPINE AND RICHARD HUNTER ("Porphyria and King George III") are mother and son. Dr. Macalpine, a physician who has retired from clinical work, served as psychiatrist at St. Bartholomew's Hospital in London. She has published papers on psychoanalysis and psychosomatic medicine. Hunter is physician in psychological medicine at National Hospital in London and lecturer in the history of psychiatry at the Institute of Psychiatry of the University of London. Dr. Macalpine writes: "Our common interest in the history of psychiatry is motivated by the belief that it helps to understand the complex and confused state of the specialty at the present time and will foster a development away from its preoccupation with treatments and toward investigative and causative research and so catch up with general medicine. We have published together the source book *Three Hundred Years of Psychiatry 1535–1860* and have edited a number of classic psychiatric texts."

### *Bibliography*

DISEASES OF PORPHYRIN METABOLISM. A. Goldberg and C. Rimington. Charles C Thomas, Publisher, 1962.

A CLINICAL REASSESSMENT OF THE "INSANITY" OF GEORGE III AND SOME OF ITS HISTORICAL IMPLICATIONS. Ida Macalpine and Richard Hunter in *Bulletin of the Institute of Historical Research,* Vol. 40, No. 102, pages 166–185; November, 1967.

PORPHYRIA–A ROYAL MALADY: ARTICLES PUBLISHED IN OR COMMISSIONED BY THE BRITISH MEDICAL JOURNAL. British Medical Association, 1968.

## 30. The Prevention of "Rhesus" Babies

### *The Author*

C. A. CLARKE ("The Prevention of 'Rhesus' Babies") is professor of medicine at the University of Liverpool. He is also director of the Nuffield Unit of Medical Genetics at the university and consulting physician to the United Liverpool Hospitals. Clarke, who has a doctorate in science from the University of Cambridge, received his medical qualification at Guy's Hospital in London in 1932. During World War II he served as a medical specialist in the Royal Naval Volunteer Reserve. Clarke is the author of *Genetics for the Clinician,* published in 1962.

### *Bibliography*

SUCCESSFUL PREVENTION OF EXPERIMENTAL RH SENSITIZATION IN MAN WITH AN ANTI-RH GAMMA$_2$-GLOBULIN ANTIBODY PREPARATION: A PRELIMINARY REPORT. Vincent J. Freda, John G. Gorman and William Pollack in *Transfusion,* Vol. 4, No. 1, pages 26–32; January–February, 1964.

THE PREVENTION OF RH HAEMOLYTIC DISEASE. R. B. McConnell in *Annual Review of Medicine,* Vol. 17, pages 291–306; 1966.

PREVENTION OF RH-HAEMOLYTIC DISEASE. C. A. Clarke in *British Medical Journal,* Vol. 4, No. 5570, pages 7–12; October 7, 1967.

PROPHYLAXIS OF RHESUS ISO-IMMUNIZATION. C. A. Clarke in *British Medical Bulletin,* Vol. 24, No. 1, pages 3–9; January, 1968.

TRANSFUSION, Vol. 8, No. 3. May–June, 1968.

PREVENTION OF RHESUS ISO-IMMUNISATION. C. A. Clarke in *The Lancet,* Vol. II for 1968, No. 7558, pages 1–7; July 6, 1968.

## 31. The Genetics of the Dunkers

### *The Author*

H. BENTLEY GLASS is distinguished professor and Academic Vice President, State University of New York, Stony Brook, and for many years previously was professor of biology at The Johns Hopkins University. In addition to teaching, research, and writing, his varied scientific activities have included membership in national advisory committees, editorships for science journals, and officerships in scientific societies. Glass was born in China in 1906 and had his pre-college education in mission schools there. He took his undergraduate work at Baylor University and his Ph.D. at the University of Texas.

### *Bibliography*

GENETIC DRIFT IN A RELIGIOUS ISOLATE: AN ANALYSIS OF THE CAUSES OF VARIATION IN BLOOD GROUP AND OTHER GENE FREQUENCIES IN A SMALL POPULATION. Bentley Glass, Milton S. Sacks, Elsa F. Jahn and Charles Hess in *The American Naturalist,* Vol. 86, No. 828, pages 145–159; May–June, 1952.

RACES: A STUDY OF THE PROBLEMS OF RACE FORMATION IN MAN. Carleton S. Coon, Stanley M. Garn and Joseph B. Birdsell. Charles C Thomas, 1950.

## 32. Hybrid Wheat

### *The Authors*

BYRD C. CURTIS and DAVID R. JOHNSTON ("Hybrid Wheat") are with Cargill, Incorporated; Curtis is head of the hybrid wheat development program at Cargill Research Farms in Fort Collins, Colo., and Johnston is a wheat breeder. Curtis received his bachelor's degree in agronomy at Oklahoma State University in 1950, his master's degree in agronomy at Kansas State University in 1951 and his Ph.D. in plant breeding and genetics from Oklahoma State in 1959. He was a member of the staff at Oklahoma State until 1963, working on the breeding of wheat, oats and barley; he then was appointed associate professor at Colorado State University and director of wheat research for the state of Colorado. He joined Cargill in 1967. Johnston, who was graduated from the University of Massachusetts in 1952, was a research worker at the University of Minnesota from 1956 to 1967, when he took up his present position.

### *Bibliography*

CROSS-BREEDING IN WHEAT, TRITICUM AESTIVUM, I: FREQUENCY OF THE POLLEN-RESTORING CHARACTER IN HYBRID WHEATS HAVING AEGILOPS OVATA CYTOPLASM. J. A. Wilson and W. M. Ross in *Crop Science,* Vol. 1, No. 3, pages 191–193; May–June, 1961.

CROSS-BREEDING IN WHEAT, TRITICUM AESTIVUM L., II: HYBRID SEED SET ON A CYTOPLASMIC MALE-STERILE WINTER WHEAT COMPOSITE SUBJECTED TO CROSS-POLLINATION. J. A. Wilson and W. M. Ross in *Crop Science,* Vol. 2, No. 5, pages 415–417; September–October, 1962.

HYBRID WHEATS: THEIR DEVELOPMENT AND FOOD POTENTIAL. Ricardo Rodríguez, Marco A. Quiñones L., Norman E. Borlaug and Ignacio Narváez in *Research Bulletin No. 3, International Maize and Wheat Improvement Center, Mexico;* July, 1967.

HYBRID WHEAT. V. A. Johnson and J. W. Schmidt in *Advances in Agronomy,* Vol. 20, pages 199–233; 1968.

## 33. Sex Differences in Cells

### *The Author*

URSULA MITTWOCH is a senior research assistant at University College London. Dr. Mittwoch was born in Berlin and completed her schooling in London. On leaving school she went to work at the John Innes Horticultural Institution under the direction of Kenneth Mather. There she became interested in genetics and later she took a degree in biology at University College London. For her Ph.D. Dr. Mittwoch did research on the genetics of fungi under the direction of J. B. S. Haldane. She began to study human genetics at University College London under L. S. Penrose and made a study of white blood cells, which led her into research on the sex differences in these and other cells.

### *Bibliography*

CHROMOSOMES FOR BEGINNERS. Bernard Lennox in *The Lancet,* Vol. 1, No. 7185, pages 1046–1051; May, 1961.

CYTOGENETICS OF ABNORMAL SEXUAL DEVELOPMENT IN MAN. P. G. Harnden and Patricia A. Jacobs in *British Medical Bulletin,* Vol. 17, No. 3, pages 206–212; September, 1961.

INDIRECT ASSESSMENT OF NUMBER OF X CHROMOSOMES IN MAN, USING NUCLEAR SEXING AND COLOUR VISION. Bernard Lennox in *British Medical Bulletin,* Vol. 17, No. 3, pages 196–199; September, 1961.

TURNER'S SYNDROME AND ALLIED CONDITIONS: CLINICAL FEATURES AND CHROMOSOME ABNORMALITIES. P. E. Polani in *British Medical Bulletin,* Vol. 17, No. 3, pages 200–205; September, 1961.

## 34. Introduction of Cancer by Viruses

### *The Author*

RENATO DULBECCO is resident fellow at the Salk Institute for Biological Studies. Born in Italy, he took a medical degree at the University of Torino in 1936 and remained there as a teacher and researcher until 1947. Moving to the U.S. in that year, he was at Indiana University for two years and at the California Institute of Technology for 14, including nine years as professor of biology. Dulbecco joined the Salk Institute in 1963 but spent the academic year 1963–1964 as Royal Society Visiting Professor at the University of Glasgow. Since 1964 he has served as a trustee of the Salk Institute while continuing his research activities there.

### *Bibliography*

CELL TRANSFORMATION BY DIFFERENT FORMS OF POLYOMA VIRUS DNA. Lionel Crawford, Renato Dulbecco, Mike Fried, Luc Montagnier and Michael Stoker in *Proceedings of the National Academy of Sciences,* Vol. 52, No. 1, pages 148–152; July, 1964.

IMMUNOLOGICAL DETERMINANTS OF POLYOMA VIRUS ONCOGENESIS. Karl Habel in *The Journal of Experimental Medicine,* Vol. 115, No. 1, pages 181–193; January 1, 1962.

STUDIES ON SPECIFIC TRANSPLANTATION RESISTANCE TO POLYOMA-VIRUS-INDUCED TUMORS, I: TRANSPLANTATION RESISTANCE INDUCED BY POLYOMA VIRUS INFECTION. Hans Olof Sjögren in *Journal of the National Cancer Institute,* Vol. 32, No. 2, pages 361–393; February, 1964.

TRANSFORMATION OF CELLS IN VITRO BY DNA-CONTAINING VIRUSES. Renato Dulbecco in *The Journal of the American Medical Association,* Vol. 190, No. 8, pages 721–726; November 23, 1964.

TRANSFORMATION OF PROPERTIES OF AN ESTABLISHED CELL LINE BY SV 40 AND POLYOMA VIRUS. George J. Todaro, Howard Green and Burton D. Goldberg in *Proceedings of the National Academy of Sciences,* Vol. 51, No. 1, pages 66–73; January, 1964.

VIRUS-CELL INTERACTION WITH A TUMOR-PRODUCING VIRUS. Marguerite Vogt and Renato Dulbecco in *Proceedings of the National Academy of Sciences,* Vol. 46, No. 3, pages 365–370; March 15, 1960.

## 35. Hybrid Somatic Cells

### *The Authors*

BORIS EPHRUSSI and MARY C. WEISS ("Hybrid Somatic Cells") are at the Centre de Génétique Moléculaire of the Centre National de la Recherche Scientifique in France; Ephrussi is director of the laboratory and Miss Weiss is an investigator there. Ephrussi, who was born in Moscow but has been a French citizen since 1932, is in addition professor of genetics at the University of Paris, where he obtained his D.Sc. in 1932. He has spent several extended periods in the U.S., notably at the California Institute of Technology as a Rockefeller Foundation Fellow for studies in genetics in 1927, 1934 and 1936, three years (1941–1944) as associate professor at Johns Hopkins University and five years (1962–1967) as professor at Western Reserve University. From 1928 to 1940 he was at the genetics laboratory of the Rothschild Foundation in Paris. Miss Weiss was graduated from Newcomb College for Women at Tulane University in 1962 and received her Ph.D. from Western Reserve University in 1966. She was a postdoctoral fellow of the U.S. Public Health Service at the New York University School of Medicine in 1966–1967 and at the Carnegie Institution of Washington in 1967–1968.

### *Bibliography*

HYBRIDIZATION OF SOMATIC CELLS AND PHENOTYPIC EXPRESSION. Boris Ephrussi in *Developmental and Metabolic Control Mechanisms and Neoplasia: A Collection of Papers Presented at the Nineteenth Annual Symposium on Fundamental Cancer Research, 1965.* The Williams and Wilkins Company, 1965.

ARTIFICIAL HETEROKARYONS OF ANIMAL CELLS FROM DIFFERENT SPECIES. H. Harris, J. F. Watkins, C. E. Ford and G. I. Schoefl in *Journal of Cell Science,* Vol. 1, No. 1, pages 1–31; March, 1966.

STUDIES OF INTERSPECIFIC (RAT × MOUSE) SOMATIC HYBRIDS. I: ISOLATION, GROWTH AND EVOLUTION OF THE KARYOTYPE. II: LACTATE DEHYDROGENASE AND $\beta$-GLUCURONIDASE. Mary C. Weiss and Boris Ephrussi in *Genetics,* Vol. 54, No. 5, pages 1095–1109, 1111–1122; November, 1966.

HUMAN-MOUSE HYBRID CELL LINES CONTAINING PARTIAL COMPLEMENTS OF HUMAN CHROMOSOMES AND FUNCTIONING HUMAN GENES. Mary C. Weiss and Howard Green in *Proceedings of the National Academy of Sciences,* Vol. 58, No. 3, pages 1104–1111; September 15, 1967.

# Index